GEORGIUS AGRICOLA

DE RE METALLICA

TRANSLATED FROM THE FIRST LATIN EDITION OF 1556

with

Biographical Introduction, Annotations and Appendices upon the Development of Mining Methods, Metallurgical Processes, Geology, Mineralogy & Mining Law from the earliest times to the 16th Century

BY

HERBERT CLARK HOOVER

A. B. Stanford University, Member American Institute of Mining Engineers, Mining and Metallurgical Society of America, Société des Ingéniéurs Civils de France, American Institute of Civil Engineers, Fellow Royal Geographical Society, etc., etc.

AND

LOU HENRY HOOVER

A. B. Stanford University, Member American Association for the Advancement of Science, The National Geographical Society, Royal Scottish Geographical Society, etc., etc.

1950

Dover Publications, Inc.

NEW YORK

TO

JOHN CASPAR BRANNER Ph.D.,

The inspiration of whose teaching is no less great than his contribution to science.

This New 1950 Edition of DE RE METALLICA is a complete and unchanged reprint of the translation published by The Mining Magazine, London, in 1912. It has been made available through the kind permission of Honorable Herbert C. Hoover and Mr. Edgar Rickard, Author and Publisher, respectively, of the original volume.

International Standard Book Number: 0-486-60006-8
Library of Congress Catalog Card Number: A51-8994

Manufactured in the United States of America

Dover Publications, Inc.
180 Varick Street
New York 14, N. Y.

TRANSLATORS' PREFACE.

HERE are three objectives in translation of works of this character: to give a faithful, literal translation of the author's statements ; to give these in a manner which will interest the reader ; and to preserve, so far as is possible, the style of the original text. The task has been doubly difficult in this work because, in using Latin, the author availed himself of a medium which had ceased to expand a thousand years before his subject had in many particulars come into being ; in consequence he was in difficulties with a large number of ideas for which there were no corresponding words in the vocabulary at his command, and instead of adopting into the text his native German terms, he coined several hundred Latin expressions to answer his needs. It is upon this rock that most former attempts at translation have been wrecked. Except for a very small number, we believe we have been able to discover the intended meaning of such expressions from a study of the context, assisted by a very incomplete glossary prepared by the author himself, and by an exhaustive investigation into the literature of these subjects during the sixteenth and seventeenth centuries. That discovery in this particular has been only gradual and obtained after much labour, may be indicated by the fact that the entire text has been re-typewritten three times since the original, and some parts more often ; and further, that the printer's proof has been thrice revised. We have found some English equivalent, more or less satisfactory, for practically all such terms, except those of weights, the varieties of veins, and a few minerals. In the matter of weights we have introduced the original Latin, because it is impossible to give true equivalents and avoid the fractions of reduction ; and further, as explained in the Appendix on Weights it is impossible to say in many cases what scale the Author had in mind. The English nomenclature to be adopted has given great difficulty, for various reasons ; among them, that many methods and processes described have never been practised in English-speaking mining communities, and so had no representatives in our vocabulary, and we considered the introduction of German terms undesirable ; other methods and processes have become obsolete and their descriptive terms with them, yet we wished to avoid the introduction of obsolete or unusual English ; but of the greatest importance of all has been the necessity to avoid rigorously such modern technical terms as would imply a greater scientific understanding than the period possessed.

Agricola's Latin, while mostly free from mediæval corruption, is somewhat tainted with German construction. Moreover some portions have not the continuous flow of sustained thought which others display, but the fact that the writing of the work extended over a period of twenty years, sufficiently explains the considerable variation in style. The technical descriptions in the later books often take the form of House-that-Jack-built sentences which have had to be at least partially broken up and the subject occasionally re-introduced. Ambiguities were also sometimes found which it was necessary to carry on into the translation. Despite these criticisms we must, however, emphasize that Agricola was infinitely clearer in his style than his contemporaries upon such subjects, or for that matter than his successors in almost any language for a couple of centuries. All of the illustrations and display letters of the original have been reproduced and the type as closely approximates to the original as the printers have been able to find in a modern font.

There are no footnotes in the original text, and Mr. Hoover is responsible for them all. He has attempted in them to give not only such comment as would tend to clarify the text, but also such information as we have been able to discover with regard to the previous history of the subjects mentioned. We have confined the historical notes to the time prior to Agricola, because to have carried them down to date in the briefest manner would have demanded very much more space than could be allowed. In the examination of such technical and historical material one is appalled at the flood of mis-information with regard to ancient arts and sciences which has been let loose upon the world by the hands of non-technical translators and commentators. At an early stage we considered that we must justify any divergence of view from such authorities, but to limit the already alarming volume of this work, we later felt compelled to eliminate most of such discussion. When the half-dozen most important of the ancient works bearing upon science have been translated by those of some scientific experience, such questions will, no doubt, be properly settled.

We need make no apologies for *De Re Metallica*. During 180 years it was not superseded as the text-book and guide to miners and metallurgists, for until Schlüter's great work on metallurgy in 1738 it had no equal. That it passed through some ten editions in three languages at a period when the printing of such a volume was no ordinary undertaking, is in itself sufficient evidence of the importance in which it was held, and is a record that no other volume upon the same subjects has equalled since. A large proportion of the technical data given by Agricola was either entirely new, or had not been given previously with sufficient detail and explanation to have enabled a worker in these arts himself to perform the operations without further guidance. Practically the whole of it must have been given from personal experience and observation, for the scant library at his service can be appreciated from his own Preface. Considering the part which the metallic arts have played in human history, the paucity of their literature down to Agricola's time is amazing. No doubt the arts were jealously guarded by their practitioners as a sort of stock-in-trade, and it is also probable that those who had knowledge were not usually of a literary turn of mind ; and, on the other hand, the small army of writers prior to his time were not much interested in the description of industrial pursuits. Moreover, in those thousands of years prior to printing, the tedious and expensive transcription of manuscripts by hand was mostly applied to matters of more general interest, and therefore many writings may have been lost in consequence. In fact, such was the fate of the works of Theophrastus and Strato on these subjects.

We have prepared a short sketch of Agricola's life and times, not only to give some indication of his learning and character, but also of his considerable position in the community in which he lived. As no appreciation of Agricola's stature among the founders of science can be gained without consideration of the advance which his works display over those of his predecessors, we therefore devote some attention to the state of knowledge of these subjects at the time by giving in the Appendix a short review of the literature then extant and a summary of Agricola's other writings. To serve the bibliophile we present such data as we have been able to collect it with regard to the various editions of his works. The full titles of the works quoted in the footnotes under simply authors' names will be found in this Appendix.

We feel that it is scarcely doing Agricola justice to publish *De Re Metallica* only. While it is of the most general interest of all of his works, yet, from the point of view of pure science, *De Natura Fossilium* and *De Ortu et Causis* are works which deserve an equally important place. It is unfortunate that Agricola's own countrymen have not given to the world competent translations into German, as his work has too often been judged by the German translations, the infidelity of which appears in nearly every paragraph.

We do not present *De Re Metallica* as a work of "practical" value. The methods and processes have long since been superseded ; yet surely such a milestone on the road of development of one of the two most basic of human industrial activities is more worthy of preservation than the thousands of volumes devoted to records of human destruction. To those interested in the history of their own profession we need make no apologies, except for the long delay in publication. For this we put forward the necessity of active endeavour in many directions ; as this book could be but a labour of love, it has had to find the moments for its execution in night hours, week-ends, and holidays, in all extending over a period of about five years. If the work serves to strengthen the traditions of one of the most important and least recognized of the world's professions we shall be amply repaid.

It is our pleasure to acknowledge our obligations to Professor H. R. Fairclough, of Stanford University, for perusal of and suggestions upon the first chapter; and to those whom we have engaged from time to time for one service or another, chiefly bibliographical work and collateral translation. We are also sensibly obligated to the printers, Messrs. Frost & Sons, for their patience and interest, and for their willingness to bend some of the canons of modern printing, to meet the demands of the 16th Century.

THE RED HOUSE, *July* 1, 1912.

HORNTON STREET, LONDON.

INTRODUCTION.

BIOGRAPHY.

EORGIUS AGRICOLA was born at Glauchau, in Saxony, on March 24th, 1494, and therefore entered the world when it was still upon the threshold of the Renaissance; Gutenberg's first book had been printed but forty years before; the Humanists had but begun that stimulating criticism which awoke the Reformation; Erasmus, of Rotterdam, who was subsequently to become Agricola's friend and patron, was just completing his student days. The Reformation itself was yet to come, but it was not long delayed, for Luther was born the year before Agricola, and through him Agricola's homeland became the cradle of the great movement; nor did Agricola escape being drawn into the conflict. Italy, already awake with the new classical revival, was still a busy workshop of antiquarian research, translation, study, and publication, and through her the Greek and Latin Classics were only now available for wide distribution. Students from the rest of Europe, among them at a later time Agricola himself, flocked to the Italian Universities, and on their return infected their native cities with the newly-awakened learning. At Agricola's birth Columbus had just returned from his great discovery, and it was only three years later that Vasco Da Gama rounded Cape Good Hope. Thus these two foremost explorers had only initiated that greatest period of geographical expansion in the world's history. A few dates will recall how far this exploration extended during Agricola's lifetime. Balboa first saw the Pacific in 1513; Cortes entered the City of Mexico in 1520; Magellan entered the Pacific in the same year; Pizarro penetrated into Peru in 1528; De Soto landed in Florida in 1539, and Potosi was discovered in 1546. Omitting the sporadic settlement on the St. Lawrence by Cartier in 1541, the settlement of North America did not begin for a quarter of a century after Agricola's death. Thus the revival of learning, with its train of Humanism, the Reformation, its stimulation of exploration and the re-awakening of the arts and sciences, was still in its infancy with Agricola.

We know practically nothing of Agricola's antecedents or his youth. His real name was Georg Bauer ("peasant"), and it was probably Latinized by his teachers, as was the custom of the time. His own brother, in receipts

[1]For the biographical information here set out we have relied principally upon the following works:—Petrus Albinus, *Meissnische Land Und Berg Chronica*, Dresden, 1590; Adam Daniel Richter, *Umständliche. . . . Chronica der Stadt Chemnitz*, Leipzig, 1754; Johann Gottfried Weller, *Altes Aus Allen Theilen Der Geschichte*, Chemnitz, 1766; Freidrich August Schmid, *Georg Agrikola's Bermannus*, Freiberg, 1806; Georg Heinrich Jacobi, *Der Mineralog Georgius Agricola*, Zwickau, 1881; Dr. Reinhold Hofmann, *Dr. Georg Agricola*, Gotha, 1905. The last is an exhaustive biographical sketch, to which we refer those who are interested.

preserved in the archives of the Zwickau Town Council, calls himself "Bauer," and in them refers to his brother "Agricola." He entered the University of Leipsic at the age of twenty, and after about three and one-half years' attendance there gained the degree of *Baccalaureus Artium*. In 1518 he became Vice-Principal of the Municipal School at Zwickau, where he taught Greek and Latin. In 1520 he became Principal, and among his assistants was Johannes Förster, better known as Luther's collaborator in the translation of the Bible. During this time our author prepared and published a small Latin Grammar[2]. In 1522 he removed to Leipsic to become a lecturer in the University under his friend, Petrus Mosellanus, at whose death in 1524 he went to Italy for the further study of Philosophy, Medicine, and the Natural Sciences. Here he remained for nearly three years, from 1524 to 1526. He visited the Universities of Bologna, Venice, and probably Padua, and at these institutions received his first inspiration to work in the sciences, for in a letter[3] from Leonardus Casibrotius to Erasmus we learn that he was engaged upon a revision of Galen. It was about this time that he made the acquaintance of Erasmus, who had settled at Basel as Editor for Froben's press.

In 1526 Agricola returned to Zwickau, and in 1527 he was chosen town physician at Joachimsthal. This little city in Bohemia is located on the eastern slope of the Erzgebirge, in the midst of the then most prolific metal-mining district of Central Europe. Thence to Freiberg is but fifty miles, and the same radius from that city would include most of the mining towns so frequently mentioned in *De Re Metallica*—Schneeberg, Geyer, Annaberg and Altenberg—and not far away were Marienberg, Gottesgab, and Platten. Joachimsthal was a booming mining camp, founded but eleven years before Agricola's arrival, and already having several thousand inhabitants. According to Agricola's own statement[4], he spent all the time not required for his medical duties in visiting the mines and smelters, in reading up in the Greek and Latin authors all references to mining, and in association with the most learned among the mining folk. Among these was one Lorenz Berman, whom Agricola afterward set up as the "learned miner" in his dialogue *Bermannus*. This book was first published by Froben at Basel in 1530, and was a sort of catechism on mineralogy, mining terms, and mining lore. The book was apparently first submitted to the great Erasmus, and the publication arranged by him, a warm letter of approval by him appearing at the beginning of the book[5]. In 1533 he published *De Mensuris et Ponderibus*, through Froben, this being a discussion of Roman and Greek weights and measures. At about this time he began *De Re Metallica* — not to be published for twenty-five years.

[2]*Georgii Agricolae Glaucii Libellus de Prima ac Simplici Institutione Grammatica*, printed by Melchior Lotther, Leipzig, 1520 Petrus Mosellanus refers to this work (without giving title) in a letter to Agricola, June, 1520.

[3]*Briefe an Desiderius Erasmus von Rotterdam*. Published by Joseph Förstemann and Otto Günther. XXVII. *Beiheft zum Zentralblatt für Bibliothekswesen*, Leipzig, 1904. p. 44.

[4]*De Veteribus et Novis Metallis*. Preface.

[5]A summary of this and of Agricola's other works is given in the Appendix A.

Agricola did not confine his interest entirely to medicine and mining, for during this period he composed a pamphlet upon the Turks, urging their extermination by the European powers. This work was no doubt inspired by the Turkish siege of Vienna in 1529. It appeared first in German in 1531, and in Latin—in which it was originally written—in 1538, and passed through many subsequent editions.

At this time, too, he became interested in the God's Gift mine at Albertham, which was discovered in 1530. Writing in 1545, he says[6]: "We, as a shareholder, through the goodness of God, have enjoyed the "proceeds of this God's Gift since the very time when the mine began first "to bestow such riches."

Agricola seems to have resigned his position at Joachimsthal in about 1530, and to have devoted the next two or three years to travel and study among the mines. About 1533 he became city physician of Chemnitz, in Saxony, and here he resided until his death in 1555. There is but little record of his activities during the first eight or nine years of his residence in this city. He must have been engaged upon the study of his subjects and the preparation of his books, for they came on with great rapidity soon after. He was frequently consulted on matters of mining engineering, as, for instance, we learn, from a letter written by a certain Johannes Hordeborch[7], that Duke Henry of Brunswick applied to him with regard to the method for working mines in the Upper Harz.

In 1543 he married Anna, widow of Matthias Meyner, a petty tithe official; there is some reason to believe from a letter published by Schmid,[8] that Anna was his second wife, and that he was married the first time at Joachimsthal. He seems to have had several children, for he commends his young children to the care of the Town Council during his absence at the war in 1547. In addition to these, we know that a son, Theodor, was born in 1550; a daughter, Anna, in 1552; another daughter, Irene, was buried at Chemnitz in 1555; and in 1580 his widow and three children—Anna, Valerius, and Lucretia—were still living.

In 1544 began the publication of the series of books to which Agricola owes his position. The first volume comprised five works and was finally issued in 1546; it was subsequently considerably revised, and re-issued in 1558. These works were: *De Ortu et Causis Subterraneorum*, in five "books," the first work on physical geology; *De Natura Eorum quae Effluunt ex Terra*, in four "books," on subterranean waters and gases; *De Natura Fossilium*, in ten "books," the first systematic mineralogy; *De Veteribus et Novis Metallis*, in two "books," devoted largely to the history of metals and topographical mineralogy; a new edition of *Bermannus* was included; and finally *Rerum Metallicarum Interpretatio*, a glossary of Latin and German mineralogical and metallurgical terms. Another work, *De Animantibus Subterraneis*, usually published with *De Re Metallica*, is dated 1548 in the preface. It

[6]*De Veteribus et Novis Metallis*, Book I.
[7]Printed in F. A Schmid's *Georg Agrikola's Bermannus*, p 14, Freiberg, 1806.
[8]Op. Cit., p. 8.

is devoted to animals which live underground, at least part of the time, but is not a very effective basis of either geologic or zoologic classification. Despite many public activities, Agricola apparently completed *De Re Metallica* in 1550, but did not send it to the press until 1553; nor did it appear until a year after his death in 1555. But we give further details on the preparation of this work on p. xv. During this period he found time to prepare a small medical work, *De Peste,* and certain historical studies, details of which appear in the Appendix. There are other works by Agricola referred to by sixteenth century writers, but so far we have not been able to find them although they may exist. Such data as we have, is given in the appendix.

As a young man, Agricola seems to have had some tendencies toward liberalism in religious matters, for while at Zwickau he composed some anti-Popish Epigrams; but after his return to Leipsic he apparently never wavered, and steadily refused to accept the Lutheran Reformation. To many even liberal scholars of the day, Luther's doctrines appeared wild and demagogic. Luther was not a scholarly man; his addresses were to the masses; his Latin was execrable. Nor did the bitter dissensions over hair-splitting theology in the Lutheran Church after Luther's death tend to increase respect for the movement among the learned. Agricola was a scholar of wide attainments, a deep-thinking, religious man, and he remained to the end a staunch Catholic, despite the general change of sentiment among his countrymen. His leanings were toward such men as his friend the humanist, Erasmus. That he had the courage of his convictions is shown in the dedication of *De Natura Eorum,* where he addresses to his friend, Duke Maurice, the pious advice that the dissensions of the Germans should be composed, and that the Duke should return to the bosom of the Church those who had been torn from her, and adds: "Yet "I do not wish to become confused by these turbulent waters, and be led to "offend anyone. It is more advisable to check my utterances." As he became older he may have become less tolerant in religious matters, for he did not seem to show as much patience in the discussion of ecclesiastical topics as he must have possessed earlier, yet he maintained to the end the respect and friendship of such great Protestants as Melanchthon, Camerarius, Fabricius, and many others.

In 1546, when he was at the age of 52, began Agricola's activity in public life, for in that year he was elected a Burgher of Chemnitz; and in the same year Duke Maurice appointed him Burgomaster—an office which he held for four terms. Before one can gain an insight into his political services, and incidentally into the character of the man, it is necessary to understand the politics of the time and his part therein, and to bear in mind always that he was a staunch Catholic under a Protestant Sovereign in a State seething with militant Protestantism.

Saxony had been divided in 1485 between the Princes Ernest and Albert, the former taking the Electoral dignity and the major portion of the Principality. Albert the Brave, the younger brother and Duke of Saxony, obtained the subordinate portion, embracing Meissen, but subject to the Elector. The Elector Ernest was succeeded in 1486 by Frederick the Wise, and under his support Luther made Saxony the cradle of the Reformation. This Elector was succeeded in 1525 by his brother John, who was in turn succeeded by his son John Frederick in 1532. Of more immediate interest to this subject is the Albertian line of Saxon Dukes who ruled Meissen, for in that Principality Agricola was born and lived, and his political fortunes were associated with this branch of the Saxon House. Albert was succeeded in 1505 by his son George, "The Bearded," and he in turn by his brother Henry, the last of the Catholics, in 1539, who ruled until 1541. Henry was succeeded in 1541 by his Protestant son Maurice, who was the Patron of Agricola.

At about this time Saxony was drawn into the storms which rose from the long-standing rivalry between Francis I., King of France, and Charles V. of Spain. These two potentates came to the throne in the same year (1515), and both were candidates for Emperor of that loose Confederation known as the Holy Roman Empire. Charles was elected, and intermittent wars between these two Princes arose—first in one part of Europe, and then in another. Francis finally formed an alliance with the Schmalkalden League of German Protestant Princes, and with the Sultan of Turkey, against Charles. In 1546 Maurice of Meissen, although a Protestant, saw his best interest in a secret league with Charles against the other Protestant Princes, and proceeded (the Schmalkalden War) to invade the domains of his superior and cousin, the Elector Frederick. The Emperor Charles proved successful in this war, and Maurice was rewarded, at the Capitulation of Wittenberg in 1547, by being made Elector of Saxony in the place of his cousin. Later on, the Elector Maurice found the association with Catholic Charles unpalatable, and joined in leading the other Protestant princes in war upon him, and on the defeat of the Catholic party and the peace of Passau, Maurice became acknowledged as the champion of German national and religious freedom. He was succeeded by his brother Augustus in 1553.

Agricola was much favoured by the Saxon Electors, Maurice and Augustus. He dedicates most of his works to them, and shows much gratitude for many favours conferred upon him. Duke Maurice presented to him a house and plot in Chemnitz, and in a letter dated June 14th, 1543,[9] in connection therewith, says: " . . . . that he may enjoy his life-long a "freehold house unburdened by all burgher rights and other municipal ser-"vice, to be used by him and inhabited as a free dwelling, and that he may "also, for the necessities of his household and of his wife and servants, brew "his own beer free, and that he may likewise purvey for himself and his "household foreign beer and also wine for use, and yet he shall not sell any "such beer. . . . We have taken the said Doctor under our especial "protection and care for our life-long, and he shall not be summoned before "any Court of Justice, but only before us and our Councillor. . . ."

Agricola was made Burgomaster of Chemnitz in 1546. A letter[10] from Fabricius to Meurer, dated May 19th, 1546, says that Agricola had been

[9]Archive 38, Chemnitz Municipal Archives.

[10]Baumgarten-Crusius. *Georgii Fabricii Chemnicensis Epistolae ad W. Meurerum et Alios Aequales*, Leipzig, 1845, p. 26.

made Burgomaster by the command of the Prince. This would be Maurice, and it is all the more a tribute to the high respect with which Agricola was held, for, as said before, he was a consistent Catholic, and Maurice a Protestant Prince. In this same year the Schmalkalden War broke out, and Agricola was called to personal attendance upon the Duke Maurice in a diplomatic and advisory capacity. In 1546 also he was a member of the Diet of Freiberg, and was summoned to Council in Dresden. The next year he continued, by the Duke's command, Burgomaster at Chemnitz, although he seems to have been away upon Ducal matters most of the time. The Duke addresses[11] the Chemnitz Council in March, 1547: "We hereby make known to you "that we are in urgent need of your Burgomaster, Dr. Georgius Agricola, "with us. It is, therefore, our will that you should yield him up and forward "him that he should with the utmost haste set forth to us here near Freiberg." He was sent on various missions from the Duke to the Emperor Charles, to King Ferdinand of Austria, and to other Princes in matters connected with the war—the fact that he was a Catholic probably entering into his appointment to such missions. Chemnitz was occupied by the troops of first one side, then the other, despite the great efforts of Agricola to have his own town specially defended. In April, 1547, the war came to an end in the Battle of Mühlberg, but Agricola was apparently not relieved of his Burgomastership until the succeeding year, for he wrote his friend Wolfgang Meurer, in April, 1548,[12] that he "was now relieved." His public duties did not end, however, for he attended the Diet of Leipzig in 1547 and in 1549, and was at the Diet at Torgau in 1550. In 1551 he was again installed as Burgomaster; and in 1553, for the fourth time, he became head of the Municipality, and during this year had again to attend the Diets at Leipzig and Dresden, representing his city. He apparently now had a short relief from public duties, for it is not until 1555, shortly before his death, that we find him again attending a Diet at Torgau.

Agricola died on November 21st, 1555. A letter[13] from his life-long friend, Fabricius, to Melanchthon, announcing this event, states: "We lost, on "November 21st, that distinguished ornament of our Fatherland, Georgius "Agricola, a man of eminent intellect, of culture and of judgment. He "attained the age of 62. He who since the days of childhood had enjoyed "robust health was carried off by a four-days' fever. He had previously "suffered from no disease except inflammation of the eyes, which he brought "upon himself by untiring study and insatiable reading. . . I know that "you loved the soul of this man, although in many of his opinions, more "especially in religious and spiritual welfare, he differed in many points from "our own. For he despised our Churches, and would not be with us in the "Communion of the Blood of Christ. Therefore, after his death, at the "command of the Prince, which was given to the Church inspectors and "carried out by Tettelbach as a loyal servant, burial was refused him, and not

[11]Hofmann, Op. cit., p. 99.
[12]Weber, *Virorum Clarorum Saeculi* XVI. *et* XVII. *Epistolae Selectae*, Leipzig, 1894, p. 8.
[13]Baumgarten-Crusius. Op. cit., p. 139.

"until the fourth day was he borne away to Zeitz and interred in the Cathedral. " . . . . I have always admired the genius of this man, so distinguished "in our sciences and in the whole realm of Philosophy—yet I wonder at his "religious views, which were compatible with reason, it is true, and were "dazzling, but were by no means compatible with truth. . . . He "would not tolerate with patience that anyone should discuss ecclesiastical "matters with him." This action of the authorities in denying burial to one of their most honored citizens, who had been ever assiduous in furthering the welfare of the community, seems strangely out of joint. Further, the Elector Augustus, although a Protestant Prince, was Agricola's warm friend, as evidenced by his letter of but a few months before (see p. xv). However, Catholics were then few in number at Chemnitz, and the feeling ran high at the time, so possibly the Prince was afraid of public disturbances. Hofmann[14] explains this occurrence in the following words :—" The feelings of Chemnitz "citizens, who were almost exclusively Protestant, must certainly be taken "into account. They may have raised objections to the solemn interment of "a Catholic in the Protestant Cathedral Church of St. Jacob, which had, "perhaps, been demanded by his relatives, and to which, according to the "custom of the time, he would have been entitled as Burgomaster. The "refusal to sanction the interment aroused, more especially in the Catholic "world, a painful sensation."

A brass memorial plate hung in the Cathedral at Zeitz had already disappeared in 1686, nor have the cities of his birth or residence ever shown any appreciation of this man, whose work more deserves their gratitude than does that of the multitude of soldiers whose monuments decorate every village and city square. It is true that in 1822 a marble tablet was placed behind the altar in the Church of St. Jacob in Chemnitz, but even this was removed to the Historical Museum later on.

He left a modest estate, which was the subject of considerable litigation by his descendants, due to the mismanagement of the guardian. Hofmann has succeeded in tracing the descendants for two generations, down to 1609, but the line is finally lost among the multitude of other Agricolas.

To deduce Georgius Agricola's character we need not search beyond the discovery of his steadfast adherence to the religion of his fathers amid the bitter storm of Protestantism around him, and need but to remember at the same time that for twenty-five years he was entrusted with elective positions of an increasingly important character in this same community. No man could have thus held the respect of his countrymen unless he were devoid of bigotry and possessed of the highest sense of integrity, justice, humanity, and patriotism.

[14]Hofmann, Op. cit., p. 123.

AGRICOLA'S INTELLECTUAL ATTAINMENTS AND POSITION IN SCIENCE.

Agricola's education was the most thorough that his times afforded in the classics, philosophy, medicine, and sciences generally. Further, his writings disclose a most exhaustive knowledge not only of an extraordinary range of classical literature, but also of obscure manuscripts buried in the public libraries of Europe. That his general learning was held to be of a high order is amply evidenced from the correspondence of the other scholars of his time—Erasmus, Melanchthon, Meurer, Fabricius, and others.

Our more immediate concern, however, is with the advances which were due to him in the sciences of Geology, Mineralogy, and Mining Engineering. No appreciation of these attainments can be conveyed to the reader unless he has some understanding of the dearth of knowledge in these sciences prior to Agricola's time. We have in Appendix B given a brief review of the literature extant at this period on these subjects. Furthermore, no appreciation of Agricola's contribution to science can be gained without a study of *De Ortu et Causis* and *De Natura Fossilium*, for while *De Re Metallica* is of much more general interest, it contains but incidental reference to Geology and Mineralogy. Apart from the book of Genesis, the only attempts at fundamental explanation of natural phenomena were those of the Greek Philosophers and the Alchemists. Orthodox beliefs Agricola scarcely mentions; with the Alchemists he had no patience. There can be no doubt, however, that his views are greatly coloured by his deep classical learning. He was in fine to a certain distance a follower of Aristotle, Theophrastus, Strato, and other leaders of the Peripatetic school. For that matter, except for the muddy current which the alchemists had introduced into this already troubled stream, the whole thought of the learned world still flowed from the Greeks. Had he not, however, radically departed from the teachings of the Peripatetic school, his work would have been no contribution to the development of science. Certain of their teachings he repudiated with great vigour, and his laboured and detailed arguments in their refutation form the first battle in science over the results of observation *versus* inductive speculation. To use his own words: "Those things which we see with our eyes and understand "by means of our senses are more clearly to be demonstrated than if learned "by means of reasoning."[15] The bigoted scholasticism of his times necessitated as much care and detail in refutation of such deep-rooted beliefs, as would be demanded to-day by an attempt at a refutation of the theory of evolution, and in consequence his works are often but dry reading to any but those interested in the development of fundamental scientific theory.

In giving an appreciation of Agricola's views here and throughout the footnotes, we do not wish to convey to the reader that he was in all things free from error and from the spirit of his times, or that his theories, constructed long before the atomic theory, are of the clear-cut order which that basic hypothesis has rendered possible to later scientific speculation in these branches. His statements are sometimes much confused, but we reiterate that

[15] *De Ortu et Causis*, Book III.

their clarity is as crystal to mud in comparison with those of his predecessors—and of most of his successors for over two hundred years. As an indication of his grasp of some of the wider aspects of geological phenomena we reproduce, in Appendix A, a passage from *De Ortu et Causis*, which we believe to be the first adequate declaration of the part played by erosion in mountain sculpture. But of all of Agricola's theoretical views those are of the greatest interest which relate to the origin of ore deposits, for in these matters he had the greatest opportunities of observation and the most experience. We have on page 108 reproduced and discussed his theory at considerable length, but we may repeat here, that in his propositions as to the circulation of ground waters, that ore channels are a subsequent creation to the contained rocks, and that they were filled by deposition from circulating solutions, he enunciated the foundations of our modern theory, and in so doing took a step in advance greater than that of any single subsequent authority. In his contention that ore channels were created by erosion of subterranean waters he was wrong, except for special cases, and it was not until two centuries later that a further step in advance was taken by the recognition by Van Oppel of the part played by fissuring in these phenomena. Nor was it until about the same time that the filling of ore channels in the main by deposition from solutions was generally accepted. While Werner, two hundred and fifty years after Agricola, is generally revered as the inspirer of the modern theory by those whose reading has taken them no farther back, we have no hesitation in asserting that of the propositions of each author, Agricola's were very much more nearly in accord with modern views. Moreover, the main result of the new ideas brought forward by Werner was to stop the march of progress for half a century, instead of speeding it forward as did those of Agricola.

In mineralogy Agricola made the first attempt at systematic treatment of the subject. His system could not be otherwise than wrongly based, as he could scarcely see forward two or three centuries to the atomic theory and our vast fund of chemical knowledge. However, based as it is upon such properties as solubility and homogeneity, and upon external characteristics such as colour, hardness, &c., it makes a most creditable advance upon Theophrastus, Dioscorides, and Albertus Magnus—his only predecessors. He is the first to assert that bismuth and antimony are true primary metals; and to some sixty actual mineral species described previous to his time he added some twenty more, and laments that there are scores unnamed.

As to Agricola's contribution to the sciences of mining and metallurgy, *De Re Metallica* speaks for itself. While he describes, for the first time, scores of methods and processes, no one would contend that they were discoveries or inventions of his own. They represent the accumulation of generations of experience and knowledge; but by him they were, for the first time, to receive detailed and intelligent exposition. Until Schlüter's work nearly two centuries later, it was not excelled. There is no measure by which we may gauge the value of such a work to the men who followed in this profession during centuries, nor the benefits enjoyed by humanity through them.

That Agricola occupied a very considerable place in the great awakening of learning will be disputed by none except by those who place the development of science in rank far below religion, politics, literature, and art. Of wider importance than the details of his achievements in the mere confines of the particular science to which he applied himself, is the fact that he was the first to found any of the natural sciences upon research and observation, as opposed to previous fruitless speculation. The wider interest of the members of the medical profession in the development of their science than that of geologists in theirs, has led to the aggrandizement of Paracelsus, a contemporary of Agricola, as the first in deductive science. Yet no comparative study of the unparalleled egotistical ravings of this half-genius, half-alchemist, with the modest sober logic and real research and observation of Agricola, can leave a moment's doubt as to the incomparably greater position which should be attributed to the latter as the pioneer in building the foundation of science by deduction from observed phenomena. Science is the base upon which is reared the civilization of to-day, and while we give daily credit to all those who toil in the superstructure, let none forget those men who laid its first foundation stones. One of the greatest of these was Georgius Agricola.

Agricola seems to have been engaged in the preparation of *De Re Metallica* for a period of over twenty years, for we first hear of the book in a letter from Petrus Plateanus, a schoolmaster at Joachimsthal, to the great humanist, Erasmus,[16] in September, 1529. He says: "The scientific world "will be still more indebted to Agricola when he brings to light the books "*De Re Metallica* and other matters which he has on hand." In the dedication of *De Mensuris et Ponderibus* (in 1533) Agricola states that he means to publish twelve books *De Re Metallica,* if he lives. That the appearance of this work was eagerly anticipated is evidenced by a letter from George Fabricius to Valentine Hertel: [17] "With great excitement the books *De Re Metallica* "are being awaited. If he treats the material at hand with his usual zeal, "he will win for himself glory such as no one in any of the fields of literature "has attained for the last thousand years." According to the dedication of *De Veteribus et Novis Metallis,* Agricola in 1546 already looked forward to its early publication. The work was apparently finished in 1550, for the dedication to the Dukes Maurice and August of Saxony is dated in December of that year. The eulogistic poem by his friend, George Fabricius, is dated in 1551.

The publication was apparently long delayed by the preparation of the woodcuts; and, according to Mathesius,[18] many sketches for them were prepared by Basilius Wefring. In the preface of *De Re Metallica,* Agricola does not mention who prepared the sketches, but does say: "I have hired "illustrators to delineate their forms, lest descriptions which are conveyed "by words should either not be understood by men of our own times, or "should cause difficulty to posterity." In 1553 the completed book was sent to Froben for publication, for a letter [19] from Fabricius to Meurer in March, 1553, announces its dispatch to the printer. An interesting letter[20] from the Elector Augustus to Agricola, dated January 18, 1555, reads: "Most learned, dear and faithful subject, whereas you have sent to the Press "a Latin book of which the title is said to be *De Rebus Metallicis,* which has "been praised to us and we should like to know the contents, it is our gracious "command that you should get the book translated when you have the "opportunity into German, and not let it be copied more than once or be "printed, but keep it by you and send us a copy. If you should need a "writer for this purpose, we will provide one. Thus you will fulfil our "gracious behest." The German translation was prepared by Philip Bechius, a Basel University Professor of Medicine and Philosophy. It is a wretched work, by one who knew nothing of the science, and who more especially had no appreciation of the peculiar Latin terms coined by Agricola, most of which

[16] *Briefe an Desiderius Erasmus von Rotterdam.* Published by Joseph Förstemann & Otto Günther. XXVII. *Beiheft zum Zentralblatt für Bibliothekswesen,* Leipzig, 1904, p. 125.

[17] Petrus Albinus, *Meissnische Land und Berg Chronica,* Dresden, 1590, p. 353.

[18] This statement is contained under "1556" in a sort of chronicle bound up with Mathesius's *Sarepta,* Nuremberg, 1562.

[19] Baumgarten-Crusius, p. 85, letter No. 93.

[20] Principal State Archives, Dresden, Cop. 259, folio 102.

he rendered literally. It is a sad commentary on his countrymen that no correct German translation exists. The Italian translation is by Michelangelo Florio, and is by him dedicated to Elizabeth, Queen of England. The title page of the first edition is reproduced later on, and the full titles of other editions are given in the Appendix, together with the author's other works. The following are the short titles of the various editions of *De Re Metallica,* together with the name and place of the publisher :—

Latin Editions.

De Re Metallica, Froben	Basel	Folio	1556.
,, ,, ,, ,, ...	,,	,,	1561.
,, ,, ,, Ludwig König	,,	,,	1621.
,, ,, ,, Emanuel König	,,	,,	1657.

In addition to these, Leupold,[21] Schmid,[22] and others mention an octavo edition, without illustrations, Schweinfurt, 1607. We have not been able to find a copy of this edition, and are not certain of its existence. The same catalogues also mention an octavo edition of *De Re Metallica,* Wittenberg, 1612 or 1614, with notes by Joanne Sigfrido ; but we believe this to be a confusion with Agricola's subsidiary works, which were published at this time and place, with such notes.

German Editions.

Vom Bergkwerck, Froben, Folio, 1557.
Bergwerck Buch, Sigmundi Feyrabendt, Frankfort-on-Main, folio, 1580.
,, ,, Ludwig König, Basel, folio, 1621.

There are other editions than these, mentioned by bibliographers, but we have been unable to confirm them in any library. The most reliable of such bibliographies, that of John Ferguson,[23] gives in addition to the above ; *Bergwerkbuch,* Basel, 1657, folio, and Schweinfurt, 1687, octavo.

Italian Edition.

L'Arte de Metalli, Froben, Basel, folio, 1563.

Other Languages.

So far as we know, *De Re Metallica* was never actually published in other than Latin, German, and Italian. However, a portion of the accounts of the firm of Froben were published in 1881[24], and therein is an entry under March, 1560, of a sum to one Leodigaris Grymaldo for some other work, and also for "correction of Agricola's *De Re Metallica* in French." This may of course, be an error for the Italian edition, which appeared a little later. There is also mention[25] that a manuscript of *De Re Metallica* in Spanish was

[21]Jacob Leupold, *Prodromus Bibliothecae Metallicae,* 1732, p. 11.
[22]F. A. Schmid, *Georg Agrikola's Bermannus,* Freiberg, 1806, p. 34.
[23]*Bibliotheca Chemica,* Glasgow, 1906, p. 10.
[24]*Rechnungsbuch der Froben und Episcopius Buchdrucker und Buchhändler zu Basel,* 1557-1564, published by R. Wackernagle, Basel, 1881, p. 20.
[25]*Colecion del Sr Monoz* t. 93, fol. 255 *En la Acad. de la Hist.* Madrid.

seen in the library of the town of Bejar. An interesting note appears in the glossary given by Sir John Pettus in his translation of Lazarus Erckern's work on assaying. He says[26] "but I cannot enlarge my observations upon any more words, because the printer calls for what I did write of a metallick dictionary, after I first proposed the printing of Erckern, but intending within the compass of a year to publish Georgius Agricola, *De Re Metallica* (being fully translated) in English, and also to add a dictionary to it, I shall reserve my remaining essays (if what I have done hitherto be approved) till then, and so I proceed in the dictionary." The translation was never published and extensive inquiry in various libraries and among the family of Pettus has failed to yield any trace of the manuscript.

[26]Sir John Pettus, *Fleta Minor*, The Laws of Art and Nature, &c., London, 1636, p. 121.

GEORGII AGRICOLAE

DE RE METALLICA LIBRI XII. QVIbus Officia, Instrumenta, Machinæ, ac omnia deniq; ad Metallicam spectantia, non modo luculentissimè describuntur, sed & per effigies, suis locis insertas, adiunctis Latinis, Germanicisq; appellationibus ita ob oculos ponuntur, ut clarius tradi non possint.

EIVSDEM

DE ANIMANTIBVS SVBTERRANEIS Liber, ab Autore recognitus: cum Indicibus diuersis, quicquid in opere tractatum est, pulchrè demonstrantibus.

BASILEAE M. D. LVI.

Cum Priuilegio Imperatoris in annos V.
& Galliarum Regis ad Sexennium.

GEORGIVS FABRICIVS IN LIbros Metallicos GEORGII AGRICOLAE philosophi præstantissimi.

AD LECTOREM.

Si iuuat ignita cognoscere fronte Chimæram,
  Semicanem nympham, semibouemq́ uirum:
Si centum capitum Titanem, totq́ ferentem
  Sublimem manibus tela cruenta Gygen:
Si iuuat Ætneum penetrare Cyclopis in antrum,
  Atque alios, Vates quos peperere, metus:
Nunc placeat mecum doctos euoluere libros,
  Ingenium AGRICOLAE quos dedit acre tibi.
Non hic uana tenet suspensam fabula mentem:
  Sed precium, utilitas multa, legentis erit.
Quidquid terra sinu, gremioq́ recondidit imo,
  Omne tibi multis eruit antè libris:
Siue fluens superas ultro nitatur in oras,
  Inueniat facilem seu magis arte uiam.
Perpetui proprijs manant de fontibus amnes,
  Est grauis Albuneæ sponte Mephitis odor.
Lethales sunt sponte scrobes Dicæarchidis oræ,
  Et micat è media conditus ignis humo.
Plana Nariscorum cùm tellus arsit in agro,
  Ter curua nondum falce resecta Ceres.
Nec dedit hoc damnum pastor, nec Iuppiter igne:
  Vulcani per se ruperat ira solum.
Terrifico aura foras erumpens, incita motu,
  Sæpe facit montes, antè ubi plana uia est.
Hæc abstrusa cauis, imoq́ incognita fundo,
  Cognita natura sæpe fuere duce.
Arte hominum, in lucem ueniunt quoq́ multa, manuq́
  Terræ multiplices effodiuntur opes.
Lydia sic nitrum profert, Islandia sulfur,
  Ac modò Tyrrhenus mittit alumen ager.
Succina, quâ trifido subit æquor Vistula cornu,
  Piscantur Codano corpora serua sinu.
Quid memorem regum preciosa insignia gemmas,
  Marmoraq́ excelsis structa sub astra iugis?
Nil lapides, nil saxa moror: sunt pulchra metalla,
  Crœse tuis opibus clara, Mydaq́ tuis,
Quæq́ acer Macedo terra Creneide fodit,
  Nomine permutans nomina prisca suo.
At nunc non ullis cedit GERMANIA terris,

Terra ferax hominum, terráq; diues opum.
Hic auri in uenis locupletibus aura refulget,
Non alio meſſis carior ulla loco.
Auricomum extulerit felix Campania ramum,
Nec fructu nobis deficiente cadit.
Eruit argenti ſolidas hoc tempore maſſas
Foſſor, de proprijs armáq; miles agris.
Ignotum Graijs eſt Heſperijsq; metallum,
Quod Biſemutum lingua paterna uocat.
Candidius nigro, ſed plumbo nigrius albo,
Noſtra quoq; hoc uena diuite fundit humus.
Funditur in tormenta, corus cum imitantia fulmen,
Æs, inq; hoſtiles ferrea maſſa domos.
Scribuntur plumbo libri: quis credidit antè
Quàm mirandam artem Teutonis ora dedit?
Nec tamen hoc alijs, aut illa petuntur ab oris,
Eruta Germano cuncta metalla ſolo.
Sed quid ego hæc repeto, monumentis tradita claris
AGRICOLAE, quæ nunc docta per ora uolant?
Hic cauſſis ortus, & formas uiribus addit,
Et quærenda quibus ſint meliora locis.
Quæ ſi mente prius legiſti candidus æqua:
Da reliquis quoq; nunc tempora pauca libris.
Vtilitas ſequitur cultorem: crede, uoluptas
Non iucunda minor, rara legentis, erit.
Iudicióq; prius ne quis malè damnet iniquo,
Quæ ſunt auctoris munera mira Dei:
Eripit ipſe ſuis primùm tela hoſtibus, inq;
Mittentis torquet ſpicula rapta caput.
Fertur equo latro, uehitur pirata triremi:
Ergo necandus equus, nec fabricanda ratis?
Viſceribus terræ lateant abſtruſa metalla,
Vti opibus neſcit quòd mala turba ſuis?
Quiſquis es, aut doctis pareto monentibus, aut te
Inter habere bonos ne fateare locum.
Se non in prærupta metallicus abijcit audax,
Vt quondam immiſſo Curtius acer equo:
Sed prius ediſcit, quæ ſunt noſcenda perito,
Quodq; facit, multa doctus ab arte facit.
Vtq; gubernator ſeruat cum ſidere uentos:
Sic minimè dubijs utitur ille notis.
Iaſides nauim, currus regit arte Metiſcus:
Foſſor opus peragit nec minus arte ſuum.
Indagat uenæ ſpacium, numerúmq;, modúmq;,
Siue obliqua ſuum, rectáue tendat iter.

Paſtor

Paſtor ut explorat quæ terra ſit apta colenti,
Quæ bene lanigeras, quæ malè paſcat oues.
En terræ intentus, quid uincula linea tendit?
Fungitur officio iam Ptolemæe tuo.
Vtq; ſuæ inuenit menſuram iuráq; uenæ,
In uarios operas diuidit inde uiros.
Iamq; aggreſſus opus, uiden' ut mouet omne quod obſtat,
Aſſidua ut uerſat ſtrenuus arma manu?
Ne tibi ſurdeſcant ferri tinnitibus aures,
Ad grauiora ideo conſpicienda ueni.
Inſtruit ecce ſuis nunc artibus ille minores:
Sedulitas nulli non operoſa loco.
Metiri docet hic uenæ ſpaciumq; modumq;,
Vtq; regat poſitis finibus arua lapis,
Ne quis transmiſſo uiolentus limite pergens,
Non ſibi conceſſas, in ſua uertat, opes.
Hic docet inſtrumenta, quibus Plutonia regna
Tutus adit, ſaxi permeat atq; uias.
Quanta (uides) ſolidas expugnet machina terras:
Machina non ullo tempore uiſa prius.
Cede nouis, nulla non inclyta laude uetuſtas,
Poſteritas meritis eſt quoq; grata tuis.
Tum quia Germano ſunt hæc inuenta ſub axe,
Si quis es, inuidiæ contrahe uela tuæ.
Auſonis ora tumet bellis, terra Attica cultu,
Germanum infractus tollit ad aſtra labor.
Nec tamen ingenio ſolet infeliciter uti,
Mite gerat Phœbi, ſeu graue Martis opus.
Tempus adeſt, ſtructis uenarum montibus, igne
Explorare, uſum quem ſibi uena ferat.
Non labor ingenio caret hic, non copia fructu,
Eſt adaperta bonæ prima feneſtra ſpei.
Ergo inſtat porrò grauiores ferre labores,
Intentas operi nec remouere manus.
Vrere ſiue locus poſcat, ſeu tundere uenas,
Siue lauare lacu præter euntis aquæ.
Seu flammis iterum modicis torrere neceſſe eſt,
Excoquere aut faſtis ignibus omne malum,
Cùm fluit æs riuis, auri argentiq; metallum,
Spes animo foſſor uix capit ipſe ſuas.
Argentum cupidus fuluo ſecernit ab auro,
Et plumbi lentam demit utriq; moram.
Separat argentum, lucri ſtudioſus, ab ære,
Seruatis, linquens deteriora, bonis.

Quæ

Quæ si cuncta uelim tenui percurrere uersu,
Ante alium reuehat Memnonis orta diem.
Postremus labor est, concretos discere succos,
Quos fert innumeris Teutona terra locis.
Quo sal, quo nitrum, quo pacto fiat alumen,
Vsibus artificis cùm parat illa manus:
Necnon chalcantum, sulfur, fluidumq́ue bitumen,
Massaq́ue quo uitri lenta dolanda modo.
Suscipit hæc hominum mirandos cura labores,
Pauperiem usq́ue adeo ferre famemq́ue graue est,
Tantus amor uictum paruis extundere natis,
Et patriæ ciuem non dare uelle malum.
Nec manet in terræ fossoris mersa latebris
Mens, sed fert domino uota precesq́ue Deo.
Munificæ expectat, spe plenus, munera dextræ,
Extollens animum lætus ad astra suum.
Diuitias CHRISTVS dat noticiamq́ue fruendi,
Cui memori grates pectore semper agit.
Hoc quoque laudati quondam fecere Philippi,
Qui uirtutis habent cum pietate decus.
Huc oculos, huc flecte animum, suauissime Lector,
Auctoremq́ue pia noscito mente Deum.
AGRICOLAE hinc optans operoso fausta labori,
Laudibus eximij candidus esto uiri.
Ille suum extollit patriæ cum nomine nomen,
Et uir in ore frequens posteritatis erit.
Cuncta cadunt letho, studij monumenta uigebunt,
Purpurei donec lumina solis erunt.

Misenæ M. D. LI.
è ludo illustri.

For completeness' sake we reproduce in the original Latin the laudation of Agricola by his friend, Georgius Fabricius, a leading scholar of his time. It has but little intrinsic value for it is not poetry of a very high order, and to make it acceptable English would require certain improvements, for which only poets have license. A "free" translation of the last few lines indicates its complimentary character :—

"He doth raise his country's fame with his own
"And in the mouths of nations yet unborn
"His praises shall be sung ; Death comes to all
"But great achievements raise a monument
"Which shall endure until the sun grows cold."

TO THE MOST ILLUSTRIOUS AND MOST MIGHTY DUKES OF

Saxony, Landgraves of Thuringia, Margraves of Meissen, Imperial Overlords of Saxony, Burgraves of Altenberg and Magdeburg, Counts of Brena, Lords of Pleissnerland, To MAURICE Grand Marshall and Elector of the Holy Roman Empire and to his brother AUGUSTUS,[1]

GEORGE AGRICOLA S. D.

OST illustrious Princes, often have I considered the metallic arts as a whole, as Moderatus Columella[2] considered the agricultural arts, just as if I had been considering the whole of the human body; and when I had perceived the various parts of the subject, like so many members of the body, I became afraid that I might die before I should understand its full extent, much less before I could immortalise it in writing. This book itself indicates the length and breadth of the subject, and the number and importance of the sciences of which at least some little knowledge is necessary to miners. Indeed, the subject of mining is a very extensive one, and one very difficult to explain; no part of it is fully dealt with by the Greek and Latin authors whose works survive; and since the art is one of the most ancient, the most necessary and the most profitable to mankind, I considered that I ought not to neglect it. Without doubt, none of the arts is older than agriculture, but that of the metals is not less ancient; in fact they are at least equal and coeval, for no mortal man ever tilled a field without implements. In truth, in all the works of agriculture, as in the other arts, implements are used which are made from metals, or which could not be made without the use of metals; for this reason the metals are of the greatest necessity to man. When an art is so poor that it lacks metals, it is not of much importance, for nothing is made without tools. Besides, of all ways whereby great wealth is acquired by good and honest means, none is more advantageous than mining; for although from fields which are well tilled (not to mention other things) we derive rich yields, yet we obtain richer products from mines; in fact, one mine is often much more beneficial to us than many fields. For this reason we learn from the history of nearly all ages that very many men have been made rich by the

[1]For Agricola's relations with these princes see p. ix.

[2]Lucius Junius Moderatus Columella was a Roman, a native of Cadiz, and lived during the 1st Century. He was the author of *De Re Rustica* in 12 books. It was first printed in 1472, and some fifteen or sixteen editions had been printed before Agricola's death.

mines, and the fortunes of many kings have been much amplified thereby. But I will not now speak more of these matters, because I have dealt with these subjects partly in the first book of this work, and partly in the other work entitled *De Veteribus et Novis Metallis*, where I have refuted the charges which have been made against metals and against miners. Now, though the art of husbandry, which I willingly rank with the art of mining, appears to be divided into many branches, yet it is not separated into so many as this art of ours, nor can I teach the principles of this as easily as Columella did of that. He had at hand many writers upon husbandry whom he could follow,—in fact, there are more than fifty Greek authors whom Marcus Varro enumerates, and more than ten Latin ones, whom Columella himself mentions. I have only one whom I can follow; that is C. Plinius Secundus,[3] and he expounds only a very few methods of digging ores and of making metals. Far from the whole of the art having been treated by any one writer, those who have written occasionally on any one or another of its branches have not even dealt completely with a single one of them. Moreover, there is a great scarcity even of these, since alone of all the Greeks, Strato of Lampsacus,[4] the successor of Theophrastus,[5] wrote a book on the subject, *De Machinis Metallicis*; except, perhaps a work by the poet Philo, a small part of which embraced to some degree the occupation of mining.[6] Pherecrates seems to have introduced into his comedy, which was similar in title, miners as slaves or as persons condemned to serve in the mines. Of the Latin writers, Pliny, as I have already said, has described a few methods of working. Also among the authors I must include the modern writers, whosoever they are, for no one should escape just condemnation who fails to award due recognition to persons whose writings he uses, even very slightly. Two books have been written in our tongue; the one on the assaying of mineral substances and metals, somewhat confused, whose author is unknown[7]; the other "On Veins," of which Pandulfus Anglus [8] is also said to have written, although the German book was written by Calbus of Freiberg, a well-known doctor; but neither of them accomplished the task

[3]We give a short review of Pliny's *Naturalis Historia* in the Appendix B.

[4]This work is not extant, as Agricola duly notes later on. Strato succeeded Theophrastus as president of the Lyceum, 288 B.C.

[5]For note on Theophrastus see Appendix B.

[6]It appears that the poet Philo did write a work on mining which is not extant. So far as we know the only reference to this work is in Athenæus' (200 A.D.) *Deipnosophistae*. The passage as it appears in C. D. Yonge's Translation (Bohn's Library, London, 1854, Vol. II, Book VII, p. 506) is: "And there is a similar fish produced in the Red Sea which "is called Stromateus; it has gold-coloured lines running along the whole of his body, as "Philo tells us in his book on Mines." There is a fragment of a poem of Pherecrates, entitled "Miners," but it seems to have little to do with mining.

[7]The title given by Agricola *De Materiae Metallicae et Metallorum Experimento* is difficult to identify. It seems likely to be the little *Probier Büchlein*, numbers of which were published in German in the first half of the 16th Century. We discuss this work at some length in the Appendix B on Ancient Authors.

[8]Pandulfus, "the Englishman," is mentioned by various 15th and 16th Century writers, and in the preface of Mathias Farinator's *Liber Moralitatum . . . Rerum Naturalium*, etc., printed in Augsburg, 1477, there is a list of books among which appears a reference to a work by Pandulfus on veins and minerals. We have not been able to find the book.

he had begun.[9] Recently Vannucci Biringuccio, of Sienna, a wise man experienced in many matters, wrote in vernacular Italian on the subject of the melting, separating, and alloying of metals.[10] He touched briefly on the methods of smelting certain ores, and explained more fully the methods of making certain juices; by reading his directions, I have refreshed my memory of those things which I myself saw in Italy; as for many matters on which I write, he did not touch upon them at all, or touched but lightly. This book was given me by Franciscus Badoarius, a Patrician of Venice, and a man of wisdom and of repute; this he had promised that he would do, when in the previous year he was at Marienberg, having been sent by the Venetians as an Ambassador to King Ferdinand. Beyond these books I do not find any writings on the metallic arts. For that reason, even if the book of Strato existed, from all these sources not one-half of the whole body of the science of mining could be pieced together.

Seeing that there have been so few who have written on the subject of the metals, it appears to me all the more wonderful that so many alchemists have arisen who would compound metals artificially, and who would change one into another. Hermolaus Barbarus,[11] a man of high rank and station, and distinguished in all kinds of learning, has mentioned the names of many in his writings; and I will proffer more, but only famous ones, for I will limit myself to a few. Thus Osthanes has written on χυμευτικά; and there are Hermes; Chanes; Zosimus, the Alexandrian, to his sister Theosebia; Olympiodorus, also an Alexandrian; Agathodæmon; Democritus, not the one of Abdera, but some other whom I know not; Orus Chrysorichites, Pebichius, Comerius, Joannes, Apulejus, Petasius, Pelagius, Africanus, Theophilus, Synesius, Stephanus to Heracleus Cæsar, Heliodorus to Theodosius, Geber, Callides Rachaidibus, Veradianus, Rodianus, Canides, Merlin, Raymond Lully, Arnold de Villa Nova, and Augustinus Pantheus of Venice; and three women, Cleopatra, the maiden Taphnutia, and Maria the Jewess.[12] All these alchemists employ obscure language, and Johanes Aurelius Augurellus of Rimini, alone has used the language of poetry. There are many other books on

[9]Jacobi (*Der Mineralog Georgius Agricola*, Zwickau, 1881, p. 47) says: "Calbus "Freibergius, so called by Agricola himself, is certainly no other than the Freiberg Doctor "Rühlein von Kalbe; he was, according to Möller, a doctor and burgomaster at Freiberg "at the end of the 15th and the beginning of the 16th Centuries. . . . The chronicler "describes him as a fine mathematician, who helped to survey and design the mining towns "of Annaberg in 1497 and Marienberg in 1521." We would call attention to the statement of Calbus' views, quoted at the end of Book III, *De Re Metallica* (p. 75), which are astonishingly similar to statements in the *Nützlich Bergbüchlin*, and leave little doubt that this "Calbus" was the author of that anonymous book on veins. For further discussion see Appendix B.

[10]For discussion of Biringuccio see Appendix B. The proper title is *De La Pirotechnia* (Venice, 1540).

[11]Hermolaus Barbarus, according to Watt (*Bibliotheca Britannica*, London, 1824), was a lecturer on Philosophy in Padua. He was born in 1454, died in 1493, and was the author of a number of works on medicine, natural history, etc., with commentaries on the older authors.

[12]The debt which humanity does owe to these self-styled philosophers must not be overlooked, for the science of Chemistry comes from three sources—Alchemy, Medicine and Metallurgy. However polluted the former of these may be, still the vast advance which it made by the discovery of the principal acids, alkalis, and the more common of their salts, should be constantly recognized. It is obviously impossible, within the space of a footnote, to

this subject, but all are difficult to follow, because the writers upon these things use strange names, which do not properly belong to the metals, and because some of them employ now one name and now another, invented by themselves, though the thing itself changes not. These masters teach their disciples that the base metals, when smelted, are broken up ; also they teach the methods by which they reduce them to the primary parts and remove whatever is superfluous in them, and by supplying what is wanted make out of them the precious metals—that is, gold and silver,—all of which they carry out in a crucible. Whether they can do these things or not I cannot decide ; but, seeing that so many writers assure us with all earnestness that they have reached that goal for which they aimed, it would seem that faith might be placed in them ; yet also seeing that we do not read of any of them ever having become rich by this art, nor do we now see them growing rich, although so many nations everywhere have produced, and are producing, alchemists, and all of them are straining every nerve night and day to the end that they may heap a great quantity of gold and silver, I should say the matter is dubious. But although it may be due to the carelessness of the writers that they have not transmitted to us the names of the masters who acquired great wealth through this occupation, certainly it is clear that their disciples either do not understand their precepts or, if they do understand them, do not follow them ; for if they do comprehend them, seeing that these disciples have been and are so numerous, they would have by to-day filled

give anything but the most casual notes as to the personages here mentioned and their writings. Aside from the classics and religious works, the libraries of the Middle Ages teemed with more material on Alchemy than on any other one subject, and since that date a never-ending stream of historical, critical, and discursive volumes and tracts devoted to the old Alchemists and their writings has been poured upon the world. A collection recently sold in London, relating to Paracelsus alone, embraced over seven hundred volumes.

Of many of the Alchemists mentioned by Agricola little is really known, and no two critics agree as to the commonest details regarding many of them ; in fact, an endless confusion springs from the negligent habit of the lesser Alchemists of attributing the authorship of their writings to more esteemed members of their own ilk, such as Hermes, Osthanes, etc., not to mention the palpable spuriousness of works under the names of the real philosophers, such as Aristotle, Plato, or Moses, and even of Jesus Christ. Knowledge of many of the authors mentioned by Agricola does not extend beyond the fact that the names mentioned are appended to various writings, in some instances to MSS yet unpublished. They may have been actual persons, or they may not. Agricola undoubtedly had perused such manuscripts and books in some leading library, as the quotation from Boerhaave given later shows. Shaw (A New Method of Chemistry, etc., London, 1753. Vol. I, p. 25) considers that the large number of such manuscripts in the European libraries at this time were composed or transcribed by monks and others living in Constantinople, Alexandria, and Athens, who fled westward before the Turkish invasion, bringing their works with them.

For purposes of this summary we group the names mentioned by Agricola, the first class being of those who are known only as names appended to MSS or not identifiable at all. Possibly a more devoted student of the history of Alchemy would assign fewer names to this department of oblivion. They are Maria the Jewess, Orus Chrysorichites, Chanes, Petasius, Pebichius, Theophilus, Callides, Veradianus, Rodianus, Canides, the maiden Taphnutia, Johannes, Augustinus, and Africanus. The last three are names so common as not to be possible of identification without more particulars, though Johannes may be the Johannes Rupeseissa (1375), an alchemist of some note. Many of these names can be found among the Bishops and Prelates of the early Christian Church, but we doubt if their owners would ever be identified with such indiscretions as open, avowed alchemy. The Theophilus mentioned might be the metal-working monk of the 12th Century, who is further discussed in Appendix B on Ancient Authors.

In the next group fall certain names such as Osthanes, Hermes, Zosimus, Agathodaemon, and Democritus, which have been the watchwords of authority to Alchemists of all ages. These certainly possessed the great secrets, either the philosopher's stone or the elixir.

whole towns with gold and silver. Even their books proclaim their vanity, for they inscribe in them the names of Plato and Aristotle and other philosophers, in order that such high-sounding inscriptions may impose upon simple people and pass for learning. There is another class of alchemists who do not change the substance of base metals, but colour them to represent gold or silver, so that they appear to be that which they are not, and when this appearance is taken from them by the fire, as if it were a garment foreign to them, they return to their own character. These alchemists, since they deceive people, are not only held in the greatest odium, but their frauds are a capital offence. No less a fraud, warranting capital punishment, is committed by a third sort of alchemists ; these throw into a crucible a small piece of gold or silver hidden in a coal, and after mixing therewith fluxes which have the power of extracting it, pretend to be making gold from orpiment, or silver from tin and like substances. But concerning the art of alchemy, if it be an art, I will speak further elsewhere. I will now return to the art of mining.

Since no authors have written of this art in its entirety, and since foreign nations and races do not understand our tongue, and, if they did understand it, would be able to learn only a small part of the art through the works of those authors whom we do possess, I have written these twelve books *De Re Metallica*. Of these, the first book contains the arguments which may be used against this art, and against metals and the mines, and what can be said in their favour. The second book describes the miner, and branches into

Hermes Trismegistos was a legendary Egyptian personage supposed to have flourished before 1,500 B.C., and by some considered to be a corruption of the god Thoth. He is supposed to have written a number of works, but those extant have been demonstrated to date not prior to the second Century ; he is referred to by the later Greek Alchemists, and was believed to have possessed the secret of transmutation. Osthanes was also a very shadowy personage, and was considered by some Alchemists to have been an Egyptian prior to Hermes, by others to have been the teacher of Zoroaster. Pliny mentions a magician of this name who accompanied Xerxes' army. Later there are many others of this name, and the most probable explanation is that this was a favourite pseudonym for ancient magicians ; there is a very old work, of no great interest, in MSS in Latin and Greek, in the Munich, Gotha, Vienna, and other libraries, by one of this name. Agathodaemon was still another shadowy character referred to by the older Alchemists. There are MSS in the Florence, Paris, Escurial, and Munich libraries bearing his name, but nothing tangible is known as to whether he was an actual man or if these writings are not of a much later period than claimed.

To the next group belong the Greek Alchemists, who flourished during the rise and decline of Alexandria, from 200 B.C. to 700 A.D., and we give them in order of their dates. Comerius was considered by his later fellow professionals to have been the teacher of the art to Cleopatra (1st Century B.C.), and a MSS with a title to that effect exists in the Bibliotheque Nationale at Paris. The celebrated Cleopatra seems to have stood very high in the estimation of the Alchemists ; perhaps her doubtful character found a response among them ; there are various works extant in MSS attributed to her, but nothing can be known as to their authenticity. Lucius Apulejus or Apuleius was born in Numidia about the 2nd Century ; he was a Roman Platonic Philosopher, and was the author of a romance, "The Metamorphosis, or the Golden Ass." Synesius was a Greek, but of unknown period ; there is a MSS treatise on the Philosopher's Stone in the library at Leyden under his name, and various printed works are attributed to him ; he mentions "water of saltpetre," and has, therefore, been hazarded to be the earliest recorder of nitric acid. The work here referred to as "Heliodorus to Theodosius" was probably the MSS in the Libraries at Paris, Vienna, Munich, etc., under the title of "Heliodorus the Philosopher's Poem to the Emperor Theodosius the Great on the Mystic Art of the Philosophers, etc." His period would, therefore, be about the 4th Century. The Alexandrian Zosimus is more generally known as Zosimus the Panopolite, from Panopolis, an ancient town on the Nile ; he flourished in the 5th Century, and belonged to the Alexandrian School of Alchemists ; he should not be confused with the Roman historian of the same name and period. The following statement is by Boerhaave (*Elementa Chemiae*, Paris, 1724, Chap. I.) :—"The name Chemistry written in Greek, or *Chemia*, is so ancient

a discourse on the finding of veins. The third book deals with veins and stringers, and seams in the rocks. The fourth book explains the method of delimiting veins, and also describes the functions of the mining officials. The fifth book describes the digging of ore and the surveyor's art. The sixth book describes the miners' tools and machines. The seventh book is on the assaying of ore. The eighth book lays down the rules for the work of roasting, crushing, and washing the ore. The ninth book explains the methods of smelting ores. The tenth book instructs those who are studious of the metallic arts in the work of separating silver from gold, and lead from gold and silver. The eleventh book shows the way of separating silver from copper. The twelfth book gives us rules for manufacturing salt, soda, alum, vitriol, sulphur, bitumen, and glass.

Although I have not fulfilled the task which I have undertaken, on account of the great magnitude of the subject, I have, at all events, endeavoured to fulfil it, for I have devoted much labour and care, and have even gone to some expense upon it; for with regard to the veins, tools, vessels, sluices, machines, and furnaces, I have not only described them, but have also hired illustrators to delineate their forms, lest descriptions which are conveyed by words should either not be understood by men of our own times, or should cause difficulty to posterity, in the same way as to us difficulty is often caused by many names which the Ancients (because such words were familiar to all of them) have handed down to us without any explanation.

I have omitted all those things which I have not myself seen, or have

" as perhaps to have been used in the antediluvian age. Of this opinion was Zosimus the " Panopolite, whose Greek writings, though known as long as before the year 1550 to George " Agricola, and afterwards perused . . . . by Jas. Scaliger and Olaus Borrichius, " still remain unpublished in the King of France's library. In one of these, entitled, 'The " Instruction of Zosimus the Panopolite and Philosopher, out of those written to Theosebeia, " etc. . . .' Olympiodorus was an Alexandrian of the 5th Century, whose writings were largely commentaries on Plato and Aristotle; he is sometimes accredited with being the first to describe white arsenic (arsenical oxide). The full title of the work styled "Stephanus to Heracleus Caesar," as published in Latin at Padua in 1573, was " Stephan of Alexandria, the " Universal Philosopher and Master, his nine processes on the great art of making gold and " silver, addressed to the Emperor Heraclius." He, therefore, if authentic, dates in the 7th Century.

To the next class belong those of the Middle Ages, which we give in order of date. The works attributed to Geber play such an important part in the history of Chemistry and Metallurgy that we discuss his book at length in Appendix B. Late criticism indicates that this work was not the production of an 8th Century Arab, but a compilation of some Latin scholar of the 12th or 13th Centuries. Arnold de Villa Nova, born about 1240, died in 1313, was celebrated as a physician, philosopher, and chemist; his first works were published in Lyons in 1504; many of them have apparently never been printed, for references may be found to some 18 different works. Raymond Lully, a Spaniard, born in 1235, who was a disciple of Arnold de Villa Nova, was stoned to death in Africa in 1315. There are extant over 100 works attributed to this author, although again the habit of disciples of writing under the master's name may be responsible for most of these. John Aurelio Augurello was an Italian Classicist, born in Rimini about 1453. The work referred to, *Chrysopoeia et Gerontica* is a poem on the art of making gold, etc., published in Venice, 1515, and re-published frequently thereafter; it is much quoted by Alchemists. With regard to Merlin, as satisfactory an account as any of this truly English magician may be found in Mark Twain's " Yankee at the Court of King Arthur." It is of some interest to note that Agricola omits from his list Avicenna (980–1037 A.D.), Roger Bacon (1214–1294), Albertus Magnus (1193–1280), Basil Valentine (end 15th century ?), and Paracelsus, a contemporary of his own. In *De Ortu et Causis* he expends much thought on refutation of theories advanced by Avicenna and Albertus, but of the others we have found no mention, although their work is, from a chemical point of view, of considerable importance.

not read or heard of from persons upon whom I can rely. That which I have neither seen, nor carefully considered after reading or hearing of, I have not written about. The same rule must be understood with regard to all my instruction, whether I enjoin things which ought to be done, or describe things which are usual, or condemn things which are done. Since the art of mining does not lend itself to elegant language, these books of mine are correspondingly lacking in refinement of style. The things dealt with in this art of metals sometimes lack names, either because they are new, or because, even if they are old, the record of the names by which they were formerly known has been lost. For this reason I have been forced by a necessity, for which I must be pardoned, to describe some of them by a number of words combined, and to distinguish others by new names,—to which latter class belong *Ingestor*, *Discretor*, *Lotor*, and *Excoctor*.[13] Other things, again, I have alluded to by old names, such as the *Cisium*; for when Nonius Marcellus wrote,[14] this was the name of a two-wheeled vehicle, but I have adopted it for a small vehicle which has only one wheel; and if anyone does not approve of these names, let him either find more appropriate ones for these things, or discover the words used in the writings of the Ancients.

These books, most illustrious Princes, are dedicated to you for many reasons, and, above all others, because metals have proved of the greatest value to you; for though your ancestors drew rich profits from the revenues of their vast and wealthy territories, and likewise from the taxes which were paid by the foreigners by way of toll and by the natives by way of tithes, yet they drew far richer profits from the mines. Because of the mines not a few towns have risen into eminence, such as Freiberg, Annaberg, Marienberg, Schneeberg, Geyer, and Altenberg, not to mention others. Nay, if I understand anything, greater wealth now lies hidden beneath the ground in the mountainous parts of your territory than is visible and apparent above ground. Farewell.

Chemnitz, Saxony,

December First, 1550.

[13]*Ingestor*,—Carrier; *Discretor*,—Sorter; *Lotor*,—Washer; *Excoctor*,—Smelter.

[14]Nonius Marcellus was a Roman grammarian of the 4th Century B.C. His extant treatise is entitled, *De Compendiosa Doctrina per Litteras ad Filium.*

BOOK I.

ANY persons hold the opinion that the metal industries are fortuitous and that the occupation is one of sordid toil, and altogether a kind of business requiring not so much skill as labour. But as for myself, when I reflect carefully upon its special points one by one, it appears to be far otherwise. For a miner must have the greatest skill in his work, that he may know first of all what mountain or hill, what valley or plain, can be prospected most profitably, or what he should leave alone; moreover, he must understand the veins, stringers[1] and seams in the rocks[2]. Then he must be thoroughly familiar with the many and varied species of earths, juices[3], gems, stones, marbles, rocks, metals, and compounds[4]. He must also have a

[1]*Fibrae*—" fibres." See Note 6, p. 70.

[2]*Commissurae saxorum*—" rock joints," " seams," or " cracks." Agricola and all of the old authors laid a wholly unwarranted geologic value on these phenomena. See description and footnotes, Book III., pages 43 and 72.

[3]*Succi*—" juice," or *succi concreti*—" solidified juice." Ger. Trans., ***saffte.*** The old English translators and mineralogists often use the word juices in the same sense, and we have adopted it. The words " solutions " and " salts " convey a chemical significance not warranted by the state of knowledge in Agricola's time. Instances of the former use of this word may be seen in Barba's "First Book of the Art of Metals," (Trans. Earl Sandwich, London, 1674, p. 2, etc.,) and in Pryce's *Mineralogia Cornubiensis* (London, 1778, p. 25, 32).

[4]In order that the reader should be able to grasp the author's point of view as to his divisions of the Mineral Kingdom, we introduce here his own statement from *De Natura Fossilium*, (p. 180). It is also desirable to read the footnote on his theory of ore-deposits on pages 43 to 53, and the review of *De Natura Fossilium* given in the Appendix.

" The subterranean inanimate bodies are divided into two classes, one of which, because
" it is a fluid or an exhalation, is called by those names, and the other class is called the
" minerals. Mineral bodies are solidified from particles of the same substance, such as pure
" gold, each particle of which is gold, or they are of different substances such as lumps which
" consist of earth, stone, and metal; these latter may be separated into earth, stone and
" metal, and therefore the first is not a mixture while the last is called a mixture. The first
" are again divided into simple and compound minerals. The simple minerals are of four
" classes, namely earths, solidified juices, stones and metals, while the mineral compounds
" are of many sorts, as I shall explain later."

" Earth is a simple mineral body which may be kneaded in the hands when moistened,
" or from which lute is made when it has been wetted. Earth, properly so called, is found
" enclosed in veins or veinlets, or frequently on the surface in fields and meadows. This
" definition is a general one. The harder earth, although moistened by water, does not at
" once become lute, but does turn into lute if it remains in water for some time. There are
" many species of earths, some of which have names but others are unnamed."

" Solidified juices are dry and somewhat hard (*subdurus*) mineral bodies which when
" moistened with water do not soften but liquefy instead; or if they do soften, they differ
" greatly from the earths by their unctuousness (*pingue*) or by the material of which they
" consist. Although occasionally they have the hardness of stone, yet because they preserve
" the form and nature which they had when less hard, they can easily be distinguished from
" the stones. The juices are divided into 'meagre' and unctuous (*macer et pinguis*). The
" 'meagre' juices, since they originate from three different substances, are of three species.
" They are formed from a liquid mixed with earth, or with metal, or with a
" mineral compound. To the first species belong salt and *Nitrum* (soda); to the second,
" chrysocolla, verdigris, iron-rust, and azure; to the third, vitriol, alum, and an acrid juice
" which is unnamed. The first two of these latter are obtained from pyrites, which is
" numbered amongst the compound minerals. The third of these comes from *Cadmia* (in
" this case the cobalt-zinc-arsenic minerals; the acrid juice is probably zinc sulphate). To
" the unctuous juices belong these species: sulphur, bitumen, realgar and orpiment. Vitriol
" and alum, although they are somewhat unctuous yet do not burn, and they differ in
" their origin from the unctuous juices, for the latter are forced out from the earth by heat,
" whereas the former are produced when pyrites is softened by moisture."

complete knowledge of the method of making all underground works. Lastly, there are the various systems of assaying[5] substances and of preparing them for smelting; and here again there are many altogether diverse methods. For there is one method for gold and silver, another for copper, another for quicksilver, another for iron, another for lead, and

"Stone is a dry and hard mineral body which may either be softened by remaining "for a long time in water and be reduced to powder by a fierce fire; or else it does not "soften with water but the heat of a great fire liquefies it. To the first species belong "those stones which have been solidified by heat, to the second those solidified (literally "'congealed') by cold. These two species of stones are constituted from their own material. "However, writers on natural subjects who take into consideration the quantity and quality "of stones and their value, divide them into four classes. The first of these has no name of "its own but is called in common parlance 'stone': to this class belong loadstone, jasper (or "bloodstone) and *Aetites* (geodes ?). The second class comprises hard stones, either pellucid "or ornamental, with very beautiful and varied colours which sparkle marvellously; they "are called gems. The third comprises stones which are only brilliant after they have been "polished, and are usually called marble. The fourth are called rocks; they are found in "quarries, from which they are hewn out for use in building, and they are cut into various "shapes. None of the rocks show colour or take a polish. Few of the stones sparkle; fewer "still are transparent. Marble is sometimes only distinguishable from opaque gems by its "volume; rock is always distinguishable from stones properly so-called by its volume. Both "the stones and the gems are usually to be found in veins and veinlets which traverse the "rocks and marble. These four classes, as I have already stated, are divided into many "species, which I will explain in their proper place."

"Metal is a mineral body, by nature either liquid or somewhat hard. The latter may "be melted by the heat of the fire, but when it has cooled down again and lost all heat, it "becomes hard again and resumes its proper form. In this respect it differs from the "stone which melts in the fire, for although the latter regain its hardness, yet it loses "its pristine form and properties. Traditionally there are six different kinds of metals, "namely gold, silver, copper, iron, tin and lead. There are really others, for quicksilver is a "metal, although the Alchemists disagree with us on this subject, and bismuth is also. The "ancient Greek writers seem to have been ignorant of bismuth, wherefore Ammonius rightly "states that there are many species of metals, animals, and plants which are unknown to us. "*Stibium* when smelted in the crucible and refined has as much right to be regarded as a "proper metal as is accorded to lead by writers. If when smelted, a certain portion be "added to tin, a bookseller's alloy is produced from which the type is made that is used by "those who print books on paper. Each metal has its own form which it preserves when "separated from those metals which were mixed with it. Therefore neither electrum nor "*Stannum* is of itself a real metal, but rather an alloy of two metals. Electrum is an alloy "of gold and silver, *Stannum* of lead and silver (see note 33 p 473). And yet if silver be "parted from the electrum, then gold remains and not electrum; if silver be taken away "from *Stannum*, then lead remains and not *Stannum*. Whether brass, however, is found as "a native metal or not, cannot be ascertained with any surety. We only know of the "artificial brass, which consists of copper tinted with the colour of the mineral calamine. "And yet if any should be dug up, it would be a proper metal. Black and white copper "seem to be different from the red kind. Metal, therefore, is by nature either solid, as I "have stated, or fluid, as in the unique case of quicksilver. But enough now concerning the "simple kinds."

"I will now speak of the compounds which are composed of the simple minerals "cemented together by nature, and under the word 'compound' I now discuss those "mineral bodies which consist of two or three simple minerals. They are likewise mineral "substances, but so thoroughly mixed and alloyed that even in the smallest part there is "not wanting any substance that is contained in the whole. Only by the force of the fire "is it possible to separate one of the simple mineral substances from another; either the "third from the other two, or two from the third, if there were three in the same compound. "These two, three or more bodies are so completely mixed into one new species that the "pristine form of none of these is recognisable."

"The 'mixed' minerals, which are composed of those same simple minerals, differ "from the 'compounds,' in that the simple minerals each preserves its own form so that "they can be separated one from the other not only by fire but sometimes by water and "sometimes by hand. As these two classes differ so greatly from one another I usually use "two different words in order to distinguish one from the other. I am well aware that

---

[5]*Experiendae*—"a trial." That actual assaying in its technical sense is meant, is sufficiently evident from Book VII.

even tin and bismuth[6] are treated differently from lead. Although the evaporation of juices is an art apparently quite distinct from metallurgy, yet they ought not to be considered separately, inasmuch as these juices are also often dug out of the ground solidified, or they are produced from certain kinds of earth and stones which the miners dig up, and some of the juices are not themselves devoid of metals. Again, their treatment is not simple, since there is one method for common salt, another for soda[7], another for alum, another for vitriol[8], another for sulphur, and another for bitumen.

Furthermore, there are many arts and sciences of which a miner should not be ignorant. First there is Philosophy, that he may discern the origin, cause, and nature of subterranean things; for then he will be able to dig out the veins easily and advantageously, and to obtain more abundant results from his mining. Secondly, there is Medicine, that he may be able to look after his diggers and other workmen, that they do not meet with those

" Galen calls the metallic earth a compound which is really a mixture, but he who wishes to " instruct others should bestow upon each separate thing a definite name."

For convenience of reference we may reduce the above to a diagram as follows:

1. Fluids and gases.

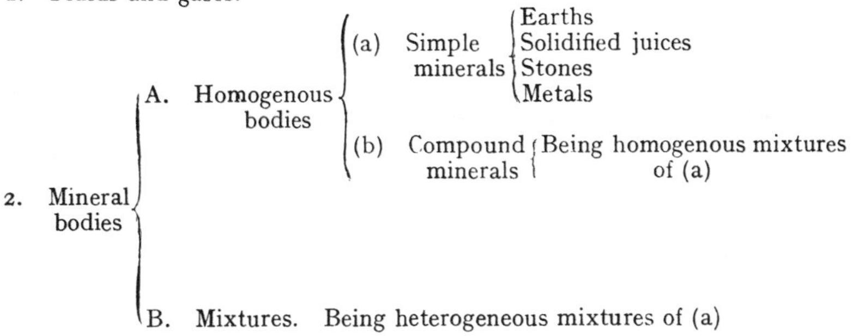

[6] . . . . *plumbum* . . . . *candidum ac cinereum vel nigrum.* "Lead ". . . white, or ash-coloured, or black." Agricola himself coined the term *plumbum cinereum* for bismuth, no doubt following the Roman term for tin—*plumbum candidum.* The following passage from *Bermannus* (p. 439) is of interest, for it appears to be the first description of bismuth, although mention of it occurs in the *Nützlich Bergbuchlin* (see Appendix B). " *Bermannus*: I will show you another kind of mineral which is numbered " amongst metals, but appears to me to have been unknown to the Ancients; we call it " *bisemutum. Naevius*: Then in your opinion there are more kinds of metals than the " seven commonly believed? *Bermannus*: More, I consider; for this which just now I " said we called *bisemutum,* cannot correctly be called *plumbum candidum* (tin), nor *nigrum* " (lead), but is different from both and is a third one. *Plumbum candidum* is whiter and " *plumbum nigrum* is darker, as you see. *Naevius*: We see that this is of the colour of " *galena. Ancon*: How then can *bisemutum,* as you call it, be distinguished from *galena*? " *Bermannus*: Easily; when you take it in your hands it stains them with black, unless " it is quite hard. The hard kind is not friable like *galena,* but can be cut. It is " blacker than the kind of *rudis* silver which we say is almost the colour of lead, and thus " is different from both. Indeed, it not rarely contains some silver. It generally indicates " that there is silver beneath the place where it is found, and because of this our miners " are accustomed to call it the 'roof of silver.' They are wont to roast this mineral, and " from the better part they make metal; from the poorer part they make a pigment of a " kind not to be despised."

[7] *Nitrum.* The Ancients comprised many salts under this head, but Agricola in the main uses it for soda, although sometimes he includes potash. He usually, however, refers to potash as *lixivium* or salt therefrom, and by other distinctive terms. For description of method of manufacture and discussion, see Book XII., p. 558.

[8] *Atramentum sutorium*—" Shoemaker's blacking." See p. 572 for description of method of manufacture and historical footnote. In the main Agricola means green vitriol, but he does describe three main varieties, green, blue, and white (*De Natura Fossilium*, p. 219). The blue was of course copper sulphate, and it is fairly certain that the white was zinc vitriol.

diseases to which they are more liable than workmen in other occupations, or if they do meet with them, that he himself may be able to heal them or may see that the doctors do so. Thirdly follows Astronomy, that he may know the divisions of the heavens and from them judge the direction of the veins. Fourthly, there is the science of Surveying that he may be able to estimate how deep a shaft should be sunk to reach the tunnel which is being driven to it, and to determine the limits and boundaries in these workings, especially in depth. Fifthly, his knowledge of Arithmetical Science should be such that he may calculate the cost to be incurred in the machinery and the working of the mine. Sixthly, his learning must comprise Architecture, that he himself may construct the various machines and timber work required underground, or that he may be able to explain the method of the construction to others. Next, he must have knowledge of Drawing, that he can draw plans of his machinery. Lastly, there is the Law, especially that dealing with metals, that he may claim his own rights, that he may undertake the duty of giving others his opinion on legal matters, that he may not take another man's property and so make trouble for himself, and that he may fulfil his obligations to others according to the law.

It is therefore necessary that those who take an interest in the methods and precepts of mining and metallurgy should read these and others of our books studiously and diligently ; or on every point they should consult expert mining people, though they will discover few who are skilled in the whole art. As a rule one man understands only the methods of mining, another possesses the knowledge of washing[9], another is experienced in the art of smelting, another has a knowledge of measuring the hidden parts of the earth, another is skilful in the art of making machines, and finally, another is learned in mining law. But as for us, though we may not have perfected the whole art of the discovery and preparation of metals, at least we can be of great assistance to persons studious in its acquisition.

But let us now approach the subject we have undertaken. Since there has always been the greatest disagreement amongst men concerning metals and mining, some praising, others utterly condemning them, therefore I have decided that before imparting my instruction, I should carefully weigh the facts with a view to discovering the truth in this matter.

So I may begin with the question of utility, which is a two-fold one, for either it may be asked whether the art of mining is really profitable or not to those who are engaged in it, or whether it is useful or not to the rest of mankind. Those who think mining of no advantage to the men who follow the occupation assert, first, that scarcely one in a hundred who dig metals or other such things derive profit therefrom; and again, that miners, because they entrust their certain and well-established wealth to dubious and slippery fortune, generally deceive themselves, and as a result, impoverished by

[9]*Lavandi*—"Washing." By this term the author includes all the operations of sluicing, buddling, and wet concentration generally. There is no English equivalent of such wide application, and there is some difficulty in interpretation without going further than the author intends. Book VIII. is devoted to the subject.

expenses and losses, in the end spend the most bitter and most miserable of lives. But persons who hold these views do not perceive how much a learned and experienced miner differs from one ignorant and unskilled in the art. The latter digs out the ore without any careful discrimination, while the former first assays and proves it, and when he finds the veins either too narrow and hard, or too wide and soft, he infers therefrom that these cannot be mined profitably, and so works only the approved ones. What wonder then if we find the incompetent miner suffers loss, while the competent one is rewarded by an abundant return from his mining? The same thing applies to husbandmen. For those who cultivate land which is alike arid, heavy, and barren, and in which they sow seeds, do not make so great a harvest as those who cultivate a fertile and mellow soil and sow their grain in that. And since by far the greater number of miners are unskilled rather than skilled in the art, it follows that mining is a profitable occupation to very few men, and a source of loss to many more. Therefore the mass of miners who are quite unskilled and ignorant in the knowledge of veins not infrequently lose both time and trouble[10]. Such men are accustomed for the most part to take to mining, either when through being weighted with the fetters of large and heavy debts, they have abandoned a business, or desiring to change their occupation, have left the reaping-hook and plough; and so if at any time such a man discovers rich veins or other abounding mining produce, this occurs more by good luck than through any knowledge on his part. We learn from history that mining has brought wealth to many, for from old writings it is well known that prosperous Republics, not a few kings, and many private persons, have made fortunes through mines and their produce. This subject, by the use of many clear and illustrious examples, I have dilated upon and explained in the first Book of my work entitled "*De Veteribus et Novis Metallis,*" from which it is evident that mining is very profitable to those who give it care and attention.

Again, those who condemn the mining industry say that it is not in the least stable, and they glorify agriculture beyond measure. But I do not see how they can say this with truth, for the silver-mines at Freiberg in Meissen remain still unexhausted after 400 years, and the lead mines of Goslar after 600 years. The proof of this can be found in the monuments of history. The gold and silver mines belonging to the communities of Schemnitz and Cremnitz have been worked for 800 years, and these latter are said to be the most ancient privileges of the inhabitants. Some then say the profit from an individual mine is unstable, as if forsooth, the miner is, or ought to be dependent on only one mine, and as if many men do not bear in common their expenses in mining, or as if one experienced in his art does not dig another vein, if fortune does not amply respond to his prayers in the first case. The New Schönberg at Freiberg has remained stable beyond the memory of man[11].

[10]*Operam et oleum perdit*—" loss of labour and oil."

[11]In *Veteribus et Novis Metallis*, and *Bermannus*, Agricola states that the mines of Schemnitz were worked 800 years before that time (1530), or about 750 A.D., and, further.

It is not my intention to detract anything from the dignity of agriculture, and that the profits of mining are less stable I will always and readily admit, for the veins do in time cease to yield metals, whereas the fields bring forth fruits every year. But though the business of mining may be less reliable it is more productive, so that in reckoning up, what is wanting in stability is found to be made up by productiveness. Indeed, the yearly profit of a lead mine in comparison with the fruitfulness of the best fields, is three times or at least twice as great. How much does the profit from gold or silver mines exceed that earned from agriculture? Wherefore truly and shrewdly does Xenophon[12] write about the Athenian silver mines: "There is land of such a nature that if you sow, it does not yield crops, but if you dig, it nourishes many more than if it had borne fruit." So let the farmers have for themselves the fruitful fields and cultivate the fertile hills for the sake of their produce; but let them leave to miners the gloomy valleys and sterile mountains, that they may draw forth from these, gems and metals which can buy, not only the crops, but all things that are sold.

The critics say further that mining is a perilous occupation to pursue, because the miners are sometimes killed by the pestilential air which they breathe; sometimes their lungs rot away; sometimes the men perish by being crushed in masses of rock; sometimes, falling from the ladders into the shafts, they break their arms, legs, or necks; and it is added there is no compensation which should be thought great enough to equalize the extreme dangers to safety and life. These occurrences, I confess, are of exceeding gravity, and moreover, fraught with terror and peril, so that I should consider that the metals should not be dug up at all, if such things were to happen very frequently to the miners, or if they could not safely guard against such risks by any means. Who would not prefer to live rather than to possess all things, even the metals? For he who thus perishes possesses nothing, but relinquishes all to his heirs. But since things like this rarely happen, and only in so far as workmen are careless, they do not deter miners from carrying on their trade any more than it would deter a carpenter from his, because one of his mates has acted incautiously and lost his life by falling from a high building. I have thus answered each argument which critics are wont to put before me when they assert that mining is an undesirable occupation, because it involves expense with uncertainty of return, because it is changeable, and because it is dangerous to those engaged in it.

Now I come to those critics who say that mining is not useful to the rest of mankind because forsooth, gems, metals, and other mineral products are worthless in themselves. This admission they try to extort from us, partly by arguments and examples, partly by misrepresentations and abuse of us. First, they make use of this argument: "The earth does not conceal and remove from our eyes those things which are useful and necessary to

that the lead mines of Goslar in the Hartz were worked by Otho the Great (936–973), and that the silver mines at Freiberg were discovered during the rule of Prince Otho (about 1170). To continue the argument to-day we could add about 360 years more of life to the mines of Goslar and Freiberg. See also Note 16, p. 36, and note 19, p. 37.

[12]Xenophon. Essay on the Revenues of Athens, I., 5.

mankind, but on the contrary, like a beneficent and kindly mother she yields in large abundance from her bounty and brings into the light of day the herbs, vegetables, grains, and fruits, and the trees. The minerals on the other hand she buries far beneath in the depth of the ground; therefore, they should not be sought. But they are dug out by wicked men who, as the poets say, are the products of the Iron Age." Ovid censures their audacity in the following lines :—

> " And not only was the rich soil required to furnish corn and due sustenance, but men even descended into the entrails of the earth, and they dug up riches, those incentives to vice, which the earth had hidden and had removed to the Stygian shades. Then destructive iron came forth, and gold, more destructive than iron ; then war came forth."[13]

Another of their arguments is this : Metals offer to men no advantages, therefore we ought not to search them out. For whereas man is composed of soul and body, neither is in want of minerals. The sweetest food of the soul is the contemplation of nature, a knowledge of the finest arts and sciences, an understanding of virtue ; and if he interests his mind in excellent things, if he exercise his body, he will be satisfied with this feast of noble thoughts and knowledge, and have no desire for other things. Now although the human body may be content with necessary food and clothing, yet the fruits of the earth and the animals of different kinds supply him in wonderful abundance with food and drink, from which the body may be suitably nourished and strengthened and life prolonged to old age. Flax, wool, and the skins of many animals provide plentiful clothing low in price ; while a luxurious kind, not hard to procure—that is the so called *seric* material, is furnished by the down of trees and the webs of the silk worm. So that the body has absolutely no need of the metals, so hidden in the depths of the earth and for the greater part very expensive. Wherefore it is said that this maxim of Euripides is approved in assemblies of learned men, and with good reason was always on the lips of Socrates :

> " Works of silver and purple are of use, not for human life, but rather for Tragedians."[14]

These critics praise also this saying from Timocreon of Rhodes :

> " O Unseeing Plutus, would that thou hadst never appeared in the earth or in the sea or on the land, but that thou didst have thy habitation in Tartarus and Acheron, for out of thee arise all evil things which overtake mankind "[15].

They greatly extol these lines from Phocylides :

> " Gold and silver are injurious to mortals ; gold is the source of crime, the plague of life, and the ruin of all things. Would that thou were not such an attractive scourge ! because of thee arise robberies,

[13]Ovid, *Metamorphoses*, I., 137 to 143.

[14]Diogenes Laertius, II., 5. The lines are assigned, however, to Philemon, not Euripides. (Kock, *Comicorum Atticorum Fragmenta* II., 512).

[15]We have not considered it of sufficient interest to cite the references to all of the minor poets and those whose preserved works are but fragmentary. The translations from the Greek into Latin are not literal and suffer again by rendering into English ; we have however considered it our duty to translate Agricola's view of the meaning.

homicides, warfare, brothers are maddened against brothers, and children against parents."

This from Naumachius also pleases them :

"Gold and silver are but dust, like the stones that lie scattered on the pebbly beach, or on the margins of the rivers."

On the other hand, they censure these verses of Euripides :

"Plutus is the god for wise men ; all else is mere folly and at the same time a deception in words."

So in like manner these lines from Theognis :

"O Plutus, thou most beautiful and placid god ! whilst I have thee, however bad I am, I can be regarded as good."

They also blame Aristodemus, the Spartan, for these words :

"Money makes the man ; no one who is poor is either good or honoured."

And they rebuke these songs of Timocles :

"Money is the life and soul of mortal men. He who has not heaped up riches for himself wanders like a dead man amongst the living."

Finally, they blame Menander when he wrote :

"Epicharmus asserts that the gods are water, wind, fire, earth, sun, and stars. But I am of opinion that the gods of any use to us are silver and gold ; for if thou wilt set these up in thy house thou mayest seek whatever thou wilt. All things will fall to thy lot ; land, houses, slaves, silver-work ; moreover friends, judges, and witnesses. Only give freely, for thus thou hast the gods to serve thee."

But besides this, the strongest argument of the detractors is that the fields are devastated by mining operations, for which reason formerly Italians were warned by law that no one should dig the earth for metals and so injure their very fertile fields, their vineyards, and their olive groves. Also they argue that the woods and groves are cut down, for there is need of an endless amount of wood for timbers, machines, and the smelting of metals. And when the woods and groves are felled, then are exterminated the beasts and birds, very many of which furnish a pleasant and agreeable food for man. Further, when the ores are washed, the water which has been used poisons the brooks and streams, and either destroys the fish or drives them away. Therefore the inhabitants of these regions, on account of the devastation of their fields, woods, groves, brooks and rivers, find great difficulty in procuring the necessaries of life, and by reason of the destruction of the timber they are forced to greater expense in erecting buildings. Thus it is said, it is clear to all that there is greater detriment from mining than the value of the metals which the mining produces.

So in fierce contention they clamour, showing by such examples as follow that every great man has been content with virtue, and despised metals. They praise Bias because he esteemed the metals merely as fortune's playthings, not as his real wealth. When his enemies had captured his native Priene, and his fellow-citizens laden with precious things had betaken themselves to flight, he was asked by one, why he carried away none of his goods with him, and he replied, "I carry all my possessions with me." And it is said that Socrates, having received twenty minae sent to him by Aristippus, a grateful disciple, refused them and sent them back to him by the command of his conscience. Aristippus, following his example in this matter, despised gold and regarded it as of no value. And once when he was making a journey with his slaves, and they, laden with the gold, went too slowly, he ordered them to keep only as much of it as they could carry without distress and to throw away the remainder[16]. Moreover, Anacreon of Teos, an ancient and noble poet, because he had been troubled about them for two nights, returned five talents which had been given him by Polycrates, saying that they were not worth the anxiety which he had gone through on their account. In like manner celebrated and exceedingly powerful princes have imitated the philosophers in their scorn and contempt for gold and silver. There was for example, Phocion, the Athenian, who was appointed general of the army so many times, and who, when a large sum of gold was sent to him as a gift by Alexander, King of Macedon, deemed it trifling and scorned it. And Marcus Curius ordered the gold to be carried back to the Samnites, as did also Fabricius Luscinus with regard to the silver and copper. And certain Republics have forbidden their citizens the use and employment of gold and silver by law and ordinance; the Lacedaemonians, by the decrees and ordinances of Lycurgus, used diligently to enquire among their citizens whether they possessed any of these things or not, and the possessor, when he was caught, was punished according to law and justice. The inhabitants of a town on the Tigris, called Babytace, buried their gold in the ground so that no one should use it. The Scythians condemned the use of gold and silver so that they might not become avaricious.

Further are the metals reviled; in the first place people wantonly abuse gold and silver and call them deadly and nefarious pests of the human race, because those who possess them are in the greatest peril, for those who have none lay snares for the possessors of wealth, and thus again and again the metals have been the cause of destruction and ruin. For example, Polymnestor, King of Thrace, to obtain possession of his gold, killed Polydorus, his noble guest and the son of Priam, his father-in-law, and old friend. Pygmalion, the King of Tyre, in order that he might seize treasures of gold and silver, killed his sister's husband, a priest, taking no account of either kinship or religion. For love of gold Eriphyle betrayed her husband Amphiaraus to his enemy. Likewise Lasthenes betrayed the city of Olynthus to Philip of Macedon. The daughter of Spurius Tarpeius, having been bribed with gold, admitted the Sabines into the citadel of Rome. Claudius Curio sold his country for gold to Cæsar, the Dictator. Gold, too, was the cause of the downfall of Aesculapius, the great physician, who it was believed was the son of Apollo. Similarly Marcus Crassus, through his eager desire for the gold of the Parthians, was completely overcome together with his son and eleven legions, and became the jest of his enemies; for they

[16]Diogenes Laertius, II.

poured liquid gold into the gaping mouth of the slain Crassus, saying: "Thou hast thirsted for gold, therefore drink gold."

But why need I cite here these many examples from history?[17] It is almost our daily experience to learn that, for the sake of obtaining gold and silver, doors are burst open, walls are pierced, wretched travellers are struck down by rapacious and cruel men born to theft, sacrilege, invasion, and robbery. We see thieves seized and strung up before us, sacrilegious persons burnt alive, the limbs of robbers broken on the wheel, wars waged for the same reason, which are not only destructive to those against whom they are waged, but to those also who carry them on. Nay, but they say that the precious metals foster all manner of vice, such as the seduction of women, adultery, and unchastity, in short, crimes of violence against the person. Therefore the Poets, when they represent Jove transformed into a golden shower and falling into the lap of Danae, merely mean that he had found for himself a safe road by the use of gold, by which he might enter the tower for the purpose of violating the maiden. Moreover, the fidelity of many men is overthrown by the love of gold and silver, judicial sentences are bought, and innumerable crimes are perpetrated. For truly, as Propertius says:

> "This is indeed the Golden Age. The greatest rewards come from gold; by gold love is won; by gold is faith destroyed; by gold is justice bought; the law follows the track of gold, while modesty will soon follow it when law is gone."

Diphilus says:

> "I consider that nothing is more powerful than gold. By it all things are torn asunder; all things are accomplished."

Therefore, all the noblest and best despise these riches, deservedly and with justice, and esteem them as nothing. And this is said by the old man in Plautus:

> "I hate gold. It has often impelled many people to many wrong acts."

In this country too, the poets inveigh with stinging reproaches against money coined from gold and silver. And especially did Juvenal:

> "Since the majesty of wealth is the most sacred thing among us; although, O pernicious money, thou dost not yet inhabit a temple, nor have we erected altars to money."

And in another place:

> "Demoralising money first introduced foreign customs, and voluptuous wealth weakened our race with disgraceful luxury."[18]

And very many vehemently praise the barter system which men used before money was devised, and which even now obtains among certain simple peoples.

And next they raise a great outcry against other metals, as iron, than

[17]An inspection of the historical incidents mentioned here and further on, indicates that Agricola relied for such information on Diogenes Laertius, Plutarch, Livy, Valerius Maximus, Pliny, and often enough on Homer, Horace, and Virgil.

[18]Juvenal. *Satires* I., l. 112, and VI., l. 298.

which they say nothing more pernicious could have been brought into the life of man. For it is employed in making swords, javelins, spears, pikes, arrows—weapons by which men are wounded, and which cause slaughter, robbery, and wars. These things so moved the wrath of Pliny that he wrote: " Iron is used not only in hand to hand fighting, but also to form the winged missiles of war, sometimes for hurling engines, sometimes for lances, sometimes even for arrows. I look upon it as the most deadly fruit of human ingenuity. For to bring Death to men more quickly we have given wings to iron and taught it to fly."[19] The spear, the arrow from the bow, or the bolt from the catapult and other engines can be driven into the body of only one man, while the iron cannon-ball fired through the air, can go through the bodies of many men, and there is no marble or stone object so hard that it cannot be shattered by the force and shock. Therefore it levels the highest towers to the ground, shatters and destroys the strongest walls. Certainly the ballistas which throw stones, the battering rams and other ancient war engines for making breaches in walls of fortresses and hurling down strongholds, seem to have little power in comparison with our present cannon. These emit horrible sounds and noises, not less than thunder, flashes of fire burst from them like the lightning, striking, crushing, and shattering buildings, belching forth flames and kindling fires even as lightning flashes. So that with more justice could it be said of the impious men of our age than of Salmoneus of ancient days, that they had snatched lightning from Jupiter and wrested it from his hands. Nay, rather there has been sent from the infernal regions to the earth this force for the destruction of men, so that Death may snatch to himself as many as possible by one stroke.

But because muskets are nowadays rarely made of iron, and the large ones never, but of a certain mixture of copper and tin, they confer more maledictions on copper and tin than on iron. In this connection too, they mention the brazen bull of Phalaris, the brazen ox of the people of Pergamus, racks in the shape of an iron dog or a horse, manacles, shackles, wedges, hooks, and red-hot plates. Cruelly racked by such instruments, people are driven to confess crimes and misdeeds which they have never committed, and innocent men are miserably tortured to death by every conceivable kind of torment.

It is claimed too, that lead is a pestilential and noxious metal, for men are punished by means of molten lead, as Horace describes in the ode addressed to the Goddess Fortune: " Cruel Necessity ever goes before thee bearing in her brazen hand the spikes and wedges, while the awful hook and molten lead are also not lacking."[20] In their desire to excite greater odium for this metal, they are not silent about the leaden balls of muskets, and they find in it the cause of wounds and death.

They contend that, inasmuch as Nature has concealed metals far within the depths of the earth, and because they are not necessary to human life, they are therefore despised and repudiated by the noblest, and should not be

[19]Pliny, XXXIV., 39.
[20]Horace. *Odes*, I., 35, ll., 17–20.

mined, and seeing that when brought to light they have always proved the cause of very great evils, it follows that mining is not useful to mankind, but on the contrary harmful and destructive. Several good men have been so perturbed by these tragedies that they conceive an intensely bitter hatred toward metals, and they wish absolutely that metals had never been created, or being created, that no one had ever dug them out. The more I commend the singular honesty, innocence, and goodness of such men, the more anxious shall I be to remove utterly and eradicate all error from their minds and to reveal the sound view, which is that the metals are most useful to mankind.

In the first place then, those who speak ill of the metals and refuse to make use of them, do not see that they accuse and condemn as wicked the Creator Himself, when they assert that He fashioned some things vainly and without good cause, and thus they regard Him as the Author of evils, which opinion is certainly not worthy of pious and sensible men.

In the next place, the earth does not conceal metals in her depths because she does not wish that men should dig them out, but because provident and sagacious Nature has appointed for each thing its place. She generates them in the veins, stringers, and seams in the rocks, as though in special vessels and receptacles for such material. The metals cannot be produced in the other elements because the materials for their formation are wanting. For if they were generated in the air, a thing that rarely happens, they could not find a firm resting-place, but by their own force and weight would settle down on to the ground. Seeing then that metals have their proper abiding place in the bowels of the earth, who does not see that these men do not reach their conclusions by good logic?

They say, "Although metals are in the earth, each located in its own proper place where it originated, yet because they lie thus enclosed and hidden from sight, they should not be taken out." But, in refutation of these attacks, which are so annoying, I will on behalf of the metals instance the fish, which we catch, hidden and concealed though they be in the water, even in the sea. Indeed, it is far stranger that man, a terrestrial animal, should search the interior of the sea than the bowels of the earth. For as birds are born to fly freely through the air, so are fishes born to swim through the waters, while to other creatures Nature has given the earth that they might live in it, and particularly to man that he might cultivate it and draw out of its caverns metals and other mineral products. On the other hand, they say that we eat fish, but neither hunger nor thirst is dispelled by minerals, nor are they useful in clothing the body, which is another argument by which these people strive to prove that metals should not be taken out. But man without metals cannot provide those things which he needs for food and clothing. For, though the produce of the land furnishes the greatest abundance of food for the nourishment of our bodies, no labour can be carried on and completed without tools. The ground itself is turned up with ploughshares and harrows, tough stalks and the tops of the roots are broken off and dug up with a mattock, the sown seed is harrowed, the corn field is hoed and weeded ; the ripe grain with part of the stalk is cut down by scythes and threshed on the floor, or its ears are cut off and stored in the barn and later beaten with flails and winnowed with fans, until finally the pure grain is stored in the granary, whence it is brought forth again when occasion demands or necessity arises. Again, if we wish to procure better and more productive fruits from trees and bushes, we must resort to cultivating, pruning, and grafting, which cannot be done without tools. Even as without vessels we cannot keep or hold liquids, such as milk, honey, wine, or oil, neither could so many living things be cared for without buildings to protect them from long-continued rain and intolerable cold. Most of the rustic instruments are made of iron, as ploughshares, share-beams, mattocks, the prongs of harrows, hoes, planes, hay-forks, straw cutters, pruning shears, pruning hooks, spades, lances, forks, and weed cutters. Vessels are also made of copper or lead. Neither are wooden instruments or vessels made without iron. Wine cellars, oil-mills, stables, or any other part of a farm building could not be built without iron tools. Then if the bull, the wether, the goat, or any other domestic animal is led away from the pasture to the butcher, or if the poulterer brings from the farm a chicken, a hen, or a capon for the cook, could any of these animals be cut up and divided without axes and knives ? I need say nothing here about bronze and copper pots for cooking, because for these purposes one could make use of earthen vessels, but even these in turn could not be made and fashioned by the potter without tools, for no instruments can be made out of wood alone, without the use of iron. Furthermore, hunting, fowling, and fishing supply man with food, but when the stag has been ensnared does not the hunter transfix him with his spear ? As he stands or runs, does he not pierce him with an arrow ? Or pierce him with a bullet ? Does not the fowler in the same way kill the moor-fowl or pheasant with an arrow ? Or does he not discharge into its body the ball from the musket ? I will not speak of the snares and other instruments with which the woodcock, wood-pecker, and other wild birds are caught, lest I pursue unseasonably and too minutely single instances. Lastly, with his fish-hook and net does not the fisherman catch the fish in the sea, in the lakes, in fish-ponds, or in rivers ? But the hook is of iron, and sometimes we see lead or iron weights attached to the net. And most fish that are caught are afterward cut up and dis-embowelled with knives and axes. But, more than enough has been said on the matter of food.

Now I will speak of clothing, which is made out of wool, flax, feathers, hair, fur, or leather. First the sheep are sheared, then the wool is combed. Next the threads are drawn out, while later the warp is suspended in the shuttle under which passes the wool. This being struck by the comb, at length cloth is formed either from threads alone or from threads and hair. Flax, when gathered, is first pulled by hooks. Then it is dipped in water and afterward dried, beaten into tow with a heavy mallet, and carded, then drawn out into threads, and finally woven into cloth. But has the artisan or weaver of the cloth any instrument not made of iron ? Can one be made of wood without the aid of iron? The cloth or web must be cut into lengths for the tailor. Can this be done without knife or scissors? Can the tailor sew together any garments without a needle? Even peoples dwelling beyond the seas cannot make a covering for their bodies, fashioned of feathers, without these same implements. Neither can the furriers do without them in sewing together the pelts of any kind of animals. The shoemaker needs a knife to cut the leather, another to scrape it, and an awl to perforate it before he can make shoes. These coverings for the body are either woven or stitched. Buildings too, which protect the same body from rain, wind, cold, and heat, are not constructed without axes, saws, and augers.

But what need of more words? If we remove metals from the service of man, all methods of protecting and sustaining health and more carefully preserving the course of life are done away with. If there were no metals, men would pass a horrible and wretched existence in the midst of wild beasts; they would return to the acorns and fruits and berries of the forest. They would feed upon the herbs and roots which they plucked up with their nails. They would dig out caves in which to lie down at night, and by day they would rove in the woods and plains at random like beasts, and inasmuch as this condition is utterly unworthy of humanity, with its splendid and glorious natural endowment, will anyone be so foolish or obstinate as not to allow that metals are necessary for food and clothing and that they tend to preserve life?

Moreover, as the miners dig almost exclusively in mountains otherwise unproductive, and in valleys invested in gloom, they do either slight damage to the fields or none at all. Lastly, where woods and glades are cut down, they may be sown with grain after they have been cleared from the roots of shrubs and trees. These new fields soon produce rich crops, so that they repair the losses which the inhabitants suffer from increased cost of timber. Moreover, with the metals which are melted from the ore, birds without number, edible beasts and fish can be purchased elsewhere and brought to these mountainous regions.

I will pass to the illustrations I have mentioned. Bias of Priene, when his country was taken, carried away out of the city none of his valuables. So strong a man with such a reputation for wisdom had no need to fear personal danger from the enemy, but this in truth cannot be said of him because he hastily took to flight; the throwing away of his goods does not seem to me so great a matter, for he had lost his house, his estates, and even his country, than which nothing is more precious. Nay, I should be convinced of Bias's contempt and scorn for possessions of this kind, if before his country was captured he had bestowed them freely on relations and friends, or had distributed them to the very poor, for this he could have done freely and without question. Whereas his conduct, which the Greeks admire so greatly, was due, it would seem, to his being driven out by the enemy and stricken with fear. Socrates in truth did not despise gold, but would not accept money for his teaching. As for Aristippus of Cyrene, if he had gathered and saved the gold which he ordered his slaves to throw away, he might have bought the things which he needed for the necessaries of life, and he would not, by reason of his poverty, have then been obliged to flatter the tyrant Dionysius, nor would he ever have been called by him a King's dog. For this reason Horace, speaking of Damasippus when reviling Staberus for valuing riches very highly, says:

> "What resemblance has the Grecian Aristippus to this fellow? He who commanded his slaves to throw away the gold in the midst of Libya because they went too slowly, impeded by the weight of their burden—which of these two men is the more insane?"[21]

Insane indeed is he who makes more of riches than of virtue. Insane also is he who rejects them and considers them as worth nothing, instead of using them with reason. Yet as to the gold which Aristippus on another occasion flung into the sea from a boat, this he did with a wise and prudent mind. For learning that it was a pirate boat in which he was sailing, and fearing for his life, he counted his gold and then throwing it of his own will into the sea, he groaned as if he had done it unwillingly. But afterward, when he escaped the peril, he said: "It is better that this gold itself should be lost than that I should have perished because of it." Let it be granted that some philosophers, as well as Anacreon of Teos, despised gold and silver. Anaxagoras of Clazomenae also gave up his sheep-farms and became a shepherd. Crates the Theban too, being annoyed that his estate and other kinds of wealth caused him worry, and that in his contemplations his mind was thereby distracted, resigned a property valued at ten talents, and taking a cloak and wallet, in poverty devoted all his thought and efforts to philosophy. Is it true that because these philosophers despised money, all others declined wealth in cattle? Did they refuse to cultivate lands or to dwell in houses? There were certainly many, on the other hand, who, though affluent, became famous in the pursuit of learning and in the knowledge of divine and human laws, such as Aristotle, Cicero, and Seneca. As for Phocion, he did not deem it honest to accept the gold sent to him by Alexander. For if he had consented to use it, the king as much as himself would have incurred the hatred and aversion of the Athenians, and these very people were afterward so ungrateful toward this excellent man that they compelled him to drink hemlock. For what would have been less becoming to Marcus Curius and Fabricius Luscinus than to accept gold from their enemies, who hoped that by these means those leaders could be corrupted or would become odious to their fellow citizens, their purpose being to cause dissentions among the Romans and destroy the Republic utterly. Lycurgus, however, ought to have given instructions to the Spartans as to the use of gold and silver, instead of abolishing things good in themselves. As to the Babytacenses, who does not see that they were senseless and envious? For with their gold they might have bought things of which they were in need, or even given it to neighbouring peoples to bind them more closely to themselves with gifts and favours. Finally, the Scythians, by condemning the use of gold and silver

[21]Horace. *Satires*, II., 3, ll., 99–102.

alone, did not free themselves utterly from avarice, because although he is not enjoying them, one who can possess other forms of property may also become avaricious.

Now let us reply to the attacks hurled against the products of mines. In the first place, they call gold and silver the scourge of mankind because they are the cause of destruction and ruin to their possessors. But in this manner, might not anything that we possess be called a scourge to human kind,—whether it be a horse, or a garment, or anything else? For, whether one rides a splendid horse, or journeys well clad, he would give occasion to a robber to kill him. Are we then not to ride on horses, but to journey on foot, because a robber has once committed a murder in order that he may steal a horse? Or are we not to possess clothing, because a vagabond with a sword has taken a traveller's life that he may rob him of his garment? The possession of gold and silver is similar. Seeing then that men cannot conveniently do all these things, we should be on our guard against robbers, and because we cannot always protect ourselves from their hands, it is the special duty of the magistrate to seize wicked and villainous men for torture, and, if need be, for execution.

Again, the products of the mines are not themselves the cause of war. Thus, for example, when a tyrant, inflamed with passion for a woman of great beauty, makes war on the inhabitants of her city, the fault lies in the unbridled lust of the tyrant and not in the beauty of the woman. Likewise, when another man, blinded by a passion for gold and silver, makes war upon a wealthy people, we ought not to blame the metals but transfer all blame to avarice. For frenzied deeds and disgraceful actions, which are wont to weaken and dishonour natural and civil laws, originate from our own vices. Wherefore Tibullus is wrong in laying the blame for war on gold, when he says: "This is the fault of a rich man's gold; there were no wars when beech goblets were used at banquets." But Virgil, speaking of Polymnestor, says that the crime of the murderer rests on avarice:

> "He breaks all law; he murders Polydorus, and obtains gold by violence. To what wilt thou not drive mortal hearts, thou accursed hunger for gold?"

And again, justly, he says, speaking of Pygmalion, who killed Sichaeus:

> "And blinded with the love of gold, he slew him unawares with stealthy sword."[22]

For lust and eagerness after gold and other things make men blind, and this wicked greed for money, all men in all times and places have considered dishonourable and criminal. Moreover, those who have been so addicted to avarice as to be its slaves have always been regarded as mean and sordid. Similarly, too, if by means of gold and silver and gems men can overcome the chastity of women, corrupt the honour of many people, bribe the course of justice and commit innumerable wickednesses, it is not the metals which are to be blamed, but the evil passions of men which become inflamed and ignited; or it is due to the blind and impious desires of their minds. But

[22]Virgil. *Æneid*, III., l. 55, and I, l. 349.

although these attacks against gold and silver may be directed especially against money, yet inasmuch as the Poets one after another condemn it, their criticism must be met, and this can be done by one argument alone. Money is good for those who use it well; it brings loss and evil to those who use it ill. Hence, very rightly, Horace says:

"Dost thou not know the value of money; and what uses it serves?
It buys bread, vegetables, and a pint of wine."

And again in another place:

"Wealth hoarded up is the master or slave of each possessor; it should follow rather than lead, the 'twisted rope.'"[23]

When ingenious and clever men considered carefully the system of barter, which ignorant men of old employed and which even to-day is used by certain uncivilised and barbarous races, it appeared to them so troublesome and laborious that they invented money. Indeed, nothing more useful could have been devised, because a small amount of gold and silver is of as great value as things cumbrous and heavy; and so peoples far distant from one another can, by the use of money, trade very easily in those things which civilised life can scarcely do without.

The curses which are uttered against iron, copper, and lead have no weight with prudent and sensible men, because if these metals were done away with, men, as their anger swelled and their fury became unbridled, would assuredly fight like wild beasts with fists, heels, nails, and teeth. They would strike each other with sticks, hit one another with stones, or dash their foes to the ground. Moreover, a man does not kill another with iron alone, but slays by means of poison, starvation, or thirst. He may seize him by the throat and strangle him; he may bury him alive in the ground; he may immerse him in water and suffocate him; he may burn or hang him; so that he can make every element a participant in the death of men. Or, finally, a man may be thrown to the wild beasts. Another may be sewn up wholly except his head in a sack, and thus be left to be devoured by worms; or he may be immersed in water until he is torn to pieces by sea-serpents. A man may be boiled in oil; he may be greased, tied with ropes, and left exposed to be stung by flies and hornets; he may be put to death by scourging with rods or beating with cudgels, or struck down by stoning, or flung from a high place. Furthermore, a man may be tortured in more ways than one without the use of metals; as when the executioner burns the groins and armpits of his victim with hot wax; or places a cloth in his mouth gradually, so that when in breathing he draws it slowly into his gullet, the executioner draws it back suddenly and violently; or the victim's hands are fastened behind his back, and he is drawn up little by little with a rope and then let down suddenly. Or similarly, he may be tied to a beam and a heavy stone fastened by a cord to his feet, or finally his limbs may be torn asunder. From these examples we see that it is not metals that are to be condemned, but our vices, such as anger, cruelty, discord, passion for power, avarice, and lust.

[23]Horace. *Satires*, I., l. 73; and Epistle, I., 10, l. 47.

The question next arises, whether we ought to count metals amongst the number of good things or class them amongst the bad. The Peripatetics regarded all wealth as a good thing, and merely spoke of externals as having to do with neither the mind nor the body. Well, let riches be an external thing. And, as they said, many other things may be classed as good if it is in one's power to use them either well or ill. For good men employ them for good, and to them they are useful. The wicked use them badly, and to them they are harmful. There is a saying of Socrates, that just as wine is influenced by the cask, so the character of riches is like their possessors. The Stoics, whose custom it is to argue subtly and acutely, though they did not put wealth in the category of good things, they did not count it amongst the evil ones, but placed it in that class which they term neutral. For to them virtue alone is good, and vice alone evil. The whole of what remains is indifferent. Thus, in their conviction, it matters not whether one be in good health or seriously ill; whether one be handsome or deformed. In short:

> "Whether, sprung from Inachus of old, and thus hast lived beneath the sun in wealth, or hast been poor and despised among men, it matters not."

For my part, I see no reason why anything that is in itself of use should not be placed in the class of good things. At all events, metals are a creation of Nature, and they supply many varied and necessary needs of the human race, to say nothing about their uses in adornment, which are so wonderfully blended with utility. Therefore, it is not right to degrade them from the place they hold among the good things. In truth, if there is a bad use made of them, should they on that account be rightly called evils? For of what good things can we not make an equally bad or good use? Let me give examples from both classes of what we term good. Wine, by far the best drink, if drunk in moderation, aids the digestion of food, helps to produce blood, and promotes the juices in all parts of the body. It is of use in nourishing not only the body but the mind as well, for it disperses our dark and gloomy thoughts, frees us from cares and anxiety, and restores our confidence. If drunk in excess, however, it injures and prostrates the body with serious disease. An intoxicated man keeps nothing to himself; he raves and rants, and commits many wicked and infamous acts. On this subject Theognis wrote some very clever lines, which we may render thus:

> "Wine is harmful if taken with greedy lips, but if drunk in moderation it is wholesome."[25]

But I linger too long over extraneous matters. I must pass on to the gifts of body and mind, amongst which strength, beauty, and genius occur to me. If then a man, relying on his strength, toils hard to maintain himself and his family in an honest and respectable manner, he uses the gift aright, but if he makes a living out of murder and robbery, he uses it wrongly. Likewise, too, if a lovely woman is anxious to please her husband

[25]Theognis. Maxims, II., l. 210.

alone she uses her beauty aright, but if she lives wantonly and is a victim of passion, she misuses her beauty. In like manner, a youth who devotes himself to learning and cultivates the liberal arts, uses his genius rightly. But he who dissembles, lies, cheats, and deceives by fraud and dishonesty, misuses his abilities. Now, the man who, because they are abused, denies that wine, strength, beauty, or genius are good things, is unjust and blasphemous towards the Most High God, Creator of the World ; so he who would remove metals from the class of blessings also acts unjustly and blasphemously against Him. Very true, therefore, are the words which certain Greek poets have written, as Pindar :

> " Money glistens, adorned with virtue ; it supplies the means by which thou mayest act well in whatever circumstances fate may have in store for thee."[26]

And Sappho :

> " Without the love of virtue gold is a dangerous and harmful guest, but when it is associated with virtue, it becomes the source and height of good."

And Callimachus :

> " Riches do not make men great without virtue ; neither do virtues themselves make men great without some wealth."

And Antiphanes :

> " Now, by the gods, why is it necessary for a man to grow rich ? Why does he desire to possess much money unless that he may, as much as possible, help his friends, and sow the seeds of a harvest of gratitude, sweetest of the goddesses."[27]

Having thus refuted the arguments and contentions of adversaries, let us sum up the advantages of the metals. In the first place, they are useful to the physician, for they furnish liberally the ingredients for medicines, by which wounds and ulcers are cured, and even plagues ; so that certainly if there were no other reasons why we should explore the depths of the earth, we should for the sake of medicine alone dig in the mines. Again, the metals are of use to painters, because they yield certain pigments which, when united with the painter's slip, are injured less than others by the moisture from without. Further, mining is useful to the architects, for thus is found marble, which is suitable not only for strengthening large buildings, but also for decoration. It is, moreover, helpful to those whose ambition urges them toward immortal glory, because it yields metals from which are made coins, statues, and other monuments, which, next to literary records, give men in a sense immortality. The metals are useful to merchants with very great cause, for, as I have stated elsewhere, the use of money which is made from metals is much more convenient to mankind than the old system of exchange of commodities. In short, to whom are the metals not of use ? In very truth, even the works of art, elegant, embellished, elaborate, useful, are fashioned in various shapes by the artist from the metals gold, silver, brass, lead, and iron. How few artists

[26]Pindar. *Olymp.* II., 58–60.
[27]Antiphanes, 4.

could make anything that is beautiful and perfect without using metals? Even if tools of iron or brass were not used, we could not make tools of wood and stone without the help of metal. From all these examples are evident the benefits and advantages derived from metals. We should not have had these at all unless the science of mining and metallurgy had been discovered and handed down to us. Who then does not understand how highly useful they are, nay rather, how necessary to the human race? In a word, man could not do without the mining industry, nor did Divine Providence will that he should.

Further, it has been asked whether to work in metals is honourable employment for respectable people or whether it is not degrading and dishonourable. We ourselves count it amongst the honourable arts. For that art, the pursuit of which is unquestionably not impious, nor offensive, nor mean, we may esteem honourable. That this is the nature of the mining profession, inasmuch as it promotes wealth by good and honest methods, we shall show presently. With justice, therefore, we may class it amongst honourable employments. In the first place, the occupation of the miner, which I must be allowed to compare with other methods of acquiring great wealth, is just as noble as that of agriculture; for, as the farmer, sowing his seed in his fields injures no one, however profitable they may prove to him, so the miner digging for his metals, albeit he draws forth great heaps of gold or silver, hurts thereby no mortal man. Certainly these two modes of increasing wealth are in the highest degree both noble and honourable. The booty of the soldier, however, is frequently impious, because in the fury of the fighting he seizes all goods, sacred as well as profane. The most just king may have to declare war on cruel tyrants, but in the course of it wicked men cannot lose their wealth and possessions without dragging into the same calamity innocent and poor people, old men, matrons, maidens, and orphans. But the miner is able to accumulate great riches in a short time, without using any violence, fraud, or malice. That old saying is, therefore, not always true that "Every rich man is either wicked himself, or is the heir to wickedness."

Some, however, who contend against us, censure and attack miners by saying that they and their children must needs fall into penury after a short time, because they have heaped up riches by improper means. According to them nothing is truer than the saying of the poet Naevius:

"Ill gotten gains in ill fashion slip away."

The following are some of the wicked and sinful methods by which they say men obtain riches from mining. When a prospect of obtaining metals shows itself in a mine, either the ruler or magistrate drives out the rightful owners of the mines from possession, or a shrewd and cunning neighbour perhaps brings a law-suit against the old possessors in order to rob them of some part of their property. Or the mine superintendent imposes on the owners such a heavy contribution on shares, that if they cannot pay, or will not, they lose their rights of possession; while the superintendent, contrary to all that is right, seizes upon all that they have lost. Or, finally, the mine foreman may conceal the vein by plastering over with clay that part where the metal abounds, or by covering it with earth, stones, stakes, or poles, in the hope that after several years the proprietors, thinking the mine exhausted, will abandon it, and the foreman can then excavate that remainder of the ore and keep it for himself. They even state that the scum of the miners exist wholly by fraud, deceit, and lying. For to speak of nothing else, but only of those deceits which are practised in buying and selling, it is said they either advertise the veins with false and imaginary praises, so that they can sell the shares in the mines at one-half more than they are worth, or on the contrary, they sometimes detract from the estimate of them so that they can buy shares for a small price. By exposing such frauds our critics suppose all good opinion of miners is lost. Now, all wealth, whether it has been gained by good or evil means, is liable by some adverse chance to vanish away. It decays and is dissipated by the fault and carelessness of the owner, since he loses it through laziness and neglect, or wastes and squanders it in luxuries, or he consumes and exhausts it in gifts, or he dissipates and throws it away in gambling :

"Just as though money sprouted up again, renewed from an exhausted coffer, and was always to be obtained from a full heap."

It is therefore not to be wondered at if miners do not keep in mind the counsel given by King Agathocles : "Unexpected fortune should be held in reverence," for by not doing so they fall into penury ; and particularly when the miners are not content with moderate riches, they not rarely spend on new mines what they have accumulated from others. But no just ruler or magistrate deprives owners of their possessions ; that, however, may be done by a tyrant, who may cruelly rob his subjects not only of their goods honestly obtained, but even of life itself. And yet whenever I have inquired into the complaints which are in common vogue, I always find that the owners who are abused have the best of reasons for driving the men from the mines ; while those who abuse the owners have no reason to complain about them. Take the case of those who, not having paid their contributions, have lost the right of possession, or those who have been expelled by the magistrate out of another man's mine : for some wicked men, mining the small veins branching from the veins rich in metal, are wont to invade the property of another person. So the magistrate expels these men accused of wrong, and drives them from the mine. They then very frequently spread unpleasant rumours concerning this amongst the populace. Or, to take another case: when, as often happens, a dispute arises between neighbours, arbitrators appointed by the magistrate settle it, or the regular judges investigate and give judgment. Consequently, when the judgment is given, inasmuch as each party has consented to submit to it, neither side should complain of injustice ; and when the controversy is adjudged, inasmuch as the decision is in accordance with the laws concerning mining, one of the parties cannot be injured by the law. I do not vigorously contest the point, that at times a mine superintendent may exact a larger contribution from the owners than necessity demands. Nay, I will admit that a foreman may plaster over, or hide with a structure, a vein where it is rich in metals. Is the wickedness of one or two to brand the many honest with fraud and trickery? What body is supposed to be more pious and virtuous in the Republic than the Senate? Yet some Senators have been detected in peculations, and have been punished. Is this any reason that so honourable a house should lose its good name and fame? The superintendent cannot exact contributions from the owners without the knowledge and permission of the Bergmeister or the deputies; for this reason deception of this kind is impossible. Should the foremen be convicted of fraud, they are beaten with rods; or of theft, they are hanged. It is complained that some sellers and buyers of the shares in mines are fraudulent. I concede it. But can they deceive anyone except a stupid, careless man, unskilled in mining matters? Indeed, a wise and prudent man, skilled in this art, if he doubts the trustworthiness of a seller or buyer, goes at once to the mine that he may for himself examine the vein which has been so greatly praised or disparaged, and may consider whether he will buy or sell the shares or not. But people say, though such an one can be on his guard against fraud, yet a simple man and one who is easily credulous, is deceived. But we frequently see a man who is trying to mislead another in this way deceive himself, and deservedly become a laughing-stock for everyone; or very often the defrauder as well as the dupe is entirely ignorant of mining. If, for instance, a vein has been found to be abundant in ore, contrary to the idea of the would-be deceiver, then he who was to have been cheated gets a profit, and he who has been the deceiver loses. Nevertheless, the miners themselves rarely buy or sell shares, but generally they have *jurati venditores*[28] who buy and sell at such prices as they have been instructed to give or accept. Seeing therefore, that magistrates decide disputes on fair and just principles, that honest men deceive nobody, while a dishonest one cannot deceive easily, or if he does he cannot do so with impunity, the criticism of those who wish to disparage the honesty of miners has therefore no force or weight.

In the next place, the occupation of the miner is objectionable to nobody. For who, unless he be naturally malevolent and envious, will hate the man who gains wealth as it were from heaven? Or who will hate a man who to amplify his fortune, adopts a method which is free from reproach? A moneylender, if he demands an excessive interest, incurs the hatred of men. If he demands a moderate and lawful rate, so that he is not injurious to the public generally and does not impoverish them, he fails to become very rich from his business. Further, the gain derived from mining is not sordid, for how can it be such, seeing that it is so great, so plentiful, and of so innocent a nature. A merchant's profits are mean and base when he sells counterfeit and spurious merchandise, or puts far too high a price on goods that he has purchased for little; for this reason the merchant

[28]*Jurati Venditores*—"Sworn brokers." (?)

would be held in no less odium amongst good men than is the usurer, did they not take account of the risk he runs to secure his merchandise. In truth, those who on this point speak abusively of mining for the sake of detracting from its merits, say that in former days men convicted of crimes and misdeeds were sentenced to the mines and were worked as slaves. But to-day the miners receive pay, and are engaged like other workmen in the common trades.

Certainly, if mining is a shameful and discreditable employment for a gentleman because slaves once worked mines, then agriculture also will not be a very creditable employment, because slaves once cultivated the fields, and even to-day do so among the Turks; nor will architecture be considered honest, because some slaves have been found skilful in that profession; nor medicine, because not a few doctors have been slaves; nor will any other worthy craft, because men captured by force of arms have practised it. Yet agriculture, architecture, and medicine are none the less counted amongst the number of honourable professions; therefore, mining ought not for this reason to be excluded from them. But suppose we grant that the hired miners have a sordid employment. We do not mean by miners only the diggers and other workmen, but also those skilled in the mining arts, and those who invest money in mines. Amongst them can be counted kings, princes, republics, and from these last the most esteemed citizens. And finally, we include amongst the overseers of mines the noble Thucydides, the historian, whom the Athenians placed in charge of the mines of Thasos.[29] And it would not be unseemly for the owners themselves to work with their own hands on the works or ore, especially if they themselves have contributed to the cost of the mines. Just as it is not undignified for great men to cultivate their own land. Otherwise the Roman Senate would not have created Dictator L. Quintius Cincinnatus, as he was at work in the fields, nor would it have summoned to the Senate House the chief men of the State from their country villas. Similarly, in our day, Maximilian Cæsar would not have enrolled Conrad in the ranks of the nobles known as Counts; Conrad was really very poor when he served in the mines of Schneeberg, and for that reason he was nicknamed the "poor man"; but

[29]There is no doubt that Thucydides had some connection with gold mines; he himself is the authority for the statement that he worked mines in Thrace. Agricola seems to have obtained his idea that Thucydides held an appointment from the Athenians in charge of mines in Thasos, from Marcellinus (*Vita*, Thucydides, 30), who also says that Thucydides obtained possession of mines in Thrace through his marriage with a Thracian woman, and that it was while residing on the mines at Scapte-Hyle that he wrote his history. Later scholars, however, find little warrant for these assertions. The gold mines of Thasos—an island off the mainland of Thrace—are frequently mentioned by the ancient authors. Herodotus, VI., 46-47, says:—"Their (the Thasians') revenue was derived partly from "their possessions upon the mainland, partly from the mines which they owned. They "were masters of the gold mines of Scapte-Hyle, the yearly produce of which amounted to "eighty talents. Their mines in Thasos yielded less, but still were so prolific that besides "being entirely free from land-tax they had a surplus of income derived from the two "sources of their territory on the mainland and their mines, in common years two hundred "and in best years three hundred talents. I myself have seen the mines in question. By "far the most curious of them are those which the Phoenicians discovered at the time "when they went with Thasos and colonized the island, which took its name from him.

not many years after, he attained wealth from the mines of Fürst, which is a city in Lorraine, and took his name from "Luck."[30] Nor would King Vladislaus have restored to the Assembly of Barons, Tursius, a citizen of Cracow, who became rich through the mines in that part of the kingdom of Hungary which was formerly called Dacia.[31] Nay, not even the common worker in the mines is vile and abject. For, trained to vigilance and work by night and day, he has great powers of endurance when occasion demands, and easily sustains the fatigues and duties of a soldier, for he is accustomed to keep long vigils at night, to wield iron tools, to dig trenches, to drive tunnels, to make machines, and to carry burdens. Therefore, experts in military affairs prefer the miner, not only to a commoner from the town, but even to the rustic.

But to bring this discussion to an end, inasmuch as the chief callings are those of the moneylender, the soldier, the merchant, the farmer, and the miner, I say, inasmuch as usury is odious, while the spoil cruelly captured from the possessions of the people innocent of wrong is wicked in the sight of God and man, and inasmuch as the calling of the miner excels in honour and dignity that of the merchant trading for lucre, while it is not less noble though far more profitable than agriculture, who can fail to realize that mining is a calling of peculiar dignity? Certainly, though it is but one of ten important and excellent methods of acquiring wealth in an honourable way, a careful and diligent man can attain this result in no easier way than by mining.

"These Phoenician workings are in Thasos itself, between Coenyra and a place called "Aenyra over against Samothrace; a high mountain has been turned upside down in "the search for ores." (Rawlinson's Trans.). The occasion of this statement of Herodotus was the relations of the Thasians with Darius (521–486 B.C.). The date of the Phoenician colonization of Thasos is highly nebular—anywhere from 1200 to 900 B.C.

[30]Agricola, *De Veteribus et Novis Metallis*, Book I., p. 392, says:—"Conrad, whose "nickname in former years was 'pauper,' suddenly became rich from the silver mines of "Mount Jura, known as the *Firstum*." He was ennobled with the title of Graf Cuntz von Glück by the Emperor Maximilian (who was Emperor of the Holy Roman Empire, 1493-1519). Conrad was originally a working miner at Schneeberg where he was known as Armer Cuntz (poor Cuntz or Conrad) and grew wealthy from the mines of Fürst in Leberthal. This district is located in the Vosges Mountains on the borders of Lorraine and Upper Alsace. The story of Cuntz or Conrad von Glück is mentioned by Albinus (*Meissnische Land und Berg Chronica*, Dresden, 1589, p. 116), Mathesius (*Sarepta*, Nuremberg, 1578, fol. XVI.), and by others.

[31]Vladislaus III. was King of Poland, 1434–44, and also became King of Hungary in 1440. Tursius seems to be a Latinized name and cannot be identified.

END OF BOOK I.

BOOK II.

UALITIES which the perfect miner should possess and the arguments which are urged for and against the arts of mining and metallurgy, as well as the people occupied in the industry, I have sufficiently discussed in the first Book. Now I have determined to give more ample information concerning the miners.

In the first place, it is indispensable that they should worship God with reverence, and that they understand the matters of which I am going to speak, and that they take good care that each individual performs his duties efficiently and diligently. It is decreed by Divine Providence that those who know what they ought to do and then take care to do it properly, for the most part meet with good fortune in all they undertake; on the other hand, misfortune overtakes the indolent and those who are careless in their work. No person indeed can, without great and sustained effort and labour, store in his mind the knowledge of every portion of the metallic arts which are involved in operating mines. If a man has the means of paying the necessary expense, he hires as many men as he needs, and sends them to the various works. Thus formerly Sosias, the Thracian, sent into the silver mines a thousand slaves whom he had hired from the Athenian Nicias, the son of Niceratus[1]. But if a man cannot afford the expenditure he chooses of the various kinds of mining that work which he himself can most easily and efficiently do. Of these kinds, the two most important are the making prospect trenches and the washing of the sands of rivers, for out of these sands are often collected gold dust, or certain black stones from which tin is smelted, or even gems are sometimes found in them; the trenching occasionally lays bare at the grass-roots veins which are found rich in metals. If therefore by skill or by luck, such sands or veins shall fall into his hands, he will be able to establish his fortune without expenditure, and from poverty rise to wealth. If on the contrary, his hopes are not realised, then he can desist from washing or digging.

When anyone, in an endeavour to increase his fortune, meets the expenditure of a mine alone, it is of great importance that he should attend to his works and personally superintend everything that he has ordered to be done. For this reason, he should either have his dwelling at the mine,

[1]Xenophon. Essay on the Revenues of Athens, IV., 14.

"But we cannot but feel surprised that the State, when it sees many private individuals "enriching themselves from its resources, does not imitate their proceedings; for we heard "long ago, indeed, at least such of us as attended to these matters, that Nicias the son of "Niceratus kept a thousand men employed in the silver mines, whom he let on hire to "Sosias of Thrace on condition that he should give him for each an obolus a day, free of all "charges; and this number he always supplied undiminished." (See also Note 6). An obolus a day each, would be about 23 oz. Troy of silver per day for the whole number. In modern value this would, of course, be but about 50s. per day, but in purchasing power the value would probably be 100 to 1 (see Note on p 28). Nicias was estimated to have a fortune of 100 talents—about 83,700 Troy ounces of silver, and was one of the wealthiest of the Athenians. (Plutarch, Life of Nicias).

where he may always be in sight of the workmen and always take care that none neglect their duties, or else he should live in the neighbourhood, so that he may frequently inspect his mining works. Then he may send word by a messenger to the workmen that he is coming more frequently than he really intends to come, and so either by his arrival or by the intimation of it, he so frightens the workmen that none of them perform their duties otherwise than diligently. When he inspects the mines he should praise the diligent workmen and occasionally give them rewards, that they and the others may become more zealous in their duties; on the other hand, he should rebuke the idle and discharge some of them from the mines and substitute industrious men in their places. Indeed, the owner should frequently remain for days and nights in the mine, which, in truth, is no habitation for the idle and luxurious; it is important that the owner who is diligent in increasing his wealth, should frequently himself descend into the mine, and devote some time to the study of the nature of the veins and stringers, and should observe and consider all the methods of working, both inside and outside the mine. Nor is this all he ought to do, for sometimes he should undertake actual labour, not thereby demeaning himself, but in order to encourage his workmen by his own diligence, and to teach them their art; for that mine is well conducted in which not only the foreman, but also the owner himself, gives instruction as to what ought to be done. A certain barbarian, according to Xenophon, rightly remarked to the King of Persia that "the eye of the master feeds the horse,"[2] for the master's watchfulness in all things is of the utmost importance.

When several share together the expenditure on a mine, it is convenient and useful to elect from amongst their own number a mine captain, and also a foreman. For, since men often look after their own interests but neglect those of others, they cannot in this case take care of their own without at the same time looking after the interests of the others, neither can they neglect the interests of the others without neglecting their own. But if no man amongst them be willing or able to undertake and sustain the burdens of these offices, it will be to the common interest to place them in the hands of most diligent men. Formerly indeed, these things were looked after by the mining prefect[3], because the owners were kings, as Priam, who owned the gold mines round Abydos, or as Midas, who was the owner of those situated in Mount Bermius, or as Gyges, or as Alyattes, or as Crœsus, who was the owner of those mines near a deserted town between Atarnea and Pergamum[4]; sometimes the mines belonged to a Republic, as, for

[2]Xenophon. *Oeconomicus* XII., 20. "'I approve,' said Ischomachus, 'of the bar-"barian's answer to the King who found a good horse, and, wishing to fatten it as soon as "possible, asked a man with a good reputation for horsemanship what would do it?' The "man's reply was: 'Its master's eye.'"

[3]*Praefectus Metallorum*. In Saxony this official was styled the *Berghauptmann*. For further information see page 94 and note on page 78.

[4]This statement is either based upon Apollodorus, whom Agricola does not mention among his authorities, or on Strabo, whom he does so include. The former in his work on Mythology makes such a statement, for which Strabo (XIV., 5, 28) takes him to task as follows: "With this vain intention they collected the stories related by the Scepsian

instance, the prosperous silver mines in Spain which belonged to Carthage[5]; sometimes they were the property of great and illustrious families, as were the Athenian mines in Mount Laurion[6].

When a man owns mines but is ignorant of the art of mining, then it is advisable that he should share in common with others the expenses, not of one only, but of several mines. When one man alone meets the expense for a long time of a whole mine, if good fortune bestows on him a vein abundant in metals, or in other products, he becomes very wealthy; if, on the contrary, the mine is poor and barren, in time he will lose everything which he has expended on it. But the man who, in common with others, has laid out his money on several mines in a region renowned for its wealth of metals, rarely spends it in vain, for fortune usually responds to his hopes in part. For when out of twelve veins in which he has a joint interest

"(Demetrius), and taken from Callisthenes and other writers, who did not clear them from "false notions respecting the Halizones; for example, that the wealth of Tantalus and of the "Pelopidae was derived, it is said, from the mines about Phrygia and Sipylus; that of Cadmus "from the mines of Thrace and Mount Pangaeum; that of Priam from the gold mines of "Astyra, near Abydos (of which at present there are small remains, yet there is a large "quantity of matter ejected, and the excavations are proofs of former workings); that of "Midas from the mines about Mount Bermium; that of Gyges, Alyattes, and Croesus, from "the mines in Lydia and the small deserted city between Atarneus and Pergamum, where "are the sites of exhausted mines." (Hamilton's Trans., Vol. III., p. 66).

In adopting this view, Agricola apparently applied a wonderful realism to some Greek mythology—for instance, in the legend of Midas, which tells of that king being rewarded by the god Dionysus, who granted his request that all he touched might turn to gold; but the inconvenience of the gift drove him to pray for relief, which he obtained by bathing in the Pactolus, the sands of which thereupon became highly auriferous. Priam was, of course, King of Troy, but Homer does not exhibit him as a mine-owner. Gyges, Alyattes, and Croesus were successively Kings of Lydia, from 687 to 546 B.C., and were no doubt possessed of great treasure in gold. Some few years ago we had occasion to inquire into extensive old workings locally reputed to be Croesus' mines, at a place some distance north of Smyrna, which would correspond very closely to the locality here mentioned.

[5]There can be no doubt that the Carthaginians worked the mines of Spain on an extensive scale for a very long period anterior to their conquest by the Romans, but whether the mines were worked by the Government or not we are unable to find any evidence.

[6]The silver mines of Mt. Laurion formed the economic mainstay of Athens for the three centuries during which the State had the ascendency in Greece, and there can be no doubt that the dominance of Athens and its position as a sea-power were directly due to the revenues from the mines. The first working of the mines is shrouded in mystery. The scarcity of silver in the time of Solon (638–598 B.C.) would not indicate any very considerable output at that time. According to Xenophon (Essay on Revenue of Athens, IV., 2), written about 355 B.C., "they were wrought in very ancient times." The first definite discussion of the mines in Greek record begins about 500 B.C., for about that time the royalties began to figure in the Athenian Budget (Aristotle, Constitution of Athens, 47). There can be no doubt that the mines reached great prosperity prior to the Persian invasion. In the year 484 B.C. the mines returned 100 Talents (about 83,700 oz. Troy) to the Treasury, and this, on the advice of Themistocles, was devoted to the construction of the fleet which conquered the Persians at Salamis (480 B.C.). The mines were much interfered with by the Spartan invasions from 431 to 425 B.C., and again by their occupation in 413 B.C.; and by 355 B.C., when Xenophon wrote the "Revenues," exploitation had fallen to a low ebb, for which he proposes the remedies noted by Agricola on p. 28. By the end of the 4th Century, B.C., the mines had again reached considerable prosperity, as is evidenced by Demosthenes' orations against Pantaenetus and against Phaenippus, and by Lycurgus' prosecution of Diphilos for robbing the supporting pillars. The domination of the Macedonians under Philip and Alexander at the end of the 4th and beginning of the 3rd Centuries B.C., however, so flooded Greece with money from the mines of Thrace, that this probably interfered with Laurion, at this time, in any event, began the decadence of these mines. Synchronous also was the decadence of Athens, and, but for fitful displays, the State was not able to maintain even its own independence, not to mention its position as a dominant State. Finally, Strabo, writing about 30 B.C. gives the epitaph of every mining district—reworking the dumps. He says (IX., I, 23): "The silver mines in Attica were at first of importance, but

one yields an abundance of metals, it not only gives back to the owner the money he has spent, but also gives a profit besides; certainly there will be for him rich and profitable mining, if of the whole number, three, or four, or more veins should yield metal. Very similar to this is the advice which Xenophon gave to the Athenians when they wished to prospect for new veins of silver without suffering loss. "There are," he said, "ten tribes of Athenians; if, therefore, the State assigned an equal number of slaves to each tribe, and the tribes participated equally in all the new veins, undoubtedly by this method, if a rich vein of silver were found by one tribe, whatever profit were made from it would assuredly be shared by the whole number. And if two, three, or four tribes, or even half the whole number find veins, their works would then become more profitable; and it is not "probable that the work of all the tribes will be disappointing"[7] Although this advice of Xenophon is full of prudence, there is no opportunity for it except in free and wealthy States; for those people who are under the authority of kings and princes, or are kept in subjection by tyranny, do not dare, without permission, to incur such expenditure; those who are endowed with little wealth and resources cannot do so on account of insufficient funds. Moreover, amongst our race it is not customary for Republics to have slaves whom they can hire out for the benefit of the people[8]; but, instead, nowadays those who are in authority administer the funds for mining in the name of the State, not unlike private individuals.

"are now exhausted. The workmen, when the mines yielded a bad return to their labour, "committed to the furnace the old refuse and scoria, and hence obtained very pure silver, "for the former workmen had carried on the process in the furnace unskilfully."

Since 1860, the mines have been worked with some success by a French Company, thus carrying the mining history of this district over a period of twenty-seven centuries. The most excellent of many memoirs upon the mines at Laurion, not only for its critical, historical, and archæological value, but also because of its author's great insight into mining and metallurgy, is that of Edouard Ardaillon (*Les Mines du Laurion dans l'Antiquité*. Paris, 1897). We have relied considerably upon this careful study for the following notes, and would refer others to it for a short bibliography on the subject. We would mention in passing that Augustus Boeckh's "Silver Mines of Laurion," which is incorporated with his "Public Economy of Athens" (English Translation by Lewis, London, 1842) has been too much relied upon by English students. It is no doubt the product of one acquainted with written history, but without any special knowledge of the industry and it is based on no antiquarian research. The Mt. Laurion mining district is located near the southern end of the Attic Peninsula. The deposits are silver-lead, and they occur along the contact between approximately horizontal limestones and slates. There are two principal beds of each, thus forming three principal contacts. The most metalliferous of these contacts are those at the base of the slates, the lowest contact of the series being the richest. The ore-bodies were most irregular, varying greatly in size, from a thin seam between schist planes, to very large bodies containing as much as 200,000 cubic metres. The ores are argentiferous galena, accompanied by considerable amounts of blende and pyrites, all oxidized near the surface. The ores worked by the Ancients appear to have been fairly rich in lead, for the discards worked in recent years by the French Company, and the pillars left behind, ran 8% to 10% lead. The ratio of silver was from 40 to 90 ounces per ton of lead. The upper contacts were exposed by erosion and could be entered by tunnels, but the lowest and most prolific contact line was only to be reached by shafts. The shafts were ordinarily from four to six feet square, and were undoubtedly cut by hammer and chisel; they were as much as 380 feet deep. In some cases long inclines for travelling roads join the vertical shafts in depth. The drives, whether tunnels or from shafts, were not level, but followed every caprice of the sinuous contact. They were from two to two and a half feet wide, often driven in parallels with cross-cuts between, in order to exploit every corner of the contact. The stoping of ore-bodies discovered was undertaken quite systematically, the methods depending in the main on the shape of the ore-body. If the body was large, its dimensions were first determined by drives, crosscuts, rises, and

Some owners prefer to buy shares[9] in mines abounding in metals, rather than to be troubled themselves to search for the veins; these men employ an easier and less uncertain method of increasing their property. Although their hopes in the shares of one or another mine may be frustrated, the buyers of shares should not abandon the rest of the mines, for all the money expended will be recovered with interest from some other mine. They should not buy only high priced shares in those mines producing metals, nor should they buy too many in neighbouring mines where metal has not yet been found, lest, should fortune not respond, they may be exhausted by their losses and have nothing with which they may meet their expenses or buy other shares which may replace their losses. This calamity overtakes those who wish to grow suddenly rich from mines, and instead, they become very much poorer than before. So then, in the buying of shares, as in other matters, there should be a certain limit of expenditure which miners should set themselves, lest blinded by the desire for excessive wealth, they throw all their money away. Moreover, a prudent owner, before he buys shares, ought to go to the mine and carefully examine the nature of the vein, for it is very important that he should be on his guard lest fraudulent sellers of shares should deceive him. Investors in shares may perhaps become less wealthy, but they are more certain of some gain than those who mine for metals at their own expense, as they are more cautious in trusting to fortune. Neither ought miners to be altogether distrustful of fortune, as we see some are, who as soon as the shares of any mine begin to go up in

winzes, as the case might require. If the ore was mainly overhead it was overhand-stoped, and the stopes filled as work progressed, inclined winzes being occasionally driven from the stopes to the original entry drives. If the ore was mainly below, it was underhand-stoped, pillars being left if necessary—such pillars in some cases being thirty feet high. They also employed timber and artificial pillars. The mines were practically dry. There is little evidence of breaking by fire. The ore was hand-sorted underground and carried out by the slaves, and in some cases apparently the windlass was used. It was treated by grinding in mills and concentrating upon a sort of buddle. These concentrates—mostly galena—were smelted in low furnaces and the lead was subsequently cupelled. Further details of metallurgical methods will be found in Notes on p. 391 and p. 465, on metallurgical subjects.

The mines were worked by slaves. Even the overseers were at times apparently slaves, for we find (Xenophon, *Memorabilia*, II., 5) that Nicias paid a whole talent for a good overseer. A talent would be about 837 Troy ounces of silver. As wages of skilled labour were about two and one half pennyweights of silver per diem, and a family income of 100 ounces of silver per annum was affluence, the ratio of purchasing power of Attic coinage to modern would be about 100 to 1. Therefore this mine manager was worth in modern value roughly £8,000. The mines were the property of the State. The areas were defined by vertical boundaries, and were let on lease for definite periods for a fixed annual rent. More ample discussion of the law will be found on p. 83.

[7]Xenophon. (Essay on The Revenues, IV., 30). "I think, however, that I am "able to give some advice with regard to this difficulty also (the risk of opening new mines), "and to show how new operations may be conducted with the greatest safety. There are ten "tribes at Athens, and if to each of these the State should assign an equal number of slaves, "and the tribes should all make new cuttings, sharing their fortunes in common, then if but "one tribe should make any useful discovery it would point out something profitable to the "whole; but if two, three, or four, or half the number should make some discovery, it is "plain that the works would be more profitable in proportion, and that they should all fail "is contrary to all experience in past times." (Watson's Trans. p. 258).

[8]Agricola here refers to the proposal of Xenophon for the State to collect slaves and hire them to work the mines of Laurion. There is no evidence that this recommendation was ever carried out.

[9]*Partes.* Agricola, p. 89–91, describes in detail the organization and management of these share companies. See Note 8, p. 90.

value, sell them, on which account they seldom obtain even moderate wealth. There are some people who wash over the dumps from exhausted and abandoned mines, and those dumps which are derived from the drains of tunnels ; and others who smelt the old slags ; from all of which they make an ample return.

Now a miner, before he begins to mine the veins, must consider seven things, namely :—the situation, the conditions, the water, the roads, the climate, the right of ownership, and the neighbours. There are four kinds of situations—mountain, hill, valley, and plain. Of these four, the first two are the most easily mined, because in them tunnels can be driven to drain off the water, which often makes mining operations very laborious, if it does not stop them altogether. The last two kinds of ground are more troublesome, especially because tunnels cannot be driven in such places. Nevertheless, a prudent miner considers all these four sorts of localities in the region in which he happens to be, and he searches for veins in those places where some torrent or other agency has removed and swept the soil away ; yet he need not prospect everywhere, but since there is a great variety, both in mountains and in the three other kinds of localities, he always selects from them those which will give him the best chance of obtaining wealth.

In the first place, mountains differ greatly in position, some being situated in even and level plains, while others are found in broken and elevated regions, and others again seem to be piled up, one mountain upon another. The wise miner does not mine in mountains which are situated on open plains, neither does he dig in those which are placed on the summits of mountainous regions, unless by some chance the veins in those mountains have been denuded of their surface covering, and abounding in metals and other products, are exposed plainly to his notice,—for with regard to what I have already said more than once, and though I never repeat it again, I wish to emphasize this exception as to the localities which should not be selected. All districts do not possess a great number of mountains crowded together ; some have but one, others two, others three, or perhaps a few more. In some places there are plains lying between them ; in others the mountains are joined together or separated only by narrow valleys. The miner should not dig in those solitary mountains, dispersed through the plains and open regions, but only in those which are connected and joined with others. Then again, since mountains differ in size, some being very large, others of medium height, and others more like hills than mountains, the miner rarely digs in the largest or the smallest of them, but generally only in those of medium size. Moreover, mountains have a great variety of shapes ; for with some the slopes rise gradually, while others, on the contrary, are all precipitous ; in some others the slopes are gradual on one side, and on the other sides precipitous ; some are drawn out in length ; some are gently curved ; others assume different shapes. But the miner may dig in all parts of them, except where there are precipices, and he should not neglect even these latter if metallic veins are exposed before his eyes. There are just as great differences in hills as there are in mountains, yet the miner does not dig except in those situated in mountainous districts, and even very rarely in those. It is however very little to be wondered at that the hill in the Island of Lemnos was excavated, for the whole is of a reddish-yellow colour, which furnishes for the inhabitants that valuable clay so especially beneficial to mankind[10]. In like manner, other hills are excavated if chalk or other varieties of earth are exposed, but these are not prospected for.

There are likewise many varieties of valleys and plains. One kind is enclosed on the sides with its outlet and entrance open; another has either its entrance or its outlet open and the rest of it is closed in; both of these are properly called valleys. There is a third variety which is surrounded on all sides by mountains, and these are called *convalles*. Some valleys again, have recesses, and others have none; one is wide, another narrow; one is long, another short; yet another kind is not higher than the neighbouring plain, and others are lower than the surrounding flat country. But the miner does not dig in those surrounded on all sides by mountains, nor in those that are open, unless there be a low plain close at hand, or unless a vein of metal descending from the mountains should extend into the valley. Plains differ from one another, one being situated at low elevation, and others higher, one being level and another with a slight incline. The miner should never excavate the low-lying plain, nor one which is perfectly level, unless it be in some mountain, and rarely should he mine in the other kinds of plains.

With regard to the conditions of the locality the miner should not contemplate mining without considering whether the place be covered with trees or is bare. If it be a wooded place, he who digs there has this advantage, besides others, that there will be an abundant supply of wood for his underground timbering, his machinery, buildings, smelting, and other necessities. If there is no forest he should not mine there unless there is a river near, by which he can carry down the timber. Yet wherever there is a hope that pure gold or gems may be found, the ground can be turned up, even though there is no forest, because the gems need only to be polished and the gold to be purified. Therefore the inhabitants of hot regions obtain these substances from rough and sandy places, where sometimes there are not even shrubs, much less woods.

The miner should next consider the locality, as to whether it has a perpetual supply of running water, or whether it is always devoid of water except when a torrent supplied by rains flows down from the summits of the mountains. The place that Nature has provided with a river or stream can

[10]This island in the northern Ægean Sea has produced this "earth" from before Theophrastus' time (372–287 B.C.) down to the present day. According to Dana (System of Mineralogy 689), it is cimolite, a hydrous silicate of aluminium. The Ancients distinguished two kinds,—one sort used as a pigment, and the other for medicinal purposes. This latter was dug with great ceremony at a certain time of the year, moulded into cubes, and stamped with a goat,—the symbol of Diana. It thus became known as *terra sigillata*, and was an article of apothecary commerce down to the last century. It is described by Galen (XII., 12), Dioscorides (V., 63), and Pliny (XXXV., 14), as a remedy for ulcers and snake bites.

be made serviceable for many things ; for water will never be wanting and can be carried through wooden pipes to baths in dwelling-houses ; it may be carried to the works, where the metals are smelted ; and finally, if the conditions of the place will allow it, the water can be diverted into the tunnels, so that it may turn the underground machinery. Yet on the other hand, to convey a constant supply of water by artificial means to mines where Nature has denied it access, or to convey the ore to the stream, increases the expense greatly, in proportion to the distance the mines are away from the river.

The miner also should consider whether the roads from the neighbouring regions to the mines are good or bad, short or long. For since a region which is abundant in mining products very often yields no agricultural produce, and the necessaries of life for the workmen and others must all be imported, a bad and long road occasions much loss and trouble with porters and carriers, and this increases the cost of goods brought in, which, therefore, must be sold at high prices. This injures not so much the workmen as the masters ; since on account of the high price of goods, the workmen are not content with the wages customary for their labour, nor can they be, and they ask higher pay from the owners. And if the owners refuse, the men will not work any longer in the mines but will go elsewhere. Although districts which yield metals and other mineral products are generally healthy, because, being often situated on high and lofty ground, they are fanned by every wind, yet sometimes they are unhealthy, as has been related in my other book, which is called "*De Natura Eorum Quae Effluunt ex Terra.*" Therefore, a wise miner does not mine in such places, even if they are very productive, when he perceives unmistakable signs of pestilence. For if a man mines in an unhealthy region he may be alive one hour and dead the next.

Then, the miner should make careful and thorough investigation concerning the lord of the locality, whether he be a just and good man or a tyrant, for the latter oppresses men by force of his authority, and seizes their possessions for himself ; but the former governs justly and lawfully and serves the common good. The miner should not start mining operations in a district which is oppressed by a tyrant, but should carefully consider if in the vicinity there is any other locality suitable for mining and make up his mind if the overlord there be friendly or inimical. If he be inimical the mine will be rendered unsafe through hostile attacks, in one of which all of the gold or silver, or other mineral products, laboriously collected with much cost, will be taken away from the owner and his workmen will be struck with terror ; overcome by fear, they will hastily fly, to free themselves from the danger to which they are exposed. In this case, not only are the fortunes of the miner in the greatest peril but his very life is in jeopardy, for which reason he should not mine in such places.

Since several miners usually come to mine the veins in one locality, a settlement generally springs up, for the miner who began first cannot keep it exclusively for himself. The *Bergmeister* gives permits to some to mine the superior and some the inferior parts of the veins; to some he gives the cross veins, to others the inclined veins. If the man who first starts work finds the vein to be metal-bearing or yielding other mining products, it will not be to his advantage to cease work because the neighbourhood may be evil, but he will guard and defend his rights both by arms and by the law. When the *Bergmeister*[11] delimits the boundaries of each owner, it is the duty of a good miner to keep within his bounds, and of a prudent one to repel encroachments of his neighbours by the help of the law. But this is enough about the neighbourhood.

The miner should try to obtain a mine, to which access is not difficult, in a mountainous region, gently sloping, wooded, healthy, safe, and not far distant from a river or stream by means of which he may convey his mining products to be washed and smelted. This indeed, is the best position. As for the others, the nearer they approximate to this position the better they are; the further removed, the worse.

Now I will discuss that kind of minerals for which it is not necessary to dig, because the force of water carries them out of the veins. Of these there are two kinds, minerals—and their fragments[12]—and juices. When there are springs at the outcrop of the veins from which, as I have already said, the above-mentioned products are emitted, the miner should consider these first, to see whether there are metals or gems mixed with the sand, or whether the waters discharged are filled with juices. In case metals or gems have settled in the pool of the spring, not only should the sand from it be washed, but also that from the streams which flow from these springs, and even from the river itself into which they again discharge. If the springs discharge water containing some juice, this also should be collected; the further such a stream has flowed from the source, the more it receives plain water and the more diluted does it become, and so much the more deficient in strength. If the stream receives no water of another kind, or scarcely any, not only the rivers, but likewise the lakes which receive these waters, are of the same nature as the springs, and serve the same uses; of this kind is the lake which the Hebrews call the Dead Sea, and which is quite full of bituminous fluids[13]. But I must return to the subject of the sands.

Springs may discharge their waters into a sea, a lake, a marsh, a river, or a stream; but the sand of the sea-shore is rarely washed, for although the water flowing down from the springs into the sea carries some metals or gems with it, yet these substances can scarcely ever be reclaimed, because they are dispersed through the immense body of waters and mixed up with

[11]*Magister Metallorum.* See Note 1, p. 78, for the reasons of the adoption of the term *Bergmeister* and page 95 for details of his duties.

[12]*Ramenta.* "Particles." The author uses this term indifferently for fragments, particles of mineral, concentrates, gold dust, black tin, etc., in all cases the result of either natural or artificial concentration. As in technical English we have no general term for both natural and artificial "concentrates," we have rendered it as the context seemed to demand.

[13]A certain amount of bitumen does float ashore in the Dead Sea; the origin of it is, however, uncertain. Strabo (XVI., 2, 42), Pliny (V., 15 and 16), and Josephus (IV., 8), all mention this fact. The lake for this reason is often referred to by the ancient writers by the name *Asphaltites.*

other sand, and scattered far and wide in different directions, or they sink down into the depths of the sea. For the same reasons, the sands of lakes can very rarely be washed successfully, even though the streams rising from the mountains pour their whole volume of water into them. The particles of metals and gems from the springs are very rarely carried into the marshes, which are generally in level and open places. Therefore, the miner, in the first place, washes the sand of the spring, then of the stream which flows from it, then finally, that of the river into which the stream discharges. It is not worth the trouble to wash the sands of a large river which is on a level plain at a distance from the mountains. Where several springs carrying metals discharge their waters into one river, there is more hope of productive results from washing. The miner does not neglect even the sands of the streams in which excavated ores have been washed.

The waters of springs taste according to the juice they contain, and they differ greatly in this respect. There are six kinds of these tastes which the worker[14] especially observes and examines; there is the salty kind, which shows that salt may be obtained by evaporation; the nitrous, which indicates soda; the aluminous kind, which indicates alum; the vitrioline, which indicates vitriol; the sulphurous kind, which indicates sulphur; and as for the bituminous juice, out of which bitumen is melted down, the colour itself proclaims it to the worker who is evaporating it. The sea-water however, is similar to that of salt springs, and may be drawn into low-lying pits, and, evaporated by the heat of the sun, changes of itself into salt; similarly the water of some salt-lakes turns to salt when dried by the heat of summer. Therefore an industrious and diligent man observes and makes use of these things and thus contributes something to the common welfare.

The strength of the sea condenses the liquid bitumen which flows into it from hidden springs, into amber and jet, as I have described already in my books "*De Subterraneorum Ortu et Causis*"[15]. The sea, with certain

[14]*Excoctor*,—literally, "Smelter" or "Metallurgist."

[15]This reference should be to the *De Natura Fossilium* (p. 230), although there is a short reference to the matter in *De Ortu et Causis* (p. 59). Agricola maintained that not only were jet and amber varieties of bitumen, but also coal and camphor and obsidian. As jet (*gagates*) is but a compact variety of coal, the ancient knowledge of this substance has more interest than would otherwise attach to the gem, especially as some materials described in this connection were no doubt coal. The Greeks often refer to a series of substances which burned, contained earth, and which no doubt comprised coal. Such substances are mentioned by Aristotle (*De Mirabilibus*. 33, 41, 125), Nicander (*Theriaca*. 37), and others, previous to the 2nd Century B.C., but the most ample description is that of Theophrastus (23–28): "Some "of the more brittle stones there also are, which become as it were burning coals when put into "a fire, and continue so a long time; of this kind are those about Bena, found in mines and "washed down by the torrents, for they will take fire on burning coals being thrown on them, "and will continue burning as long as anyone blows them; afterward they will deaden, and "may after that be made to burn again. They are therefore of long continuance, but their "smell is troublesome and disagreeable. That also which is called the *spinus*, is found in "mines. This stone, cut in pieces and thrown together in a heap, exposed to the sun, burns; "and that the more, if it be moistened or sprinkled with water (a pyritiferous shale ?). But "the *Lipara* stone empties itself, as it were, in burning, and becomes like the *pumice*, "changing at once both its colour and density; for before burning it is black, smooth, and "compact. This stone is found in the Pumices, separately in different places, as it were, in

directions of the wind, throws both these substances on shore, and for this reason the search for amber demands as much care as does that for coral.

Moreover, it is necessary that those who wash the sand or evaporate the water from the springs, should be careful to learn the nature of the locality, its roads, its salubrity, its overlord, and the neighbours, lest on account of difficulties in the conduct of their business they become either impoverished by exhaustive expenditure, or their goods and lives are imperilled. But enough about this.

The miner, after he has selected out of many places one particular spot adapted by Nature for mining, bestows much labour and attention on the veins. These have either been stripped bare of their covering by chance and thus lie exposed to our view, or lying deeply hidden and concealed they are found after close search; the latter is more usual, the former more rarely happens, and both of these occurrences must be explained. There is more than one force which can lay bare the veins unaided by the industry or toil of man; since either a torrent might strip off the surface, which happened in the case of the silver mines of Freiberg (concerning which I have

"cells, nowhere continuous to the matter of them. It is said that in Melos the pumice "is produced in this manner in some other stone, as this is on the contrary in it; but the "stone which the pumice is found in is not at all like the *Lipara* stone which is found in it. "Certain stones there are about Tetras, in Sicily, which is over against Lipara, which "empty themselves in the same manner in the fire. And in the promontory called Erineas, "there is a great quantity of stone like that found about Bena, which, when "burnt, emits a bituminous smell, and leaves a matter resembling calcined earth. Those "fossil substances that are called coals, and are broken for use, are earthy; they kindle, "however, and burn like wood coals. These are found in Liguria, where there also is amber, "and in Elis, on the way to Olympia over the mountains. These are used by smiths." (Based on Hill's Trans.). Dioscorides and Pliny add nothing of value to this description.

Agricola (*De Nat. Fos.*, p. 229–230) not only gives various localities of jet, but also records its relation to coal. As to the latter, he describes several occurrences, and describes the deposits as *vena dilatata*. Coal had come into considerable use all over Europe, particularly in England, long before Agricola's time; the oft-mentioned charter to mine sea-coal given to the Monks of Newbottle Abbey, near Preston, was dated 1210.

Amber was known to the Greeks by the name *electrum*, but whether the alloy of the same name took its name from the colour of amber or *vice versa* is uncertain. The gum is supposed to be referred to by Homer (Od. XV. 460), and Thales of Miletus (640–546 B.C.) is supposed to have first described its power of attraction. It is mentioned by many other Greek authors, Æschylus, Euripides, Aristotle, and others. The latter (*De Mirabilibus*, 81) records of the amber islands in the Adriatic, that the inhabitants tell the story that on these islands amber falls from poplar trees. "This, they say, resembles gum and hardens "like stone, the story of the poets being that after Phaeton was struck by lightning his sisters "turned to poplar trees and shed tears of amber." Theophrastus (53) says: "Amber is "also a stone; it is dug out of the earth in Liguria and has, like the before-mentioned (lode- "stone), a power of attraction." Pliny (XXXVII., 11) gives a long account of both the substance, literature, and mythology on the subject. His view of its origin was: "Certainly amber is obtained from the islands of the Northern Ocean, and is called by the "Germans *glaesum*. For this reason the Romans, when Germanicus Cæsar commanded in "those parts, called one of them *Glaesaria*, which was known to the barbarians as "*Austeravia*. Amber originates from gum discharged by a kind of pine tree, like gum from "cherry and resin from the ordinary pine. It is liquid at first, and issues abundantly and "hardens in time by cold, or by the sea when the rising tides carry off the fragments from "the shores of those islands. Certainly it is thrown on the coasts, and is so light that it "appears to roll in the water. Our forefathers believed that it was the juice of a tree, for "they called it *succinum*. And that it belongs to a kind of pine tree is proved by the odour "of the pine tree which it gives when rubbed, and that it burns when ignited like a pitch "pine torch." The term amber is of Arabic origin—from *Ambar*—and this term was adopted by the Greeks after the Christian era. Agricola uses the Latin term *succinum* and (*De Nat. Fos.*, p. 231–5) disputes the origin from tree gum, and contends for submarine bitumen springs.

written in Book I. of my work "*De Veteribus et Novis Metallis*")[16]; or they may be exposed through the force of the wind, when it uproots and destroys the trees which have grown over the veins; or by the breaking away of the rocks; or by long-continued heavy rains tearing away the mountain; or by an earthquake; or by a lightning flash; or by a snowslide; or by the violence of the winds: "Of such a nature are the rocks hurled down from the mountains by the force of the winds aided by the ravages of time." Or the plough may uncover the veins, for Justin relates in his history that nuggets of gold had been turned up in Galicia by the plough; or this may occur through a fire in the forest, as Diodorus Siculus tells us happened in the silver mines in Spain; and that saying of Posidonius is appropriate enough: "The earth violently moved by the fires consuming the forest sends forth new products, namely, gold and silver."[17]. And indeed, Lucretius has explained the same thing more fully in the following lines: "Copper and gold and iron were discovered, and at the same time weighty silver and the substance of lead, when fire had burned up vast forests on the great hills, either by a discharge of heaven's lightning, or else because, when men were waging war with one another, forest fires had carried fire among the enemy in order to strike terror to them, or because, attracted by the goodness of the soil, they wished to clear rich fields and bring the country into pasture, or else to destroy wild beasts and enrich themselves with the game; for hunting with pitfalls and with fire came into use before the practice of enclosing the wood with toils and rousing the game with dogs. Whatever the fact is, from

[16]The statement in *De Veteribus et Novis Metallis* (p. 394) is as follows:—

"It came about by chance and accident that the silver mines were discovered at "Freiberg in Meissen. By the river Sala, which is not unknown to Strabo, is Hala, which "was once country, but is now a large town; the site, at any rate, even from Roman times "was famous and renowned for its salt springs, for the possession of which the Hermunduri "fought with the Chatti. When people carried the salt thence in wagons, as they now do "straight through Meissen (Saxony) into Bohemia—which is lacking in that seasoning to-day "no less than formerly—they saw galena in the wheel tracks, which had been uncovered by "the torrents. This lead ore, since it was similar to that of Goslar, they put into their carts "and carried to Goslar, for the same carriers were accustomed to carry lead from that city. "And since much more silver was smelted from this galena than from that of Goslar, certain "miners betook themselves to that part of Meissen in which is now situated Freiberg, a "great and wealthy town; and we are told by consistent stories and general report that "they grew rich out of the mines." Agricola places the discovery of the mines at Freiberg at about 1170. See Note 11, p. 5.

[17]Diodorus Siculus (v., 35). "These places being covered with woods, it is said that "in ancient times these mountains were set on fire by shepherds, and continued burning for "many days, and parched the earth, so that an abundance of silver ore was melted, and "the metal flowed in streams of pure silver like a river." Aristotle, nearly three centuries before Diodorus, mentions this same story (*De Mirabilibus*, 87): "They say that in Ibernia "the woods were set on fire by certain shepherds, and the earth thus heated, the country "visibly flowed silver; and when some time later there were earthquakes, and the earth "burst asunder at different places, a large amount of silver was collected." As the works of Posidonius are lost, it is probable that Agricola was quoting from Strabo (III., 2, 9), who says, in describing Spain: "Posidonius, in praising the amount and excellence of the "metals, cannot refrain from his accustomed rhetoric, and becomes quite enthusiastic in "exaggeration. He tells us we are not to disbelieve the fable that formerly the forests "having been set on fire, the earth, which was loaded with silver and gold, melted and "threw up these metals to the surface, for inasmuch as every mountain and wooded hill "seemed to be heaped up with money by a lavish fortune." (Hamilton's Trans. I., p. 220). Or he may have been quoting from the *Deipnosophistae* of Athenaeus (VI.), where Posidonius is quoted: "And the mountains . . . when once the woods upon them had caught fire, spontaneously ran with liquid silver."

whatever cause the heat of flame had swallowed up the forests with a frightful crackling from their very roots, and had thoroughly baked the earth with fire, there would run from the boiling veins and collect into the hollows of the grounds a stream of silver and gold, as well as of copper and lead."[18] But yet the poet considers that the veins are not laid bare in the first instance so much by this kind of fire, but rather that all mining had its origin in this. And lastly, some other force may by chance disclose the veins, for a horse, if this tale can be believed, disclosed the lead veins at Goslar by a blow from his hoof[19]. By such methods as these does fortune disclose the veins to us.

But by skill we can also investigate hidden and concealed veins, by observing in the first place the bubbling waters of springs, which cannot be very far distant from the veins because the source of the water is from them; secondly, by examining the fragments of the veins which the torrents break off from the earth, for after a long time some of these fragments are again buried in the ground. Fragments of this kind lying about on the ground, if they are rubbed smooth, are a long distance from the veins, because the torrent, which broke them from the vein, polished them while it rolled them a long distance; but if they are fixed in the ground, or if they are rough, they are nearer to the veins. The soil also should be considered, for this is often the cause of veins being buried more or less deeply under the earth; in this case the fragments protrude more or less widely apart, and miners are wont to call the veins discovered in this manner "*fragmenta*."[20]

Further, we search for the veins by observing the hoar-frosts, which whiten all herbage except that growing over the veins, because the veins emit a warm and dry exhalation which hinders the freezing of the moisture, for which reason such plants appear rather wet than whitened by the frost. This may be observed in all cold places before the grass has grown to its full size, as in the months of April and May; or when the late crop of

[18]Lucretius *De Rerum Natura* v. 1241.

[19]Agricola's account of this event in *De Veteribus et Novis Metallis* is as follows (p. 393): "Now veins are not always first disclosed by the hand and labour of man, nor has art "always demonstrated them; sometimes they have been disclosed rather by chance or by "good fortune. I will explain briefly what has been written upon this matter in history, "what miners tell us, and what has occurred in our times. Thus the mines at Goslar are "said to have been found in the following way. A certain noble, whose name is not recorded, "tied his horse, which was named Ramelus, to the branch of a tree which grew on the "mountain. This horse, pawing the earth with its hoofs, which were iron shod, and thus "turning it over, uncovered a hidden vein of lead, not unlike the winged Pegasus, who in the "legend of the poets opened a spring when he beat the rock with his hoof. So just as that "spring is named Hipprocrene after that horse, so our ancestors named the mountain "Rammelsberg. Whereas the perennial water spring of the poets would long ago have dried "up, the vein even to-day exists, and supplies an abundant amount of excellent lead. That "a horse can have opened a vein will seem credible to anyone who reflects in how many ways "the signs of veins are shown by chance, all of which are explained in my work *De Re* "*Metallica*. Therefore, here we will believe the story, both because it may happen that a "horse may disclose a vein, and because the name of the mountain agrees with the story." Agricola places the discovery of Goslar in the Hartz at prior to 936. See Note 11, p. 5.

[20]*Fragmenta*. The glossary gives "*Geschube*." This term is defined in the *Bergwerks' Lexicon* (Chemnitz, 1743, p 250) as the pieces of stone, especially tin-stone, broken from the vein and washed out by the water—the croppings.

hay, which is called the *cordum*, is cut with scythes in the month of September. Therefore in places where the grass has a dampness that is not congealed into frost, there is a vein beneath; also if the exhalation be excessively hot, the soil will produce only small and pale-coloured plants. Lastly, there are trees whose foliage in spring-time has a bluish or leaden tint, the upper branches more especially being tinged with black or with any other unnatural colour, the trunks cleft in two, and the branches black or discoloured. These phenomena are caused by the intensely hot and dry exhalations which do not spare even the roots, but scorching them, render the trees sickly; wherefore the wind will more frequently uproot trees of this kind than any others. Verily the veins do emit this exhalation. Therefore, in a place where there is a multitude of trees, if a long row of them at an unusual time lose their verdure and become black or discoloured, and frequently fall by the violence of the wind, beneath this spot there is a vein. Likewise along a course where a vein extends, there grows a certain herb or fungus which is absent from the adjacent space, or sometimes even from the neighbourhood of the veins. By these signs of Nature a vein can be discovered.

There are many great contentions between miners concerning the forked twig[21], for some say that it is of the greatest use in discovering veins, and others deny it. Some of those who manipulate and use the twig, first cut a fork from a hazel bush with a knife, for this bush they consider more efficacious than any other for revealing the veins, especially if the hazel

[21]So far as we are able to discover, this is the first published description of the divining rod as applied to minerals or water. Like Agricola, many authors have sought to find its origin among the Ancients. The magic rods of Moses and Homer, especially the rod with which the former struck the rock at Horeb, the rod described by Ctesias (died 398 B.C.) which attracted gold and silver, and the ***virgula divina*** of the Romans have all been called up for proof. It is true that the Romans are responsible for the name ***virgula divina***, "divining rod," but this rod was used for taking auguries by casting bits of wood (Cicero, *De Divinatione*). Despite all this, while the ancient naturalists all give detailed directions for finding water, none mention anything akin to the divining rod of the Middle Ages. It is also worth noting that the Monk Theophilus in the 12th Century also gives a detailed description of how to find water, but makes no mention of the rod. There are two authorities sometimes cited as prior to Agricola, the first being Basil Valentine in his "Last Will and Testament" (XXIV–VIII.), and while there may be some reason (see Appendix) for accepting the authenticity of the "Triumphal Chariot of Antimony" by this author, as dating about 1500, there can be little doubt that the "Last Will and Testament" was spurious and dated about 50 years after Agricola. Paracelsus (*De Natura Rerum* IX.), says: "These (divinations) are vain and misleading, and among the first of them are divining rods, which have "deceived many miners. If they once point rightly they deceive ten or twenty times." In his *De Origine Morborum Invisibilium* (Book I.) he adds that the "faith turns the rod." These works were no doubt written prior to *De Re Metallica*—Paracelsus died in 1541—but they were not published until some time afterward. Those interested in the strange persistence of this superstition down to the present day—and the files of the patent offices of the world are full of it—will find the subject exhaustively discussed in M. E. Chevreul's "*De la Baguette Divinatoire*," Paris, 1845; L. Figuier, "*Histoire du Merveilleux dans les temps moderne II.*", Paris, 1860; W. F. Barrett, Proceedings of the Society of Psychical Research, part 32, 1897, and 38, 1900; R. W. Raymond, American Inst. of Mining Engineers, 1883, p. 411. Of the descriptions by those who believed in it there is none better than that of William Pryce (*Mineralogia Cornubiensis*, London, 1778, pp. 113–123), who devotes much pains to a refutation of Agricola. When we consider that a century later than Agricola such an advanced mind as Robert Boyle (1626–1691), the founder of the Royal Society, was convinced of the genuineness of the divining rod, one is more impressed with the clarity of Agricola's vision. In fact, there were few indeed, down to the 19th Century, who did not believe implicitly in the effectiveness of this instrument, and while science has long since abandoned it, not a year passes but some new manifestation of its hold on the popular mind breaks out.

bush grows above a vein. Others use a different kind of twig for each metal, when they are seeking to discover the veins, for they employ hazel twigs for veins of silver; ash twigs for copper; pitch pine for lead and especially tin, and rods made of iron and steel for gold. All alike grasp the forks of the twig with their hands, clenching their fists, it being necessary that the clenched fingers should be held toward the sky in order that the twig should be raised at that end where the two branches meet. Then they wander hither and thither at random through mountainous regions. It is said that the moment they place their feet on a vein the twig immediately turns and twists, and so by its action discloses the vein; when they move their feet again and go away from that spot the twig becomes once more immobile.

The truth is, they assert, the movement of the twig is caused by the power of the veins, and sometimes this is so great that the branches of trees growing near a vein are deflected toward it. On the other hand, those who say that the twig is of no use to good and serious men, also deny that the motion is due to the power of the veins, because the twigs will not move for everybody, but only for those who employ incantations and craft. Moreover, they deny the power of a vein to draw to itself the branches of trees, but they say that the warm and dry exhalations cause these contortions. Those who advocate the use of the twig make this reply to these objections: when one of the miners or some other person holds the twig in his hands, and it is not turned by the force of a vein, this is due to some peculiarity of the individual, which hinders and impedes the power of the vein, for since the power of the vein in turning and twisting the twig may be not unlike that of a magnet attracting and drawing iron toward itself, this hidden quality of a man weakens and breaks the force, just the same as garlic weakens and overcomes the strength of a magnet. For a magnet smeared with garlic juice cannot attract iron; nor does it attract the latter when rusty. Further, concerning the handling of the twig, they warn us that we should not press the fingers together too lightly, nor clench them too firmly, for if the twig is held lightly they say that it will fall before the force of the vein can turn it; if however, it is grasped too firmly the force of the hands resists the force of the veins and counteracts it. Therefore, they consider that five things are necessary to insure that the twig shall serve its purpose: of these the first is the size of the twig, for the force of the veins cannot turn too large a stick; secondly, there is the shape of the twig, which must be forked or the vein cannot turn it; thirdly, the power of the vein which has the nature to turn it; fourthly, the manipulation of the twig; fifthly, the absence of impeding peculiarities. These advocates of the twig sum up their conclusions as follows: if the rod does not move for everybody, it is due to unskilled manipulation or to the impeding peculiarities of the man which oppose and resist the force of the veins, as we said above, and those who search for veins by means of the twig need not necessarily make incantations, but it is sufficient that they handle it suitably and are devoid of impeding power; therefore, the twig may be of use to good and serious

A—Twig. B—Trench.

men in discovering veins. With regard to deflection of branches of trees they say nothing and adhere to their opinion.

Since this matter remains in dispute and causes much dissention amongst miners, I consider it ought to be examined on its own merits. The wizards, who also make use of rings, mirrors and crystals, seek for veins with a divining rod shaped like a fork; but its shape makes no difference in the matter,—it might be straight or of some other form—for it is not the form of the twig that matters, but the wizard's incantations which it would not become me to repeat, neither do I wish to do so. The Ancients, by means of the divining rod, not only procured those things necessary for a livelihood or for luxury, but they were also able to alter the forms of things by it; as when the magicians changed the rods of the Egyptians into serpents, as the writings of the Hebrews relate[22]; and as in Homer, Minerva with a divining rod turned the aged Ulysses suddenly into a youth, and then restored him back again to old age; Circe also changed Ulysses' companions into beasts, but afterward gave them back again their human form[23]; moreover by his rod, which was called "Caduceus," Mercury gave

[22]Exodus VII., 10, 11, 12.
[23]Odyssey XVI., 172, and X., 238.

sleep to watchmen and awoke slumberers[24]. Therefore it seems that the divining rod passed to the mines from its impure origin with the magicians. Then when good men shrank with horror from the incantations and rejected them, the twig was retained by the unsophisticated common miners, and in searching for new veins some traces of these ancient usages remain.

But since truly the twigs of the miners do move, albeit they do not generally use incantations, some say this movement is caused by the power of the veins, others say that it depends on the manipulation, and still others think that the movement is due to both these causes. But, in truth, all those objects which are endowed with the power of attraction do not twist things in circles, but attract them directly to themselves; for instance, the magnet does not turn the iron, but draws it directly to itself, and amber rubbed until it is warm does not bend straws about, but simply draws them to itself. If the power of the veins were of a similar nature to that of the magnet and the amber, the twig would not so much twist as move once only, in a semi-circle, and be drawn directly to the vein, and unless the strength of the man who holds the twig were to resist and oppose the force of the vein, the twig would be brought to the ground; wherefore, since this is not the case, it must necessarily follow that the manipulation is the cause of the twig's twisting motion. It is a conspicuous fact that these cunning manipulators do not use a straight twig, but a forked one cut from a hazel bush, or from some other wood equally flexible, so that if it be held in the hands, as they are accustomed to hold it, it turns in a circle for any man wherever he stands. Nor is it strange that the twig does not turn when held by the inexperienced, because they either grasp the forks of the twig too tightly or hold them too loosely. Nevertheless, these things give rise to the faith among common miners that veins are discovered by the use of twigs, because whilst using these they do accidentally discover some; but it more often happens that they lose their labour, and although they might discover a vein, they become none the less exhausted in digging useless trenches than do the miners who prospect in an unfortunate locality. Therefore a miner, since we think he ought to be a good and serious man, should not make use of an enchanted twig, because if he is prudent and skilled in the natural signs, he understands that a forked stick is of no use to him, for as I have said before, there are the natural indications of the veins which he can see for himself without the help of twigs. So if Nature or chance should indicate a locality suitable for mining, the miner should dig his trenches there; if no vein appears he must dig numerous trenches until he discovers an outcrop of a vein.

A vena dilatata is rarely discovered by men's labour, but usually some force or other reveals it, or sometimes it is discovered by a shaft or a tunnel on a *vena profunda*[25].

[24]Odyssey XXIV., 1, etc. The *Caduceus* of Hermes had also the power of turning things to gold, and it is interesting to note that in its oldest form, as the insignia of heralds and of ambassadors, it had two prongs.

[25]In a general way *venae profundae* were fissure veins and *venae dilatatae* were sheeted deposits. For description see Book III.

The veins after they have been discovered, and likewise the shafts and tunnels, have names given them, either from their discoverers, as in the case at Annaberg of the vein called "Kölergang," because a charcoal burner discovered it; or from their owners, as the Geyer, in Joachimstal, because part of the same belonged to Geyer; or from their products, as the "Pleygang" from lead, or the "Bissmutisch" at Schneeberg from bismuth[26]; or from some other circumstances, such as the rich alluvials from the torrent by which they were laid bare in the valley of Joachim. More often the first discoverers give the names either of persons, as those of German Kaiser, Apollo, Janus; or the name of an animal, as that of lion, bear, ram, or cow; or of things inanimate, as "silver chest" or "ox stalls"; or of something ridiculous, as "glutton's nightshade"; or finally, for the sake of a good omen, they call it after the Deity. In ancient times they followed the same custom and gave names to the veins, shafts and tunnels, as we read in Pliny: "It is wonderful that the shafts begun by Hannibal in Spain are still worked, their names being derived from their discoverers. One of these at the present day, called Baebelo, furnished Hannibal with three hundred pounds weight (of silver) per day."[27]

[26]These mines are in the Erzgebirge. We have adopted the names given in the German translation.

[27] The quotation from Pliny (XXXIII., 31) as a whole reads as follows:—

"Silver is found in nearly all the provinces, but the finest of all in Spain; where it "is found in the barren lands, and in the mountains. Wherever one vein of silver has been "found, another is sure to be found not far away. This is the case of nearly all the metals, "whence it appears that the Greeks derived *metalla*. It is wonderful that the shafts begun "by Hannibal in Spain still remain, their names being derived from their makers. One of "these at the present day called Baebelo, furnished Hannibal with three hundred pounds' "weight (of silver) per day. This mountain is excavated for a distance of fifteen hundred "paces; and for this distance there are waterbearers lighted by torches standing night and "day baling out the water in turns, thus making quite a river." Hannibal dates 247–183 B.C. and was therefore dead 206 years when Pliny was born. According to a footnote in Bostock and Riley's translation of Pliny, these workings were supposed to be in the neighbourhood of Castulo, now Cazlona, near Linares. It was at Castulo that Hannibal married his rich wife Himilce; and in the hills north of Linares there are ancient silver mines still known as Los Pozos de Anibal.

END OF BOOK II.

BOOK III.

PREVIOUSLY I have given much information concerning the miners, also I have discussed the choice of localities for mining, for washing sands, and for evaporating waters; further, I described the method of searching for veins. With such matters I was occupied in the second book; now I come to the third book, which is about veins and stringers, and the seams in the rocks[1]. The term "vein" is sometimes used to indicate *canales* in the earth, but very often elsewhere by this name I have described that which may be put in vessels[2]; I now attach a second significance to these words, for by them I mean to designate any mineral substances which the earth keeps hidden within her own deep receptacles.

[1]Modern nomenclature in the description of ore-deposits is so impregnated with modern views of their origin, that we have considered it desirable in many instances to adopt the Latin terms used by the author, for we believe this method will allow the reader greater freedom of judgment as to the author's views. The Latin names retained are usually expressive even to the non-Latin student. In a general way, a *vena profunda* is a fissure vein, a *vena dilatata* is a bedded deposit, and a *vena cumulata* an impregnation, or a replacement or a *stockwerk.* The *canales,* as will appear from the following footnote, were ore channels. "The seams of the rocks" (*commissurae saxorum*) are very puzzling. The author states, as appears in the following note, that they are of two kinds,—contemporaneous with the formation of the rocks, and also of the nature of veinlets. However, as to their supposed relation to the strike of veins, we can offer no explanation. There are passages in this chapter where if the word "ore-shoot" were introduced for "seams in the rocks" the text would be intelligible. That is, it is possible to conceive the view that the determination of whether an east-west vein ran east or ran west was dependent on the dip of the ore-shoot along the strike. This view, however, is utterly impossible to reconcile with the description and illustration of *commissurae saxorum* given on page 54, where they are defined as the finest stringers. The following passage from the *Nützliche Bergbüchlin* (see Appendix), reads very much as though the dip of ore-shoots was understood at this time in relation to the direction of veins. "Every vein (*gang*) has two (outcrops) *ausgehen,* one of the "*ausgehen* is toward daylight along the whole length of the vein, which is called the *ausgehen* "of the whole vein. The other *ausgehen* is contrary to or toward the strike (*streichen*) of "the vein, according to its rock (*gestein*), that is called the *gesteins ausgehen*; for instance, "every vein that has its strike from east to west has its *gesteins ausgehen* to the east, and "*vice-versa.*"

Agricola's classification of ore-deposits, after the general distinction between alluvial and *in situ* deposits, is based entirely upon form, as will be seen in the quotation below relating to the origin of *canales.* The German equivalents in the Glossary are as follows:—

Fissure vein (*vena profunda*) ...... *Gang.*
Bedded deposit (*vena dilatata*) ...... *Schwebender gang oder fletze.*
Stockwerk or impregnation (*vena cumulata*) ...... *Geschute oder stock.*
Stringer (*fibra*) ...... *Klufft.*
Seams or joints (*commissurae saxorum*) ...... *Absetzen des gesteins.*

It is interesting to note that in *De Natura Fossilium* he describes coal and salt, and later in *De Re Metallica* he describes the Mannsfeld copper schists, as all being *venae dilatatae.* This nomenclature and classification is not original with Agricola. Pliny (XXXIII, 21) uses the term *vena* with no explanations, and while Agricola coined the Latin terms for various kinds of veins, they are his transliteration of German terms already in use. The *Nützliche Bergbüchlin* gives this same classification.

HISTORICAL NOTE ON THE THEORY OF ORE DEPOSITS. Prior to Agricola there were three schools of explanation of the phenomena of ore deposits, the orthodox followers of the Genesis, the Greek Philosophers, and the Alchemists. The geology of the Genesis—the contemporaneous formation of everything—needs no comment other than that for anyone to have proposed an alternative to the dogma of the orthodox during the Middle Ages, required

---

[2]The Latin *vena,* "vein," is also used by the author for ore; hence this descriptive warning as to its intended double use.

First I will speak of the veins, which, in depth, width, and length, differ very much one from another. Those of one variety descend from the surface of the earth to its lowest depths, which on account of this characteristic, I am accustomed to call "*venae profundae.*"

much independence of mind. Of the Greek views—which are meagre enough—that of the Peripatetics greatly dominated thought on natural phenomena down to the 17th century. Aristotle's views may be summarized: The elements are earth, water, air, and fire; they are transmutable and never found pure, and are endowed with certain fundamental properties which acted as an "efficient" force upon the material cause—the elements. These properties were dryness and dampness and heat and cold, the latter being active, the former passive. Further, the elements were possessed of weight and lightness, for instance earth was absolutely heavy, fire absolutely light. The active and passive properties existed in binary combinations, one of which is characteristic, *i.e.*, "earth" is cold and dry, water damp and cold, fire hot and dry, air hot and wet; transmutation took place, for instance, by removing the cold from water, when air resulted (really steam), and by removing the dampness from water, when "earth" resulted (really any dissolved substance). The transmutation of the elements in the earth (meaning the globe) produces two "exhalations," the one fiery (probably meaning gases), the other damp (probably meaning steam). The former produces stones, the latter the metals. Theophrastus (On Stones, I to VII.) elaborates the views of Aristotle on the origin of stones, metals, etc.: "Of things "formed in the earth some have their origin from water, others from earth. Water is the "basis of metals, silver, gold, and the rest; 'earth' of stones, as well the more precious "as the common. . . . All these are formed by solidification of matter pure and "equal in its constituent parts, which has been brought together in that state by mere "afflux or by means of some kind of percolation, or separated. . . . The solidification "is in some of these substances due to heat and in others to cold." (Based on Hill's Trans., pp. 3-11). That is, the metals inasmuch as they become liquid when heated must be in a large part water, and, like water, they solidify with cold. Therefore, the "metals are cold and damp." Stones, on the other hand, solidify with heat and do not liquefy, therefore, they are "dry and hot" and partake largely of "earth." This "earth" was something indefinite, but purer and more pristine than common clay. In discussing the ancient beliefs with regard to the origin of deposits, we must not overlook the import of the use of the word "vein" (*vena*) by various ancient authors including Pliny (XXXIII, 21), although he offers no explanation of the term.

During the Middle Ages there arose the horde of Alchemists and Astrologers, a review of the development of whose muddled views is but barren reading. In the main they held more or less to the Peripatetic view, with additions of their own. Geber (13th (?) century, see Appendix B) propounded the conception that all metals were composed of varying proportions of "spiritual" sulphur and quicksilver, and to these Albertus Magnus added salt. The Astrologers contributed the idea that the immediate cause of the metals were the various planets. The only work devoted to description of ore-deposits prior to Agricola was the *Bergbüchlin* (about 1,520, see Appendix B), and this little book exhibits the absolute apogee of muddled thought derived from the Peripatetics, the Alchemists, and the Astrologers. We believe it is of interest to reproduce the following statement, if for no other reason than to indicate the great advance in thought shown by Agricola.

"The first chapter or first part; on the common origin of ore, whether silver, gold, "tin, copper, iron, or lead ore, in which they all appear together, and are called by the common "name of metallic ore. It must be noticed that for the washing or smelting of metallic ore, "there must be the one who works and the thing that is worked upon, or the material upon "which the work is expended. The general worker (efficient force) on the ore and on all "things that are born, is the heavens, its movement, its light and influences, as the "philosophers say. The influence of the heavens is multiplied by the movement of the "firmaments and the movements of the seven planets. Therefore, every metallic ore "receives a special influence from its own particular planet, due to the properties of the "planet and of the ore, also due to properties of heat, cold, dampness, and dryness. Thus "gold is of the Sun or its influence, silver of the Moon, tin of Jupiter, copper of Venus, iron "of Mars, lead of Saturn, and quicksilver of Mercury. Therefore, metals are often called by "these names by hermits and other philosophers. Thus gold is called the Sun, in Latin *Sol*, "silver is called the Moon, in Latin *Luna*, as is clearly stated in the special chapters on each "metal. Thus briefly have we spoken of the 'common worker' of metal and ore. But the "thing worked upon, or the common material of all metals, according to the opinion of "the learned, is sulphur and quicksilver, which through the movement and influence of the "heavens must have become united and hardened into one metallic body or one ore. "Certain others hold that through the movement and the influence of the heavens, vapours "or *braden*, called mineral exhalations, are drawn up from the depths of the earth, from "sulphur and quicksilver, and the rising fumes pass into the veins and stringers and are

A. C.—THE MOUNTAIN. B—*Vena profunda.*

Another kind, unlike the *venae profundae,* neither ascend to the surface of the earth nor descend, but lying under the ground, expand over a large area; and on that account I call them "*venae dilatatae.*"

A. D.—THE MOUNTAIN. B. C—*Vena dilatata.*

Another occupies a large extent of space in length and width; therefore I usually call it "*vena cumulata,*" for it is nothing else than an accumulation of some certain kind of mineral, as I have described in the book

"united through the effect of the planets and made into ore. Certain others hold that "metal is not formed from quicksilver, because in many places metallic ore is found and "no quicksilver. But instead of quicksilver they maintain a damp and cold and slimy "material is set up on all sulphur which is drawn out from the earth, like your perspiration, "and from that mixed with sulphur all metals are formed. Now each of these opinions is "correct according to a good understanding and right interpretation; the ore or metal is "formed from the fattiness of the earth as the material of the first degree (primary element), "also the vapours or *braden* on the one part and the materials on the other part, both of which "are called quicksilver. Likewise in the mingling or union of the quicksilver and the "sulphur in the ore, the sulphur is counted the male and quicksilver the female, as in the "bearing or conception of a child. Also the sulphur is a special worker in ore or metal.

"The second chapter or part deals with the general capacity of the mountain. "Although the influence of the heavens and the fitness of the material are necessary to the "formation of ore or metal, yet these are not enough thereto. But there must be adapt-"ability of the natural vessel in which the ore is formed, such are the veins, namely "*steinendegange, flachgange, schargange, creutzgange,* or as these may be termed in provincial "names. Also the mineral force must have easy access to the natural vessel such as "through the *kluffte* (stringers), namely *hengkluft, querklufte, flacheklufte, creutzkluft,* and "other occasional *flotzwerk,* according to their various local names. Also there must be a "suitable place in the mountain which the veins and stringers can traverse."

Agricola's Views on the Origin of Ore Deposits. Agricola rejected absolutely the Biblical view which, he says, was the opinion of the vulgar; further, he repudiates the alchemistic and astrological view with great vigour. There can be no doubt, however, that he was greatly influenced by the Peripatetic philosophy. He accepted absolutely the four elements—earth, fire, water, and air, and their "binary" properties, and the theory that every substance had a material cause operated upon by an efficient force. Beyond this he did not go, and a large portion of *De Ortu et Causis* is devoted to disproof of the origin of metals and stones from the Peripatetic "exhalations."

No one should conclude that Agricola's theories are set out with the clarity of Darwin or Lyell. However, the matter is of such importance in the history of the theory of ore-deposits, and has been either so ignored or so coloured by the preconceptions of narrators, that we consider it justifiable to devote the space necessary to a reproduction of his own statements in *De Ortu et Causis* and other works. Before doing so we believe it will be of service to readers to summarize these views, and in giving quotations from the Author's other works, to group them under special headings, following the outline of his theory given below. His theory was:—

(1) Openings in the earth (*canales*) were formed by the erosion of subterranean waters.

(2) These ground waters were due (*a*) to the infiltration of the surface waters, rain, river, and sea water; (*b*) to the condensation of steam (*halitus*) arising from the penetration of the surface waters to greater depths,—the production of this *halitus* being due to subterranean heat, which in his view was in turn due in the main to burning bitumen (a comprehensive genera which embraced coal).

(3) The filling of these *canales* is composed of "earth," "solidified juices," "stone," metals, and "compounds," all deposited from water and "juices" circulating in the *canales.* (See also note 4, page 1).

"Earth" comprises clay, mud, ochre, marl, and "peculiar earths" generally. The origin of these "earths" was from rocks, due to erosion, transportation, and deposition by water. "Solidified juices" (*succi concreti*) comprised salt, soda, vitriol, bitumen, etc., being generally those substances which he conceived were soluble in and deposited from water. "Stones" comprised precious, semi-precious, and unusual stones, such as quartz, fluor-spar, etc., as distinguished from country rock; the origin of these he attributed in minor proportion to transportation of fragments of rock, but in the main to deposits from ordinary mineral juice and from "stone juice" (*succus lapidescens*). Metals comprised the seven traditional metals; the "compounds" comprised the metallic minerals; and both were due to deposition from juices, the compounds being due to a mixture of juices. The "juices" play the most important part in Agricola's theory. Each substance had its own particular juice, and in his theory every substance had a material and an efficient cause, the first being the juice, the second being heat or cold. Owing to the latter the juices fell into two categories—those solidified by heat (*i.e.*, by evaporation, such as salt), and those solidified by cold, (*i.e*, because metals melt and flow by heat, therefore their solidification was due to cold, and the juice underwent similar treatment). As to the origin of these juices, some were generated by the solution of their own particular substance, but in the

entitled *De Subterraneorum Ortu et Causis.* It occasionally happens, though it is unusual and rare, that several accumulations of this kind are found in one place, each one or more fathoms in depth and four or five in

main their origin was due to the combination of "dry things," such as "earth," with water, the mixture being heated, and the resultant metals depended upon the proportions of "earth" and water. In some cases we have been inclined to translate *succus* (juice) as "solution," but in other cases it embraced substances to which this would not apply, and we feared implying in the text a chemical understanding not warranted prior to the atomic theory. In order to distinguish between earths, (clays, etc.,) the Peripatetic "earth" (a pure element) and the earth (the globe) we have given the two former in quotation marks. There is no doubt some confusion between earth (clays, etc.) and the Peripatetic "earth," as the latter was a pure substance not found in its pristine form in nature; it is, however, difficult to distinguish between the two.

ORIGIN OF CANALES (*De Ortu*, p. 35). "I now come to the *canales* in the earth. "These are veins, veinlets, and what are called 'seams in the rocks.' These serve as "vessels or receptacles for the material from which minerals (*res fossiles*) are formed. "The term *vena* is most frequently given to what is contained in the *canales*, but likewise "the same name is applied to the *canales* themselves. The term vein is borrowed from "that used for animals, for just as their veins are distributed through all parts of the "body, and just as by means of the veins blood is diffused from the liver throughout the "whole body, so also the veins traverse the whole globe, and more particularly the "mountainous districts; and water runs and flows through them. With regard to veinlets "or stringers and 'seams in the rocks,' which are the thinnest stringers, the following is the "mode of their arrangement. Veins in the earth, just like the veins of an animal, have certain "veinlets of their own, but in a contrary way. For the larger veins of animals pour blood "into the veinlets, while in the earth the humours are usually poured from the veinlets into "the larger veins, and rarely flow from the larger into the smaller ones. As for the seams in "the rocks (*commissurae saxorum*) we consider that they are produced by two methods: by "the first, which is peculiar to themselves, they are formed at the same time as the rocks, "for the heat bakes the refractory material into stone and the non-refractory material "similarly heated exhales its humours and is made into 'earth,' generally friable. The "other method is common also to veins and veinlets, when water is collected into one "place it softens the rock by its liquid nature, and by its weight and pressure breaks and "divides it. Now, if the rock is hard, it makes seams in the rocks and veinlets, and if it is "not too hard it makes veins. However, if the rocks are not hard, seams and veinlets are "created as well as veins. If these do not carry a very large quantity of water, or if they "are pressed by a great volume of it, they soon discharge themselves into the nearest veins. "The following appears to be the reason why some veinlets or stringers and veins are "*profundae* and others *dilatatae*. The force of the water crushes and splits the brittle rocks; "and when they are broken and split, it forces its way through them and passes on, at one "time in a downward direction, making small and large *venae profundae*, at another time "in a lateral direction, in which way *venae dilatatae* are formed. Now since in each "class there are found some which are straight, some inclined, and some crooked, it should "be explained that the water makes the *vena profunda* straight when it runs straight "downward, inclined when it runs in an inclined direction; and that it makes a *vena* "*dilatata* straight when it runs horizontally to the right or left, and in a similar way inclined "when it runs in a sloping direction. Stringers and large veins of the *profunda* sort, extending "for considerable lengths, become crooked from two causes. In one case when narrow "veins are intersected by wide ones, then the latter bend or drag the former a little. In "the other case, when the water runs against very hard rock, being unable to break through, "it goes around the nearest way, and the stringers and veins are formed bent and crooked. "This last is also the reason we sometimes see crooked small and large *venae dilatatae*, not "unlike the gentle rise and fall of flowing water. Next, *venae profundae* are wide, either "because of abundant water or because the rock is fragile. On the other hand, they are "narrow, either because but little water flows and trickles through them, or because the "rock is very hard. The *venae dilatatae*, too, for the same reasons, are either thin or thick. "There are other differences, too, in stringers and veins, which I will explain in my work "*De Re Metallica*. . . . There is also a third kind of vein which, as it cannot be "described as a wide *vena profunda*, nor as a thick *vena dilatata*, we will call a *vena cumulata*. "These are nothing else than places where some species of mineral is accumulated; "sometimes exceeding in depth and also in length and breadth 600 feet; sometimes, or "rather generally, not so deep nor so long, nor so wide. These are created when water "has broken away the rock for such a length, breadth, and thickness, and has flung aside "and ejected the stones and sand from the great cavern which is thus made; and afterward "when the mouth is obstructed and closed up, the whole cavern is filled with material "from which there is in time produced some one or more minerals. Now I have stated

width, and one is distant from another two, three, or more fathoms. When the excavation of these accumulations begins, they at first appear in the shape of a disc; then they open out wider; finally from each of such

"when discoursing on the origin of subterranean humours, that water erodes away "substances inside the earth, just as it does those on the surface, and least of all does it "shun minerals; for which reason we may daily see veinlets and veins sometimes filled with "air and water, but void and empty of mining products, and sometimes full of these same "materials. Even those which are empty of minerals become finally obstructed, and when "the rock is broken through at some other point the water gushes out. It is certain that "old springs are closed up in some way and new ones opened in others. In the same "manner, but much more easily and quickly than in the solid rock, water produces stringers "and veins in surface material, whether it be in plains, hills, or mountains. Of this kind are "the stringers in the banks of rivers which produce gold, and the veins which produce "peculiar earth. So in this manner in the earth are made *canales* which bear minerals."

ORIGIN OF GROUND WATERS. (*De Ortu* p. 5). " . . . . Besides rain there is "another kind of water by which the interior of the earth is soaked, so that being heated "it can continually give off *halitus*, from which arises a great and abundant force of waters." In description of the *modus operandi* of *halitum*, he says (p. 6): " . . . . *Halitus* "rises to the upper parts of the *canales*, where the congealing cold turns it into water, which "by its gravity and weight again runs down to the lowest parts and increases the flow of "water if there is any. If any finds its way through a *canales dilatata* the same thing "happens, but it is carried a long way from its place of origin. The first phase of distillation "teaches us how this water is produced, for when that which is put into the ampulla is "warmed it evaporates (*expirare*), and this *halitus* rising into the operculum is converted "by cold into water, which drips through the spout. In this way water is being continually "created underground." (*De Ortu*, p. 7): "And so we know from all this that of the waters "which are under the earth, some are collected from rain, some arise from *halitus* (steam), some "from river-water, some from sea-water; and we know that the *halitum* is produced within "the earth partly from rain-water, partly from river-water, and partly from sea-water." It would require too much space to set out Agricola's views upon the origin of the subterranean heat which produced this steam. It is an involved theory embracing clashing winds, burning bitumen, coal, etc., and is fully set out in the latter part of Book II, *De Ortu et Causis*.

ORIGIN OF GANGUE MINERALS. It is necessary to bear in mind that Agricola divided minerals (*res fossiles*—"Things dug up," see note 4, p. 1) into "earths," "solidified juices," "stones," "metals," and "compounds;" and, further, to bear in mind that in his conception of the origin of things generally, he was a disciple of the Peripatetic logic of a "material substance" and an "efficient force," as mentioned above.

As to the origin of "earths," he says (*De Ortu*, *p.* 38): "Pure and simple 'earth' "originates in the *canales* in the following way: rain water, which is absorbed by the surface "of the earth, first of all penetrates and passes into the inner parts of the earth and "mixes with it; next, it is collected from all sides into stringers and veins, where it, "and sometimes water of other origin, erodes the 'earth' away,—a great quantity of it if the "stringers and veins are in 'earth,' a small quantity if they are in rock. The softer the "rock is, the more the water wears away particles by its continual movement. To this "class of rock belongs limestone, from which we see chalk, clay, and marl, and other unctuous "'earths' made; also sandstone, from which are made those barren 'earths' which we may "see in ravines and on bare rocks. For the rain softens limestone or sandstone and carries "particles away with it, and the sediment collects together and forms mud, which afterward "solidifies into some kind of 'earth.' In a similar way under the ground the power of water "softens the rock and dissolves the coarser fragments of stone. This is clearly shown by "the following circumstance, that frequently the powder of rock or marble is found in a "soft state and as if partly dissolved. Now, the water carries this mixture into the course "of some underground *canalis*, or dragging it into narrow places, filters away. And in each "case the water flows away and a pure and uniform material is left from which 'earth' "is made. . . . Particles of rock, however, are only by force of long time so softened "by water as to become similar to particles of 'earth.' It is possible to see 'earth' being "made in this way in underground *canales* in the earth, when drifts or tunnels are driven into "the mountains, or when shafts are sunk, for then the *canales* are laid bare; also it can be "seen above ground in ravines, as I have said, or otherwise disclosed. For in both cases "it is clear to the eye that they are made out of the 'earth' or rocks, which are often of the "same colour. And in just the same way they are made in the springs which the veins "discharge. Since all those things which we see with our eyes and which are perceived "with our senses, are more clearly understood than if they were learnt by means of reasoning, "we deem it sufficient to explain by this argument our view of the origin of 'earth.' In "the manner which I have described, 'earths' originate in veins and veinlets, seams in the "rocks, springs, ravines, and other openings, therefore all 'earths' are made in this way.

A, B, C, D—THE MOUNTAIN. E, F, G, H, I, K—*Vena cumulata.*

accumulations is usually formed a "*vena cumulata.*"

" As to those that are found in underground *canales* which do not appear to have been derived " from the earth or rock adjoining, these have undoubtedly been carried by the water for a " greater distance from their place of origin ; which may be made clear to anyone who seeks " their source."

On the origin of solidified juices he states (*De Ortu*, p. 43) : " I will now speak of " solidified juices (*succi concreti*). I give this name to those minerals which are without " difficulty resolved into liquids (*humore*). Some stones and metals, even though they are " themselves composed of juices, have been compressed so solidly by the cold that they can only " be dissolved with difficulty or not at all. . . . For juices, as I said above, are either " made when dry substances immersed in moisture are cooked by heat, or else they are " made when water flows over 'earth,' or when the surrounding moisture corrodes metallic " material ; or else they are forced out of the ground by the power of heat alone. There- " fore, solidified juices originate from liquid juices, which either heat or cold have condensed. " But that which heat has dried, fire reduces to dust, and moisture dissolves. Not only " does warm or cold water dissolve certain solidified juices, but also humid air ; and a juice " which the cold has condensed is liquefied by fire and warm water. A salty juice is con- " densed into salt ; a bitter one into soda ; an astringent and sharp one into alum or into " vitriol. Skilled workmen in a similar way to nature, evaporate water which contains " juices of this kind until it is condensed ; from salty ones they make salt, from " aluminous ones alum, from one which contains vitriol they make vitriol. These workmen " imitate nature in condensing liquid juices with heat, but they cannot imitate nature in " condensing them by cold. From an astringent juice not only is alum made and vitriol, but " also *sory*, *chalcitis*, and *misy*, which appears to be the 'flower' of vitriol, just as *melanteria* " is of *sory*. (See note on p. 573 for these minerals.) When humour corrodes pyrites so that " it is friable, an astringent juice of this kind is obtained."

ON THE ORIGIN OF STONES (*De Ortu*, p. 50), he states : " It is now necessary to " review in a few words what I have said as to all of the material from which stones are " made ; there is first of all mud ; next juice which is solidified by severe cold ; then frag- " ments of rock ; afterward stone juice (*succus lapidescens*), which also turns to stone when " it comes out into the air ; and lastly, everything which has pores capable of receiving a " stony juice." As to an " efficient force," he states (p. 54) : " But it is now necessary " that I should explain my own view, omitting the first and antecedent causes. Thus the

A—*Vena profunda*. B—*Intervenium*. C—ANOTHER *vena profunda*.

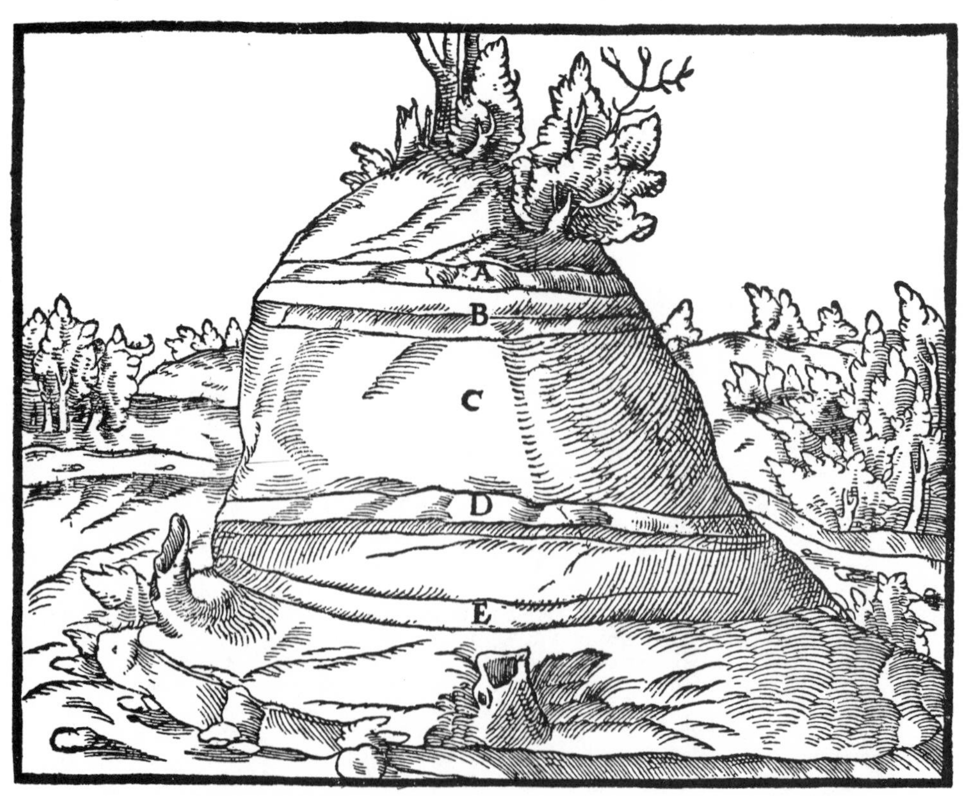

A & B—*Venae dilatatae*. C—*Intervenium*. D & E—OTHER *venae dilatatae*.

The space between two veins is called an *interventum* ; this interval between the veins, if it is between *venae dilatatae* is entirely hidden underground. If, however, it lies between *venae profundae* then the top is plainly in sight, and the remainder is hidden.

Venae profundae differ greatly one from another in width, for some of them are one fathom wide, some are two cubits, others one cubit; others again are a foot wide, and some only half a foot; all of which our miners call wide veins. Others on the contrary, are only a palm wide, others three digits,

"immediate causes are heat and cold; next in some way a stony juice. For we know that "stones which water has dissolved, are solidified when dried by heat; and on the contrary, "we know that stones which melt by fire, such as quartz, solidify by cold. For solidification "and the conditions which are opposite thereto, namely, dissolving and liquefying, spring "from causes which are the opposite to each other. Heat, driving the water (*humorem*) out of "a substance, makes it hard; and cold, by withdrawing the air, solidifies the same stone "firmly. But if a stony juice, either alone or mixed with water, finds its way into the pores "either of plants or animals . . . . it creates stones. . . . If stony juice is "obtained in certain stony places and flows through the veins, for this reason certain springs, "brooks, streams, and lakes, have the power of turning things to stone."

ON THE ORIGIN OF METALS, he says (*De Ortu*, p. 71): "Having now refuted the "opinions of others, I must explain what it really is from which metals are produced. "The best proof that there is water in their materials is the fact that they flow when "melted, whereas they are again solidified by the cold of air or water. This, however, "must be understood in the sense that there is more water in them and less 'earth'; for it "is not simply water that is their substance but water mixed with 'earth.' And such a "proportion of 'earth' is in the mixture as may obscure the transparency of the water, but "not remove the brilliance which is frequently in unpolished things. Again, the purer the "mixture, the more precious the metal which is made from it, and the greater its resistance "to fire. But what proportion of 'earth' is in each liquid from which a metal is made "no mortal can ever ascertain, or still less explain, but the one God has known it, Who has "given certain sure and fixed laws to nature for mixing and blending things together. It "is a juice (*succus*) then, from which metals are formed; and this juice is created by various "operations. Of these operations the first is a flow of water which softens the 'earth' or "carries the 'earth' along with it, thus there is a mixture of 'earth' and water, then the "power of heat works upon the mixtures so as to produce that kind of a juice. We have "spoken of the substance of metals; we must now speak of their efficient cause. . . . "(p. 75): We do not deny the statement of Albertus Magnus that the mixture of 'earth' "and water is baked by subterranean heat to a certain denseness, but it is our opinion that "the juice so obtained is afterward solidified by cold so as to become a metal. . . . "We grant, indeed, that heat is the efficient cause of a good mixture of elements, and also "cooks this same mixture into a juice, but until this juice is solidified by cold it is not a "metal." . . . (p. 76): This view of Aristotle is the true one. For metals melt "through the heat and somehow become softened; but those which have become softened "through heat are again solidified by the influence of cold, and, on the contrary, those "which become softened by moisture are solidified by heat."

ON THE ORIGIN OF COMPOUNDS, he states (*De Ortu*, p. 80): "There now remain "for our consideration the compound minerals (*mistae*), that is to say, minerals which "contain either solidified juice (*succus concretus*) and 'stone,' or else metal or metals and "'stone,' or else metal-coloured 'earth,' of which two or more have so grown together "by the action of cold that one body has been created. By this sign they are distin-"guished from mixed minerals (*composita*), for the latter have not one body. For "example, pyrites, galena, and ruby silver are reckoned in the category of compound "minerals, whereas we say that metallic 'earths' or stony 'earths' or 'earths' mingled with "juices, are mixed minerals; or similarly, stones in which metal or solidified juices adhere, "or which contain 'earth.' But of both these classes I will treat more fully in my book *De* "*Natura Fossilium*. I will now discuss their origin in a few words. A compound mineral "is produced when either a juice from which some metal is obtained, or a *humour* and some "other juice from which stone is obtained, are solidified by cold, or when two or more juices "of different metals mixed with the juice from which stone is made, are condensed by the same "cold, or when a metallic juice is mixed with 'earth' whose whole mass is stained with its "colour, and in this way they form one body. To the first class belongs *galena*, composed "of lead juice and of that material which forms the substance of opaque stone. Similarly, "transparent ruby silver is made out of silver juice and the juice which forms the

or even two; these they call narrow. But in other places where there are very wide veins, the widths of a cubit, or a foot, or half a foot, are said to be narrow; at Cremnitz, for instance, there is a certain vein which measures in one place fifteen fathoms in width, in another eighteen, and in another twenty; the truth of this statement is vouched for by the inhabitants.

"substance of transparent stone; when it is smelted into pure silver, since from it is "separated the transparent juice, it is no longer transparent. Then too, there is pyrites, "or *lapis fissilis*, from which sulphur is melted. To the second kind belongs that kind of "pyrites which contains not only copper and stone, but sometimes copper, silver, and stone; "sometimes copper, silver, gold, and stone; sometimes silver, lead, tin, copper and silver "glance. That compound minerals consist of stone and metal is sufficiently proved by "their hardness; that some are made of 'earth' and metal is proved from brass, which is "composed of copper and calamine; and also proved from white brass, which is coloured "by artificial white arsenic. Sometimes the heat bakes some of them to such an extent that "they appear to have flowed out of blazing furnaces, which we may see in the case of "*cadmia* and pyrites. A metallic substance is produced out of 'earth' when a metallic "juice impregnating the 'earth' solidifies with cold, the 'earth' not being changed. A "stony substance is produced when viscous and non-viscous 'earth' are accumulated in "one place and baked by heat; for then the viscous part turns into stone and the non-"viscous is only dried up."

THE ORIGIN OF JUICES. The portion of Agricola's theory surrounding this subject is by no means easy to follow in detail, especially as it is difficult to adjust one's point of view to the Peripatetic elements, fire, water, earth, and air, instead of to those of the atomic theory which so dominates our every modern conception. That Agricola's 'juice' was in most cases a solution is indicated by the statement (*De Ortu*, p. 48): "Nor is juice "anything but water, which on the other hand has absorbed 'earth' or has corroded or "touched metal and somehow become heated." That he realized the difference between mechanical suspension and solution is evident from (*De Ortu*, p. 50): "A stony juice differs "from water which has abraded something from rock, either because it has more of that which "deposits, or because heat, by cooking water of that kind, has thickened it, or because there "is something in it which has powerful astringent properties." Much of the author's notion of juices has already been given in the quotations regarding various minerals, but his most general statement on the subject is as follows:—(*De Ortu*, p. 9): "Juices, however, are "distinguished from water by their density (*crassitudo*), and are generated in various ways— "either when dry things are soaked with moisture and the mixture is heated, in which way "by far the greatest part of juices arise, not only inside the earth, but outside it: or when "water running over the earth is made rather dense, in which way, for the most "part the juice becomes salty and bitter; or when the moisture stands upon metal, "especially copper, and corrodes it, and in this way is produced the juice from which "chrysocolla originates. Similarly, when the moisture corrodes friable cupriferous pyrites "an acrid juice is made from which is produced vitriol and sometimes alum; or, finally, "juices are pressed out by the very force of the heat from the earth. If the force is great "the juice flows like pitch from burning pine . . . . in this way we know a kind of "bitumen is made in the earth. In the same way different kinds of moisture are generated "in living bodies, so also the earth produces waters differing in quality, and in the same "way juices."

CONCLUSION. If we strip his theory of the necessary influence of the state of knowledge of his time, and of his own deep classical learning, we find two propositions original with Agricola, which still to-day are fundamentals:

(1) That ore channels were of origin subsequent to their containing rocks; (2) That ores were deposited from solutions circulating in these openings. A scientist's work must be judged by the advancement he gave to his science, and with this gauge one can say unhesitatingly that the theory which we have set out above represents a much greater step from what had gone before than that of almost any single observer since. Moreover, apart from any tangible proposition laid down, the deduction of these views from actual observation instead of from fruitless speculation was a contribution to the very foundation of natural science. Agricola was wrong in attributing the creation of ore channels to erosion alone, and it was not until Von Oppel (*Anleitung zur Markscheidekunst*, Dresden, 1749 and other essays), two centuries after Agricola, that the positive proposition that ore channels were due to fissuring was brought forward. Von Oppel, however, in neglecting channels due to erosion (and in this term we include solution) was not altogether sound. Nor was it until late in the 18th century that the filling of ore channels by deposition from solutions was generally accepted. In the meantime, Agricola's successors in the study of ore deposits exhibited positive retrogression from the true fundamentals advocated by him. Gesner, Utman, Meier, Lohneys, Barba,

A—Wide *vena profunda*. B—Narrow *vena profunda*.

Venae dilatatae, in truth, differ also in thickness, for some are one fathom thick, others two, or even more ; some are a cubit thick, some a foot, some only half a foot ; and all these are usually called thick veins. Some on the other hand, are but a palm thick, some three digits, some two, some one ; these are called thin veins.

Rössler, Becher, Stahl, Henckel, and Zimmerman, all fail to grasp the double essentials. Other writers of this period often enough merely quote Agricola, some not even acknowledging the source, as, for instance, Pryce (*Mineralogia Cornubiensis*, London, 1778) and Williams (Natural History of the Mineral Kingdom, London, 1789). After Von Oppel, the two fundamental principles mentioned were generally accepted, but then arose the complicated and acrimonious discussion of the origin of solutions, and nothing in Agricola's view was so absurd as Werner's contention (*Neue Theorie von der Entstehung der Gänge*, Freiberg, 1791) of the universal chemical deluge which penetrated fissures open at the surface. While it is not the purpose of these notes to pursue the history of these subjects subsequent to the author's time, it is due to him and to the current beliefs as to the history of the theory of ore deposits, to call the attention of students to the perverse representation of Agricola's views by Werner (op. cit.) upon which most writers have apparently relied. Why this author should be (as, for instance, by Posepny, Amer. Inst. Mining Engineers, 1901) so generally considered the father of our modern theory, can only be explained by a general lack of knowledge of the work of previous writers on ore deposition. Not one of the propositions original with Werner still holds good, while his rejection of the origin of solutions within the earth itself halted the march of advance in thought on these subjects for half a century. It is our hope to discuss exhaustively at some future time the development of the history of this, one of the most far-reaching of geologic hypotheses.

A—Thin *vena dilatata.* B—Thick *vena dilatata.*

Venae profundae vary in direction ; for some run from east to west.

South

East

West

North

A, B, C—Vein. D, E, F—Seams in the Rock (*Commissurae Saxorum*).

Others, on the other hand, run from west to east.

SOUTH.

EAST.

WEST.

NORTH.

A, B, C—VEIN. D, E, F—*Seams in the Rocks.*

Others run from south to north.

SOUTH.

EAST.

WEST.

NORTH.

A, B, C—VEIN. D, E, F—*Seams in the Rocks.*

Others, on the contrary, run from north to south.

A, B, C—Vein. D, E, F—*Seams in the Rocks.*

The seams in the rocks indicate to us whether a vein runs from the east or from the west. For instance, if the rock seams incline toward the westward as they descend into the earth, the vein is said to run from east to west; if they incline toward the east, the vein is said to run from west to east; in a similar manner, we determine from the rock seams whether the veins run north or south.

Now miners divide each quarter of the earth into six divisions; and by this method they apportion the earth into twenty-four directions, which they divide into two parts of twelve each. The instrument which indicates these directions is thus constructed. First a circle is made; then at equal intervals on one half portion of it right through to the other, twelve straight lines called by the Greeks διάμετροι, and in the Latin *dimetientes*, are drawn through a central point which the Greeks call κέντρον, so that the circle is thus divided into twenty-four divisions, all being of an equal size. Then, within the circle are inscribed three other circles, the outermost of which has cross-lines dividing it into twenty-four equal parts; the space between it and the next circle contains two sets of twelve numbers, inscribed on the lines called "diameters"; while within the innermost circle it is hollowed out to contain a magnetic needle[3]. The needle lies directly

[3]The endeavour to discover the origin of the compass with the Chinese, Arabs, or other Orientals having now generally ceased, together with the idea that the knowledge of the lodestone involved any acquaintance with the compass, it is permissible to take a rational

over that one of the twelve lines called " diameters " on which the number XII is inscribed at both ends.

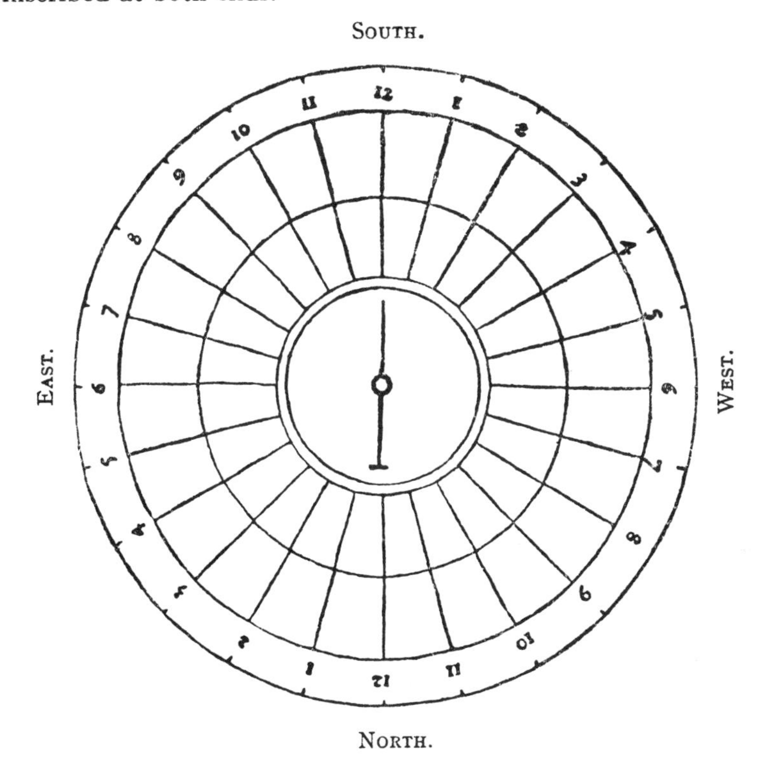

When the needle which is governed by the magnet points directly from the north to the south, the number XII at its tail, which is forked, signifies the north, that number XII which is at its point indicates the south. The sign VI superior indicates the east, and VI inferior the west. Further, between each two cardinal points there are always five others which are not so important. The first two of these directions are called the prior directions ; the last two are called the posterior, and the fifth direction lies immediately between the former and the latter ; it is halved, and one half is attributed to one cardinal point and one half to the other. For example, between the northern number XII and the eastern number VI, are points numbered I, II, III, IV, V, of which I and

view of the subject. The lodestone was well known even before Plato and Aristotle, and is described by Theophrastus (see Note 10, p. 115.) The first authentic and specific mention of the compass appears to be by Alexander Neckam (an Englishman who died in 1217), in his works *De Utensilibus* and *De Naturis Rerum*. The first tangible description of the instrument was in a letter to Petrus Peregrinus de Maricourt, written in 1269, a translation of which was published by Sir Sylvanus Thompson (London, 1902). His circle was divided into four quadrants and these quarters divided into 90 degrees each. The first mention of a compass in connection with mines so far as we know is in the *Nützlich Bergbüchlin*, a review of which will be found in Appendix B. This book, which dates from 1500, gives a compass much like the one described above by Agricola. It is divided in like manner into two halves of 12 divisions each. The four cardinal points being marked *Mitternacht*, *Morgen*, *Mittag*, and *Abend*. Thus the directions read were referred to as II. after midnight, etc. According to Joseph Carne (Trans. Roy. Geol. Socy. of Cornwall, Vol. II, 1814), the Cornish miners formerly referred to North-South veins as 12 o'clock veins ; South-East North-West veins as 9 o'clock veins, etc.

II are northern directions lying toward the east, IV and V are eastern directions lying toward the north, and III is assigned, half to the north and half to the east.

One who wishes to know the direction of the veins underground, places over the vein the instrument just described; and the needle, as soon as it becomes quiet, will indicate the course of the vein. That is, if the vein proceeds from VI to VI, it either runs from east to west, or from west to east; but whether it be the former or the latter, is clearly shown by the seams in the rocks. If the vein proceeds along the line which is between V and VI toward the opposite direction, it runs from between the fifth and sixth divisions of east to the west, or from between the fifth and sixth divisions of west to the east; and again, whether it is the one or the other is clearly shown by the seams in the rocks. In a similar manner we determine the other directions.

Now miners reckon as many points as the sailors do in reckoning up the number of the winds. Not only is this done to-day in this country, but it was also done by the Romans who in olden times gave the winds partly Latin names and partly names borrowed from the Greeks. Any miner who pleases may therefore call the directions of the veins by the names of the winds. There are four principal winds, as there are four cardinal points: the *Subsolanus*, which blows from the east; and its opposite the *Favonius*, which blows from the west; the latter is called by the Greeks Ζέφυρος, and the former Ἀπηλιώτης. There is the *Auster*, which blows from the south; and opposed to it is the *Septentrio*, from the north; the former the Greeks called Νότος, and the latter Ἀπαρκτίας. There are also subordinate winds, to the number of twenty, as there are directions, for between each two principal winds there are always five subordinate ones. Between the *Subsolanus* (east wind) and the *Auster* (south wind) there is the *Ornithiae* or the Bird wind, which has the first place next to the *Subsolanus*; then comes *Caecias*; then *Eurus*, which lies in the midway of these five; next comes *Vulturnus*; and lastly, *Euronotus*, nearest the *Auster* (south wind). The Greeks have given these names to all of these, with the exception of *Vulturnus*, but those who do not distinguish the winds in so precise a manner say this is the same as the Greeks called Εὖρος. Between the *Auster* (south wind) and the *Favonius* (west wind) is first *Altanus*, to the right of the *Auster* (south wind); then *Libonotus*; then *Africus*, which is the middle one of these five; after that comes *Subvesperus*; next *Argestes*, to the left of *Favonius* (west wind). All these, with the exception of *Libonotus* and *Argestes*, have Latin names; but *Africus* also is called by the Greeks Λίψ. In a similar manner, between *Favonius* (west wind) and *Septentrio* (north wind), first to the right of *Favonius* (west wind), is the *Etesiae*; then *Circius*; then *Caurus*, which is in the middle of these five; then *Corus*; and lastly *Thrascias* to the left of *Septentrio* (north wind). To all of these, except that of *Caurus*, the Greeks gave the names, and those who do not distinguish the winds by so exact a plan, assert that the wind which the Greeks called Κόρος and the Latins *Caurus* is one and the same.

Again, between *Septentrio* (north wind) and the *Subsolanus* (east wind), the first to the right of *Septentrio* (north wind) is *Gallicus*; then *Supernas*; then *Aquilo*, which is the middle one of these five; next comes *Boreas*; and lastly *Carbas*, to the left of *Subsolanus* (east wind). Here again, those who do not consider the winds to be in so great a multitude, but say there are but twelve winds in all, or at the most fourteen, assert that the wind called

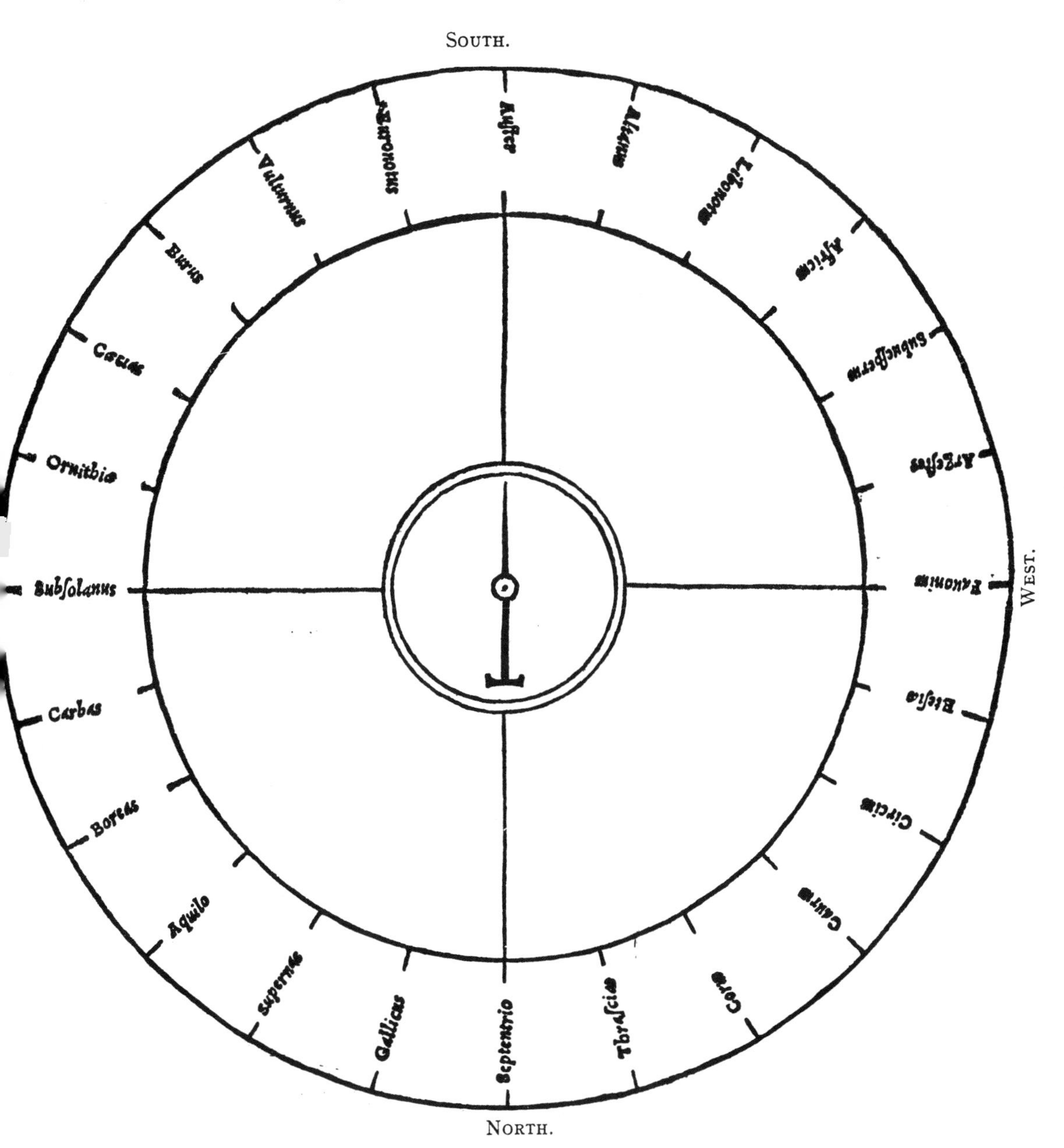

by the Greeks Βορέας and the Latins *Aquilo* is one and the same. For our purpose it is not only useful to adopt this large number of winds, but even to double it, as the German sailors do. They always reckon that between each two there is one in the centre taken from both. By this method we also are able to signify the intermediate directions by means of the names of the winds. For instance, if a vein runs from VI east to VI west, it is said to proceed from *Subsolanus* (east wind) to *Favonius* (west wind) ; but one which proceeds from between V and VI of the east to between V and VI west is said to proceed out of the middle of *Carbas* and *Subsolanus* to between *Argestes* and *Favonius* ; the remaining directions, and their intermediates are similarly designated. The miner, on account of the natural properties of a magnet, by which the needle points to the south, must fix the instrument already described so that east is to the left and west to the right.

In a similar way to *venae profundae*, the *venae dilatatae* vary in their lateral directions, and we are able to understand from the seams in the rocks in which direction they extend into the ground. For if these incline toward the west in depth, the vein is said to extend from east to west ; if on the contrary, they incline toward the east, the vein is said to go from west to east. In the same way, from the rock seams we can determine veins running south and north, or the reverse, and likewise to the subordinate directions and their intermediates.

A, B—*Venae dilatatae.* C—*Seams in the Rocks.*

Further, as regards the question of direction of a *vena profunda*, one runs straight from one quarter of the earth to that quarter which is opposite, while another one runs in a curve, in which case it may happen that a vein proceeding from the east does not turn to the quarter opposite, which is the west, but twists itself and turns to the south or the north.

A—STRAIGHT *vena profunda*. B—CURVED *vena profunda* [should be *vena dilatata*(?)].

Similarly some *venae dilatatae* are horizontal, some are inclined, and some are curved.

A—HORIZONTAL *vena dilatata*. B—INCLINED *vena dilatata*. C—CURVED *vena dilatata*.

Also the veins which we call *profundae* differ in the manner in which they descend into the depths of the earth ; for some are vertical (A), some are inclined and sloping (B), others crooked (C).

Moreover, *venae profundae* (B) differ much among themselves regarding the kind of locality through which they pass, for some extend along the slopes of mountains or hills (A-C) and do not descend down the sides.

Other *Venae Profundae* (D, E, F) from the very summit of the mountain or hill descend the slope (A) to the hollow or valley (B), and they again ascend the slope or the side of the mountain or hill opposite (C)

Other *Venae Profundae* (C, D) descend the mountain or hill (A) and extend out into the plain (B).

Some veins run straight along on the plateaux, the hills, or plains.

A—Mountainous Plain. B—*Vena profunda.*

A—Principal vein. B—Transverse vein. C—Vein cutting principal one obliquely.

In the next place, *venae profundae* differ not a little in the manner in which they intersect, since one may cross through a second transversely, or one may cross another one obliquely as if cutting it in two.

If a vein which cuts through another principal one obliquely be the harder of the two, it penetrates right through it, just as a wedge of beech or iron can be driven through soft wood by means of a tool. If it be softer, the principal vein either drags the soft one with it for a distance of three feet, or perhaps one, two, three, or several fathoms, or else throws it forward along the principal vein; but this latter happens very rarely. But that the vein which cuts the principal one is the same vein on both sides, is shown by its having the same character in its foot walls and hanging walls.

A—PRINCIPAL VEIN. B—VEIN WHICH CUTS A OBLIQUELY. C—PART CARRIED AWAY. D—THAT PART WHICH HAS BEEN CARRIED FORWARD.

Sometimes *venae profundae* join one with another, and from two or more outcropping veins[4], one is formed; or from two which do not outcrop one is made, if they are not far distant from each other, and the one dips into the other, or if each dips toward the other, and they thus join when they have descended in depth. In exactly the same way, out of three or more veins, one may be formed in depth.

[4]*Crudariis.* Pliny (XXXIII., 31), says:—"*Argenti vena in summo reperta crudaria appellatur.*" "Silver veins discovered at the surface are called *crudaria.*" The German translator of Agricola uses the term *sylber gang*—silver vein, obviously misunderstanding the author's meaning.

A, B—Two veins descend inclined and dip toward each other.
C—Junction. Likewise two veins. D—Indicates one descending vertically.
E—Marks the other descending inclined, which dips toward D. F—Their junction.

However, such a junction of veins sometimes disunites and in this way it happens that the vein which was the right-hand vein becomes the left; and again, the one which was on the left becomes the right.

Furthermore, one vein may be split and divided into parts by some hard rock resembling a beak, or stringers in soft rock may sunder the vein and make two or more. These sometimes join together again and sometimes remain divided.

A, B—Veins dividing. C—The same joining.

Whether a vein is separating from or uniting with another can be determined only from the seams in the rocks. For example, if a principal vein runs from the east to the west, the rock seams descend in depth likewise from the east toward the west, and the associated vein which joins with the principal vein, whether it runs from the south or the north, has its rock seams extending in the same way as its own, and they do not conform with the seams in the rock of the principal vein—which remain the same after the junction—unless the associated vein proceeds in the same direction as the principal vein. In that case we name the broader vein the principal one, and the narrower the associated vein. But if the principal vein splits, the rock seams which belong respectively to the parts, keep the same course when descending in depth as those of the principal vein.

But enough of *venae profundae*, their junctions and divisions. Now we come to *venae dilatatae*. A *vena dilatata* may either cross a *vena profunda*, or join with it, or it may be cut by a *vena profunda*, and be divided into parts.

A, C—*Vena dilatata* CROSSING A *vena profunda*. B—*Vena profunda*. D, E—*Vena dilatata* WHICH JUNCTIONS WITH A *vena profunda*. F—*Vena profunda*. G—*Vena dilatata*. H, I—ITS DIVIDED PARTS. K—*Vena profunda* WHICH DIVIDES THE *vena dilatata*.

Finally, a *vena profunda* has a " beginning " (*origo*), an " end " (*finis*), a " head " (*caput*), and a " tail " (*cauda*). That part whence it takes its rise is said to be its " beginning," that in which it terminates the " end." Its " head "[5] is that part which emerges into daylight ; its " tail " that part which is hidden in the earth. But miners have no need to seek the " beginning " of veins, as formerly the kings of Egypt sought for the source of the Nile, but it is enough for them to discover some other part of the vein and to recognise its direction, for seldom can either the " beginning " or the " end " be found. The direction in which the head of the vein comes into the light, or the direction toward which the tail extends, is indicated by its footwall and hangingwall. The latter is said to hang, and the former to lie. The vein rests on the footwall, and the hangingwall overhangs it ; thus, when we descend a shaft, the part to which we turn the face is the footwall and seat of the vein, that to which we turn the back is the hangingwall. Also in another way, the head accords with the footwall and the tail with the hangingwall, for if the footwall is toward the south, the vein extends its head into the light toward the south ; and the hangingwall, because it is always opposite to the footwall, is then toward the north. Consequently the vein extends its tail toward the north if it is an inclined *vena profunda*. Similarly, we can determine with regard to east and west and the subordinate and their intermediate directions. A *vena profunda* which descends into the earth may be either vertical, inclined, or crooked , the footwall of an inclined vein is easily distinguished from the hangingwall, but it is not so with a vertical vein ; and again, the footwall of a crooked vein is inverted and changed into the hangingwall, and contrariwise the hangingwall is twisted into the footwall, but very many of these crooked veins may be turned back to vertical or inclined ones.

[5]It might be considered that the term " outcrop " could be used for " head," but it will be noticed that a *vena dilatata* would thus be stated to have no outcrop.

A—THE "BEGINNING" (*origo*). B—THE "END" (*finis*). C—THE "HEAD" (*caput*). D—THE "TAIL" (*cauda*).

A *vena dilatata* has only a "beginning" and an "end," and in the place of the "head" and "tail" it has two sides.

A—THE "BEGINNING." B—THE "END." C, D—THE "SIDES."

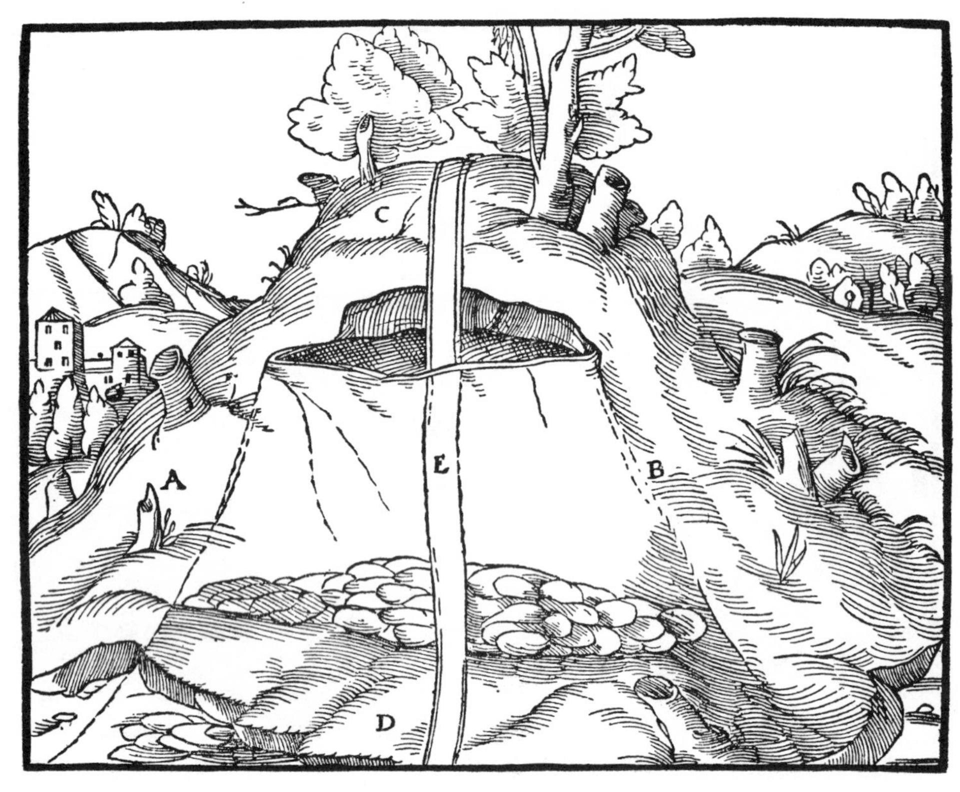

A—The "beginning." B—The "end." C—The "head." D—The "tail."
E—Transverse vein.

A *vena cumulata* has a "beginning," an "end," a "head," and a "tail," just as a *vena profunda*. Moreover, a *vena cumulata*, and likewise a *vena dilatata*, are often cut through by a transverse *vena profunda*.

Stringers (*fibrae*)[6], which are little veins, are classified into *fibrae transversae, fibrae obliquae* which cut the vein obliquely, *fibrae sociae, fibrae dilatatae,* and *fibrae incumbentes*. The *fibra transversa* crosses the vein; the *fibra obliqua* crosses the vein obliquely; the *fibra socia* joins with the vein itself; the *fibra dilatata*, like the *vena dilatata*, penetrates through it; but the *fibra dilatata*, as well as the *fibra profunda*, is usually found associated with a vein.

The *fibra incumbens* does not descend as deeply into the earth as the other stringers, but lies on the vein, as it were, from the surface to the hangingwall or footwall, from which it is named *Subdialis*.[7]

In truth, as to direction, junctions, and divisions, the stringers are not different from the veins.

[6]It is possible that "veinlets" would be preferred by purists, but the word "stringer" has become fixed in the nomenclature of miners and we have adopted it. The old English term was "stringe," and appears in Edward Manlove's "Rhymed Chronicle," London, 1653; Pryce's, *Mineralogia Cornubiensis*, London, 1778, pp. 103 and 329; Mawe's "Mineralogy of Devonshire," London, 1802, p. 210, etc., etc.

[7]*Subdialis*. "In the open air." The Glossary gives the meaning as *Ein tag klufft oder tag gehenge*—a surface stringer.

A, B—VEINS. C—TRANSVERSE STRINGER. D—OBLIQUE STRINGER. E—ASSOCIATED STRINGER. F—*Fibra dilatata*

A—VEIN. B—*Fibra incumbens* FROM THE SURFACE OF THE HANGINGWALL. C—SAME FROM THE FOOTWALL.

Lastly, the seams, which are the very finest stringers (*fibrae*), divide the rock, and occur sometimes frequently, sometimes rarely. From whatever direction the vein comes, its seams always turn their heads toward the light in the same direction. But, while the seams usually run from one point of the compass to another immediately opposite it, as for instance, from east to west, if hard stringers divert them, it may happen that these very seams, which before were running from east to west, then contrariwise proceed from west to east, and the direction of the rocks is thus inverted. In such a case, the direction of the veins is judged, not by the direction of the seams which occur rarely, but by those which constantly recur.

A—Seams which proceed from the east. B—The inverse.

Both veins or stringers may be solid or drusy, or barren of minerals, or pervious to water. Solid veins contain no water and very little air. The drusy veins rarely contain water; they often contain air. Those which are barren of minerals often carry water. Solid veins and stringers consist sometimes of hard materials, sometimes of soft, and sometimes of a kind of medium between the two.

A—SOLID VEIN. B—SOLID STRINGER. C—CAVERNOUS VEIN. D—CAVERNOUS STRINGER. E—BARREN VEIN. F—BARREN STRINGER.

But to return to veins. A great number of miners consider[8] that the best veins in depth are those which run from the VI or VII direction of the east to the VI or VII direction of the west, through a mountain slope which inclines to the north ; and whose hangingwalls are in the south, and whose footwalls are in the north, and which have their heads rising to the north, as explained before, always like the footwall, and finally, whose rock seams turn their heads to the east. And the veins which are the next

[8]The following from Chapter IV of the *Nützlich Bergbüchlin* (see Appendix B) may indicate the source of the theory which Agricola here discards :—" As to those veins which " are most profitable to work, it must be remarked that the most suitable location for the vein " is on the slope of the mountain facing south, so its strike is from VII or VI east to VI or " VII west. According to the above-mentioned directions, the outcrop of the whole vein " should face north, its *gesteins ausgang* toward the east, its hangingwall toward the south, " and its footwall toward the north, for in such mountains and veins the influence of the " planets is conveniently received to prepare the matter out of which the silver is to be made " or formed. . . . The other strikes of veins from between east and south to the region " between west and north are esteemed more or less valuable, according to whether they are " nearer or further away from the above-mentioned strikes, but with the same hanging- " wall, footwall, and outcrops. But the veins having their strike from north to south, " their hangingwall toward the west, their footwall and their outcrops toward the east, " are better to work than veins which extend from south to north, whose hangingwalls " are toward the east, and footwalls and outcrops toward the west. Although the latter " veins sometimes yield solid and good silver ore, still it is not sure and certain, because " the whole mineral force is completely scattered and dispersed through the outcrop, etc."

best are those which, on the contrary, extend from the VI or VII direction of the west to the VI or VII direction of the east, through the slope of a mountain which similarly inclines to the north, whose hangingwalls are also in the south, whose footwalls are in the north, and whose heads rise toward the north; and lastly, whose rock seams raise their heads toward the west. In the third place, they recommend those veins which extend from XII north to XII south, through the slope of a mountain which faces east; whose hangingwalls are in the west, whose footwalls are in the east; whose heads rise toward the east; and whose rock seams raise their heads toward the north. Therefore they devote all their energies to those veins, and give very little or nothing to those whose heads, or the heads of whose rock seams rise toward the south or west. For although they say these veins sometimes show bright specks of pure metal adhering to the stones, or they come upon lumps of metal, yet these are so few and far between that despite them it is not worth the trouble to excavate such veins; and miners who persevere in digging in the hope of coming upon a quantity of metal, always lose their time and trouble. And they say that from veins of this kind, since the sun's rays draw out the metallic material, very little metal is gained. But in this matter the actual experience of the miners who thus judge of the veins does not always agree with their opinions, nor is their reasoning sound; since indeed the veins which run from east to west through the slope of a mountain which inclines to the south, whose heads rise likewise to the south, are not less charged with metals, than those to which miners are wont to accord the first place in productiveness; as in recent years has been proved by the St. Lorentz vein at Abertham, which our countrymen call Gottsgaab, for they have dug out of it a large quantity of pure silver; and lately a vein in Annaberg, called by the name of Himmelsch hoz[9], has made it

[9]The names in the Latin are given as *Donum Divinum*—"God's Gift," and *Coelestis Exercitus*—"Heavenly Host." The names given in the text are from the German Translation. The former of these mines was located in the valley of Joachim, where Agricola spent many years as the town physician at Joachimsthal. It is of further interest, as Agricola obtained an income from it as a shareholder. He gives the history of the mine (*De Veteribus et Novis Metallis*, Book I.), as follows:—"The mines at Abertham were discovered, partly "by chance, partly by science. In the eleventh year of Charles V. (1530), on the 18th of "February, a poor miner, but one skilled in the art of mining, dwelt in the middle of the "forest in a solitary hut, and there tended the cattle of his employer. While digging a little "trench in which to store milk, he opened a vein. At once he washed some in a bowl and saw "particles of the purest silver settled at the bottom. Overcome with joy he informed his "employer, and went to the *Bergmeister* and petitioned that official to give him a head "mining lease, which in the language of our people he called *Gottsgaab*. Then he proceeded "to dig the vein, and found more fragments of silver, and the miners were inspired with "great hopes as to the richness of the vein. Although such hopes were not frustrated, "still a whole year was spent before they received any profits from the mine; whereby "many became discouraged and did not persevere in paying expenses, but sold their shares "in the mine; and for this reason, when at last an abundance of silver was being drawn "out, a great change had taken place in the ownership of the mine; nay, even the first "finder of the vein was not in possession of any share in it, and had spent nearly all the "money which he had obtained from the selling of his shares. Then this mine yielded such "a quantity of pure silver as no other mine that has existed within our own or our "fathers' memories, with the exception of the St. George at Schneeberg. We, as a share-"holder, through the goodness of God, have enjoyed the proceeds of this 'God's Gift' "since the very time when the mine began first to bestow such riches." Later on in the

plain by the production of much silver that veins which extend from the north to the south, with their heads rising toward the west, are no less rich in metals than those whose heads rise toward the east.

It may be denied that the heat of the sun draws the metallic material out of these veins; for though it draws up vapours from the surface of the ground, the rays of the sun do not penetrate right down to the depths; because the air of a tunnel which is covered and enveloped by solid earth to the depth of only two fathoms is cold in summer, for the intermediate earth holds in check the force of the sun. Having observed this fact, the inhabitants and dwellers of very hot regions lie down by day in caves which protect them from the excessive ardour of the sun. Therefore it is unlikely that the sun draws out from within the earth the metallic bodies. Indeed, it cannot even dry the moisture of many places abounding in veins, because they are protected and shaded by the trees. Furthermore, certain miners, out of all the different kinds of metallic veins, choose those which I have described, and others, on the contrary, reject copper mines which are of this sort, so that there seems to be no reason in this. For what can be the reason if the sun draws no copper from copper veins, that it draws silver from silver veins, and gold from gold veins?

Moreover, some miners, of whose number was Calbus[10], distinguish between the gold-bearing rivers and streams. A river, they say, or a stream, is most productive of fine and coarse grains of gold when it comes from the east and flows to the west, and when it washes against the foot of mountains which are situated in the north, and when it has a level plain toward the south or west. In the second place, they esteem a river or a stream which flows in the opposite course from the west toward the east, and which has the mountains to the north and the level plain to the south. In the third place, they esteem the river or the stream which flows from the north to the south and washes the base of the mountains which are situated in the east. But they say that the river or stream is least productive of gold which flows in a contrary direction from the south to the north, and washes the base of

same book he gives the following further information with regard to these mines:—"Now "if all the individual mines which have proved fruitful in our own times are weighed in "the balance, the one at Annaberg, which is known as the *Himmelsch hoz*, surpasses all "others. For the value of the silver which has been dug out has been estimated at 420,000 "Rhenish gulden. Next to this comes the lead mine in Joachimsthal, whose name is the "*Sternen*, from which as much silver has been dug as would be equivalent to 350,000 Rhenish "gulden; from the Gottsgaab at Abertham, explained before, the equivalent of 300,000. "But far before all others within our fathers' memory stands the St. George of Schneeberg, "whose silver has been estimated as being equal to two million Rhenish gulden." A Rhenish gulden was about 6.9 shillings, or, say, \$1.66. However, the ratio value of silver to gold at this period was about 11.5 to one, or in other words an ounce of silver was worth about a gulden, so that, for purposes of rough calculation, one might say that the silver product mentioned in gulden is practically of the same number of ounces of silver. Moreover, it must be remembered that the purchasing power of money was vastly greater then.

[10]The following passage occurs in the *Nützlich Bergbüchlin* (Chap. V.), which is interesting on account of the great similarity to Agricola's quotation:—"The best position of the stream is "when it has a cliff beside it on the north and level ground on the south, but its current should "be from east to west—that is the most suitable. The next best after this is from west to "east, with the same position of the rocks as already stated. The third in order is when the "stream flows from north to south with rocks toward the east, but the worst flow of water "for the preparation of gold is from south to north if a rock or hill rises toward the west." Calbus was probably the author of this booklet.

mountains which are situated in the west. Lastly, of the streams or rivers which flow from the rising sun toward the setting sun, or which flow from the northern parts to the southern parts, they favour those which approach the nearest to the lauded ones, and say they are more productive of gold, and the further they depart from them the less productive they are. Such are the opinions held about rivers and streams. Now, since gold is not generated in the rivers and streams, as we have maintained against Albertus[11] in the book entitled "*De Subterraneorum Ortu et Causis,*" Book V, but is torn away from the veins and stringers and settled in the sands of torrents and water-courses, in whatever direction the rivers or streams flow, therefore it is reasonable to expect to find gold therein; which is not opposed by experience. Nevertheless, we do not deny that gold is generated in veins and stringers which lie under the beds of rivers or streams, as in other places.

[11]Albertus Magnus.

END OF BOOK III.

BOOK IV.

HE third book has explained the various and manifold varieties of veins and stringers. This fourth book will deal with mining areas and the method of delimiting them, and will then pass on to the officials who are connected with mining affairs[1].

Now the miner, if the vein he has uncovered is to his liking, first of all goes to the *Bergmeister* to request to be granted a right to mine, this official's special function and office being to adjudicate in respect of the mines. And so to the first man who has discovered the vein the *Bergmeister* awards the head meer, and to others the remaining meers, in the order in which each makes his application. The size of a meer is measured by fathoms, which for miners are reckoned at six feet each. The length, in fact, is that of a man's extended arms and hands measured across his chest; but different peoples assign to it different lengths,

[1]The nomenclature in this chapter has given unusual difficulty, because the organisation of mines, either past or present, in English-speaking countries provides no exact equivalents for many of these offices and for many of the legal terms. The Latin terms in the text were, of course, coined by the author, and have no historical basis to warrant their adoption, while the introduction of the original German terms is open to much objection, as they are not only largely obsolete, but also in the main would convey no meaning to the majority of readers. We have, therefore, reached a series of compromises, and in the main give the nearest English equivalent. Of much interest in this connection is a curious exotic survival in mining law to be found in the High Peak of Derbyshire. We believe (see note on p. 85) that the law of this district was of Saxon importation, for in it are not only many terms of German origin, but the character of the law is foreign to the older English districts and shows its near kinship to that of Saxony. It is therefore of interest in connection with the nomenclature to be adopted in this book, as it furnishes about the only English precedents in many cases. The head of the administration in the Peak was the Steward, who was the chief judicial officer, with functions somewhat similar to the *Berghauptmann*. However, the term Steward has come to have so much less significance that we have adopted a literal rendering of the Latin. Under the Steward was the Barmaster, Barghmaster, or Barmar, as he was variously called, and his duties were similar to those of the *Bergmeister*. The English term would seem to be a corruption of the German, and as the latter has come to be so well understood by the English-speaking mining class, we have in this case adopted the German. The Barmaster acted always by the consent and with the approval of a jury of from 12 to 24 members. In this instance the English had functions much like a modern jury, while the *Geschwornen* of Saxony had much more widely extended powers. The German *Geschwornen* were in the main Inspectors; despite this, however, we have not felt justified in adopting any other than the literal English for the Latin and German terms. We have vacillated a great deal over the term *Praefectus Fodinae*, the German *Steiger* having, like the Cornish "Captain," in these days degenerated into a foreman, whereas the duties as described were not only those of the modern Superintendent or Manager, but also those of Treasurer of the Company, for he made the calls on shares and paid the dividends. The term Purser has been used for centuries in English mining for the Accountant or Cashier, but his functions were limited to paying dividends, wages, etc., therefore we have considered it better not to adopt the latter term, and have compromised upon the term Superintendent or Manager, although it has a distinctly modern flavor. The word for *area* has also caused much hesitation, and the "meer" has finally been adopted with some doubt. The title described by Agricola has a very close equivalent in the meer of old Derbyshire. As will be seen later, the mines of Saxony were Regal property, and were held subject to two essential conditions, *i.e.*, payment of a tithe, and continuous operation. This form of title thus approximates more closely to the "lease" of Australia than to the old Cornish *sett*, or the American *claim*. The *fundgrube* of Saxony and Agricola's equivalent, the *area capitis*—head lease—we have rendered literally as "head meer," although in some ways "founders' meer" might be better, for, in Derbyshire, this was called the "finder's" or founder's meer, and was awarded under similar circumstances. It has also an analogy in Australian law in the "reward" leases. The term "measure" has the merit of being a literal rendering of the Latin, and also of being the identical term in the same

for among the Greeks, who called it an ὀργυιά, it was six feet, among the Romans five feet. So this measure which is used by miners seems to have come down to the Germans in accordance with the Greek mode of reckoning. A miner's foot approaches very nearly to the length of a Greek foot, for it exceeds it by only three-quarters of a Greek digit, but like that of the Romans it is divided into twelve *unciae*[2].

Now square fathoms are reckoned in units of one, two, three, or more "measures", and a "measure" is seven fathoms each way. Mining meers are for the most part either square or elongated ; in square meers all the sides are of equal length, therefore the numbers of fathoms on the two sides multiplied together produce the total in square fathoms. Thus, if the shape of a "measure" is seven fathoms on every side, this number multiplied by itself makes forty-nine square fathoms.

The sides of a long meer are of equal length, and similarly its ends are equal ; therefore, if the number of fathoms in one of the long sides be multiplied by the number of fathoms in one of the ends, the total produced by the

use in the High Peak. The following table of the principal terms gives the originals of the Latin text, their German equivalents according in the Glossary and other sources, and those adopted in the translation :—

AGRICOLA.	GERMAN GLOSSARY.	TERM ADOPTED.
Praefectus Metallorum	*Bergamptmann*	Mining Prefect.
Magister Metallicorum	*Bergmeister*	Bergmeister
Scriba Magister Metallicorum	*Bergmeister's schreiber*	Bergmeister's clerk.
Jurati	*Geschwornen*	Jurates or Jurors.
Publicus Signator	*Gemeiner sigler*	Notary..
Decumanus	*Zehender*	Tithe gatherer.
Distributor	*Aussteiler*	Cashier.
Scriba partium	*Gegenschreiber*	Share clerk.
Scriba fodinarum	*Bergschreiber*	Mining clerk.
Praefectus fodinae	*Steiger*	Manager of the Mine.
Praefectus cuniculi		Manager of the Tunnel.
Praeses fodinae	*Schichtmeister*	Foreman of the Mine.
Praeses cuniculi		Foreman of the Tunnel.
Fossores	*Berghauer*	Miners or diggers.
Ingestores	*Berganschlagen*	Shovellers.
Vectarii	*Hespeler*	Lever workers (windlass men).
Discretores	*Ertzpucher*	Sorters.
Lotores	*Wescher und seiffner*	Washers, buddlers, sifters, etc.
Excoctores	*Schmeltzer*	Smelters.
Purgator Argenti	*Silber brenner*	Silver refiner.
Magister Monetariorum	*Müntzmeister*	Master of the Mint.
Monetarius	*Müntzer*	Coiner.
Area fodinarum	*Masse*	Meer.
Area Capitis Fodinarum	*Fundgrube*	Head meer.
Demensum	*Lehen*	Measure.

[2]The following are the equivalents of the measures mentioned in this book. It is not always certain which "foot" or "fathom" Agricola actually had in mind although they were probably the German.

GREEK—								
Dactylos	= .76 inches	16	= *Pous*	= 12.13 inches	6	= *Orguia*	= 72.81 inches.
ROMAN							
Uncia	= .97 "	12	= *Pes*	= 11.6 "	5	= *Passus*	= 58.1 "
GERMAN—							
Zoll	= .93 "	12	= *Werckschuh*	= 11.24 "	6	= *Lachter*	= 67.5 "
ENGLISH—							
Inch	= 1.0 "	12	= Foot	= 12.00 "	6	= Fathom	= 72.0 "

The discrepancies are due to variations in authorities and to decimals dropped. The *werckschuh* taken is the Chemnitz foot deduced from Agricola's statement in his *De Mensuris et Ponderibus*, Basel, 1533, p. 29. For further notes see Appendix C.

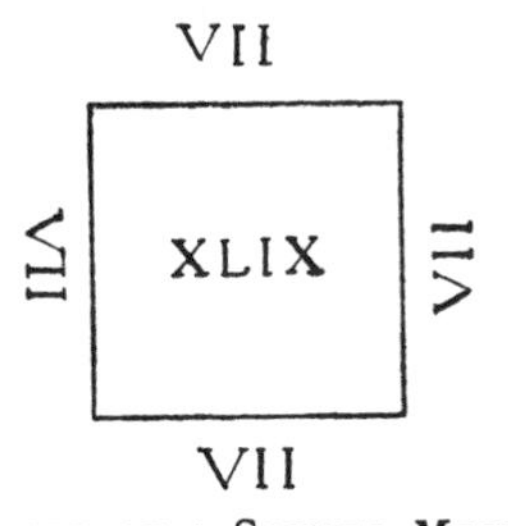

SHAPE OF A SQUARE MEER.

multiplication is the total number of square fathoms in the long meer. For example, the double measure is fourteen fathoms long and seven broad, which two numbers multiplied together make ninety-eight square fathoms.

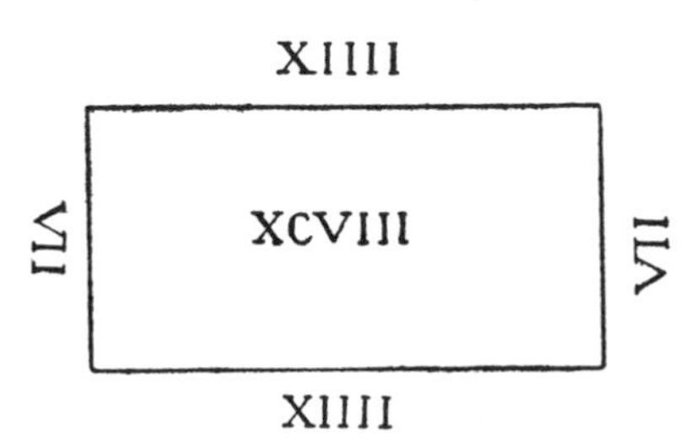

SHAPE OF A LONG MEER OR DOUBLE MEASURE.

Since meers vary in shape according to the different varieties of veins it is necessary for me to go more into detail concerning them and their measurements. If the vein is a *vena profunda*, the head meer is composed of three double measures, therefore it is forty-two fathoms in length and seven in width, which numbers multiplied together give two hundred and ninety-four square fathoms, and by these limits the *Bergmeister* bounds the owner's rights in a head-meer.

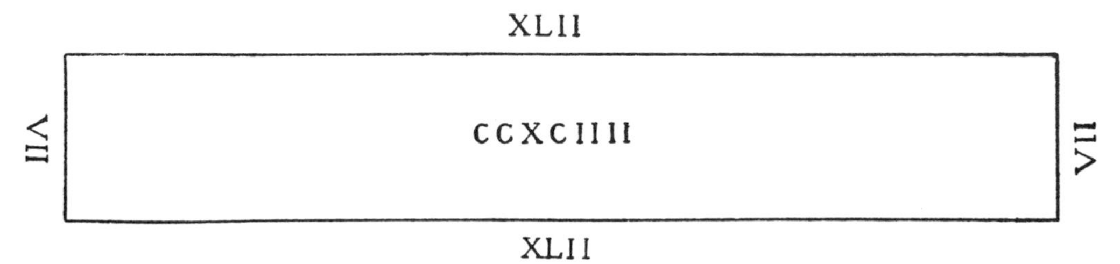

SHAPE OF A HEAD MEER.

The area of every other meer consists of two double measures, on whichever side of the head meer it lies, or whatever its number in order may be, that is to say, whether next to the head meer, or second, third, or any later number. Therefore, it is twenty-eight fathoms long and seven wide, so multiplying the length by the width we get one hundred and ninety-six square fathoms, which is the extent of the meer, and by these boundaries the *Bergmeister* defines the right of the owner or company over each mine.

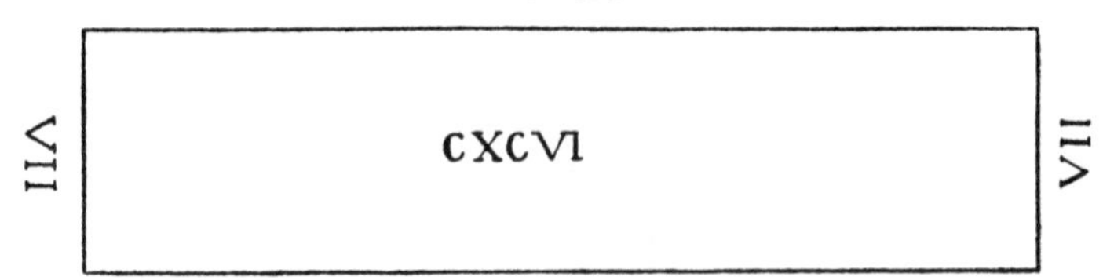

Shape of a Meer.

Now we call that part of the vein which is first discovered and mined, the head-meer, because all the other meers run from it, just as the nerves from the head. The *Bergmeister* begins his measurements from it, and the reason why he apportions a larger area to the head-meer than to the others, is that he may give a suitable reward to the one who first found the vein and may encourage others to search for veins. Since meers often reach to a torrent, or river, or stream, if the last meer cannot be completed it is called a fraction[3]. If it is the size of a double measure, the *Bergmeister* grants the right of mining it to him who makes the first application, but if it is the size of a single measure or a little over, he divides it between the nearest meers on either side of it. It is the custom among miners that the first meer beyond a stream on that part of the vein on the opposite side is a new head-meer, and they call it the " opposite,"[4] while the other meers beyond are only ordinary meers. Formerly every head-meer was composed of three double measures and one single one, that is, it was forty-nine fathoms long and seven wide, and so if we multiply these two together we have three hundred and forty-three square fathoms, which total gives us the area of an ancient head-meer.

XLIX

VII

CCCXLIIII

XLIX

Shape of an ancient Head-Meer.

Every ancient meer was formed of a single measure, that is to say, it was seven fathoms in length and width, and was therefore square. In memory of which miners even now call the width of every meer which is located on a *vena profunda* a " square "[5]. The following was formerly the

[3] *Subcisivum*—" Remainder." German Glossary, *Ueberschar*. The term used in Mendip and Derbyshire was *primgap* or *primegap*. It did not, however, in this case belong to adjacent mines, but to the landlord.

[4] *Adversum*. Glossary, *gegendrumb*. The *Bergwerk Lexicon*, Chemnitz, 1743, gives *gegendrom* or *gegentramm*, and defines it as the *masse* or lease next beyond a stream.

[5] *Quadratum*. Glossary, *vierung*. The *vierung* in old Saxon title meant a definite zone on either side of the vein, 3½ *lachter* (*lachter* = 5ft. 7.5 inches) into the hanging-wall and the same into the footwall, the length of one *vierung* being 7 *lachter* along the strike. It

usual method of delimiting a vein: as soon as the miner found metal, he gave information to the *Bergmeister* and the tithe-gatherer, who either proceeded personally from the town to the mountains, or sent thither men of good repute, at least two in number, to inspect the metal-bearing vein. Thereupon, if they thought it of sufficient importance to survey, the *Bergmeister* again having gone forth on an appointed day, thus questioned him who first found the vein, concerning the vein and the diggings: "Which is your vein?" "Which digging carried metal?" Then the discoverer, pointing his finger to his vein and diggings, indicated them, and next the *Bergmeister* ordered him to approach the windlass and place two fingers of his right hand upon his head, and swear this oath in a clear voice: "I swear by God and all the Saints, and I call them all to witness, that this is my vein; and moreover if it is not mine, may neither this my head nor these my hands henceforth perform their functions." Then the *Bergmeister*, having started from the centre of the windlass, proceeded to measure the vein with a cord, and to give the measured portion to the discoverer,—in the first instance a half and then three full measures; afterward one to the King or Prince, another to his Consort, a third to the Master of the Horse, a fourth to the Cup-bearer, a fifth to the Groom of the Chamber, a sixth to himself. Then, starting from the other side of the windlass, he proceeded to measure the vein in a similar manner. Thus the discoverer of the vein obtained the head-meer, that is, seven single measures; but the King or Ruler, his Consort, the leading dignitaries, and lastly, the *Bergmeister*, obtained two measures each, or two ancient meers. This is the reason there are to be found at Freiberg in Meissen so many shafts with so many intercommunications on a single vein—which are to a great extent destroyed by age. If, however, the *Bergmeister* had already fixed the boundaries of the meers on one side of the shaft for the benefit of some other discoverer, then for those dignitaries I have just mentioned, as many meers as he was unable to award on that side he duplicated on the other. But if on both sides of the shaft he had already defined the boundaries of meers, he proceeded to measure out only that part of the vein which remained free, and thus it sometimes happened that some of those persons I have mentioned obtained no meer at all. To-day, though that old-established custom is observed, the method of allotting the vein and granting title has been changed. As I have explained above, the head-meer consists of three double measures, and each other meer of two measures, and the *Bergmeister* grants one each of the meers to him who makes the first application. The King or Prince, since all metal is taxed, is himself content with that, which is usually one-tenth.

Of the width of every meer, whether old or new, one-half lies on the footwall side of a *vena profunda* and one half on the hangingwall side. If the vein descends vertically into the earth, the boundaries similarly descend

must be borne in mind that the form of rights here referred to entitled the miner to follow his vein, carrying the side line with him in depth the same distance from the vein, in much the same way as with the Apex Law of the United States. From this definition as given in the *Bergwerk Lexicon*, p. 585, it would appear that the vein itself was not included in the measurements, but that they started from the walls.

vertically ; but if the vein inclines, the boundaries likewise will be inclined. The owner always holds the mining right for the width of the meer, however far the vein descends into the depth of the earth.[6] Further, the *Bergmeister*, on application being made to him, grants to one owner or company a right

[6]Historical Note on the Development of Mining Law.—There is no branch of the law of property, of which the development is more interesting and illuminating from a social point of view than that relating to minerals. Unlike the land, the minerals have ever been regarded as a sort of fortuitous property, for the title of which there have been four principal claimants—that is, the Overlord, as represented by the King, Prince, Bishop, or what not; the Community or the State, as distinguished from the Ruler; the Landowner; and the Mine Operator, to which class belongs the Discoverer. The one of these that possessed the dominant right reflects vividly the social state and sentiment of the period. The Divine Right of Kings; the measure of freedom of their subjects; the tyranny of the land-owning class; the rights of the Community as opposed to its individual members; the rise of individualism; and finally, the modern return to more communal view, have all been reflected promptly in the mineral title. Of these parties the claims of the Overlord have been limited only by the resistance of his subjects; those of the State limited by the landlord; those of the landlord by the Sovereign or by the State; while the miner, ever in a minority in influence as well as in numbers, has been buffeted from pillar to post, his only protection being the fact that all other parties depended upon his exertion and skill.

The conception as to which of these classes had a right in the title have been by no means the same in different places at the same time, and in all it varies with different periods; but the whole range of legislation indicates the encroachment of one factor in the community over another, so that their relative rights have been the cause of never-ending contention, ever since a record of civil and economic contentions began. In modern times, practically over the whole world, the State has in effect taken the rights from the Overlord, but his claims did not cease until his claims over the bodies of his subjects also ceased. However, he still remains in many places with his picture on the coinage. The Landlord has passed through many vicissitudes; his complete right to minerals was practically never admitted until the doctrine of *laissez-faire* had become a matter of faith, and this just in time to vest him with most of the coal and iron deposits in the world; this, no doubt, being also partially due to the little regard in which such deposits were generally held at that time, and therefore to the little opposition to his ever-ready pretentions. Their numbers, however, and their prominence in the support of the political powers *de jure* have usually obtained them some recognition. In the rise of individualism, the apogee of the *laissez-faire* fetish came about the time of the foundation of the United States, and hence the relaxation in the claims of the State in that country and the corresponding position attained by the landlord and miner. The discoverer and the operator—that is, the miner himself—has, however, had to be reckoned with by all three of the other claimants, because they have almost universally sought to escape the risks of mining, to obtain the most skilful operation, and to stimulate the productivity of the mines; thereupon the miner has secured at least partial consideration. This stands out in all times and all places, and while the miner has had to take the risks of his fortuitous calling, the Overlord, State, or Landlord have all made for complacent safety by demanding some kind of a tithe on his exertions. Moreover, there has often been a low cunning displayed by these powers in giving something extra to the first discoverer. In these relations of the powers to the mine operator, from the very first we find definite records of the imposition of certain conditions with extraordinary persistence—so fixed a notion that even the United States did not quite escape it. This condition was, no doubt, designed as a stimulus to productive activity, and was the requirement that the miner should continuously employ himself digging in the piece of ground allotted to him. The Greeks, Romans, Mediæval Germans, old and modern Englishmen, modern Australians, all require the miner to keep continuously labouring at his mines, or lose his title. The American, as his inauguration of government happened when things were easier for individuals, allows him a vacation of 11 months in the year for a few years, and finally a holiday altogether. There are other points where the Overlord, the State, or the Landlord have always considered that they had a right to interfere, principally as to the way the miner does his work, lest he should miss, or cause to be missed, some of the mineral; so he has usually been under pains and penalties as to his methods—these quite apart from the very proper protection to human life, which is purely a modern invention, largely of the miner himself. Somebody has had to keep peace and settle disputes among the usually turbulent miners (for what other sort of operators would undertake the hazards and handicaps ?), and therefore special officials and codes, or Courts, for his benefit are of the oldest and most persistent of institutions.

Between the Overlord and the Landowner the fundamental conflict of view as to their respective rights has found its interpretation in the form of the mineral title. The Overlord claimed the metals as distinguished from the land, while the landowner claimed all beneath his

over not only the head meer, or another meer, but also the head meer and the next meer or two adjoining meers. So much for the shape of meers and their dimensions in the case of a *vena profunda*.

I now come to the case of *venae dilatatae*. The boundaries of the areas

soil. Therefore, we find two forms of title—that in which the miner could follow the ore regardless of the surface (the "apex" conception), and that in which the boundaries were vertical from the land surface. Lest the Americans think that the Apex Law was a sin original to themselves, we may mention that it was made use of in Europe a few centuries before Agricola, who will be found to set it out with great precision.

From these points of view, more philosophical than legal, we present a few notes on various ancient laws of mines, though space forbids a discussion of a tithe of the amount it deserves at some experienced hand.

Of the Ancient Egyptian, Lydian, Assyrian, Persian, Indian, and Chinese laws as to mines we have no record, but they were of great simplicity, for the bodies as well as the property of subjects were at the abject disposition of the Overlord. We are informed on countless occasions of Emperors, Kings, and Princes of various degree among these races, owning and operating mines with convicts, soldiers, or other slaves, so we may take it for certain that continuous labour was enforced, and that the boundaries, inspection, and landlords did not cause much anxiety. However, herein lies the root of regalian right.

Our first glimpse of a serious right of the subject to mines is among some of the Greek States, as could be expected from their form of government. With republican ideals, a rich mining district at Mount Laurion, an enterprising and contentious people, it would be surprising indeed if Athenian Literature was void on the subject. While we know that the active operation of these mines extended over some 500 years, from 700 to 200 B.C., the period of most literary reference was from 400 to 300 B.C. Our information on the subject is from two of Demosthenes' orations—one against Pantaenetus, the other against Phaenippis—the first mining lawsuit in which the address of counsel is extant. There is also available some information in Xenophon's Essay upon the Revenues, Aristotle's Constitution of Athens, Lycurgus' prosecution of Diphilos, the Tablets of the Poletae, and many incidental references and inscriptions of minor order. The minerals were the property of the State, a conception apparently inherited from the older civilizations. Leases for exploitation were granted to individuals for terms of three to ten years, depending upon whether the mines had been previously worked, thus a special advantage was conferred upon the pioneer. The leases did not carry surface rights, but the boundaries at Mt. Laurion were vertical, as necessarily must be the case everywhere in horizontal deposits. What they were elsewhere we do not know. The landlord apparently got nothing. The miner must continuously operate his mine, and was required to pay a large tribute to the State, either in the initial purchase of his lease or in annual rent. There were elaborate regulations as to interference and encroachment, and proper support of the workings. Diphilos was condemned to death and his fortune confiscated for robbing pillars. The mines were worked with slaves.

The Romans were most intensive miners and searchers after metallic wealth already mined. The latter was obviously the objective of most Roman conquest, and those nations rich in these commodities, at that time necessarily possessed their own mines. Thus a map showing the extensions of Empire coincides in an extraordinary manner with the metal distribution of Europe, Asia, and North Africa. Further, the great indentations into the periphery of the Imperial map, though many were rich from an agricultural point of view, had no lure to the Roman because they had no mineral wealth. On the Roman law of mines the student is faced with many perplexities. With the conquest of the older States, the plunderers took over the mines and worked them, either by leases from the State to public companies or to individuals; or even in some cases worked them directly by the State. There was thus maintained the concept of State ownership of the minerals which, although apparently never very specifically defined, yet formed a basis of support to the contention of regalian rights in Europe later on. Parallel with this system, mines were discovered and worked by individuals under tithe to the State, and in Pliny (XXXIV, 49) there is reference to the miners in Britain limiting their own output. Individual mining appears to have increased with any relaxation of central authority, as for instance under Augustus. It appears, as a rule, that the mines were held on terminable leases, and that the State did at times resume them; the labour was mostly slaves. As to the detailed conditions under which the mine operator held his title, we know less than of the Greeks—in fact, practically nothing other than that he paid a tithe. The Romans maintained in each mining district an official—the *Procurator Metallorum*—who not only had general charge of the leasing of the mines on behalf of the State, but was usually the magistrate of the district. A bronze tablet found near Aljustrel, in Portugal, in 1876, generally known as the Aljustrel Tablet, appears to be the third of a series setting out the regulations of the mining district. It refers mostly to the regulation of public auctions, the baths, barbers, and tradesmen; but one clause (VII.) is devoted to the regulation of those

on such veins are not all measured by one method. For in some places the *Bergmeister* gives them shapes similar to the shapes of the meers on *venae profundae*, in which case the head-meer is composed of three double measures, and the area of every other mine of two measures, as I have

who work dumps of scoria, etc., and provides for payment to the administrator of the mines of a *capitation* on the slaves employed. It does not, however, so far as we can determine, throw any light upon the actual regulations for working the mines. (Those interested will find ample detail in Jacques Flach, "*La Table de Bronze d'Aljustrel* : *Nouvelle Revue Historique de Droit Francais et Etranger*, 1878, p. 655 ; *Estacio da Veiga, Memorias da Acad. Real das Ciencias de Lisbon, Nova Serie, Tome V, Part II*, Lisbon, 1882.) Despite the systematic law of property evolved by the Romans, the codes contain but small reference to mines, and this in itself is indirect evidence of the concept that they were the property of the State. Any general freedom of the metals would have given rise to a more extensive body of law. There are, of course, the well-known sections in the Justinian and Theodosian Codes, but the former in the main bears on the collection of the tithe and the stimulation of mining by ordering migrant miners to return to their own hearths. There is also some intangible prohibition of mining near edifices. There is in the Theodosian code evident extension of individual right to mine or quarry, and this "freeing" of the mines was later considerably extended. The Empire was, however, then on the decline ; and no doubt it was hoped to stimulate the taxable commodities. There is nothing very tangible as to the position of the landlord with regard to minerals found on his property ; the metals were probably of insufficient frequency on the land of Italian landlords to matter much, and the attitude toward subject races was not usually such as to require an extensive body of law.

In the chaos of the Middle Ages, Europe was governed by hundreds of potentates, great and small, who were unanimous on one point, and this that the minerals were their property. In the bickerings among themselves, the stronger did not hesitate to interpret the Roman law in affirming regalian rights as an excuse to dispossess the weaker. The rights to the mines form no small part of the differences between these Potentates and the more important of their subjects ; and with the gradual accretion of power into a few hands, we find only the most powerful of vassals able to resist such encroachment. However, as to what position the landlord or miner held in these rights, we have little indication until about the beginning of the 13th century, after which there appear several well-known charters, which as time went on were elaborated into practical codes of mining law. The earliest of these charters are those of the Bishop of Trent, 1185 ; that of the Harz Miners, 1219 ; of the town of Iglau in 1249. Many such in connection with other districts appear throughout the 13th, 14th, and 15th centuries. (References to the most important of such charters may be found in Sternberg, *Umrisse der Geschichte des Bergbaues*, Prague, 1838 ; Eisenhart, *De Regali Metalli Fodinarium*, Helmestadt, 1681 ; Gmelin, *Beyträge zur Geschichte des Teutschen Bergbaus*, Halle, 1783 ; Inama-Strenegg, *Deutsche Wirthschaftsgeschichte*, Leipzig, 1879–1901 ; Transactions, Royal Geol. Soc. Cornwall VI, 155 ; Lewis, The Stannaries, New York 1908.) By this time a number of mining communities had grown up, and the charters in the main are a confirmation to them of certain privileges ; they contain, nevertheless, rigorous reservation of the regalian right. The landlord, where present, was usually granted some interest in the mine, but had to yield to the miner free entry. The miner was simply a sort of tributer to the Crown, loaded with an obligation when upon private lands to pay a further portion of his profits to the landlord. He held tenure only during strenuous operation. However, it being necessary to attract skilled men, they were granted many civil privileges not general to the people ; and from many of the principal mining towns "free cities" were created, possessing a measure of self-government. There appear in the Iglau charter of 1249 the first symptoms of the "apex" form of title, this being the logical development of the conception that the minerals were of quite distinct ownership from the land. The law, as outlined by Agricola, is much the same as set out in the Iglavian Charter of three centuries before, and we must believe that such fully developed conceptions as that charter conveys were but the confirmation of customs developed over generations.

In France the landlord managed to maintain a stronger position *vis-à-vis* with the Crown, despite much assertion of its rights ; and as a result, while the landlord admitted the right to a tithe for the Crown, he maintained the actual possession, and the boundaries were defined with the land.

In England the law varied with special mining communities, such as Cornwall, Devon, the Forest of Dean, the Forest of Mendip, Alston Moor, and the High Peak, and they exhibit a curious complex of individual growth, of profound interest to the student of the growth of institutions. These communities were of very ancient origin, some of them at least pre-Roman ; but we are, except for the reference in Pliny, practically without any idea of their legal doings until after the Norman occupation (1066 A.D.). The genius of these conquerors for systematic government soon led them to inquire into the doings of these communities, and while gradually systematising their customs into law, they lost no occasion to assert the

explained more fully above. In this case, however, he measures the meers with a cord, not only forward and backward from the ends of the head-meer, as he is wont to do in the case where the owner of a *vena profunda* has a meer granted him, but also from the sides. In this way meers are marked

regalian right to the minerals. In the two centuries subsequent to their advent there are on record numerous inquisitions, with the recognition and confirmation of "the customs and liberties which had existed from time immemorial," always with the reservation to the Crown of some sort of royalty. Except for the High Peak in Derbyshire, the period and origin of these "customs and liberties" are beyond finding out, as there is practically no record of English History between the Roman withdrawal and the Norman occupation. There may have been "liberties" under the Romans, but there is not a shred of evidence on the subject, and our own belief is that the forms of self-government which sprang up were the result of the Roman evacuation. The miner had little to complain of in the Norman treatment in these matters; but between the Crown and the landlord as represented by the Barons, Lords of the Manor, etc., there were wide differences of opinion on the regalian rights, for in the extreme interpretation of the Crown it tended greatly to curtail the landlord's position in the matter, and the success of the Crown on this subject was by no means universal. In fact, a considerable portion of English legal history of mines is but the outcropping of this conflict, and one of the concessions wrung from King John at Runnymede in 1215 was his abandonment of a portion of such claims.

The mining communities of Cornwall and Devon were early in the 13th century definitely chartered into corporations—"The Stannaries"—possessing definite legislative and executive functions, judicial powers, and practical self-government; but they were required to make payment of the tithe in the shape of "coinage" on the tin. Such recognition, while but a ratification of prior custom, was not obtained without struggle, for the Norman Kings early asserted wide rights over the mines. Tangible record of mining in these parts, from a legal point of view, practically begins with a report by William de Wrotham in 1198 upon his arrangements regarding the coinage. A charter of King John in 1201, while granting free right of entry to the miners, thus usurped the rights of the landlords—a claim which he was compelled by the Barons to moderate; the Crown, as above mentioned did maintain its right to a royalty, but the landlord held the minerals. It is not, however, until the time of Richard Carew's "Survey of Cornwall" (London, 1602) that we obtain much insight into details of miners' title, and the customs there set out were maintained in broad principle down to the 19th century. At Carew's time the miner was allowed to prospect freely upon "Common" or wastrel lands (since mostly usurped by landlords), and upon mineral discovery marked his boundaries, within which he was entitled to the vertical contents. Even upon such lands, however, he must acknowledge the right of the lord of the manor to a participation in the mine. Upon "enclosed" lands he had no right of entry without the consent of the landlord; in fact, the minerals belonged to the land as they do to-day except where voluntarily relinquished. In either case he was compelled to "renew his bounds" once a year, and to operate more or less continuously to maintain the right once obtained. There thus existed a "labour condition" of variable character, usually imposed more or less vigorously in the bargains with landlords. The regulations in Devonshire differed in the important particular that the miner had right of entry to private lands, although he was not relieved of the necessity to give a participation of some sort to the landlord. The Forests of Dean, Mendip, and other old mining communities possessed a measure of self-government, which do not display any features in their law fundamentally different from those of Cornwall and Devon. The High Peak lead mines of Derbyshire, however, exhibit one of the most profoundly interesting of these mining communities. As well as having distinctively Saxon names for some of the mines, the customs there are of undoubted Saxon origin, and as such their ratification by the Normans caused the survival of one of the few Saxon institutions in England—a fact which, we believe, has been hitherto overlooked by historians. Beginning with inquisitions by Edward I. in 1288, there is in the Record Office a wealth of information, the bare titles of which form too extensive a list to set out here. (Of published works, the most important are Edward Manlove's "The Liberties and Customs of the Lead Mines within the Wapentake of Wirksworth," London, 1653, generally referred to as the "Rhymed Chronicle"; Thomas Houghton, "Rara Avis in Terra," London, 1687; William Hardy, "The Miner's Guide," Sheffield, 1748; Thomas Tapping, "High Peak Mineral Customs," London, 1851.) The miners in this district were presided over by a "Barmaster," "Barghmaster," or "Barmar," as he was variously spelled, all being a corruption of the German Bergmeister, with precisely the same functions as to the allotment of title, settlement of disputes, etc., as his Saxon progenitor had, and, like him, he was advised by a jury. The miners had entry to all lands except churchyards (this regulation waived upon death), and a few similar exceptions, and was subject to royalty to the Crown and the landlord. The discoverer was entitled to a finder's "meer" of extra size, and his title was to the vein within the end lines, *i.e.*, the "apex" law. This title was held subject to rigorous labour con-out when a torrent or some other force of Nature has laid open a *vena dilatata* in a valley, so that it appears either on the slope of a mountain or hill or on a plain. Elsewhere the *Bergmeister* doubles the width of the head-meer and it is made fourteen fathoms wide, while the width of each of the other meers remains single, that is seven fathoms, but the length is not defined by boundaries. In some places the head-meer consists of three double measures, but has a width of fourteen fathoms and a length of twenty-one.

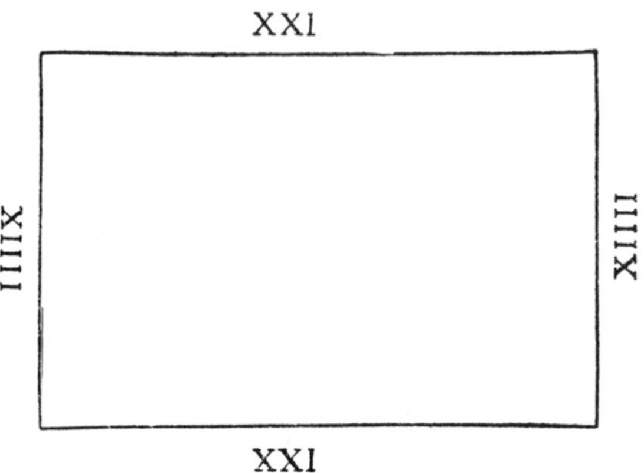

SHAPE OF A HEAD-MEER.

In the same way, every other meer is composed of two measures, doubled in the same fashion, so that it is fourteen fathoms in width and of the same length.

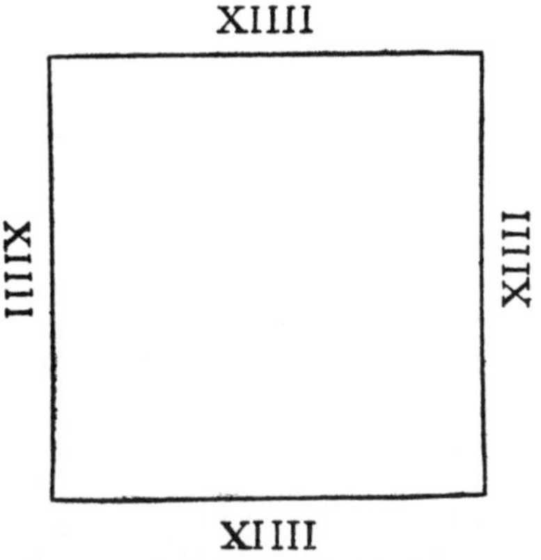

SHAPE OF EVERY OTHER MEER.

ditions, amounting to forfeiture for failure to operate the mine for a period of nine weeks. Space does not permit of the elaboration of the details of this subject, which we hope to pursue elsewhere in its many historical bearings. Among these we may mention that if the American "Apex law" is of English descent, it must be laid to the door of Derbyshire, and not of Cornwall, as is generally done. Our own belief, however, is that the American "apex" conception came straight from Germany.

It is not our purpose to follow these inquiries into mining law beyond the 15th century, but we may point out that with the growth of the sentiment of individualism the miners and landlords obtained steadily wider and wider rights at the cost of the State, until well within the 19th century. The growth of stronger communal sentiment since the middle of the last century has already found its manifestation in the legislation with regard to mines, for the laws of South Africa, Australia, and England, and the agitation in the United States are all toward greater restrictions on the mineral ownership in favour of the State.

Elsewhere every meer, whether a head-meer or other meer, comprises forty-two fathoms in width and as many in length.

In other places the *Bergmeister* gives the owner or company all of some locality defined by rivers or little valleys as boundaries. But the boundaries of every such area of whatsoever shape it be, descend vertically into the earth; so the owner of that area has a right over that part of any *vena dilatata* which lies beneath the first one, just as the owner of the meer on a *vena profunda* has a right over so great a part of all other *venae profundae* as lies within the boundaries of his meer; for just as wherever one *vena profunda* is found, another is found not far away, so wherever one *vena dilatata* is found, others are found beneath it.

Finally, the *Bergmeister* divides *vena cumulata* areas in different ways, for in some localities the head-meer is composed of three measures, doubled in such a way that it is fourteen fathoms wide and twenty-one long; and every other meer consists of two measures doubled, and is square, that is, fourteen fathoms wide and as many long. In some places the head-meer is composed of three single measures, and its width is seven fathoms and its length twenty-one, which two numbers multiplied together make one hundred and forty-seven square fathoms.

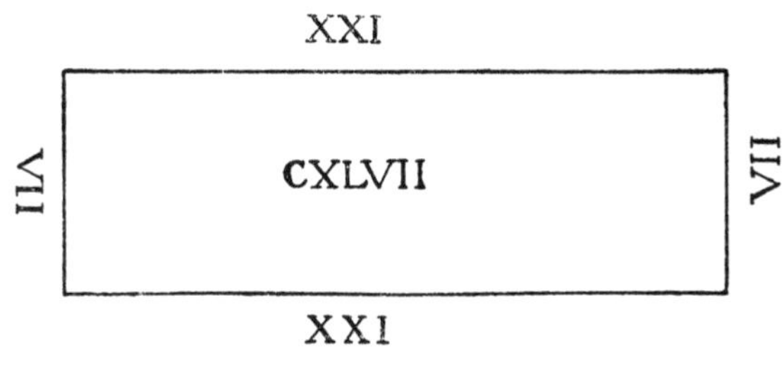

SHAPE OF A HEAD-MEER.

Each other meer consists of one double measure. In some places the head-meer is given the shape of a double measure, and every other meer that of a single measure. Lastly, in other places the owner or a company is given a right over some complete specified locality bounded by little streams, valleys, or other limits. Furthermore, all meers on *venae cumulatae*, as in the case of *dilatatae*, descend vertically into the depths of the earth, and each meer has the boundaries so determined as to prevent disputes arising between the owners of neighbouring mines.

The boundary marks in use among miners formerly consisted only of stones, and from this their name was derived, for now the marks of a boundary are called "boundary stones." To-day a row of posts, made either of oak or pine, and strengthened at the top with iron rings to prevent them from being damaged, is fixed beside the boundary stones to make them more conspicuous. By this method in former times the boundaries of the fields were marked by stones or posts, not only as written of in the book "*De Limitibus Agrorum*,"[7] but also as testified to by the songs of the poets. Such

[7] ?*De Limitibus et de Re Agraria* of Sextus Julius Frontinus (about 50–90 A.D.)

then is the shape of the meers, varying in accordance with the different kinds of veins.

Now tunnels are of two sorts, one kind having no right of property, the other kind having some limited right. For when a miner in some particular locality is unable to open a vein on account of a great quantity of water, he runs a wide ditch, open at the top and three feet deep, starting on the slope and running up to the place where the vein is found. Through it the water flows off, so that the place is made dry and fit for digging. But if it is not sufficiently dried by this open ditch, or if a shaft which he has now for the first time begun to sink is suffering from overmuch water, he goes to the *Bergmeister* and asks that official to give him the right for a tunnel. Having obtained leave, he drives the tunnel, and into its drains all the water is diverted, so that the place or shaft is made fit for digging. If it is not seven fathoms from the surface of the earth to the bottom of this kind of tunnel, the owner possesses no rights except this one: namely, that the owners of the mines, from whose leases the owner of the tunnel extracts gold or silver, themselves pay him the sum he expends within their meer in driving the tunnel through it.

To a depth or height of three and a half fathoms above and below the mouth of the tunnel, no one is allowed to begin another tunnel. The reason for this is that this kind of a tunnel is liable to be changed into the other kind which has a complete right of property, when it drains the meers to a depth of seven fathoms, or to ten, according as the old custom in each place acquires the force of law. In such case this second kind of tunnel has the following right; in the first place, whatever metal the owner, or company owning it, finds in any meer through which it is driven, all belongs to the tunnel owner within a height or depth of one and a quarter fathoms. In the years which are not long passed, the owner of a tunnel possessed all the metal which a miner standing at the bottom of the tunnel touched with a bar, whose handle did not exceed the customary length; but nowadays a certain prescribed height and width is allowed to the owner of the tunnel, lest the owners of the mines be damaged, if the length of the bar be longer than usual. Further, every metal-yielding mine which is drained and supplied with ventilation by a tunnel, is taxed in the proportion of one-ninth for the benefit of the owner of the tunnel. But if several tunnels of this kind are driven through one mining area which is yielding metals, and all drain it and supply it with ventilation, then of the metal which is dug out from above the bottom of each tunnel, one-ninth is given to the owner of that tunnel; of that which is dug out below the bottom of each tunnel, one-ninth is in each case given to the owner of the tunnel which follows next in order below. But if the lower tunnel does not yet drain the shaft of that meer nor supply it with ventilation, then of the metal which is dug out below the bottom of the higher tunnel, one-ninth part is given to the owner of such upper tunnel. Moreover, no one tunnel deprives another of its right to one-ninth part, unless it be a lower one, from the bottom of which to the bottom of the one above must not be less than seven or ten fathoms, according as the king or prince has decreed. Further, of all the money which the owner of the tunnel has spent on his tunnel while driving it through a meer, the owner of that meer pays one-fourth part. If he does not do so he is not allowed to make use of the drains.

Finally, with regard to whatever veins are discovered by the owner at whose expense the tunnel is driven, the right of which has not been already awarded to anyone, on the application of such owner the *Bergmeister* grants him a right of a head-meer, or of a head-meer together with the next meer. Ancient custom gives the right for a tunnel to be driven in any direction for an unlimited length. Further, to-day he who commences a tunnel is given, on his application, not only the right over the tunnel, but even the head and sometimes the next meer also. In former days the owner of the tunnel obtained only so much ground as an arrow shot from the bow might cover, and he was allowed to pasture cattle therein. In a case where the shafts of several meers on some vein could not be worked on account of the great quantity of water, ancient custom also allowed the *Bergmeister* to grant the right of a large meer to anyone who would drive a tunnel. When, however, he had driven a tunnel as far as the old shafts and had found metal, he used to return to the *Bergmeister* and request him to bound and mark off the extent of his right to a meer. Thereupon, the *Bergmeister*, together with a certain number of citizens of the town—in whose place Jurors have now succeeded—used to proceed to the mountain and mark off with boundary stones a large meer, which consisted of seven double measures, that is to say, it was ninety-eight fathoms long and seven wide, which two numbers multiplied together make six hundred and eighty-six square fathoms.

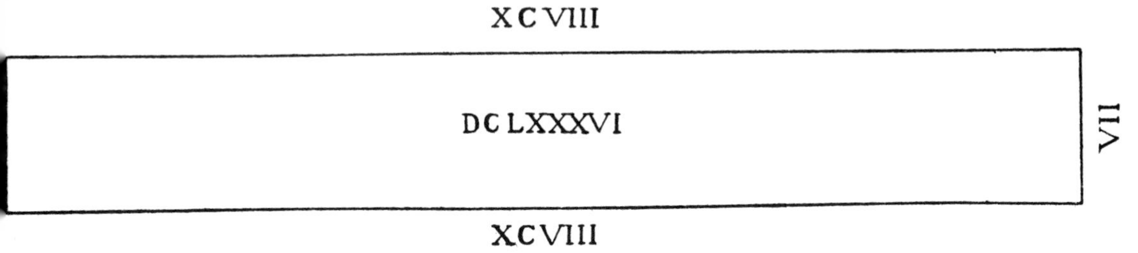

XCVIII

DCLXXXVI	VII

XCVIII

LARGE AREA.

But each of these early customs has been changed, and we now employ the new method.

I have spoken of tunnels ; I will now speak about the division of ownership in mines and tunnels. One owner is allowed to possess and to work one, two, three, or more whole meers, or similarly one or more separate tunnels, provided he conforms to the decrees of the laws relating to metals, and to the orders of the *Bergmeister*. And because he alone provides the expenditure of money on the mines, if they yield metal he alone obtains the product from them. But when large and frequent expenditures are necessary in mining, he to whom the *Bergmeister* first gave the right often admits others to share with him, and they join with him in forming a company, and they each lay out a part of the expense and share with him the profit or loss of the mine. But the title of the mines or tunnels remains undivided, although for the purpose of dividing the expense and profit it may be said each mine or tunnel is divided into parts[8].

This division is made in various ways. A mine, and the same thing must be understood with regard to a tunnel, may be divided into two halves, that is into two similar portions, by which method two owners spend an equal amount on it and draw an equal profit from it, for each possesses one half. Sometimes it is divided into four shares, by which compact four persons can be owners, so that each possesses one-fourth, or also two persons, so that one possesses three-fourths, and the other only one-fourth; or three owners, so that the first has two-fourths, and the second and third one-fourth each. Sometimes it is divided into eight shares, by which plan there may be eight owners, so that each is possessor of one-eighth; sometimes there are two owners, so that one has five-sixths[9] together with one twenty-fourth, and the other one-eighth; or there may be three owners, in which one has three-quarters and the second and third each one-eighth; or it may be divided so that one owner has seven-twelfths, together with one twenty-fourth, a second owner has one-quarter, and a third owner has one-eighth; or so that the first has one-half, the second one-third and one twenty-fourth, and the third one-eighth; or so that the first has one-half, as before, and the second and third each one-quarter; or so that the first and second each have one-third and one twenty-fourth, and the third one-quarter; and in the same way the divisions may be adjusted in all the other proportions. The different ways of dividing the shares originate from the different proportions of ownership. Sometimes a mine is divided into sixteen parts, each of which is a twenty-fourth and a forty-eighth; or it may be divided into thirty-two parts, each of which is a forty-eighth and half a seventy-second and a two hundred and eighty-eighth; or into sixty-four parts of which each share is one seventy-second and one five hundred and seventy-sixth; or finally, into one hundred and twenty-eight parts, any one of which is half a seventy-second and half of one five hundred and seventy-sixth.

Now an iron mine either remains undivided or is divided into two, four, or occasionally more shares, which depends on the excellence of the veins. But a lead, bismuth, or tin mine, and likewise one of copper or even quicksilver, is also divided into eight shares, or into sixteen or thirty-two, and less commonly into sixty-four. The number of the divisions of the silver mines at Freiberg in Meissen did not formerly progress beyond this; but

[8]Such a form of ownership is very old. Apparently upon the instigation of Xenophon (see Note 7, p. 29) the Greeks formed companies to work the mines of Laurion, further information as to which is given in note 6, p. 27. Pliny (Note 7, p. 232) mentions the Company working the quicksilver mines in Spain. In fact, company organization was very common among the Romans, who speculated largely in the shares, especially in those companies which farmed the taxes of the provinces, or leased public lands, or took military and civil contracts.

[9]The Latin text gives one-sixth, obviously an error.

within the memory of our fathers, miners have divided a silver mine, and similarly the tunnel at Schneeberg, first of all into one hundred and twenty-eight shares, of which one hundred and twenty-six are the property of private owners in the mines or tunnels, one belongs to the State and one to the Church; while in Joachimsthal only one hundred and twenty-two shares of the mines or tunnels are the property of private owners, four are proprietary shares, and the State and Church each have one in the same way. To these there has lately been added in some places one share for the most needy of the population, which makes one hundred and twenty-nine shares. It is only the private owners of mines who pay contributions. A proprietary holder, though he holds as many as four shares such as I have described, does not pay contributions, but gratuitously supplies the owners of the mines with sufficient wood from his forests for timbering, machinery, buildings, and smelting; nor do those belonging to the State, Church, and the poor pay contributions, but the proceeds are used to build or repair public works and sacred buildings, and to support the most needy with the profits which they draw from the mines. Furthermore, in our State, the one hundred and twenty-eighth share has begun to be divided into two, four, or eight parts, or even into three, six, twelve, or smaller parts. This is done when one mine is created out of two, for then the owner who formerly possessed one-half becomes owner of one-fourth; he who possessed one-fourth, of one-eighth; he who possessed one-third, of one-sixth; he who possessed one-sixth, of one-twelfth. Since our countrymen call a mine a *symposium*, that is, a drinking bout, we are accustomed to call the money which the owners subscribe a *symbolum*, or a contribution[10]. For, just as those who go to a banquet (*symposium*) give contributions (*symbola*), so those who purpose making large profits from mining are accustomed to contribute toward the expenditure. However, the manager of the mine assesses the contributions of the owners annually, or for the most part quarterly, and as often he renders an account of receipts and expenses. At Freiberg in Meissen the old practice was for the manager to exact a contribution from the owners every week, and every week to distribute among them the profits of the mines, but this practice during almost the last fifteen years has been so far changed that contribution and distribution are made four[11] times each year. Large or small contributions are imposed according to the number of workmen which the mine or tunnel requires; as a result, those who possess many shares provide many contributions. Four times a year the owners contribute to the cost, and four times during the year the profits of the mines are distributed among them; these are sometimes large, sometimes small, according as there is more or less gold or silver or other metal dug out. Indeed, from the St. George mine in Schneeberg the miners extracted so much silver in a quarter of a year that silver cakes, which were worth

[10]A *symposium* is a banquet, and a *symbola* is a contribution of money to a banquet. This sentence is probably a play on the old German *Zeche*, mine, this being also a term for a drinking bout.

[11]In the Latin text this is " three "—obviously an error.

1,100 Rhenish guldens, were distributed to each one hundred and twenty-eighth share. From the Annaberg mine which is known as the Himmelich Höz, they had a dole of eight hundred thaler; from a mine in Joachimsthal which is named the Sternen, three hundred thaler; from the head mine at Abertham, which is called St. Lorentz, two hundred and twenty-five thaler[12]. The more shares of which any individual is owner the more profits he takes.

I will now explain how the owners may lose or obtain the right over a mine, or a tunnel, or a share. Formerly, if anyone was able to prove by witnesses that the owners had failed to send miners for three continuous shifts[13], the *Bergmeister* deprived them of their right over the mine, and gave the right over it to the informer, if he desired it. But although miners preserve this custom to-day, still mining share owners who have paid their contributions do not lose their right over their mines against their will. Formerly, if water which had not been drawn off from the higher shaft of some mine percolated through a vein or stringer into the shaft of another mine and impeded their work, then the owners of the mine which suffered the damage went to the *Bergmeister* and complained of the loss, and he sent to the shafts two Jurors. If they found that matters were as claimed, the right over the mine which caused the injury was given to the owners who suffered the injury. But this custom in certain places has been changed, for the *Bergmeister*, if he finds this condition of things proved in the case of two shafts, orders the owners of the shaft which causes the injury to contribute part of the expense to the owners of the shaft which receives the injury; if they fail to do so, he then deprives them of their right over their mine; on the other hand, if the owners send men to the workings to dig and draw off the water from the shafts, they keep their right over their mine. Formerly owners used to obtain a right over any tunnel, firstly, if in its bottom they made drains and cleansed them of mud and sand so that the water might flow out without any hindrance, and restored those drains which had been damaged; secondly, if they provided shafts or openings to supply the miners with air, and restored those which had fallen in; and finally, if three miners were employed continuously in driving the tunnel. But the principal reason for losing the title to a tunnel was that for a period of eight days no miner was employed upon it; therefore, when anyone was able to prove by witnesses that the owners of a tunnel had not done these things, he brought his accusation before the *Bergmeister*, who, after going out from the town to the tunnel and inspecting the drains and the ventilating machines and everything else, and finding the charge to be true, placed the witness under oath, and asked him: "Whose tunnel is this at the present time?" The witness would reply: "The King's" or "The

[12]See Note 9, p. 74, for further information with regard to these mines. The Rhenish gulden was about 6.9 shillings, or $1.66. Silver was worth about this amount per Troy ounce at this period, so that roughly, silver of a value of 1,100 gulden would be about 1,100 Troy ounces. The Saxon thaler was worth about 4.64 shillings or about $1.11. The thaler, therefore, represented about .65 Troy ounces of silver, so that 300 thalers were about 195 Troy ounces, and 225 thalers about 146 Troy ounces.

[13]*Opera continens.* The Glossary gives *schicht,*—the origin of the English "shift."

Prince's." Thereupon the *Bergmeister* gave the right over the tunnel to the first applicant. This was the severe rule under which the owners at one time lost their rights over a tunnel; but its severity is now considerably mitigated, for the owners do not now forthwith lose their right over a tunnel through not having cleaned out the drains and restored the shafts or ventilation holes which have suffered damage; but the *Bergmeister* orders the tunnel manager to do it, and if he does not obey, the authorities fine the tunnel. Also it is sufficient for one miner to be engaged in driving the tunnel. Moreover, if the owner of a tunnel sets boundaries at a fixed spot in the rocks and stops driving the tunnel, he may obtain a right over it so far as he has gone, provided the drains are cleaned out and ventilation holes are kept in repair. But any other owner is allowed to start from the established mark and drive the tunnel further, if he pays the former owners of the tunnel as much money every three months as the *Bergmeister* decides ought to be paid.

There remain for discussion, the shares in the mines and tunnels. Formerly if anybody conveyed these shares to anyone else, and the latter had once paid his contribution, the seller[14] was bound to stand by his bargain, and this custom to-day has the force of law. But if the seller denied that the contribution had been paid, while the buyer of the shares declared that he could prove by witnesses that he had paid his contribution to the other proprietors, and a case arose for trial, then the evidence of the other proprietors carried more weight than the oath of the seller. To-day the buyer of the shares proves that he has paid his contribution by a document which the mine or tunnel manager always gives each one; if the buyer has contributed no money there is no obligation on the seller to keep his bargain. Formerly, as I have said above, the proprietors used to contribute money weekly, but now contributions are paid four times each year. To-day, if for the space of a month anyone does not take proceedings against the seller of the shares for the contribution, the right of taking proceedings is lost. But when the Clerk has already entered on the register the shares which had been conveyed or bought, none of the owners loses his right over the share unless the money is not contributed which the manager of the mine or tunnel has demanded from the owner or his agent. Formerly, if on the application of the manager the owner or his agent did not pay, the matter was referred to the *Bergmeister*, who ordered the owner or his agent to make his contribution; then if he failed to contribute for three successive weeks, the *Bergmeister* gave the right to his shares to the first applicant. To-day this custom is unchanged, for if owners fail for the space of a month to pay the contributions which the manager of the mine has imposed on them, on a stated day their names are proclaimed aloud and struck off the list of owners, in the presence of the *Bergmeister*, the Jurors, the Mining Clerk, and the Share Clerk, and each of such shares is entered on the proscribed list. If, how-

[14]The terms in the Latin text are *donator*, a giver of a gift, and *donatus*, a receiver. It appears to us, however, that some consideration passed, and we have, therefore, used "seller" and "buyer."

ever, on the third, or at latest the fourth day, they pay their contributions to the manager of the mine or tunnel, and pay the money which is due from them to the Share Clerk, he removes their shares from the proscribed list. They are not thereupon restored to their former position unless the other owners consent; in which respect the custom now in use differs from the old practice, for to-day if the owners of shares constituting anything over half the mine consent to the restoration of those who have been proscribed, the others are obliged to consent whether they wish to or not. Formerly, unless such restoration had been sanctioned by the approval of the owners of one hundred shares, those who had been proscribed were not restored to their former position.

The procedure in suits relating to shares was formerly as follows: he who instituted a suit and took legal proceedings against another in respect of the shares, used to make a formal charge against the accused possessor before the *Bergmeister*. This was done either at his house or in some public place or at the mines, once each day for three days if the shares belonged to an old mine, and three times in eight days if they belonged to a head-meer. But if he could not find the possessor of the shares in these places, it was valid and effectual to make the accusation against him at the house of the *Bergmeister*. When, however, he made the charge for the third time, he used to bring with him a notary, whom the *Bergmeister* would interrogate: "Have I earned the fee?" and who would respond: "You have earned it"; thereupon the *Bergmeister* would give the right over the shares to him who made the accusation, and the accuser in turn would pay down the customary fee to the *Bergmeister*. After these proceedings, if the man whom the *Bergmeister* had deprived of his shares dwelt in the city, one of the proprietors of the mine or of the head-mine was sent to him to acquaint him with the facts, but if he dwelt elsewhere proclamation was made in some public place, or at the mine, openly and in a loud voice in the hearing of numbers of miners. Nowadays a date is defined for the one who is answerable for the debt of shares or money, and information is given the accused by an official if he is near at hand, or if he is absent, a letter is sent him; nor is the right over his shares taken from anyone for the space of one and a half months. So much for these matters.

Now, before I deal with the methods which must be employed in working, I will speak of the duties of the Mining Prefect, the *Bergmeister*, the Jurors, the Mining Clerk, the Share Clerk, the manager of the mine or tunnel, the foreman of the mine or tunnel, and the workmen.

To the Mining Prefect, whom the King or Prince appoints as his deputy, all men of all races, ages, and rank, give obedience and submission. He governs and regulates everything at his discretion, ordering those things which are useful and advantageous in mining operations, and prohibiting those which are to the contrary. He levies penalties and punishes offenders; he arranges disputes which the *Bergmeister* has been unable to settle, and if even he cannot arrange them, he allows the owners who are at variance over some point to proceed to litigation; he even lays down the law, gives orders as a magistrate, or bids them leave their rights in abeyance, and he determines the pay of persons who hold any post or office. He is present in person when the mine managers present their quarterly accounts of profits and expenses, and generally represents the King or Prince and upholds his dignity. The Athenians in this way set Thucydides, the famous historian, over the mines of Thasos[15].

Next in power to the Mining Prefect comes the *Bergmeister*, since he has jurisdiction over all who are connected with mines, with a few exceptions, which are the Tithe Gatherer, the Cashier, the Silver Refiner, the Master of the Mint, and the Coiners themselves. Fraudulent, negligent, or dissolute men he either throws into prison, or deprives of promotion, or fines; of these fines, part is given as a tribute to those in power. When the mine owners have a dispute over boundaries he arbitrates it; or if he cannot settle the dispute, he pronounces judgment jointly with the Jurors; from them, however, an appeal lies to the Mining Prefect. He transcribes his decrees in a book and sets up the records in public. It is also his duty to grant the right over the mines to those who apply, and to confirm their rights; he also must measure the mines, and fix their boundaries, and see that the mine workings are not allowed to become dangerous. Some of these duties he observes on fixed days; for on Wednesday in the presence of the Jurors he confirms the rights over the mines which he has granted, settles disputes about boundaries, and pronounces judgments. On Mondays, Tuesdays, Thursdays, and Fridays, he rides up to the mines, and dismounting at some of them explains what is required to be done, or considers the boundaries which are under controversy. On Saturday all the mine managers and mine foremen render an account of the money which they have spent on the mines during the preceding week, and the Mining Clerk transcribes this account into the register of expenses. Formerly, for one Principality there was one *Bergmeister*, who used to create all the judges and exercise jurisdiction and control over them; for every mine had its own judge, just as to-day each locality has a *Bergmeister* in his place, the name alone being changed. To this ancient *Bergmeister*, who used to dwell at Freiberg in Meissen, disputes were referred; hence right up to the present time the one at Freiberg still has the power of pronouncing judgment when mine owners who are engaged in disputes among themselves appeal to him. The old *Bergmeister* could try everything which was presented to him in any mine whatsoever; whereas the judge could only try the things which were done in his own district, in the same way that every modern *Bergmeister* can.

To each *Bergmeister* is attached a clerk, who writes out a schedule signifying to the applicant for a right over a mine, the day and hour on which the right is granted, the name of the applicant, and the location of the mine. He also affixes at the entrance to the mine, quarterly, at the appointed time, a sheet of paper on which is shown how much contribution must be paid to the manager of the mine. These notices are prepared jointly with the

[15]See Note 29, p. 23.

Mining Clerk, and in common they receive the fee rendered by the foremen of the separate mines.

I now come to the Jurors, who are men experienced in mining matters and of good repute. Their number is greater or less as there are few or more mines; thus if there are ten mines there will be five pairs of Jurors, like a *decemviral college*[16]. Into however many divisions the total number of mines has been divided, so many divisions has the body of Jurors; each pair of Jurors usually visits some of the mines whose administration is under their supervision on every day that workmen are employed; it is usually so arranged that they visit all the mines in the space of fourteen days. They inspect and consider all details, and deliberate and consult with the mine foreman on matters relating to the underground workings, machinery, timbering, and everything else. They also jointly with the mine foreman from time to time make the price per fathom to the workmen for mining the ore, fixing it at a high or low price, according to whether the rock is hard or soft; if, however, the contractors find that an unforeseen and unexpected hardness occurs, and for that reason have difficulty and delay in carrying out their work, the Jurors allow them something in excess of the price fixed; while if there is a softness by reason of water, and the work is done more easily and quickly, they deduct something from the price. Further, if the Jurors discover manifest negligence or fraud on the part of any foreman or workman, they first admonish or reprimand him as to his duties and obligations, and if he does not become more diligent and improve, the matter is reported to the *Bergmeister*, who by right of his authority deprives such persons of their functions and office, or, if they have committed a crime, throws them into prison. Lastly, because the Jurors have been given to the *Bergmeister* as councillors and advisors, in their absence he does not confirm the right over any mine, nor measure the mines, nor fix their boundaries, nor settle disputes about boundaries, nor pronounce judgment, nor, finally, does he without them listen to any account of profits and expenditure.

Now the Mining Clerk enters each mine in his books, the new mines in one book, the old mines which have been re-opened in another. This is done in the following way: first is written the name of the man who has applied for the right over the mine, then the day and hour on which he made his application, then the vein and the locality in which it is situated, next the conditions on which the right has been given, and lastly, the day on which the *Bergmeister* confirmed it. A document containing all these particulars is also given to the person whose right over a mine has been confirmed. The Mining Clerk also sets down in another book the names of the owners of each mine over which the right has been confirmed; in another any intermission of work permitted to any person for cer-

[16]*Decemviri*—"The Ten Men." The original *Decemviri* were a body appointed by the Romans in 452 B.C., principally to codify the law. Such commissions were afterward instituted for other purposes, but the analogy of the above paragraph is a little remote.

tain reasons by the *Bergmeister* ; in another the money which one mine supplies to another for drawing off water or making machinery ; and in another the decisions of the *Bergmeister* and the Jurors, and the disputes settled by them as honorary arbitrators. All these matters he enters in the books on Wednesday of every week ; if holidays fall on that day he does it on the following Thursday. Every Saturday he enters in another book the total expenses of the preceding week, the account of which the mine manager has rendered ; but the total quarterly expenses of each mine manager, he enters in a special book at his own convenience. He enters similarly in another book a list of owners who have been proscribed. Lastly, that no one may be able to bring a charge of falsification against him, all these books are enclosed in a chest with two locks, the key of one of which is kept by the Mining Clerk, and of the other by the *Bergmeister.*

The Share Clerk enters in a book the owners of each mine whom the first finder of the vein names to him, and from time to time replaces the names of the sellers with those of the buyers of the shares. It sometimes happens that twenty or more owners come into the possession of some particular share. Unless, however, the seller is present, or has sent a letter to the Mining Clerk with his seal, or better still with the seal of the Mayor of the town where he dwells, his name is not replaced by that of anyone else ; for if the Share Clerk is not sufficiently cautious, the law requires him to restore the late owner wholly to his former position. He writes out a fresh document, and in this way gives proof of possession. Four times a year, when the accounts of the quarterly expenditure are rendered, he names the new proprietors to the manager of each mine, that the manager may know from whom he should demand contributions and among whom to distribute the profits of the mines. For this work the mine manager pays the Clerk a fixed fee.

I will now speak of the duties of the mine manager. In the case of the owners of every mine which is not yielding metal, the manager announces to the proprietors their contributions in a document which is affixed to the doors of the town hall, such contributions being large or small, according as the *Bergmeister* and two Jurors determine. If anyone fails to pay these contributions for the space of a month, the manager removes their names from the list of owners, and makes their shares the common property of the other proprietors. And so, whomsoever the mine manager names as not having paid his contribution, that same man the Mining Clerk designates in writing, and so also does the Share Clerk. Of the contribution, the mine manager applies part to the payment of the foreman and workmen, and lays by a part to purchase at the lowest price the necessary things for the mine, such as iron tools, nails, firewood, planks, buckets, drawing-ropes, or grease. But in the case of a mine which is yielding metal, the Tithe-gatherer pays the mine manager week by week as much money as suffices to discharge the workmen's wages and to provide the necessary implements for mining. The mine manager of each mine also, in the presence of its foreman, on Saturday in each week renders an account of his expenses to the *Bergmeister* and the Jurors, he renders an account of his receipts, whether the money has been contributed by the owners or taken from the Tithe-gatherer ; and of his quarterly expenditure in the same way to them and to the Mining Prefect and to the Mining Clerk, four times a year at the appointed time ; for just as there are four seasons of the year, namely, Spring, Summer, Autumn, and Winter, so there are fourfold accounts of profits and expenses. In the beginning of the first month of each quarter an account is rendered of the money which the manager has spent on the mine during the previous quarter, then of the profit which he has taken from it during the same period ; for example, the account which is rendered at the beginning of spring is an account of all the profits and expenses of each separate week of winter, which have been entered by the Mining Clerk in the book of accounts. If the manager has spent the money of the proprietors advantageously in the mine and has faithfully looked after it, everyone praises him as a diligent and honest man ; if through ignorance in these matters he has caused loss, he is generally deprived of his office ; if by his carelessness and negligence the owners have suffered loss, the *Bergmeister* compels him to make good the loss ; and finally, if he has been guilty of fraud or theft, he is punished with fine, prison, or death. Further, it is the business of the manager to see that the foreman of the mine is present at the beginning and end of the shifts, that he digs the ore in an advantageous manner, and makes the required timbering, machines, and drains. The manager also makes the deductions from the pay of the workmen whom the foreman has noted as negligent. Next, if the mine is rich in metal, the manager must see that its ore-house is closed on those days on which no work is performed ; and if it is a rich vein of gold or silver, he sees that the miners promptly transfer the output from the shaft or tunnel into a chest or into the strong room next to the house where the foreman dwells, that no opportunity for theft may be given to dishonest persons. This duty he shares in common with the foreman, but the one which follows is peculiarly his own. When ore is smelted he is present in person, and watches that the smelting is performed carefully and advantageously. If from it gold or silver is melted out, when it is melted in the cupellation furnace he enters the weight of it in his books and carries it to the Tithe-gatherer, who similarly writes a note of its weight in his books ; it is then conveyed to the refiner. When it has been brought back, both the Tithe-gatherer and manager again enter its weight in their books. Why again ? Because he looks after the goods of the owners just as if they were his own. Now the laws which relate to mining permit a manager to have charge of more than one mine, but in the case of mines yielding gold or silver, to have charge of only two. If, however, several mines following the head-mine begin to produce metal, he remains in charge of these others until he is freed from the duty of looking after them by the *Bergmeister*. Last of all, the manager, the *Bergmeister*, and the two Jurors, in agreement with the owners, settle the remuneration for the labourers. Enough of the duties and occupation of the manager.

I will now leave the manager, and discuss him who controls the workmen of the mine, who is therefore called the foreman, although some call him the watchman. It is he who distributes the work among the labourers, and sees diligently that each faithfully and usefully performs his duties. He also discharges workmen on account of incompetence, or negligence, and supplies others in their places if the two Jurors and manager give their consent. He must be skilful in working wood, that he may timber shafts, place posts, and make underground structures capable of supporting an undermined mountain, lest the rocks from the hangingwall of the veins, not being supported, become detached from the mass of the mountain and overwhelm the workmen with destruction. He must be able to make and lay out the drains in the tunnels, into which the water from the veins, stringers, and seams in the rocks may collect, that it may be properly guided and can flow away. Further, he must be able to recognize veins and stringers, so as to sink shafts to the best advantage, and must be able to discern one kind of material which is mined from another, or to train his subordinates that they may separate the materials correctly. He must also be well acquainted with all methods of washing, so as to teach the washers how the metalliferous earth or sand is washed. He supplies the miners with iron tools when they are about to start to work in the mines, and apportions a certain weight of oil for their lamps, and trains them to dig to the best advantage, and sees that they work faithfully. When their shift is finished, he takes back the oil which has been left. On account of his numerous and important duties and labours, only one mine is entrusted to one foreman, nay, rather sometimes two or three foremen are set over one mine.

Since I have mentioned the shifts, I will briefly explain how these are carried on. The twenty-four hours of a day and night are divided into three shifts, and each shift consists of seven hours. The three remaining hours are intermediate between the shifts, and form an interval during which the workmen enter and leave the mines. The first shift begins at the fourth hour in the morning and lasts till the eleventh hour; the second begins at the twelfth and is finished at the seventh; these two are day shifts in the morning and afternoon. The third is the night shift, and commences at the eighth hour in the evening and finishes at the third in the morning. The *Bergmeister* does not allow this third shift to be imposed upon the workmen unless necessity demands it. In that case, whether they draw water from the shafts or mine the ore, they keep their vigil by the night lamps, and to prevent themselves falling asleep from the late hours or from fatigue, they lighten their long and arduous labours by singing, which is neither wholly untrained nor unpleasing. In some places one miner is not allowed to undertake two shifts in succession, because it often happens that he either falls asleep in the mine, overcome by exhaustion from too much labour, or arrives too late for his shift, or leaves sooner than he ought. Elsewhere he is allowed to do so, because he cannot subsist on the pay of one shift, especially if provisions grow dearer. The *Bergmeister* does not, however, forbid an extraordinary shift when he concedes only one ordinary shift.

When it is time to go to work the sound of a great bell, which the foreigners call a "campana," gives the workmen warning, and when this is heard they run hither and thither through the streets toward the mines. Similarly, the same sound of the bell warns the foreman that a shift has just been finished; therefore as soon as he hears it, he stamps on the woodwork of the shaft and signals the workmen to come out. Thereupon, the nearest as soon as they hear the signal, strike the rocks with their hammers, and the sound reaches those who are furthest away. Moreover, the lamps show that the shift has come to an end when the oil becomes almost consumed and fails them. The labourers do not work on Saturdays, but buy those things which are necessary to life, nor do they usually work on Sundays or annual festivals, but on these occasions devote the shift to holy things. However, the workmen do not rest and do nothing if necessity demands their labour; for sometimes a rush of water compels them to work, sometimes an impending fall, sometimes something else, and at such times it is not considered irreligious to work on holidays. Moreover, all workmen of this class are strong and used to toil from birth.

The chief kinds of workmen are miners, shovelers, windlass men, carriers, sorters, washers, and smelters, as to whose duties I will speak in the following books, in their proper place. At present it is enough to add this one fact, that if the workmen have been reported by the foreman for negligence, the *Bergmeister*, or even the foreman himself, jointly with the manager, dismisses them from their work on Saturday, or deprives them of part of their pay; or if for fraud, throws them into prison. However, the owners of works in which the metals are smelted, and the master of the smelter, look after their own men. As to the government and duties of miners, I have now said enough; I will explain them more fully in another work entitled *De Jure et Legibus Metallicis*[17].

[17]This work was apparently never published; see Appendix A.

END OF BOOK IV.

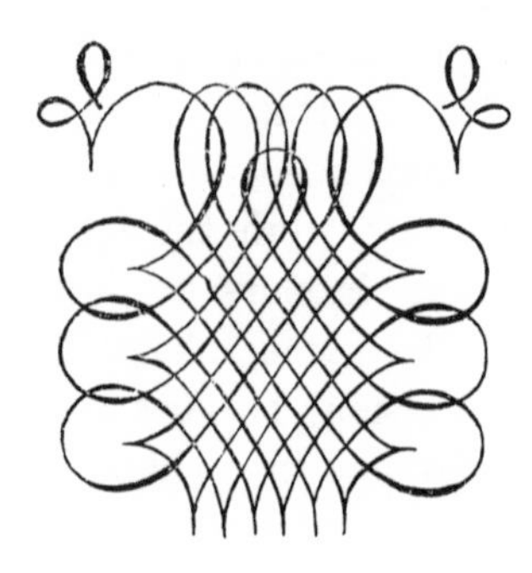

BOOK V.

N the last book I have explained the methods of delimiting the meers along each kind of vein, and the duties of mine officials. In this book[1] I will in like manner explain the principles of underground mining and the art of surveying. First then, I will proceed to deal with those matters which pertain to the former heading, since both the subject and methodical arrangement require it. And so I will describe first of all the digging of shafts, tunnels, and drifts on *venae profundae;* next I will discuss the good indications shown by *canales*[2], by the materials which are dug out, and by the rocks; then I will speak of the tools by which veins and rocks are broken down and excavated; the method by which fire shatters the hard veins; and further, of the machines with which water is drawn from the shafts and air is forced into deep shafts and long tunnels, for digging is impeded by the inrush of the former or the failure of the latter; next I will deal with the two kinds of shafts, and with the making of them and of tunnels; and finally, I will describe the method of mining *venae dilatatae, venae cumulatae,* and stringers.

[1]It has been suggested that we should adopt throughout this volume the mechanical and mining terms used in English mines at Agricola's time. We believe, however, that but a little inquiry would illustrate the undesirability of this course as a whole. Where there is choice in modern miner's nomenclature between an old and a modern term, we have leaned toward age, if it be a term generally understood. But except where the subject described has itself become obsolete, we have revived no obsolete terms. In substantiation of this view, we append a few examples of terms which served the English miner well for centuries, some of which are still extant in some local communities, yet we believe they would carry as little meaning to the average reader as would the reproduction of the Latin terms coined by Agricola.

Rake	= A perpendicular vein.	Slough	= Drainage tunnel.
Woughs	= Walls of the vein.	Sole	= Lowest drift.
Shakes	= Cracks in the walls.	Stool	= Face of a drift or stope.
Flookan	= Gouge.	Winds	
Bryle	= Outcrop.	Turn	= Winze.
Hade	= Incline or underlay of the vein.	Dippas	
		Grove	= Shaft.
Dawling	= Impoverishment of the vein.	Dutins	= Set of timber.
Rither	= A "horse" in a vein.	Stemple	= Post or stull.
Twitches	= "Pinching" of a vein.	Laths	= Lagging.

As examples of the author's coinage and adaptations of terms in this book we may cite:—

Fossa latens	= Drift.
Fossa latens transversa	= Crosscut.
Tectum	= Hangingwall.
Fundamentum	= Footwall.
Tigna per intervalla posita	= Wall plate.
Arbores dissectae	= Lagging.
Formae	= Hitches.

We have adopted the term "tunnel" for openings by way of outlet to the mine. The word in this narrow sense is as old as "adit," a term less expressive and not so generally used in the English-speaking mining world. We have for the same reason adopted the word "drift" instead of the term "level" so generally used in America, because that term always leads to confusion in discussion of mine surveys. We may mention, however, that the term "level" is a heritage from the Derbyshire mines, and is of an equally respectable age as "drift."

[2]See note on p. 46-47. The *canales*, as here used, were the openings in the earth, in which minerals were deposited.

Now when a miner discovers a *vena profunda* he begins sinking a shaft and above it sets up a windlass, and builds a shed over the shaft to prevent the rain from falling in, lest the men who turn the windlass be numbed by the cold or troubled by the rain. The windlass men also place their barrows in it, and the miners store their iron tools and other implements therein. Next to the shaft-house another house is built, where the mine foreman and the other workmen dwell, and in which are stored the ore and other things which are dug out. Although some persons build only one house, yet because sometimes boys and other living things fall into the shafts, most miners deliberately place one house apart from the other, or at least separate them by a wall.

Now a shaft is dug, usually two fathoms long, two-thirds of a fathom wide, and thirteen fathoms deep; but for the purpose of connecting with a tunnel which has already been driven in a hill, a shaft may be sunk to a depth of only eight fathoms, at other times to fourteen, more or less[3]. A shaft may be made vertical or inclined, according as the vein which the miners follow in the course of digging is vertical or inclined. A tunnel is a subterranean ditch driven lengthwise, and is nearly twice as high as it is broad, and wide enough that workmen and others may be able to pass and carry their loads. It is usually one and a quarter fathoms high, while its width is about three and three-quarters feet. Usually two workmen are required to drive it, one of whom digs out the upper and the other the lower part, and the one goes forward, while the other follows closely after. Each sits upon small boards fixed securely from the footwall to the hangingwall, or if the vein is a soft one, sometimes on a wedge-shaped plank fixed on to the vein itself. Miners sink more inclined shafts than vertical, and some of each kind do not reach to tunnels, while some connect with them. But as for some shafts, though they have already been sunk to the required depth, the tunnel which is to pierce the mountain may not yet have been driven far enough to connect with them.

It is advantageous if a shaft connects with a tunnel, for then the miners and other workmen carry on more easily the work they have undertaken; but if the shaft is not so deep, it is usual to drift from one or both sides of it. From these openings the owner or foreman becomes acquainted with the veins and stringers that unite with the principal vein, or cut across it, or

[3]This statement, as will appear by the description later on, refers to the depth of winzes or to the distance between drifts, that is "the lift." We have not, however, been justified in using the term "winze," because some of these were openings to the surface. As showing the considerable depth of shafts in Agricola's time, we may quote the following from *Bermannus* (p. 442): "The depths of our shafts "forced us to invent hauling machines suitable for them. There are some of them "larger and more ingenious than this one, for use in deep shafts, as, for instance, "those in my native town of Geyer, but more especially at Schneeberg, where the "shaft of the mine from which so much treasure was taken in our memory has reached the "depth of about 200 fathoms (feet ?), wherefore the necessity of this kind of machinery. "*Naevius:* What an enormous depth! Have you reached the Inferno? *Bermannus:* Oh, "at Kuttenberg there are shafts more than 500 fathoms (feet ?) deep. *Naevius:* And "not yet reached the Kingdom of Pluto?" It is impossible to accept these as fathoms, as this would in the last case represent 3,000 feet vertically. The expression used, however, for fathoms is *passus*, presumably the Roman measure equal to 58·1 inches.

divide it obliquely; however, my discourse is now concerned mainly with *vena profunda*, but most of all with the metallic material which it contains.

Three vertical shafts, of which the first, A, does not reach the tunnel; the second, B, reaches the tunnel; to the third, C, the tunnel has not yet been driven. D—Tunnel.

Excavations of this kind were called by the Greeks κρυπται for, extending along after the manner of a tunnel, they are entirely hidden within the

THREE INCLINED SHAFTS, OF WHICH A DOES NOT YET REACH THE TUNNEL; B REACHES THE TUNNEL; TO THE THIRD, C, THE TUNNEL HAS NOT YET BEEN DRIVEN. D—TUNNEL.

ground. This kind of an opening, however, differs from a tunnel in that it is dark throughout its length, whereas a tunnel has a mouth open to daylight.

A—Shaft. B, C—Drift. D—Another shaft. E—Tunnel. F—Mouth of tunnel.

I have spoken of shafts, tunnels, and drifts. I will now speak of the indications given by the *canales*, by the materials which are dug out, and by the rocks. These indications, as also many others which I will explain, are to a great extent identical in *venae dilatatae* and *venae cumulatae* with *venae profundae.*

When a stringer junctions with a main vein and causes a swelling, a shaft should be sunk at the junction. But when we find the stringer intersecting the main vein crosswise or obliquely, if it descends vertically down to the depths of the earth, a second shaft should be sunk to the point where the stringer cuts the main vein; but if the stringer cuts it obliquely the shaft should be two or three fathoms back, in order that the junction may be pierced lower down. At such junctions lies the best hope of finding the ore for the sake of which we explore the ground, and if ore has already been found, it is usually found in much greater abundance at that spot. Again, if several stringers descend into the earth, the miner, in order to pierce through the point of contact, should sink the shaft in the midst of these stringers, or else calculate on the most prominent one.

Since an inclined vein often lies near a vertical vein, it is advisable to sink a shaft at the spot where a stringer or cross-vein cuts them both; or where a *vena dilatata* or a stringer *dilatata* passes through, for minerals are usually found there. In the same way we have a good prospect of finding metal at the point where an inclined vein joins a vertical one; this is why miners cross-cut the hangingwall or footwall of a main vein, and in these openings seek for a vein which may junction with the principal vein a few fathoms below. Nay, further, these same miners, if no stringer or cross-vein intersects the main vein so that they can follow it in their workings, even cross-cut through the solid rock of the hangingwall or footwall. These cross-cuts are likewise called "κρυπταί," whether the beginning of the opening which has to be undertaken is made from a tunnel or from a drift. Miners have some hope when only a cross vein cuts a main vein. Further, if a vein which cuts the main vein obliquely does not appear anywhere beyond it, it is advisable to dig into that side of the main vein toward which the oblique vein inclines, whether the right or left side, that we may ascertain if the main vein has absorbed it; if after cross-cutting six fathoms it is not found, it is advisable to dig on the other side of the main vein, that we may know for certain whether it has carried it forward. The owners of a main vein can often dig no less profitably on that side where the vein which cuts the main vein again appears, than where it first cuts it; the owners of the intersecting vein, when that is found again, recover their title, which had in a measure been lost.

The common miners look favourably upon the stringers which come from the north and join the main vein; on the other hand, they look unfavourably upon those which come from the south, and say that these do much harm to the main vein, while the former improve it. But I think that miners should not neglect either of them: as I showed in Book III, experience does not confirm those who hold this opinion about veins, so now again I could furnish examples of each kind of stringers rejected by the common miners which have proved good, but I know this could be of little or no benefit to posterity.

If the miners find no stringers or veins in the hangingwall or footwall of the main vein, and if they do not find much ore, it is not worth while to undertake the labour of sinking another shaft. Nor ought a shaft to be sunk where a vein is divided into two or three parts, unless the indications are satisfactory that those parts may be united and joined together a little later. Further, it is a bad indication for a vein rich in mineral to bend and turn hither and thither, for unless it goes down again into the ground vertically or inclined, as it first began, it produces no more metal; and even though it does go down again, it often continues barren. Stringers which in their outcrops bear metals, often disappoint miners, no metal being found in depth. Further, inverted seams in the rocks are counted among the bad indications.

The miners hew out the whole of solid veins when they show clear evidence of being of good quality; similarly they hew out the drusy[4] veins, especially if the cavities are plainly seen to have formerly borne metal, or if the cavities are few and small. They do not dig barren veins through which water flows, if there are no metallic particles showing; occasionally, however, they dig even barren veins which are free from water, because of the pyrites which is devoid of all metal, or because of a fine black soft substance which is like wool. They dig stringers which are rich in metal, or sometimes, for the purpose of searching for the vein, those that are devoid of ore which lie near the hangingwall or footwall of the main vein. This then, generally speaking, is the mode of dealing with stringers and veins.

Let us now consider the metallic material which is found in the *canales* of *venae profundae*, *venae dilatatae*, and *venae cumulatae*, being in all these either cohesive and continuous, or scattered and dispersed among them, or swelling out in bellying shapes, or found in veins or stringers which originate from the main vein and ramify like branches; but these latter veins and stringers are very short, for after a little space they do not appear again. If we come across a small quantity of metallic material it is an indication; but if a large quantity, it is not an "indication," but the very thing for which we explore the earth. As soon as a miner who searches for veins discovers pure metal or minerals, or rich metallic material, or a great abundance of material which is poor in metal, let him sink a shaft on the spot without any delay. If the material appears more abundant or of better quality on the one side, he will incline his digging in that direction.

Gold, silver, copper, and quicksilver are often found native[5]; less often iron and bismuth; almost never tin and lead. Nevertheless tin-stone is not far removed from the pure white tin which is melted out of them, and galena, from which lead is obtained, differs little from that metal itself.

Now we may classify gold ores. Next after native gold, we come to the

[4]*Cavernos*. The Glossary gives *drusen*, our word *drusy* having had this origin.
[5]*Purum*,—"pure." *Interpretatio* gives the German as *gedigen*,—"native."

rudis[6], of yellowish green, yellow, purple, black, or outside red and inside gold colour. These must be reckoned as the richest ores, because the gold exceeds the stone or earth in weight. Next come all gold ores of which each one hundred *librae* contains more than three *unciae* of gold[7]; for although but a small proportion of gold is found in the earth or stone, yet it equals in value other metals of greater weight.[8] All other gold ores are considered poor, because

[6]*Rudis*,—" Crude." By this expression the author really means ores very rich in any designated metal. In many cases it serves to indicate the minerals of a given metal, as distinguished from the metal itself. Our system of mineralogy obviously does not afford an acceptable equivalent. Agricola (*De Nat. Foss.*, p. 360) says: " I find it necessary to call " each genus (of the metallic minerals) by the name of its own metal, and to this I add a " word which differentiates it from the pure (*puro*) metal, whether the latter has been mined " or smelted; so I speak of *rudis* gold, silver, quicksilver, copper, tin, bismuth, lead, or iron. " This is not because I am unaware that Varro called silver *rudis* which had not yet been " refined and stamped, but because a word which will distinguish the one from the other is " not to be found."

[7]The reasons for retaining the Latin weights are given in the Appendix on Weights and Measures. A *centumpondium* weighs 70.6 lbs. avoirdupois, an *uncia* 412.2 Troy grains, therefore, this value is equal to 72 ounces 18 pennyweights per short ton.

[8]Agricola mentions many minerals in *De Re Metallica*, but without such description as would make possible a hazard at their identity. From his *De Natura Fossilium*, however, and from other mineralogies of the 16th Century, some can be fully identified and others surmised. While we consider it desirable to set out the probable composition of these minerals, on account of the space required, the reasons upon which our opinion has been based cannot be given in detail, as that would require extensive quotations. In a general way, we have throughout the text studiously evaded the use of modern mineralogical terms—unless the term used to-day is of Agricola's age—and have adopted either old English terms of pre-chemistry times or more loose terms used by common miners. Obviously modern mineralogic terms imply a precision of knowledge not existing at that period. It must not be assumed that the following is by any means a complete list of the minerals described by Agricola, but they include most of those referred to in this chapter. His system of mineralogy we have set out in note 4, p. 1, and it requires no further comment here. The grouping given below is simply for convenience and does not follow Agricola's method. Where possible, we tabulate in columns the Latin term used in *De Re Metallica*; the German equivalent given by the Author in either the *Interpretatio* or the Glossary; our view of the probable modern equivalent based on investigation of his other works and other ancient mineralogies, and lastly the terms we have adopted in the text. The German spelling is that given in the original. As an indication of Agricola's position as a mineralogist, we mark with an asterisk the minerals which were first specifically described by him. We also give some notes on matters of importance bearing on the nomenclature used in *De Re Metallica*. Historical notes on the chief metals will be found elsewhere, generally with the discussion of smelting methods. We should not omit to express our indebtedness to Dana's great " System of Mineralogy," in the matter of correlation of many old and modern minerals.

Gold Minerals. Agricola apparently believed that there were various gold minerals, green, yellow, purple, black, etc. There is nothing, however, in his works that permits of any attempt to identify them, and his classification seems to rest on gangue colours.

Silver Minerals.

Argentum purum in venis reperitur	*Gedigen silber* ..	.. ..	*Native silver
Argentum rude	*Gedigen silber ertz* ..	.. ..	*Rudis* silver, or pure silver minerals
Argentum rude plumbei coloris	*Glas ertz*	Argentite (Ag_2S)	*Silver glance
Argentum rude rubrum ..	*Rot gold ertz* ..	Pyrargyrite (Ag_3SbS_3)	*Red silver
Argentum rude rubrum translucidum	*Durchsichtig rod gulden ertz*	Proustite (Ag_3AsS_3)	*Ruby silver
Argentum rude album ..	*Weis rod gulden ertz: Dan es ist frisch wie offtmals rod gulden ertz pfleget zusein* ..	.. ..	White silver

the earth or stone too far outweighs the gold. A vein which contains a larger proportion of silver than of gold is rarely found to be a rich one. Earth, whether it be dry or wet, rarely abounds in gold; but in dry earth there is more often found a greater quantity of gold, especially if it has the

Argentum rude jecoris colore	*Gedigen leberfarbig ertz*	Part Bromyrite (Ag Br)	Liver-coloured silver
Argentum rude luteum ..	*Gedigen geelertz* ..	.. ..	Yellow silver
Argentum rude cineraceum	*Gedigen graw ertz* ..	Part Cerargurite (Ag Cl) (Horn Silver) Part Stephanite (Ag_5SbS_4)	*Grey silver
Argentum rude nigrum ..	*Gedigen schwartz ertz*		*Black silver
Argentum rude purpureum	*Gedigen braun ertz* ..		*Purple silver

The last six may be in part also alteration products from all silver minerals.

The reasons for indefiniteness in determination usually lie in the failure of ancient authors to give sufficient or characteristic descriptions. In many cases Agricola is sufficiently definite as to assure certainty, as the following description of what we consider to be silver glance, from *De Natura Fossilium* (p. 360), will indicate: "Lead-coloured *rudis* silver is "called by the Germans from the word glass (*glasertz*), not from lead. Indeed, it has "the colour of the latter or of galena (*plumbago*), but not of glass, nor is it transparent "like glass, which one might indeed expect had the name been correctly derived. This "mineral is occasionally so like galena in colour, although it is darker, that one who is not "experienced in minerals is unable to distinguish between the two at sight, but in substance "they differ greatly from one another. Nature has made this kind of silver out of a little "earth and much silver. Whereas galena consists of stone and lead containing some silver. "But the distinction between them can be easily determined, for galena may be ground "to powder in a mortar with a pestle, but this treatment flattens out this kind of *rudis* silver. "Also galena, when struck by a mallet or bitten or hacked with a knife, splits and breaks to "pieces; whereas this silver is malleable under the hammer, may be dented by the teeth, "and cut with a knife."

COPPER MINERALS.

Aes purum fossile	*Gedigen kupfer* ..	Native copper ..	Native copper
Aes rude plumbei coloris ..	*Kupferglas ertz* ..	Chalcocite (Cu_2S) ..	*Copper glance
Chalcitis ..	*Rodt atrament* ..	A decomposed copper or iron sulphide ..	*Chalcitis* (see notes on p. 573)
Pyrites aurei colore .. / *Pyrites aerosus* ..	*Geelkis oder kupferkis* ..	Part chalcopyrite (Cu Fe S) part bornite (Cu_3FeS_3)	Copper pyrites
Caeruleum ..	*Berglasur* ..	Azurite	Azure
Chrysocolla ..	*Berggrün und schifergrün* ..	Part chrysocolla .. / Part Malachite ..	Chrysocolla (see note 7, p. 560)
Molochites ..	*Molochit* ..	Malachite	Malachite
Lapis aerarius ..	*Kupfer ertz* ..	.. .. ..	Copper ore
Aes caldarium rubrum fuscum or *Aes sui coloris* ..	*Lebeter kupfer* .. / *Rotkupfer* ..	When used for an ore, is probably cuprite ..	*Ruby copper ore
Aes nigrum ..	*Schwartz kupfer* ..	Probably CuO from oxidation of other minerals	*Black copper

In addition to the above the Author uses the following, which were in the main artificial products:

Aerugo	*Grünspan oder Spanschgrün* ..	Verdigris	Verdigris
Aes luteum ..	*Gelfarkupfer* ..	Impure blister copper	Unrefined copper (see note 16, p. 511)
Aes caldarium ..	*Lebeterkupfer* ..		
Aeris flos ..	*Kupferbraun* ..	Cupric oxide scales ..	Copper flower
Aeris squama ..	*Kupferhammerschlag*		Copper scale (see note 9, p. 233)
Atramentum sutorium caeruleum or *chalcanthum* ..	*Blaw kupfer wasser*	Chalcanthite	Native blue vitriol (see note on p. 572)

appearance of having been melted in a furnace, and if it is not lacking in scales resembling mica. The solidified juices, azure, chrysocolla, orpiment, and realgar, also frequently contain gold. Likewise native or *rudis* gold is found sometimes in large, and sometimes in small quantities in quartz,

Blue and green copper minerals were distinguished by all the ancient mineralogists. Theophrastus, Dioscorides, Pliny, etc., all give sufficient detail to identify their *cyanus* and *caeruleum* partly with modern azurite, and their *chrysocolla* partly with the modern mineral of the same name. However, these terms were also used for vegetable pigments, as well as for the pigments made from the minerals. The Greek origin of *chrysocolla* (*chrusos*, gold and *kolla*, solder) may be blamed with another and distinct line of confusion, in that this term has been applied to soldering materials, from Greek down to modern times, some of the ancient mineralogists even asserting that the copper mineral *chrysocolla* was used for this purpose. Agricola uses *chrysocolla* for borax, but is careful to state in every case (see note xx., p. x): " *Chrysocolla* made from *nitrum*," or " *Chrysocolla* which the Moors call Borax." Dioscorides and Pliny mention substances which were evidently copper sulphides, but no description occurs prior to Agricola that permits a hazard as to different species.

LEAD MINERALS.

Plumbarius lapis	*Glantz*	Galena	Galena
Galena	*Glantz und pleiertz*	Galena	Galena
Plumbum nigrum lutei coloris .. / *Plumbago metallica*	*Pleiertz oder pleischweis*	Cerussite ($Pb\ CO_3$) ..	Yellow lead ore
Cerussa ..	*Pleiweis*	Artificial White-lead..	White-lead (see note 4, p. 440)
Ochra facticia or ochra plumbaria	*Pleigeel*	Massicot (Pb O) ..	*Lead-ochre (see note 8, p. 232)
Molybdaena .. / *Plumbago fornacis*	*Herdplei*	Part litharge	Hearth-lead (see note 37, p. 476)
Spuma argenti .. / *Lithargyrum* ..	*Glett*	Litharge	Litharge (see note on p. 465)
Minium secundarium	*Menning*	Minium (Pb_3O_4) ..	Red-lead (see note 7, p. 232)

So far as we can determine, all of these except the first three were believed by Agricola to be artificial products. Of the first three, galena is certain enough, but while he obviously was familiar with the alteration lead products, his descriptions are inadequate and much confused with the artificial oxides. Great confusion arises in the ancient mineralogies over the terms *molybdaena*, *plumbago*, *plumbum*, *galena*, and *spuma argenti*, all of which, from Roman mineralogists down to a century after Agricola, were used for lead in some form. Further discussion of such confusion will be found in note 37, p. 476. Agricola in *Bermannus* and *De Natura Fossilium*, devotes pages to endeavouring to reconcile the ancient usages of these terms, and all the confusion existing in Agricola's time was thrice confounded when the names *molybdaena* and *plumbago* were assigned to non-lead minerals.

TIN. Agricola knew only one tin mineral : *Lapilli nigri ex quibus conflatur plumbum candidum, i.e.*, " Little black stones from which tin is smelted," and he gives the German equivalent as *zwitter*, " tinstone." He describes them as being of different colours, but probably due to external causes.

ANTIMONY. (*Interpretatio,—spiesglas.*) The *stibi* or *stibium* of Agricola was no doubt the sulphide, and he follows Dioscorides in dividing it into male and female species. This distinction, however, is impossible to apply from the inadequate descriptions given. The mineral and metal known to Agricola and his predecessors was almost always the sulphide, and we have not felt justified in using the term antimony alone, as that implies the refined product, therefore, we have adopted either the Latin term or the old English term " grey antimony." The smelted antimony of commerce sold under the latter term was the sulphide. For further notes see p. 428.

BISMUTH*. *Plumbum cinereum* (*Interpretatio,—bismut*). Agricola states that this mineral occasionally occurs native, " but more often as a mineral of another colour " (*De Nat. Fos.*, p. 337), and he also describes its commonest form as black or grey. This, considering his localities, would indicate the sulphide, although he assigns no special name to it. Although bismuth is mentioned before Agricola in the *Nützliche Bergbüchlin*, he was the first to describe it (see p. 433).

QUICKSILVER. Apart from native quicksilver, Agricola adequately describes cinnabar only. The term used by him for the mineral is *minium nativum* (*Interpretatio,— bergzinober* or *cinnabaris*). He makes the curious statement (*De Nat. Fos.* p. 335) that *rudis* quicksilver also occurs liver-coloured and blackish,—probably gangue colours. (See p. 432).

schist, marble, and also in stone which easily melts in fire of the second degree, and which is sometimes so porous that it seems completely decomposed. Lastly, gold is found in pyrites, though rarely in large quantities.

When considering silver ores other than native silver, those ores are

ARSENICAL MINERALS. Metallic arsenic was unknown, although it has been maintained that a substance mentioned by Albertus Magnus (*De Rebus Metallicis*) was the metallic form. Agricola, who was familiar with all Albertus's writings, makes no mention of it, and it appears to us that the statement of Albertus referred only to the oxide from sublimation. Our word "arsenic" obviously takes root in the Greek for orpiment, which was also used by Pliny (XXXIV, 56) as *arrhenicum*, and later was modified to *arsenicum* by the Alchemists, who applied it to the oxide. Agricola gives the following in *Bermannus* (p. 448), who has been previously discussing realgar and orpiment:—"*Ancon:* Avicenna "also has a white variety. *Bermannus:* I cannot at all believe in a mineral of a white "colour; perhaps he was thinking of an artificial product; there are two which the Alchemists "make, one yellow and the other white, and they are accounted the most powerful poisons "to-day, and are called only by the name *arsenicum*." In *De Natura Fossilium* (p. 219) is described the making of "the white variety" by sublimating orpiment, and also it is noted that realgar can be made from orpiment by heating the latter for five hours in a sealed crucible. In *De Re Metallica* (Book X.), he refers to *auripigmentum facticum*, and no doubt means the realgar made from orpiment. The four minerals of arsenic base mentioned by Agricola were:—

Auripigmentum ..	*Operment* ..	Orpiment (As_2S_3) ..	Orpiment
Sandaraca	*Rosgeel* ..	Realgar (As S) ..	Realgar
Arsenicum	*Arsenik* ..	Artificial arsenical oxide	White arsenic
Lapis subrutilus atque .. splendens	*Mistpuckel* ..	Arsenopyrite (Fe As S)	*Mispickel

We are somewhat uncertain as to the identification of the last. The yellow and red sulphides, however, were well known to the Ancients, and are described by Aristotle, Theophrastus (71 and 89), Dioscorides (V, 81), Pliny (XXXIII, 22, etc.); and Strabo (XII, 3, 40) mentions a mine of them near Pompeiopolis, where, because of its poisonous character none but slaves were employed. The Ancients believed that the yellow sulphide contained gold—hence the name *auripigmentum*, and Pliny describes the attempt of the Emperor Caligula to extract the gold from it, and states that he did obtain a small amount, but unprofitably. So late a mineralogist as Hill (1750) held this view, which seemed to be general. Both realgar and orpiment were important for pigments, medicinal purposes, and poisons among the Ancients. In addition to the above, some arsenic-cobalt minerals are included under *cadmia*.

IRON MINERALS.

Ferrum purum ..	*Gedigen eisen* ..	Native iron	*Native iron
Terra ferria	*Eisen ertz* ..		
Ferri vena	*Eisen ertz* ..		
Galenae genus tertium omnis metalli inanissimi ..	*Eisen glantz* ..	Various soft and hard iron ores, probably mostly hematite ..	Ironstone
Schistos	*Glasköpfe oder blütstein* ..		
Ferri vena jecoris colore	*Leber ertz* ..		
Ferrugo	*Rüst*	Part limonite ..	Iron rust
Magnes	*Siegelstein oder magnet* ..	Magnetite	Lodestone
Ochra nativa	*Berg geel* ..	Limonite	Yellow ochre or ironstone
Haematites	*Blüt stein* ..	Part hematite .. / Part jasper	Bloodstone or ironstone
Schistos	*Glas köpfe* ..	Part limonite	Ironstone
Pyrites	*Kis* ..	Pyrites	Pyrites
Pyrites argenti coloris	*wasser oder weisser kis* ..	Marcasite	*White iron pyrites
Misy	*Gel atrament* ..	Part copiapite ..	*Misy* (see note on p. 573)
Sory	*Graw und schwartz atrament*	Partly a decomposed iron pyrite	*Sory* (see note on p. 573)
Melanteria	*Schwartz und grau atrament*	Melanterite (native vitriol)	*Melanteria* (see note on p. 573)

The classification of iron ores on the basis of exterior characteristics, chiefly hardness and

classified as rich, of which each one hundred *librae* contains more than three *librae* of silver. This quality comprises *rudis* silver, whether silver glance or ruby silver, or whether white, or black, or grey, or purple, or yellow, or liver-

brilliancy, does not justify a more narrow rendering than "ironstone." Agricola (*De Nat. Fos.*, Book V.) gives elaborate descriptions of various iron ores, but the descriptions under any special name would cover many actual minerals. The subject of pyrites is a most confused one; the term originates from the Greek word for fire, and referred in Greek and Roman times to almost any stone that would strike sparks. By Agricola it was a generic term in somewhat the same sense that it is still used in mineralogy, as, for instance, iron pyrite, copper pyrite, etc. So much was this the case later on, that Henckel, the leading mineralogist of the 18th Century, entitled his large volume *Pyritologia*, and in it embraces practically all the sulphide minerals then known. The term *marcasite*, of mediæval Arabic origin, seems to have had some vogue prior and subsequent to Agricola. He, however, puts it on one side as merely a synonym for pyrite, nor can it be satisfactorily defined in much better terms. Agricola apparently did not recognise the iron base of pyrites, for he says (*De Nat. Fos.*, p. 366): "Sometimes, however, pyrites do not contain any gold, silver, copper, "or lead, and yet it is not a pure stone, but a compound, and consists of stone and a substance "which is somewhat metallic, which is a species of its own." Many varieties were known to him and described, partly by their other metal association, but chiefly by their colour.

CADMIA. The minerals embraced under this term by the old mineralogists form one of the most difficult chapters in the history of mineralogy. These complexities reached their height with Agricola, for at this time various new minerals classed under this heading had come under debate. All these minerals were later found to be forms of zinc, cobalt, or arsenic, and some of these minerals were in use long prior to Agricola. From Greek and Roman times down to long after Agricola, brass was made by cementing zinc ore with copper. Aristotle and Strabo mention an earth used to colour copper, but give no details. It is difficult to say what zinc mineral the *cadmium* of Dioscorides (v, 46) and Pliny (XXXIV, 2), really was. It was possibly only furnace calamine, or perhaps blende, for it was associated with copper. They amply describe *cadmia* produced in copper furnaces, and *pompholyx* (zinc oxide). It was apparently not until Theophilus (1150) that the term *calamina* appears for that mineral. Precisely when the term "zinc," and a knowledge of the metal, first appeared in Europe is a matter of some doubt; it has been attributed to Paracelsus, a contemporary of Agricola (see note on p. 409), but we do not believe that author's work in question was printed until long after. The quotations from Agricola given below, in which *zincum* is mentioned in an obscure way, do not appear in the first editions of these works, but only in the revised edition of 1559. In other words, Agricola himself only learned of a substance under this name a short period before his death in 1555. The metal was imported into Europe from China prior to this time. He however does describe actual metallic zinc under the term *conterfei*, and mentions its occurrence in the cracks of furnace walls. (See also notes on p. 409).

The word cobalt (German *kobelt*) is from the Greek word *cobalos*, "mime," and its German form was the term for gnomes and goblins. It appears that the German miners, finding a material (Agricola's "corrosive material") which injured their hands and feet, connected it with the goblins, or used the term as an epithet, and finally it became established for certain minerals (see note 21, p. 214, on this subject). The first written appearance of the term in connection with minerals, appears in Agricola's *Bermannus* (1530). The first practical use of cobalt was in the form of *zaffre* or cobalt blue. There seems to be no mention of the substance by the Greek or Roman writers, although analyses of old colourings show some traces of cobalt, but whether accidental or not is undetermined. The first mention we know of, was by Biringuccio in 1540 (*De La Pirotechnica*, Book II, Chap. IX.), who did not connect it with the minerals then called *cobalt* or *cadmia*. "*Zaffera* is another mineral "substance, like a metal of middle weight, which will not melt alone, but accompanied "by vitreous substances it melts into an azure colour so that those who colour glass, or "paint vases or glazed earthenware, make use of it. Not only does it serve for the above-"mentioned operations, but if one uses too great a quantity of it, it will be black and all other "colours, according to the quantity used." Agricola, although he does not use the word *zaffre*, does refer to a substance of this kind, and in any event also missed the relation between *zaffre* and cobalt, as he seems to think (*De Nat. Fos.*, p. 347) that *zaffre* came from bismuth, a belief that existed until long after his time. The cobalt of the Erzgebirge was of course, intimately associated with this mineral. He says, "the slag of bismuth, mixed "together with metalliferous substances, which when melted make a kind of glass, will tint "glass and earthenware vessels blue." *Zaffre* is the roasted mineral ground with sand, while *smalt*, a term used more frequently, is the fused mixture with sand.

The following are the substances mentioned by Agricola, which, we believe, relate to cobalt and zinc minerals, some of them arsenical compounds. Other arsenical minerals we give above.

coloured, or any other. Sometimes quartz, schist, or marble is of this quality also, if much native or *rudis* silver adheres to it. But that ore is considered of poor quality if three *librae* of silver at the utmost are found in each one hundred *librae* of it[9]. Silver ore usually contains a greater quantity

Cadmia fossilis	*Calmei*; *lapis calaminaris*	Calamine	Calamine
Cadmia metallica	*Kobelt*	Part cobalt	**Cadmia metallica*
Cadmia fornacis	*Mitlere und obere offenbrüche*	Furnace accretions or furnace calamine	Furnace accretions
Bituminosa cadmia	*Kobelt des bergwacht*	(Mansfeld copper schists)	*Bituminosa cadmia* (see note 4, p. 273)
Galena inanis	*Blende*	Sphalerite* ($Zn\ S$)	*Blende
Cobaltum cineraceum	..	Smallite* ($CoAs_2$)	*Cadmia metallica*
Cobaltum nigrum	..	Abolite*	
Cobaltum ferri colore	..	Cobaltite (CoAsA)	
Zincum	*Zinck*	Zinc	Zinc
Liquor Candidus ex fornace . . . etc	*Conterfei*	Zinc	See note 48, p. 408
Atramentum sutorium, candidum, potissimum reperitur Goselariae	..	Goslarite ($Zn\ SO_4$)	*Native white vitriol
Spodos subterranea cinerea	*Geeler zechen rauch*	Either natural or artificial zinc oxides, no doubt containing arsenical oxides	Grey *spodos*
Spodos subterranea nigra	*Schwartzer zechen rauch, auff dem, Altenberge nennet man in kis*		Black *spodos*
Spodos subterranea viridis	*Grauer zechen rauch*		Green *spodos*
Pompholyx	*Hüttenrauch*		*Pompholyx* (see note 26, p. 394)

As seen from the following quotations from Agricola, on *cadmia* and cobalt, there was infinite confusion as to the zinc, cobalt, and arsenic minerals; nor do we think any good purpose is served by adding to the already lengthy discussion of these passages, the obscurity of which is natural to the state of knowledge; but we reproduce them as giving a fairly clear idea of the amount of confusion then existing. It is, however, desirable to bear in mind that the mines familiar to Agricola abounded in complex mixtures of cobalt, nickel, arsenic, bismuth, zinc, and antimony. Agricola frequently mentions the garlic odour from *cadmia metallica*, which, together with the corrosive qualities mentioned below, would obviously be due to arsenic. *Bermannus* (p. 459). "This kind of pyrites miners call *cobaltum*, if it be allowed "to me to use our German name. The Greeks call it *cadmia*. The juices, however, out "of which pyrites and silver are formed, appear to solidify into one body, and thus is produced "what they call *cobaltum*. There are some who consider this the same as pyrites, because "it is almost the same. There are some who distinguish it as a species, which pleases me, "for it has the distinctive property of being extremely corrosive, so that it consumes the "hands and feet of the workmen, unless they are well protected, which I do not believe that "pyrites can do. Three kinds are found, and distinguished more by the colour than by other "properties; they are black (abolite ?), grey (smallite ?), and iron colour (cobalt glance ?). "Moreover, it contains more silver than does pyrites. . ." *Bermannus* (p. 431). "It (a "sort of pyrites) is so like the colour of galena that not without cause might anybody have "doubt in deciding whether it be pyrites or galena. . . . . Perhaps this kind is neither "pyrites nor galena, but has a genus of its own. For it has not the colour of pyrites, nor the "hardness. It is almost the colour of galena, but of entirely different components. From "it there is made gold and silver, and a great quantity is dug out from Reichenstein which "is in Silesia, as was lately reported to me. Much more is found at Raurici, which they call "*zincum*; which species differs from pyrites, for the latter contains more silver than gold, "the former only gold, or hardly any silver."

(*De Natura Fossilium*, p. 170). "*Cadmia fossilis* has an odour like garlic" . . (p. 367). "We now proceed with *cadmia*, not the *cadmia fornacis* (furnace accretions) of "which I spoke in the last book, nor the *cadmia fossilis* (calamine) devoid of metal, which "is used to colour copper, whose nature I explained in Book V, but the metallic mineral "(*fossilis metallica*), which Pliny states to be an ore from which copper is made. The "Ancients have left no record that another metal could be smelted from it. Yet it is a fact

[9]Three *librae* of silver per *centumpondium* would be equal to 875 ounces per short ton.

than this, because Nature bestows quantity in place of quality; such ore is mixed with all kinds of earth and stone compounds, except the various kinds of *rudis* silver; especially with pyrites, *cadmia metallica fossilis*, galena, *stibium*, and others.

"that not only copper but also silver may be smelted from it, and indeed occasionally both "copper and silver together. Sometimes, as is the case with pyrites, it is entirely devoid "of metal. It is frequently found in copper mines, but more frequently still in silver mines. "And there are likewise veins of *cadmia* itself. . . . There are several species of the "*cadmia fossilis* just as there were of *cadmia fornacum*. For one kind has the form of grapes "and another of broken tiles, a third seems to consist of layers. But the *cadmia fossilis* "has much stronger properties than that which is produced in the furnaces. Indeed, it often "possesses such highly corrosive power that it corrodes the hands and feet of the miners. "It, therefore, differs from pyrites in colour and properties. For pyrites, if it does not "contain vitriol, is generally either of a gold or silver colour, rarely of any other. *Cadmia* "is either black or brown or grey, or else reddish like copper when melted in the furnace. ". . . . For this *cadmia* is put in a suitable vessel, in the same way as quicksilver, so "that the heat of the fire will cause it to sublimate, and from it is made a black or brown or "grey body which the Alchemists call "sublimated *cadmia*" (*cadmiam sublimatam*). This "possesses corrosive properties of the highest degree. Cognate with *cadmia* and pyrites "is a compound which the Noricians and Rhetians call *zincum*. This contains gold and "silver, and is either red or white. It is likewise found in the Sudetian mountains, and is "devoid of those metals. . . . With this *cadmia* is naturally related mineral *spodos*, "known to the Moor Serapion, but unknown to the Greeks; and also *pompholyx*—for both "are produced by fire where the miners, breaking the hard rocks in drifts, tunnels, and "shafts, burn the *cadmia* or pyrites or galena or other similar minerals. From *cadmia* is "made black, brown, and grey *spodos*; from pyrites, white *pompholyx* and *spodos*; from "galena is made yellow or grey *spodos*. But *pompholyx* produced from copper stone (*lapide* "*aeroso*) after some time becomes green. The black *spodos*, similar to soot, is found at "Altenberg in Meissen. The white *pompholyx*, like wool which floats in the air in summer, "is found in Hildesheim in the seams in the rocks of almost all quarries except in the sand-"stone. But the grey and the brown and the yellow *pompholyx* are found in those silver "mines where the miners break up the rocks by fire. All consist of very fine particles which "are very light, but the lightest of all is white *pompholyx*."

QUARTZ MINERALS.			
Quarzum ("which Latins call *silex*")	*Quertz oder kiselstein* ..	Quartz	Quartz (see note 15, p. 380)
Silex	*Hornstein oder feurstein* ..	Flinty or jaspery quartz	Hornstone
Crystallum ..	*Crystal*	Clear crystals.. ..	Crystal
Achates	*Achat*	Agate	Agate
Sarda	*Carneol*	Carnelian	Carnelian
Jaspis	*Jaspis*	Part coloured quartz, part jade	*Jaspis*
Murrhina	*Chalcedonius* ..	Chalcedony	Chalcedony
Coticula	*Goldstein* ..	A black silicious stone	Touchstone (see note 37, p. 252)
Amethystus ..	*Amethyst* ..	Amethyst	Amethyst
LIME MINERALS.			
Lapis specularis .. / *Gypsum*	*Gips*	Gypsum	Gypsum
Marmor	*Marmelstein* ..	Marble	Marble
Marmor alabastrites	*Alabaster* ..	Alabaster	Alabaster
Marmor glarea ..	.. .. ..	Calcite (?)	Calc spar(?)
Saxum calcis ..	*Kalchstein* ..	Limestone	Limestone
Marga	*Mergel*	Marl	Marl
Tophus	*Toffstein oder topstein* ..	Sintry limestones, stalagmites, etc. ..	*Tophus* (see note 13, p. 233)
MISCELLANEOUS.			
Amiantus	*Federwis, pliant salamanderhar* ..	Usually asbestos	Asbestos
Magnetis	*Silberweis oder katzensilber* ..	Mica	*Mica
Bracteolae magnetidi simile	.. .. ..		
Mica	*Katzensilber oder glimmer* ..		

As regards other kinds of metal, although some rich ores are found, still, unless the veins contain a large quantity of ore, it is very rarely worth while to dig them. The Indians and some other races do search for gems in veins hidden deep in the earth, but more often they are noticed from their clearness, or rather their brilliancy, when metals are mined. When they outcrop, we follow veins of marble by mining in the same way as is done with rock or building-stones when we come upon them. But gems, properly so called, though they sometimes have veins of their own, are still for the most part found in mines and rock quarries, as the lodestone in iron mines, the emery in silver mines, the *lapis judaicus*, *trochites*, and the like in stone quarries where the diggers, at the bidding of the owners, usually collect them from the seams in the rocks.[10] Nor does the miner neglect the digging of "extraordinary earths,"[11] whether they are found

Silex ex eo ictu ferri facile ignis elicitur. . . . excubus figuris	.. ..	Feldspar	*Feldspar
Medulla saxorum ..	*Steinmarck*	Kaolinite	Porcelain clay
Fluores (lapides gemmarum simili) ..	*Flusse*	Fluorspar ..	*Fluorspar (see note 15, p. 380)
Marmor in metallis repertum	*Spat*	Barite	*Heavy spar

Apart from the above, many other minerals are mentioned in other chapters, and some information is given with regard to them in the footnotes.

[10]As stated in note on p. 2, Agricola divided "stones so called" into four kinds; the first, common stones in which he included lodestone and jasper or bloodstone; the second embraced gems; the third were decorative stones, such as marble, porphyry, etc.; the fourth were rocks, such as sandstone and limestone.

LODESTONE. (*Magnes*; *Interpretatio* gives *Siegelstein oder magnet*). The lodestone was well-known to the Ancients under various names—*magnes*, *magnetis*, *heraclion*, and *sideritis*. A review of the ancient opinions as to its miraculous properties would require more space than can be afforded. It is mentioned by many Greek writers, including Hippocrates (460–372 B.C.) and Aristotle; while Theophrastus (53), Dioscorides (v, 105), and Pliny (XXXIV, 42, XXXVI 25) describe it at length. The Ancients also maintained the existence of a stone, *theamedes*, having repellant properties, and the two were supposed to exist at times in the same stone.

EMERY. (*Smiris*; *Interpretatio* gives *smirgel*). Agricola (*De Natura Fossilium.*, p. 265) says: "The ring-makers polish and clean their hard gems with *smiris*. The glaziers "use it to cut their glass into sheets. It is found in the silver mines of Annaberg in Meissen "and elsewhere." Stones used for polishing gems are noted by the ancient authors, and Dana (Syst. of Mineralogy, p. 211) considers the stone of Armenia, of Theophrastus (77), to be emery, although it could quite well be any hard stone, such as Novaculite—which is found in Armenia. Dioscorides (v, 166) describes a stone with which the engravers polish gems.

LAPIS JUDAICUS. (*Interpretatio* gives *Jüden stein*). This was undoubtedly a fossil, possibly a *pentremites*. Agricola (*De Natura Fosilium*, p. 256) says: "It is shaped like an "acorn, from the obtuse end to the point proceed raised lines, all equidistant, etc." Many fossils were included among the semi-precious stones by the Ancients. Pliny (XXXVII, 55, 66, 73) describes many such stones, among them the *balanites*, *phoenicitis* and the *pyren*, which resemble the above.

TROCHITIS. (*Interpretatio* gives *spangen oder rederstein*). This was also a fossil, probably crinoid stems. Agricola (*De Natura Fosilium*, p. 256) describes it: "*Trochites* is so "called from a wheel, and is related to *lapis judaicus*. Nature has indeed given it the shape "of a drum (*tympanum*). The round part is smooth, but on both ends as it were there is a "module from which on all sides there extend radii to the outer edge, which corresponds with "the radii. These radii are so much raised that it is fluted. The size of these *trochites* "varies greatly, for the smallest is so little that the largest is ten times as big, and the largest "are a digit in length by a third of a digit in thickness . . . when immersed in vinegar "they make bubbles."

[11]The "extraordinary earths" of Agricola were such substances as ochres, tripoli, fullers earth, potters' clay, clay used for medicinal purposes, etc., etc.

in gold mines, silver mines, or other mines; nor do other miners neglect them if they are found in stone quarries, or in their own veins; their value is usually indicated by their taste. Nor, lastly, does the miner fail to give attention to the solidified juices which are found in metallic veins, as well as in their own veins, from which he collects and gathers them. But I will say no more on these matters, because I have explained more fully all the metals and mineral substances in the books "*De Natura Fossilium.*"

But I will return to the indications. If we come upon earth which is like lute, in which there are particles of any sort of metal, native or *rudis*, the best possible indication of a vein is given to miners, for the metallic material from which the particles have become detached is necessarily close by. But if this kind of earth is found absolutely devoid of all metallic material, but fatty, and of white, green, blue, and similar colours, they must not abandon the work that has been started. Miners have other indications in the veins and stringers, which I have described already, and in the rocks, about which I will speak a little later. If the miner comes across other dry earths which contain native or *rudis* metal, that is a good indication; if he comes across yellow, red, black, or some other "extraordinary" earth, though it is devoid of mineral, it is not a bad indication. Chrysocolla, or azure, or verdigris, or orpiment, or realgar, when they are found, are counted among the good indications. Further, where underground springs throw up metal we ought to continue the digging we have begun, for this points to the particles having been detached from the main mass like a fragment from a body. In the same way the thin scales of any metal adhering to stone or rock are counted among the good indications. Next, if the veins which are composed partly of quartz, partly of clayey or dry earth, descend one and all into the depths of the earth together, with their stringers, there is good hope of metal being found; but if the stringers afterward do not appear, or little metallic material is met with, the digging should not be given up until there is nothing remaining. Dark or black or horn or liver-coloured quartz is usually a good sign; white is sometimes good, sometimes no sign at all. But calc-spar, showing itself in a *vena profunda*, if it disappears a little lower down is not a good indication; for it did not belong to the vein proper, but to some stringer. Those kinds of stone which easily melt in fire, especially if they are translucent (fluorspar ?), must be counted among the medium indications, for if other good indications are present they are good, but if no good indications are present, they give no useful significance. In the same way we ought to form our judgment with regard to gems. Veins which at the hangingwall and footwall have horn-coloured quartz or marble, but in the middle clayey earth, give some hope; likewise those give hope in which the hangingwall or footwall shows iron-rust coloured earth, and in the middle greasy and sticky earth; also there is hope for those which have at the hanging or footwall that kind of earth which we call "soldiers' earth," and in the middle black earth or earth which looks as if burnt. The special indication of gold is orpiment; of silver is bismuth and *stibium*; of copper is verdigris, *melanteria*, *sory*, *chalcitis*, *misy*, and vitriol; of tin is the large pure black stones of which the tin itself is made, and a material they dig up resembling litharge; of iron, iron rust. Gold and copper are equally indicated by chrysocolla and azure; silver and lead, by the lead. But, though miners rightly call bismuth "the roof of silver," and though copper pyrites is the common parent of vitriol and *melanteria*, still these sometimes have their own peculiar minerals, just as have orpiment and *stibium*.

Now, just as certain vein materials give miners a favourable indication, so also do the rocks through which the *canales* of the veins wind their way, for sand discovered in a mine is reckoned among the good indications, especially if it is very fine. In the same way schist, when it is of a bluish or blackish colour, and also limestone, of whatever colour it may be, is a good sign for a silver vein. There is a rock of another kind that is a good sign; in it are scattered tiny black stones from which tin is smelted; especially when the whole space between the veins is composed of this kind of rock. Very often indeed, this good kind of rock in conjunction with valuable stringers contains within its folds the *canales* of mineral bearing veins: if it descends vertically into the earth, the benefit belongs to that mine in which it is seen first of all; if inclined, it benefits the other neighbouring mines[12]. As a result the miner who is not ignorant of geometry can calculate from the other mines the depth at which the *canales* of a vein bearing rich metal will wind its way through the rock into his mine. So much for these matters.

I now come to the mode of working, which is varied and complex, for in some places they dig crumbling ore, in others hard ore, in others a harder ore, and in others the hardest kind of ore. In the same way, in some places the hangingwall rock is soft and fragile, in others hard, in others harder, and in still others of the hardest sort. I call that ore "crumbling" which is composed of earth, and of soft solidified juices; that ore "hard" which is composed of metallic minerals and moderately hard stones, such as for the most part are those which easily melt in a fire of the first and second orders, like lead and similar materials. I call that ore "harder" when with those I have already mentioned are combined various sorts of quartz, or stones which easily melt in fire of the third degree, or pyrites, or *cadmia*, or very hard marble. I call that ore hardest, which is composed throughout the whole vein of these hard stones and compounds. The hanging or footwalls of a vein are hard, when composed of rock in which there are few stringers or seams; harder, in which they are fewer; hardest, in which they are fewest or none at all. When these are absent, the rock is quite devoid of water which softens it. But the hardest rock of the hanging or footwall, however, is seldom as hard as the harder class of ore.

Miners dig out crumbling ore with the pick alone. When the metal has not yet shown itself, they do not discriminate between the hangingwall and the veins; when it has once been found, they work with the utmost care. For first of all they tear away the hangingwall rock separately from the vein, afterward with a pick they dislodge the crumbling vein from the footwall

[12]Presumably the ore-body dips into a neighbouring property.

into a dish placed underneath to prevent any of the metal from falling to the ground. They break a hard vein loose from the footwall by blows with a hammer upon the first kind of iron tool[13], all of which are designated by appropriate names, and with the same tools they hew away the hard hangingwall rock. They hew out the hangingwall rock in advance more frequently, the rock of the footwall more rarely; and indeed, when the rock of the footwall resists iron tools, the rock of the hangingwall certainly cannot be broken unless it is allowable to shatter it by fire. With regard to the harder veins which are tractable to iron tools, and likewise with regard to the harder and hardest kind of hangingwall rock, they generally attack them with more powerful iron tools, in fact, with the fourth kind of iron tool, which are called by their appropriate names; but if these are not ready to hand, they use two or three iron tools of the first kind together. As for the hardest kind of metal-bearing vein, which in a measure resists iron tools, if the owners of the neighbouring mines give them permission, they break it with fires. But if these owners refuse them permission, then first of all they hew out the rock of the hangingwall, or of the footwall if it be less hard; then they place timbers set in hitches in the hanging or footwall, a little above the vein, and from the front and upper part, where the vein is seen to be seamed with small cracks, they drive into one of the little cracks one of the iron tools which I have mentioned; then in each fracture they place four thin iron blocks, and in order to hold them more firmly, if necessary, they place as many thin iron plates back to back; next they place thinner iron plates between each two iron blocks, and strike and drive them by turns with hammers, whereby the vein rings with a shrill sound; and the moment when it begins to be detached from the hangingwall or footwall rock, a tearing sound is heard. As soon as this grows distinct the miners hastily flee away; then a great crash is heard as the vein is broken and torn, and falls down. By this method they throw down a portion of a vein weighing a hundred pounds more or less. But if the miners by any other method hew the hardest kind of vein which is rich in metal, there remain certain cone-shaped portions which can be cut out afterward only with difficulty. As for this knob of hard ore, if it is devoid of metal, or if they are not allowed to apply fire to it, they proceed round it by digging to the right or left, because it cannot be broken into by iron wedges without great expense. Meantime, while the workmen are carrying out the task they have undertaken, the depths of the earth often resound with sweet singing, whereby they lighten a toil which is of the severest kind and full of the greatest dangers.

As I have just said, fire shatters the hardest rocks, but the method of its application is not simple[14]. For if a vein held in the rocks cannot be hewn

[13]The various kinds of iron tools are described in great detail in Book VI.

[14]Fire-setting as an aid to breaking rock is of very ancient origin, and moreover it persisted in certain German and Norwegian mines down to the end of the 19th century—270 years after the first application of explosives to mining. The first specific reference to fire-setting in mining is by Agatharchides (2nd century B.C.) whose works are not extant, but who is quoted by both Diodorus Siculus and Photius, for which statement see note 8, p. 279. Pliny (XXXIII, 21) says: "Occasionally a kind of silex is met with, which must be "broken with fire and vinegar, or as the tunnels are filled with suffocating fumes and smoke,

out because of the hardness or other difficulty, and the drift or tunnel is low, a heap of dried logs is placed against the rock and fired; if the drift or tunnel is high, two heaps are necessary, of which one is placed above the other, and both burn until the fire has consumed them. This force does not generally soften a large portion of the vein, but only some of the surface. When the rock in the hanging or footwall can be worked by the iron tools and the vein is so hard that it is not tractable to the same tools, then the walls are hollowed out; if this be in the end of the drift or tunnel or above or below, the vein is then broken by fire, but not by the same method; for if the hollow is wide, as many logs are piled into it as possible, but if narrow, only a few. By the one method the greater fire separates the vein more completely from the footwall or sometimes from the hangingwall, and by the other, the smaller fire breaks away less of the vein from the rock, because in that case the fire is confined and kept in check by portions of the rock which surround the wood held in such a narrow excavation. Further, if the excavation is low, only one pile of logs is placed in it, if high, there are two, one placed above the other, by which plan the lower bundle being kindled sets alight the upper one; and the fire being driven by the draught into the vein, separates it from the rock which, however hard it may be, often becomes so softened as to be the most easily breakable of all. Applying this principle, Hannibal, the Carthaginian General, imitating the Spanish miners,

"they frequently use bruising machines, carrying 150 *librae* of iron." This combination of fire and vinegar he again refers to (XXIII, 27), where he dilates in the same sentence on the usefulness of vinegar for breaking rock and for salad dressing. This myth about breaking rocks with fire and vinegar is of more than usual interest, and its origin seems to be in the legend that Hannibal thus broke through the Alps. Livy (59 B.C., 17 A.D.) seems to be the first to produce this myth in writing; and, in any event, by Pliny's time (23–79 A.D.) it had become an established method—in literature. Livy (XXI, 37) says, in connection with Hannibal's crossing of the Alps: "They set fire to it (the timber) when a wind had arisen suitable to "excite the fire, then when the rock was hot it was crumbled by pouring on vinegar (*infuso* "*aceto*). In this manner the cliff heated by the fire was broken by iron tools, and the "declivities eased by turnings, so that not only the beasts of burden but also the elephants "could be led down." Hannibal crossed the Alps in 218 B.C. and Livy's account was written 200 years later, by which time Hannibal's memory among the Romans was generally surrounded by Herculean fables. Be this as it may, by Pliny's time the vinegar was generally accepted, and has been ceaselessly debated ever since. Nor has the myth ceased to grow, despite the remarks of Gibbon, Lavalette, and others. A recent historian (Hennebert, *Histoire d' Annibal* II, p. 253) of that famous engineer and soldier, soberly sets out to prove that inasmuch as literal acceptance of ordinary vinegar is impossible, the Phoenecians must have possessed some mysterious high explosive. A still more recent biographer swallows this argument *in toto*. (Morris, "Hannibal," London, 1903, p. 103). A study of the commentators of this passage, although it would fill a volume with sterile words, would disclose one generalization: That the real scholars have passed over the passage with the comment that it is either a corruption or an old woman's tale, but that hosts of soldiers who set about the biography of famous generals and campaigns, almost to a man take the passage seriously, and seriously explain it by way of the rock being limestone, or snow, or by the use of explosives, or other foolishness. It has been proposed, although there are grammatical objections, that the text is slightly corrupt and read *infosso acuto*, instead of *infuso aceto*, in which case all becomes easy from a mining point of view. If so, however, it must be assumed that the corruption occurred during the 20 years between Livy and Pliny.

By the use of fire-setting in recent times at Königsberg (Arthur L. Collins, "Fire-setting," Federated Inst. of Mining Engineers, Vol. V, p. 82) an advance of from 5 to 20 feet per month in headings was accomplished, and on the score of economy survived the use of gunpowder, but has now been abandoned in favour of dynamite. We may mention that the use of gunpowder for blasting was first introduced at Schemnitz by Caspar Weindle, in 1627, but apparently was not introduced into English mines for nearly 75 years afterward, as the late 17th century English writers continue to describe fire-setting.

overcame the hardness of the Alps by the use of vinegar and fire. Even if a vein is a very wide one, as tin veins usually are, miners excavate into the small streaks, and into those hollows they put dry wood and place amongst them at frequent intervals sticks, all sides of which are shaved down fan-shaped, which easily take light, and when once they have taken fire communicate it to the other bundles of wood, which easily ignite.

A—Kindled logs. B—Sticks shaved down fan-shaped. C—Tunnel.

While the heated veins and rock are giving forth a foetid vapour and the shafts or tunnels are emitting fumes, the miners and other workmen do not go down in the mines lest the stench affect their health or actually kill them, as I will explain in greater detail when I come to speak of the evils which affect miners. The *Bergmeister*, in order to prevent workmen from being suffocated, gives no one permission to break veins or rock by fire in shafts or tunnels where it is possible for the poisonous vapour and smoke to permeate the veins or stringers and pass through into the neighbouring mines, which have no hard veins or rock. As for that part of a vein or the surface of the rock which the fire has separated from the remaining mass, if it is overhead, the miners dislodge it with a crowbar, or if it still has some degree of hardness, they thrust a smaller crowbar into the cracks and so break it down, but if it is on the sides they break it with hammers. Thus broken off, the rock tumbles down ; or if it still remains, they break it off with picks. Rock and earth on the one hand, and metal and ore on the other, are filled into buckets separately and drawn up to the open air or to the nearest tunnel. If the shaft is not deep, the buckets are drawn up by a machine turned by men ; if it is deep, they are drawn by machines turned by horses.

It often happens that a rush of water or sometimes stagnant air hinders the mining ; for this reason miners pay the greatest attention to these matters, just as much as to digging, or they should do so. The water of the veins and stringers and especially of vacant workings, must be drained out through the shafts and tunnels. Air, indeed, becomes stagnant both in tunnels and in shafts ; in a deep shaft, if it be by itself, this occurs if it is neither reached by a tunnel nor connected by a drift with another shaft ; this occurs in a tunnel if it has been driven too far into a mountain and no shaft has yet been sunk deep enough to meet it ; in neither case can the air move or circulate. For this reason the vapours become heavy and resemble mist, and they smell of mouldiness, like a vault or some underground chamber which has been completely closed for many years. This suffices to prevent miners from continuing their work for long in these places, even if the mine is full of silver or gold, or if they do continue, they cannot breathe freely and they have headaches ; this more often happens if they work in these places in great numbers, and bring many lamps, which then supply them with a feeble light, because the foul air from both lamps and men make the vapours still more heavy.

A small quantity of water is drawn from the shafts by machines of different kinds which men turn or work. If so great a quantity has flowed into one shaft as greatly to impede mining, another shaft is sunk some fathoms distant from the first, and thus in one of them work and labour are carried on without hindrance, and the water is drained into the other, which is sunk lower than the level of the water in the first one ; then by these machines or by those worked by horses, the water is drawn up into the drain and flows out of the shaft-house or the mouth of the nearest tunnel. But when into the shaft of one mine, which is sunk more deeply, there flows all the water of all the neighbouring mines, not only from that vein in which the shaft is sunk, but also from other veins, then it becomes necessary for a large sump to be made to collect the water ; from this sump the water is drained by machines which draw it through pipes, or by ox-hides, about which I will say more in the next book. The water which pours into the tunnels from the veins and stringers and seams in the rocks is carried away in the drains.

Air is driven into the extremities of deep shafts and long tunnels by powerful blowing machines, as I will explain in the following book, which will deal with these machines also. The outer air flows spontaneously into the caverns of the earth, and when it can pass through them comes out again. This, however, comes about in different ways, for in spring and summer it flows into the deeper shafts, traverses the tunnels or drifts, and finds its way out of the shallower shafts ; similarly at the same season it pours into the lowest tunnel and, meeting a shaft in its course, turns aside to a higher tunnel and passes out therefrom ; but in autumn and winter, on the other hand, it enters the upper tunnel or shaft and comes out at the deeper ones. This change in the flow of air currents occurs in temperate regions at the beginning of spring and the end of autumn, but in cold regions at the end of spring and the beginning of autumn. But at each period, before the air regularly assumes its own accustomed course, generally for a space of fourteen days it undergoes frequent variations, now blowing into an upper shaft or tunnel, now into a lower one. But enough of this, let us now proceed to what remains.

There are two kinds of shafts, one of the depth already described, of which kind there are usually several in one mine ; especially if the mine is entered by a tunnel and is metal-bearing. For when the first tunnel is connected with the first shaft, two new shafts are sunk ; or if the inrush of water hinders sinking, sometimes three are sunk ; so that one may take the place of a sump and the work of sinking which has been begun may be continued by means of the remaining two shafts ; the same is done in the case of the second tunnel and the third, or even the fourth, if so many are driven into a mountain. The second kind of shaft is very deep, sometimes as much as sixty, eighty, or one hundred fathoms. These shafts continue vertically toward the depths of the earth, and by means of a hauling-rope the broken rock and metalliferous ores are drawn out of the mine ; for which reason miners call them vertical shafts. Over these shafts are erected machines by which water is extracted ; when they are above ground the machines are usually worked by horses, but when they are in tunnels, other kinds are used which are turned by water-power. Such are the shafts which are sunk when a vein is rich in metal.

Now shafts, of whatever kind they may be, are supported in various ways. If the vein is hard, and also the hanging and footwall rock, the shaft does not require much timbering, but timbers are placed at intervals, one end of each of which is fixed in a hitch cut into the rock of the hangingwall and the other fixed into a hitch cut in the footwall. To these timbers are fixed small timbers along the footwall, to which are fastened the lagging and ladders. The lagging is also fixed to the timbers, both to those which screen off the shaft on the ends from the vein, and to those which screen off the rest of the shaft from that part in which the ladders are placed. The lagging on the sides of the shaft confine the vein, so as to prevent fragments of it which have become loosened by water from dropping into the shaft and terrifying, or injuring, or knocking off the miners and other workmen who are going up or down the ladders from one part of the mine to another. For the same reason, the lagging between the ladders and the haulage-way on the other hand, confine and shut off from the ladders the fragments of rock which fall from the buckets or baskets while they are being drawn up ; moreover, they make the arduous and difficult descent and ascent to appear less terrible, and in fact to be less dangerous.

If a vein is soft and the rock of the hanging and footwalls is weak, a closer structure is necessary; for this purpose timbers are joined together in rectangular shapes and placed one after the other without a break. These

A—Wall plates. B—Dividers. C—Long end posts. D—End plates.

are arranged on two different systems; for either the square ends of the timbers, which reach from the hangingwall to the footwall, are fixed into corresponding square holes in the timbers which lie along the hanging or footwall, or the upper part of the end of one and the lower part of the end of the other are cut out and one laid on the other. The great weight of these joined timbers is sustained by stout beams placed at intervals, which are deeply set into hitches in the footwall and hangingwall, but are inclined. In order that these joined timbers may remain stationary, wooden wedges or poles cut from trees are driven in between the timbers and the vein and the hanging wall and the footwall; and the space which remains empty is filled with loose dirt. If the hanging and footwall rock is sometimes hard and sometimes soft, and the vein likewise, solid joined timbers are not used, but timbers are placed at intervals; and where the rock is soft and the vein crumbling, carpenters put in lagging between them and the wall rocks, and behind these they fill with loose dirt; by this means they fill up the void.

When a very deep shaft, whether vertical or inclined, is supported by joined timbers, then, since they are sometimes of bad material and a fall is threatened, for the sake of greater firmness three or four pairs of strong end posts are placed between these, one pair on the hangingwall side, the other on the footwall side. To prevent them from falling out of position and to make them firm and substantial, they are supported by frequent end plates, and in order that these may be more securely fixed they are mortised into the posts. Further, in whatever way the shaft may be timbered, dividers are placed upon the wall plates, and to these is fixed lagging, and this marks off and separates the ladder-way from the remaining part of the shaft. If a vertical shaft is a very deep one, planks are laid upon the timbers by the side of the ladders and fixed on to the timbers, in order that the men who are going up or down may sit or stand upon them and rest when they are tired. To prevent danger to the shovellers from rocks which, after being drawn up from so deep a shaft fall down again, a little above the bottom of the shaft small rough sticks are placed close together on the timbers, in such a way as to cover the whole space of the shaft except the ladder-way. A hole, however, is left in this structure near the footwall, which is kept open so that there may be one opening to the shaft from the bottom, that the buckets full of the materials which have been dug out may be drawn from the shaft through it by machines, and may be returned to the same place again empty; and so the shovellers and other workmen, as it were hiding beneath this structure, remain perfectly safe in the shaft.

In mines on one vein there are driven one, two, or sometimes three or more tunnels, always one above the other. If the vein is solid and hard, and likewise the hanging and footwall rock, no part of the tunnel needs support, beyond that which is required at the mouth, because at that spot there is not yet solid rock; if the vein is soft, and the hanging and footwall rock are likewise soft, the tunnel requires frequent strong timbering, which is provided in the following way. First, two dressed posts are erected and set into the tunnel floor, which is dug out a little; these are of medium thickness, and high enough that their ends, which are cut square, almost touch the top of the tunnel; then upon them is placed a smaller dressed cap, which is mortised into the heads of the posts; at the bottom, other small timbers, whose ends are similarly squared, are mortised into the posts. At each interval of one and a half fathoms, one of these sets is erected; each one of these the miners call a "little doorway," because it opens a certain amount of passage way; and indeed, when necessity requires it, doors are fixed to the timbers of each little doorway so that it can be closed. Then lagging of planks or of poles is placed upon the caps lengthwise, so as to reach from one set of timbers to another, and is laid along the sides, in case some portion of the body of the mountain may fall, and by its bulk impede passage or crush persons coming in or out. Moreover, to make the timbers remain stationary, wooden pegs are driven between them and the sides of the tunnel. Lastly, if rock or earth are carried out in wheelbarrows, planks joined together are laid upon the sills; if the rock is hauled out in trucks, then two timbers three-quarters of a foot thick and wide are laid on the sills, and, where they join, these are usually hollowed out so that in the hollow, as in a road, the iron pin of the truck may be pushed along; indeed, because of this pin in the groove, the truck does not leave the worn track to the left or right. Beneath the sills are the drains through which the water flows away.

A—Posts. B—Caps. C—Sills. D—Doors. E—Lagging. F—Drains.

Miners timber drifts in the same way as tunnels. These do not, however, require sill-pieces, or drains; for the broken rock is not hauled very far, nor does the water have far to flow. If the vein above is metal-bearing, as it sometimes is for a distance of several fathoms, then from the upper part of tunnels or even drifts that have already been driven, other drifts are driven again and again until that part of the vein is reached which does not yield metal. The timbering of these openings is done as follows: stulls are set at intervals into hitches in the hanging and footwall, and upon them smooth poles are laid continuously; and that they may be able to bear the weight, the stulls are generally a foot and a half thick. After the ore has been taken out and the mining of the vein is being done elsewhere, the rock then broken, especially if it cannot be taken away without great difficulty, is thrown into these openings among the timber, and the carriers of the ore are saved toil, and the owners save half the expense. This then, generally speaking, is the method by which everything relating to the timbering of shafts, tunnels, and drifts is carried out.

All that I have hitherto written is in part peculiar to *venae profundae*, and in part common to all kinds of veins; of what follows, part is specially applicable to *venae dilatatae*, part to *venae cumulatae*. But first I will describe how *venae dilatatae* should be mined. Where torrents, rivers, or streams have by inundations washed away part of the slope of a mountain or a hill, and have disclosed a *vena dilatata*, a tunnel should be driven first straight and narrow, and then wider, for nearly all the vein should be hewn away; and when this tunnel has been driven further, a shaft which supplies air should be sunk in the mountain or hill, and through it from time to time the ore, earth, and rock can be drawn up at less expense than if they be drawn out through the very great length of the tunnel; and even in those places to which the tunnel does not yet reach, miners dig shafts in order to open a *vena dilatata* which they conjecture must lie beneath the soil. In this way, when the upper layers are removed, they dig through rock sometimes of one kind and colour, sometimes of one kind but different colours, sometimes of different kinds but of one colour, and, lastly, of different kinds and different colours. The thickness of rock, both of each single stratum and of all combined, is uncertain, for the whole of the strata are in some places twenty fathoms deep, in others more than fifty; individual strata are in some places half a foot thick; in others, one, two, or more feet; in others, one, two, three, or more fathoms. For example, in those districts which lie at the foot of the Harz mountains, there are many different coloured strata, covering a copper *vena dilatata*. When the soil has been stripped, first of all is disclosed a stratum which is red, but of a dull shade and of a thickness of twenty, thirty, or five and thirty fathoms. Then there is another stratum, also red, but of a light shade, which has usually a thickness of about two fathoms. Beneath this is a stratum of ash-coloured clay nearly a fathom thick, which, although it is not metalliferous, is reckoned a vein. Then follows a third stratum, which is ashy, and about three fathoms thick. Beneath this lies a vein of ashes to the thickness of five fathoms, and these ashes are mixed with rock of the same colour. Joined to the last, and underneath, comes a stratum, the fourth in number, dark in colour and a foot thick. Under this comes the fifth stratum, of a pale or yellowish colour, two feet thick; underneath which is the sixth stratum, likewise dark, but rough and three feet thick. Afterward occurs the seventh stratum, likewise of dark colour, but still darker than the last, and two feet thick. This is followed by an eighth stratum, ashy, rough, and a foot thick. This kind, as also the others, is sometimes distinguished by stringers of the stone which easily melts in fire of the second order. Beneath this is another ashy rock, light in weight, and five feet thick. Next to this comes a lighter ash-coloured one, a foot thick; beneath this lies the eleventh stratum, which is dark and very much like the seventh, and two feet thick. Below the last is a twelfth stratum of a whitish colour and soft, also two feet thick; the weight of this rests on a thirteenth stratum, ashy and one foot thick, whose weight is in turn supported by a fourteenth stratum, which is blackish and half a foot thick. There follows this, another stratum of black colour, likewise half a foot thick, which is again followed by a sixteenth stratum still blacker in colour, whose thickness is also the same. Beneath this, and last of all, lies the cupriferous stratum, black coloured and schistose, in which there sometimes glitter scales of gold-coloured pyrites in the very thin sheets, which, as I said elsewhere, often take the forms of various living things.[15]

The miners mine out a *vena dilatata* laterally and longitudinally by driving a low tunnel in it, and if the nature of the work and place permit, they sink also a shaft in order to discover whether there is a second vein beneath the first one; for sometimes beneath it there are two, three, or more similar metal-bearing veins, and these are excavated in the same way laterally and longitudinally. They generally mine *venæ dilatatæ* lying down; and to

[15]The strata here enumerated are given in the Glossary of *De Re Metallica* as follows:—

Corium terrae	*Die erd oder leim.*
Saxum rubrum	*Rot gebirge.*
Alterum item rubrum	*Roterkle.*
Argilla cinerea	*Thone.*
Tertium saxum	*Gerhulle.*
Cineris vena	*Asche.*
Quartum saxum	*Gniest.*
Quintum saxum	*Schwehlen.*
Sextum saxum	*Oberrauchstein.*
Septimum saxum	*Zechstein.*
Octavum saxum	*Underrauchstein.*
Nonum saxum	*Blitterstein.*
Decimum saxum	*Oberschuelen.*
Undecimum saxum	*Mittelstein.*
Duodecimum saxum	*Underschuelen.*
Decimumtertium saxum	*Dach.*
Decimumquartum saxum	*Norweg.*
Decimumquintum saxum	*Lotwerg.*
Decimumsextum saxum	*Kamme.*
Lapis aerosus fissilis	*Schifer*

The description is no doubt that of the Mannsfeld cupriferous slates. It is of some additional interest as the first attempt at stratigraphic distinctions, although this must not be taken too literally, for we have rendered the different numbered "*saxum*" in this connection as "stratum." The German terms given by Agricola above, can many of them be identified in the miners' terms to-day for the various strata at Mannsfeld. Over the *kupferschiefer* the names to-day are *kammschale*, *dach*, *faule*, *zechstein*, *rauchwacke*, *rauchstein*, *asche*. The relative thickness of these beds is much the same as given by Agricola. The stringers in the 8th stratum of stone, which fuse in the fire of the second order, were possibly calcite. The *rauchstein* of the modern section is distinguished by stringers of calcite, which give it at times a brecciated appearance.

avoid wearing away their clothes and injuring their left shoulders they usually bind on themselves small wooden cradles. For this reason, this particular class of miners, in order to use their iron tools, are obliged to bend their necks to the left, not infrequently having them twisted. Now these veins also sometimes divide, and where these parts re-unite, ore of a richer and a better quality is generally found; the same thing occurs where the stringers, of which they are not altogether devoid, join with them, or cut them cross-wise, or divide them obliquely. To prevent a mountain or hill, which has in this way been undermined, from subsiding by its weight, either some natural pillars and arches are left, on which the pressure rests as on a foundation, or timbering is done for support. Moreover, the materials which are dug out and which are devoid of metal are removed in bowls, and are thrown back, thus once more filling the caverns.

Next, as to *venæ cumulatæ*. These are dug by a somewhat different method, for when one of these shows some metal at the top of the ground, first of all one shaft is sunk; then, if it is worth while, around this one many shafts are sunk and tunnels are driven into the mountain. If a torrent or spring has torn fragments of metal from such a vein, a tunnel is first driven into the mountain or hill for the purpose of searching for the ore; then when it is found, a vertical shaft is sunk in it. Since the whole mountain, or more especially the whole hill, is undermined, seeing that the whole of it is composed of ore, it is necessary to leave the natural pillars and arches, or the place is timbered. But sometimes when a vein is very hard it is broken by fire, whereby it happens that the soft pillars break up, or the timbers are burnt away, and the mountain by its great weight sinks into itself, and then the shaft buildings are swallowed up in the great subsidence. Therefore, about a *vena cumulata* it is advisable to sink some shafts which are not subject to this kind of ruin, through which the materials that are excavated may be carried out, not only while the pillars and underpinnings still remain whole and solid, but also after the supports have been destroyed by fire and have fallen. Since ore which has thus fallen must necessarily be broken by fire, new shafts through which the smoke can escape must be sunk in the abyss. At those places where stringers intersect, richer ore is generally obtained from the mine; these stringers, in the case of tin mines, sometimes have in them black stones the size of a walnut. If such a vein is found in a plain, as not infrequently happens in the case of iron, many shafts are sunk, because they cannot be sunk very deep. The work is carried on by this method because the miners cannot drive a tunnel into a level plain of this kind.

There remain the stringers in which gold alone is sometimes found, in the vicinity of rivers and streams, or in swamps. If upon the soil being removed, many of these are found, composed of earth somewhat baked and burnt, as may sometimes be seen in clay pits, there is some hope that gold may be obtained from them, especially if several join together. But the very point of junction must be pierced, and the length and width searched for ore, and in these places very deep shafts cannot be sunk.

I have completed one part of this book, and now come to the other, in which I will deal with the art of surveying. Miners measure the solid mass of the mountains in order that the owners may lay out their plans, and that their workmen may not encroach on other people's possessions. The surveyor either measures the interval not yet wholly dug through, which lies between the mouth of a tunnel and a shaft to be sunk to that depth, or between the mouth of a shaft and the tunnel to be driven to that spot which lies under the shaft, or between both, if the tunnel is neither so long as to reach to the shaft, nor the shaft so deep as to reach to the tunnel; and thus on both sides work is still to be done. Or in some cases, within the tunnels and drifts, are to be fixed the boundaries of the meers, just as the *Bergmeister* has determined the boundaries of the same meers above ground.[16]

Each method of surveying depends on the measuring of triangles. A small triangle should be laid out, and from it calculations must be made regarding a larger one. Most particular care must be taken that we do not deviate at all from a correct measuring; for if, at the beginning, we are drawn

[16]The history of surveying and surveying instruments, and in a subsidiary way their application to mine work, is a subject upon which there exists a most extensive literature. However, that portion of such history which relates to the period prior to Agricola represents a much less proportion of the whole than do the citations to this chapter in *De Re Metallica*, which is the first comprehensive discussion of the mining application. The history of such instruments is too extensive to be entered upon in a footnote, but there are some fundamental considerations which, if they had been present in the minds of historical students of this subject, would have considerably abridged the literature on it. First, there can be no doubt that measuring cords or rods and boundary stones existed almost from the first division of land. There is, therefore, no need to try to discover their origins. Second, the history of surveying and surveying instruments really begins with the invention of instruments for taking levels, or for the determination of angles with a view to geometrical calculation. The meagre facts bearing upon this subject do not warrant the endless expansion they have received by argument as to what was probable, in order to accomplish assumed methods of construction among the Ancients. For instance, the argument that in carrying the Grand Canal over watersheds with necessary reservoir supply, the Chinese must have had accurate levelling and surveying instruments before the Christian Era, and must have conceived in advance a completed work, does not hold water when any investigation will demonstrate that the canal grew by slow accretion from the lateral river systems, until it joined almost by accident. Much the same may be said about the preconception of engineering results in several other ancient works. There can be no certainty as to who first invented instruments of the order mentioned above; for instance, the invention of the dioptra has been ascribed to Hero, *vide* his work on the *Dioptra*. He has been assumed to have lived in the 1st or 2nd Century B.C. Recent investigations, however, have shown that he lived about 100 A.D. (Sir Thomas Heath, Encyc. Brit. 11th Ed., XIII, 378). As this instrument is mentioned by Vitruvius (50 – 0 B.C.) the myth that Hero was the inventor must also disappear. Incidentally Vitruvius (VIII, 5) describes a levelling instrument called a *chorobates*, which was a frame levelled either by a groove of water or by plumb strings. Be the inventor of the *dioptra* who he may, Hero's work on that subject contains the first suggestion of mine surveys in the problems (XIII, XIV, XV, XVI), where geometrical methods are elucidated for determining the depths required for the connection of shafts and tunnels. On the compass we give further notes on p. 56. It was probably an evolution of the 13th Century. As to the application of angle- and level-determining instruments to underground surveys, so far as we know there is no reference prior to Agricola, except that of Hero. Mr. Bennett Brough (Cantor Lecture, London, 1892) points outthat the *Nützliche Bergbüchlin* (see Appendix) describes a mine compass, but there is not the slightest reference to its use for anything but surface direction of veins.

Although map-making of a primitive sort requires no instruments, except legs, the oldest map in the world possesses unusual interest because it happens to be a map of a mining region. This well-known Turin papyrus dates from Seti I. (about 1300 B.C.), and it represents certain gold mines between the Nile and the Red Sea. The best discussion is by Chabas (*Inscriptions des Mines d'Or*, Chalons-sur-Saone, Paris, 1862, p. 30-36). Fragments of another papyrus, in the Turin Museum, are considered by Lieblein (*Deux Papyras Hiėratiques*, Christiania, 1868) also to represent a mine of the time of Rameses I. If so, this one dates from about 1400 B.C. As to an actual map of underground workings (disregarding illustrations) we know of none until after Agricola's time. At his time maps were not made, as will be gathered from the text.

by carelessness into a slight error, this at the end will produce great errors. Now these triangles are of many shapes, since shafts differ among themselves and are not all sunk by one and the same method into the depths of the earth, nor do the slopes of all mountains come down to the valley or plain in the same manner. For if a shaft is vertical, there is a triangle with a right angle, which the Greeks call ὀρθογώνιον and this, according to the inequalities of the mountain slope, has either two equal sides or three unequal sides. The Greeks call the former τρίγωνον ἰσοσκελές the latter σκαληνόν for a right angle triangle cannot have three equal sides. If a shaft is inclined and sunk in the same vein in which the tunnel is driven, a triangle is likewise made with a right angle, and this again, according to the various inequalities of the mountain slope, has either two equal or three unequal sides. But if a shaft is inclined and is sunk in one vein, and a tunnel is driven in another vein, then a triangle comes into existence which has either an obtuse angle or all acute angles. The former the Greeks call ἀμβλυγώνιον, the latter ὀξυγώνιον. That triangle which has an obtuse angle cannot have three equal sides, but in accordance with the different mountain slopes has either two equal sides or three unequal sides. That triangle which has all acute angles in accordance with the different mountain slopes has either three equal sides, which the Greeks call τρίγωνον ἰσόπλευρον or two equal sides or three unequal sides.

The surveyor, as I said, employs his art when the owners of the mines desire to know how many fathoms of the intervening ground require to be dug; when a tunnel is being driven toward a shaft and does not yet reach it; or when the shaft has not yet been sunk to the depth of the bottom of the tunnel which is under it; or when neither the tunnel reaches to that point, nor has the shaft been sunk to it. It is of importance that miners should know how many fathoms remain from the tunnel to the shaft, or from the shaft to the tunnel, in order to calculate the expenditure; and in order that the owners of a metal-bearing mine may hasten the sinking of a shaft and the excavation of the metal, before the tunnel reaches that point and the tunnel owners excavate part of the metal by any right of their own; and on the other hand, it is important that the owners of a tunnel may similarly hasten their driving before a shaft can be sunk to the depth of a tunnel, so that they may excavate the metal to which they will have a right.

The surveyor, first of all, if the beams of the shaft-house do not give him the opportunity, sets a pair of forked posts by the sides of the shaft in such a manner that a pole may be laid across them. Next, from the pole he lets down into the shaft a cord with a weight attached to it. Then he stretches a second cord, attached to the upper end of the first cord, right down along the slope of the mountain to the bottom of the mouth of the tunnel, and fixes it to the ground. Next, from the same pole not far from the first cord, he lets down a third cord, similarly weighted, so that it may intersect the second cord, which descends obliquely. Then, starting from that point where the third cord cuts the second cord which descends obliquely to the mouth of the tunnel, he measures the second cord upward to where it reaches the end of

A—Upright forked posts. B—Pole over the posts. C—Shaft. D—First cord. E—Weight of first cord. F—Second cord. G—Same fixed ground. H—Head of first cord. I—Mouth of tunnel. K—Third cord. L—Weight of third cord. M—First side minor triangle. N—Second side minor triangle. O—Third side minor triangle. P—The minor triangle.

the first cord, and makes a note of this first side of the minor triangle[17]. Afterward, starting again from that point where the third cord intersects the second cord, he measures the straight space which lies between that point and the opposite point on the first cord, and in that way forms the minor triangle, and he notes this second side of the minor triangle in the same way as before. Then, if it is necessary, from the angle formed by the first cord and the second side of the minor triangle, he measures upward to the end of the first cord and also makes a note of this third side of the minor triangle. The third side of the minor triangle, if the shaft is vertical or inclined and is sunk on the same vein in which the tunnel is driven, will necessarily be the same length as the third cord above the point where it intersects the second cord; and so, as often as the first side of the minor triangle is contained in the length of the whole cord which descends obliquely, so many times the length of the second side of the minor triangle indicates the distance between the mouth of the tunnel and the point to which the shaft must be sunk; and similarly, so many times the length of the third side of the minor triangle gives the distance between the mouth of the shaft and the bottom of the tunnel.

When there is a level bench on the mountain slope, the surveyor first measures across this with a measuring-rod; then at the edges of this bench he sets up forked posts, and applies the principle of the triangle to the two sloping parts of the mountain; and to the fathoms which are the length of that part of the tunnel determined by the triangles, he adds the number of fathoms which are the width of the bench. But if sometimes the mountain side stands up, so that a cord cannot run down from the shaft to the mouth of the tunnel, or, on the other hand, cannot run up from the mouth of the tunnel to the shaft, and, therefore, one cannot connect them in a straight line, the surveyor, in order to fix an accurate triangle, measures the mountain; and going downward he substitutes for the first part of the cord a pole one fathom long, and for the second part a pole half a fathom long. Going upward, on the contrary, for the first part of the cord he substitutes a pole half a fathom long, and for the next part, one a whole fathom long; then where he requires to fix his triangle he adds a straight line to these angles.

To make this system of measuring clear and more explicit, I will proceed by describing each separate kind of triangle. When a shaft is vertical or inclined, and is sunk in the same vein on which the tunnel is driven, there is created, as I said, a triangle containing a right angle. Now if the minor triangle has the two sides equal, which, in accordance with the numbering used by surveyors, are the second and third sides, then the second and third sides of the major triangle will be equal; and so also the intervening distances will be equal which lie between the mouth of the tunnel and the bottom of the shaft, and which lie between the mouth of the shaft and the bottom of the tunnel. For example, if the first side of the minor triangle is seven feet long and the second and likewise the third sides are five feet, and

[17]For greater clarity we have in a few places interpolated the terms "major" and "minor" triangles.

the length shown by the cord for the side of the major triangle is 101 times seven feet, that is 117 fathoms and five feet, then the intervening space, of course, whether the whole of it has been already driven through or has yet to be driven, will be one hundred times five feet, which makes eighty-three fathoms and two feet. Anyone with this example of proportions will be able to construct the major and minor triangles in the same way as I have done, if there be the necessary upright posts and cross-beams. When a shaft is vertical the triangle is absolutely upright; when it is inclined and is sunk on the same vein in which the tunnel is driven, it is inclined toward one side.

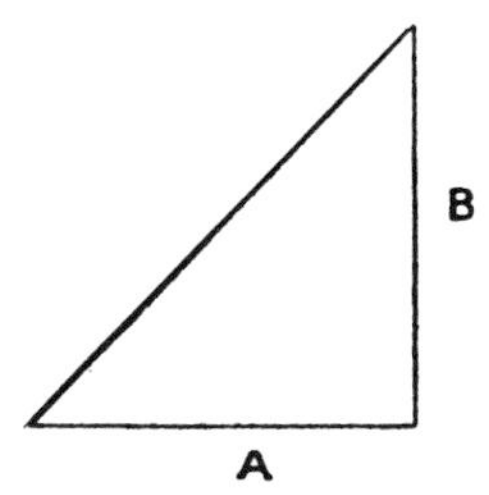

A TRIANGLE HAVING A RIGHT ANGLE AND TWO EQUAL SIDES.

Therefore, if a tunnel has been driven into the mountain for sixty fathoms, there remains a space of ground to be penetrated twenty-three fathoms and two feet long; for five feet of the second side of the major triangle, which lies above the mouth of the shaft and corresponds with the first side of the minor triangle, must not be added. Therefore, if the shaft has been sunk in the middle of the head meer, a tunnel sixty fathoms long will reach to the boundary of the meer only when the tunnel has been extended a further two fathoms and two feet; but if the shaft is located in the middle of an ordinary meer, then the boundary will be reached when the tunnel has been driven a further length of nine fathoms and two feet. Since a tunnel, for every one hundred fathoms of length, rises in grade one fathom, or at all events, ought to rise as it proceeds toward the shaft, one more fathom must always be taken from the depth allowed to the shaft, and one added to the length allowed to the tunnel. Proportionately, because a tunnel fifty fathoms long is raised half a fathom, this amount must be taken from the depth of the shaft and added to the length of the tunnel. In the same way if a tunnel is one hundred or fifty fathoms shorter or longer, the same proportion also must be taken from the depth of the one and added to the length of the other. For this reason, in the case mentioned above, half a fathom and a little more must be added to the distance to be driven through, so that there remain twenty-three fathoms, five feet, two palms, one and a half digits and a fifth of a digit; that is, if even the minutest proportions are carried out; and surveyors do not neglect these without good cause. Similarly, if the shaft is seventy fathoms deep, in order that it may reach to the bottom of the tunnel, it still must be sunk a further depth of thirteen fathoms and two feet, or rather twelve fathoms and a half, one foot, two digits, and four-fifths of half a digit. And in this instance five feet must be deducted from the reckoning, because these five feet complete the third side of the minor triangle, which is above the mouth of the shaft, and from its depth there must be deducted half a fathom, two palms, one and a half digits and the fifth part of half a digit. But if the tunnel has been driven to a point where it is under the shaft, then to reach the roof of the tunnel the shaft must still be sunk a depth of eleven fathoms, two and a half feet, one palm, two digits, and four-fifths of half a digit.

If a minor triangle is produced of the kind having three unequal sides, then the sides of the greater triangle cannot be equal; that is, if the first side of the minor triangle is eight feet long, the second six feet long, and the third five feet long, and the cord along the side of the greater triangle, not to go too far from the example just given, is one hundred and one times eight feet, that is, one hundred and thirty-four fathoms and four feet, the distance which lies between the mouth of the tunnel and the bottom of the shaft will occupy one hundred times six feet in length, that is, one hundred fathoms. The distance between the mouth of the shaft and the bottom of the tunnel is one hundred times five feet, that is, eighty-three fathoms and two feet. And so, if the tunnel is eighty-five fathoms long, the remainder to be driven into the mountain is fifteen fathoms long, and here, too, a correction in measurement must be taken from the depth of the shaft and added to the length of the tunnel; what this is precisely, I will pursue no further, since everyone having a small knowledge of arithmetic can work it out. If the shaft is sixty-seven fathoms deep, in order that it may reach the bottom of the tunnel, the further distance required to be sunk amounts to sixteen fathoms and two feet.

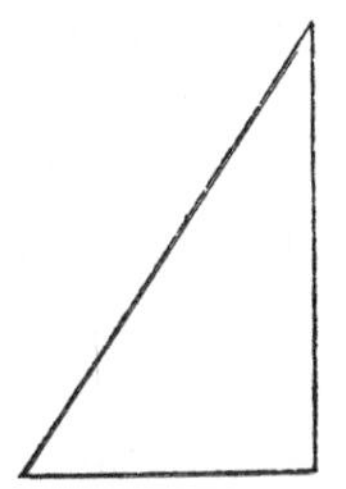

A TRIANGLE HAVING A RIGHT ANGLE AND THREE UNEQUAL SIDES.

The surveyor employs this same method in measuring the mountain, whether the shaft and tunnel are on one and the same vein, whether the vein is vertical or inclined, or whether the shaft is on the principal vein and the tunnel on a transverse vein descending vertically to the depths of the earth; in the latter case the excavation is to be made where the transverse vein cuts the vertical vein. If the principal vein descends on an incline and the cross-vein descends vertically, then a minor triangle is created having one obtuse angle or all three angles acute. If the minor triangle has one angle obtuse and the two sides which are the second and third are equal, then the second and third sides of the major triangle will be equal, so that if the first side of the minor triangle is nine feet, the second, and likewise the third, will be five feet. Then the first side of the major triangle will be one hundred and one times nine feet, or one hundred and fifty-one and one-half fathoms, and each of the other sides of the major triangle will be one hundred times five feet, that is, eighty-three fathoms and two feet. But when the first shaft is inclined, generally speaking, it is not deep; but there are usually several, all inclined, and one always following the other. Therefore, if a tunnel is seventy-seven fathoms long, it will reach to the middle of the bottom of a shaft when six fathoms and two feet further have been sunk. But if all such inclined shafts are seventy-six fathoms deep, in order that the last one may reach the bottom of the tunnel, a depth of seven fathoms and two feet remains to be sunk.

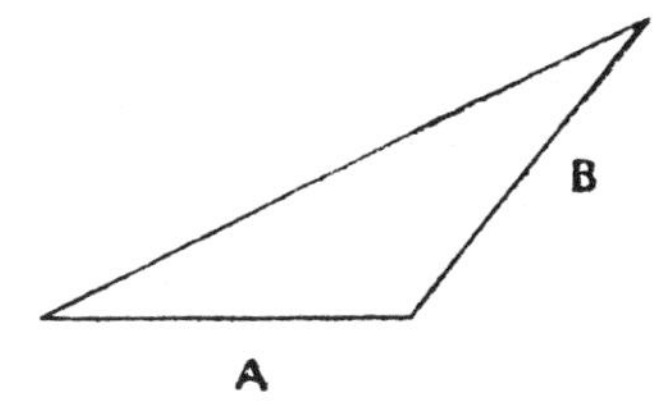

TRIANGLE HAVING AN OBTUSE ANGLE AND TWO EQUAL SIDES.

If a minor triangle is made which has an obtuse angle and three unequal sides, then again the sides of the large triangle cannot be equal. For example, if the first side of the minor triangle is six feet long, the second three feet, and the third four feet, and the cord along the side of the greater triangle one hundred and one times six feet, that is, one hundred and one fathoms, the distance between the mouth of the tunnel and the bottom of the last shaft will be a length one hundred times three feet, or fifty fathoms; but the depth that lies between the mouth of the first shaft and the bottom of the tunnel is one hundred times four feet, or sixty-six fathoms and four feet. Therefore, if a tunnel is forty-four fathoms long, the remaining distance to be driven is six fathoms. If the shafts are fifty-eight fathoms deep, the newest will touch the bottom of the tunnel when eight fathoms and four feet have been sunk.

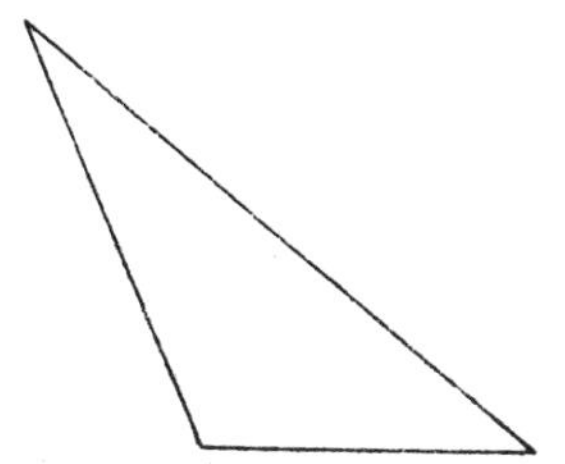

TRIANGLE HAVING AN OBTUSE ANGLE AND THREE UNEQUAL SIDES.

If a minor triangle is produced which has all its angles acute and its three sides equal, then necessarily the second and third sides of the minor triangle will be equal, and likewise the sides of the major triangle frequently referred to will be equal. Thus if each side of the minor triangle is six feet long, and the cord measurement for the side of the major triangle is one hundred and one times six feet, that is, one hundred and one fathoms, then both the distances to be dug will be one hundred fathoms. And thus if the tunnel is ninety fathoms long, it will reach the middle of the bottom of the last shaft when ten fathoms further have been driven. If the shafts are ninety-five fathoms deep, the last will reach the bottom of the tunnel when it is sunk a further depth of five fathoms.

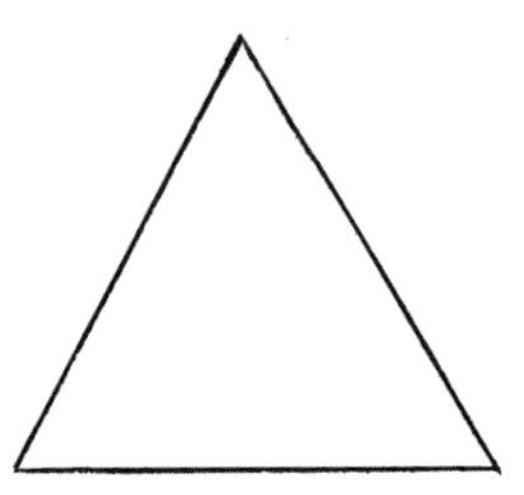

A TRIANGLE HAVING ALL ITS ANGLES ACUTE AND ITS THREE SIDES EQUAL.

If a triangle is made which has all its angles acute, but only two sides equal, namely, the first and third, then the second and third sides are not equal; therefore the distances to be dug cannot be equal. For example, if the first side of the minor triangle is six feet long, and the second is four feet, and the third is six feet, and the cord measurement for the side of the major triangle is one hundred and one times six feet, that is, one hundred and one fathoms, then the distance between the mouth of the tunnel and the bottom of the last shaft will be sixty-six fathoms and four feet. But the distance from the mouth of the first shaft to the bottom of the tunnel is one hundred fathoms. So if the tunnel is sixty fathoms long, the remaining distance to be driven into the mountain is six fathoms and four feet. If the shaft is ninety-seven fathoms deep, the last one will reach the bottom of the tunnel when a further depth of three fathoms has been sunk.

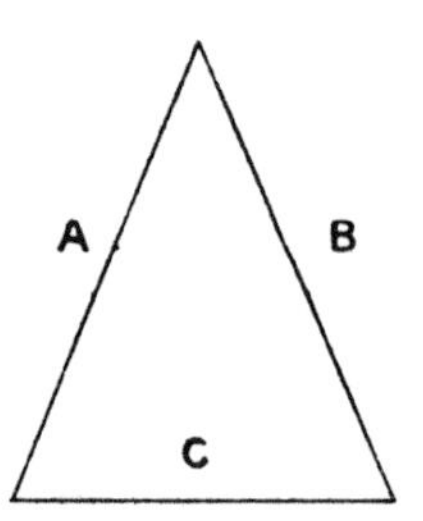

TRIANGLE HAVING ALL ITS ANGLES ACUTE AND TWO SIDES EQUAL, A, B, UNEQUAL SIDE C.

If a minor triangle is produced which has all its angles acute, but its three sides unequal, then again the distances to be dug cannot be equal. For example, if the first side of the minor triangle is seven feet long, the second side is four feet, and the third side is six feet, and the cord measurement for the side of the major triangle is one hundred and one times seven feet or one hundred and seventeen fathoms and four feet, the distance between the mouth of the tunnel and the bottom of the last shaft will be four hundred feet or sixty-six fathoms, and the depth between the mouth of the first shaft and the bottom of the tunnel will be one hundred fathoms. Therefore, if a tunnel is fifty fathoms long, it will reach the middle of the bottom of the newest shaft when it has been driven sixteen fathoms and four feet further. But if the shafts are then ninety-two fathoms deep, the last shaft will reach the bottom of the tunnel when it has been sunk a further eight fathoms.

A TRIANGLE HAVING ALL ITS ANGLES ACUTE AND ITS THREE SIDES UNEQUAL.

This is the method of the surveyor in measuring the mountain, if the principal vein descends inclined into the depths of the earth or the transverse vein is vertical. But if they are both inclined, the surveyor uses the same method, or he measures the slope of the mountain separately from the slope of the shaft. Next, if a transverse vein in which a tunnel is driven does not cut the principal vein in that spot where the shaft is sunk, then it is necessary for the starting point of the survey to be in the other shaft in which the transverse vein cuts the principal vein. But if there be no shaft on that spot where the outcrop of the transverse vein cuts the outcrop of the principal vein, then the surface of the ground which lies between the shafts must be measured, or that between the shaft and the place where the outcrop of the one vein intersects the outcrop of the other.

Some surveyors, although they use three cords, nevertheless ascertain only the length of a tunnel by that method of measuring, and determine the depth of a shaft by another method ; that is, by the method by which cords are re-stretched on a level part of the mountain or in a valley, or in flat fields, and are measured again. Some, however, do not employ this method in surveying the depth of a shaft and the length of a tunnel, but use only two cords, a graduated hemicycle[18] and a rod half a fathom long. They suspend in the shaft one cord, fastened from the upper pole and weighted, just as the others do. Fastened to the upper end of this cord, they stretch another right down the slope of the mountain to the bottom of the mouth of the tunnel and fix it to the ground. Then to the upper part of this second cord they apply on its lower side the broad part of a hemicycle. This consists of half a circle, the outer margin of which is covered with wax, and within this are six semi-circular lines. From the

[18]The names of the instruments here described in the original text, their German equivalents in the Glossary, and the terms adopted in translation are given below :—

LATIN TEXT.	GLOSSARY.	TERMS ADOPTED.
Funiculus	..	Cord
Pertica	*Stab*	Rod
Hemicyclium	*Donlege bretlein*	Hemicycle
Tripus	*Stul*	Tripod
Instrumentum cui index	*Compass*	Compass
Orbis	*Scheube*	Orbis
Libra stativa	*Auffsatz*	Standing plummet level
Libra pensilis	*Wage*	Suspended plummet level
Instrumentum cui index Alpinum	*Der schiner compass*	Swiss compass

waxed margin through the first semi-circular line, and reaching to the second, there proceed straight lines converging toward the centre of the hemicycle; these mark the middles of intervening spaces lying between other straight lines which extend to the fourth semi-circular line. But all lines whatsoever, from the waxed margin up to the fourth line, whether they go beyond it or not, correspond with the graduated lines which mark the minor spaces of a rod. Those which go beyond the fourth line correspond with the lines marking

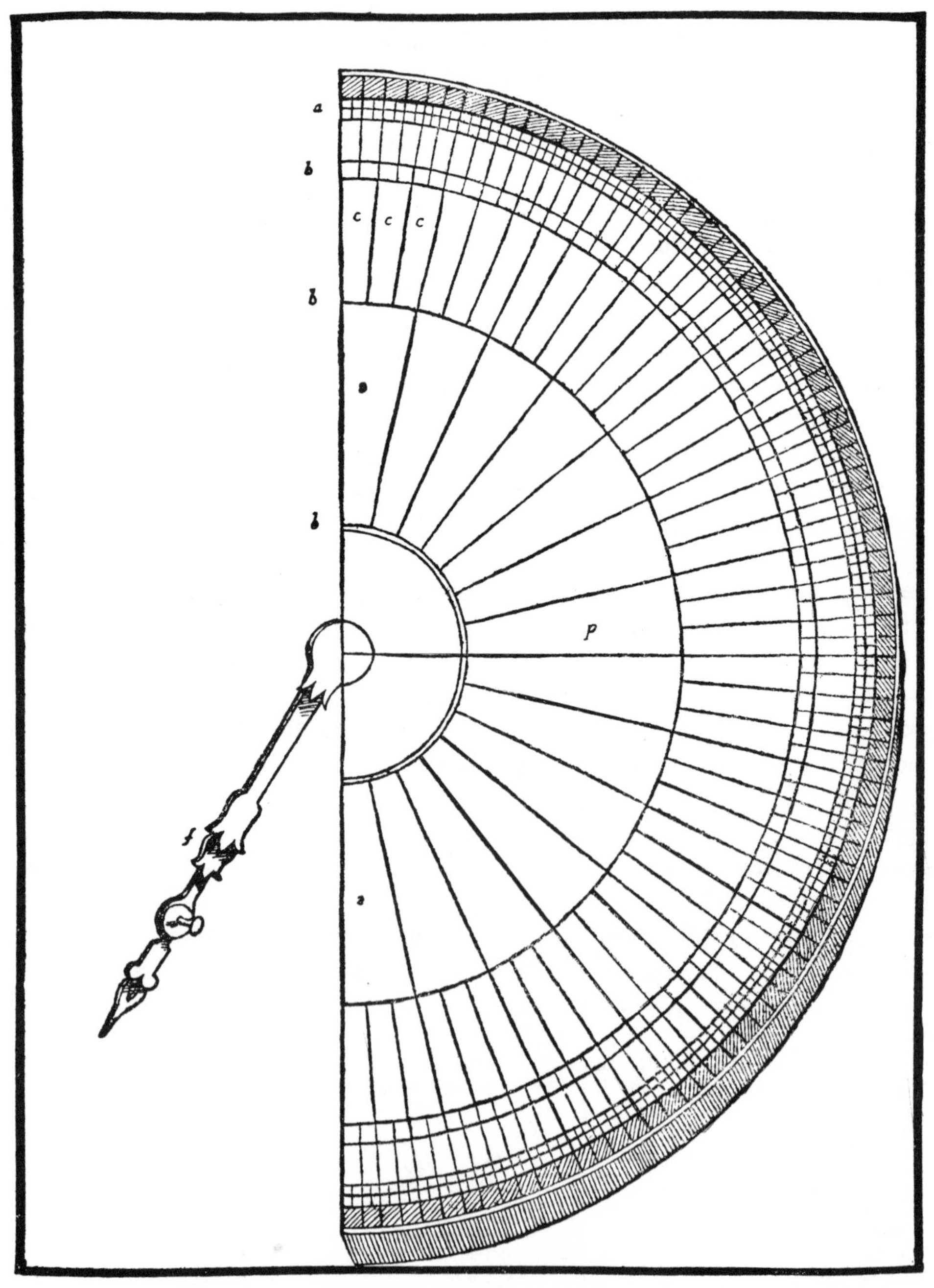

A—Waxed semicircle of the hemicycle. B—Semicircular lines. C—Straight lines. D—Line measuring the half. E—Line measuring the whole. F—Tongue.

the major spaces on the rod, and those which proceed further, mark the middle of the intervening space which lies between the others. The straight lines, which run from the fifth to the sixth semi-circular line, show nothing further. Nor does the line which measures the half, show anything when it has already passed from the sixth straight line to the base of the hemicycle. When the hemicycle is applied to the cord, if its tongue indicates the sixth straight line which lies between the second and third semi-circular lines, the surveyor counts on the rod six lines which separate the minor spaces, and if the length of this portion of the rod be taken from the second cord, as many times as the cord itself is half-fathoms long, the remaining length of cord shows the distance the tunnel must be driven to reach under the shaft. But if he sees that the tongue has gone so far that it marks the sixth line between the fourth and fifth semi-circular lines, he counts six lines which separate the major spaces on the rod ; and this entire space is deducted from the length of the second cord, as many times as the number of whole fathoms which the cord contains ; and then, in like manner, the remaining length of cord shows us the distance the tunnel must be driven to reach under the shaft.[19]

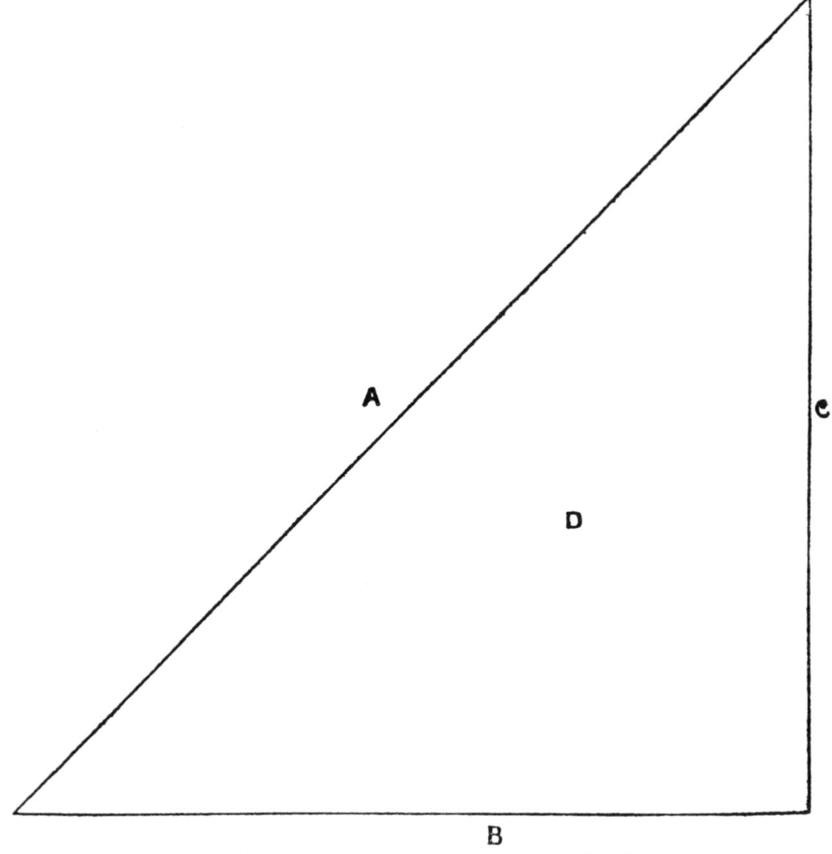

Stretched cords : A—First cord. B—Second cord. C—Third cord. D—Triangle.

[19]It is interesting to note that the ratio of any length so obtained, to the whole length of the staff, is practically equal to the cosine of the angle represented by the corresponding gradation on the hemicycle ; the gradations on the rod forming a fairly accurate table of cosines.

Both these surveyors, as well as the others, in the first place make use of the haulage rope. These they measure by means of others made of linden bark, because the latter do not stretch at all, while the former become very slack. These cords they stretch on the surveyor's field, the first one to represent the parts of mountain slopes which descend obliquely. Then the second cord, which represents the length of the tunnel to be driven to reach the shaft, they place straight, in such a direction that one end of it can touch the lower end of the first cord ; then they similarly lay the third cord straight, and in such a direction that its upper end may touch the upper end of the first cord, and its lower end the other extremity of the second cord, and thus a triangle is formed. This third cord is measured by the instrument with the index, to determine its relation to the perpendicular ; and the length of this cord shows the depth of the shaft.

Some surveyors, to make their system of measuring the depth of a shaft more certain, use five stretched cords : the first one descending obliquely ; two, that is to say the second and third, for ascertaining the length of the tunnel ; two for the depth of the shaft ; in which way they form a quadrangle divided into two equal triangles, and this tends to greater accuracy.

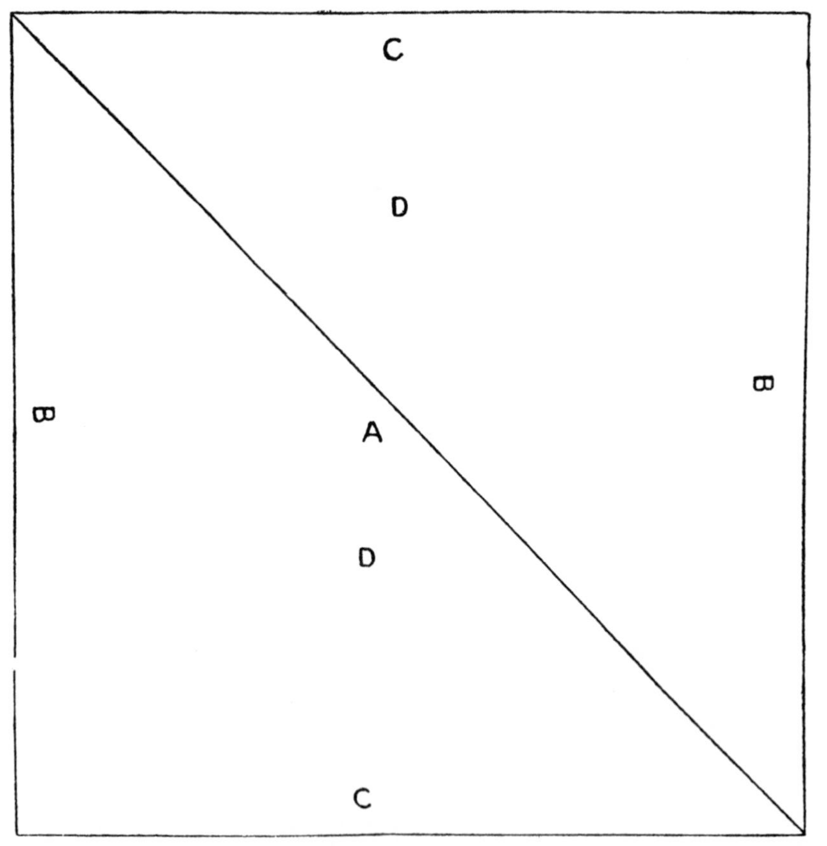

Stretched cords : A—First. B—Second. B—Third. C—Fourth. C—Fifth. D—Quadrangle.

These systems of measuring the depth of a shaft and the length of a tunnel, are accurate when the vein and also the shaft or shafts go down to the tunnel vertically or inclined, in an uninterrupted course. The same is true when a tunnel runs straight on to a shaft. But when each of them bends now in this, now in that direction, if they have not been completely driven and sunk, no living man is clever enough to judge how far they are deflected from a straight course. But if the whole of either one of the two has been excavated its full distance, then we can estimate more easily the length of one, or the depth of the other; and so the location of the tunnel, which is below a newly-started shaft, is determined by a method of surveying which I will describe. First of all a tripod is fixed at the mouth of the tunnel, and likewise at the mouth of the shaft which has been started, or at the place where the shaft will be started. The tripod is made of three stakes fixed to the ground, a small rectangular board being placed upon the stakes and fixed to them, and on this is set a compass. Then from the lower tripod a weighted cord is let down perpendicularly to the earth, close to which cord a stake is fixed in the ground. To this stake another cord is tied and drawn straight into the tunnel to a point as far as it can go without being bent by the hangingwall or the footwall of the vein. Next, from the cord which hangs from the lower tripod, a third cord likewise fixed is brought straight up the sloping side of the mountain to the stake of the upper tripod, and fastened to it. In order that the measuring of the depth of the shaft may be more certain, the third cord should touch one and the same side of the cord hanging from the lower tripod which is touched by the second cord—the one which is drawn into the tunnel. All this having been correctly carried out, the surveyor, when at length the cord which has been drawn straight into the tunnel is about to be bent by the hangingwall or footwall, places a plank in the bottom of the tunnel and on it sets the orbis, an instrument which has an indicator peculiar to itself. This instrument, although it also has waxed circles, differs from the other, which I have described in the third book. But by both these instruments, as well as by a rule and a square, he determines whether the stretched cords reach straight to the extreme end of the tunnel, or whether they sometimes reach straight, and are sometimes bent by the footwall or hangingwall. Each instrument is divided into parts, but the compass into twenty-four parts, the orbis into sixteen parts; for first of all it is divided into four principal parts, and then each of these is again divided into four. Both have waxed circles, but the compass has seven circles, and the orbis only five circles. These waxed circles the surveyor marks, whichever instrument he uses, and by the succession of these same marks he notes any change in the direction in which the cord extends. The orbis has an opening running from its outer edge as far as the centre, into which opening he puts an iron screw, to which he binds the second cord, and by screwing it into the plank, fixes it so that the orbis may be immovable. He takes care to prevent the second cord, and afterward the others which are put up, from being pulled off the screw, by employing a heavy iron, into an opening of which he fixes the head of the screw. In the case of the compass, since it has no opening, he merely places it by the side of the screw. That the instrument does not incline forward or backward, and in that way the measurement become a greater length than it should be, he sets upon the instrument a standing plummet level, the tongue of which, if the instrument is level, indicates no numbers, but the point from which the numbers start.

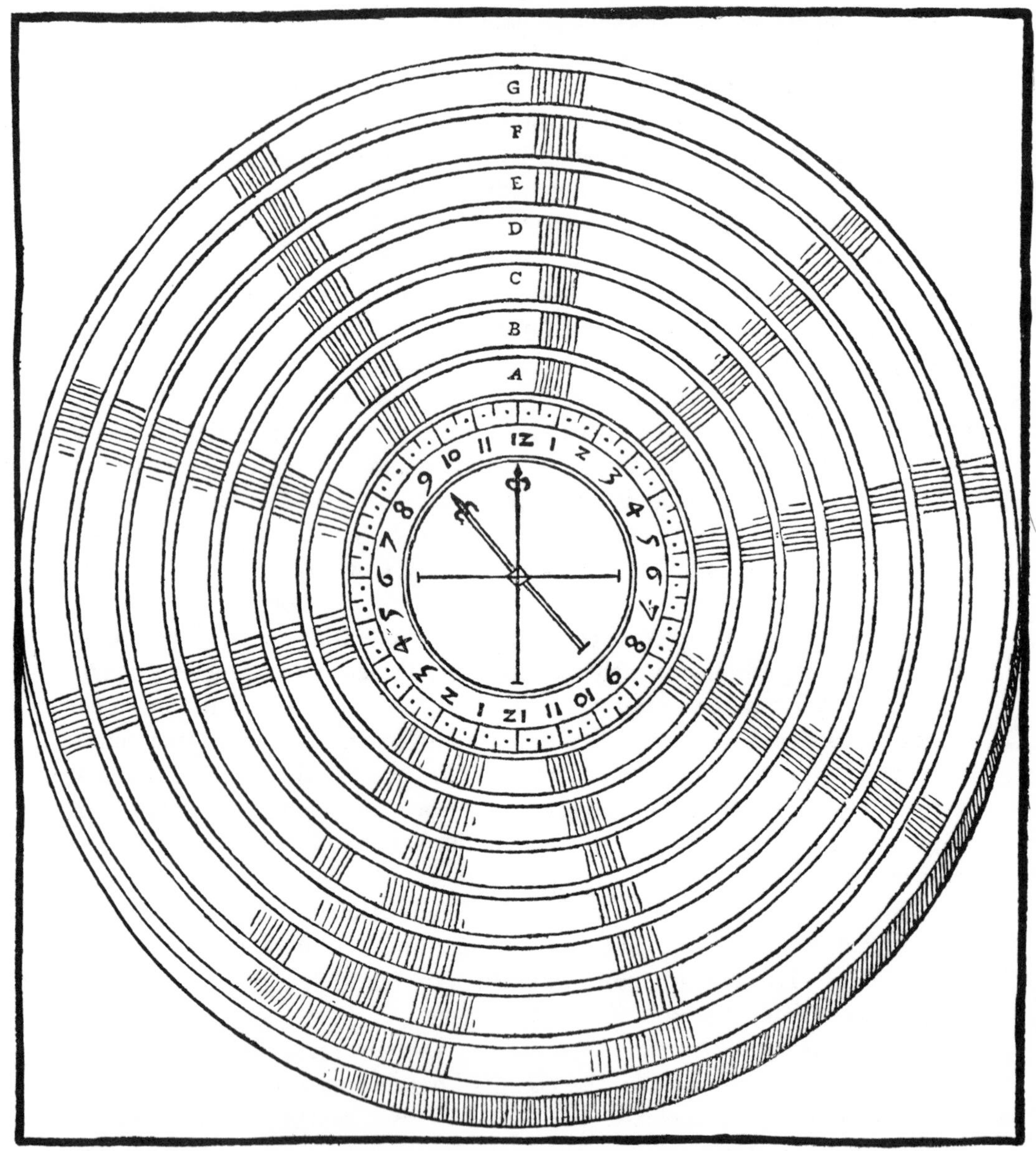

COMPASS. A, B, C, D, E, F, G ARE THE SEVEN WAXED CIRCLES.

When the surveyor has carefully observed each separate angle of the tunnel and has measured such parts as he ought to measure, then he lays them out in the same way on the surveyor's field[20] in the open air, and again no less carefully observes each separate angle and measures them. First of all, to each angle, according as the calculation of his triangle and his art require it, he lays out a straight cord as a line. Then he stretches a cord at

[20]It must be understood that instead of "plotting" a survey on a reduced scale on paper, as modern surveyors do, the whole survey was reproduced in full scale on the "surveyor's field."

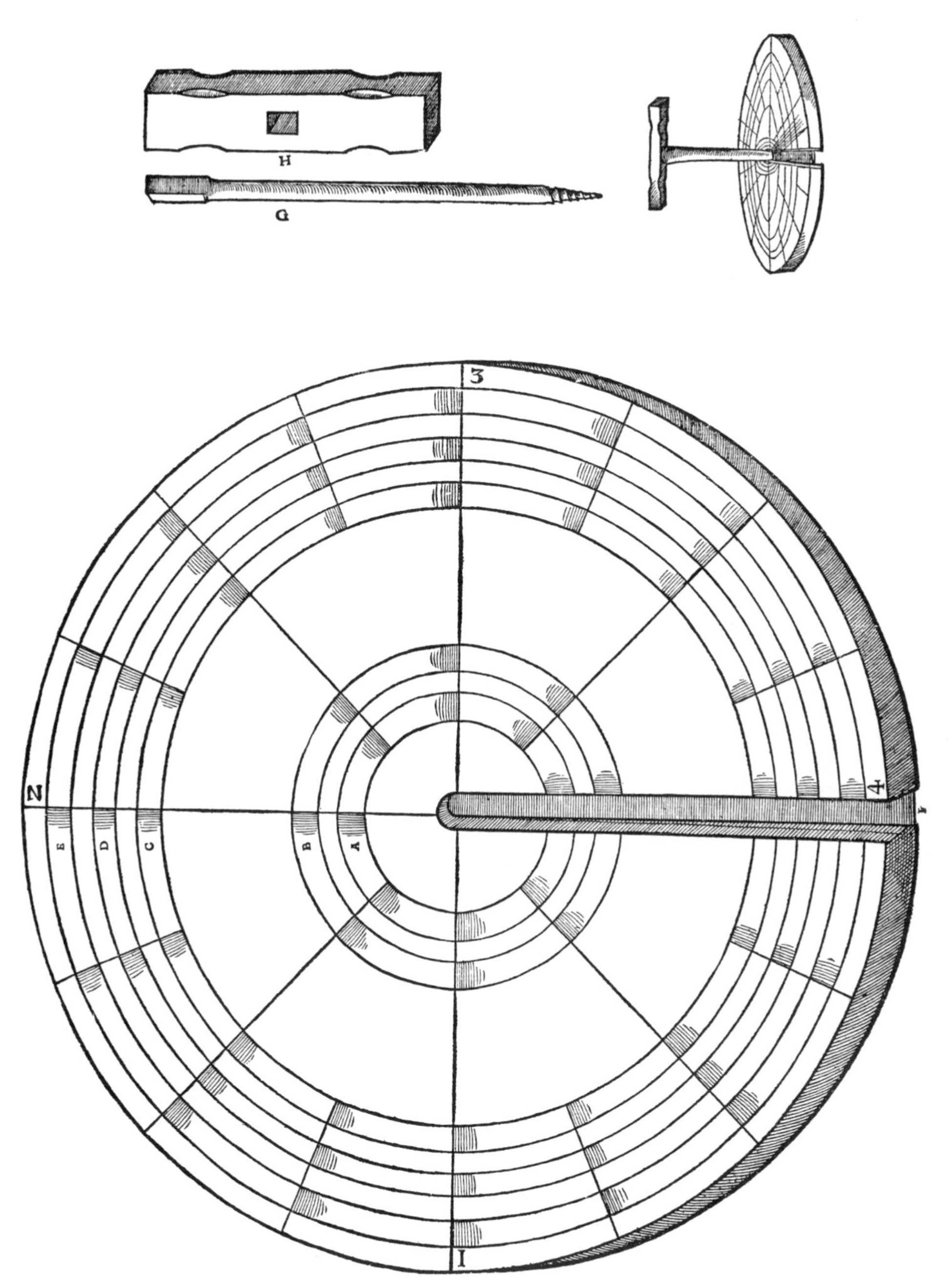

A, B, C, D, E—Five waxed circles of the *orbis.*
F—Opening of same. G—Screw. H—Perforated iron.

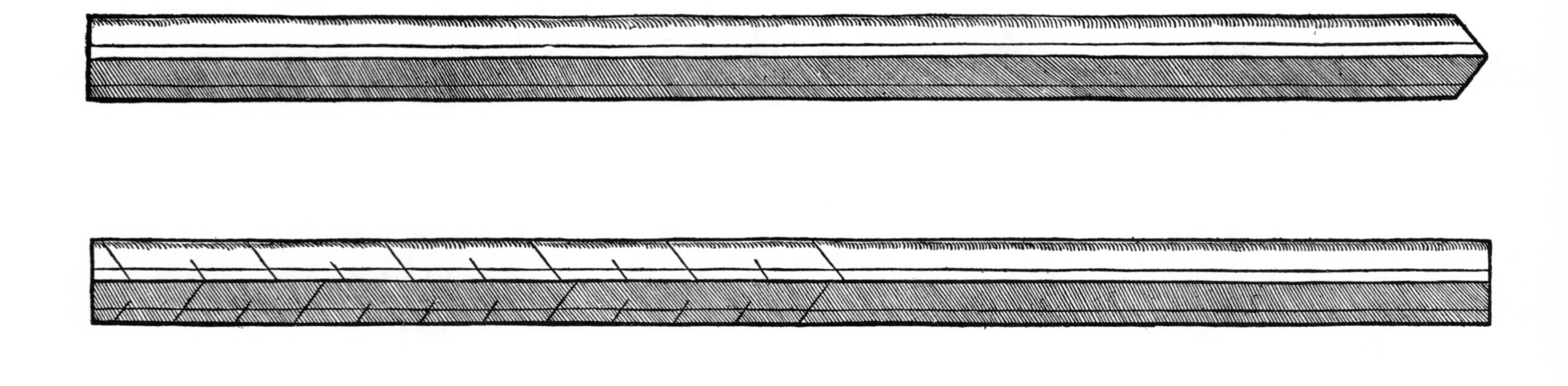

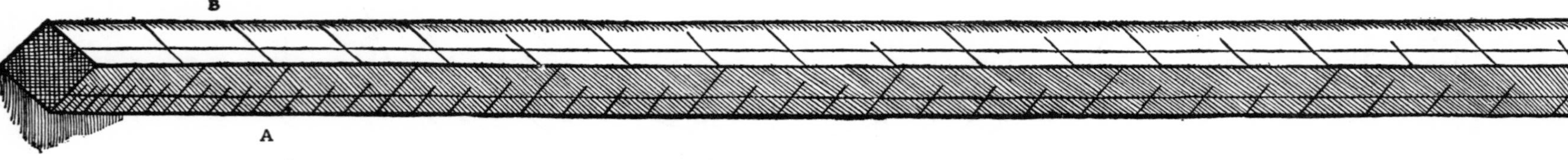

A—LINES OF THE ROD WHICH SEPARATE MINOR SPACES. B—LINES OF THE ROD WHICH SEPARATE MAJOR SPACES.

such an angle as represents the slope of the mountain, so that its lower end may reach the end of the straight cord; then he stretches a third cord

A—Standing plummet level. B—Tongue. C—Level and tongue.

similarly straight and at such an angle, that with its upper end it may reach the upper end of the second cord, and with its lower end the last end of the first cord. The length of the third cord shows the depth of the shaft, as I said before, and at the same time that point on the tunnel to which the shaft will reach when it has been sunk.

If one or more shafts reach the tunnel through intermediate drifts and shafts, the surveyor, starting from the nearest which is open to the air, measures in a shorter time the depth of the shaft which requires to be sunk, than if he starts from the mouth of the tunnel. First of all he measures that space on the surface which lies between the shaft which has been sunk and the one which requires to be sunk. Then he measures the incline of all the shafts which it is necessary to measure, and the length of all the drifts with which they are in any way connected to the tunnel. Lastly, he measures part of the tunnel; and when all this is properly done, he demonstrates the depth of the shaft and the point in the tunnel to which the shaft will reach. But sometimes a very deep straight shaft requires to be sunk at the same place where there is a previous inclined shaft, and to the same depth, in order that loads may be raised and drawn straight up by machines. Those machines on the surface are turned by horses; those inside the earth, by the same means, and also by water-power. And so, if it becomes necessary to sink such a shaft, the surveyor first of all fixes an iron screw in the upper part of the old shaft, and from the screw he lets down a cord as far as the first angle, where again he fixes a screw, and again lets down the cord as far as the second angle; this he repeats again and again until the cord reaches to the bottom of the shaft. Then to each angle of the cord he applies a hemicycle, and marks the waxed semi-circle according to the lines which the tongue indicates, and designates it by a number, in case it should be moved; then he measures the separate parts of the cord with another cord made of linden bark. Afterward, when he has come back out of the shaft, he goes away and transfers the markings from the waxed semi-circle of the hemicycle to an orbis similarly waxed. Lastly, the cords are stretched on the surveyor's field, and he measures the angles, as the system of measuring by triangles requires, and ascertains which part of the footwall and which part of the hangingwall rock must be cut away in order that the shaft may descend straight. But if the surveyor is required to show the owners of the mine, the spot in a drift or a tunnel in which a shaft needs to be raised from the bottom upward, that it should cut through more quickly, he begins measuring from the bottom of the drift or tunnel, at a point beyond the spot at which the bottom of the shaft will arrive, when it has been sunk. When he has measured the part of the drift or tunnel up to the first shaft which connects with an upper drift, he measures the incline of this shaft by applying a hemicycle or orbis to the cord. Then in a like manner he measures the upper drift and the incline shaft which is sunk therein toward which a raise is being dug, then again all the cords are stretched in the surveyor's field, the last cord in such a way that it reaches the first, and then he measures them. From this measurement is known in what part of the drift or tunnel the raise should be made, and how many fathoms of vein remain to be broken through in order that the shaft may be connected.

I have described the first reason for surveying; I will now describe another. When one vein comes near another, and their owners are different persons who have late come into possession, whether they drive a tunnel or a drift, or sink a shaft, they may encroach, or seem to encroach, without any lawful right, upon the boundaries of the older owners, for which reason the latter very often seek redress, or take legal proceedings. The surveyor either himself settles the dispute between the owners, or by his art gives evidence to the judges for making their decision, that one shall not encroach on the mine of the other. Thus, first of all he measures the mines of each party with a basket rope and cords of linden bark; and having applied to the cords an orbis or a compass, he notes the directions in which they extend. Then he stretches the cords on the surveyor's field; and starting from that point whose owners are in possession of the old meer toward the other, whether it is in the hanging or footwall of the vein, he stretches a cross-cord in a straight line, according to the sixth division of the compass, that is, at a right angle to the vein, for a distance of three and a half fathoms, and assigns to the older owners that which belongs to them. But if both ends of one vein are being dug out in two tunnels, or drifts from opposite directions, the surveyor first of all considers the lower tunnel or drift and afterward the upper one, and judges how much each of them has risen little by little. On each side strong men take in their hands a stretched cord and hold it so that there is no point where it is not strained tight; on each side the surveyor supports the cord with a rod half a fathom long, and stays the rod at the end with a short stick as often as he thinks it necessary. But some fasten cords to the rods to make them steadier. The surveyor attaches a suspended plummet level to the middle of the cord to enable him to calculate more accurately on both sides, and from this he ascertains whether one tunnel has risen more than another, or in like manner one drift more than another. Afterward he measures the incline of the shafts on both sides, so that he can estimate their position on each side. Then he easily sees how many fathoms remain in the space which must be broken through. But the grade of each tunnel, as I said, should rise one fathom in the distance of one hundred fathoms.

The Swiss surveyors, when they wish to measure tunnels driven into the highest mountains, also use a rod half a fathom long, but composed of three parts, which screw together, so that they may be shortened. They use a cord made of linden bark to which are fastened slips of paper showing the number of fathoms. They also employ an instrument peculiar to them, which has a needle; but in place of the waxed circles they carry in their hands a chart on which they inscribe the readings of the instrument. The instrument is placed on the back part of the rod so that the tongue, and the extended cord which runs through the three holes in the tongue, demonstrates the direction, and they note the number of fathoms. The tongue shows whether the cord inclines forward or backward. The tongue does not hang, as in the case of the suspended plummet level, but is fixed to the instrument in a half-lying position. They measure the tunnels for the purpose of knowing how many fathoms they have been increased in elevation ; how many fathoms the lower is distant from the upper one ; how many fathoms of interval is

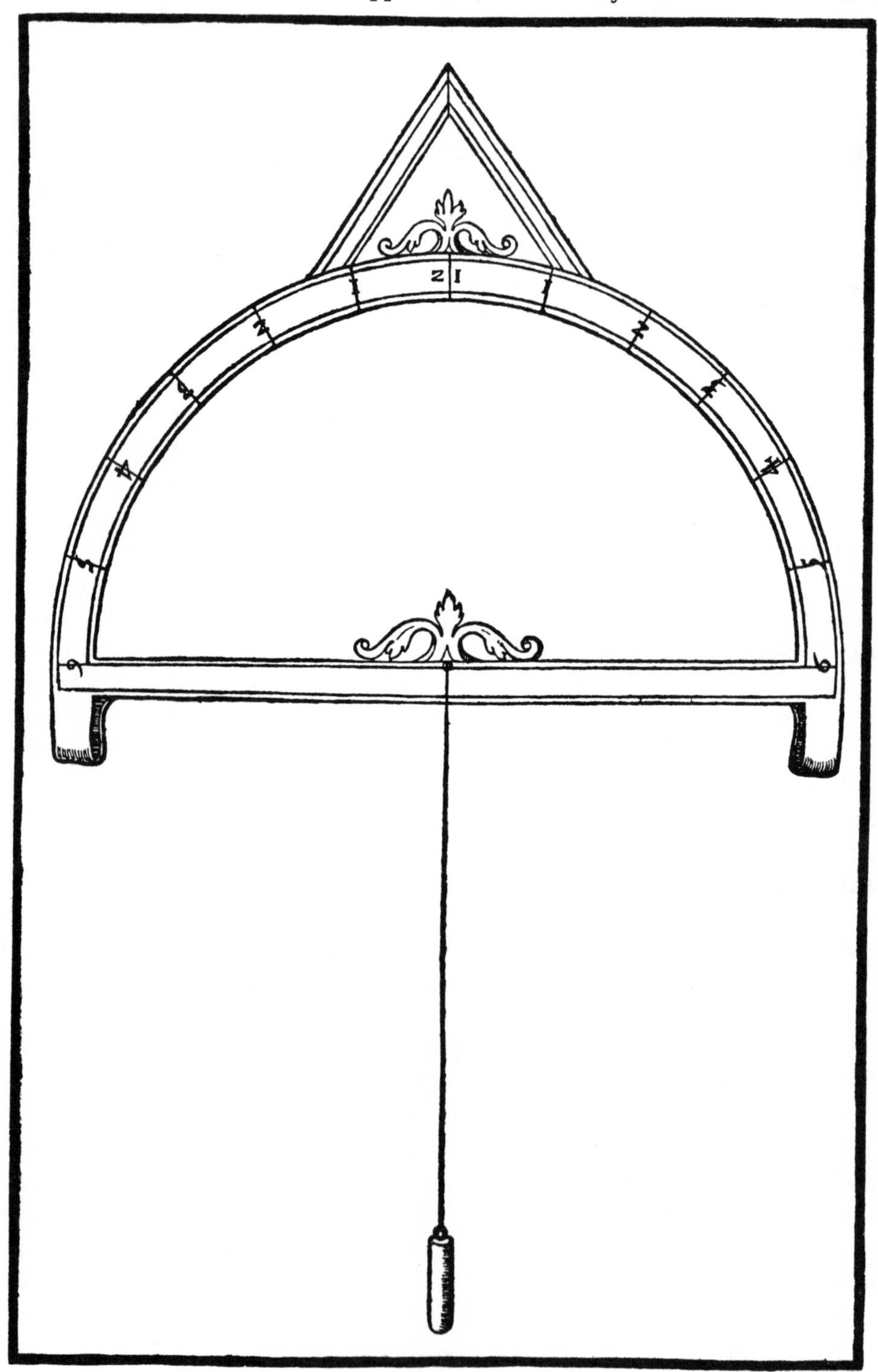

INDICATOR OF A SUSPENDED PLUMMET LEVEL.

not yet pierced between the miners who on opposite sides are digging on the same vein, or cross-stringers, or two veins which are approaching one another.

But I return to our mines. If the surveyor desires to fix the boundaries of the meer within the tunnels or drifts, and mark to them with a sign cut in the rock, in the same way that the *Bergmeister* has marked these boundaries above ground, he first of all ascertains, by measuring in the manner which I have explained above, which part of the tunnel or drift lies beneath the surface boundary mark, stretching the cords along the drifts to a point beyond that spot in the rock where he judges the mark should be cut. Then, after the same cords have been laid out on the surveyor's field, he starts from that upper cord at a point which shows the boundary mark, and stretches another cross-cord straight downward according to the sixth

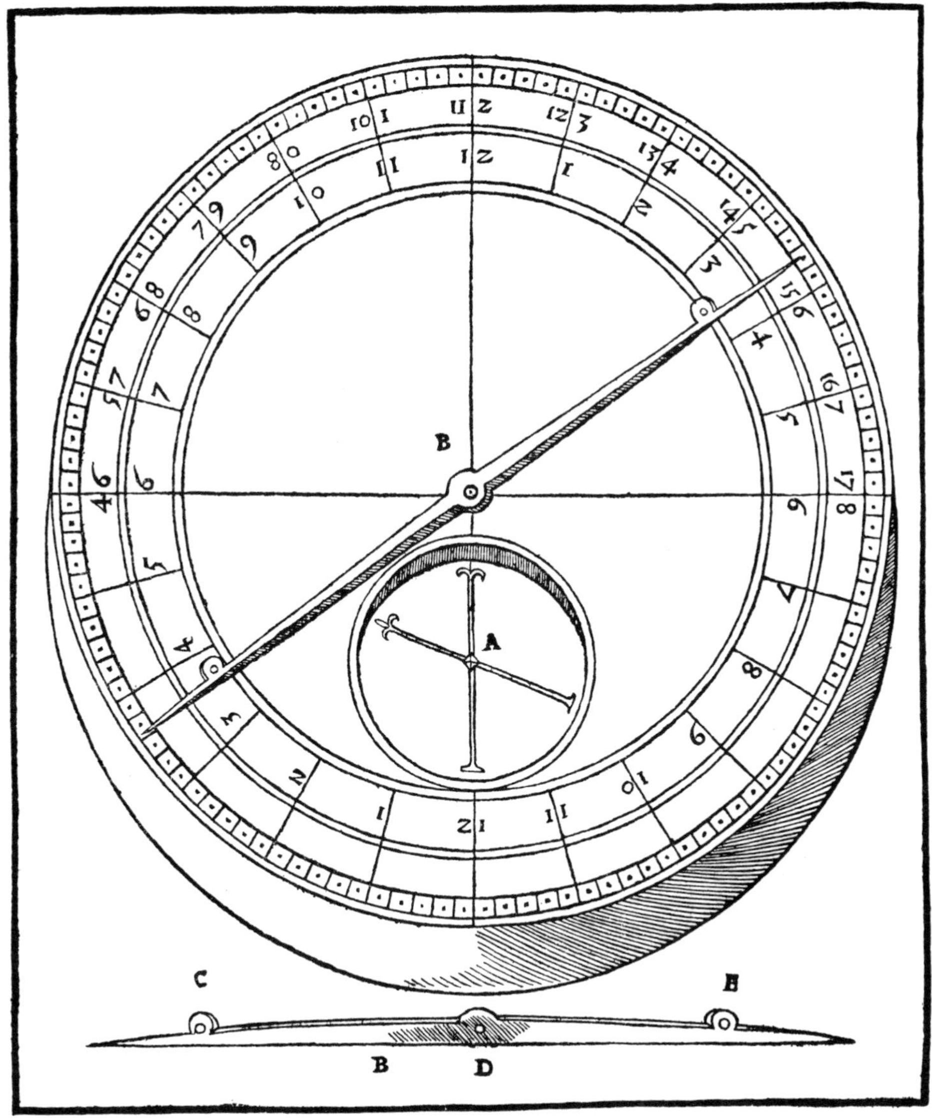

A—NEEDLE OF THE INSTRUMENT. B—ITS TONGUE. C, D, E—HOLES IN THE TONGUE.

division of the compass—that is at a right angle. Then that part of the lowest cord which lies beyond the part to which the cross-cord runs being removed, it shows at what point the boundary mark should be cut into the rock of the tunnel or drift. The cutting is made in the presence of the two Jurors and the manager and the foreman of each mine. For as the *Bergmeister* in the presence of these same persons sets the boundary stones on the surface, so the surveyor cuts in the rock a sign which for this reason is called the boundary rock. If he fixes the boundary mark of a meer in which a shaft has recently begun to be sunk on a vein, first of all he measures and notes the incline of that shaft by the compass or by another way with the applied cords; then he measures all the drifts up to that one in whose rock the boundary mark has to be cut. Of these drifts he measures each angle; then the cords, being laid out on the surveyor's field, in a similar way he stretches a cross-cord, as I said, and cuts the sign on the rock. But if the underground boundary rock has to be cut in a drift which lies beneath the first drift, the surveyor starts from the mark in the first drift, notes the different angles, one by one, takes his measurements, and in the lower drift stretches a cord beyond that place where he judges the mark ought to be cut; and then, as I said before, lays out the cords on the surveyor's field. Even if a vein runs differently in the lower drift from the upper one, in which the first boundary mark has been cut in the rock, still, in the lower drift the mark must be cut in the rock vertically beneath. For if he cuts the lower mark obliquely from the upper one some part of the possession of one mine is taken away to its detriment, and given to the other. Moreover, if it happens that the underground boundary mark requires to be cut in an angle, the surveyor, starting from that angle, measures one fathom toward the front of the mine and another fathom toward the back, and from these measurements forms a triangle, and dividing its middle by a cross-cord, makes his cutting for the boundary mark.

Lastly, the surveyor sometimes, in order to make more certain, finds the boundary of the meers in those places where many old boundary marks are cut in the rock. Then, starting from a stake fixed on the surface, he first of all measures to the nearest mine; then he measures one shaft after another; then he fixes a stake on the surveyors' field, and making a beginning from it stretches the same cords in the same way and measures them, and again fixes in the ground a stake which for him will signify the end of his measuring. Afterward he again measures underground from that spot at which he left off, as many shafts and drifts as he can remember. Then he returns to the surveyor's field, and starting again from the second stake, makes his measurements; and he does this as far as the drift in which the boundary mark must be cut in the rock. Finally, commencing from the stake first fixed in the ground, he stretches a cross-cord in a straight line to the last stake, and this shows the length of the lowest drift. The point where they touch, he judges to be the place where the underground boundary mark should be cut.

END OF BOOK V.

BOOK VI.

IGGING of veins I have written of, and the timbering of shafts, tunnels, drifts, and other excavations, and the art of surveying. I will now speak first of all, of the iron tools with which veins and rocks are broken, then of the buckets into which the lumps of earth, rock, metal, and other excavated materials are thrown, in order that they may be drawn, conveyed, or carried out. Also, I will speak of the water vessels and drains, then of the machines of different kinds,[1] and lastly of the maladies of miners. And while all these matters are being described accurately, many methods of work will be explained.

There are certain iron tools which the miners designate by names of their own, and besides these, there are wedges, iron blocks, iron plates, hammers, crowbars, pikes, picks, hoes, and shovels. Of those which are especially referred to as "iron tools" there are four varieties, which are different from one another in length or thickness, but not in shape, for the upper end of all of them is broad and square, so that it can be struck by the

[1]This Book is devoted in the main to winding, ventilating, and pumping machinery. Their mechanical principles are very old. The block and pulley, the windlass, the use of water-wheels, the transmission of power through shafts and gear-wheels, chain-pumps, piston-pumps with valves, were all known to the Greeks and Romans, and possibly earlier. Machines involving these principles were described by Ctesibius, an Alexandrian of 250 B.C., by Archimedes (287–212 B.C.), and by Vitruvius (1st Century B.C.) As to how far these machines were applied to mining by the Ancients we have but little evidence, and this largely in connection with handling water. Diodorus Siculus (1st Century B.C.) referring to the Spanish mines, says (Book V.): "Sometimes at great depths they meet great rivers underground, "but by art give check to the violence of the streams, for by cutting trenches they divert the "current, and being sure to gain what they aim at when they have begun, they never leave "off till they have finished it. And they admirably pump out the water with those instru- "ments called Egyptian pumps, invented by Archimedes, the Syracusan, when he was in "Egypt. By these, with constant pumping by turns they throw up the water to the mouth of "the pit and thus drain the mine; for this engine is so ingeniously contrived that a vast "quantity of water is strangely and with little labour cast out."

Strabo (63 B.C.—24 A.D., III., 2, 9), also referring to Spanish mines, quoting from Posidonius (about 100 B.C.), says: "He compares with these (the Athenians) the activity "and diligence of the Turdetani, who are in the habit of cutting tortuous and deep tunnels, "and draining the streams which they frequently encounter by means of Egyptian screws." (Hamilton's Tran., Vol. I., p. 221). The "Egyptian screw" was Archimedes' screw, and was thus called because much used by the Egyptians for irrigation. Pliny (XXXIII., 31) also says, in speaking of the Spanish silver-lead mines: "The mountain has been excavated for a distance of 1,500 paces, and along this distance there are water-carriers standing by torch-light night and day steadily baling the water (thus) making quite a river." The re-opening of the mines at Rio Tinto in the middle of the 18th Century disclosed old Roman stopes, in which were found several water-wheels. These were about 15 feet in diameter, lifting the water by the reverse arrangement to an overshot water-wheel. A wooden Archimedian screw was also found in the neighbourhood. (Nash, The Rio Tinto Mine, its History and Romance, London, 1904).

Until early in the 18th Century, water formed the limiting factor in the depth of mines. To the great devotion to this water problem we owe the invention of the steam engine. In 1705 Newcomen—no doubt inspired by Savery's unsuccessful attempt—invented his engine, and installed the first one on a colliery at Wolverhampton, in Staffordshire. With its success, a new era was opened to the miner, to be yet further extended by Watts's improvements sixty years later. It should be a matter of satisfaction to mining engineers that not only was the steam engine the handiwork of their profession, but that another mining engineer, Stephenson, in his effort to further the advance of his calling, invented the locomotive.

hammer. The lower end is pointed so as to split the hard rocks and veins with its point. All of these have eyes except the fourth. The first, which is in daily use among miners, is three-quarters of a foot long, a digit and a half wide, and a digit thick. The second is of the same width as the first, and the same thickness, but one and one half feet long, and is used to shatter the hardest veins in such a way that they crack open. The third is the same length as the second, but is a little wider and thicker; with this one they dig the bottoms of those shafts which slowly accumulate water. The fourth is nearly three palms and one digit long, two digits thick, and in the upper end it is three digits wide, in the middle it is one palm wide, and at the lower end it is pointed like the others; with this they cut out the harder veins. The eye in the first tool is one palm distant from the upper end, in the second and third it is seven digits distant; each swells out around the eye on both sides, and into it they fit a wooden handle, which they hold with one hand, while they strike the iron tool with a hammer, after placing it against the rock. These tools are made larger or smaller as necessary. The smiths, as far as possible, sharpen again all that become dull.

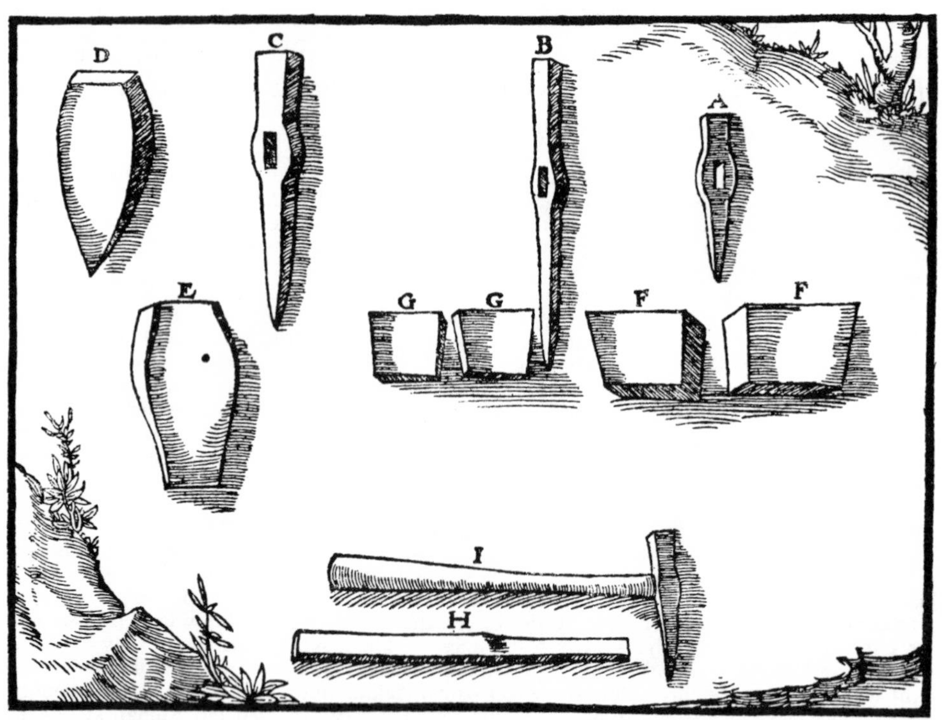

A—First "iron tool." B—Second. C—Third. D—Fourth.[2] E—Wedge. F—Iron block. G—Iron plate. H—Wooden handle. I—Handle inserted in first tool.

A wedge is usually three palms and two digits long and six digits wide; at the upper end, for a distance of a palm, it is three digits thick, and beyond that point it becomes thinner by degrees, until finally it is quite sharp.

[2]While these particular tools serve the same purpose as the "gad" and the "moil," the latter are not fitted with handles, and we have, therefore, not felt justified in adopting these terms, but have given a literal rendering of the Latin.

The iron block is six digits in length and width ; at the upper end it is two digits thick, and at the bottom a digit and a half. The iron plate is the same length and width as the iron block, but it is very thin. All of these, as I explained in the last book, are used when the hardest kind of veins are hewn out. Wedges, locks, and plates, are likewise made larger or smaller.

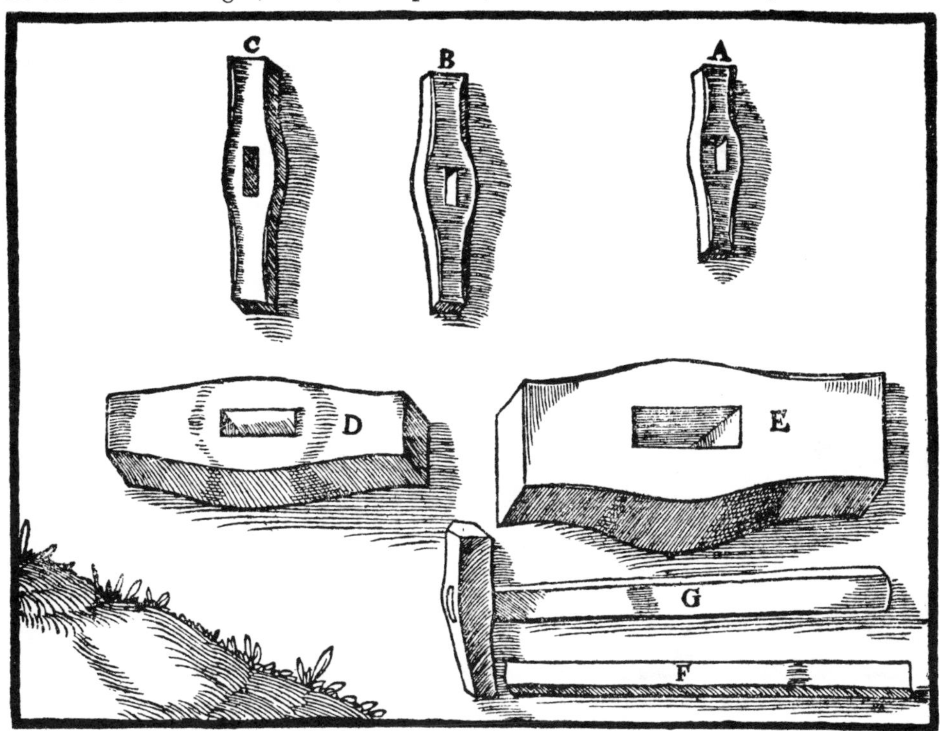

A—Smallest of the smaller hammers. B—Intermediate. C—Largest. D—Small kind of the larger hammer. E—Large kind. F—Wooden handle. G—Handle fixed in the smallest hammer.

Hammers are of two kinds, the smaller ones the miners hold in one hand, and the larger ones they hold with both hands. The former, because of their size and use, are of three sorts. With the smallest, that is to say, the lightest, they strike the second "iron tool;" with the intermediate one the first "iron tool;" and with the largest the third "iron tool"; this one is two digits wide and thick. Of the larger sort of hammers there are two kinds ; with the smaller they strike the fourth "iron tool;" with the larger they drive the wedges into the cracks ; the former are three, and the latter five digits wide and thick, and a foot long. All swell out in their middle, in which there is an eye for a handle, but in most cases the handles are somewhat light, in order that the workmen may be able to strike more powerful blows by the hammer's full weight being thus concentrated.

[2](Continued)—The Latin and old German terms for these tools were :—

First Iron tool	=	*Ferramentum primum*	=	*Bergeisen.*
Second ,,	=	,, *secundum*	=	*Rutzeisen.*
Third ,,	=	,, *tertium*	=	*Sumpffeisen.*
Fourth ,,	=	,, *quartum*	=	*Fimmel.*
Wedge	=	*Cuneus*	=	*Keil.*
Iron block	=	*Lamina*	=	*Plôtz.*
Iron plate	=	*Bractea*	=	*Feder.*

The German words obviously had local value and do not bear translation literally.

The iron crowbars are likewise of two kinds, and each kind is pointed at one end. One is rounded, and with this they pierce to a shaft full of water when a tunnel reaches to it; the other is flat, and with this they knock out of the stopes on to the floor, the rocks which have been softened by the fire, and which cannot be dislodged by the pike. A miner's pike, like a sailor's, is a long rod having an iron head.

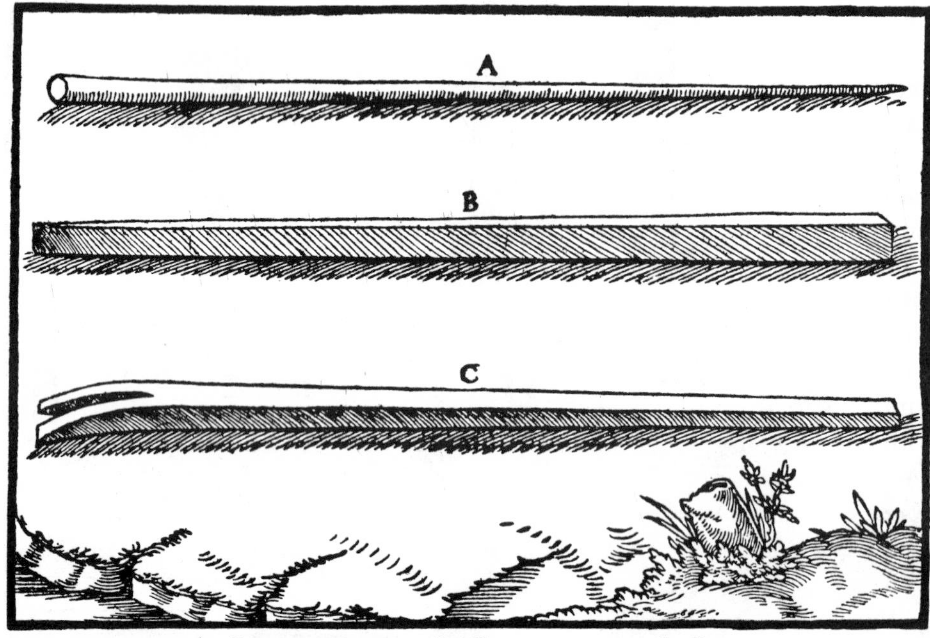

A—Round crowbar. B—Flat crowbar. C—Pike.

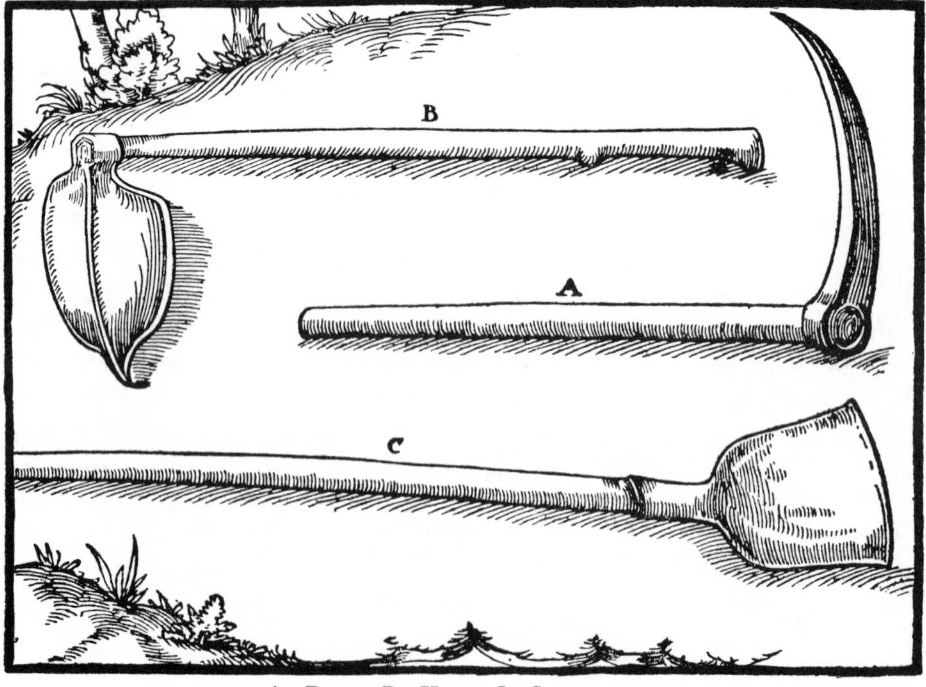

A—Pick. B—Hoe. C—Shovel.

The miner's pick differs from a peasant's pick in that the latter is wide at the bottom and sharp, but the former is pointed. It is used to dig out ore which is not hard, such as earth. Likewise a hoe and shovel are in no way different from the common articles, with the one they scrape up earth and sand, with the other they throw it into vessels.

Now earth, rock, mineral substances and other things dug out with the pick or hewn out with the "iron tools" are hauled out of the shaft in buckets, or baskets, or hide buckets; they are drawn out of tunnels in wheelbarrows or open trucks, and from both they are sometimes carried in trays.

Buckets are of two kinds, which differ in size, but not in material or shape. The smaller for the most part hold only about one *metreta*; the larger are generally capable of carrying one-sixth of a *congius*; neither is of unchangeable capacity, but they often vary.[3] Each is made of staves circled with hoops, one of which binds the top and the other the bottom. The hoops are sometimes made of hazel and oak, but these are easily broken by dashing against the shaft, while those made of iron are more durable. In the larger buckets the staves are thicker and wider, as also are both hoops, and in order that the buckets may be more firm and strong, they have eight iron straps, somewhat broad, four of which run from the upper hoop downwards, and four from the lower hoop upwards, as if to meet each other. The bottom of each bucket, both inside and outside, is furnished with two or three straps of iron, which run from one side of the lower hoop to the other, but the straps which are on the outside are fixed crosswise. Each bucket has two iron hafts which project above the edge, and it has an iron semi-circular bail whose lower ends are fixed directly into the hafts, that the bucket may be handled more easily. Each kind of bucket is much deeper than it is wide, and each is wider at the top, in order that the material which is dug out may be the more easily poured in and poured out again. Into the smaller buckets strong boys, and into larger ones men, fill earth from the bottom of the shaft with hoes; or the other material dug up is shovelled into them or filled in with their hands, for which reason these men are called "shovellers.[4]" Afterward they fix the hook of the drawing-rope into the bale; then the buckets are drawn up by machines—the smaller ones, because of their lighter weight, by machines turned by men, and the larger ones, being heavier, by the machines turned by horses. Some, in place of these buckets, substitute baskets which hold just as much, or even more, since they are lighter than the buckets; some use sacks made of ox-hide instead of buckets, and the drawing-rope hook is fastened to their iron bale, usually three of these filled with excavated material are drawn up at the same time as three are being lowered and three are being filled by boys. The latter are generally used at Schneeberg and the former at Freiberg.

[3]One *metreta*, a Greek measure, equalled about nine English gallons, and a *congius* contained about six pints.

[4]*Ingestores*. This is a case of Agricola coining a name for workmen from the work, the term being derived from *ingero*, to pour or to throw in, used in the previous clause—hence the "reason." See p. xxxi.

A—Small bucket. B—Large bucket. C—Staves. D—Iron hoops. E—Iron straps. F—Iron straps on the bottom. G—Hafts. H—Iron bale. I—Hook of drawing-rope. K—Basket. L—Hide bucket or sack.

That which we call a *cisium*[5] is a vehicle with one wheel, not with two, such as horses draw. When filled with excavated material it is pushed

[5]*Cisium.* A two-wheeled cart. In the preface Agricola gives this as an example of his intended adaptations. See p. xxxi.

by a workman out of tunnels or sheds. It is made as follows: two planks are chosen about five feet long, one foot wide, and two digits thick ; of each of these the lower side is cut away at the front for a length of one foot, and at the back for a length of two feet, while the middle is left whole. Then in the front parts are bored circular holes, in order that the ends of an axle may revolve in them. The intermediate parts of the planks are perforated twice near the bottom, so as to receive the heads of two little cleats on which the planks are fixed; and they are also perforated in the middle, so as to receive the heads of two end-boards, while keys fixed in these projecting heads strengthen the whole structure. The handles are made out of the extreme ends of the long planks, and they turn downward at the ends that they may be grasped more firmly in the hands. The small wheel, of which there is only one, neither has a nave nor does it revolve around the axle, but turns around with it. From the felloe, which the Greeks called ἀψῖδες, two transverse spokes fixed into it pass through the middle of the axle toward the opposite felloe ; the axle is square, with the exception of the ends, each of which is rounded so as to turn in the opening. A workman draws out this barrow full of earth and rock and draws it back empty. Miners also have another wheelbarrow, larger than this one, which they use when they wash earth mixed with tin-stone on to which a stream has been turned. The front end-board of this one is deeper, in order that the earth which has been thrown into it may not fall out.

A—Small wheelbarrow. B—Long planks thereof. C—End-boards. D—Small wheel. E—Larger barrow. F—Front end-board thereof.

A—Rectangular iron bands on truck. B—Its iron straps. C—Iron axle. D—Wooden rollers. E—Small iron keys. F—Large blunt iron pin. G—Same truck upside down.

The open truck has a capacity half as large again as a wheelbarrow; it is about four feet long and about two and a half feet wide and deep; and since its shape is rectangular, it is bound together with three rectangular iron bands, and besides these there are iron straps on all sides. Two small iron axles are fixed to the bottom, around the ends of which wooden rollers revolve on either side; in order that the rollers shall not fall off the immovable axles, there are small iron keys. A large blunt pin fixed to the bottom of the truck runs in a groove of a plank in such a way that the truck does not leave the beaten track. Holding the back part with his hands, the carrier pushes out the truck laden with excavated material, and pushes it back again empty. Some people call it a "dog"[6], because when it moves it makes a noise which seems to them not unlike the bark of a dog. This truck is used when they draw loads out of the longest tunnels, both because it is moved more easily and because a heavier load can be placed in it.

Bateas[7] are hollowed out of a single block of wood; the smaller kind are generally two feet long and one foot wide. When they have been filled with ore, especially when but little is dug from the shafts and tunnels, men either carry them out on their shoulders, or bear them away hung from

[6]*Canis.* The Germans in Agricola's time called a truck a *hundt*—a hound.

[7]*Alveus,*—"Tray." The Spanish term *batea* has been so generally adopted into the mining vocabulary for a wooden bowl for these purposes, that we introduce it here.

A—Small batea. B—Rope. C—Large batea.

their necks. Pliny[8] is our authority that among the ancients everything which was mined was carried out on men's shoulders, but in truth this method of carrying forth burdens is onerous, since it causes great fatigue to a great number of men, and involves a large expenditure for labour ; for this reason it has been rejected and abandoned in our day. The length of the larger batea is as much as three feet, the width up to a foot and a palm. In these bateas the metallic earth is washed for the purpose of testing it.

Water-vessels differ both in the use to which they are put and in the material of which they are made; some draw the water from the shafts and pour it into other things, as dippers ; while some of the vessels filled with water are drawn out by machines, as buckets and bags ; some are made of wood, as the dippers and buckets, and others of hides, as the bags. The water-buckets, just like the buckets which are filled with dry material, are of two kinds, the smaller and the larger , but these are unlike the other buckets at the top, as in this case they are narrower, in order that the water may not be spilled by being bumped against the timbers when they are being drawn out of the shafts, especially those considerably inclined. The water is poured into these buckets by dippers, which are small wooden buckets, but unlike the water-buckets, they are neither narrow at the top nor bound with iron hoops, but with hazel,—because there is no necessity for either. The smaller buckets are drawn up by machines turned by men, the larger ones by those turned by horses.

[8]Pliny (xxxiii., 21). "The fragments are carried on workmen's shoulders ; night "and day each passes the material to his neighbour, only the last of them seeing the daylight."

A—Smaller water-bucket. B—Larger water-bucket. C—Dipper

A—Water-bag which takes in water by itself. B—Water-bag into which water pours when it is pushed with a shovel.

Our people give the name of water-bags to those very large skins for carrying water which are made of two, or two and a half, ox-hides. When these water-bags have undergone much wear and use, first the hair comes off them and they become bald and shining ; after this they become torn. If the tear is but a small one, a piece of smooth notched stick is put into the broken part, and the broken bag is bound into its notches on either side and sewn together ; but if it is a large one, they mend it with a piece of ox-hide. The water-bags are fixed to the hook of a drawing-chain and let down and dipped into the water, and as soon as they are filled they are drawn up by the largest machine. They are of two kinds ; the one kind take in the water by themselves ; the water pours into the other kind when it is pushed in a certain way by a wooden shovel.

When the water has been drawn out from the shafts, it is run off in troughs, or into a hopper, through which it runs into the trough. Likewise the water which flows along the sides of the tunnels is carried off in drains. These are composed of two hollowed beams joined firmly together, so as to hold the water which flows through them, and they are covered by planks all along their course, from the mouth of the tunnel right up to the extreme end of it, to prevent earth or rock falling into them and obstructing the flow of the water. If much mud gradually settles in them the planks are raised and the drains are cleaned out, for they would otherwise become stopped up and obstructed by this accident. With regard to the trough lying above

A—Trough. B—Hopper.

ground, which miners place under the hoppers which are close by the shaft houses, these are usually hollowed out of single trees. Hoppers are generally made of four planks, so cut on the lower side and joined together that the top part of the hopper is broader and the bottom part narrower.

I have sufficiently indicated the nature of the miners' iron tools and their vessels. I will now explain their machines, which are of three kinds, that is, hauling machines, ventilating machines, and ladders. By means of the hauling machines loads are drawn out of the shafts ; the ventilating machines receive the air through their mouths and blow it into shafts or tunnels, for if this is not done, diggers cannot carry on their labour without great difficulty in breathing ; by the steps of the ladders the miners go down into the shafts and come up again.

Hauling machines are of varied and diverse forms, some of them being made with great skill, and if I am not mistaken, they were unknown to the Ancients. They have been invented in order that water may be drawn from the depths of the earth to which no tunnels reach, and also the excavated material from shafts which are likewise not connected with a tunnel, or if so, only with very long ones. Since shafts are not all of the same depth, there is a great variety among these hauling machines. Of those by which dry loads are drawn out of the shafts, five sorts are in the most common use, of which I will now describe the first. Two timbers a little longer than the shaft are placed beside it, the one in the front of the shaft, the other at the back. Their extreme ends have holes through which stakes, pointed at the bottom like wedges, are driven deeply into the ground, so that the timbers may remain stationary. Into these timbers are mortised the ends of two cross-timbers, one laid on the right end of the shaft, while the other is far enough from the left end that between it and that end there remains suitable space for placing the ladders. In the middle of the cross-timbers, posts are fixed and secured with iron keys. In hollows at the top of these posts thick iron sockets hold the ends of the barrel, of which each end projects beyond the hollow of the post, and is mortised into the end of another piece of wood a foot and a half long, a palm wide and three digits thick ; the other end of these pieces of wood is seven digits wide, and into each of them is fixed a round handle, likewise a foot and a half long. A winding-rope is wound around the barrel and fastened to it at the middle part. The loop at each end of the rope has an iron hook which is engaged in the bale of a bucket, and so when the windlass revolves by being turned by the cranks, a loaded bucket is always being drawn out of the shaft and an empty one is being sent down into it. Two robust men turn the windlass, each having a wheelbarrow near him, into which he unloads the bucket which is drawn up nearest to him ; two buckets generally fill a wheelbarrow ; therefore when four buckets have been drawn up, each man runs his own wheelbarrow out of the shed and empties it. Thus it happens that if shafts are dug deep, a hillock rises around the shed of the windlass. If a vein is not metal-bearing, they pour out the earth and rock without discriminating ; whereas if it is metal-bearing, they preserve these materials, which they unload separately and crush and wash. When they draw up buckets of water they empty the water through the hopper into a trough, through which it flows away.

A—Timber placed in front of the shaft. B—Timber placed at the back of the shaft. C—Pointed stakes. D—Cross-timbers. E—Posts or thick planks. F—Iron sockets. G—Barrel. H—Ends of barrel. I—Pieces of wood. K—Handle. L—Drawing-rope. M—Its hook. N—Bucket. O—Bale of the bucket.

The next kind of machine, which miners employ when the shaft is deeper, differs from the first in that it possesses a wheel as well as cranks. This windlass, if the load is not being drawn up from a great depth, is turned by one windlass man, the wheel taking the place of the other man. But if the depth is greater, then the windlass is turned by three men, the wheel being substituted for a fourth, because the barrel having been once set in motion, the rapid revolutions of the wheel help, and it can be turned more easily. Sometimes masses of lead are hung on to this wheel, or are fastened to the spokes, in order that when it is turned they depress the spokes by their weight and increase the motion ; some persons for the same reason fasten into the barrel two, three, or four iron rods, and weight their ends with lumps of lead. The windlass wheel differs from the wheel of a carriage and from the one

A—Barrel. B—Straight levers. C—Usual crank. D—Spokes of wheel. E—Rim of the same wheel.

which is turned by water power, for it lacks the buckets of a water-wheel and it lacks the nave of a carriage wheel. In the place of the nave it has a thick barrel, in which are mortised the lower ends of the spokes, just as their upper ends are mortised into the rim. When three windlass men turn this machine, four straight levers are fixed to the one end of the barrel, and to the other the crank which is usual in mines, and which is composed of two limbs, of which the rounded horizontal one is grasped by the hands; the rectangular limb, which is at right angles to the horizontal one, has mortised in its lower end the round handle, and in the upper end the end of the barrel. This crank is worked by one man, the levers by two men, of whom one pulls while the other pushes; all windlass workers, whatsoever kind of a machine they may turn, are necessarily robust that they can sustain such great toil.

The third kind of machine is less fatiguing for the workman, while it raises larger loads; even though it is slower, like all other machines which have drums, yet it reaches greater depths, even to a depth of 180 feet. It consists of an upright axle with iron journals at its extremities, which turn in two iron sockets, the lower of which is fixed in a block set in the ground and the upper one in the roof beam. This axle has at its lower end a

A—UPRIGHT AXLE. B—BLOCK. C—ROOF BEAM. D—WHEEL. E—TOOTHED-DRUM. F—HORIZONTAL AXLE. G—DRUM COMPOSED OF RUNDLES. H—DRAWING ROPE. I—POLE. K—UPRIGHT POSTS. L—CLEATS ON THE WHEEL.

wheel made of thick planks joined firmly together, and at its upper end a toothed drum ; this toothed drum turns another drum made of rundles, which is on a horizontal axle. A winding-rope is wound around this latter axle, which turns in iron bearings set in the beams. So that they may not fall, the two workmen grasp with their hands a pole fixed to two upright posts, and then pushing the cleats of the lower wheel backward with their feet, they revolve the machine ; as often as they have drawn up and emptied one bucket full of excavated material, they turn the machine in the opposite direction and draw out another.

The fourth machine raises burdens once and a half as large again as the two machines first explained. When it is made, sixteen beams are erected each forty feet long, one foot thick and one foot wide, joined at the top with clamps and widely separated at the bottom. The lower ends of all of them are mortised into separate sills laid flat upon the ground ; these sills are five feet long, a foot and a half wide, and a foot thick. Each beam is also connected with its sill by a post, whose upper end is mortised into the beam and its lower end mortised into the sill ; these posts are four feet long, one foot thick, and one foot wide. Thus a circular area is made, the diameter of which is fifty feet ; in the middle of this area a hole is sunk to a depth of ten feet, and rammed down tight, and in order to give it sufficient firmness, it is strengthened with contiguous small timbers, through which pins are driven, for by them the earth around the hole is held so that it cannot fall in. In the bottom of the hole is planted a sill, three or four feet long and a foot and a half thick and wide ; in order that it may remain fixed, it is set into the small timbers ; in the middle of it is a steel socket in which the pivot of the axle turns. In like manner a timber is mortised into two of the large beams, at the top beneath the clamps ; this has an iron bearing in which the other iron journal of the axle revolves. Every axle used in mining, to speak of them once for all, has two iron journals, rounded off on all sides, one fixed with keys in the centre of each end. That part of this journal which is fixed to the end of the axle is as broad as the end itself and a digit thick ; that which projects beyond the axle is round and a palm thick, or thicker if necessity requires ; the ends of each miner's axle are encircled and bound by an iron band to hold the journal more securely. The axle of this machine, except at the ends, is square, and is forty feet long, a foot and a half thick and wide. Mortised and clamped into the axle above the lower end are the ends of four inclined beams ; their outer ends support two double cross-beams similarly mortised into them ; the inclined beams are eighteen feet long, three palms thick, and five wide. The two cross-beams are fixed to the axle and held together by wooden keys so that they will not separate, and they are twenty-four feet long. Next, there is a drum which is made of three wheels, of which the middle one is seven feet distant from the upper one and from the lower one ; the wheels have four spokes which are supported by the same number of inclined braces, the lower ends of which are joined together round the axle by a clamp ; one end of each spoke is mortised into the axle and the other into the rim. There are rundles all round the wheels, reaching from the rim of the lowest one to the rim of the middle one, and likewise from the rim of the middle wheel to the rim of the top one ; around these rundles are wound the drawing-ropes, one between the lowest wheel and the middle one, the other between the middle and top wheels. The whole of this construction is shaped like a cone, and is covered with a shingle roof, with the exception of that square part which faces the shaft. Then cross-beams, mortised at both ends, connect a double row of upright posts ; all of these are eighteen feet long, but the posts are one foot thick and one foot wide, and the cross-beams are three palms thick and wide. There are sixteen posts and eight cross-beams, and upon these cross-beams are laid two timbers a foot wide and three palms thick, hollowed out to a width of half a foot and to a depth of five digits ; the one is laid upon the upper cross-beams and the other upon the lower ; each is long enough to reach nearly from the drum of the whim to the shaft. Near the same drum each timber has a small round wooden roller six digits thick, whose ends are

A—Upright beams. B—Sills laid flat upon the ground. C—Posts. D—Area. E—Sill set at the bottom of the hole. F—Axle. G—Double cross-beams. H—Drum. I—Winding-ropes. K—Bucket. L—Small pieces of wood hanging from double cross-beams. M—Short wooden block. N—Chain. O—Pole bar. P—Grappling hook. (Some members mentioned in the text are not shown).

covered with iron bands and revolve in iron rings. Each timber also has a wooden pulley, which together with its iron axle revolves in holes in the timber. These pulleys are hollowed out all round, in order that the drawing-rope may not slip out of them, and thus each rope is drawn tight and turns over its own roller and its own pulley. The iron hook of each rope is engaged with the bale of the bucket. Further, with regard to the double cross-beams which are mortised to the lower part of the main axle, to each end of them there is mortised a small piece of wood four feet long. These appear to hang from the double cross-beams, and a short wooden block is fixed to the lower part of them, on which a driver sits. Each of these blocks has an iron clavis which holds a chain, and that in turn a pole-bar. In this way it is possible for two horses to draw this whim, now this way and now that ; turn by turn one bucket is drawn out of the shaft full and another is let down into it empty ; if, indeed, the shaft is very deep four horses turn the whim. When a bucket has been drawn up, whether filled with dry or wet materials, it must be emptied, and a workman inserts a grappling hook and overturns it ; this hook hangs on a chain made of three or four links, fixed to a timber.

The fifth machine is partly like the whim, and partly like the third rag and chain pump, which draws water by balls when turned by horse power, as I will explain a little later. Like this pump, it is turned by horse power and has two axles, namely, an upright one—about whose lower end, which decends into an underground chamber, there is a toothed drum—and a horizontal one, around which there is a drum made of rundles. It has indeed two drums around its horizontal axle, similar to those of the big machine, but smaller, because it draws buckets from a shaft almost two hundred and forty feet deep. One drum is made of hubs to which cleats are fixed, and the other is made of rundles ; and near the latter is a wheel two feet deep, measured on all sides around the axle, and one foot wide ; and against this impinges a brake,[10] which holds the whim when occasion demands that it be stopped. This is necessary when the hide buckets are emptied after being drawn up full of rock fragments or earth, or as often as water is poured out of buckets similarly drawn up ; for this machine not only raises dry loads, but also wet ones, just like the other four machines which I have already described. By this also, timbers fastened on to its winding-chain are let down into a shaft. The brake is made of a piece of wood one foot thick and half a foot long, projecting from a timber that is suspended by a chain from one end of a beam which oscillates on an iron pin, this in turn being supported in the claws of an upright post ; and from the other end of this oscillating beam a long timber is suspended by a chain, and from this long timber again a short beam is suspended. A workman sits on the short beam when the machine needs to be stopped, and lowers it ; he then inserts a plank or small stick so that the two timbers are held down and cannot be raised. In this way the brake is raised, and seizing the drum, presses it so tightly that sparks often fly from it ; the suspended timber to which the short beam is attached, has several holes in which the chain is

[10]*Harpago*,—A " grapple " or " hook."

A—Toothed drum which is on the upright axle. B—Horizontal axle. C—Drum which is made of rundles. D—Wheel near it. E—Drum made of hubs. F—Brake. G—Oscillating beam. H—Short beam. I—Hook.

fixed, so that it may be raised as much as is convenient. Above this wheel there are boards to prevent the water from dripping down and wetting it, for if it becomes wet the brake will not grip the machine so well. Near the other drum is a pin from which hangs a chain, in the last link of which there is an iron hook three feet long ; a ring is fixed to the bottom of the bucket, and this hook, being inserted into it, holds the bucket back so that the water may be poured out or the fragments of rock emptied.

The miners either carry, draw, or roll down the mountains the ore which is hauled out of the shafts by these five machines or taken out of the tunnels. In the winter time our people place a box on a sledge and draw it down the low mountains with a horse ; and in this season they also fill sacks made of hide and load them on dogs, or place two or three of them on a small sledge which is higher in the fore part and lower at the back. Sitting on these sacks, not without risk of his life, the bold driver guides the sledge as it rushes down the mountain into the valleys with a stick, which he carries in his hand ; when it is rushing down too quickly he arrests it with the stick, or with the same stick brings it back to the track when it is turning aside from its proper course. Some of the

A—Sledge with box placed on it. B—Sledge with sacks placed on it. C—Stick. D—Dogs with pack-saddles. E—Pig-skin sacks tied to a rope.

Noricians[11] collect ore during the winter into sacks made of bristly pigskins, and drag them down from the highest mountains, which neither horses, mules nor asses can climb. Strong dogs, that are trained to bear pack saddles, carry these sacks when empty into the mountains. When they are filled with ore, bound with thongs, and fastened to a rope, a man, winding the rope round his arm or breast, drags them down through the snow to a place where horses, mules, or asses bearing pack-saddles can climb. There the ore is removed from the pigskin sacks and put into other sacks made of double or triple twilled linen thread, and these placed on the pack-saddles of the beasts are borne down to the works where the ores are washed or smelted. If, indeed, the horses, mules, or asses are able to climb the mountains, linen sacks filled with ore are placed on their saddles, and they carry these down the narrow mountain paths, which are passable neither by wagons nor sledges, into the valleys lying below the steeper portions of the mountains. But on the declivity of cliffs which beasts cannot climb, are placed long open boxes made of planks, with transverse cleats to hold them together ; into these boxes is thrown the ore which has been brought in wheelbarrows, and when it has run down to the level it is gathered into sacks, and the beasts either carry it away on their backs or drag it away after it has been thrown into sledges or wagons. When the drivers bring ore down steep mountain slopes they use two-wheeled carts, and they drag behind them on the ground the trunks of two trees, for these by their weight hold back the heavily-laden carts, which contain ore in their boxes, and check their descent, and but for these the driver would often be obliged to bind chains to the wheels. When these men bring down ore from mountains which do not have such declivities, they use wagons whose beds are twice as long as those of the carts. The planks of these are so put together that, when the ore is unloaded by the drivers, they can be raised and taken apart, for they are only held together by bars. The drivers employed by the owners of the ore bring down thirty or sixty wagon-loads, and the master of the works marks on a stick the number of loads for each driver. But some ore, especially tin, after being taken from the mines, is divided into eight parts, or into nine, if the owners of the mine give "ninth parts" to the owners of the tunnel. This is occasionally done by measuring with a bucket, but more frequently planks are put together on a spot where, with the addition of the level ground as a base, it forms a hollow box. Each owner provides for removing, washing, and smelting that portion which has fallen to him. (Illustration p. 170).

Into the buckets, drawn by these five machines, the boys or men throw the earth and broken rock with shovels, or they fill them with their hands ; hence they get their name of shovellers. As I have said, the same machines raise not only dry loads, but also wet ones, or water ; but before I explain the varied and diverse kinds of machines by which miners are wont

[11]Ancient Noricum covered the region of modern Tyrol, with parts of Bavaria, Salzburg, etc.

A—HORSES WITH PACK-SADDLES. B—LONG BOX PLACED ON THE SLOPE OF THE CLIFF. C—CLEATS THEREOF. D—WHEELBARROW. E—TWO-WHEELED CART. F—TRUNKS OF TREES. G—WAGON. H—ORE BEING UNLOADED FROM THE WAGON. I—BARS. K—MASTER OF THE WORKS MARKING THE NUMBER OF CARTS ON A STICK. L—BOXES INTO WHICH ARE THROWN THE ORE WHICH HAS TO BE DIVIDED.

to draw water alone, I will explain how heavy bodies, such as axles, iron chains, pipes, and heavy timbers, should be lowered into deep vertical shafts. A windlass is erected whose barrel has on each end four straight levers ; it is fixed into upright beams and around it is wound a rope, one end of which is fastened to the barrel and the other to those heavy bodies which are slowly lowered down by workmen ; and if these halt at any part of the shaft they are drawn up a little way. When these bodies are very heavy, then behind this windlass another is erected just like it, that their combined strength may be equal to the load, and that it may be lowered slowly. Sometimes for the same reason, a pulley is fastened with cords to the roof-beam, and the rope descends and ascends over it.

A—Windlass. B—Straight levers. C—Upright beams. D—Rope. E—Pulley. F—Timbers to be lowered.

Water is either hoisted or pumped out of shafts. It is hoisted up after being poured into buckets or water-bags ; the water-bags are generally brought up by a machine whose water-wheels have double paddles, while the buckets are brought up by the five machines already described, although in certain localities the fourth machine also hauls up water-bags of moderate size. Water is drawn up also by chains of dippers, or by suction pumps, or by "rag and chain" pumps.[12] When there is but a small quantity, it is either brought up in buckets or drawn up by chains of dippers or suction pumps, and when there is much water it is either drawn up in hide bags or by rag and chain pumps.

First of all, I will describe the machines which draw water by chains of dippers, of which there are three kinds. For the first, a frame is made entirely of iron bars; it is two and a half feet high, likewise two and a half feet long, and in addition one-sixth and one-quarter of a digit long, one-fourth and one-twenty-fourth of a foot wide. In it there are three little horizontal iron axles, which revolve in bearings or wide pillows of steel, and also four iron wheels, of which two are made with rundles and the same number are toothed. Outside the frame, around the lowest axle, is a wooden fly-wheel, so that it can be more readily turned, and inside the frame is a smaller drum which is made of eight rundles, one-sixth and one twenty-fourth of a foot long. Around the second axle, which does not project beyond the frame, and is therefore only two and a half feet and one-twelfth and one-third part of a digit long, there is on the one side, a smaller toothed wheel, which has forty-eight teeth, and on the other side a larger drum, which is surrounded by twelve rundles one-quarter of a foot long. Around the third axle, which is one inch and one-third thick, is a larger toothed wheel projecting one foot from the axle in all directions, which has seventy-two teeth. The teeth of each wheel are fixed in with screws, whose threads are screwed into threads in the wheel, so that those teeth which are broken can be replaced by others; both the teeth and rundles are steel. The upper axle projects beyond the frame, and is so skilfully mortised into the body of another axle that it has the appearance of being one; this axle proceeds through a frame made of beams which stands around the shaft, into an iron fork set in a stout oak timber, and turns on a roller made of pure steel. Around this axle is a drum of the kind possessed by those machines which draw water by rag and chain; this drum has triple curved iron clamps, to which the links of an iron chain hook themselves, so that a great weight cannot tear them away. These links are not whole like the links of other chains, but each one being curved in the upper part on each side catches the one which comes next, whereby it presents the appearance of a double chain. At the point where one catches the other, dippers made of iron or brass plates and holding half a *congius*[13] are bound to them with thongs; thus, if there are one hundred links there will be the same number of dippers pouring out water. When the shafts are inclined, the mouths of the dippers project and are covered on the top that they may not spill out the water, but when the shafts are vertical the dippers do not require a cover. By fitting the end of the lowest small axle into the crank, the man who works the crank turns the axle, and at the same time the drum whose rundles turn the toothed wheel of the second axle; by this wheel is driven the one that is made of rundles, which

[12] *Machina quae pilis aquas haurit.* "Machine which draws water with balls." This apparatus is identical with the Cornish "rag and chain pump" of the same period, and we have therefore adopted that term.

[13] A *congius* contained about six pints.

A—Iron frame. B—Lowest axle. C—Fly-wheel. D—Smaller drum made of rundles. E—Second axle. F—Smaller toothed wheel. G—Larger drum made of rundles. H—Upper axle. I—Larger toothed wheel. K—Bearings. L—Pillow. M—Framework. N—Oak timber. O—Support of iron bearing. P—Roller. Q—Upper drum. R—Clamps. S—Chain. T—Links. V—Dippers. X—Crank. Y—Lower drum or balance weight.

again turns the toothed wheel of the upper small axle and thus the drum to which the clamps are fixed. In this way the chain, together with the empty dippers, is slowly let down, close to the footwall side of the vein, into the sump to the bottom of the balance drum, which turns on a little iron axle, both ends of which are set in a thick iron bearing. The chain is rolled round the drum and the dippers fill with water; the chain being drawn up close to the hanging-wall side, carries the dippers filled with water above the drum of the upper axle. Thus there are always three of the dippers inverted and pouring water into a lip, from which it flows away into the drain of the tunnel. This machine is less useful, because it cannot be constructed without great expense, and it carries off but little water and is somewhat slow, as also are other machines which possess a great number of drums.

A—Wheel which is turned by treading. B—Axle. C—Double chain. D—Link of double chain. E—Dippers. F—Simple clamps. G—Clamp with triple curves.

The next machine of this kind, described in a few words by Vitruvius,[14] more rapidly brings up dippers, holding a *congius*; for this reason, it is

[14]Vitruvius (x., 9). "But if the water is to be supplied to still higher places, a double "chain of iron is made to revolve on the axis of the wheel, long enough to reach to the lower "level. This is furnished with brazen buckets, each holding about a *congius*. Then by turning "the wheel, the chain also turns upon the axis and brings the buckets to the top thereof, on "passing which they are inverted and pour into the conduits the water they have raised."

more useful than the first one for drawing water out of shafts, into which much water is continually flowing. This machine has no iron frame nor drums, but has around its axle a wooden wheel which is turned by treading; the axle, since it has no drum, does not last very long. In other respects this pump resembles the first kind, except that it differs from it by having a double chain. Clamps should be fixed to the axle of this machine, just as to the drum of the other one; some of these are made simple and others with triple curves, but each kind has four barbs.

The third machine, which far excels the two just described, is made when a running stream can be diverted to a mine; the impetus of the stream striking the paddles revolves a water-wheel in place of the wheel turned by treading. With regard to the axle, it is like the second machine,

A—Wheel whose paddles are turned by the force of the stream. B—Axle. C—Drum of axle, to which clamps are fixed. D—Chain. E—Link. F—Dippers. G—Balance drum.

but the drum which is round the axle, the chain, and the balance drum, are like the first machine. It has much more capacious dippers than even the second machine, but since the dippers are frequently broken, miners rarely use these machines; for they prefer to lift out small quantities of water by the first five machines or to draw it up by suction pumps, or, if there is much water, to drain it by the rag and chain pump or to bring it up in water-bags.

Enough, then, of the first sort of pumps. I will now explain the other, that is the pump which draws, by means of pistons, water which has been raised by suction. Of these there are seven varieties, which though they differ from one another in structure, nevertheless confer the same benefits upon miners, though some to a greater degree than others. The first pump is made as follows. Over the sump is placed a flooring, through which a pipe—or two lengths of pipe, one of which is joined into the other—are let down to the bottom of the sump; they are fastened with pointed iron clamps driven in straight on both sides, so that the pipes may remain fixed. The lower end of the lower pipe is enclosed in a trunk two feet deep ; this trunk, hollow like the pipe, stands at the bottom of the sump, but the lower opening of it is blocked with a round piece of wood ; the trunk has perforations round about, through which water flows into it. If there is one length of pipe, then in the upper part of the trunk which has been hollowed out there is enclosed a box of iron, copper, or brass, one palm deep, but without a bottom, and a rounded valve so tightly closes it that the water, which has been drawn up by suction, cannot run back ; but if there are two lengths of pipe, the box is enclosed in the lower pipe at the point of junction. An opening or a spout in the upper pipe reaches to the drain of the tunnel. Thus the workman, eager at his labour, standing on the flooring boards, pushes the piston down into the pipe and draws it out again. At the top of the piston-rod is a hand-bar and the bottom is fixed in a shoe ; this is the name given to the leather covering, which is almost cone-shaped, for it is so stitched that it is tight at the lower end, where it is fixed to the piston-rod which it surrounds, but in the upper end where it draws the water it is wide open. Or else an iron disc one digit thick is used, or one of wood six digits thick, each of which is far superior to the shoe. The disc is fixed by an iron key which penetrates through the bottom of the piston-rod, or it is screwed on to the rod ; it is round, with its upper part protected by a cover, and has five or six openings, either round or oval, which taken together present a star-like appearance ; the disc has the same diameter as the inside of the pipe, so that it can be just drawn up and down in it. When the workman draws the piston up, the water which has passed in at the openings of the disc, whose cover is then closed, is raised to the hole or little spout, through which it flows away ; then the valve of the box opens, and the water which has passed into the trunk is drawn up by the suction and rises into the pipe ; but when the workman pushes down the piston, the valve closes and allows the disc again to draw in the water.

The piston of the second pump is more easily moved up and down. When this pump is made, two beams are placed over the sump, one near the right side of it, and the other near the left. To one beam a pipe is fixed with iron clamps ; to the other is fixed either the forked branch of a tree or a timber cut out at the top in the shape of a fork, and through the prongs of the fork a round hole is bored. Through a wide round hole in the middle of a sweep passes

A—Sump. B—Pipes. C—Flooring. D—Trunk. E—Perforations of trunk. F—Valve. G—Spout. H—Piston-rod. I—Hand-bar of piston. K—Shoe. L—Disc with round openings. M—Disc with oval openings. N—Cover. O—This man is boring logs and making them into pipes. P—Borer with auger. Q—Wider borer.

A—Erect timber. B—Axle. C—Sweep which turns about the axle. D—Piston rod. E—Cross-bar. F—Ring with which two pipes are generally joined.

an iron axle, so fastened in the holes in the fork that it remains fixed, and the sweep turns on this axle. In one end of the sweep the upper end of a piston-rod is fastened with an iron key ; at the other end a cross-bar is also fixed, to the extreme ends of which are handles to enable it to be held more firmly in the hands. And so when the workman pulls the cross-bar upward, he forces the piston into the pipe ; when he pushes it down again he draws the piston out of the pipe ; and thus the piston carries up the water which has been drawn in at the openings of the disc, and the water flows away through the spout into the drains. This pump, like the next one, is identical with the first in all that relates to the piston, disc, trunk, box, and valve.

The third pump is not unlike the one just described, but in place of one upright, posts are erected with holes at the top, and in these holes the ends of an axle revolve. To the middle of this axle are fixed two wooden bars, to the end of one of which is fixed the piston, and to the end of the other a heavy piece of wood, but short, so that it can pass between the two posts and may move backward and forward. When the workman pushes this piece of wood, the piston is drawn out of the pipe ; when it returns by its

A—Posts. B—Axle. C—Wooden bars. D—Piston rod. E—Short piece of wood. F—Drain. G—This man is diverting the water which is flowing out of the drain, to prevent it from flowing into the trenches which are being dug.

own weight, the piston is pushed in. In this way, the water which the pipe contains is drawn through the openings in the disc and emptied by the piston through the spout into the drain. There are some who place a hand-bar underneath in place of the short piece of wood. This pump, as also the last before described, is less generally used among miners than the others.

The fourth kind is not a simple pump but a duplex one. It is made as follows. A rectangular block of beechwood, five feet long, two and a half feet wide, and one and a half feet thick, is cut in two and hollowed out wide and deep enough so that an iron axle with cranks can revolve in it. The axle is placed between the two halves of this box, and the first part of the axle, which is in contact with the wood, is round and the straight end forms a journal. Then the axle is bent down the depth of a foot and again bent so as to continue straight, and at this point a round piston-rod hangs from it; next it is bent up as far as it was bent down; then it continues a little way straight again, and then it is bent up a foot and again continues straight, at which point a second round piston-rod is hung from it; afterward it

A—Box B—Lower part of box. C—Upper part of same. D—Clamps. E—Pipes below the box. F—Column pipe fixed above the box. G—Iron axle. H—Piston-rods. I—Washers to protect the bearings. K—Leathers. L—Eyes in the axle. M—Rods whose ends are weighted with lumps of lead. N—Crank.
(*This plate is unlettered in the first edition but corrected in those later.*)

is bent down the same distance as it was bent up the last time ; the other end of it, which also acts as a journal, is straight. This part which protrudes through the wood is protected by two iron washers in the shape of discs, to which are fastened two leather washers of the same shape and size, in order to prevent the water which is drawn into the box from gushing out. These discs are around the axle ; one of them is inside the box and the other outside. Beyond this, the end of the axle is square and has two eyes, in which are fixed two iron rods, and to their ends are weighted lumps of lead, so that the axle may have a greater propensity to revolve ; this axle can easily be turned when its end has been mortised in a crank. The upper part of the box is the shallower one, and the lower part the deeper ; the upper part is bored out once straight down through the middle, the diameter of the opening being the same as the outside diameter of the column pipe ; the lower box has, side by side, two apertures also bored straight down ; these are for two pipes, the space of whose openings therefore is twice as great as that of the upper part ; this lower part of the box is placed upon the two pipes, which are fitted into it at their upper ends, and the lower ends of these pipes penetrate into trunks which stand in the sump. These trunks have perforations through which the water flows into them. The iron axle is placed in the inside of the box, then the two iron piston-rods which hang from it are let down through the two pipes to the depth of a foot. Each piston has a screw at its lower end which holds a thick iron plate, shaped like a disc and full of openings, covered with a leather, and similarly to the other pump it has a round valve in a little box. Then the upper part of the box is placed upon the lower one and properly fitted to it on every side, and where they join they are bound by wide thick iron plates, and held with small wide iron wedges, which are driven in and are fastened with clamps. The first length of column pipe is fixed into the upper part of the box, and another length of pipe extends it, and a third again extends this one, and so on, another extending on another, until the uppermost one reaches the drain of the tunnel. When the crank worker turns the axle, the pistons in turn draw the water through their discs ; since this is done quickly, and since the area of openings of the two pipes over which the box is set, is twice as large as the opening of the column pipe which rises from the box, and since the pistons do not lift the water far up, the impetus of the water from the lower pipes forces it to rise and flow out of the column pipe into the drain of the tunnel. Since a wooden box frequently cracks open, it is better to make it of lead or copper or brass.

The fifth kind of pump is still less simple, for it is composed of two or three pumps whose pistons are raised by a machine turned by men, for each piston-rod has a tappet which is raised, each in succession, by two cams on a barrel ; two or four strong men turn it. When the pistons descend into the pipes their discs draw the water ; when they are raised these force the water out through the pipes. The upper part of each of these piston-rods, which is half a foot square, is held in a slot in a cross-beam ; the lower part, which drops down into the pipes, is made of another piece of wood and is round. Each of these three pumps is composed of two lengths of pipe fixed

A—Tappets of piston-rods. B—Cams of the barrel. C—Square upper parts of piston-rods. D—Lower rounded parts of piston-rods. E—Cross-beams. F—Pipes. G—Apertures of pipes. H—Trough. (Fifth kind of pump—see p. 181).

A—Water-wheel. B—Axle. C—Trunk on which the lowest pipe stands. D—Basket surrounding trunk. (Sixth kind of pump—see p. 184.)

to the shaft timbers. This machine draws the water higher, as much as twenty-four feet. If the diameter of the pipes is large, only two pumps are made; if smaller, three, so that by either method the volume of water is the same. This also must be understood regarding the other machines and their pipes. Since these pumps are composed of two lengths of pipe, the little iron box having the iron valve which I described before, is not enclosed in a trunk, but is in the lower length of pipe, at that point where it joins the upper one; thus the rounded part of the piston-rod is only as long as the upper length of pipe; but I will presently explain this more clearly.

The sixth kind of pump would be just the same as the fifth were it not that it has an axle instead of a barrel, turned not by men but by a water-wheel, which is revolved by the force of water striking its buckets. Since water-power far exceeds human strength, this machine draws water through its pipes by discs out of a shaft more than one hundred feet deep. The bottom of the lowest pipe, set in the sump, not only of this pump but also of the others, is generally enclosed in a basket made of wicker-work, to prevent wood shavings and other things being sucked in. (See p. 183.)

The seventh kind of pump, invented ten years ago, which is the most ingenious, durable, and useful of all, can be made without much expense. It is composed of several pumps, which do not, like those last described, go down into the shaft together, but of which one is below the other, for if there are three, as is generally the case, the lower one lifts the water of the sump and pours it out into the first tank; the second pump lifts again from that tank into a second tank, and the third pump lifts it into the drain of the tunnel. A wheel fifteen feet high raises the piston-rods of all these pumps at the same time and causes them to drop together. The wheel is made to revolve by paddles, turned by the force of a stream which has been diverted to the mountain. The spokes of the water-wheel are mortised in an axle six feet long and one foot thick, each end of which is surrounded by an iron band, but in one end there is fixed an iron journal; to the other end is attached an iron like this journal in its posterior part, which is a digit thick and as wide as the end of the axle itself. Then the iron extends horizontally, being rounded and about three digits in diameter, for the length of a foot, and serves as a journal; thence, it bends to a height of a foot in a curve, like the horn of the moon, after which it again extends straight out for one foot; thus it comes about that this last straight portion, as it revolves in an orbit becomes alternately a foot higher and a foot lower than the first straight part. From this round iron crank there hangs the first flat pump-rod, for the crank is fixed in a perforation in the upper end of this flat pump-rod just as the iron key of the first set of "claws" is fixed into the lower end. In order to prevent the pump-rod from slipping off it, as it could easily do, and that it may be taken off when necessary, its opening is wider than the corresponding part of the crank, and it is fastened on both sides by iron keys. To prevent friction, the ends of the pump-rods are protected by iron plates or intervening leathers. This first pump-rod is about twelve feet long, the other two are twenty-six feet, and each is a palm

A—Shaft. B—Bottom pump. C—First tank. D—Second pump. E—Second tank. F—Third pump. G—Trough. H—The iron set in the axle. I—First pump rod. K—Second pump rod. L—Third pump rod. M—First piston rod. N—Second piston rod. O—Third piston rod. P—Little axles. Q—"Claws."

wide and three digits thick. The sides of each pump-rod are covered and protected by iron plates, which are held on by iron screws, so that a part which has received damage can be repaired. In the "claws" is set a small round axle, a foot and a half long and two palms thick. The ends are encircled by iron bands to prevent the iron journals which revolve in the iron bearings of the wood from slipping out of it.[15] From this little axle the wooden "claws" extend two feet, with a width and thickness of six digits; they are three palms distant from each other, and both the inner and outer sides are covered with iron plates. Two rounded iron keys two digits thick are immovably fixed into the claws. The one of these keys perforates the lower end of the first pump-rod, and the upper end of the second pump-rod which is held fast. The other key, which is likewise immovable, perforates the iron end of the first piston-rod, which is bent in a curve and is immovable. Each such piston-rod is thirteen feet long and three digits thick, and descends into the first pipe of each pump to such depth that its disc nearly reaches the valve-box. When it descends into the pipe, the water, penetrating through the openings of the disc, raises the leather, and when the piston-rod is raised the water presses down the leather, and this supports its weight; then the valve closes the box as a door closes an entrance. The pipes are joined by two iron bands, one palm wide, one outside the other, but the inner one is sharp all round that it may fit into each pipe and hold them together. Although at the present time pipes lack the inner band, still they have nipples by which they are joined together, for the lower end of the upper one holds the upper end of the lower one, each being hewn away for a length of seven digits, the former inside, the latter outside, so that the one can fit into the other. When the piston-rod descends into the first pipe, that valve which I have described is closed; when the piston-rod is raised, the valve is opened so that the water can run in through the perforations. Each one of such pumps is composed of two lengths of pipe, each of which is twelve feet long, and the inside diameter is seven digits. The lower one is placed in the sump of the shaft, or in a tank, and its lower end is blocked by a round piece of wood, above which there are six perforations around the pipe through which the water flows into it. The upper part of the upper pipe has a notch one foot deep and a palm wide, through which the water flows away into a tank or trough. Each tank is two feet long and one foot wide and deep. There is the same number of axles, "claws," and rods of each kind as there are pumps; if there are three pumps, there are only two tanks, because the sump of the shaft and the drain of the tunnel take the place of two. The following is the way this machine draws water from a shaft. The wheel being turned raises the first pump-rod, and the pump-rod raises the first "claw," and thus also the second pump-rod, and the first piston-rod; then the second pump-rod raises the second "claw," and thus the third pump-rod and the second piston-rod; then the third pump-rod raises the third "claw" and the third piston-rod,

[15]This description certainly does not correspond in every particular with the illustration.

for there hangs no pump-rod from the iron key of these claws, for it can be of no use in the last pump. In turn, when the first pump-rod descends, each set of "claws" is lowered, each pump-rod and each piston-rod. And by this system, at the same time the water is lifted into the tanks and drained out of them; from the sump at the bottom of the shaft it is drained out, and it is poured into the trough of the tunnel. Further, around the main axle there may be placed two water wheels, if the river supplies enough water to turn them, and from the back part of each round iron crank, one or two pump-rods can be hung, each of which can move the piston-rods of three pumps. Lastly, it is necessary that the shafts from which the water is pumped out in pipes should be vertical, for as in the case of the hauling machines, all pumps which have pipes do not draw the water so high if the pipes are inclined in inclined shafts, as if they are placed vertically in vertical shafts.

If the river does not supply enough water-power to turn the last-described pump, which happens because of the nature of the locality or occurs during the summer season when there are daily droughts, a machine is built with a wheel so low and light that the water of ever so little a

A—WATER WHEEL OF UPPER MACHINE. B—ITS PUMP. C—ITS TROUGH. D—WHEEL OF LOWER MACHINE. E—ITS PUMP. F—RACE.

stream can turn it. This water, falling into a race, runs therefrom on to a second high and heavy wheel of a lower machine, whose pump lifts the water out of a deep shaft. Since, however, the water of so small a stream cannot alone revolve the lower water-wheel, the axle of the latter is turned at the start with a crank worked by two men, but as soon as it has poured out into a pool the water which has been drawn up by the pumps, the upper wheel draws up this water by its own pump, and pours it into the race, from which it flows on to the lower water-wheel and strikes its buckets. So both this water from the mine, as well as the water of the stream, being turned down the races on to that subterranean wheel of the lower machine, turns it, and water is pumped out of the deeper part of the shaft by means of two or three pumps.[16]

If the stream supplies enough water straightway to turn a higher and heavier water-wheel, then a toothed drum is fixed to the other end of the axle, and this turns the drum made of rundles on another axle set below it. To each end of this lower axle there is fitted a crank of round iron curved like the horns of the moon, of the kind employed in machines of this description. This machine, since it has rows of pumps on each side, draws great quantities of water.

Of the rag and chain pumps there are six kinds known to us, of which the first is made as follows: A cave is dug under the surface of earth or in a tunnel, and timbered on all sides by stout posts and planks, to prevent either the men from being crushed or the machine from being broken by its collapse. In this cave, thus timbered, is placed a water-wheel fitted to an angular axle. The iron journals of the axle revolve in iron pillows, which are held in timbers of sufficient strength. The wheel is generally twenty-four feet high, occasionally thirty, and in no way different from those which are made for grinding corn, except that it is a little narrower. The axle has on one side a drum with a groove in the middle of its circumference, to which are fixed many four-curved iron clamps. In these clamps catch the links of the chain, which is drawn through the pipes out of the sump, and which again falls, through a timbered opening, right down to the bottom into the sump to a balancing drum. There is an iron band around the small axle of the balancing drum, each journal of which revolves in an iron bearing fixed to a timber. The chain turning about this drum brings up the water by the balls through the pipes. Each length of pipe is encircled and protected by five iron bands, a palm wide and a digit thick, placed at equal distances from each other; the first band on the pipe is shared in common with the preceding length of pipe into which it is fitted, the last band with the succeeding length of pipe which is fitted into it. Each length of pipe, except the first, is bevelled on the outer circumference of the upper end to a distance of seven digits and for a depth of three digits, in order that it may be inserted into the length of pipe which goes before it; each, except the last, is reamed out on the inside of the lower end to a like distance, but to the depth

[16]There is a certain deficiency in the hydraulics of this machine.

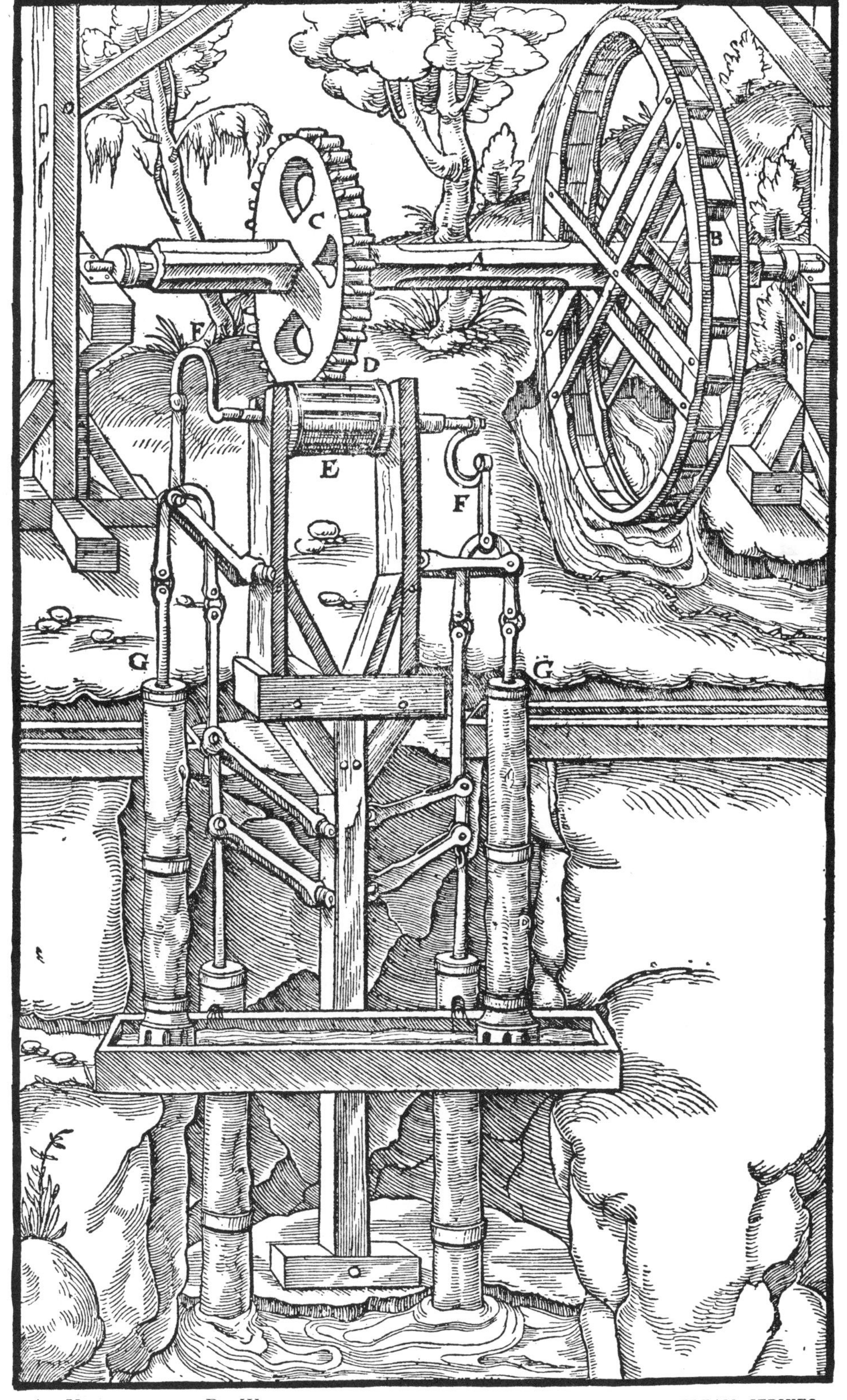

A—Upper axle. B—Wheel whose buckets the force of the stream strikes. C—Toothed drum. D—Second axle. E—Drum composed of rundles. F—Curved round irons. G—Rows of pumps.

of a palm, that it may be able to take the end of the pipe which follows. And each length of pipe is fixed with iron clamps to the timbers of the shaft, that it may remain stationary. Through this continuous series of pipes, the water is drawn by the balls of the chain up out of the sump as far as the tunnel, where it flows out into the drains through an aperture in the highest pipe. The balls which lift the water are connected by the iron links of the chain, and are six feet distant from one another ; they are made of the hair of a horse's tail sewn into a covering to prevent it from being pulled out by the iron clamps on the drum ; the balls are of such size that one can be held in each hand. If this machine is set up on the surface of the earth, the stream which turns the water-wheel is led away through open-air ditches ; if in a tunnel, the water is led away through the subterranean drains. The buckets of the water-wheel, when struck by the impact of the stream, move forward and turn the wheel, together with the drum, whereby the chain is wound up and the balls expel the water through the pipes. If the wheel of this machine is twenty-four feet in diameter, it draws water from a shaft two hundred and ten feet deep ; if thirty feet in diameter, it will draw water from a shaft two hundred and forty feet deep. But such work requires a stream with greater water-power.

The next pump has two drums, two rows of pipes and two drawing-chains whose balls lift out the water ; otherwise they are like the last pump. This pump is usually built when an excessive amount of water flows into the sump. These two pumps are turned by water-power ; indeed, water draws water.

The following is the way of indicating the increase or decrease of the water in an underground sump, whether it is pumped by this rag and chain pump or by the first pump, or the third, or some other. From a beam which is as high above the shaft as the sump is deep, is hung a cord, to one end of which there is fastened a stone, the other end being attached to a plank. The plank is lowered down by an iron wire fastened to the other end ; when the stone is at the mouth of the shaft the plank is right down the shaft in the sump, in which water it floats. This plank is so heavy that it can drag down the wire and its iron clasp and hook, together with the cord, and thus pull the stone upwards. Thus, as the water decreases, the plank decends and the stone is raised ; on the contrary, when the water increases the plank rises and the stone is lowered. When the stone nearly touches the beam, since this indicates that the water has been exhausted from the sump by the pump, the overseer in charge of the machine closes the water-race and stops the water-wheel : when the stone nearly touches the ground at the side of the shaft, this indicates that the sump is full of water which has again collected in it, because the water raises the plank and thus the stone drags back both the rope and the iron wire ; then the overseer opens the water-race, whereupon the water of the stream again strikes the buckets of the water-wheel and turns the pump. As workmen generally cease from their labours on the yearly holidays, and

A—Wheel. B—Axle. C—Journals. D—Pillows. E—Drum. F—Clamps. G—Drawing-chain. H—Timbers. I—Balls. K—Pipe. L—Race of stream.

sometimes on working days, and are thus not always near the pump, and as the pump, if necessary, must continue to draw water all the time, a bell rings aloud continuously, indicating that this pump, or any other kind, is uninjured and nothing is preventing its turning. The bell is hung by a cord from a small wooden axle held in the timbers which stand over the shaft, and a second long cord whose upper end is fastened to the small axle is lowered into the shaft; to the lower end of this cord is fastened a piece of wood; and as often as a cam on the main axle strikes it, so often does the bell ring and give forth a sound.

The third pump of this kind is employed by miners when no river capable of turning a water-wheel can be diverted, and it is made as follows. They first dig a chamber and erect strong timbers and planks to prevent the sides from falling in, which would overwhelm the pump and kill the men. The roof of the chamber is protected with contiguous timbers, so arranged that the horses which pull the machine can travel over it. Next they again set up sixteen beams forty feet long and one foot wide and thick, joined by clamps at the top and spreading apart at the bottom, and they fit the lower end of each beam into a separate sill laid flat on the ground, and join these by a post; thus there is created a circular area of which the diameter is fifty feet. Through an opening in the centre of this area there descends an upright square axle, forty-five feet long and a foot and a half wide and thick; its lower pivot revolves in a socket in a block laid flat on the ground in the chamber, and the upper pivot revolves in a bearing in a beam which is mortised into two beams at the summit beneath the clamps; the lower pivot is seventeen feet distant from either side of the chamber, *i.e.*, from its front and rear. At the height of a foot above its lower end, the axle has a toothed wheel, the diameter of which is twenty-two feet. This wheel is composed of four spokes and eight rim pieces; the spokes are fifteen feet long and three-quarters of a foot wide and thick[17]; one end of them is mortised in the axle, the other in the two rims where they are joined together. These rims are three-quarters of a foot thick and one foot wide, and from them there rise and project upright teeth three-quarters of a foot high, half a foot wide, and six digits thick. These teeth turn a second horizontal axle by means of a drum composed of twelve rundles, each three feet long and six digits wide and thick. This drum, being turned, causes the axle to revolve, and around this axle there is a drum having iron clamps with four-fold curves in which catch the links of a chain, which draws water through pipes by means of balls. The iron journals of this horizontal axle revolve on pillows which are set in the centre of timbers. Above the roof of the chamber there are mortised into the upright axle the ends of two beams which rise obliquely; the upper ends of these beams support double cross-beams, likewise mortised to the axle. In the outer end of each cross-beam there is mortised a small wooden piece which appears to hang down; in this wooden piece there is similarly

[17]The dimensions given in this description for the various members do not tally.

A—Upright axle. B—Toothed wheel. C—Teeth. D—Horizontal axle. E—Drum which is made of rundles. F—Second drum. G—Drawing-chain. H—The balls.

mortised at the lower end a short board ; this has an iron key which engages a chain, and this chain again a pole-bar. This machine, which draws water from a shaft two hundred and forty feet deep, is worked by thirty-two horses ; eight of them work for four hours, and then these rest for twelve hours, and the same number take their place. This kind of machine is employed at the foot of the Harz[18] mountains and in the neighbourhood. Further, if necessity arises, several pumps of this kind are often built for the purpose of mining one vein, but arranged differently in different localities varying according to the depth. At Schemnitz, in the Carpathian mountains, there are three pumps, of which the lowest lifts water from the lowest sump to the first drains, through which it flows into the second sump ; the intermediate one lifts from the second sump to the second drain, from which it flows into the third sump ; and the upper one lifts it to the drains of the tunnel, through which it flows away. This system of three machines of this kind is turned by ninety-six horses ; these horses go down to the machines by an inclined

A—Axle. B—Drum. C—Drawing-chain. D—Balls. E—Clamps.

[18]*Melibocian*,—the Harz.

shaft, which slopes and twists like a screw and gradually descends. The lowest of these machines is set in a deep place, which is distant from the surface of the ground 660 feet.

The fourth species of pump belongs to the same genera, and is made as follows. Two timbers are erected, and in openings in them, the ends of a barrel revolve. Two or four strong men turn the barrel, that is to say, one or two pull the cranks, and one or two push them, and in this way help the others ; alternately another two or four men take their place. The barrel of this machine, just like the horizontal axle of the other machines, has a drum whose iron clamps catch the links of a drawing-chain. Thus water is drawn through the pipes by the balls from a depth of forty-eight feet. Human strength cannot draw water higher than this, because such very heavy labour exhausts not only men, but even horses ; only water-power can drive continuously a drum of this kind. Several pumps of this kind, as of the last, are often built for the purpose of mining on a single vein, but they are arranged differently for different positions and depths.

A—Axles. B—Levers. C—Toothed drum. D—Drum made of rundles. E—Drum in which iron clamps are fixed.

The fifth pump of this kind is partly like the third and partly like the fourth, because it is turned by strong men like the last, and like the third it has two axles and three drums, though each axle is horizontal. The journals of each axle are so fitted in the pillows of the beams that they cannot fly out ; the lower axle has a crank at one end and a toothed drum at the other end ; the upper axle has at one end a drum made of rundles, and at the other end, a drum to which are fixed iron clamps, in which the links of a chain catch in the same way as before, and from the same depth, draw water through pipes by means of balls. This revolving machine is turned by two pairs of men alternately, for one pair stands working while the other sits taking a rest ; while they are engaged upon the task of turning, one pulls the crank and the other pushes, and the drums help to make the pump turn more easily.

The sixth pump of this kind likewise has two axles. At one end of the lower axle is a wheel which is turned by two men treading, this is twenty-three feet high and four feet wide, so that one man may stand alongside the other. At the other end of this axle is a toothed wheel. The upper[19] axle has two drums and one wheel ; the first drum is made of rundles, and to the other there are fixed the iron clamps. The wheel is like the one on the second machine which is chiefly used for drawing earth and broken rock out of shafts. The treaders, to prevent themselves from falling, grasp in their hands poles which are fixed to the inner sides of the wheel. When they turn this wheel, the toothed drum being made to revolve, sets in motion the other drum which is made of rundles, by which means again the links of the chain catch to the cleats of the third drum and draw water through pipes by means of balls,—from a depth of sixty-six feet.

But the largest machine of all those which draw water is the one which follows. First of all a reservoir is made in a timbered chamber ; this reservoir is eighteen feet long and twelve feet wide and high. Into this reservoir a stream is diverted through a water-race or through the tunnel ; it has two entrances and the same number of gates. Levers are fixed to the upper part of these gates, by which they can be raised and let down again, so that by one way the gates are opened and in the other way closed. Beneath the openings are two plank troughs which carry the water flowing from the reservoir, and pour it on to the buckets of the water-wheel, the impact of which turns the wheel. The shorter trough carries the water, which strikes the buckets that turn the wheel toward the reservoir, and the longer trough carries the water which strikes those buckets that turn the wheel in the opposite direction. The casing or covering of the wheel is made of joined boards to which strips are affixed on the inner side. The wheel itself is thirty-six feet in diameter, and is mortised to an axle, and it has, as I have already said, two rows of buckets, of which one is set the opposite way to the other, so that the wheel may be turned toward the reservoir or in the opposite

[19]In the original text this is given as " lower," and appears to be an error.

A—Axles. B—Wheel which is turned by treading. C—Toothed wheel. D—Drum made of rundles. E—Drum to which are fixed iron clamps. F—Second wheel. G—Balls.

direction. The axle is square and is thirty-five feet long and two feet thick and wide. Beyond the wheel, at a distance of six feet, the axle has four hubs, one foot wide and thick, each one of which is four feet distant from the next; to these hubs are fixed by iron nails as many pieces of wood as are necessary to cover the hubs, and, in order that the wood pieces may fit tight, they are broader on the outside and narrower on the inside; in this way a drum is made, around which is wound a chain to whose ends are hooked leather bags. The reason why a drum of this kind is made, is that the axle may be kept in good condition, because this drum when it becomes worn away by use can be repaired easily. Further along the axle, not far from the end, is another drum one foot broad, projecting two feet on all sides around the axle. And to this, when occasion demands, a brake is applied forcibly and holds back the machine; this kind of brake I have explained before. Near the axle, in place of a hopper, there is a floor with a considerable slope, having in front of the shaft a width of fifteen feet and the same at the back; at each side of it there is a stout post carrying an iron chain which has a large hook. Five men operate this machine; one lets down the doors which close the reservoir gates, or by drawing down the levers, opens the water-races; this man, who is the director of this machine, stands in a hanging cage beside the reservoir. When one bag has been drawn out nearly as far as the sloping floor, he closes the water gate in order that the wheel may be stopped; when the bag has been emptied he opens the other water gate, in order that the other set of buckets may receive the water and drive the wheel in the opposite direction. If he cannot close the water-gate quickly enough, and the water continues to flow, he calls out to his comrade and bids him raise the brake upon the drum and stop the wheel. Two men alternately empty the bags, one standing on that part of the floor which is in front of the shaft, and the other on that part which is at the back. When the bag has been nearly drawn up—of which fact a certain link of the chain gives warning—the man who stands on the one part of the floor, catches a large iron hook in one link of the chain, and pulls out all the subsequent part of the chain toward the floor, where the bag is emptied by the other man. The object of this hook is to prevent the chain, by its own weight, from pulling down the other empty bag, and thus pulling the whole chain from its axle and dropping it down the shaft. His comrade in the work, seeing that the bag filled with water has been nearly drawn out, calls to the director of the machine and bids him close the water of the tower so that there will be time to empty the bag; this being emptied, the director of the machine first of all slightly opens the other water-gate of the tower to allow the end of the chain, together with the empty bag, to be started into the shaft again, and then opens entirely the water-gates. When that part of the chain which has been pulled on to the floor has been wound up again, and has been let down over the shaft from the drum, he takes out the large hook which was fastened into a link of the chain. The fifth man stands in a sort of cross-cut beside the sump, that he may not be hurt, if it should happen that a link

A—Reservoir. B—Race. C, D—Levers. E, F—Troughs under the water gates. G, H—Double rows of buckets. I—Axle. K—Larger drum. L—Drawing-chain. M—Bag. N—Hanging cage. O—Man who directs the machine. P, Q—Men emptying bags.

is broken and part of the chain or anything else should fall down ; he guides the bag with a wooden shovel, and fills it with water if it fails to take in the water spontaneously. In these days, they sew an iron band into the top of each bag that it may constantly remain open, and when lowered into the sump may fill itself with water, and there is no need for a man to act as governor of the bags. Further, in these days, of those men who stand on the floor the one empties the bags, and the other closes the gates of the reservoir and opens them again, and the same man usually fixes the large hook in the link of the chain. In this way, three men only are employed in working this machine; or even—since sometimes the one who empties the bag presses the brake which is raised against the other drum and thus stops the wheel—two men take upon themselves the whole labour.

But enough of haulage machines ; I will now speak of ventilating machines. If a shaft is very deep and no tunnel reaches to it, or no drift from another shaft connects with it, or when a tunnel is of great length and no shaft reaches to it, then the air does not replenish itself. In such a case it weighs heavily on the miners, causing them to breathe with difficulty, and sometimes they are even suffocated, and burning lamps are also extinguished. There is, therefore, a necessity for machines which the Greeks call πνευματικάι and the Latins *spiritales*—though they do not give forth any sound—which enable the miners to breathe easily and carry on their work.

These devices are of three genera. The first receives and diverts into the shaft the blowing of the wind, and this genus is divided into three species, of which the first is as follows. Over the shaft—to which no tunnel connects—are placed three sills a little longer than the shaft, the first over the front, the second over the middle, and the third over the back of the shaft. Their ends have openings, through which pegs, sharpened at the bottom, are driven deeply into the ground so as to hold them immovable, in the same way that the sills of the windlass are fixed. Each of these sills is mortised into each of three cross-beams, of which one is at the right side of the shaft, the second at the left, and the third in the middle. To the second sill and the second cross-beam—each of which is placed over the middle of the shaft—planks are fixed which are joined in such a manner that the one which precedes always fits into the groove of the one which follows. In this way four angles and the same number of intervening hollows are created, which collect the winds that blow from all directions. The planks are roofed above with a cover made in a circular shape, and are open below, in order that the wind may not be diverted upward and escape, but may be carried downward ; and thereby the winds of necessity blow into the shafts through these four openings. However, there is no need to roof this kind of machine in those localities in which it can be so placed that the wind can blow down through its topmost part.

A—Sills. B—Pointed stakes. C—Cross-beams. D—Upright Planks. E—Hollows. F—Winds. G—Covering disc. H—Shafts. I—Machine without a covering.

The second machine of this genus turns the blowing wind into a shaft through a long box-shaped conduit, which is made of as many lengths of planks, joined together, as the depth of the shaft requires ; the joints are smeared with fat, glutinous clay moistened with water. The mouth of this conduit either projects out of the shaft to a height of three or four feet, or it does not project ; if it projects, it is shaped like a rectangular funnel, broader and wider at the top than the conduit itself, that it may the more easily gather the wind ; if it does not project, it is not broader than the conduit, but planks are fixed to it away from the direction in which the wind is blowing, which catch the wind and force it into the conduit.

The third of this genus of machine is made of a pipe or pipes and a barrel. Above the uppermost pipe there is erected a wooden barrel, four

A—Projecting mouth of conduit. B—Planks fixed to the mouth of the conduit which does not project.

feet high and three feet in diameter, bound with wooden hoops ; it has a square blow-hole always open, which catches the breezes and guides them down either by a pipe into a conduit or by many pipes into the shaft. To the top of the upper pipe is attached a circular table as thick as the bottom of the barrel, but of a little less diameter, so that the barrel may be turned around on it ; the pipe projects out of the table and is fixed in a round opening in the centre of the bottom of the barrel. To the end of the pipe a perpendicular axle is fixed which runs through the centre of the barrel into a hole in the cover, in which it is fastened, in the same way as at the bottom. Around this fixed axle and the table on the pipe, the movable barrel is easily turned by a zephyr, or much more by a wind, which govern the wing on it. This wing is made of thin boards and fixed to the upper part of the barrel on the side furthest away from the blow-hole ; this, as I have said, is square and always open. The wind, from whatever quarter of the world it blows, drives the wing straight toward the opposite direction, in which way the barrel turns the blow-hole towards the wind itself; the blow-hole receives the wind, and it is guided down into the shaft by means of the conduit or pipes.

A—WOODEN BARRELS. B—HOOPS. C—BLOW-HOLES. D—PIPE. E—TABLE. F—AXLE. G—OPENING IN THE BOTTOM OF THE BARREL. H—WING.

The second genus of blowing machine is made with fans, and is likewise varied and of many forms, for the fans are either fitted to a windlass barrel or to an axle. If to an axle, they are either contained in a hollow drum, which is made of two wheels and a number of boards joining them together, or else in a box-shaped casing. The drum is stationary and closed on the sides, except for round holes of such size that the axle may turn in them; it has two square blow-holes, of which the upper one receives the air, while the lower one empties into the conduit through which the air is led down the shaft. The ends of the axle, which project on each side of the drum, are supported by forked posts or hollowed beams plated with thick iron; one end of the axle has a crank, while in the other end are fixed four rods with thick heavy ends, so that they weight the axle, and when turned, make it

A—Drum. B—Box-shaped casing. C—Blow-hole. D—Second hole.
E—Conduit. F—Axle. G—Lever of axle. H—Rods.

prone to motion as it revolves. And so, when the workman turns the axle by the crank, the fans, the description of which I will give a little later, draw in the air by the blow-hole, and force it through the other blow-hole which leads to the conduit, and through this conduit the air penetrates into the shaft.

The one with the box-shaped casing is furnished with just the same things as the drum, but the drum is far superior to the box ; for the fans so fill the drum that they almost touch it on every side, and drive into the conduit all the air that has been accumulated ; but they cannot thus fill the box-shaped casing, on account of its angles, into which the air partly retreats ; therefore it cannot be as useful as the drum. The kind with a box-shaped casing is not only placed on the ground, but is also set up on timbers like a windmill, and its axle, in place of a crank, has four sails outside, like the sails of a windmill. When these are struck by the wind they turn the axle, and in this way its fans—which are placed within the casing—drive

A—Box-shaped casing placed on the ground. B—Its blow-hole. C—Its axle with fans. D—Crank of the axle. E—Rods of same. F—Casing set on timbers. G—Sails which the axle has outside the casing.

the air through the blow-hole and the conduit into the shaft. Although this machine has no need of men whom it is necessary to pay to work the crank, still when the sky is devoid of wind, as it often is, the machine does not turn, and it is therefore less suitable than the others for ventilating a shaft.

In the kind where the fans are fixed to an axle, there is generally a hollow stationary drum at one end of the axle, and on the other end is fixed a drum made of rundles. This rundle drum is turned by the toothed wheel of a lower axle, which is itself turned by a wheel whose buckets receive the impetus of water. If the locality supplies an abundance of water this machine is most useful, because to turn the crank does not need men who require pay, and because it forces air without cessation through the conduit into the shaft.

A—Hollow drum. B—Its blow-hole. C—Axle with fans. D—Drum which is made of rundles. E—Lower axle. F—Its toothed wheel. G—Water wheel.

Of the fans which are fixed on to an axle contained in a drum or box, there are three sorts. The first sort is made of thin boards of such length and width as the height and width of the drum or box require ; the second sort is made of boards of the same width, but shorter, to which are bound long thin blades of poplar or some other flexible wood ; the third sort has boards like the last, to which are bound double and triple rows of goose feathers. This last is less used than the second, which in turn is less used than the first. The boards of the fan are mortised into the quadrangular parts of the barrel axle.

A—First kind of fan. B—Second kind of fan. C—Third kind of fan. D—Quadrangular part of axle. E—Round part of same. F—Crank.

Blowing machines of the third genus, which are no less varied and of no fewer forms than those of the second genus, are made with bellows, for by its blasts the shafts and tunnels are not only furnished with air through conduits or pipes, but they can also be cleared by suction of their heavy and pestilential vapours. In the latter case, when the bellows is opened it draws the vapours from the conduits through its blow-hole and sucks these vapours into itself ; in the former case, when it is compressed, it drives the air through its nozzle into the conduits or pipes. They are compressed either by a man, or by a horse or by water-power; if by a man, the lower board of a large bellows is fixed to the timbers above the conduit which projects out of the shaft, and so placed that when the blast is blown through the conduit, its nozzle is set in the conduit. When it is desired to suck out heavy or pestilential vapours, the blow-hole of the bellows is fitted all round the mouth of the conduit. Fixed to the upper bellows board is a lever which couples with another running downward from a little axle, into which it is mortised so that it may remain immovable; the iron journals of this little axle revolve in openings of upright posts; and so when the workman pulls down the lever the upper board of the bellows is raised, and at the same time the flap of the blow-hole is dragged open by the force of the wind. If the nozzle of the bellows is enclosed in the conduit it draws pure air into itself, but if its blow-hole is fitted all round the mouth of the conduit it exhausts the heavy and pestilential vapours out of the conduit and thus from the shaft, even if it is one hundred and twenty feet deep. A stone placed on the upper board of the bellows depresses it and then the flap of the blow-hole is

A—Smaller part of shaft. B—Square conduit. C—Bellows. D—Larger part of shaft.

closed. The bellows, by the first method, blows fresh air into the conduit through its nozzle, and by the second method blows out through the nozzle the heavy and pestilential vapours which have been collected. In this latter case fresh air enters through the larger part of the shaft, and the miners getting the benefit of it can sustain their toil. A certain smaller part of the shaft which forms a kind of estuary, requires to be partitioned off from the other larger part by uninterrupted lagging, which reaches from the top of the shaft to the bottom ; through this part the long but narrow conduit reaches down nearly to the bottom of the shaft.

When no shaft has been sunk to such depth as to meet a tunnel driven far into a mountain, these machines should be built in such a manner that the workman can move them about. Close by the drains of the tunnel through which the water flows away, wooden pipes should be placed and joined tightly together in such a manner that they can hold the air ; these should reach from the mouth of the tunnel to its furthest end. At the mouth of the tunnel the bellows should be so placed that through its nozzle it can blow its accumulated blasts into the pipes or the conduit ; since one blast

A—Tunnel. B—Pipe. C—Nozzle of double bellows.

always drives forward another, they penetrate into the tunnel and change the air, whereby the miners are enabled to continue their work.

If heavy vapours need to be drawn off from the tunnels, generally three double or triple bellows, without nozzles and closed in the forepart, are placed upon benches. A workman compresses them by treading with his feet, just as persons compress those bellows of the organs which give out varied and sweet sounds in churches. These heavy vapours are thus drawn along the air-pipes and through the blow-hole of the lower bellows board, and are expelled through the blow-hole of the upper bellows board into the open air, or into some shaft or drift. This blow-hole has a flap-valve, which the noxious blast opens, as often as it passes out. Since one volume of air constantly rushes in to take the place of another which has been drawn out by the bellows, not only is the heavy air drawn out of a tunnel as great as 1,200 feet long, or even longer, but also the wholesome air is naturally drawn in through that part of the tunnel which is open outside the conduits. In this way the air is changed, and the miners are enabled to carry on the work they have begun. If machines of this kind had not been invented, it would be necessary for miners to drive two tunnels into a mountain, and continually, at every two hundred feet at most, to sink a shaft from the upper tunnel to the lower one, that the air passing into the one, and descending by the shafts into the other, would be kept fresh for the miners ; this could not be done without great expense.

There are two different machines for operating, by means of horses, the above described bellows. The first of these machines has on its axle a wooden wheel, the rim of which is covered all the way round by steps ; a horse is kept continually within bars, like those within which horses are held to be shod with iron, and by treading these steps with its feet it turns the wheel, together with the axle ; the cams on the axle press down the sweeps which compress the bellows. The way the instrument is made which raises the bellows again, and also the benches on which the bellows rest, I will explain more clearly in Book IX. Each bellows, if it draws heavy vapours out of a tunnel, blows them out of the hole in the upper board ; if they are drawn out of a shaft, it blows them out through its nozzle. The wheel has a round hole, which is transfixed with a pole when the machine needs to be stopped.

The second machine has two axles ; the upright one is turned by a horse, and its toothed drum turns a drum made of rundles on a horizontal axle ; in other respects this machine is like the last. Here, also, the nozzles of the bellows placed in the conduits blow a blast into the shaft or tunnel.

In the same way that this last machine can refresh the heavy air of a shaft or tunnel, so also could the old system of ventilating by the constant shaking of linen cloths, which Pliny [20] has explained ; the air not only grows

[20]Pliny (xxxi, 28). "In deep wells, the occurrence of *sulphurata* or *aluminosa* "vapor is fatal to the diggers. The presence of this peril is shown if a lighted lamp let down "into the well is extinguished. If so, other wells are sunk to the right and left, which carry "off these noxious gases. Apart from these evils, the air itself becomes noxious with depth, "which can be remedied by constantly shaking linen cloths, thus setting the air in motion."

A—Machine first described. B—This workman, treading with his feet, is compressing the bellows. C—Bellows without nozzles. D—Hole by which heavy vapours or blasts are blown out. E—Conduits. F—Tunnel. G—Second machine described. H—Wooden wheel. I—Its steps. K—Bars. L—Hole in same wheel. M—Pole. N—Third machine described. O—Upright axle. P—Its toothed drum. Q—Horizontal axle. R—Its drum which is made of rundles.

A—TUNNEL. B—LINEN CLOTH.

heavier with the depth of a shaft, of which fact he has made mention, but also with the length of a tunnel.

The climbing machines of miners are ladders, fixed to one side of the shaft, and these reach either to the tunnel or to the bottom of the shaft. I need not describe how they are made, because they are used everywhere, and need not so much skill in their construction as care in fixing them. However, miners go down into mines not only by the steps of ladders, but they are also lowered into them while sitting on a stick or a wicker basket, fastened to the rope of one of the three drawing machines which I described at first. Further, when the shafts are much inclined, miners and other workmen sit in the dirt which surrounds their loins and slide down in the same way that boys do in winter-time when the water on some hillside has congealed with the cold, and to prevent themselves from falling, one arm is wound about a rope, the upper end of which is fastened to a beam at the mouth of the shaft, and the lower end to a stake fixed in the bottom of the shaft. In these three ways miners descend into the shafts. A fourth way may be mentioned which is employed when men and horses go down to the underground

A—Descending into the shaft by ladders. B—By sitting on a stick. C—By sitting on the dirt. D—Descending by steps cut in the rock.

machines and come up again, that is by inclined shafts which are twisted like a screw and have steps cut in the rock, as I have already described.

It remains for me to speak of the ailments and accidents of miners, and of the methods by which they can guard against these, for we should always devote more care to maintaining our health, that we may freely perform our bodily functions, than to making profits. Of the illnesses, some affect the joints, others attack the lungs, some the eyes, and finally some are fatal to men.

Where water in shafts is abundant and very cold, it frequently injures the limbs, for cold is harmful to the sinews. To meet this, miners should make themselves sufficiently high boots of rawhide, which protect their legs from the cold water ; the man who does not follow this advice will suffer much ill-health, especially when he reaches old age. On the other hand, some mines are so dry that they are entirely devoid of water, and this dryness causes the workmen even greater harm, for the dust which is stirred and beaten up by digging penetrates into the windpipe and lungs, and produces difficulty in breathing, and the disease which the Greeks call ἄσθμα. If the dust has corrosive qualities, it eats away the lungs, and implants consumption in the body ; hence in the mines of the Carpathian Mountains women are found who have married seven husbands, all of whom this terrible consumption has carried off to a premature death. At Altenberg in Meissen there is found in the mines black *pompholyx*, which eats wounds and ulcers to the bone ; this also corrodes iron, for which reason the keys of their sheds are made of wood. Further, there is a certain kind of *cadmia* [21] which eats away the feet of the workmen when they have become wet, and similarly their hands, and injures their lungs and eyes. Therefore, for their

[21]This is given in the German translation as *kobelt.* The *kobelt* (or *cobaltum* of Agricola) was probably arsenical-cobalt, a mineral common in the Saxon mines. The origin of the application of the word cobalt to a mineral appears to lie in the German word for the gnomes and goblins (*kobelts*) so universal to Saxon miners' imaginations,—this word in turn probably being derived from the Greek *cobali* (mimes). The suffering described above seems to have been associated with the malevolence of demons, and later the word for these demons was attached to this disagreeable ore. A quaint series of mining " sermons," by Johann Mathesius, entitled *Sarepta oder Bergpostill,* Nürnberg, 1562, contains the following passage (p. 154) which bears out this view. We retain the original and varied spelling of cobalt and also add another view of Mathesius, involving an experience of Solomon and Hiram of Tyre with some mines containing cobalt.

" Sometimes, however, from dry hard veins a certain black, greenish, grey or ash-" coloured earth is dug out, often containing good ore, and this mineral being burnt gives strong " fumes and is extracted like ' tutty.' It is called *cadmia fossilis.* You miners call it *cobelt.* " Germans call the Black Devil and the old Devil's furies, old and black *cobel,* who injure people " and their cattle with their witchcrafts. Now the Devil is a wicked, malicious spirit, who " shoots his poisoned darts into the hearts of men, as sorcerers and witches shoot at the limbs " of cattle and men, and work much evil and mischief with *cobalt* or *hipomane* or horses' " poison. After quicksilver and *rotgültigen* ore, are *cobalt* and *wismuth* fumes ; these are the " most poisonous of the metals, and with them one can kill flies, mice, cattle, birds, and men. " So, fresh *cobalt* and *kisswasser* (vitriol ?) devour the hands and feet of miners, and the dust " and fumes of *cobalt* kill many mining people and workpeople who do much work among the " fumes of the smelters. Whether or not the Devil and his hellish crew gave their name to " *cobelt,* or *kobelt,* nevertheless, *cobelt* is a poisonous and injurious metal even if it contains " silver. I find in I. Kings 9, the word *Cabul.* When Solomon presented twenty towns in " Galilee to the King of Tyre, Hiram visited them first, and would not have them, and said the " land was well named *Cabul* as Joshua had christened it. It is certain from Joshua that these

digging they should make for themselves not only boots of rawhide, but gloves long enough to reach to the elbow, and they should fasten loose veils over their faces ; the dust will then neither be drawn through these into their windpipes and lungs, nor will it fly into their eyes. Not dissimilarly, among the Romans[22] the makers of vermilion took precautions against breathing its fatal dust.

Stagnant air, both that which remains in a shaft and that which remains in a tunnel, produces a difficulty in breathing ; the remedies for this evil are the ventilating machines which I have explained above. There is another illness even more destructive, which soon brings death to men who work in those shafts or levels or tunnels in which the hard rock is broken by fire. Here the air is infected with poison, since large and small veins and seams in the rocks exhale some subtle poison from the minerals, which is driven out by the fire, and this poison itself is raised with the smoke not unlike *pompholyx*,[23] which clings to the upper part of the walls in the works in which ore is smelted. If this poison cannot escape from the ground, but falls down into the pools and floats on their surface, it often causes danger, for if at any time the water is disturbed through a stone or anything else, these fumes rise again from the pools and thus overcome the men, by being drawn in with their breath ; this is even much worse if the fumes of the fire have not yet all escaped. The bodies of living creatures who are infected with this poison generally swell immediately and lose all movement and feeling, and they die without pain ; men even in the act of climbing from the shafts by the steps of ladders fall back into the shafts when the poison overtakes them, because their hands do not perform their office, and seem to them to be round and spherical, and likewise their feet. If by good fortune the injured ones escape these evils, for a little while they are pale and look like dead men. At such times, no one should descend into the mine or into the neighbouring mines, or if he is in them he should come out quickly. Prudent and skilled miners burn the piles of wood on Friday, towards evening, and

" twenty towns lay in the Kingdom of Aser, not far from our *Sarepta*, and that there had been " iron and copper mines there, as Moses says in another place. Inasmuch, then, as these twenty " places were mining towns, and *cobelt* is a metal, it appears quite likely that the mineral took " its name from the land of Cabul. History and circumstances bear out the theory that Hiram " was an excellent and experienced miner, who obtained much gold from Ophir, with which he " honoured Solomon. Therefore, the Great King wished to show his gratitude to his good " neighbour by honouring a miner with mining towns. But because the King of Tyre was " skilled in mines, he first inspected the new mines, and saw that they only produced poor " metal and much wild *cobelt* ore, therefore he preferred to find his gold by digging the gold " and silver in India rather than by getting it by the *cobelt* veins and ore. For truly, *cobelt* " ores are injurious, and are usually so embedded in other ore that they rob them in the " fire and consume (*madtet und frist*) much lead before the silver is extracted, and when this " happens it is especially *speysig*. Therefore Hiram made a good reckoning as to the mines " and would not undertake all the expense of working and smelting, and so returned Solomon " the twenty towns."

[22]Pliny (XXXIII, 40). " Those employed in the works preparing vermilion, cover " their faces with a bladder-skin, that they may not inhale the pernicious powder, yet they " can see through the skin."

[23]*Pompholyx* was a furnace deposit, usually mostly zinc oxide, but often containing arsenical oxide, and to this latter quality this reference probably applies. The symptoms mentioned later in the text amply indicate arsenical poisoning, of which a sort of spherical effect on the hands is characteristic. See also note on p. 112 for discussion of " corrosive " *cadmia* ; further information on *pompholyx* is given in Note 26, p. 394.

they do not descend into the shafts nor enter the tunnels again before Monday, and in the meantime the poisonous fumes pass away.

There are also times when a reckoning has to be made with Orcus,[24] for some metalliferous localities, though such are rare, spontaneously produce poison and exhale pestilential vapour, as is also the case with some openings in the ore, though these more often contain the noxious fumes. In the towns of the plains of Bohemia there are some caverns which, at certain seasons of the year, emit pungent vapours which put out lights and kill the miners if they linger too long in them. Pliny, too, has left a record that when wells are sunk, the sulphurous or aluminous vapours which arise kill the well-diggers, and it is a test of this danger if a burning lamp which has been let down is extinguished. In such cases a second well is dug to the right or left, as an air-shaft, which draws off these noxious vapours. On the plains they construct bellows which draw up these noxious vapours and remedy this evil ; these I have described before.

Further, sometimes workmen slipping from the ladders into the shafts break their arms, legs, or necks, or fall into the sumps and are drowned ; often, indeed, the negligence of the foreman is to blame, for it is his special work both to fix the ladders so firmly to the timbers that they cannot break away, and to cover so securely with planks the sumps at the bottom of the shafts, that the planks cannot be moved nor the men fall into the water ; wherefore the foreman must carefully execute his own work. Moreover, he must not set the entrance of the shaft-house toward the north wind, lest in winter the ladders freeze with cold, for when this happens the men's hands become stiff and slippery with cold, and cannot perform their office of holding. The men, too, must be careful that, even if none of these things happen, they do not fall through their own carelessness.

Mountains, too, slide down and men are crushed in their fall and perish. In fact, when in olden days Rammelsberg, in Goslar, sank down, so many men were crushed in the ruins that in one day, the records tell us, about 400 women were robbed of their husbands. And eleven years ago, part of the mountain of Altenberg, which had been excavated, became loose and sank, and suddenly crushed six miners ; it also swallowed up a hut and one mother and her little boy. But this generally occurs in those mountains which contain *venae cumulatae*. Therefore, miners should leave numerous arches under the mountains which need support, or provide underpinning. Falling pieces of rock also injure their limbs, and to prevent this from happening, miners should protect the shafts, tunnels, and drifts.

The venomous ant which exists in Sardinia is not found in our mines. This animal is, as Solinus[25] writes, very small and like a spider in shape ; it is called *solifuga*, because it shuns (*fugit*) the light (*solem*). It is very common

[24]Orcus, the god of the infernal regions,—otherwise Pluto.

[25]Caius Julius Solinus was an unreliable Roman Grammarian of the 3rd Century. There is much difference of opinion as to the precise animal meant by *solifuga*. The word is variously spelled *solipugus*, *solpugus*, *solipuga*, *solipunga*, etc., and is mentioned by Pliny (VIII., 43), and other ancient authors all apparently meaning a venomous insect, either an ant or a spider. The term in later times indicated a scorpion.

in silver mines; it creeps unobserved and brings destruction upon those who imprudently sit on it. But, as the same writer tells us, springs of warm and salubrious waters gush out in certain places, which neutralise the venom inserted by the ants.

In some of our mines, however, though in very few, there are other pernicious pests. These are demons of ferocious aspect, about which I have spoken in my book *De Animantibus Subterraneis*. Demons of this kind are expelled and put to flight by prayer and fasting.[26]

Some of these evils, as well as certain other things, are the reason why pits are occasionally abandoned. But the first and principal cause is that they do not yield metal, or if, for some fathoms, they do bear metal they become barren in depth. The second cause is the quantity of water which flows in; sometimes the miners can neither divert this water into the tunnels, since tunnels cannot be driven so far into the mountains, or they cannot draw it out with machines because the shafts are too deep; or if they could draw it out with machines, they do not use them, the reason undoubtedly being that the expenditure is greater than the profits of a moderately poor vein. The third cause is the noxious air, which the owners sometimes cannot overcome either by skill or expenditure, for which reason the digging is sometimes abandoned, not only of shafts, but also of tunnels. The fourth cause is the poison produced in particular places, if it is not in our power either completely to remove it or to moderate its effects. This is the reason why the caverns in the Plain known as Laurentius[27] used not to be

[26]The presence of demons or gnomes in the mines was so general a belief that Agricola fully accepted it. This is more remarkable, in view of our author's very general scepticism regarding the supernatural. He, however, does not classify them all as bad—some being distinctly helpful. The description of gnomes of kindly intent, which is contained in the last paragraph in *De Animantibus* is of interest:—

"Then there are the gentle kind which the Germans as well as the Greeks call *cobalos*, "because they mimic men. They appear to laugh with glee and pretend to do much, but "really do nothing. They are called little miners, because of their dwarfish stature, which "is about two feet. They are venerable looking and are clothed like miners in a filleted "garment with a leather apron about their loins. This kind does not often trouble the miners, "but they idle about in the shafts and tunnels and really do nothing, although they pretend to "be busy in all kinds of labour, sometimes digging ore, and sometimes putting into buckets "that which has been dug. Sometimes they throw pebbles at the workmen, but they rarely "injure them unless the workmen first ridicule or curse them. They are not very dissimilar "to Goblins, which occasionally appear to men when they go to or from their day's work, or "when they attend their cattle. Because they generally appear benign to men, the Germans "call them *guteli*. Those called *trulli*, which take the form of women as well as men, actually "enter the service of some people, especially the *Suions*. The mining gnomes are especially "active in the workings where metal has already been found, or where there are hopes of "discovering it, because of which they do not discourage the miners, but on the contrary "stimulate them and cause them to labour more vigorously."

The German miners were not alone in such beliefs, for miners generally accepted them—even to-day the faith in "knockers" has not entirely disappeared from Cornwall. Neither the sea nor the forest so lends itself to the substantiation of the supernatural as does the mine. The dead darkness, in which the miners' lamps serve only to distort every shape, the uncanny noises of restless rocks whose support has been undermined, the approach of danger and death without warning, the sudden vanishing or discovery of good fortune, all yield a thousand corroborations to minds long steeped in ignorance and prepared for the miraculous through religious teaching.

[27]The Plains of Laurentius extend from the mouth of the Tiber southward—say twenty miles south of Rome. What Agricola's authority was for silver mines in this region we cannot discover. This may, however, refer to the lead-silver district of the Attic Peninsula, Laurion being sometimes Latinized as *Laurium* or *Laurius*.

worked, though they were not deficient in silver. The fifth cause are the fierce and murderous demons, for if they cannot be expelled, no one escapes from them. The sixth cause is that the underpinnings become loosened and collapse, and a fall of the mountain usually follows ; the underpinnings are then only restored when the vein is very rich in metal. The seventh cause is military operations. Shafts and tunnels should not be re-opened unless we are quite certain of the reasons why the miners have deserted them, because we ought not to believe that our ancestors were so indolent and spiritless as to desert mines which could have been carried on with profit. Indeed, in our own days, not a few miners, persuaded by old women's tales, have re-opened deserted shafts and lost their time and trouble. Therefore, to prevent future generations from being led to act in such a way, it is advisable to set down in writing the reason why the digging of each shaft or tunnel has been abandoned, just as it is agreed was once done at Freiberg, when the shafts were deserted on account of the great inrush of water.

END OF BOOK VI.

BOOK VII.

INCE the Sixth Book has described the iron tools, the vessels and the machines used in mines, this Book will describe the methods of assaying[1] ores; because it is desirable to first test them in order that the material mined may be advantageously smelted, or that the dross may be purged away and the metal made pure. Although writers have mentioned such tests, yet none of them have set down the directions for performing them, wherefore it is no wonder that those who come later have written nothing on the subject. By tests of this kind miners can determine with certainty whether ores contain any metal in them or not; or if it has already been indicated that the ore contains one or more metals, the tests show whether it is much or little; the miners also ascertain by such tests the method by which the metal can be separated from that part of the ore devoid of it; and further, by these tests, they determine that part in which there is much metal from that part in which there is little. Unless these tests have been carefully applied before the metals are melted out, the ore cannot be smelted without great loss to the owners, for the parts which do not easily melt in the fire carry the metals off with them or consume them. In the last case, they pass off with the fumes; in the other case they are mixed with the slag and furnace accretions, and in such event the owners lose the labour which they have spent in preparing the furnaces and the crucibles, and further, it is necessary for them to incur fresh expenditure for fluxes and other things. Metals, when they have been melted out, are usually assayed in order that we may ascertain what proportion of silver is in a *centumpondium* of copper or lead, or what quantity of gold is in one *libra* of silver; and, on the other hand, what proportion of copper or lead is contained in a *centumpondium* of silver, or what quantity of silver is contained in one *libra* of gold. And from this we can calculate whether it will be worth while to separate the precious metals from the base metals, or not. Further, a test of this kind shows whether coins are good or are debased; and readily detects silver, if the coiners have mixed more than is lawful with the gold; or copper, if the coiners have alloyed with the gold or silver more of it than is allowable. I will explain all these methods with the utmost care that I can.

[1]We have but little record of anything which could be called "assaying" among the Greeks and Romans. The fact, however, that they made constant use of the touchstone (see note 37, p. 252) is sufficient proof that they were able to test the purity of gold and silver. The description of the touchstone by Theophrastus contains several references to "trial" by fire (see note 37, p. 252). They were adepts at metal working, and were therefore familiar with melting metals on a small scale, with the smelting of silver, lead, copper, and tin ores (see note 1, p. 353) and with the parting of silver and lead by cupellation. Consequently, it would not require much of an imaginative flight to conclude that there existed some system of tests of ore and metal values by fire. Apart from the statement of Theophrastus referred to, the first references made to anything which might fill the *rôle* of assaying are from the Alchemists, particularly Geber (prior to 1300), for they describe methods of solution, precipitation, distillation, fusing in crucibles, cupellation, and of the parting of gold and silver by acid and by sulphur, antimony, or cementation. However, they were not bent on

The method of assaying ore used by mining people, differs from smelting only by the small amount of material used. Inasmuch as, by smelting a small quantity, they learn whether the smelting of a large

determining quantitative values, which is the fundamental object of the assayer's art, and all their discussion is shrouded in an obscure cloak of gibberish and attempted mysticism. Nevertheless, therein lies the foundation of many cardinal assay methods, and even of chemistry itself.

The first explicit records of assaying are the anonymous booklets published in German early in the 16th Century under the title *Probierbüchlein*. Therein the art is disclosed well advanced toward maturity, so far as concerns gold and silver, with some notes on lead and copper. We refer the reader to Appendix B for fuller discussion of these books, but we may repeat here that they are a collection of disconnected recipes lacking in arrangement, the items often repeated, and all apparently the inheritance of wisdom passed from father to son over many generations. It is obviously intended as a sort of reminder to those already skilled in the art, and would be hopeless to a novice. Apart from some notes in Biringuccio (Book III, Chaps. 1 and 2) on assaying gold and silver, there is nothing else prior to *De Re Metallica*. Agricola was familiar with these works and includes their material in this chapter. The very great advance which his account represents can only be appreciated by comparison, but the exhaustive publication of other works is foreign to the purpose of these notes. Agricola introduces system into the arrangement of his materials, describes implements, and gives a hundred details which are wholly omitted from the previous works, all in a manner which would enable a beginner to learn the art. Furthermore, the assaying of lead, copper, tin, quicksilver, iron, and bismuth, is almost wholly new, together with the whole of the argument and explanations. We would call the attention of students of the history of chemistry to the general oversight of these early 16th Century attempts at analytical chemistry, for in them lie the foundations of that science. The statement sometimes made that Agricola was the first assayer, is false if for no other reason than that science does not develop with such strides at any one human hand. He can, however, fairly be accounted as the author of the first proper text-book upon assaying. Those familiar with the art will be astonished at the small progress made since his time, for in his pages appear most of the reagents and most of the critical operations in the dry analyses of gold, silver, lead, copper, tin, bismuth, quicksilver, and iron of to-day. Further, there will be recognised many of the "kinks" of the art used even yet, such as the method of granulation, duplicate assays, the "assay ton" method of weights, the use of test lead, the introduction of charges in leaf lead, and even the use of beer instead of water to damp bone-ash.

The following table is given of the substances mentioned requiring some comment, and the terms adopted in this book, with notes for convenience in reference. The German terms are either from Agricola's Glossary of *De Re Metallica*, his *Interpretatio*, or the German Translation. We have retained the original German spelling. The fifth column refers to the page where more ample notes are given :—

Terms adopted.	Latin.	German.	Remarks.	Further Notes.
Alum	*Alumen*	*Alaun*	Either potassium or ammonia alum	p. 564
Ampulla	*Ampulla*	*Kolb*	A distillation jar	
Antimony	*Stibium*	*Spiesglas*	Practically always antimony sulphide	p. 428
Aqua valens or *aqua*	*Aqua valens*	*Scheidewasser*	Mostly nitric acid	p. 439
Argol	*Feces vini siccae*	*Die weinheffen*	Crude tartar	p. 234
Ash of lead	*Nigrum plumbum cinereum*		Artificial lead sulphide	p. 237
Ash of musk ivy (Salt made from)	*Sal ex anthyllidis cinere factus*	*Salalkali*	Mostly potash	p. 560
Ashes which wool-dyers use	*Cineres quo infectores lanarum utuntur*		Mostly potash	p. 559
Assay	*Venas experiri*	*Probiren*		
Assay furnace	*Fornacula*	*Probir ofen*	"Little" furnace	
Azure	*Caeruleum*	*Lasur*	Partly copper carbonate (azurite) partly silicate	p. 110

quantity will compensate them for their expenditure; hence, if they are not particular to employ assays, they may, as I have already said, sometimes smelt the metal from the ore with a loss or sometimes without any profit; for they

Terms adopted.	Latin.	German.	Remarks.	Further Notes.
Bismuth	*Plumlum Cinereum*	*Wismut*	*Bismuth*	p. 433
Bitumen	*Bitumen*	*Bergwachs*		p. 581
Blast furnace	*Prima fornax*	*Schmeltzofen*		
Borax	*Chrysocolla ex nitro confecta; chrysocolla quam boracem nominant*	*Borras; Tincar*		p. 560
Burned alum	*Alumen coctum*	*Gesottener alaun*	Probably de hydrated alum	p. 565
Cadmia (see note 8, p. 112)			(1) Furnace accretions (2) Calamine (3) Zinc blende (4) Cobalt arsenical sulphides	p. 112
Camphor	*Camphora*	*Campffer*		p. 238
Chrysocolla called borax (see borax)				
Chrysocolla (copper mineral)	*Chrysocolla*	*Berggrün und Schifergrün*	Partly chrysocolla, partly malachite	p. 110
Copper filings	*Aeris scobs elimata*	*Kupferfeilich*	Apparently finely divided copper metal	p. 233
Copper flowers	*Aeris flos*	*Kupferbraun*	Cupric oxide	p. 538
Copper scales	*Aeris squamae*	*Kupfer hammerschlag oder kessel braun*	Probably cupric oxide	
Copper minerals (see note 8, p. 109)				
Crucible (triangular)	*Catillus triangularis*	*Dreieckichtschirbe*	See illustration	p. 229
Cupel	*Catillus cinereus*	*Capelle*		
Cupellation furnace	*Secunda fornax*	*Treibherd*		
Flux	*Additamentum*	*Zusetze*		p. 232
Furnace accretions	*Cadmia fornacum*	*Mitlere und obere offenbrüche*		
Galena	*Lapis plumbarius*	*Glantz*	Lead sulphide	p. 110
Glass-gall	*Recrementum vitri*	*Glassgallen*	Skimmings from glass melting	p. 235
Grey antimony or stibium	*Stibi* or *stibium*	*Spiesglas*	Antimony sulphide, stibnite	p. 428
Hearth-lead	*Molybdaena*	*Herdplei*	The saturated furnace bottoms from cupellation	p. 476
Hoop (iron)	*Circulus ferreus*	*Ring*	A forge for crucibles	p. 226
Iron filings	*Ferri scobs elimata*	*Eisen feilich*	Metallic iron	
Iron scales	*Squamae ferri*	*Eisen hammerschlag*	Partly iron oxide	
Iron slag	*Recrementum ferri*	*Sinder*		
Lead ash	*Cinis plumbi nigri*	*Pleiasche*	Artificial lead sulphide	p. 237
Lead granules	*Globuli plumbei*	*Gekornt plei*	Granulated lead	
Lead ochre	*Ochra plumbaria*	*Pleigeel*	Modern massicot (PbO)	p. 232
Lees of *aqua* which separates gold from silver	*Feces aquarum quae aurum ab argento secernunt*	*Scheidewasser heffe*	Uncertain	p. 234
Dried lees of vinegar	*Siccae feces aceti*	*Heffe des essigs*	Argol	p. 234
Dried lees of wine	*Feces vini siccae*	*Wein heffen*	Argol	p. 234

can assay the ore at a very small expense, and smelt it only at a great expense. Both processes, however, are carried out in the same way, for just as we assay ore in a little furnace, so do we smelt it in the large furnace. Also in both cases charcoal and not wood is burned. Moreover, in the crucible when metals are tested, be they gold, silver, copper, or lead, they are mixed in precisely the same way as they are mixed in the blast furnace when they are smelted. Further, those who assay ores with fire, either pour out the metal in a liquid state, or, when it has cooled, break the crucible and clean

Terms adopted.	Latin.	German.	Remarks.	Further Notes.
Limestone	*Saxum calcis*	*Kalchstein*		
Litharge	*Spuma argenti*	*Glette*		
Lye	*Lixivium*	*Lauge durch asschen gemacht*	Mostly potash	p. 233
Muffle	*Tegula*	*Muffel*	Latin, literally "Roof-tile"	
Operculum	*Operculum*	*Helm oder alembick*	Helmet or cover for a distillation jar	
Orpiment	*Auripigmentum*	*Operment*	Yellow sulphide of arsenic (As_2S_3)	p. 111
Pyrites	*Pyrites*	*Kis*	Rather a genus of sulphides, than iron pyrite in particular	p. 112
Pyrites (Cakes from)	*Panes ex pyrite conflati*	*Stein*	Iron or copper matte	p. 350
Realgar	*Sandaraca*	*Rosgeel*	Red sulphide of arsenic (AsS)	p. 111
Red lead	*Minium*	*Menning*	Pb_3O_4	p. 232
Roasted copper	*Aes ustum*	*Gebrandt kupffer*	Artificial copper sulphide (?)	p. 233
Salt	*Sal*	*Saltz*	NaCl	p. 233
Salt (Rock)	*Sal fossilis*	*Berg saltz*	NaCl	p. 233
Sal artificiosus	*Sal artificiosus*		A stock flux ?	p. 236
Sal ammoniac	*Sal ammoniacus*	*Salarmoniac*	NH_4Cl	p. 560
Saltpetre	*Halinitrum*	*Salpeter*	KNO_3	p. 561
Salt (refined)	*Sal facticius purgatus*		NaCl	
Sal tostus	*Sal tostus*	*Geröst saltz*	Apparently simply heated or melted common salt	p. 233
Sal torrefactus	*Sal torrefactus*	*Geröst saltz*		p. 233
Salt (melted)	*Sal liquefactus*	*Geflossen saltz*	Melted salt or salt glass	p. 233
Scorifier	*Catillus fictilis*	*Scherbe*		
Schist	*Saxum fissile*	*Schifer*		
Silver minerals (see note 8, p. 108)				
Slag	*Recrementum*	*Schlacken*		
Soda	*Nitrum*		Mostly soda from Egypt, Na_2Co_3	p. 558
Stones which easily melt	*Lapides qui facile igni liquescunt*	*Flüs*	Quartz and fluorspar	p. 380
Sulphur	*Sulfur*	*Schwefel*		p. 579
Tophus	*Tophus*	*Topstein*	Marl (?)	p. 233
Touchstone	*Coticula*	*Goldstein*		
Venetian glass	*Venetianum vitrum*			
Verdigris	*Aerugo*	*Grünspan oder Spanschgrün*	Copper sub-acetate	p. 440
Vitriol	*Atramentum sutorium*	*Kupferwasser*	Mostly $FeSO_4$	p. 572
White schist	*Saxum fissile album*	*Weisser schifer*		p. 234
Weights (see Appendix).				

the metal from slag; and in the same way the smelter, as soon as the metal flows from the furnace into the forehearth, pours in cold water and takes the slag from the metal with a hooked bar. Finally, in the same way that gold and silver are separated from lead in a cupel, so also are they separated in the cupellation furnace.

It is necessary that the assayer who is testing ore or metals should be prepared and instructed in all things necessary in assaying, and that he should close the doors of the room in which the assay furnace stands, lest

ROUND ASSAY FURNACE.

RECTANGULAR ASSAY FURNACE.

anyone coming at an inopportune moment might disturb his thoughts when they are intent on the work. It is also necessary for him to place his balances in a case, so that when he weighs the little buttons of metal the scales may not be agitated by a draught of air, for that is a hindrance to his work.

Now I will describe the different things which are necessary in assaying, beginning with the assay furnace, of which one differs from another in shape, material, and the place in which it is set. In shape, they may be round or rectangular, the latter shape being more suited to assaying ores. The materials of the assay furnaces differ, in that one is made of bricks, another of iron, and certain ones of clay. The one of bricks is built on a chimney-hearth which is three and a half feet high; the iron one is placed in the same position, and also the one of clay. The brick one is a cubit high, a foot wide on the inside, and one foot two digits long; at a point five digits above the hearth—which is usually the thickness of an unbaked[2] brick—an iron plate is laid, and smeared over with lute on the upper side to prevent it from being injured by the fire; in front of the furnace above the plate is a mouth a palm high, five digits wide, and rounded at the top. The iron plate

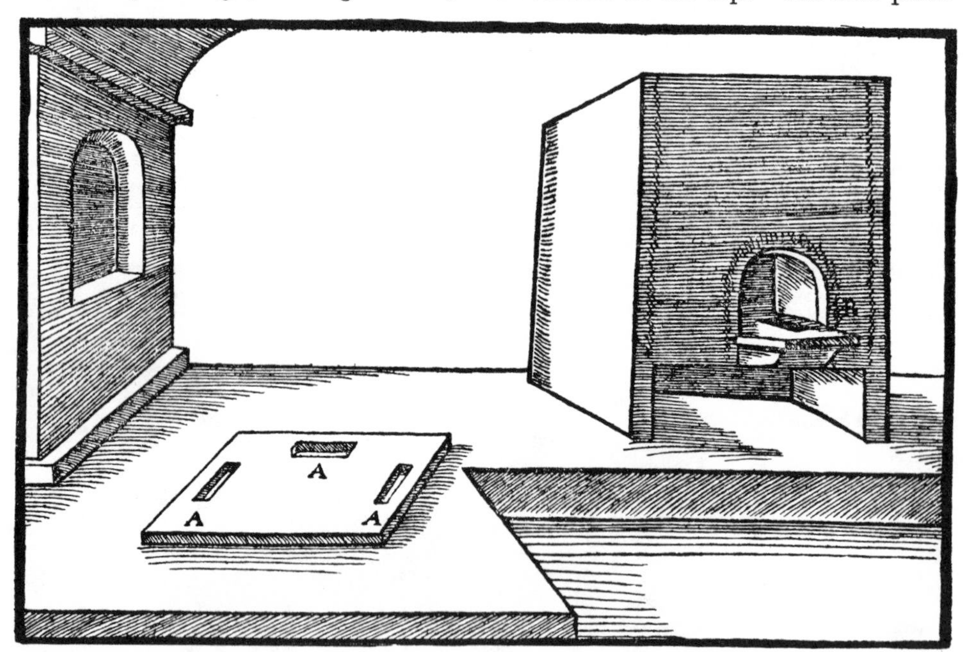

A—Openings in the plate. B—Part of plate which projects beyond the furnace.

has three openings which are one digit wide and three digits long, one is at each side and the third at the back; through them sometimes the ash falls from the burning charcoal, and sometimes the draught blows through the chamber which is below the iron plate, and stimulates the fire. For this reason this furnace when used by metallurgists is named from assaying, but when used by the alchemists it is named from the wind[3]. The part of the iron plate which projects from the furnace is generally three-quarters of a

[2]*Crudorum*,—unbaked ?

[3]This reference is not very clear. Apparently the names refer to the German terms *probier ofen* and *windt ofen*.

palm long and a palm wide; small pieces of charcoal, after being laid thereon, can be placed quickly in the furnace through its mouth with a pair of tongs, or again, if necessary, can be taken out of the furnace and laid there.

The iron assay furnace is made of four iron bars a foot and a half high; which at the bottom are bent outward and broadened a short distance to enable them to stand more firmly; the front part of the furnace is made from two of these bars, and the back part from two of them; to these bars on both sides are joined and welded three iron cross-bars, the first at a height of a palm from the bottom, the second at a height of a foot, and the third at the top. The upright bars are perforated at that point where the side cross-bars are joined to them, in order that three similar iron bars on the remaining sides can be engaged in them; thus there are twelve cross-bars, which make three stages at unequal intervals. At the lower stage, the upright bars are distant from each other one foot and five digits; and at the middle stage the front is distant from the back three palms and one digit, and the sides are distant from each other three palms and as many digits; at the highest stage from the front to the back there is a distance of two palms, and between the sides three palms, so that in this way the furnace becomes narrower at the top. Furthermore, an iron rod, bent to the shape of the mouth, is set into the lowest bar of the front; this mouth, just like that of the brick furnace, is a palm high and five digits wide. Then the front cross-bar of the lower stage is perforated on each side of the mouth, and likewise the back one; through these perforations there pass two iron rods, thus making altogether four bars in the lower stage, and these support an iron plate smeared with lute; part of this plate also projects outside the furnace. The outside of the furnace from the lower stage to the upper, is covered with iron plates, which are bound to the bars by iron wires, and smeared with lute to enable them to bear the heat of the fire as long as possible.

As for the clay furnace, it must be made of fat, thick clay, medium so far as relates to its softness or hardness. This furnace has exactly the same height as the iron one, and its base is made of two earthenware tiles, one foot and three palms long and one foot and one palm wide. Each side of the fore part of both tiles is gradually cut away for the length of a palm, so that they are half a foot and a digit wide, which part projects from the furnace; the tiles are about a digit and a half thick. The walls are similarly of clay, and are set on the lower tiles at a distance of a digit from the edge, and support the upper tiles; the walls are three digits high and have four openings, each of which is about three digits high; those of the back part and of each side are five digits wide, and of the front, a palm and a half wide, to enable the freshly made cupels to be conveniently placed on the hearth, when it has been thoroughly warmed, that they may be dried there. Both tiles are bound on the outer edge with iron wire, pressed into them, so that they will be less easily broken; and the tiles, not unlike the iron bed-plate, have three openings three digits long and a digit wide, in order that when the upper one on account of the heat of the fire or for some other reason has become damaged, the lower one may be exchanged and take its place. Through these holes, the ashes from the burning charcoal, as I have stated, fall down, and air blows into the furnace after passing through the openings in the walls of the chamber. The furnace is rectangular, and inside at the lower part it is three palms and one digit wide and three palms and as many digits long. At the upper part it is two palms and three digits wide, so that it also grows narrower; it is one foot high; in the middle of the back it is cut out at the bottom in the shape of a semicircle, of half a digit radius. Not unlike the furnace before described, it has in its forepart a mouth which is rounded at the top, one palm high and a palm and a digit wide. Its door is also made of clay, and this has a window and a handle; even the lid of the furnace which is made of clay has its own handle, fastened on with iron wire. The outer parts and sides of this furnace are bound with iron wires, which are usually pressed in, in the shape of triangles. The brick furnaces must remain stationary; the clay and iron ones can be carried from one place to another. Those of brick can be prepared more quickly, while those of iron are more lasting, and those of clay are more suitable. Assayers also make temporary furnaces in another way; they stand three bricks on a hearth, one on each side and a third one at the back, the fore-part lies open to the draught, and on these bricks is placed an iron plate, upon which they again stand three bricks, which hold and retain the charcoal.

The setting of one furnace differs from another, in that some are placed higher and others lower; that one is placed higher, in which the man who is assaying the ore or metals introduces the scorifier through the mouth with the tongs; that one is placed lower, into which he introduces the crucible through its open top.

In some cases the assayer uses an iron hoop[4] in place of a furnace; this is placed upon the hearth of a chimney, the lower edge being daubed with lute to prevent the blast of the bellows from escaping under it. If the blast is given slowly, the ore will be smelted and the copper will melt in the triangular crucible, which is placed in it and taken away again with the tongs. The hoop is two palms high and half a digit thick; its diameter is generally one foot and one palm, and where the blast from the bellows enters into it, it is notched out. The bellows is a double one, such as goldworkers use, and sometimes smiths. In the middle of the bellows there is a board in which there is an air-hole, five digits wide and seven long, covered by a little flap which is fastened over the air-hole on the lower side of the board; this flap is of equal length and width. The bellows, without its head, is three feet long, and at the back is one foot and one palm wide and somewhat rounded, and it is three palms wide at the head; the head itself is three palms long and two palms and a digit wide at the part where it joins the boards, then it gradually becomes narrower. The nozzle, of which there is only one, is one foot and two digits long; this nozzle, and one-half of the head in which the nozzle is fixed, are placed in an opening of the wall, this being one foot and one palm thick; it reaches only to the iron hoop on the

[4]*Circulus.* This term does not offer a very satisfactory equivalent, as such a furnace has no distinctive name in English. It is obviously a sort of forge for fusing in crucibles.

hearth, for it does not project beyond the wall. The hide of the bellows is fixed to the bellows-boards with its own peculiar kind of iron nails. It joins both bellows-boards to the head, and over it there are cross strips of hide fixed to the bellows-boards with broad-headed nails, and similarly fixed to the head. The middle board of the bellows rests on an iron bar, to which it is fastened with iron nails clinched on both ends, so that it cannot move; the iron bar is fixed between two upright posts, through which it penetrates. Higher up on these upright posts there is a wooden axle, with iron journals which revolve in the holes in the posts. In the middle of this axle there is mortised a lever, fixed with iron nails to prevent it from flying out; the lever is five and a half feet long, and its posterior end is engaged in the iron ring of an iron rod which reaches to the "tail" of the lowest bellows-board, and there engages another similar ring. And so when the workman pulls down the lever, the lower part of the bellows is raised and drives the wind into the nozzle; then the wind, penetrating through the hole in the middle bellows-board, which is called the air-hole, lifts up the upper part of the bellows, upon whose upper board is a piece of lead, heavy enough to press down that part of the bellows again, and this being pressed down blows a blast through the nozzle. This is the principle of the double bellows, which is peculiar to the iron hoop where are placed the triangular crucibles in which copper ore is smelted and copper is melted.

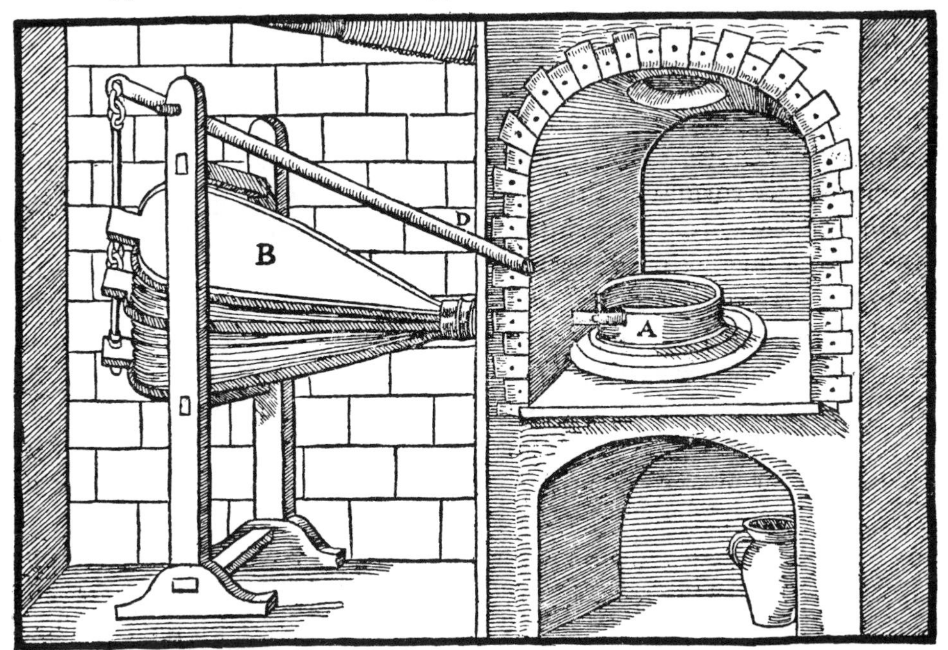

A—Iron hoop. B—Double bellows. C—Its nozzle. D—Lever.

I have spoken of the furnaces and the iron hoop; I will now speak of the muffles and the crucibles. The muffle is made of clay, in the shape of an inverted gutter tile; it covers the scorifiers, lest coal dust fall into them and interfere with the assay. It is a palm and a half broad, and the height, which corresponds with the mouth of the furnace, is generally a palm, and it is nearly as long as the furnace ; only at the front end does it touch the mouth of the furnace, everywhere else on the sides and at the back there is a space of three digits, to allow the charcoal to lie in the open space between it and the furnace. The muffle is as thick as a fairly thick earthen jar ; its upper part is entire ; the back has two little windows, and each side has two or three or even four, through which the heat passes into the scorifiers and melts the ore. In place of little windows, some muffles have small holes, ten in the back and more on each side. Moreover, in the back below the little windows, or small holes, there are cut away three semi-circular notches half a digit high, and on each side there are four. The back of the muffle is generally a little lower than the front.

A—Broad little windows of muffle. B—Narrow ones. C—Openings in the back thereof.

The crucibles differ in the materials from which they are made, because they are made of either clay or ashes ; and those of clay, which we also call "earthen," differ in shape and size. Some are made in the shape of a moderately thick salver (scorifiers), three digits wide, and of a capacity of an *uncia* measure ; in these the ore mixed with fluxes is melted, and they are used by those who assay gold or silver ore. Some are triangular and much thicker and more capacious, holding five, or six, or even more *unciae* ; in these copper is melted, so that it can be poured out, expanded, and tested with fire, and in these copper ore is usually melted.

The cupels are made of ashes ; like the preceding scorifiers they are tray-shaped, and their lower part is very thick but their capacity is less. In these lead is separated from silver, and by them assays are concluded. Inasmuch as the assayers themselves make the cupels, something must be said about the material from which they are made, and the method of making them. Some make them out of all kinds of ordinary ashes ; these are not good, because ashes of this kind contain a certain amount of fat, whereby such cupels are easily broken when they are hot. Others make them likewise out of any kind of ashes which have been previously leached ; of this kind are the ashes into which warm water has been infused for the purpose of making lye. These ashes, after being dried in the sun or a furnace, are sifted in a hair sieve; and although warm water washes away the

A—SCORIFIER. B—TRIANGULAR CRUCIBLE. C—CUPEL.

fat from the ashes, still the cupels which are made from such ashes are not very good because they often contain charcoal dust, sand, and pebbles. Some make them in the same way out of any kind of ashes, but first of all pour water into the ashes and remove the scum which floats thereon; then, after it has become clear, they pour away the water, and dry the ashes; they then sift them and make the cupels from them. These, indeed, are good, but not of the best quality, because ashes of this kind are also not devoid of small pebbles and sand. To enable cupels of the best quality to be made, all the impurities must be removed from the ashes. These impurities are of two kinds; the one sort light, to which class belong charcoal dust and fatty material and other things which float in water, the other sort heavy, such as small stones, fine sand, and any other materials which settle in the bottom of a vessel. Therefore, first of all, water should be poured into the ashes and the light impurities removed; then the ashes should be kneaded with the hands, so that they will become properly mixed with the water. When the water has become muddy and turbid, it should be poured into a second vessel. In this way the small stones and fine sand, or any other heavy substance which may be there, remain in the first vessel, and should be thrown away. When all the ashes have settled in this second vessel, which will be shown if the water has become clear and does not taste of the flavour of lye, the water should be thrown away, and the ashes which have settled in the vessel should be dried in the sun or in a furnace. This material is suitable for the cupels, especially if it is the ash of beech wood or other wood which has a small annual growth; those ashes made from twigs and limbs of vines, which have rapid annual growth, are not so good, for the cupels made from them, since they are not sufficiently dry, frequently crack and break in the fire and absorb the metals. If ashes of beech or similar wood are not to be had, the assayer makes little balls of such ashes as he can get, after they have been cleared of impurities in the manner before described, and puts them in a baker's or potter's oven to burn, and from these the cupels are made, because the fire consumes whatever fat or damp there may be. As to all kinds of ashes, the older they are the better, for it is necessary that they should have the greatest possible dryness. For this reason ashes obtained from burned bones, especially from the bones of the heads of animals, are the most suitable for cupels, as are also those ashes obtained from the horns of deer and the spines of fishes. Lastly, some take the ashes which are obtained from burnt scrapings of leather, when the tanners scrape the hides to clear them from hair. Some prefer to use compounds, that one being recommended which has one and a half parts of ashes from the bones of animals or the spines of fishes, and one part of beech ashes, and half a part of ashes of burnt hide scrapings. From this mixture good cupels are made, though far better ones are obtained from equal portions of ashes of burnt hide scrapings, ashes of the bones of heads of sheep and calves, and ashes of deer horns. But the best of all are produced from deer horns alone, burnt to powder ; this kind, by reason of its extreme dryness, absorbs metals least of all. Assayers of our own day, however, generally make the cupels from beech ashes. These ashes, after being prepared in the manner just described, are first of all sprinkled with beer or water, to make them stick together, and are then ground in a small mortar. They are ground again after being mixed with the ashes obtained from the skulls of beasts or from the spines of fishes ; the more the ashes are ground the better they are. Some rub bricks and sprinkle the dust so obtained, after sifting it, into the beech ashes, for dust of this kind does not allow the hearth-lead to absorb the gold or silver by eating away the cupels. Others, to guard against the same thing, moisten the cupels with white of egg after they have been made, and when they have been dried in the sun, again crush them ; especially if they want to assay in it an ore or copper which contains iron. Some moisten the ashes again and again with cow's milk, and dry them, and grind them in a small mortar, and then mould the cupels. In the works in which silver is separated from copper, they make cupels from two parts of the ashes of the crucible of the cupellation furnace, for these ashes are very dry, and from one part of bone-ash. Cupels which have been made in these ways also need to be placed in the sun or in a furnace ; afterward, in whatever way they have been made, they must be kept a long time in dry places, for the older they are, the dryer and better they are.

Not only potters, but also the assayers themselves, make scorifiers and triangular crucibles. They make them out of fatty clay, which is dry[5], and neither hard nor soft. With this clay they mix the dust of old broken crucibles, or of burnt and worn bricks ; then they knead with a pestle the clay thus mixed with dust, and then dry it. As to these crucibles,

[5] *Spissa*,—"Dry." This term is used in contra-distinction to *pingue*, unctuous or "fatty."

the older they are, the dryer and better they are. The moulds in which the cupels are moulded are of two kinds, that is, a smaller size and a larger size. In the smaller ones are made the cupels in which silver or gold is purged from the lead which has absorbed it ; in the larger ones are made cupels in which silver is separated from copper and lead. Both moulds are made out of brass and have no bottom, in order that the cupels can be taken out of them whole. The pestles also are of two kinds, smaller and larger, each likewise of brass, and from the lower end of them there projects a round knob, and this alone is pressed into the mould and makes the hollow part of the cupel. The part which is next to the knob corresponds to the upper part of the mould.

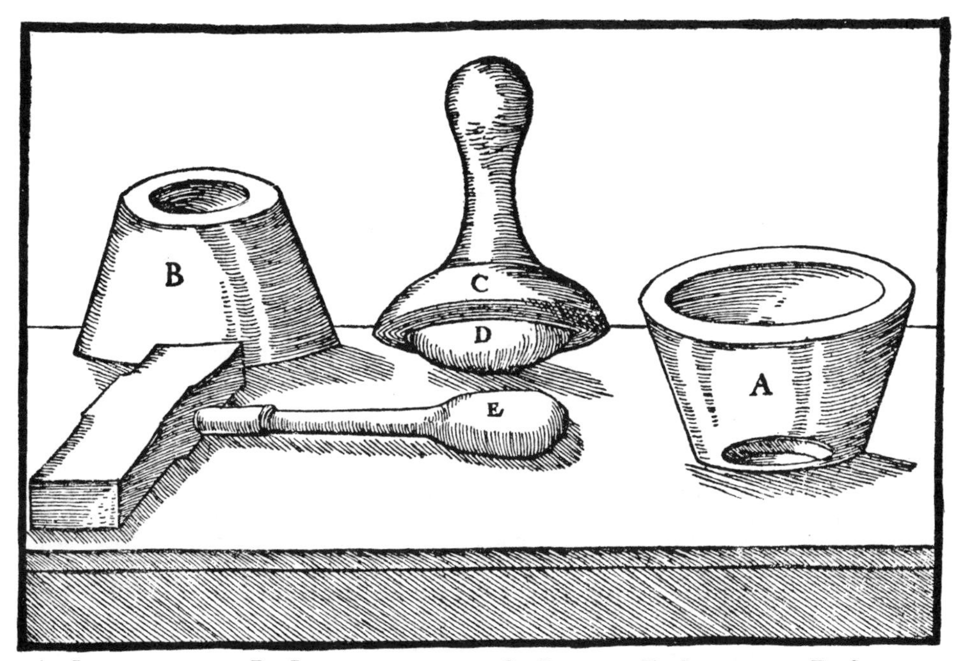

A—Little mould. B—Inverted mould. C—Pestle. D—Its knob. E—Second pestle.

So much for these matters. I will now speak of the preparation of the ore for assaying. It is prepared by roasting, burning, crushing, and washing. It is necessary to take a fixed weight of ore in order that one may determine how great a portion of it these preparations consume. The hard stone containing the metal is burned in order that, when its hardness has been overcome, it can be crushed and washed ; indeed, the very hardest kind, before it is burned, is sprinkled with vinegar, in order that it may more rapidly soften in the fire. The soft stone should be broken with a hammer, crushed in a mortar and reduced to powder ; then it should be washed and then dried again. If earth is mixed with the mineral, it is washed in a basin, and that which settles is assayed in the fire after it is dried. All mining products which are washed must again be dried. But ore which is rich in metal is neither burned nor crushed nor washed, but is roasted, lest that method of preparation should lose some of the metal. When the fires have been kindled, this kind of ore is roasted in an enclosed pot, which is stopped up with lute. A less valuable ore is even burned on a hearth, being placed upon the charcoal; for we do not make a great expenditure upon metals, if they are not worth it. However, I will go into fuller details as to all these methods of preparing ore, both a little later, and in the following Book.

For the present, I have decided to explain those things which mining people usually call fluxes[6] because they are added to ores, not only for assaying, but also for smelting. Great power is discovered in all these fluxes, but we do not see the same effects produced in every case; and some are of a very complicated nature. For when they have been mixed with the ore and are melted in either the assay or the smelting furnace, some of them, because they melt easily, to some extent melt the ore; others, because they either make the ore very hot or penetrate into it, greatly assist the fire in separating the impurities from the metals, and they also mix the fused part with the lead, or they partly protect from the fire the ore whose metal contents would be either consumed in the fire, or carried up with the fumes and fly out of the furnace; some fluxes absorb the metals. To the first order belongs lead, whether it be reduced to little granules or resolved into ash by fire, or red-lead[7], or ochre made from lead[8], or litharge, or hearth-lead, or

[6]*Additamenta,*—" Additions." Hence the play on words.

We have adopted " flux " because the old English equivalent for all these materials was " flux," although in modern nomenclature the term is generally restricted to those substances which, by chemical combination in the furnace, lower the melting point of some of the charge. The " additions " of Agricola, therefore, include reducing, oxidizing, sulphurizing, desulphurizing, and collecting agents as well as fluxes. A critical examination of the fluxes mentioned in the next four pages gives point to the Author's assertion that " some are of a very complicated nature." However, anyone of experience with home-taught assayers has come in contact with equally extraordinary combinations. The four orders of " additions " enumerated are quite impossible to reconcile from a modern metallurgical point of view.

[7]*Minium secundarium.* (*Interpretatio,—menning.* Pb_3O_4). Agricola derived his Latin term from Pliny. There is great confusion in the ancient writers on the use of the word *minium*, for prior to the Middle Ages it was usually applied to vermilion derived from cinnabar. Vermilion was much adulterated with red-lead, even in Roman times, and finally in later centuries the name came to be appropriated to the lead product. Theophrastus (103) mentions a substitute for vermilion, but, in spite of commentators, there is no evidence that it was red-lead. The first to describe the manufacture of real red-lead was apparently Vitruvius (VII, 12), who calls it *sandaraca* (this name was usually applied to red arsenical sulphide), and says: " White-lead is heated in a furnace and by the force of the " fire becomes red lead. This invention was the result of observation in the case of an " accidental fire, and by the process a much better material is obtained than from the mines." He describes *minium* as the product from cinnabar. Dioscorides (V, 63), after discussing white-lead, says it may be burned until it becomes the colour of *sandaracha*, and is called *sandyx*. He also states (V, 69) that those are deceived who consider cinnabar to be the same as *minium*, for *minium* is made in Spain out of stone mixed with silver sands. Therefore he is not in agreement with Vitruvius and Pliny on the use of the term. Pliny (XXXIII, 40) says: " These barren stones (apparently lead ores barren of silver) may be " recognised by their colour; it is only in the furnace that they turn red. After being " roasted it is pulverized and is *minium secundarium*. It is known to few and is very " inferior to the natural kind made from those sands we have mentioned (*cinnabar*). It is " with this that the genuine *minium* is adulterated in the works of the Company." This proprietary company who held a monopoly of the Spanish quicksilver mines, " had many " methods of adulterating it (*minium*)—a source of great plunder to the Company." Pliny also describes the making of red lead from white.

[8]*Ochra plumbaria,* (*Interpretatio,—pleigeel*; modern German,—*Bleigelb*). The German term indicates that this " Lead Ochre," a form of PbO, is what in the English trade is known as *massicot*, or *masticot*. This material can be a partial product from almost any cupellation where oxidation takes place below the melting point of the oxide. It may have been known to the Ancients among the various species into which they divided

galena; also copper, the same either roasted or in leaves or filings[9]; also the slags of gold, silver, copper, and lead; also soda[10], its slags, saltpetre, burned alum, vitriol, *sal tostus*, and melted salt[11]; stones which easily melt in hot furnaces, the sand which is made from them[12]; soft *tophus*[13],

litharge, but there is no valid reason for assigning to it any special one of their terms, so far as we can see.

[9]There are four forms of copper named as re-agents by Agricola:

Copper filings — *Aeris scobs elimata.*
Copper scales — *Aeris squamae.*
Copper flowers — *Aeris flos.*
Roasted copper — *Aes ustum.*

The first of these was no doubt finely divided copper metal; the second, third, and fourth were probably all cupric oxide. According to Agricola (*De Nat. Fos.*, p. 352), the scales were the result of hammering the metal; the flowers came off the metal when hot bars were quenched in water, and a third kind were obtained from calcining the metal. "Both flowers (*flos*) and hammer-scales (*squama*) have the same properties as *crematum* copper. ". . . The particles of flower copper are finer than scales or *crematum* copper." If we assume that the verb *uro* used in *De Re Metallica* is of the same import as *cremo* in the *De Natura Fossilium*, we can accept this material as being merely cupric oxide, but the *aes ustum* of Pliny—Agricola's usual source of technical nomenclature—is probably an artificial sulphide. Dioscorides (v, 47), who is apparently the source of Pliny's information, says:—"Of *chalcos cecaumenos*, the best is red, and pulverized resembles the colour of cinnabar; "if it turns black, it is over-burnt. It is made from broken ship nails put into a rough "earthen pot, with alternate layers of equal parts of sulphur and salt. The opening should "be smeared with potter's clay and the pot put in the furnace until it is thoroughly heated," etc. Pliny (xxxiv, 23) states: "Moreover Cyprian copper is roasted in crude earthen "pots with an equal amount of sulphur; the apertures of the pots are well luted, and they "are kept in the furnace until the pot is thoroughly heated. Some add salt, others use "*alumen* instead of sulphur, others add nothing, but only sprinkle it with vinegar."

[10]The reader is referred to note 6, p. 558, for more ample discussion of the alkalis. Agricola gives in this chapter four substances of that character:

Soda (*nitrum*). Lye. "Ashes which wool-dyers use."
"Salt made from the ashes of musk ivy."

The last three are certainly potash, probably impure. While the first might be either potash or soda, the fact that the last three are mentioned separately, together with other evidence, convinces us that by the first is intended the *nitrum* so generally imported into Europe from Egypt during the Middle Ages. This imported salt was certainly the natural bicarbonate, and we have, therefore, used the term "soda."

[11]In this chapter are mentioned seven kinds of common salt:

Salt — *Sal.*
Rock salt — *Sal fossilis.*
"Made" salt — *Sal factictius.*
Refined salt — *Sal purgatius.*
Melted salt — *Sal liquefactus.*

And in addition *sal tostus* and *sal torrefactus*. *Sal facticius* is used in distinction from rock-salt. The melted salt would apparently be salt-glass. What form the *sal tostus* and *sal torrefactus* could have we cannot say, however, but they were possibly some form of heated salt; they may have been combinations after the order of *sal artificiosus* (see p. 236).

[12] "Stones which easily melt in hot furnaces and sand which is made from them" (*lapides qui in ardentibus fornacibus facile liquescunt arenae ab eis resolutae*). These were probably quartz in this instance, although fluorspar is also included in this same genus. For fuller discussion see note on p. 380.

[13]*Tophus.* (*Interpretatio*; *Toffstein oder topstein*). According to Dana (Syst. of Min., p. 678), the German *topfstein* was English potstone or soapstone, a magnesian silicate. It is scarcely possible, however, that this is what Agricola meant by this term, for such a substance would be highly infusible. Agricola has a good deal to say about this mineral in *De Natura Fossilium* (p. 189 and 313), and from these descriptions it would seem to be a tufaceous limestone of various sorts, embracing some marls, stalagmites, calcareous sinter, etc. He states: "Generally fire does not melt it, but makes it harder and breaks it into "powder. Tophus is said to be a stone found in caverns, made from the dripping of stone "juice solidified by cold . . . . sometimes it is found containing many shells, and "likewise the impressions of alder leaves; our people make lime by burning it." Pliny, upon whom Agricola depends largely for his nomenclature, mentions such a substance (xxxvi, 48): "Among the multitude of stones there is *tophus*. It is unsuitable for

and a certain white schist[14]. But lead, its ashes, red-lead, ochre, and litharge, are more efficacious for ores which melt easily; hearth-lead for those which melt with difficulty; and galena for those which melt with greater difficulty. To the second order belong iron filings, their slag, *sal artificiosus*, argol, dried lees of vinegar[15], and the lees of the *aqua* which separates gold from silver[16]; these lees and *sal artificiosus* have the power of penetrating into ore, the argol to a considerable degree, the lees of vinegar to a greater degree, but most of all those of the *aqua* which separates gold from silver; filings and slags of iron, since they melt more slowly, have the power of heating the ore. To the third order belong pyrites, the cakes which are melted from them, soda, its slags, salt, iron, iron scales, iron filings, iron slags, vitriol, the sand which is resolved from stones which easily melt in the fire, and *tophus*; but first of all are pyrites and the cakes which are melted from it, for they absorb the metals of the ore and guard them from the fire which consumes them. To the fourth order belong lead and copper, and their relations. And so with regard to fluxes, it is manifest that some are natural, others fall in the category of slags, and the rest are purged from slag. When we

"buildings, because it is perishable and soft. Still, however, there are some places which "have no other, as Carthage, in Africa. It is eaten away by the emanations from the "sea, crumbled to dust by the wind, and washed away by the rain." In fact, *tophus* was a wide genus among the older mineralogists, Wallerius (*Meditationes Physico—Chemicae De Origine Mundi*, Stockholm, 1776, p. 186), for instance, gives 22 varieties. For the purposes for which it is used we believe it was always limestone of some form.

[14]*Saxum fissile album.* (*The Interpretatio* gives the German as *schifer*) Agricola mentions it in *Bermannus* (459), in *De Natura Fossilium* (p. 319), but nothing definite can be derived from these references. It appears to us from its use to have been either a quartzite or a fissile limestone.

[15]Argol (*Feces vini siccae*,—"Dried lees of wine." Germ. trans. gives *die wein heffen*, although the usual German term of the period was *weinstein*). The lees of wine were the crude tartar or argols of commerce and modern assayers. The argols of white wine are white, while they are red from red wine. The white argol which Agricola so often specifies would have no special excellence, unless it may be that it is less easily adulterated. Agricola (*De Nat. Fos.*, p. 344) uses the expression "*Fex vini sicca* called *tartarum*"—one of the earliest appearances of the latter term in this connection. The use of argol is very old, for Dioscorides (1st Century A.D.) not only describes argol, but also its reduction to impure potash. He says (v, 90): "The lees (*tryx*) are to be selected from old Italian wine; if not, "from other similar wine. Lees of vinegar are much stronger. They are carefully dried and "then burnt. There are some who burn them in a new earthen pot on a large fire until they "are thoroughly incinerated. Others place a quantity of the lees on live coals and pursue "the same method. The test as to whether it is completely burned, is that it becomes white "or blue, and seems to burn the tongue when touched. The method of burning lees of "vinegar is the same. . . . It should be used fresh, as it quickly grows stale; it should "be placed in a vessel in a secluded place." Pliny (xxiii, 31) says: "Following these, come "the lees of these various liquids. The lees of wine (*vini faecibus*) are so powerful as to be "fatal to persons on descending into the vats. The test for this is to let down a lamp, which, "if extinguished, indicates the peril. . . . Their virtues are greatly increased by the "action of fire." Matthioli, commenting on this passage from Dioscorides in 1565, makes the following remark (p. 1375): "The precipitate of the wine which settles in the casks of the winery forms stone-like crusts, and is called by the works-people by the name *tartarum*." It will be seen above that these lees were rendered stronger by the action of fire, in which case the tartar was reduced to potassium carbonate. The *weinstein* of the old German metallurgists was often the material lixiviated from the incinerated tartar.

Dried lees of vinegar (*siccae feces aceti; Interpretaltio, die heffe des essigs*). This would also be crude tartar. Pliny (xxiii, 32) says: "The lees of vinegar (*faex aceti*); owing to the "more acrid material are more aggravating in their effects. . . . When combined with "*melanthium* it heals the bites of dogs and crocodiles."

[16]Dried lees of *aqua* which separates gold and silver. (*Siccae feces aquarum quae aurum ab argento secernunt.* German translation, *Der scheidwasser heffe*). There is no pointed description in Agricola's works, or in any other that we can find, as to what this material was. The "separating *aqua*" was undoubtedly nitric acid (see p. 439, Book X). There

assay ores, we can without great expense add to them a small portion of any sort of flux, but when we smelt them we cannot add a large portion without great expense. We must, therefore, consider how great the cost is, to avoid incurring a greater expense on smelting an ore than the profit we make out of the metals which it yields.

The colour of the fumes which the ore emits after being placed on a hot shovel or an iron plate, indicates what flux is needed in addition to the lead, for the purpose of either assaying or smelting. If the fumes have a purple tint, it is best of all, and the ore does not generally require any flux whatever. If the fumes are blue, there should be added cakes melted out of pyrites or other cupriferous rock ; if yellow, litharge and sulphur should be added ; if red, glass-galls[17] and salt ; if green, then cakes melted from cupriferous stones, litharge, and glass-galls ; if the fumes are black, melted salt or iron slag, litharge and white lime rock. If they are white, sulphur and iron which is eaten with rust ; if they are white with green patches, iron slag and sand obtained from stones which easily melt ; if the middle part of the fumes are yellow and thick, but the outer parts green, the same sand and iron slag. The colour of the fumes not only gives us information as to the proper remedies which should be applied to each ore, but also more or less indication as to the solidified juices which are mixed with it, and which give forth such fumes. Generally, blue fumes signify that the ore contains azure ; yellow, orpiment ; red, realgar ; green, chrysocolla ; black, black bitumen ; white, tin[18] ; white with green patches, the same mixed with chrysocolla ; the middle part yellow and other parts green show that it contains sulphur. Earth, however, and other things dug up which contain metals, sometimes emit similarly coloured fumes.

If the ore contains any *stibium*, then iron slag is added to it ; if pyrites, then are added cakes melted from a cupriferous stone and sand made from stones which easily melt. If the ore contains iron, then pyrites and sulphur are added ; for just as iron slag is the flux for an ore mixed with sulphur, so on the contrary, to a gold or silver ore containing iron, from which they are

are two precipitates possible, both referred to as *feces*,—the first, a precipitate of silver chloride from clarifying the *aqua valens*, and the second, the residues left in making the acid by distillation. It is difficult to believe that silver chloride was the *feces* referred to in the text, because such a precipitate would be obviously misleading when used as a flux through the addition of silver to the assays, too expensive, and of no merit for this purpose. Therefore one is driven to the conclusion that the *feces* must have been the residues left in the retorts when nitric acid was prepared. It would have been more in keeping with his usual mode of expression, however, to have referred to this material as a *residuus*. The materials used for making acid varied greatly, so there is no telling what such a *feces* contained. A list of possibilities is given in note 8, p. 443. In the main, the residue would be undigested vitriol, alum, saltpetre, salt, etc., together with potassium, iron, and alum sulphates. The *Probierbüchlin* (p. 27) also gives this re-agent under the term *Toden kopff das ist schlam oder feces auss dem scheydwasser*.

[17] *Recrementum vitri*. (*Interpretatio Glassgallen*). Formerly, when more impure materials were employed than nowadays, the surface of the mass in the first melting of glass materials was covered with salts, mostly potassium and sodium sulphates and chlorides which escaped perfect vitrification. This "slag" or "*glassgallen*" of Agricola was also termed *sandiver*.

[18] The whole of this expression is "*candidus, candido*." It is by no means certain that this is tin, for usually tin is given as *plumbum candidum*.

not easily separated, is added sulphur and sand made from stones which easily melt.

Sal artificiosus[19] suitable for use in assaying ore is made in many ways. By the first method, equal portions of argol, lees of vinegar, and urine, are all boiled down together till turned into salt. The second method is from equal portions of the ashes which wool-dyers use, of lime, of argol purified, and of melted salt; one *libra* of each of these ingredients is thrown into twenty *librae* of urine; then all are boiled down to one-third and strained, and afterward there is added to what remains one *libra* and four *unciae* of unmelted salt, eight pounds of lye being at the same time poured into the pots, with litharge smeared around on the inside, and the whole is boiled till the salt becomes thoroughly dry. The third method follows. Unmelted salt, and iron which is eaten with rust, are put into a vessel, and after urine has been poured in, it is covered with a lid and put in a warm place for thirty days; then the iron is washed in the urine and taken out, and the residue is boiled until it is turned into salt. In the fourth method by which *sal artificiosus* is prepared, the lye made from equal portions of lime and the ashes which wool-dyers use, together with equal portions of salt, soap, white argol, and saltpetre, are boiled until in the end the mixture evaporates and becomes salt. This salt is mixed with the concentrates from washing, to melt them.

Saltpetre is prepared in the following manner, in order that it may be suitable for use in assaying ore. It is placed in a pot which is smeared on the inside with litharge, and lye made of quicklime is repeatedly poured over it, and it is heated until the fire consumes it. Wherefore the saltpetre does not kindle with the fire, since it has absorbed the lime which preserves it, and thus it is prepared[20].

The following compositions[21] are recommended to smelt all ores which the heat of fire breaks up or melts only with difficulty. Of these, one is made from stones of the third order, which easily melt when thrown into hot furnaces. They are crushed into pure white powder, and with half an *uncia*

[19]*Sal artificiosus.* These are a sort of stock fluxes. Such mixtures are common in all old assay books, from the *Probierbüchlin* to later than John Cramer in 1737 (whose Latin lectures on Assaying were published in English under the title of "Elements of the Art of Assaying Metals," London, 1741). Cramer observes (p. 51) that: "Artificers compose a "great many fluxes with the above-mentioned salts and with the reductive ones; nay, "some use as many different fluxes as there are different ores and metals; all which, however, "we think needless to describe. It is better to have explained a few of the simpler ones, "which serve for all the others, and are very easily prepared, than to tire the reader with "confused compositions: and this chiefly because unskilled artificers sometimes attempt "to obtain with many ingredients of the same nature heaped up beyond measure, and with "much labour, though not more properly and more securely, what might have been easily "effected, with one only and the same ingredient, thus increasing the number, not at all "the virtue of the things employed. Nevertheless, if anyone loves variety, he may, according "to the proportions and cautions above prescribed, at his will chuse among the simpler kinds "such as will best suit his purpose, and compose a variety of fluxes with them."

[20]This operation apparently results in a coating to prevent the deflagration of the saltpetre—in fact, it might be permitted to translate *inflammatur* "deflagrate," instead of kindle.

[21]The results which would follow from the use of these "fluxes" would obviously depend upon the ore treated. They can all conceivably be successful. Of these, the first is the lead-glass of the German assayers—a flux much emphasized by all old authorities,

of this powder there are mixed two *unciae* of yellow litharge, likewise crushed. This mixture is put into a scorifier large enough to hold it, and placed under the muffle of a hot furnace; when the charge flows like water, which occurs after half an hour, it is taken out of the furnace and poured on to a stone, and when it has hardened it has the appearance of glass, and this is likewise crushed. This powder is sprinkled over any metalliferous ore which does not easily melt when we are assaying it, and it causes the slag to exude.

Others, in place of litharge, substitute lead ash,[22] which is made in the following way: sulphur is thrown into lead which has been melted in a crucible, and it soon becomes covered with a sort of scum; when this is removed, sulphur is again thrown in, and the skin which forms is again taken off; this is frequently repeated, in fact until all the lead is turned into powder. There is a powerful flux compound which is made from one *uncia* each of prepared saltpetre, melted salt, glass-gall, and argol, and one-third of an *uncia* of litharge and a *bes* of glass ground to powder; this flux, being added to an equal weight of ore, liquefies it. A more powerful flux is made by placing together in a pot, smeared on the inside with litharge, equal portions of white argol, common salt, and prepared saltpetre, and these are heated until a white powder is obtained from them, and this is mixed with as much litharge; one part of this compound is mixed with two parts of the ore which is to be assayed. A still more powerful flux than this is made out of ashes of black lead, saltpetre, orpiment, *stibium*, and dried lees of the *aqua* with which gold workers separate gold from silver. The ashes of lead[23] are made from one pound of lead and one pound of sulphur; the lead is flattened out into sheets by pounding with a hammer, and placed alternately with sulphur in a crucible or pot, and they are heated together until the fire consumes the sulphur and the lead turns to ashes. One *libra* of crushed saltpetre is mixed with one *libra* of orpiment similarly ground to powder, and the two are cooked in an iron pan until they liquefy; they are then poured out, and after cooling are again ground to powder. A *libra* of *stibium* and a *bes* of the dried lees (*of what* ?) are placed alternately in a crucible and heated to the point at which they form a button, which is similarly reduced to powder. A *bes* of this powder and one *libra* of the ashes of lead, as well as a *libra* of powder made out of the saltpetre and orpiment, are mixed together and a

including Loehneys, Ercker and Cramner, and used even yet. The "powerful flux" would be a reducing, desulphurizing, and an acid flux. The "more powerful" would be a basic flux in which the reducing action of the argols would be largely neutralized by the nitre. The "still more powerful" would be a strongly sulphurizing basic flux, while the "most powerful" would be a still more sulphurizing flux, but it is badly mixed as to its oxidation and basic properties. (See also note 19 on *sal artificiosus*).

[22]Lead ash (*Cinis Plumbi*. Glossary, *Pleyasch*).—This was obviously, from the method of making, an artificial lead sulphide.

[23]Ashes of lead (*Nigri plumbi cinis*). This, as well as lead ash, was also an artificial lead sulphide. Such substances were highly valued by the Ancients for medicinal purposes. Dioscorides (v, 56) says: "Burned lead (*Molybdos cecaumenos*) is made in this "way: Sprinkle sulphur over some very thinnest lead plates and put them into a new "earthen pot, add other layers, putting sulphur between each layer until the pot is full; set "it alight and stir the melted lead with an iron rod until it is entirely reduced to ashes and "until none of the lead remains unburned. Then take it off, first stopping up your nose, "because the fumes of burnt lead are very injurious. Or burn the lead filings in a pot with "sulphur as aforesaid." Pliny (XXXIV., 50) gives much the same directions.

powder is made from them, one part of which added to two parts of ore liquefies it and cleanses it of dross. But the most powerful flux is one which has two *drachmae* of sulphur and as much glass-galls, and half an *uncia* of each of the following,—*stibium*, salt obtained from boiled urine, melted common salt, prepared saltpetre, litharge, vitriol, argol, salt obtained from ashes of musk ivy, dried lees of the *aqua* by which gold-workers separate gold from silver, alum reduced by fire to powder, and one *uncia* of camphor[24] combined with sulphur and ground into powder. A half or whole portion of this mixture, as the necessity of the case requires, is mixed with one portion of the ore and two portions of lead, and put in a scorifier; it is sprinkled with powder of crushed Venetian glass, and when the mixture has been heated for an hour and a half or two hours, a button will settle in the bottom of the scorifier, and from it the lead is soon separated.

There is also a flux which separates sulphur, orpiment and realgar from metalliferous ore. This flux is composed of equal portions of iron slag, white *tophus*, and salt. After these juices have been secreted, the ores themselves are melted, with argol added to them. There is one flux which preserves *stibium* from the fire, that the fire may not consume it, and which preserves the metals from the *stibium*; and this is composed of equal portions of sulphur, prepared saltpetre, melted salt, and vitriol, heated together in lye until no odour emanates from the sulphur, which occurs after a space of three or four hours.[25]

It is also worth while to substitute certain other mixtures. Take two portions of ore properly prepared, one portion of iron filings, and likewise one portion of salt, and mix; then put them into a scorifier and place them in a muffle furnace; when they are reduced by the fire and run together, a button will settle in the bottom of the scorifier. Or else take equal portions of ore and of lead ochre, and mix with them a small quantity of iron filings, and put them into a scorifier, then scatter iron filings over the mixture. Or else take ore which has been ground to powder and sprinkle it in a crucible, and then sprinkle over it an equal quantity of salt that has been three or four times moistened with urine and dried; then, again and again alternately, powdered ore and salt; next, after the crucible has been covered with a lid and sealed, it is placed upon burning charcoal. Or else take one portion of ore, one portion of minute lead granules, half a portion of Venetian glass, and the same quantity of glass-galls. Or else take one portion of ore, one portion of lead granules, half a portion of salt, one-fourth of a portion of argol, and the same quantity of lees of the *aqua* which separates gold from silver. Or else take equal portions of prepared ore and a powder in which there

[24]Camphor (*camphora*). This was no doubt the well-known gum. Agricola, however, believed that camphor (*De Nat. Fossilium*, p. 224) was a species of bitumen, and he devotes considerable trouble to the refutation of the statements by the Arabic authors that it was a gum. In any event, it would be a useful reducing agent.

[25]Inasmuch as orpiment and realgar are both arsenical sulphides, the use of iron "slag," if it contains enough iron, would certainly matte the sulphur and arsenic. Sulphur and arsenic are the "juices" referred to (see note 4, p. 1). It is difficult to see the object of preserving the antimony with such a sulphurizing "addition," unless it was desired to secure a regulus of antimony alone from a given antimonial ore.

are equal portions of very minute lead granules, melted salt, *stibium* and iron slag. Or else take equal portions of gold ore, vitriol, argol, and of salt. So much for the fluxes.

In the assay furnace, when it has been prepared in the way in which I have described, is first placed a muffle. Then selected pieces of live charcoals are laid on it, for, from pieces of inferior quality, a great quantity of ash collects around the muffle and hinders the action of the fire. Then the scorifiers are placed under the muffle with tongs, and glowing coals are placed under the fore part of the muffle to warm the scorifiers more quickly; and when the lead or ore is to be placed in the scorifiers, they are taken out again with the tongs. When the scorifiers glow in the heat, first of all the ash or small charcoals, if any have fallen into them, should be blown away with an iron pipe two feet long and a digit in diameter; this same thing must be done if ash or small coal has fallen into the cupels. Next, put in a small ball of lead with the tongs, and when this lead has begun to be turned into fumes and consumed, add to it the prepared ore wrapped in paper. It is preferable that the assayer should wrap it in paper, and in this way put it in the scorifier, than that he should drop it in with a copper ladle; for when the scorifiers are small, if he uses a ladle he frequently spills some part of the ore. When the paper is burnt, he stirs the ore with a small charcoal held in the tongs, so that the lead may absorb the metal which is mixed in the ore; when this mixture has taken place, the slag partly adheres by its circumference to the scorifier and makes a kind of black ring, and partly floats on the lead in which is mixed the gold or silver; then the slag must be removed from it.

The lead used must be entirely free from every trace of silver, as is that which is known as *Villacense*.[26] But if this kind is not obtainable, the lead must be assayed separately, to determine with certainty that proportion of silver it contains, so that it may be deducted from the calculation of the ore, and the result be exact; for unless such lead be used, the assay will be false and misleading. The lead balls are made with a pair of iron tongs, about one foot long; its iron claws are so formed that when pressed together they are egg-shaped; each claw contains a hollow cup, and when the claws are closed there extends upward from the cup a passage, so there are two openings, one of which leads to each hollow cup. And so when the molten lead is poured in through the openings, it flows down into the hollow cup, and two balls are formed by one pouring.

In this place I ought not to omit mention of another method of assaying employed by some assayers. They first of all place prepared ore in the scorifiers and heat it, and afterward they add the lead. Of this method I cannot approve, for in this way the ore frequently becomes cemented, and for this reason it does not stir easily afterward, and is very slow in mixing with the lead.

[26]The lead free from silver, called *villacense*, was probably from Bleyberg, not far from Villach in Upper Austria, this locality having been for centuries celebrated for its pure lead. These mines were worked prior to, and long after, Agricola's time.

If the whole space of the furnace covered by the muffle is not filled with scorifiers, cupels are put in the empty space, in order that they may become warmed in the meantime. Sometimes, however, it is filled with scorifiers, when we are assaying many different ores, or many portions of one ore at the same time. Although the cupels are usually dried in one hour, yet smaller ones are done more quickly, and the larger ones more slowly. Unless the cupels are heated before the metal mixed with lead is placed in them, they

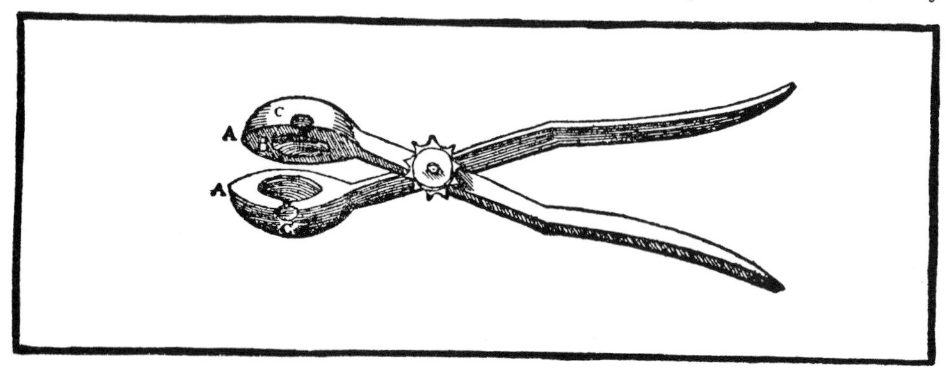

A—Claws of the tongs. B—Iron, giving form of an egg. C—Opening.

frequently break, and the lead always sputters and sometimes leaps out of them; if the cupel is broken or the lead leaps out of it, it is necessary to assay another portion of ore; but if the lead only sputters, then the cupels should be covered with broad thin pieces of glowing charcoal, and when the lead strikes these, it falls back again, and thus the mixture is slowly exhaled. Further, if in the cupellation the lead which is in the mixture is not consumed, but remains fixed and set, and is covered by a kind of skin, this is a sign that it has not been heated by a sufficiently hot fire; put into the mixture, therefore, a dry pine stick, or a twig of a similar tree, and hold it in the hand in order that it can be drawn away when it has been heated. Then take care that the heat is sufficient and equal; if the heat has not passed all round the charge, as it should when everything is done rightly, but causes it to have a lengthened shape, so that it appears to have a tail, this is a sign that the heat is deficient where the tail lies. Then in order that the cupel may be equally heated by the fire, turn it around with a small iron hook, whose handle is likewise made of iron and is a foot and a half long.

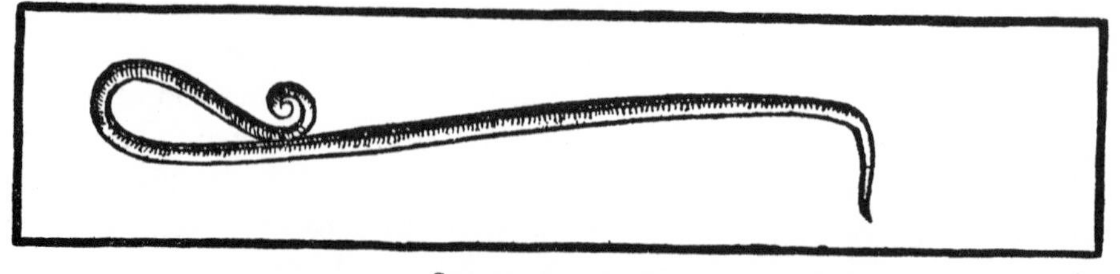

Small iron hook.

Next, if the mixture has not enough lead, add as much of it as is required with the iron tongs, or with the brass ladle to which is fastened a very long handle. In order that the charge may not be cooled, warm the lead beforehand.

But it is better at first to add as much lead as is required to the ore which needs melting, rather than afterward when the melting has been half finished, that the whole quantity may not vanish in fumes, but part of it remain fast. When the heat of the fire has nearly consumed the lead, then is the time when the gold and silver gleam in their varied colours, and when all the lead has been consumed the gold or silver settles in the cupel. Then as soon as possible remove the cupel out of the furnace, and take the button out of it while it is still warm, in order that it does not adhere to the ashes. This generally happens if the button is already cold when it is taken out. If the ashes do adhere to it, do not scrape it with a knife, lest some of it be lost and the assay be erroneous, but squeeze it with the iron tongs, so that the ashes drop off through the pressure. Finally, it is of advantage to make two or three assays of the same ore at the same time, in order that if by chance one is not successful, the second, or in any event the third, may be certain.

While the assayer is assaying the ore, in order to prevent the great heat of the fire from injuring his eyes, it will be useful for him always to have ready a thin wooden tablet, two palms wide, with a handle by which it may be held, and with a slit down the middle in order that he may look through it as through a crack, since it is necessary for him to look frequently within and carefully to consider everything.

A—Handle of tablet. B—Its crack.

Now the lead which has absorbed the silver from a metallic ore is consumed in the cupel by the heat in the space of three quarters of an hour. When the assays are completed the muffle is taken out of the furnace, and the ashes removed with an iron shovel, not only from the brick and iron furnaces, but also from the earthen one, so that the furnace need not be removed from its foundation.

From ore placed in the triangular crucible a button is melted out, from which metal is afterward made. First of all, glowing charcoal is put into the iron hoop, then is put in the triangular crucible, which contains the ore together with those things which can liquefy it and purge it of its dross; then the fire is blown with the double bellows, and the ore is heated until the button settles in the bottom of the crucible. We have explained that there are two methods of assaying ore,—one, by which the lead is mixed with ore in the scorifier and afterward again separated from it in the cupel; the other, by which it is first melted in the triangular earthen crucible and afterward mixed with lead in the scorifier, and later separated from it in the cupel. Now let us consider which is more suitable for each ore, or, if neither is suitable, by what other method in one way or another we can assay it.

We justly begin with a gold ore, which we assay by both methods, for if it is rich and seems not to be strongly resistant to fire, but to liquefy easily, one *centumpondium* of it (known to us as the lesser weights),[27] together with one and a half, or two *unciae* of lead of the larger weights, are mixed together and placed in the scorifier, and the two are heated in the fire until they are well mixed. But since such an ore sometimes resists melting, add a little salt to it, either *sal torrefactus* or *sal artificiosus*, for this will subdue it, and prevent the alloy from collecting much dross; stir it frequently with an iron rod, in order that the lead may flow around the gold on every side, and absorb it and cast out the waste. When this has been done, take out the alloy and cleanse it of slag; then place it in the cupel and heat it until it exhales all the lead, and a bead of gold settles in the bottom.

If the gold ore is seen not to be easily melted in the fire, roast it and extinguish it with brine. Do this again and again, for the more often you roast it and extinguish it, the more easily the ore can be crushed fine, and the more quickly does it melt in the fire and give up whatever dross it possesses.

[27]This method of proportionate weights for assay charges is simpler than the modern English "assay ton," both because of the use of 100 units in the standard of weight (the *centumpondium*), and because of the lack of complication between the Avoirdupois and Troy scales. For instance, an ore containing a *libra* of silver to the *centumpondium* would contain 1/100th part, and the same ratio would obtain, no matter what the actual weight of a *centumpondium* of the "lesser weight" might be. To follow the matter still further, an *uncia* being 1/1,200 of a *centumpondium*, if the ore ran one "*uncia* of the lesser weight" to the "*centumpondium* of the lesser weight," it would also run one actual *uncia* to the actual *centumpondium*; it being a matter of indifference what might be the actual weight of the *centumpondium* upon which the scale of lesser weights is based. In fact Agricola's statement (p. 261) indicates that it weighed an actual *drachma*. We have, in some places, interpolated the expressions "lesser" and "greater" weights for clarity.

This is not the first mention of this scheme of lesser weights, as it appears in the *Probierbüchlein* (1500? see Appendix B) and Biringuccio (1540). For a more complete discussion of weights and measures see Appendix C. For convenience, we repeat here the Roman scale, although, as will be seen in the Appendix, Agricola used the Latin terms in many places merely as nomenclature equivalents of the old German scale.

		Troy Grains.		Ozs. dwts. gr. per short ton.		
1 *Siliqua*	.. .. ..	2.87	Per *Centumpondium* ..	0	3	9
6 *Siliquae*	= 1 *Scripulum* ..	17.2	,, ,, ..	1	0	6
4 *Scripula*	= 1 *Sextula* ..	68.7	,, ,, ..	4	1	0
6 *Sextulae*	= 1 *Uncia* ..	412.2	,, ,, ..	24	6	2
12 *Unciae*	= 1 *Libra* ..	4946.4	,, ,, ..	291	13	8
100 *Librae*	= 1 *Centumpondium*	494640.0				
However Agricola may occasionally use						
16 *Unciae*	= 1 *Libra* ..	6592.0 (?)				
100 *Librae*	= 1 *Centumpondium*	659200.0 (?)				

Also

				Oz. dwts. gr. per short ton.		
1 *Scripulum*	.. .. ..	17.2	Per *Centumpondium* ..	1	0	6
3 *Scripula*	= 1 *Drachma* ..	51.5	,, ,, ..	3	0	19
2 *Drachmae*	= 1 *Sicilicus* ..	103.0	,, ,, ..	6	1	15
4 *Sicilici*	= 1 *Uncia* ..	412.2	,, ,, ..	24	6	12
8 *Unciae*	= 1 *Bes*	3297.6	,, ,, ..	194	12	0

Mix one part of this ore, when it has been roasted, crushed, and washed, with three parts of some powder compound which melts ore, and six parts of lead. Put the charge into the triangular crucible, place it in the iron hoop to which the double bellows reaches, and heat first in a slow fire, and afterward gradually in a fiercer fire, till it melts and flows like water. If the ore does not melt, add to it a little more of these fluxes, mixed with an equal portion of yellow litharge, and stir it with a hot iron rod until it all melts. Then take the crucible out of the hoop, shake off the button when it has cooled, and when it has been cleansed, melt first in the scorifier and afterward in the cupel. Finally, rub the gold which has settled in the bottom of the cupel, after it has been taken out and cooled, on the touchstone, in order to find out what proportion of silver it contains. Another method is to put a *centumpondium* (of the lesser weights) of gold ore into the triangular crucible, and add to it a *drachma* (of the larger weights) of glass-galls. If it resists melting, add half a *drachma* of roasted argol, and if even then it resists, add the same quantity of roasted lees of vinegar, or lees of the *aqua* which separates gold from silver, and the button will settle in the bottom of the crucible. Melt this button again in the scorifier and a third time in the cupel.

We determine in the following way, before it is melted in the muffle furnace, whether pyrites contains gold in it or not: if, after being three times roasted and three times quenched in sharp vinegar, it has not broken nor changed its colour, there is gold in it. The vinegar by which it is quenched should be mixed with salt that is put in it, and frequently stirred and dissolved for three days. Nor is pyrites devoid of gold, when, after being roasted and then rubbed on the touchstone, it colours the touchstone in the same way that it coloured it when rubbed in its crude state. Nor is gold lacking in that, whose concentrates from washing, when heated in the fire, easily melt, giving forth little smell and remaining bright; such concentrates are heated in the fire in a hollowed piece of charcoal covered over with another charcoal.

We also assay gold ore without fire, but more often its sand or the concentrates which have been made by washing, or the dust gathered up by some other means. A little of it is slightly moistened with water and heated until it begins to exhale an odour, and then to one portion of ore are placed two portions of quicksilver[28] in a wooden dish as deep as a basin. They are mixed together with a little brine, and are then ground with a wooden pestle for the space of two hours, until the mixture becomes of the thickness of dough, and the quicksilver can no longer be distinguished from the concentrates made by the washing, nor the concentrates from the quicksilver. Warm, or at least tepid, water is poured into the dish and the material is washed until the water runs out clear. Afterward cold water is poured into the same dish, and soon the quicksilver, which has absorbed all the gold, runs together into a separate place away from the rest of the concentrates made by washing. The quicksilver is afterward separated from the gold by means of a pot covered with soft leather, or with canvas made of woven threads of cotton; the amalgam is poured into the middle of the cloth or

[28]The amalgamation of gold ores is fully discussed in note 12, p. 297.

leather, which sags about one hand's breadth; next, the leather is folded over and tied with a waxed string, and the dish catches the quicksilver which is squeezed through it. As for the gold which remains in the leather, it is placed in a scorifier and purified by being placed near glowing coals. Others do not wash away the dirt with warm water, but with strong lye and vinegar, for they pour these liquids into the pot, and also throw into it the quicksilver mixed with the concentrates made by washing. Then they set the pot in a warm place, and after twenty-four hours pour out the liquids with the dirt, and separate the quicksilver from the gold in the manner which I have described. Then they pour urine into a jar set in the ground, and in the jar place a pot with holes in the bottom, and in the pot they place the gold; then the lid is put on and cemented, and it is joined with the jar; they afterward heat it till the pot glows red. After it has cooled, if there is copper in the gold they melt it with lead in a cupel, that the copper may be separated from it; but if there is silver in the gold they separate them by means of the *aqua* which has the power of parting these two metals. There are some who, when they separate gold from quicksilver, do not pour the amalgam into a leather, but put it into a gourd-shaped earthen vessel, which they place in the furnace and heat gradually over burning charcoal; next, with an iron plate, they cover the opening of the operculum, which exudes vapour, and as soon as it has ceased to exude, they smear it with lute and heat it for a short time; then they remove the operculum from the pot, and wipe off the quicksilver which adheres to it with a hare's foot, and preserve it for future use. By the latter method, a greater quantity of quicksilver is lost, and by the former method, a smaller quantity.

If an ore is rich in silver, as is *rudis* silver[29], frequently silver glance, or rarely ruby silver, gray silver, black silver, brown silver, or yellow silver, as soon as it is cleansed and heated, a *centumpondium* (of the lesser weights) of it is placed in an *uncia* of molten lead in a cupel, and is heated until the lead exhales. But if the ore is of poor or moderate quality, it must first be dried, then crushed, and then to a *centumpondium* (of the lesser weights) an *uncia* of lead is added, and it is heated in the scorifier until it melts. If it is not soon melted by the fire, it should be sprinkled with a little powder of the first order of fluxes, and if then it does not melt, more is added little by little until it melts and exudes its slag; that this result may be reached sooner, the powder which has been sprinkled over it should be stirred in with an iron rod. When the scorifier has been taken out of the assay furnace, the alloy should be poured into a hole in a baked brick; and when it has cooled and been cleansed of the slag, it should be placed in a cupel and heated until it exhales all its lead; the weight of silver which remains in the cupel indicates what proportion of silver is contained in the ore.

We assay copper ore without lead, for if it is melted with it, the copper usually exhales and is lost. Therefore, a certain weight of such an ore

[29]For discussion of the silver ores, see note 8, p. 108. *Rudis* silver was a fairly pure silver mineral, the various coloured silvers were partly horn-silver and partly alteration products.

is first roasted in a hot fire for about six or eight hours ; next, when it has cooled, it is crushed and washed ; then the concentrates made by washing are again roasted, crushed, washed, dried, and weighed. The portion which it has lost whilst it is being roasted and washed is taken into account, and these concentrates by washing represent the cake which will be melted out of the copper ore. Place three *centumpondia* (lesser weights) of this, mixed with three *centumpondia* (lesser weights) each of copper scales[30], saltpetre, and Venetian glass, mixed, into the triangular crucible, and place it in the iron hoop which is set on the hearth in front of the double bellows. Cover the crucible with charcoal in such a way that nothing may fall into the ore which is to be melted, and so that it may melt more quickly. At first blow a gentle blast with the bellows in order that the ore may be heated gradually in the fire ; then blow strongly till it melts, and the fire consumes that which has been added to it, and the ore itself exudes whatever slag it possesses. Next, cool the crucible which has been taken out, and when this is broken you will find the copper ; weigh this, in order to ascertain how great a portion of the ore the fire has consumed. Some ore is only once roasted, crushed, and washed ; and of this kind of concentrates, three *centumpondia* (lesser weights) are taken with one *centumpondium* each of common salt, argol and glass-galls. Heat them in the triangular crucible, and when the mixture has cooled a button of pure copper will be found, if the ore is rich in this metal. If, however, it is less rich, a stony lump results, with which the copper is intermixed ; this lump is again roasted, crushed, and, after adding stones which easily melt and saltpetre, it is again melted in another crucible, and there settles in the bottom of the crucible a button of pure copper. If you wish to know what proportion of silver is in this copper button, melt it in a cupel after adding lead. With regard to this test I will speak later.

Those who wish to know quickly what portion of silver the copper ore contains, roast the ore, crush and wash it, then mix a little yellow litharge with one *centumpondium* (lesser weights) of the concentrates, and put the mixture into a scorifier, which they place under the muffle in a hot furnace for the space of half an hour. When the slag exudes, by reason of the melting force which is in the litharge, they take the scorifier out ; when it has cooled, they cleanse it of slag and again crush it, and with one *centumpondium* of it they mix one and a half *unciae* of lead granules. They then put it into another scorifier, which they place under the muffle in a hot furnace, adding to the mixture a little of the powder of some one of the fluxes which cause ore to melt ; when it has melted they take it out, and after it has cooled, cleanse it of slag ; lastly, they heat it in the cupel till it has exhaled all of the lead, and only silver remains.

Lead ore may be assayed by this method : crush half an *uncia* of pure lead-stone and the same quantity of the *chrysocolla* which they call borax, mix them together, place them in a crucible, and put a glowing coal

[30]It is difficult to see why copper scales (*squamae aeris*—copper oxide ?) are added, unless it be to collect a small ratio of copper in the ore. This additional copper is not mentioned again, however. The whole of this statement is very confused.

in the middle of it. As soon as the borax crackles and the lead-stone melts, which soon occurs, remove the coal from the crucible, and the lead will settle to the bottom of it; weigh it out, and take account of that portion of it which the fire has consumed. If you also wish to know what portion of silver is contained in the lead, melt the lead in the cupel until all of it exhales.

Another way is to roast the lead ore, of whatsoever quality it be, wash it, and put into the crucible one *centumpondium* of the concentrates, together with three *centumpondia* of the powdered compound which melts ore, mixed together, and place it in the iron hoop that it may melt; when it has cooled, cleanse it of its slag, and complete the test as I have already said. Another way is to take two *unciae* of prepared ore, five *drachmae* of roasted copper, one *uncia* of glass, or glass-galls reduced to powder, a *semi-uncia* of salt, and mix them. Put the mixture into the triangular crucible, and heat it over a gentle fire to prevent it from breaking; when the mixture has melted, blow the fire vigorously with the bellows; then take the crucible off the live coals and let it cool in the open air; do not pour water on it, lest the lead button being acted upon by the excessive cold should become mixed with the slag, and the assay in this way be erroneous. When the crucible has cooled, you will find in the bottom of it the lead button. Another way is to take two *unciae* of ore, a *semi-uncia* of litharge, two *drachmae* of Venetian glass and a *semi-uncia* of saltpetre. If there is difficulty in melting the ore, add to it iron filings, which, since they increase the heat, easily separate the waste from lead and other metals. By the last way, lead ore properly prepared is placed in the crucible, and there is added to it only the sand made from stones which easily melt, or iron filings, and then the assay is completed as formerly.

You can assay tin ore by the following method. First roast it, then crush, and afterward wash it; the concentrates are again roasted, crushed, and washed. Mix one and a half *centumpondia* of this with one *centumpondium* of the *chrysocolla* which they call borax; from the mixture, when it has been moistened with water, make a lump. Afterwards, perforate a large round piece of charcoal, making this opening a palm deep, three digits wide on the upper side and narrower on the lower side; when the charcoal is put in its place the latter should be on the bottom and the former uppermost. Let it be placed in a crucible, and let glowing coal be put round it on all sides; when the perforated piece of coal begins to burn, the lump is placed in the upper part of the opening, and it is covered with a wide piece of glowing coal, and after many pieces of coal have been put round it, a hot fire is blown up with the bellows, until all the tin has run out of the lower opening of the charcoal into the crucible. Another way is to take a large piece of charcoal, hollow it out, and smear it with lute, that the ore may not leap out when white hot. Next, make a small hole through the middle of it, then fill up the large opening with small charcoal, and put the ore upon this; put fire in the small hole and blow the fire with the nozzle of a hand bellows; place the piece of charcoal in a small crucible, smeared with lute, in which, when the melting is finished, you will find a button of tin.

In assaying bismuth ore, place pieces of ore in the scorifier, and put it under the muffle in a hot furnace ; as soon as they are heated, they drip with bismuth, which runs together into a button.

Quicksilver ore is usually tested by mixing one part of broken ore with three-parts of charcoal dust and a handful of salt. Put the mixture into a crucible or a pot or a jar, cover it with a lid, seal it with lute, place it on glowing charcoal, and as soon as a burnt cinnabar colour shows in it, take out the vessel ; for if you continue the heat too long the mixture exhales the quicksilver with the fumes. The quicksilver itself, when it has become cool, is found in the bottom of the crucible or other vessel. Another way is to place broken ore in a gourd-shaped earthen vessel, put it in the assay furnace, and cover with an operculum which has a long spout ; under the spout, put an ampulla to receive the quicksilver which distills. Cold water should be poured into the ampulla, so that the quicksilver which has been heated by the fire may be continuously cooled and gathered together, for the quicksilver is borne over by the force of the fire, and flows down through the spout of the operculum into the ampulla. We also assay quicksilver ore in the very same way in which we smelt it. This I will explain in its proper place.

Lastly, we assay iron ore in the forge of a blacksmith. Such ore is burned, crushed, washed, and dried ; a magnet is laid over the concentrates, and the particles of iron are attracted to it ; these are wiped off with a brush, and are caught in a crucible, the magnet being continually passed over the concentrates and the particles wiped off, so long as there remain any particles which the magnet can attract to it. These particles are heated in the crucible with saltpetre until they melt, and an iron button is melted out of them. If the magnet easily and quickly attracts the particles to it, we infer that the ore is rich in iron ; if slowly, that it is poor ; if it appears actually to repel the ore, then it contains little or no iron. This is enough for the assaying of ores.

I will now speak of the assaying of the metal alloys. This is done both by coiners and merchants who buy and sell metal, and by miners, but most of all by the owners and mine masters, and by the owners and masters of the works in which the metals are smelted, or in which one metal is parted from another.

First I will describe the way assays are usually made to ascertain what portion of precious metal is contained in base metal. Gold and silver are now reckoned as precious metals and all the others as base metals. Once upon a time the base metals were burned up, in order that the precious metals should be left pure ; the Ancients even discovered by such burning what portion of gold was contained in silver, and in this way all the silver was consumed, which was no small loss. However, the famous mathematician, Archimedes[31], to gratify King Hiero, invented a method of testing the silver,

[31]This old story runs that Hiero, King of Syracuse, asked Archimedes to tell him whether a crown made for him was pure gold or whether it contained some proportion of silver. Archimedes is said to have puzzled over it until he noticed the increase in water-level upon entering his bath. Whereupon he determined the matter by immersing bars of pure gold and pure silver, and thus determining the relative specific weights. The best

which was not very rapid, and was more accurate for testing a large mass than a small one. This I will explain in my commentaries. The alchemists have shown us a way of separating silver from gold by which neither of them is lost[32].

Gold which contains silver,[33] or silver which contains gold, is first rubbed on the touchstone. Then a needle in which there is a similar amount of gold or silver is rubbed on the same touchstone, and from the lines which are produced in this way, is perceived what portion of silver there is in the gold, or what portion of gold there is in the silver. Next there is added to the silver which is in the gold, enough silver to make it three times as much as the gold. Then lead is placed in a cupel and melted; a little later, a small amount of copper is put in it, in fact, half an *uncia* of it, or half an *uncia* and a *sicilicus* (of the smaller weights) if the gold or silver does not contain any copper. The cupel, when the lead and copper are wanting, attracts the particles of gold and silver, and absorbs them. Finally, one-third of a *libra* of the gold, and one *libra*[34] of the silver must be placed together in the same cupel and melted; for if the gold and silver were first placed in the cupel and melted, as I have already said, it absorbs particles of them, and the gold, when separated from the silver, will not be found pure. These metals are heated until the lead and the copper are consumed, and again, the same weight of each is melted in the same manner in another cupel. The buttons are pounded with a hammer and flattened out, and each little leaf is shaped in the form of a tube, and each is put into a small glass ampulla. Over these there is poured one *uncia* and one *drachma* (of the large weight) of the third quality *aqua valens*, which I will describe in the Tenth Book. This is heated over a slow fire, and small bubbles, resembling pearls in shape, will be seen to adhere to the tubes. The redder the *aqua* appears, the better it is judged to be; when the redness has vanished, small white bubbles are seen to be resting on the tubes, resembling pearls not only in shape, but also in colour. After a short time the *aqua* is poured off and other is poured on; when this has again raised six or eight small white bubbles, it is poured off and the tubes are taken out and washed four or five times with spring water; or if they are heated with the same water, when it is boiling, they will shine more brilliantly. Then they are placed in a saucer, which is held in the hand and gradually dried by the gentle heat of the fire; afterward the saucer is placed over glowing charcoal and covered with a charcoal, and a moderate blast is blown upon it

ancient account of this affair is to be found in Vitruvius, IX, Preface. The story does not seem very probable, seeing that Theophrastus, who died the year Archimedes was born, described the touchstone in detail, and that it was of common knowledge among the Greeks before (see note 37). In any event, there is not sufficient evidence in this story on which to build the conclusion of Meyer (Hist. of Chemistry, p. 14) and others, that, inasmuch as Archimedes was unable to solve the problem until his discovery of specific weights, therefore the Ancients could not part gold and silver. The probability that he did not want to injure the King's jewellery would show sufficient reason for his not parting these metals. It seems probable that the Ancients did part gold and silver by cementation. (See note on p. 458).

[32]The Alchemists (with whose works Agricola was familiar—*vide* preface) were the inventors of nitric acid separation. (See note on p. 460).

[33]Parting gold and silver by nitric acid is more exhaustively discussed in Book X. and notes 10, p. 443.

[34]The lesser weights, probably.

with the mouth and then a blue flame will be emitted. In the end the tubes are weighed, and if their weights prove equal, he who has undertaken this work has not laboured in vain. Lastly, both are placed in another balance-pan and weighed; of each tube four grains must not be counted, on account of the silver which remains in the gold and cannot be separated from it. From the weight of the tubes we learn the weight both of the gold and of the silver which is in the button. If some assayer has omitted to add so much silver to the gold as to make it three times the quantity, but only double, or two and a half times as much, he will require the stronger quality of *aqua* which separates gold from silver, such as the fourth quality. Whether the *aqua* which he employs for gold and silver is suitable for the purpose, or whether it is more or less strong than is right, is recognised by its effect. That of medium strength raises the little bubbles on the tubes and is found to colour the ampulla and the operculum a strong red; the weaker one is found to colour them a light red, and the stronger one to break the tubes. To pure silver in which there is some portion of gold, nothing should be added when they are being heated in the cupel prior to their being parted, except a *bes* of lead and one-fourth or one-third its amount of copper of the lesser weights. If the silver contains in itself a certain amount of copper, let it be weighed, both after it has been melted with the lead, and after the gold has been parted from it; by the former we learn how much copper is in it, by the latter how much gold. Base metals are burnt up even to-day for the purpose of assay, because to lose so little of the metal is small loss, but from a large mass of base metal, the precious metal is always extracted, as I will explain in Books X. and XI.

We assay an alloy of copper and silver in the following way. From a few cakes of copper the assayer cuts out portions, small samples from small cakes, medium samples from medium cakes, and large samples from large cakes; the small ones are equal in size to half a hazel nut, the large ones do not exceed the size of half a chestnut, and those of medium size come between the two. He cuts out the samples from the middle of the bottom of each cake. He places the samples in a new, clean, triangular crucible and fixes to them pieces of paper upon which are written the weight of the cakes of copper, of whatever size they may be; for example, he writes, "These samples have been cut from copper which weighs twenty *centumpondia*." When he wishes to know how much silver one *centumpondium* of copper of this kind has in it, first of all he throws glowing coals into the iron hoop, then adds charcoal to it. When the fire has become hot, the paper is taken out of the crucible and put aside, he then sets that crucible on the fire and gradually heats it for a quarter of an hour until it becomes red hot. Then he stimulates the fire by blowing with a blast from the double bellows for half an hour, because copper which is devoid of lead requires this time to become hot and to melt; copper not devoid of lead melts quicker. When he has blown the bellows for about the space of time stated, he removes the glowing charcoal with the tongs, and stirs the copper with a splinter of wood, which he grasps with the tongs. If it does not stir easily, it is a sign that the copper is not wholly liquefied ; if he finds this is the case, he again places a large piece of charcoal in the crucible, and replaces the glowing charcoal which had been removed, and again blows the bellows for a short time. When all the copper has melted he stops using the bellows, for if he were to continue to use them, the fire would consume part of the copper, and then that which remained would be richer than the cake from which it had been cut ; this is no small mistake. Therefore, as soon as the copper has become sufficiently liquified, he pours it out into a little iron mould, which may be large or small, according as more or less copper is melted in the crucible for the purpose of the assay. The mould has a handle, likewise made of iron, by which it is held when the copper is poured in, after which, he plunges it into a tub of water placed near at hand, that the copper may be cooled. Then he again dries the copper by the fire, and cuts off its point with an iron wedge ; the portion nearest the point he hammers on an anvil and makes into a leaf, which he cuts into pieces.

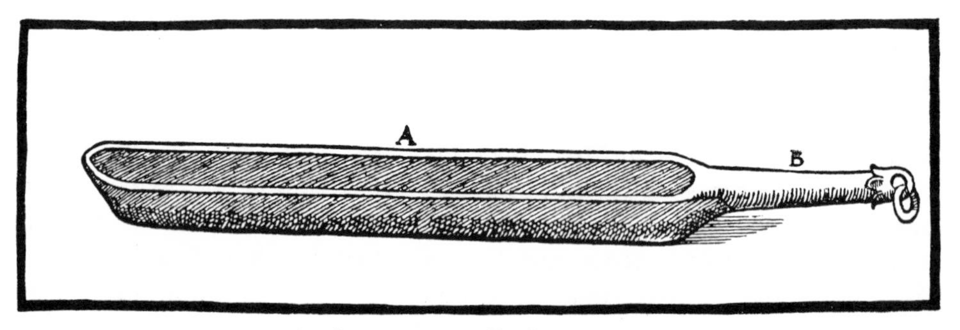

A—Iron mould. B—Its handle.

Others stir the molten copper with a stick of linden tree charcoal, and then pour it over a bundle of new clean birch twigs, beneath which is placed a wooden tub of sufficient size and full of water, and in this manner the copper is broken up into little granules as small as hemp seeds. Others employ straw in place of twigs. Others place a broad stone in a tub and pour in enough water to cover the stone, then they run out the molten copper from the crucible on to the stone, from which the minute granules roll off ; others pour the molten copper into water and stir it until it is resolved into granules. The fire does not easily melt the copper in the cupel unless it has been poured and a thin leaf made of it, or unless it has been resolved into granules or made into filings ; and if it does not melt, all the labour has been undertaken in vain. In order that they may be accurately weighed out, silver and lead are resolved into granules in the same manner as copper. But to return to the assay of copper. When the copper has been prepared by these methods, if it is free of lead and iron, and rich in silver, to each *centumpondium* (lesser weights) add one and a half *unciae* of lead (larger weights). If, however, the copper contains some lead, add one *uncia* of lead ; if it contains iron, add two *unciae.* First put the lead into a cupel, and after it begins to smoke, add the copper ; the fire generally consumes the copper, together with the lead, in about one hour and a quarter. When this is done, the silver will be found in the bottom of the cupel. The fire consumes both of those metals more quickly if they are heated in that furnace which draws in air. It is better to cover the upper half of it with a lid, and not only to put on the muffle door, but also to close the window of the muffle door with a piece of charcoal, or with a piece of brick. If the copper be such that the silver can only be separated from it with difficulty, then before it is tested with fire in the cupel, lead should first be put into the scorifier, and then the copper should be added with a moderate quantity of melted salt, both that the lead may absorb the copper and that the copper may be cleansed of the dross which abounds in it.

Tin which contains silver should not at the beginning of the assay be placed in a cupel, lest the silver, as often happens, be consumed and converted into fumes, together with the tin. As soon as the lead[35] has begun to fume in the scorifier, then add that[36] to it. In this way the lead will take the silver and the tin will boil and turn into ashes, which may be removed with a wooden splinter. The same thing occurs if any alloy is melted in which there is tin. When the lead has absorbed the silver which was in the tin, then, and not till then, it is heated in the cupel. First place the lead with which the silver is mixed, in an iron pan, and stand it on a hot furnace and let it melt; afterward pour this lead into a small iron mould, and then beat it out with a hammer on an anvil and make it into leaves in the same way as the copper. Lastly, place it in the cupel, which assay can be carried out in the space of half an hour. A great heat is harmful to it, for which reason there is no necessity either to cover the half of the furnace with a lid or to close up its mouth.

The minted metal alloys, which are known as money, are assayed in the following way. The smaller silver coins which have been picked out from the bottom and top and sides of a heap are first carefully cleansed; then, after they have been melted in the triangular crucible, they are either resolved into granules, or made into thin leaves. As for the large coins which weigh a *drachma*, a *sicilicus*, half an *uncia*, or an *uncia*, beat them into leaves. Then take a *bes* of the granules, or an equal weight of the leaves, and likewise take another *bes* in the same way. Wrap each sample separately in paper, and afterwards place two small pieces of lead in two cupels which have first been heated. The more precious the money is, the smaller portion of lead do we require for the assay, the more base, the larger is the portion required; for if a *bes* of silver is said to contain only half an *uncia* or one *uncia* of copper, we add to the *bes* of granules half an *uncia* of lead. If it is composed of equal parts of silver and copper, we add an *uncia* of lead, but if in a *bes* of copper there is only half an *uncia* or one *uncia* of silver, we add an *uncia* and a half of lead. As soon as the lead has begun to fume, put into each cupel one of the papers in which is wrapped the sample of silver alloyed with copper, and close the mouth of the muffle with charcoal. Heat them with a gentle fire until all the lead and copper are consumed, for a hot fire by its heat forces the

[35] Lead and Tin seem badly mixed in this paragraph.
[36] It is not clear what is added.

silver, combined with a certain portion of lead, into the cupel, in which way the assay is rendered erroneous. Then take the beads out of the cupel and clean them of dross. If neither depresses the pan of the balance in which it is placed, but their weight is equal, the assay has been free from error; but if one bead depresses its pan, then there is an error, for which reason the assay must be repeated. If the *bes* of coin contains but seven *unciae* of pure silver it is because the King, or Prince, or the State who coins the money, has taken one *uncia*, which he keeps partly for profit and partly for the expense of coining, he having added copper to the silver. Of all these matters I have written extensively in my book *De Precio Metallorum et Monetis*.

We assay gold coins in various ways. If there is copper mixed with the gold, we melt them by fire in the same way as silver coins; if there is silver mixed with the gold, they are separated by the strongest *aqua valens*; if there is copper and silver mixed with the gold, then in the first place, after the addition of lead, they are heated in the cupel until the fire consumes the copper and the lead, and afterward the gold is parted from the silver.

It remains to speak of the touchstone[37] with which gold and silver are tested, and which was also used by the Ancients. For although the assay made by fire is more certain, still, since we often have no furnace, nor muffle, nor crucibles, or some delay must be occasioned in using them, we can always rub gold or silver on the touchstone, which we can have in readiness. Further, when gold coins are assayed in the fire, of what use are they afterward? A touchstone must be selected which is thoroughly black and free of sulphur, for the blacker it is and the more devoid of sulphur, the better it

[37]Historical Note on Touchstone (*Coticula. Interpretatio,—Goldstein*). Theophrastus is, we believe, the first to describe the touchstone, although it was generally known to the Greeks, as is evidenced by the metaphors of many of the poets,—Pindar, Theognis, Euripides, etc. The general knowledge of the constituents of alloys which is implied, raises the question as to whether the Greeks did not know a great deal more about parting metals, than has been attributed to them. Theophrastus says (78–80): "The nature of the stone which tries "gold is also very wonderful, as it seems to have the same power with fire; which is also "a test of that metal. Some people have for this reason questioned the truth of this power "in the stone, but their doubts are ill-founded, for this trial is not of the same nature or "made in the same manner as the other. The trial by fire is by the colour and by the "quantity lost by it; but that by the stone is made only by rubbing the metal on it; the "stone seeming to have the power to receive separately the distinct particles of different "metals. It is said also that there is a much better kind of this stone now found out, than "that which was formerly used; insomuch that it now serves not only for the trial of refined "gold, but also of copper or silver coloured with gold; and shows how much of the "adulterating matter by weight is mixed with gold; this has signs which it yields from "the smallest weight of the adulterating matter, which is a grain, from thence a colybus, "and thence a quadrans or semi-obolus, by which it is easy to distinguish if, and in what "degree, that metal is adulterated. All these stones are found in the River Tmolus; their "texture is smooth and like that of pebbles; their figure broad, not round; and their "bigness twice that of the common larger sort of pebbles. In their use in the trial of metals "there is a difference in power between their upper surface, which has lain toward the sun, "and their under, which has been to the earth; the upper performing its office the more "nicely; and this is consonant to reason, as the upper part is dryer; for the humidity of "the other surface hinders its receiving so well the particles of metals; for the same reason "also it does not perform its office as well in hot weather as in colder, for in the hot it emits "a kind of humidity out of its substance, which runs all over it. This hinders the metalline "particles from adhering perfectly, and makes mistakes in the trials. This exudation of a "humid matter is also common to many other stones, among others, to those of which "statues are made; and this has been looked on as peculiar to the statue." (Based on

generally is; I have written elsewhere of its nature[38]. First the gold is rubbed on the touchstone, whether it contains silver or whether it is obtained from the mines or from the smelting; silver also is rubbed in the same way. Then one of the needles, that we judge by its colour to be of similar composition, is rubbed on the touchstone; if this proves too pale, another needle which has a stronger colour is rubbed on the touchstone; and if this proves too deep in colour, a third which has a little paler colour is used. For this will show us how great a proportion of silver or copper, or silver and copper together, is in the gold, or else how great a proportion of copper is in silver.

These needles are of four kinds.[39] The first kind are made of gold and silver, the second of gold and copper, the third of gold, silver, and copper, and the fourth of silver and copper. The first three kinds of needles are used principally for testing gold, and the fourth for silver. Needles of this kind are prepared in the following ways. The lesser weights correspond proportionately to the larger weights, and both of them are used, not only by mining people, but by coiners also. The needles are made in accordance with the lesser weights, and each set corresponds to a *bes*, which, in our own vocabulary, is called a *mark*. The *bes*, which is employed by those who coin gold, is divided into twenty-four double *sextulae*, which

Hill's trans.) This humid "exudation of fine-grained stones in summer" would not sound abnormal if it were called condensation. Pliny (XXXIII, 43) says: "The mention of "gold and silver should be accompanied by that of the stone called *coticula*. Formerly, "according to Theophrastus, it was only to be found in the river Tmolus but now found in "many parts, it was found in small pieces never over four inches long by two broad. That "side which lay toward the sun is better than that toward the ground. Those experienced "with the *coticula* when they rub ore (*vena*) with it, can at once say how much gold it contains, "how much silver or copper. This method is so accurate that they do not mistake it to a "scruple." This purported use for determining values of *ore* is of about Pliny's average accuracy. The first detailed account of touchneedles and their manner of making, which we have been able to find, is that of the *Probierbüchlein* (1527? see Appendix) where many of the tables given by Agricola may be found.

[38]*De Natura Fossilium* (p. 267) and *De Ortu et Causis Subterraneorum* (p. 59). The author does not add any material mineralogical information to the quotations from Theophrastus and Pliny given above.

[39]In these tables Agricola has simply adopted Roman names as equivalents of the old German weights, but as they did not always approximate in proportions, he coined terms such as "units of 4 *siliquae*," etc. It might seem more desirable to have introduced the German terms into this text, but while it would apply in this instance, as we have discussed on p. 259, the actual values of the Roman weights are very different from the German, and as elsewhere in the book actual Roman weights are applied, we have considered it better to use the Latin terms consistently throughout. Further, the obsolete German would be to most readers but little improvement upon the Latin. For convenience of readers we set out the various scales as used by Agricola, together with the German:—

Roman Scale.				Old German Scale.		
6 *Siliquae*	=	1 *Scripulum*	..	3 *Grenlin*	=	1 *Gran*
4 *Scripula*	=	1 *Sextula*	..	4 *Gran*	=	1 *Krat*
2 *Sextulae*	=	1 *Duella*	..	24 *Kratt*	=	1 *Mark*
24 *Duellae*	=	1 *Bes*		or		
				24 *Grenlin*	=	1 "*Nummus*"
				12 "*Nummi*"	=	1 *Mark*.

Also the following scales are applied to fineness by Agricola:—

3 *Scripula*	=	1 *Drachma*	..	4 *Pfennige*	=	1 *Quintlein*
2 *Drachmae*	=	1 *Sicilicus*	..	4 *Quintlein*	=	1 *Loth*
2 *Sicilici*	=	1 *Semuncia*	..	16 *Loth*	=	1 *Mark*
16 *Semunciae*	=	1 *Bes*				

The term "*nummus*," a coin, given above and in the text, appears in the German translation as *pfennig* as applied to both German scales, but as they are of different values, are now called after the Greek name *ceratia;* and each double *sextula* is divided into four *semi-sextulae,* which are called *granas ;* and each *semi-sextula* is divided into three units of four *siliquae* each, of which each unit is called a *grenlin.* If we made the needles to be each four *siliquae,* there would be two hundred and eighty-eight in a *bes,* but if each were made to be a *semi-sextula* or a double *scripula,* then there would be ninety-six in a *bes.* By these two methods too many needles would be made, and the majority of them, by reason of the small difference in the proportion of the gold, would indicate nothing, therefore it is advisable to make them each of a double *sextula* ; in this way twenty-four needles are made, of which the first is made of twenty-three *duellae* of silver and one of gold. Fannius is our authority that the Ancients called the double *sextula* a *duella.* When a bar of silver is rubbed on the touchstone and colours it just as this needle does, it contains one *duella* of gold. In this manner we determine by the other needles what proportion of gold there is, or when the gold exceeds the silver in weight, what proportion of silver.

The needles are made[40] :—

The	1st	needle	of	23	*duellae*	of silver and	1	*duella*	of gold.
,,	2nd		,,	22	,,	,,	2	*duellae*	of gold.
,,	3rd		,,	21	,,	,,	3	,,	,,
,,	4th		,,	20	,,	,,	4	,,	,,
,,	5th		,,	19	,,	,,	5	,,	,,
,,	6th		,,	18	,,	,,	6	,,	,,
,,	7th		,,	17	,,	,,	7	,,	,,
,,	8th		,,	16	,,	,,	8	,,	,,

we have left Agricola's adaptation in one scale to avoid confusion. The Latin terms adopted by Agricola are given below, together with the German :—

Roman Term.	German Term.	Number in one Mark or Bes.	Value in *Siliquae*.
Siliqua	.. ..	1152	1
" Unit of 4 *Siliquae* "	*Grenlin*	288	4
	Pfennig	256	—
Scripulum	*Scruple* (?)	192	6
Semi-sextula	*Gran*	96	12
Drachma	*Quintlein*	64	18
Sextula	*Halb Krat*	48	24
Sicilicus	*Halb Loth*	32	36
Duella	*Krat*	24	48
Semuncia	*Loth*	16	72
" *Unit of* 5 *Drachmae & 1 Scripulum* "	" *Nummus* "	12	96
Uncia	*Untzen*	8	144
Bes	*Mark*	1	1152

While the proportions in a *bes* or *mark* are the same in both scales, the actual weight values are vastly different—for instance, the *mark* contained about 3609.6, and the *bes* 3297 Troy Grains. Agricola also uses :

Selibra	*Halb-pfundt*
Libra	*Pfundt*
Centumpondium	*Centner.*

As the Roman *libra* contains 12 *unciae* and the German *pfundt* 16 *untzen,* the actual weights of these latter quantities are still further apart—the former 4946 and the latter 7219 Troy grains.

[40]There are no tables in the Latin text, the whole having been written out *in extenso,* but they have now been arranged as above, as being in a much more convenient and expressive form.

The	9th	needle of	15	*duellae* of silver and	9	*duellae* of gold.
,,	10th	,,	14	,, ,,	10	,, ,,
,,	11th	,,	13	,, ,,	11	,, ,,
,,	12th	,,	12	,, ,,	12	,, ,,
,,	13th	,,	11	,, ,,	13	,, ,,
,,	14th	,,	10	,, ,,	14	,, ,,
,,	15th	,,	9	,, ,,	15	,, ,,
,,	16th	,,	8	,, ,,	16	,, ,,
,,	17th	,,	7	,, ,,	17	,, ,,
,,	18th	,,	6	,, ,,	18	,, ,,
,,	19th	,,	5	,, ,,	19	,, ,,
,,	20th	,,	4	,, ,,	20	,, ,,
,,	21st	,,	3	,, ,,	21	,, ,,
,,	22nd	,,	2	,, ,,	22	,, ,,
,,	23rd	,,	1	,, ,,	23	,, ,,
,,	24th	,,	pure gold			

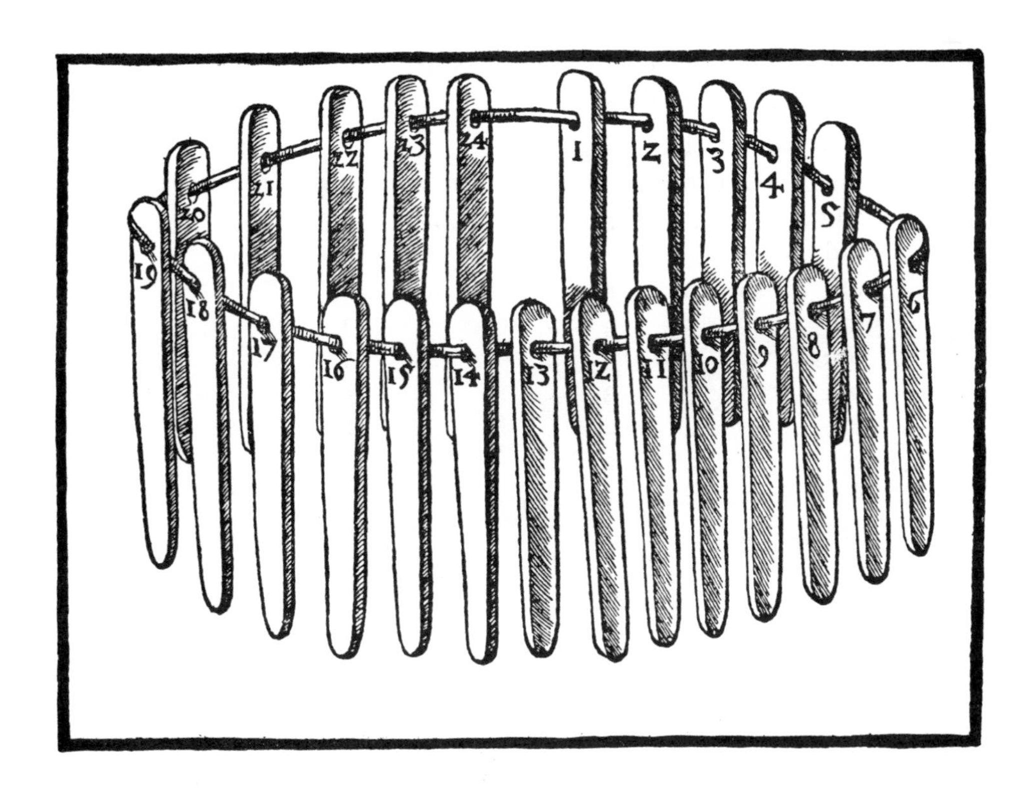

By the first eleven needles, when they are rubbed on the touchstone, we test what proportion of gold a bar of silver contains, and with the remaining thirteen we test what proportion of silver is in a bar of gold; and also what proportion of either may be in money.

Since some gold coins are composed of gold and copper, thirteen needles of another kind are made as follows :—

The	1st	of	12	*duellae*	of gold and	12	*duellae*	of copper.
,,	2nd	,,	13	,,	,,	11	,,	,,
,,	3rd	,,	14	,,	,,	10	,,	,,
,,	4th	,,	15	,,	,,	9	,,	,,
,,	5th	,,	16	,,	,,	8	,,	,,
,,	6th	,,	17	,,	,,	7	,,	,,
,,	7th	,,	18	,,	,,	6	,,	,,
,,	8th	,,	19	,,	,,	5	,,	,,
,,	9th	,,	20	,,	,,	4	,,	,,
,,	10th	,,	21	,,	,,	3	,,	,,
,,	11th	,,	22	,,	,,	2	,,	,,
,,	12th	,,	23	,,	,,	1	,,	,,
,,	13th	,,	pure gold.					

These needles are not much used, because gold coins of that kind are somewhat rare; the ones chiefly used are those in which there is much copper. Needles of the third kind, which are composed of gold, silver, and copper, are more largely used, because such gold coins are common. But since with the gold there are mixed equal or unequal portions of silver and copper, two sorts of needles are made. If the proportion of silver and copper is equal, the needles are as follows :—

			Gold.		Silver.				Copper.			
The	1st	of	12	*duellae*	6	*duellae*	0	*sextula*	6	*duellae*	0	*sextula*
,,	2nd	,,	13	,,	5	,,	1	,,	5	,,	1	,,
,,	3rd	,,	14	,,	5	,,			5	,,		
,,	4th	,,	15	,,	4	,,	1	,,	4	,,	1	,,
,,	5th	,,	16	,,	4	,,			4	,,		
,,	6th	,,	17	,,	3	,,	1	,,	3	,,	1	,,
,,	7th	,,	18	,,	3	,,			3	,,		
,,	8th	,,	19	,,	2	,,	1	,,	2	,,	1	,,
,,	9th	,,	20	,,	2	,,			2	,,		
,,	10th	,,	21	,,	1	,,	1	,,	1	,,	1	,,
,,	11th	,,	22	,,	1	,,			1	,,		
,,	12th	,,	23		1	,,						
,,	13th	,,	pure gold.									

Some make twenty-five needles, in order to be able to detect the two *scripula* of silver or copper which are in a *bes* of gold. Of these needles, the first is composed of twelve *duellae* of gold and six of silver, and the same number of copper. The second, of twelve *duellae* and one *sextula* of gold and five *duellae* and one and a half *sextulae* of silver, and the same number of *duellae* and one and a half *sextulae* of copper. The remaining needles are made in the same proportion.

Pliny is our authority that the Romans could tell to within one *scripulum* how much gold was in any given alloy, and how much silver or copper.

Needles may be made in either of two ways, namely, in the ways of which I have spoken, and in the ways of which I am now about to speak. If unequal portions of silver and copper have been mixed with the gold, thirty-seven needles are made in the following way :—

	Gold.	Silver.			Copper.		
	Duellae.	*Duellae*	*Sextulae*	*Siliquae.*	*Duellae*	*Sextulae*	*Siliquae.*
The 1st of	12	9	0	0	3	0	0
,, 2nd ,,	12	8	0	0	4	0	0
,, 3rd ,,	12	7			5		
,, 4th ,,	13	8	½		2	½	
,, 5th ,,	13	7	½	4	3	1	8
,, 6th ,,	13	6	½	8	4	1	4
,, 7th ,,	14	7	1		2	1	
,, 8th ,,	14	6	1	8	3	½	4
,, 9th ,,	14	5	1½	4	4		8
,, 10th ,,	15	6	1½		2	½	
,, 11th ,,	15	6			3		
,, 12th ,,	15	5	½		3	1½	
,, 13th ,,	16	6			2		
,, 14th ,,	16	5	½	4	2	1	8
,, 15th ,,	16	4	1	8	3	½	4
,, 16th ,,	17	5	½	0	1	1½	
,, 17th ,,	17	4	1	8	2	½	4
,, 18th ,,	17	4	4		2	1½	8
,, 19th ,,	18	4	1		1	1	
,, 20th ,,	18	4	0		2	1	
,, 21st ,,	18	3	1		2		
,, 22nd ,,	19	2	1½		1	½	
,, 23rd ,,	19	3	½	4	1	1	8
,, 24th ,,	19	2	1½	8	2		4
,, 25th ,,	20	3			1		
,, 26th ,,	20	2	1	8	1	½	4
,, 27th ,,	20	2	½	4	1	1	8
,, 28th ,,	21	2	½		1½		
,, 29th ,,	21	2			1		
,, 30th ,,	21	1	1½		1	½	
,, 31st ,,	22	1	1		1		
,, 32nd ,,	22	1	½	4	0	1	8
,, 33rd ,,	22	1		8		1½	4
,, 34th ,,	23		1½			½	
,, 35th ,,	23		1	8		½	4
,, 36th ,,	23		1	4		½	8
,, 37th ,,	pure gold.						

Since it is rarely found that gold, which has been coined, does not amount to at least fifteen *duellae* of gold in a *bes*, some make only twenty-eight needles, and some make them different from those already described, inasmuch as the alloy of gold with silver and copper is sometimes differently proportioned.

These needles are made :—

	Gold.	Silver.			Copper.		
	Duellae.	*Duellae*	*Sextulae*	*Siliquae.*	*Duellae*	*Sextulae*	*Siliquae.*
The 1st of	15	6	1	8	2	½	4
,, 2nd ,,	15	6		4	2	1½	8
,, 3rd ,,	15	5	½		3	1½	
,, 4th ,,	16	6	½		1	1½	
,, 5th ,,	16	5	1	8	2	½	4
,, 6th ,,	16	4	1½	8	3		4
,, 7th ,,	17	5	1	4	1	½	8
,, 8th ,,	17	5		4	1	1½	8
,, 9th ,,	17	4	1	4	2	½	8
,, 10th ,,	18	4	1		1	1	
,, 11th ,,	18	4			2		
,, 12th ,,	18	3	1		2	1	
,, 13th ,,	19	3	1½	4	1	8	
,, 14th ,,	19	3	½	4	1	1	8
,, 15th ,,	19	2	1½	4	2		8
,, 16th ,,	20	3			1		
,, 17th ,,	20	2			1	1	
,, 18th ,,	20	2			2		
,, 19th ,,	21	2	½	4		1	8
,, 20th ,,	21	1	1½	4	1		8
,, 21st ,,	21	1	1	8	1	½	4
,, 22nd ,,	22	1	1	8	½	4	
,, 23rd ,,	22	1	1			1	
,, 24th ,,	22	1	½	4	1	8	
,, 25th ,,	23		1½	4			8
,, 26th ,,	23		1½			½	
,, 27th ,,	23		1	8		½	4
,, 28th ,,	pure gold						

Next follows the fourth kind of needles, by which we test silver coins which contain copper, or copper coins which contain silver. The *bes* by which we weigh the silver is divided in two different ways. It is either divided twelve times, into units of five *drachmae* and one *scripulum* each, which the ordinary people call *nummi*[41]; each of these units we again divide into twenty-four units of four *siliquae* each, which the same ordinary people call a *grenlin*; or else the *bes* is divided into sixteen *semunciae* which are called *loths*, each of which is again divided into eighteen units of four *siliquae* each, which they call *grenlin*. Or else the *bes* is divided into sixteen *semunciae*, of which each is divided into four *drachmae*, and each *drachma* into four *pfennige*. Needles are made in accordance with each method of dividing the *bes*. According to the first method, to the number of twenty-four half *nummi*; according to the second method, to the number of thirty-one half *semunciae*, that is to say a *sicilicus*; for if the needles were made to the number of the smaller weights, the number of needles would again be too large, and not a few of them, by reason of the small difference in proportion of silver or copper, would have no significance. We test both bars and coined money composed of silver and copper by both scales. The one is as follows: the first needle is made of twenty-three parts of copper and one part silver; whereby, whatsoever bar or coin, when rubbed on the touchstone, colours it just as this needle does, in that bar or money there is one twenty-fourth part of silver, and so also, in accordance with the proportion of silver, is known the remaining proportion of the copper.

The	1st	needle is	made of	23	parts of copper	and	1	of silver.
,,	2nd	,,	,,	22	,,	,,	2	,,
,,	3rd	,,	,,	21	,,	,,	3	,,
,,	4th	,,	,,	20	,,	,,	4	,,
,,	5th	,,	,,	19	,,	,,	5	,,
,,	6th	,,	,,	18	,,	,,	6	,,
,,	7th	,,	,,	17	,,	,,	7	,,
,,	8th	,,	,,	16	,,	,,	8	,,
,,	9th	,,	,,	15	,,	,,	9	,,
,,	10th	,,	,,	14	,,	,,	10	,,
,,	11th	,,	,,	13	,,	,,	11	,,
,,	12th	,,	,,	12	,,	,,	12	,,
,,	13th	,,	,,	11	,,	,,	13	,,
,,	14th	,,	,,	10	,,	,,	14	,,
,,	15th	,,	,,	9	,,	,,	15	,,
,,	16th	,,	,,	8	,,	,,	16	,,
,,	17th	,,	,,	7	,,	,,	17	,,
,,	18th	,,	,,	6	,,	,,	18	,,
,,	19th	,,	,,	5	,,	,,	19	,,
,,	20th	,,	,,	4	,,	,,	20	,,
,,	21st	,,	,,	3	,,	,,	21	,,
,,	22nd	,,	,,	2	,,	,,	22	,,
,,	23rd	,,	,,	1	,,	,,	23	,,
,,	24th of pure silver.							

[41]See note 39 above.

The other method of making needles is as follows :—

				Copper.		Silver.	
				Semunciae	*Sicilici.*	*Semunciae*	*Sicilici*
The	1st	is	of	15		1	
,,	2nd	,,	,,	14	1	1	1
,,	3rd	,,	,,	14		2	
,,	4th	,,	,,	13	1	2	1
,,	5th	,,	,,	13		3	
,,	6th	,,	,,	12	1	3	1
,,	7th	,,	,,	12		4	
,,	8th	,,	,,	11	1		1
,,	9th	,,	,,	11		5	
,,	10th	,,	,,	10	1	5	1
,,	11th	,,	,,	10		6	
,,	12th	,,	,,	9	1	6	1
,,	13th	,,	,,	9		7	
,,	14th	,,	,,	8	1	7	1
,,	15th	,,	,,	8		8	
,,	16th	,,	,,	7	1	8	1
,,	17th	,,	,,	7		9	
,,	18th	,,	,,	6	1	9	1
,,	19th	,,	,,	6		10	
,,	20th	,,	,,	5	1	10	1
,,	21st	,,	,,	5		11	
,,	22nd	,,	,,	4	1	11	1
,,	23rd	,,	,,	4		12	
,,	24th	,,	,,	3	1	12	1
,,	25th	,,	,,	3		13	
,,	26th	,,	,,	2	1	13	1
,,	27th	,,	,,	2		14	
,,	28th	,,	,,	1	1	14	1
,,	29th	,,	,,	1		15	
,,	30th	,,	,,		1	15	1
,,	31st of pure silver.						

So much for this. Perhaps I have used more words than those most highly skilled in the art may require, but it is necessary for the understanding of these matters.

I will now speak of the weights, of which I have frequently made mention. Among mining people these are of two kinds, that is, the greater weights and the lesser weights. The *centumpondium* is the first and largest weight, and of course consists of one hundred *librae*, and for that reason is called a hundred weight.

The various weights are :—

1st = 100 *librae* = *centumpondium*.
2nd = 50 ,,
3rd = 52 ,,
4th = 16 ,,
5th = 8 ,,
6th = 4 ,,
7th = 2 ,,
8th = 1 *libra*.

This *libra* consists of sixteen *unciae*, and the half part of the *libra* is the *selibra*, which our people call a *mark*, and consists of eight *unciae*, or, as they divide it, of sixteen *semunciae* :—

9th = 8 *unciae*.
10th = 8 *semunciae*.
11th = 4 ,,
12th = 2 ,,
13th = 1 *semuncia*.
14th = 1 *sicilicus*.
15th = 1 *drachma*.
16th = 1 *dimidi-drachma*.

The above is how the "greater" weights are divided. The "lesser" weights are made of silver or brass or copper. Of these, the first and largest generally weighs one *drachma*, for it is necessary for us to weigh, not only ore, but also metals to be assayed, and smaller quantities of lead. The first of these weights is called a *centumpondium* and the number of *librae* in it corresponds to the larger scale, being likewise one hundred[42].

The 1st is called 1 *centumpondium*.
,, 2nd ,, 50 *librae*.
,, 3rd ,, 25 ,,
,, 4th ,, 16 ,,
,, 5th ,, 8 ,,
,, 6th ,, 4 ,,
,, 7th ,, 2 ,,
,, 8th ,, 1 ,,
,, 9th ,, 1 *selibra*.
,, 10th ,, 8 *semunciae*.
,, 11th ,, 4 ,,
,, 12th ,, 2 ,,
,, 13th ,, 1 ,,
,, 14th ,, 1 *sicilicus*.

The fourteenth is the last, for the proportionate weights which correspond with a *drachma* and half a *drachma* are not used. On all these weights of the lesser scale, are written the numbers of *librae* and of *semunciae*. Some

[42]See note 27, p. 242, for discussion of this "Assay ton" arrangement.

copper assayers divide both the lesser and greater scale weights into divisions of a different scale. Their largest weight of the greater scale weighs one hundred and twelve *librae*, which is the first unit of measurement.

1st	=	112 *librae.*
2nd	=	64 „
3rd	=	32 „
4th	=	16 „
5th	=	8 „
6th	=	4 „
7th	=	2 „
8th	=	1 „
9th	=	1 *selibra* or sixteen *semunciae.*
10th	=	8 *semunciae.*
11th	=	4 „
12th	=	2 „
13th	=	1 „

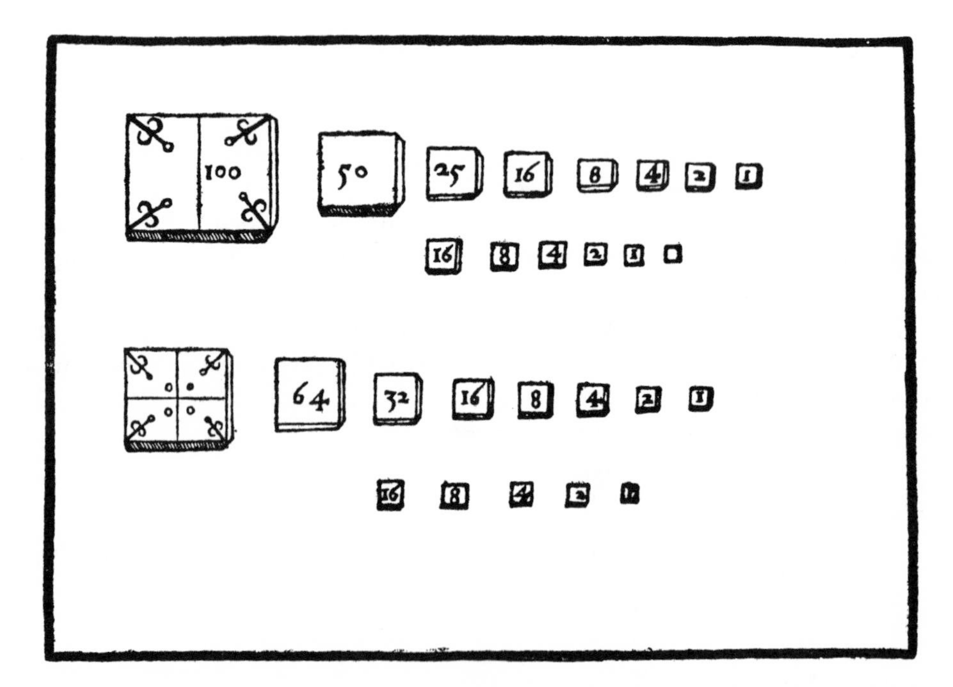

As for the *selibra* of the lesser weights, which our people, as I have often said, call a *mark*, and the Romans call a *bes*, coiners who coin gold, divide it just like the greater weights scale, into twenty-four units of two *sextulae* each, and each unit of two *sextulae* is divided into four *semi-sextulae* and each *semi-sextula* into three units of four *siliquae* each. Some also divide the separate units of four *siliquae* into four individual *siliquae*, but most, omitting the *semi-sextulae*, then divide the double *sextula* into twelve units of four *siliquae* each, and do not divide these into four individual *siliquae*. Thus the first and greatest unit of measurement, which is the *bes*, weighs twenty-four double *sextulae*.

The 2nd = 12 double *sextulae.*
,, 3rd = 6 ,, ,,
,, 4th = 3 ,, ,,
,, 5th = 2 ,, ,,
,, 6th = 1 ,, ,,
,, 7th = 2 *semi-sextulae* or four *semi-sextulae.*
,, 8th = 1 *semi-sextula* or 3 units of 4 *siliquae* each.
,, 9th = 2 units of four *siliquae* each.
,, 10th = 1 ,, ,, ,,

Coiners who mint silver also divide the *bes* of the lesser weights in the same way as the greater weights; our people, indeed, divide it into sixteen *semunciae,* and the *semuncia* into eighteen units of four *siliquae* each.

There are ten weights which are placed in the other pan of the balance, when they weigh the silver which remains from the copper that has been consumed, when they assay the alloy with fire.

The 1st = 16 *semunciae* = 1 *bes.*
,, 2nd = 8 ,,
,, 3rd = 4 ,,
,, 4th = 2 ,,
,, 5th = 1 ,, or 18 units of 4 *siliquae* each.
,, 6th = 9 units of 4 *siliquae* each.
,, 7th = 6 ,, ,,
,, 8th = 3 ,, ,,
,, 9th = 2 ,, ,,
,, 10th = 1 ,, ,,

The coiners of Nuremberg who mint silver, divide the *bes* into sixteen *semunciae,* but divide the *semuncia* into four *drachmae,* and the *drachma* into four *pfennige.* They employ nine weights.

The 1st = 16 *semunciae.*
,, 2nd = 8 ,,
,, 3rd = 4 ,,
,, 4th = 2 ,,
,, 5th = 1 ,,

For they divide the *bes* in the same way as our own people, but since they divide the *semuncia* into four *drachmae,*

the 6th weight = 2 *drachmae.*
,, 7th ,, = 1 *drachma* or 4 *pfennige.*
,, 8th ,, = 2 *pfennige.*
,, 9th ,, = 1 *pfennig*

The men of Cologne and Antwerp[43] divide the *bes* into twelve units of five *drachmae* and one *scripulum,* which weights they call *nummi.* Each of these they again divide into twenty-four units of four *siliquae* each, which they call *grenlins.* They have ten weights, of which

[43] *Agrippinenses* and *Antuerpiani.*

the 1st	= 12 *nummi*	= 1 *bes*.	
„ 2nd	= 6 „		
„ 3rd	= 3 „		
„ 4th	= 2 „		
„ 5th	= 1 „	= 24 units of 4 *siliquae* each.	
„ 6th	= 12 units of 4 *siliquae* each.		
„ 7th	= 6 „	„	
„ 8th	= 3 „	„	
„ 9th	= 2 „	„	
„ 10th	= 1 „	„	

And so with them, just as with our own people, the *mark* is divided into two hundred and eighty-eight *grenlins*, and by the people of Nuremberg it is divided into two hundred and fifty-six *pfennige*. Lastly, the Venetians divide the *bes* into eight *unciae*. The *uncia* into four *sicilici*, the *sicilicus* into thirty-six *siliquae*. They make twelve weights, which they use whenever they wish to assay alloys of silver and copper. Of these

the 1st	= 8 *unciae*	= 1 *bes*.
„ 2nd	= 4 „	
„ 3rd	= 2 „	
„ 4th	= 1 „	or 4 *sicilici*.
„ 5th	= 2 *sicilici*.	
„ 6th	= 1 *sicilicus*.	
„ 7th	= 18 *siliquae*.	
„ 8th	= 9 „	
„ 9th	= 6 „	
„ 10th	= 3 „	
„ 11th	= 2 „	
„ 12th	= 1 „	

Since the Venetians divide the *bes* into eleven hundred and fifty-two *siliquae*, or two hundred and eighty-eight units of 4 *siliquae* each, into which number our people also divide the *bes*, they thus make the same number of *siliquae*, and both agree, even though the Venetians divide the *bes* into smaller divisions.

This, then, is the system of weights, both of the greater and the lesser kinds, which metallurgists employ, and likewise the system of the lesser weights which coiners and merchants employ, when they are assaying metals and coined money. The *bes* of the larger weight with which they provide themselves when they weigh large masses of these things, I have explained in my work *De Mensuris et Ponderibus*, and in another book, *De Precio Metallorum et Monetis*.

There are three small balances by which we weigh ore, metals, and fluxes. The first, by which we weigh lead and fluxes, is the largest among these smaller balances, and when eight *unciae* (of the greater weights) are placed in one of its pans, and the same number in the other, it sustains no damage. The second is more delicate, and by this we weigh the ore or the metal, which is to be assayed ; this is well able to carry one *centumpondium* of the lesser weights in one pan, and in the other, ore or metal as heavy as that weight. The third is the most delicate, and by this we weigh the beads of gold or silver, which, when the assay is completed, settle in the bottom of the cupel. But if anyone weighs lead in the second balance, or an ore in the third, he will do them much injury.

Whatsoever small amount of metal is obtained from a *centumpondium* of the lesser weights of ore or metal alloy, the same greater weight of metal is smelted from a *centumpondium* of the greater weight of ore or metal alloy.

A—First small balance. B—Second. C—Third, placed in a case.

END OF BOOK VII.

BOOK VIII.

UESTIONS of assaying were explained in the last Book, and I have now come to a greater task, that is, to the description of how we extract the metals. First of all I will explain the method of preparing the ore[1]; for since Nature usually creates metals in an impure state, mixed with earth, stones, and solidified juices, it is necessary to separate most of these impurities from the ores as far as can be, before they are smelted, and therefore I will now describe the methods by which the ores are sorted, broken with hammers, burnt, crushed with stamps, ground into powder, sifted, washed, roasted, and calcined[2].

[1]As would be expected, practically all the technical terms used by Agricola in this chapter are adaptations. The Latin terms, *canalis, area, lacus, vasa, cribrum,* and *fossa,* have had to be pressed into service for many different devices, largely by extemporised combinations. Where the devices described have become obsolete, we have adopted the nomenclature of the old works on Cornish methods. The following examples may be of interest:—

Simple buddle	= *Canalis simplex*	Short strake	= *Area curta*
Divided buddle	= *Canalis tabellis distinctus*	Canvas strake	= *Area linteis extensis contecta*
Ordinary strake	= *Canalis devexus*	Limp	= *Radius.*

The strake (or streke) when applied to alluvial tin, would have been termed a "tye" in some parts of Cornwall, and the "short strake" a "gounce." In the case of the stamp mill, inasmuch as almost every mechanical part has its counterpart in a modern mill, we have considered the reader will have less difficulty if the modern designations are used instead of the old Cornish. The following are the essential terms in modern, old Cornish, and Latin:—

Stamp	..Stamper	..*Pilum*	Cams	..Caps	..*Dentes*
Stamp-stem	..Lifter	..*Pilum*	Tappets	..Tongues	..*Pili dentes*
Shoes	..Stamp-heads	..*Capita*	Screens	..Crate	..*Laminae foraminum plenae*
Mortar-box	..Box	..*Capsa*	Settling pit	..Catchers	..*Lacus*
Cam-shaft	..Barrell	..*Axis*	Jigging sieve	..Dilleugher	..*Cribrum angustum*

[2]Agricola uses four Latin verbs in connection with heat operations at temperatures under the melting point: *Calefacio, uro, torreo,* and *cremo.* The first he always uses in the sense of "to warm" or "to heat," but the last three he uses indiscriminately in much the same way as the English verbs burn, roast, and calcine are used; but in general he uses the Latin verbs in the order given to indicate degrees of heat. We have used the English verbs in their technical sense as indicated by the context.

It is very difficult to say when roasting began as a distinct and separate metallurgical step in sulphide ore treatment. The Greeks and Romans worked both lead and copper sulphides (see note on p. 391, and note on p. 403), but neither in the remains of old works nor in their literature is there anything from which satisfactory details of such a step can be obtained. The Ancients, of course, understood lime-burning, and calcined several salts to purify them or to render them more caustic. Practically the only specific mention is by Pliny regarding lead ores (see p. 391). Even the statement of Theophilus (1050-1100, A.D.), may refer simply to rendering ore more fragile, for he says (p. 305) in regard to copper ore: "This stone dug up in abundance is placed upon a pile and burned (*comburitur*) after the "manner of lime. Nor does it change colour, but loses its hardness and can be broken up, "and afterward it is smelted." The *Probierbüchlein* casually mentions roasting prior to assaying, and Biringuccio (III, 2) mentions incidentally that "dry and ill-disposed ores "before everything must be roasted in an open oven so that the air can get in." He gives no further information; and therefore this account of Agricola's becomes practically the first. Apparently roasting, as a preliminary to the treatment of copper sulphides, did not come into use in England until some time later than Agricola, for in Col. Grant Francis' "Smelting of Copper in the Swansea District" (London, 1881, p. 29), a report is set of the "Doeinges of "Jochim Ganse"—an imported German—at the "Mynes by Keswicke in Cumberland, "A.D., 1581," wherein the delinquencies of the then current practice are described: "Thei "never coulde, nether yet can make (copper) under XXII. tymes passinge thro the fire, and "XXII. weekes doeing thereof ane sometyme more. But now the nature of these IX. hurtfull "humors abovesaid being discovered and opened by Jochim's way of doeing, we can, by his "order of workeinge, so correct theim, that parte of theim beinge by nature hurtfull to the

A—Long table. B—Tray. C—Tub.

I will start at the beginning with the first sort of work. Experienced miners, when they dig the ore, sort the metalliferous material from earth, stones, and solidified juices before it is taken from the shafts and tunnels, and they put the valuable metal in trays and the waste into buckets. But if some miner who is inexperienced in mining matters has omitted to do this, or even if some experienced miner, compelled by some unavoidable necessity, has been unable to do so, as soon as the material which has been dug out has been removed from the mine, all of it should be examined, and that part of the ore which is rich in metal sorted from that part of it which is devoid of metal, whether such part be earth, or solidified juices, or stones. To smelt waste together with an ore involves a loss, for some expenditure is thrown away, seeing that out of earth and stones only empty and useless slags are

"copper in wasteinge of it, ar by arte maide freindes, and be not onely an encrease to the "copper, but further it in smeltinge; and the rest of the other evill humors shalbe so "corrected, and their humors so taken from them, that by once rosteinge and once smeltinge "the ure (which shalbe done in the space of three dayes), the same copper ure shall yeeld us "black copper." Jochim proposed by 'rostynge' to be rid of "sulphur, arsineque, and "antimony."

melted out, and further, the solidified juices also impede the smelting of the metals and cause loss. The rock which lies contiguous to rich ore should also be broken into small pieces, crushed, and washed, lest any of the mineral should be lost. When, either through ignorance or carelessness, the miners while excavating have mixed the ore with earth or broken rock, the work of sorting the crude metal or the best ore is done not only by men, but also by boys and women. They throw the mixed material upon a long table, beside which they sit for almost the whole day, and they sort out the ore; when it has been sorted out, they collect it in trays, and when collected they throw it into tubs, which are carried to the works in which the ores are smelted.

The metal which is dug out in a pure or crude state, to which class belong native silver, silver glance, and gray silver, is placed on a stone by the mine foreman and flattened out by pounding with heavy square hammers. These masses, when they have been thus flattened out like plates, are placed either on the stump of a tree, and cut into pieces by pounding an iron chisel into them with a hammer, or else they are cut with an iron tool similar to a pair of shears. One blade of these shears is three feet long, and is firmly fixed in a stump, and the other blade which cuts the metal is six feet long.

A—MASSES OF METAL. B—HAMMER. C—CHISEL. D—TREE STUMPS. E—IRON TOOL SIMILAR TO A PAIR OF SHEARS.

These pieces of metal are afterward heated in iron basins and smelted in the cupellation furnace by the smelters.

Although the miners, in the shafts or tunnels, have sorted over the material which they mine, still the ore which has been broken down and carried out must be broken into pieces by a hammer or minutely crushed, so that the more valuable and better parts can be distinguished from the inferior and worthless portions. This is of the greatest importance in smelting ore, for if the ore is smelted without this separation, the valuable part frequently receives great damage before the worthless part melts in the fire, or else the one consumes the other; this latter difficulty can, however, be partly avoided by the exercise of care and partly by the use of fluxes. Now, if a vein is of poor quality, the better portions which have been broken down and carried out should be thrown together in one place, and the inferior portion and the rock thrown away. The sorters place a hard broad stone on a table; the tables are generally four feet square and made of joined planks, and to the edge of the sides and back are fixed upright planks, which rise about a foot from the table; the front, where the sorter sits, is left open. The

A—Tables. B—Upright planks. C—Hammer. D—Quadrangular hammer. E—Deeper vessel. F—Shallower vessel. G—Iron rod.

lumps of ore, rich in gold or silver, are put by the sorters on the stone and broken up with a broad, but not thick, hammer ; they either break them into pieces and throw them into one vessel, or they break and sort—whence they get their name—the more precious from the worthless, throwing and collecting them separately into different vessels. Other men crush the lumps of ore less rich in gold or silver, which have likewise been put on the stone, with a broad thick hammer, and when it has been well crushed, they collect it and throw it into one vessel. There are two kinds of vessels ; one is deeper, and a little wider in the centre than at the top or bottom ; the other is not so deep though it is broader at the bottom, and becomes gradually a little narrower toward the top. The latter vessel is covered with a lid, while the former is not covered ; an iron rod through the handles, bent over on either end, is grasped in the hand when the vessel is carried. But, above all, it behooves the sorters to be assiduous in their labours.

By another method of breaking ore with hammers, large hard fragments of ore are broken before they are burned. The legs of the workmen —at all events of those who crush pyrites in this manner with large hammers in Goslar—are protected with coverings resembling leggings, and their hands

A—Pyrites. B—Leggings. C—Gloves. D—Hammer.

are protected with long gloves, to prevent them from being injured by the chips which fly away from the fragments.

In that district of Greater Germany which is called Westphalia and in that district of Lower Germany which is named Eifel, the broken ore which has been burned, is thrown by the workmen into a round area paved with the hardest stones, and the fragments are pounded up with iron tools, which are very much like hammers in shape and are used like threshing sledges. This tool is a foot long, a palm wide, and a digit thick, and has an opening in the middle just as hammers have, in which is fixed a wooden handle of no great thickness, but up to three and a half feet long, in order that the workmen can pound the ore with greater force by reason of its weight falling from a greater height. They strike and pound with the broad side of the tool, in the same way as corn is pounded out on a threshing floor with the threshing sledges, although the latter are made of wood and are smooth and fixed to poles. When the ore has been broken into small pieces, they sweep it together with brooms and remove it to the works, where it is washed

A—Area paved with stones. B—Broken ore. C—Area covered with broken ore. D—Iron tool. E—Its handle. F—Broom. G—Short strake. H—Wooden hoe.

in a short strake, at the head of which stands the washer, who draws the water upward with a wooden hoe. The water running down again, carries all the light particles into a trough placed underneath. I shall deal more fully with this method of washing a little later.

Ore is burned for two reasons; either that from being hard, it may become soft and more easily broken and more readily crushed with a hammer or stamps, and then can be smelted; or that the fatty things, that is to say, sulphur, bitumen, orpiment, or realgar[3] may be consumed. Sulphur is frequently found in metallic ores, and, generally speaking, is more harmful to the metals, except gold, than are the other things. It is most harmful of all to iron, and less to tin than to bismuth, lead, silver, or copper. Since very rarely gold is found in which there is not some silver, even gold ores containing sulphur ought to be roasted before they are smelted, because, in a very vigorous furnace fire, sulphur resolves metal into ashes and makes slag of it. Bitumen acts in the same way, in fact sometimes it consumes silver, which we may see in bituminous *cadmia*[4].

I now come to the methods of roasting, and first of all to that one which is common to all ores. The earth is dug out to the required extent, and thus is made a quadrangular area of fair size, open at the front, and above this, firewood is laid close together, and on it other wood is laid transversely, likewise close together, for which reason our countrymen call this pile of wood a crate; this is repeated until the pile attains a height of one or two cubits. Then there is placed upon it a quantity of ore that has been broken into small pieces with a hammer; first the largest of these pieces, next those of medium size, and lastly the smallest, and thus is built up a gently sloping cone. To prevent it from becoming scattered, fine sand of the

[3]*Orpiment* and *realgar* are the red and yellow arsenical sulphides. (See note on p. 111).

[4]*Cadmia bituminosa*. The description of this substance by Agricola, given below, indicates that it was his term for the complex copper-zinc-arsenic-cobalt minerals found in the well-known, highly bituminous, copper schists at Mannsfeld. The later Mineralogists, Wallerius (*Mineralogia*, Stockholm, 1747), Valmont De Bomare (*Mineralogie*, Paris, 1762), and others assume Agricola's *cadmia bituminosa* to be "black arsenic" or "arsenic noir," but we see no reason for this assumption. Agricola's statement (*De Nat. Foss.*, p. 369) is ". . . . the schistose stone dug up at the foot of the Melibocus Mountains, or as they are "now called the Harz (*Hercynium*), near Eisleben, Mannsfeld, and near Hettstedt, is similar "to *spinos* (a bituminous substance described by Theophrastus), if not identical with it. "This is black, bituminous, and cupriferous, and when first extracted from the mine it is thrown "out into an open space and heaped up in a mound. Then the lower part of the mound is "surrounded by faggots, on to which are likewise thrown stones of the same kind. Then "the faggots are kindled and the fire soon spreads to the stones placed upon them; by "these the fire is communicated to the next, which thus spreads to the whole heap. This "easy reception of fire is a characteristic which bitumen possesses in common with sulphur. "Yet the small, pure and black bituminous ore is distinguished from the stones as follows: "when they burn they emit the kind of odour which is usually given off by burning "bituminous coal, and besides, if while they are burning a small shower of rain should fall, they "burn more brightly and soften more quickly. Indeed, when the wind carries the fumes "so that they descend into nearby standing waters, there can be seen floating in it "something like a bituminous liquid, either black, or brown, or purple, which is sufficient to "indicate that those stones were bituminous. And that genus of stones has been recently "found in the Harz in layers, having occasionally gold-coloured specks of pyrites adhering "to them, representing various flat sea-fish or pike or perch or birds, and poultry cocks, "and sometimes salamanders."

A—AREA. B—WOOD. C—ORE. D—CONE-SHAPED PILES. E—CANAL.

same ore is soaked with water and smeared over it and beaten on with shovels; some workers, if they cannot obtain such fine sand, cover the pile with charcoal-dust, just as do charcoal-burners. But at Goslar, the pile, when it has been built up in the form of a cone, is smeared with *atramentum sutorium rubrum*[5], which is made by the leaching of roasted pyrites soaked with water. In some districts the ore is roasted once, in others twice, in others three times, as its hardness may require. At Goslar, when pyrites is roasted for the third time, that which is placed on the top of the pyre exudes a certain greenish, dry, rough, thin substance, as I have elsewhere written[6]; this is no more easily burned by the fire than is asbestos. Very often also, water is put on

[5]*Atramentum sutorium rubrum*. Literally, this would be red vitriol. The German translation gives *rot kupferwasser*, also red vitriol. We must confess that we cannot make this substance out, nor can we find it mentioned in the other works of Agricola. It may be the residue from leaching roasted pyrites for vitriol, which would be reddish oxide of iron.

[6]The statement "elsewhere" does not convey very much more information. It is (*De Nat. Fos.*, p. 253): "When Goslar pyrites and Eisleben (copper) schists are placed on "the pyre and roasted for the third time, they both exude a certain substance which is of a "greenish colour, dry, rough, and fibrous (*tenue*). This substance, like asbestos, is not "consumed by the fire. The schists exude it more plentifully than the pyrites." The *Interpretatio* gives *federwis*, as the German equivalent of *amiantus* (asbestos). This term was used for the feathery alum efflorescence on aluminous slates.

to the ore which has been roasted, while it is still hot, in order to make it softer and more easily broken; for after fire has dried up the moisture in the ore, it breaks up more easily while it is still hot, of which fact burnt limestone affords the best example.

By digging out the earth they make the areas much larger, and square; walls should be built along the sides and back to hold the heat of the fire more effectively, and the front should be left open. In these compartments tin ore is roasted in the following manner. First of all wood about twelve feet long should be laid in the area in four layers, alternately straight and transverse. Then the larger pieces of ore should be laid upon them, and on these again the smaller ones, which should also be placed around the sides; the fine sand of the same ore should also be spread over the pile and pounded with shovels, to prevent the pile from falling before it has been roasted; the wood should then be fired.

A—Lighted pyre. B—Pyre which is being constructed. C—Ore. D—Wood. E—Pile of the same wood.

Lead ore, if roasting is necessary, should be piled in an area just like the last, but sloping, and the wood should be placed over it. A tree trunk should be laid right across the front of the ore to prevent it from falling out. The ore, being roasted in this way, becomes partly melted and resembles slag.

Thuringian pyrites, in which there is gold, sulphur, and vitriol, after the last particle of vitriol has been obtained by heating it in water, is thrown into a furnace, in which logs are placed. This furnace is very similar to an oven in shape, in order that when the ore is roasted the valuable contents may not fly away with the smoke, but may adhere to the roof of the furnace. In this way sulphur very often hangs like icicles from the two openings of the roof through which the smoke escapes.

A—Burning pyre which is composed of lead ore with wood placed above it. B—Workman throwing ore into another area. C—Oven-shaped furnace. D—Openings through which the smoke escapes.

If pyrites or *cadmia*, or any other ore containing metal, possesses a good deal of sulphur or bitumen, it should be so roasted that neither is lost. For this purpose it is thrown on an iron plate full of holes, and roasted with charcoal placed on top; three walls support this plate, two on the sides and the third at the back. Beneath the plate are placed pots containing water, into which the sulphurous or bituminous vapour descends, and in the water the fat accumulates and floats on the top. If it is sulphur, it is generally of a yellow colour; if bitumen, it is black like pitch. If these were not drawn out they would do much harm to the metal, when the ore is being smelted. When they have thus been separated they prove of some service to man, especially the sulphurous kind. From the vapour which is carried down, not

A—Iron plates full of holes. B—Walls. C—Plate on which ore is placed. D—Burning charcoal placed on the ore. E—Pots. F—Furnace. G—Middle part of upper chamber. H—The other two compartments. I—Divisions of the lower chamber. K—Middle wall. L—Pots which are filled with ore. M—Lids of same pots. N—Grating.

into the water, but into the ground, there is created a sulphurous or a bituminous substance resembling *pompholyx*[7], and so light that it can be blown away with a breath. Some employ a vaulted furnace, open at the front and divided into two chambers. A wall built in the middle of the furnace divides the lower chamber into two equal parts, in which are set pots containing water, as above described. The upper chamber is again divided into three parts, the middle one of which is always open, for in it the wood is placed, and it is not broader than the middle wall, of which it forms the topmost portion. The other two compartments have iron doors which are closed, and which, together with the roof, keep in the heat when the wood is lighted. In these upper compartments are iron bars which take the place of a floor, and on these are arranged pots without bottoms, having in place of a bottom, a grating made of iron wire, fixed to each, through the openings of which the sulphurous or bituminous vapours roasted from the ore run into the lower pots. Each of the upper pots holds a hundred

A—HEAP OF CUPRIFEROUS STONES. B—KINDLED HEAP. C—STONES BEING TAKEN TO THE BEDS OF FAGGOTS.

[7]Bearing in mind that bituminous cadmia contained arsenical-cobalt minerals, this substance "resembling *pompholyx*" would probably be arsenic oxide. In *De Natura Fossilium* (p. 368), Agricola discusses the *pompholyx* from *cadmia* at length and pronounces it to be of remarkably "corrosive" quality. (See also note on p. 112.)

pounds of ore; when they are filled they are covered with lids and smeared with lute.

In Eisleben and the neighbourhood, when they roast the schistose stone from which copper is smelted, and which is not free from bitumen, they do not use piles of logs, but bundles of faggots. At one time, they used to pile this kind of stone, when extracted from the pit, on bundles of faggots and roast it by firing the faggots; nowadays, they first of all carry these same stones to a heap, where they are left to lie for some time in such a way as to allow the air and rain to soften them. Then they make a bed of faggot bundles near the heap, and carry the nearest stones to this bed; afterward they again place bundles of faggots in the empty place from which the first stones have been removed, and pile over this extended bed, the stones which lay nearest to the first lot; and they do this right up to the end, until all the stones have been piled mound-shape on a bed of faggots. Finally they fire the faggots, not, however, on the side where the wind is blowing, but on the opposite side, lest the fire blown up by the force of the wind should consume the faggots before the stones are roasted and made soft; by this method the stones which are adjacent to the faggots take fire and communicate it to the next ones, and these again to the adjoining ones, and in this way the heap very often burns continuously for thirty days or more. This schist rock when rich in copper, as I have said elsewhere, exudes a substance of a nature similar to asbestos.

Ore is crushed with iron-shod stamps, in order that the metal may be separated from the stone and the hanging-wall rock.[8] The machines which miners use for this purpose are of four kinds, and are made by the following method. A block of oak timber six feet long, two feet and a palm square, is laid on the ground. In the middle of this is fixed a mortar-box, two feet and six digits long, one foot and six digits deep; the front, which might be called a

[8]HISTORICAL NOTE ON CRUSHING AND CONCENTRATION OF ORES. There can be no question that the first step in the metallurgy of ores was direct smelting, and that this antedates human records. The obvious advantages of reducing the bulk of the material to be smelted by the elimination of barren portions of the ore, must have appealed to metallurgists at a very early date. Logically, therefore, we should find the second step in metallurgy to be concentration in some form. The question of crushing is so much involved with concentration that we have not endeavoured to keep them separate. The earliest indication of these processes appears to be certain inscriptions on monuments of the IV Dynasty (4,000 B.C. ?) depicting gold washing (Wilkinson, The Ancient Egyptians, London, 1874, II, p. 137). Certain stele of the XII Dynasty (2,400 B.C.) in the British Museum (144 Bay 1 and 145 Bay 6) refer to gold washing in the Sudan, and one of them appears to indicate the working of gold ore as distinguished from alluvial. The first written description of the Egyptian methods—and probably that reflecting the most ancient technology of crushing and concentration—is that of Agatharchides, a Greek geographer of the second Century B.C. This work is lost, but the passage in question is quoted by Diodorus Siculus (1st Century B.C.) and by Photius (died 891 A.D.). We give Booth's translation of Diodorus (London, 1700, p. 89), slightly amended: "In the confines of Egypt and the "neighbouring countries of Arabia and Ethiopia there is a place full of rich gold mines, "out of which with much cost and pains of many labourers gold is dug. The soil here "is naturally black, but in the body of the earth run many white veins, shining like "white marble, surpassing in lustre all other bright things. Out of these laborious "mines, those appointed overseers cause the gold to be dug up by the labour of a vast "multitude of people. For the Kings of Egypt condemn to these mines notorious "criminals, captives taken in war, persons sometimes falsely accused, or against "whom the King is incens'd; and not only they themselves, but sometimes all their

mouth, lies open; the bottom is covered with a plate of iron, a palm thick and two palms and as many digits wide, each end of which is wedged into the timber with broad wedges, and the front and back part of it are fixed to the timber with iron nails. To the sides of the mortar above the block are fixed two upright posts, whose upper ends are somewhat cut back and are mortised to the timbers of the building. Two and a half feet above the mortar

"kindred and relations together with them, are sent to work here, both to punish "them, and by their labour to advance the profit and gain of the Kings. There are "infinite numbers upon these accounts thrust down into these mines, all bound in fetters, "where they work continually, without being admitted any rest night or day, and so "strictly guarded that there is no possibility or way left to make an escape. For they "set over them barbarians, soldiers of various and strange languages, so that it is not "possible to corrupt any of the guard by discoursing one with another, or by the gaining "insinuations of familiar converse. The earth which is hardest and full of gold they "soften by putting fire under it, and then work it out with their hands. The rocks thus "soften'd and made more pliant and yielding, several thousands of profligate wretches "break in pieces with hammers and pickaxes. There is one artist that is the overseer of the "whole work, who marks out the stone, and shows the labourers the way and manner "how he would have it done. Those that are the strongest amongst them that are "appointed to this slavery, provided with sharp iron pickaxes, cleave the marble-shining rock "by mere force and strength, and not by arts or sleight-of-hand. They undermine not the "rock in a direct line, but follow the bright shining vein of the mine. They carry lamps "fastened to their foreheads to give them light, being otherwise in perfect darkness in the "various windings and turnings wrought in the mine; and having their bodies appearing "sometimes of one colour and sometimes of another (according to the nature of the mine "where they work) they throw the lumps and pieces of the stone cut out of the rock upon the "floor. And thus they are employed continually without intermission, at the very nod of "the overseer, who lashes them severely besides. And there are little boys who penetrate "through the galleries into the cavities and with great labour and toil gather up the lumps "and pieces hewed out of the rock as they are cast upon the ground, and carry them forth "and lay them upon the bank. Those that are over thirty years of age take a piece of the "rock of such a certain quantity, and pound it in a stone mortar with iron pestles till it be "as small as a vetch; then those little stones so pounded are taken from them by women "and older men, who cast them into mills that stand together there near at hand in a long "row, and two or three of them being employed at one mill they grind a certain measure given "to them at a time, until it is as small as fine meal. No care at all is taken of the bodies of "these poor creatures, so that they have not a rag so much as to cover their nakedness, and "no man that sees them can choose but commiserate their sad and deplorable condition. "For though they are sick, maimed, or lame, no rest nor intermission in the least is allowed "them; neither the weakness of old age, nor women's infirmities are any plea to excuse them; "but all are driven to their work with blows and cudgelling, till at length, overborne with "the intolerable weight of their misery, they drop down dead in the midst of their insufferable "labours; so that these miserable creatures always expect the future to be more terrible "than even the present, and therefore long for death as far more desirable than life.

"At length the masters of the work take the stone thus ground to powder, and carry "it away in order to perfect it. They spread the mineral so ground upon a broad board, some-"what sloping, and pouring water upon it, rub it and cleanse it; and so all the earthy and "drossy part being separated from the rest by the water, it runs off the board, and the gold "by reason of its weight remains behind. Then washing it several times again, they first rub "it lightly with their hands; afterward they draw off any earthy and drossy matter with "slender sponges gently applied to the powdered dust, till it be clean, pure gold. At last "other workmen take it away by weight and measure, and these put it into earthen pots, and "according to the quantity of the gold in every pot they mix with it some lead, grains of "salt, a little tin and barley bran. Then, covering every pot close, and carefully "daubing them over with clay, they put them in a furnace, where they abide five days and "nights together; then after a convenient time that they have stood to cool, nothing of the "other matter is to be found in the pots but only pure, refined gold, some little "thing diminished in the weight. And thus gold is prepared in the borders of Egypt, and "perfected and completed with so many and so great toils and vexations. And, therefore, "I cannot but conclude that nature itself teaches us, that as gold is got with labour and toil, "so it is kept with difficulty; it creates everywhere the greatest cares; and the use of it is "mixed both with pleasure and sorrow."

The remains at Mt. Laurion show many of the ancient mills and concentration works of the Greeks, but we cannot be absolutely certain at what period in the history of these mines crushing and concentration were introduced. While the mines were worked with

are placed two cross-beams joined together, one in front and one in the back, the ends of which are mortised into the upright posts already mentioned. Through each mortise is bored a hole, into which is driven an iron clavis; one end of the clavis has two horns, and the other end is perforated in order that a wedge driven through, binds the beams more firmly; one horn of the clavis turns up and the other down. Three and a half feet above the cross-

great activity prior to 500 B.C. (see note 6, p. 27), it was quite feasible for the ancient miner to have smelted these argentiferous lead ores direct. However, at some period prior to the decadence of the mines in the 3rd Century B.C., there was in use an extensive system of milling and concentration. For the following details we are indebted mostly to Edouard Ardaillon (*Les Mines Du Laurion dans l'Antiquité*, Chap. IV.). The ore was first hand-picked (in 1869 one portion of these rejects was estimated at 7,000,000 tons) and afterward it was apparently crushed in stone mortars some 16 to 24 inches in diameter, and thence passed to the mills. These mills, which crushed dry, were of the upper and lower millstone order, like the old-fashioned flour mills, and were turned by hand. The stones were capable of adjustment in such a way as to yield different sizes. The sand was sifted and the oversize returned to the mills. From the mills it was taken to washing plants, which consisted essentially of an inclined area, below which a canal, sometimes with riffles, lead through a series of basins, ultimately returning the water again to near the head of the area. These washing areas, constructed with great care, were made of stone cemented over smoothly, and were so efficiently done as to remain still intact. In washing, a workman brushed upward the pulp placed on the inclined upper portion of the area, thus concentrating there a considerable proportion of the galena; what escaped had an opportunity to settle in the sequence of basins, somewhat on the order of the buddle. A quotation by Strabo (III, 2, 10) from the lost work of Polybius (200–125 B.C.) also indicates concentration of lead-silver ores in Spain previous to the Christian era: "Polybius speaking of the silver mines of New Carthage, "tells us that they are extremely large, distant from the city about 20 stadia, and occupy a "circuit of 400 stadia, that there are 40,000 men regularly engaged in them, and that they "yield daily to the Roman people (a revenue of) 25,000 drachmae. The rest of the process "I pass over, as it is too long, but as for the silver ore collected, he tells us that it is broken "up, and sifted through sieves over water; that what remains is to be again broken, and the "water having been strained off, it is to be sifted and broken a third time. The dregs which "remain after the fifth time are to be melted, and the lead being poured off, the silver is "obtained pure. These silver mines still exist; however, they are no longer the property "of the state, neither these nor those elsewhere, but are possessed by private individuals. The "gold mines, on the contrary, nearly all belong to the state. Both at Castlon and other "places there are singular lead mines worked. They contain a small proportion of silver, but "not sufficient to pay for the expense of refining." (Hamilton's Translation, Vol. I., p. 222). While Pliny gives considerable information on vein mining and on alluvial washing, the following obscure passage (XXXIII, 21) appears to be the only reference to concentration of ores: "That which is dug out is crushed, washed, roasted, and ground to powder. This "powder is called *apitascudes*, while the silver (lead ?) which becomes disengaged in the "furnace is called *sudor* (sweat). That which is ejected from the chimney is called *scoria* "as with other metals. In the case of gold this *scoria* is crushed and melted again." It is evident enough from these quotations that the Ancients by "washing" and "sifting," grasped the practical effect of differences in specific gravity of the various components of an ore. Such processes are barely mentioned by other mediæval authors, such as Theophilus, Biringuccio, etc., and thus the account in this chapter is the first tangible technical description. Lead mining has been in active progress in Derbyshire since the 13th century, and concentration was done on an inclined board until the 16th century, when William Humpfrey (see below) introduced the jigging sieve. Some further notes on this industry will be found in note 1, p. 77. However, the buddle and strake which appear at that time, are but modest improvements over the board described by Agatharchides in the quotation above.

The ancient crushing appliances, as indicated by the ancient authors and by the Greek and Roman remains scattered over Europe, were hand-mortars and mill-stones of the same order as those with which they ground flour. The stamp-mill, the next advance over grinding in mill-stones, seems to have been invented some time late in the 15th or early in the 16th centuries, but who invented it is unknown. Beckmann (Hist. of Inventions, II, p. 335) says: "In the year 1519 the process of sifting and wet-stamping was established "at Joachimsthal by Paul Grommestetter, a native of Schwarz, named on that account "the Schwarzer, whom Melzer praises as an ingenious and active washer; and we are "told that he had before introduced the same improvements at Schneeberg. Soon after, "that is in 1521, a large stamping-work was erected at Joachimsthal, and the process "of washing was begun. A considerable saving was thus made, as a great many metallic "particles were before left in the washed sand, which was either thrown away or used as "mortar for building. In the year 1525, Hans Pörtner employed at Schlackenwalde the

beams, two other cross-beams of the same kind are again joined in a similar manner; these cross-beams have square openings, in which the iron-shod stamps are inserted. The stamps are not far distant from each other, and fit closely in the cross-beams. Each stamp has a tappet at the back, which requires to be daubed with grease on the lower side that it can be raised more easily. For each stamp there are on a cam-shaft, two cams, rounded on

"wet method of stamping, whereas before that period the ore there was ground. In the "Harz this invention was introduced at Wildenmann by Peter Philip, who was assay-"master there soon after the works at the Upper Harz were resumed by Duke Henry the "Younger, about the year 1524. This we learn from the papers of Herdan Hacke or "Haecke, who was preacher at Wildenmann in 1572."

In view of the great amount of direct and indirect reference to tin mining in Cornwall, covering four centuries prior to Agricola, it would be natural to expect some statement bearing upon the treatment of ore. Curiously enough, while alluvial washing and smelting of the black-tin are often referred to, there is nothing that we have been able to find, prior to Richard Carew's "Survey of Cornwall" (London, 1602, p. 12) which gives any tangible evidence on the technical phases of ore-dressing. In any event, an inspection of charters, tax-rolls, Stannary Court proceedings, etc., prior to that date gives the impression that vein mining was a very minor portion of the source of production. Although Carew's work dates 45 years after Agricola, his description is of interest: "As much almost dooth it "exceede credite, that the Tynne, for and in so small quantitie digged up with so great toyle, "and passing afterwards thorow the managing of so many hands, ere it comes to sale, should "be any way able to acquite the cost: for being once brought above ground in the stone, "it is first broken in peeces with hammers; and then carryed, either in waynes, or on horses' "backs, to a stamping mill, where three, and in some places sixe great logges of timber, "bounde at the ends with yron, and lifted up and downe by a wheele, driven with the water, "doe break it smaller. If the stones be over-moyst, they are dried by the fire in an yron "cradle or grate. From the stamping mill, it passeth to the crazing mill, which betweene "two grinding stones, turned also with a water-wheel, bruseth the same to a find sand; "howbeit, of late times they mostly use wet stampers, and so have no need of the crazing "mills for their best stuffe, but only for the crust of their tayles. The streame, after it hath "forsaken the mill, is made to fall by certayne degrees, one somewhat distant from another; "upon each of which, at every discent, lyeth a greene turfe, three or foure foote square, and "one foote thick. On this the Tinner layeth a certayne portion of the sandie Tinne, and "with his shovel softly tosseth the same to and fro, that, through this stirring, the water "which runneth over it may wash away the light earth from the Tinne, which of a heavier "substance lyeth fast on the turfe. Having so clensed one portion, he setteth the same "aside, and beginneth with another, until his labour take end with his taske. The best of "those turfes (for all sorts serve not) are fetched about two miles to the eastwards of S. "Michael's Mount, where at low water they cast aside the sand, and dig them up: they "are full of rootes of trees, and on some of them nuts have been found, which confirmeth "my former assertion of the sea's intrusion. After it is thus washed, they put the remnant "into a wooden dish, broad, flat, and round, being about two foote over, and having two "handles fastened at the sides, by which they softly shogge the same to and fro in the water "betweene their legges, as they sit over it, untill whatsoever of the earthie substance that "was yet left be flitted away. Some of later time, with a sleighter invention, and lighter "labour, doe cause certayne boyes to stir it up and down with their feete, which worketh "the same effect; the residue, after this often clensing, they call Blacke Tynne."

It will be noticed that the "wet stampers" and the buddle—worked with "boyes "feete"—are "innovations of late times." And the interesting question arises as to whether Cornwall did not derive the stamp-mill, buddle, and strake, from the Germans. The first adequate detailed description of Cornish appliances is that of Pryce (*Mineralogia Cornubiensis*, London, 1778) where the apparatus is identical with that described by Agricola 130 years before. The word "stamper" of Cornwall is of German origin, from *stampfer*, or, as it is often written in old German works, *stamper*. However, the pursuit of the subject through etymology ends here, for no derivatives in German can be found for buddle, tye, strake, or other collateral terms. The first tangible evidence of German influence is to be found in Carew who, continuing after the above quotation, states: "But sithence I gathered "stickes to the building of this poore nest, Sir Francis Godolphin (whose kind helpe hath much "advanced this my playing labour) entertained a Dutch Mynerall man, and taking light from "his experience, but building thereon farre more profitable conclusions of his owne invention, "hath practised a more saving way in these matters, and besides, made Tynne with good "profit of that refuse which Tynners rejected as nothing worth." Beyond this quotation we can find no direct evidence of the influence of "Dutch Mynerall men" in Cornish tin mining at this time. There can be no doubt, however, that in copper mining in Cornwall and elsewhere in England, the "Dutch Mynerall men" did play a large part in the latter

the outer end, which alternately raise the stamp, in order that, by its dropping into the mortar, it may with its iron head pound and crush the rock which has been thrown under it. To the cam-shaft is fixed a water-wheel whose buckets are turned by water-power. Instead of doors, the mouth of the mortar has a board, which is fitted into notches cut out of the front of the block. This board can be raised, in order that when the mouth is open, the workmen

part of the 16th Century. Pettus (*Fodinæ Regales*, London, 1670, p. 20) states that "about "the third year of Queen Elizabeth (1561) she by the advice of her Council sent over for "some Germans experienced in mines, and being supplied, she, on the tenth of October, in the "sixth of her reign, granted the mines of eight counties . . . . to Houghsetter, a "German whose name and family still continue in Cardiganshire." Elizabeth granted large mining rights to various Germans, and the opening paragraphs of two out of several Charters may be quoted in point. This grant is dated 1565, and in part reads: "ELIZABETH, "by the Grace of God, Queen of England, France, and Ireland, Defender of the Faith, &c. "To all Men to whom these Letters Patents shall come, Greeting. Where heretofore we "have granted Privileges to Cornelius de Voz, for the Mining and Digging in our Realm "of England, for Allom and Copperas, and for divers Ewers of Metals that were to be found "in digging for the said Allom and Copperas, incidently and consequently without fraud "or guile, as by the same our Privilege may appear. And where we also moved, by credible "Report to us made, of one Daniel Houghsetter, a German born, and of his Skill and Know-"ledge of and in all manner of Mines, of Metals and Minerals, have given and granted "Privilege to Thomas Thurland, Clerk, one of our Chaplains, and Master of the Hospital of "Savoy, and to the same Daniel, for digging and mining for all manner of Ewers of Gold, "Silver, Copper, and Quicksilver, within our Counties of York, Lancaster, Cumberland, "Westmorland, Cornwall, Devon, Gloucester, and Worcester, and within our Principality "of Wales; and with the same further to deal, as by our said Privilege thereof granted and "made to the said Thomas Thurland and Daniel Houghsetter may appear. *And* we now "being minded that the said Commodities, and all other Treasures of the Earth, in all other "Places of our Realm of England . . . ." On the same date another grant reads: "ELIZABETH, by the Grace of God, Queen of England, France, and Ireland, Defender of the "Faith, &c. To all Men to whom these our Letters Patents shall come, Greeting. Where "we have received credible Information that our faithful and well-beloved Subject William "Humfrey, Saymaster of our Mint within our Tower of London, by his great Endeavour, "Labour, and Charge, hath brought into this our Realm of England one Christopher Shutz, "an Almain, born at *St. Annen Berg*, under the Obedience of the Electer of Saxony; a "Workman as it is reported, of great Cunning, Knowledge, and Experience, as well in the "finding of the Calamin Stone, call'd in Latin, *lapis calaminaris*, and in the right and proper "use and commodity thereof, for the Composition of the mix'd Metal commonly call'd "*latten*, etc." Col. Grant-Francis, in his most valuable collection (Smelting of Copper in the Swansea District, London, 1881) has published a collection of correspondence relating to early mining and smelting operations in Great Britain. And among them (p. 1., etc.) are letters in the years 1583–6 from William Carnsewe and others to Thomas Smyth, with regard to the first smelter erected at Neath, which was based upon copper mines in Cornwall. He mentions "Mr. Weston's (a partner) provydence in bringynge hys Dutch myners hether "to aplye such businys in this countrye ys more to be commendyd than his ignorance of "our countrymen's actyvytyes in suche matters." The principal "Dutche Mineral Master" referred to was one Ulrick Frosse, who had charge of the mine at Perin Sands in Cornwall, and subsequently of the smelter at Neath. Further on is given (p. 25) a Report by Jochim Gaunse upon the Smelting of copper ores at Keswick in Cumberland in 1581, referred to in note 2, p. 267. The Daniel Hochstetter mentioned in the Charter above, together with other German and English gentlemen, formed the "Company of Mines Royal" and among the properties worked were those with which Gaunse's report is concerned. There is in the Record Office, London (Exchequer K.R. Com. Derby 611. Eliz.) the record of an interesting inquisition into Derbyshire methods in which a then recent great improvement was the jigging sieve, the introduction of which was due to William Humphrey (mentioned above). It is possible that he learned of it from the German with whom he was associated. Much more evidence of the activity of the Germans in English mining at this period can be adduced.

On the other hand, Cornwall has laid claims to having taught the art of tin mining and metallurgy to the Germans. Matthew Paris, a Benedictine monk, by birth an Englishman, who died in 1259, relates (*Historia Major Angliae*, London, 1571) that a Cornishman who fled to Germany on account of a murder, first discovered tin there in 1241, and that in consequence the price of tin fell greatly. This statement is recalled with great persistence by many writers on Cornwall. (Camden, *Britannia*, London, 1586; Borlase, Natural History of Cornwall, Oxford, 1758; Pryce, *Mineralogia Cornubiensis*, London, 1778, p. 70, and others).

A—Mortar. B—Upright posts. C—Cross-beams. D—Stamps. E—Their heads. F—Axle (cam-shaft). G—Tooth of the stamp (tappet). H—Teeth of axle (cams).

can remove with a shovel the fine sand, and likewise the coarse sand and broken rock, into which the rocks have been crushed; this board can be lowered, so that the mouth thus being closed, the fresh rock thrown in may be crushed with the iron-shod stamps. If an oak block is not available, two timbers are placed on the ground and joined together with iron clamps, each of the timbers being six feet long, a foot wide, and a foot and a half thick. Such depth as should be allowed to the mortar, is obtained by cutting out the first beam to a width of three-quarters of a foot and to a length of two and a third and one twenty-fourth of a foot. In the bottom of the part thus dug out, there should be laid a very hard rock, a foot thick and three-quarters of a foot wide; about it, if any space remains, earth or sand should be filled in and pounded. On the front, this bed rock is covered with a plank; this rock when it has been broken, should be taken away and replaced by another. A smaller mortar having room for only three stamps may also be made in the same manner.

The stamp-stems are made of small square timbers nine feet long and half a foot wide each way. The iron head of each is made in the following way; the lower part of the head is three palms long and the upper part the same length. The lower part is a palm square in the middle for two palms, then below this, for a length of two digits it gradually spreads until it becomes five digits square; above the middle part, for a length of two digits, it again gradually swells out until it becomes a palm and a half square. Higher up, where the head of the shoe is enclosed in the stem, it is bored through and similarly the stem itself is pierced, and through the opening of each, there passes a broad iron wedge, which prevents the head falling off the stem. To prevent the stamp head from becoming broken by the constant striking of fragments of ore or rocks, there is placed around it a quadrangular iron band a digit thick, seven digits wide, and six digits deep. Those who use three stamps, as is common, make them much larger, and they are made square and three palms broad each way; then the iron shoe of each has a total length of two feet and a palm; at the lower end, it is hexagonal, and at that point it is seven digits wide and thick. The lower part of it which projects beyond the stem is one foot and two palms long; the upper part, which is enclosed in the stem, is three palms long; the

A—STAMP. B—STEM CUT OUT IN LOWER PART. C—SHOE. D—THE OTHER SHOE, BARBED AND GROOVED. E—QUADRANGULAR IRON BAND. F—WEDGE. G—TAPPET. H—ANGULAR CAM-SHAFT. I—CAMS. K—PAIR OF COMPASSES.

lower part is a palm wide and thick ; then gradually the upper part becomes narrower and thinner, so that at the top it is three digits and a half wide and two thick. It is bored through at the place where the angles have been somewhat cut away ; the hole is three digits long and one wide, and is one digit's distance from the top. There are some who make that part of the head which is enclosed in the stem, barbed and grooved, in order that when the hooks have been fixed into the stem and wedges fitted to the grooves, it may remain tightly fixed, especially when it is also held with two quadrangular iron bands. Some divide the cam-shaft with a compass into six sides, others into nine ; it is better for it to be divided into twelve sides, in order that successively one side may contain a cam and the next be without one.

The water-wheel is entirely enclosed under a quadrangular box, in case either the deep snows or ice in winter, or storms, may impede its running and its turning around. The joints in the planks are stopped all around with moss. The cover, however, has one opening, through which there passes a race bringing down water which, dropping on the buckets of the wheel, turns it round, and flows out again in the lower race under the box. The spokes of the water-wheel are not infrequently mortised into the middle of

A—Box. Although the upper part is not open, it is shown open here, that the wheel may be seen. B—Wheel. C—Cam-shaft. D—Stamps.

the cam-shaft; in this case the cams on both sides raise the stamps, which either both crush dry or wet ore, or else the one set crushes dry ore and the other set wet ore, just as circumstances require the one or the other; further, when the one set is raised and the iron clavises in them are fixed into openings in the first cross-beam, the other set alone crushes the ore.

Broken rock or stones, or the coarse or fine sand, are removed from the mortar of this machine and heaped up, as is also done with the same materials when raked out of the dump near the mine. They are thrown by a workman into a box, which is open on the top and the front, and is three feet long and nearly a foot and a half wide. Its sides are sloping and made of planks, but its bottom is made of iron wire netting, and fastened with wire to two iron rods, which are fixed to the two side planks. This bottom has openings, through which broken rock of the size of a hazel nut cannot pass; the pieces which are too large to pass through are removed by the workman, who again places them under stamps, while those which have passed through, together with the coarse and fine sand, he collects in a large vessel and keeps for the washing. When he is performing his laborious

A—Box laid flat on the ground. B—Its bottom which is made of iron wire. C—Box inverted. D—Iron rods. E—Box suspended from a beam, the inside being visible. F—Box suspended from a beam, the outside being visible.

task he suspends the box from a beam by two ropes. This box may rightly be called a quadrangular sieve, as may also that kind which follows.

Some employ a sieve shaped like a wooden bucket, bound with two iron hoops; its bottom, like that of the box, is made of iron wire netting. They place this on two small cross-planks fixed upon a post set in the ground. Some do not fix the post in the ground, but stand it on the ground until there arises a heap of the material which has passed through the sieve, and in this the post is fixed. With an iron shovel the workman throws into this sieve broken rock, small stones, coarse and fine sand raked out of the dump; holding the handles of the sieve in his hands, he agitates it up and down in

A—Sieve. B—Small planks. C—Post. D—Bottom of sieve. E—Open box. F—Small cross-beam. G—Upright posts.

order that by this movement the dust, fine and coarse sand, small stones, and fine broken rock may fall through the bottom. Others do not use a sieve, but an open box, whose bottom is likewise covered with wire netting; this they fix on a small cross-beam fastened to two upright beams and tilt it backward and forward.

Some use a sieve made of copper, having square copper handles on both sides, and through these handles runs a pole, of which one end projects three-quarters of a foot beyond one handle; the workman then places that end in a rope which is suspended from a beam, and rapidly shakes the pole alternately backward and forward. By this movement the small particles fall through the bottom of the sieve. In order that the end of the pole may be easily placed in the rope, a stick, two palms long, holds open the lower part of the rope as it hangs double, each end of the rope being tied to the beam; part of the rope, however, hangs beyond the stick to a length of half a foot. A large box is also used for this purpose, of which the bottom is either made of a plank full of holes or of iron netting, as are the other boxes. An iron bale is fastened from the middle of the planks which form its sides; to this bale is fastened a rope which is suspended from a wooden beam, in order that the box may be moved or tilted in any direction.

A—Box. B—Bale. C—Rope. D—Beam. E—Handles. F—Five-toothed rake. G—Sieve. H—Its handles. I—Pole. K—Rope. L—Timber.

There are two handles on each end, not unlike the handles of a wheelbarrow; these are held by two workmen, who shake the box to and fro. This box is the one principally used by the Germans who dwell in the Carpathian mountains. The smaller particles are separated from the larger ones by means of three boxes and two sieves, in order that those which pass through each, being of equal size, may be washed together; for the bottoms of both the boxes and sieves have openings which do not let through broken rock of the size of a hazel nut. As for the dry remnants in the bottoms of the sieves, if they contain any metal the miners put them under the stamps. The larger pieces of broken rock are not separated from the smaller by this method until the men and boys, with five-toothed rakes, have separated them from the rock fragments, the little stones, the coarse and the fine sand and earth, which have been thrown on to the dumps.

At Neusohl, in the Carpathians, there are mines where the veins of copper lie in the ridges and peaks of the mountains, and in order to save expense being incurred by a long and difficult transport, along a rough and sometimes very precipitous road, one workman sorts over the dumps which have been thrown out from the mines, and another carries in a wheelbarrow the earth, fine and coarse sand, little stones, broken rock, and even the poorer ore, and overturns the barrow into a long open chute fixed to a steep rock. This chute is held apart by small cleats, and the material slides down a distance of about one hundred and fifty feet into a short box, whose bottom is made of a thick copper plate, full of holes. This box has two handles by which it is shaken to and fro, and at the top there are two bales made of hazel sticks, in which is fixed the iron hook of a rope hung from the branch of a tree or from a wooden beam which projects from an upright post. From time to time a sifter pulls this box and thrusts it violently against the tree or post, by which means the small particles passing through its holes descend down another chute into another short box, in whose bottom there are smaller holes. A second sifter, in like manner, thrusts this box violently against a tree or post, and a second time the smaller particles are received into a third chute, and slide down into a third box, whose bottom has still smaller holes. A third sifter, in like manner, thrusts this box violently against a tree or post, and for the third time the tiny particles fall through the holes upon a table. While the workman is bringing in the barrow, another load which has been sorted from the dump, each sifter withdraws the hooks from his bale and carries away his own box and overturns it, heaping up the broken rock or sand which remains in the bottom of it. As for the tiny particles which have slid down upon the table, the first washer—for there are as many washers as sifters—sweeps them off and in a tub nearly full of water, washes them through a sieve whose holes are smaller than the holes of the third box. When this tub has been filled with the material which has passed through the sieve, he draws out the plug to let the water run away; then he removes with a shovel that which has settled in the tub and throws it upon the table of a second washer, who washes it in a sieve with smaller holes. The sediment which has this time settled in his tub, he takes out and throws on the table of a third washer, who washes it in a sieve with the smallest holes. The copper concentrates which have settled in the last tub are taken out and smelted; the sediment which each washer has removed with a limp is washed on a canvas strake. The sifters at Altenberg, in the tin mines of the mountains bordering on Bohemia, use such boxes as I have described, hung from wooden beams. These, however, are a little larger and open in the front, through which opening the broken rock which has not gone through the sieve can be shaken out immediately by thrusting the sieve against its post.

A—Workman carrying broken rock in a barrow. B—First chute. C—First box. D—Its handles. E—Its bales. F—Rope. G—Beam. H—Post. I—Second chute. K—Second box. L—Third chute. M—Third box. N—First table. O—First sieve. P—First tub. Q—Second table. R—Second sieve. S—Second tub. T—Third table. V—Third sieve. X—Third tub. Y—Plugs.

If the ore is rich in metal, the earth, the fine and coarse sand, and the pieces of rock which have been broken from the hanging-wall, are dug out of the dump with a spade or rake and, with a shovel, are thrown into a large sieve or basket, and washed in a tub nearly full of water. The sieve is generally a cubit broad and half a foot deep; its bottom has holes of such size that the larger pieces of broken rock cannot pass through them, for this material rests upon the straight and cross iron wires, which at their points of contact are bound by small iron clips. The sieve is held together by an iron band and by two cross-rods likewise of iron; the rest of the sieve is made of staves in the shape of a little tub, and is bound with two iron hoops; some, however, bind it with hoops of hazel or oak, but in that case they use three of them. On each side it has handles, which are held in the hands by whoever washes the metalliferous material. Into this sieve a boy throws the material to be washed, and a woman shakes it up and down, turning it alternately to the

A—SIEVE. B—ITS HANDLES. C—TUB. D—BOTTOM OF SIEVE MADE OF IRON WIRES. E—HOOP. F—RODS. G—HOOPS. H—WOMAN SHAKING THE SIEVE. I—BOY SUPPLYING IT WITH MATERIAL WHICH REQUIRES WASHING. K—MAN WITH SHOVEL REMOVING FROM THE TUB THE MATERIAL WHICH HAS PASSED THROUGH THE SIEVE.

right and to the left, and in this way passes through it the smaller pieces of earth, sand, and broken rock. The larger pieces remain in the sieve, and these are taken out, placed in a heap and put under the stamps. The mud, together with fine sand, coarse sand, and broken rock, which remain after the water has been drawn out of the tub, is removed by an iron shovel and washed in the sluice, about which I will speak a little later.

The Bohemians use a basket a foot and a half broad and half a foot deep, bound together by osiers. It has two handles by which it is grasped, when they move it about and shake it in the tub or in a small pool nearly full of water. All that passes through it into the tub or pool they take out and wash in a bowl, which is higher in the back part and lower and flat in the front; it is grasped by the two handles and shaken in the water, the lighter particles flowing away, and the heavier and mineral portion sinking to the bottom.

A—Basket. B—Its handles. C—Dish. D—Its back part. E—Its front part. F—Handles of same.

Gold ore, after being broken with hammers or crushed by the stamps, and even tin ore, is further milled to powder. The upper millstone, which is turned by water-power, is made in the following way. An axle is rounded to compass measure, or is made angular, and its iron pinions turn in iron sockets which are held in beams. The axle is turned by a water-wheel, the buckets of which are fixed to the rim and are struck by the force of a stream.

A—Axle. B—Water-wheel. C—Toothed drum. D—Drum made of rundles. E—Iron axle. F—Millstone. G—Hopper. H—Round wooden plate. I—Trough.

Into the axle is mortised a toothed drum, whose teeth are fixed in the side of the rim. These teeth turn a second drum of rundles, which are made of very hard material. This drum surrounds an iron axle which has a pinion at the bottom and revolves in an iron cup in a timber. At the top of the iron axle is an iron tongue, dove-tailed into the millstone, and so when the teeth of the one drum turn the rundles of the other, the millstone is made to turn round. An overhanging machine supplies it with ore through a hopper, and the ore, being ground to powder, is discharged from a round wooden plate into a trough and flowing away through it accumulates on the floor; from there the ore is carried away and reserved for washing. Since this method of grinding requires the millstone to be now raised and now lowered, the timber in whose socket the iron of the pinion axle revolves, rests upon two beams, which can be raised and lowered.

There are three mills in use in milling gold ores, especially for quartz[11] which is not lacking in metal. They are not all turned by water-power, but some by the strength of men, and two of them even by the power of beasts of burden. The first revolving one differs from the next only in its driving wheel, which is closed in and turned by men treading it, or by horses, which are placed inside, or by asses, or even by strong goats; the eyes of these beasts are covered by linen bands. The second mill, both when pushed and turned round, differs from the two above by having an upright axle in the place of the horizontal one; this axle has at its lower end a disc, which two workmen turn by treading back its cleats with their feet, though frequently one man sustains all the labour; or sometimes there projects from the axle a pole which is turned by a horse or an ass, for which reason it is called an *asinaria*. The toothed drum which is at the upper end of the axle turns the drum which is made of rundles, and together with it the millstone.

The third mill is turned round and round, and not pushed by hand; but between this and the others there is a great distinction, for the lower millstone is so shaped at the top that it can hold within it the upper millstone, which revolves around an iron axle; this axle is fastened in the centre of the lower stone and passes through the upper stone. A workman, by grasping in his hand an upright iron bar placed in the upper millstone, moves it round. The middle of the upper millstone is bored through, and the ore, being thrown into this opening, falls down upon the lower millstone and is there ground to powder, which gradually runs out through its opening; it is washed by various methods before it is mixed with quicksilver, which I will explain presently.

Some people build a machine which at one and the same time can crush, grind, cleanse, and wash the gold ore, and mix the gold with quicksilver. This machine has one water-wheel, which is turned by a stream striking its buckets; the main axle on one side of the water-wheel has long cams, which raise the stamps that crush the dry ore. Then the crushed ore is thrown into the hopper of the upper millstone, and gradually falling through the opening, is ground to powder. The lower millstone is square, but has a round depression in which the round, upper millstone turns, and it has an outlet from which the powder falls into the first tub. A vertical iron axle is dovetailed into a cross-piece, which is in turn fixed into the upper millstone; the upper pinion of this axle is held in a bearing fixed in a beam; the drum of the vertical axle is made of rundles, and is turned by the toothed drum on the main axle, and thus turns the millstone. The powder falls continually into the first tub, together with water, and from there runs into a second tub which is set lower down, and out of the second into a third, which is the lowest; from the third, it generally flows into a small trough hewn out of a

[11]*Lapidibus liquescentibus.* (See note 15, p. 380).

A—First mill. B—Wheel turned by goats. C—Second mill. D—Disc of upright axle. E—Its toothed drum. F—Third mill. G—Shape of lower millstone. H—Small upright axle of the same. I—Its opening. K—Lever of the upper millstone. L—Its opening.

tree trunk. Quicksilver[12] is placed in each tub, across which is fixed a small plank, and through a hole in the middle of each plank there passes a small upright axle, which is enlarged above the plank to prevent it from dropping into the tub lower than it should. At the lower end of the axle three sets of paddles intersect, each made from two little boards fixed to the axle opposite each other. The upper end of this axle has a pinion held by a bearing set in a beam, and around each of these axles is a small drum made of rundles, each of which is turned by a small toothed drum on a horizontal

[12]HISTORICAL NOTE ON AMALGAMATION. The recovery of gold by the use of mercury possibly dates from Roman times, but the application of the process to silver does not seem to go back prior to the 16th Century. Quicksilver was well-known to the Greeks, and is described by Theophrastus (105) and others (see note 58, p. 432, on quicksilver). However, the Greeks made no mention of its use for amalgamation, and, in fact, Dioscorides (V, 70) says "it is kept in vessels of glass, lead, tin or silver; if kept in "vessels of any other kind it consumes them and flows away." It was used by them for medicinal purposes. The Romans amalgamated gold with mercury, but whether they took advantage of the principle to recover gold from ores we do not know. Vitruvius (VII, 8) makes the following statement:—"If quicksilver be placed in a vessel and a "stone of a hundred pounds' weight be placed on it, it will swim at the top, and will, "notwithstanding its weight, be incapable of pressing the liquid so as to break or separate "it. If this be taken out, and only a single scruple of gold be put in, that will not swim, but "immediately descend to the bottom. This is a proof that the gravity of a body does not "depend on its weight, but on its nature. Quicksilver is used for many purposes; without "it, neither silver nor brass can be properly gilt. When gold is embroidered on a garment "which is worn out and no longer fit for use, the cloth is burnt over the fire in earthen pots; "the ashes are thrown into water and quicksilver added to them; this collects all the "particles of gold and unites with them. The water is then poured off and the residuum "placed in a cloth, which, when squeezed with the hands, suffers the liquid quicksilver to "pass through the pores of the cloth, but retains the gold in a mass within it." (Gwilt's Trans., p. 217). Pliny is rather more explicit (XXXIII, 32): "All floats on it (quicksilver) "except gold. This it draws into itself, and on that account is the best means of purifying; "for, on being repeatedly agitated in earthen pots it casts out the other things and the "impurities. These things being rejected, in order that it may give up the gold, it is squeezed "in prepared skins, through which, exuding like perspiration, it leaves the gold pure." It may be noted particularly that both these authors state that gold is the only substance that does not float, and, moreover, nowhere do we find any reference to silver combining with mercury, although Beckmann (Hist. of Inventions, Vol. 1, p. 14) not only states that the above passage from Pliny refers to silver, but in further error, attributes the origin of silver amalgamation of ores to the Spaniards in the Indies.

The Alchemists of the Middle Ages were well aware that silver would amalgamate with mercury. There is, however, difficulty in any conclusion that it was applied by them to separating silver or gold from ore. The involved gibberish in which most of their utterances was couched, obscures most of their reactions in any event. The School of Geber (Appendix B) held that all metals were a compound of "spiritual" mercury and sulphur, and they clearly amalgamated silver with mercury, and separated them by distillation. The *Probierbüchlein* (1520 ?) describes a method of recovering silver from the cement used in parting gold and silver, by mixing the cement (silver chlorides) with quicksilver. Agricola nowhere in this work mentions the treatment of silver ores by amalgamation, although he was familiar with Biringuccio (*De La Pirotechnia*), as he himself mentions in the Preface. This work, published at least ten years before *De Re Metallica*, contains the first comprehensive account of silver amalgamation. There is more than usual interest in the description, because, not only did it precede *De Re Metallica*, but it is also a specific explanation of the fundamental essentials of the Patio Process long before the date when the Spaniards could possibly have invented that process in Mexico. We quote Mr. A. Dick's translation from Percy (Metallurgy of Silver and Gold, p. 560):

"He was certainly endowed with much useful and ingenious thought who invented "the short method of extracting metal from the sweepings produced by those arts which have "to do with gold and silver, every substance left in the refuse by smelters, and also the "substance from certain ores themselves, without the labour of fusing, but by the sole "means and virtue of mercury. To effect this, a large basin is first constructed of stone or "timber and walled, into which is fitted a millstone made to turn like that of a mill. Into the "hollow of this basin is placed matter containing gold (*della materia vra che tiene oro*), well "ground in a mortar and afterward washed and dried; and, with the above-mentioned

axle, one end of which is mortised into the large horizontal axle, and the other end is held in a hollow covered with thick iron plates in a beam. Thus the paddles, of which there are three sets in each tub, turn round, and agitating the powder, thoroughly mix it with water and separate the minute particles of gold from it, and these are attracted by the quicksilver and purified. The water carries away the waste. The quicksilver is poured into a bag made of leather or cloth woven from cotton, and when this bag is squeezed, as I have described elsewhere, the quicksilver drips through it into a jar placed underneath. The pure gold[13] remains in the bag. Some people substitute three broad sluices for the tubs, each of which has an angular axle on which are set six narrow spokes, and to them are fixed the same number of broad paddles ; the water that is poured in strikes these paddles and turns them round, and they agitate the powder which is mixed with the water and separate the metal from it. If the powder which is being treated contains gold particles, the first method of washing is far superior, because the quicksilver in the tubs immediately attracts the gold ; if it is powder in which are the small black stones from which tin is smelted, this latter method is not to be despised. It is very advantageous to place interlaced fir boughs in the sluices in which such tin-stuff is washed, after it has run through the launders from the mills, because the fine tin-stone is either held back by the twigs, or if the current carries them along they fall away from the water and settle down.

" millstone, it is ground while being moistened with vinegar, or water, in which has been " dissolved corrosive sublimate (*solimato*), verdigris (*verde rame*), and common salt. Over " these materials is then put as much mercury as will cover them ; they are then stirred for " an hour or two, by turning the millstone, either by hand, or horse-power, according " to the plan adopted, bearing in mind that the more the mercury and the materials are " bruised together by the millstone, the more the mercury may be trusted to have taken up " the substance which the materials contain. The mercury, in this condition, can then be " separated from the earthy matter by a sieve, or by washing, and thus you will recover " the auriferous mercury (*el vro mercurio*). After this, by driving off the mercury by " means of a flask (*i.e.*, by heating in a retort or an alembic), or by passing it through a bag, " there will remain, at the bottom, the gold, silver, or copper, or whatever metal was placed " in the basin under the millstone to be ground. Having been desirous of knowing this " secret, I gave to him who taught it to me a ring with a diamond worth 25 ducats ; he also " required me to give him the eighth part of any profit I might make by using it. This I " wished to tell you, not that you should return the ducats to me for teaching you the secret, " but in order that you should esteem it all the more and hold it dear."

In another part of the treatise Biringuccio states that washed (concentrated) ores may be ultimately reduced either by lead or mercury. Concerning these silver concentrates he writes : " Afterward drenching them with vinegar in which has been put green " copper (*i.e.*, verdigris) ; or drenching them with water in which has been dissolved vitriol " and green copper. . . ." He next describes how this material should be ground with mercury. The question as to who was the inventor of silver amalgamation will probably never be cleared up. According to Ulloa (*Relacion Historica Del Viage a la America Meridional*, Madrid, 1748) Dom Pedro Fernandes De Velasco discovered the process in Mexico in 1566. The earliest technical account is that of Father Joseph De Acosta (*Historia Natural y Moral de las Indias*, Seville, 1590, English trans. Edward Grimston, London, 1604, republished by the Hakluyt Society, 1880). Acosta was born in 1540, and spent the years 1570 to 1585 in Peru, and 1586 in Mexico. It may be noted that Potosi was discovered in 1545. He states that refining silver with mercury was introduced at Potosi by Pedro Fernandes de Velasco from Mexico in 1571, and states (Grimston's Trans., Vol. 1, p. 219) : " . . . They put the powder of the metall into the vessels upon furnaces, whereas they " anoint it and mortifie it with brine, putting to every fiftie quintalles of powder five " quintalles of salt. And this they do for that the salt separates the earth and filth, to the " end the quicksilver may the more easily draw the silver unto it. After, they put quick-

[13]*Aurum in ea remanet purum.* This same error of assuming squeezed amalgam to be pure gold occurs in Pliny: see previous footnote.

A—Water-wheel. B—Axle. C—Stamp. D—Hopper in the upper millstone. E—Opening passing through the centre. F—Lower millstone. G—Its round depression. H—Its outlet. I—Iron axle. K—Its crosspiece. L—Beam. M—Drum of rundles on the iron axle. N—Toothed drum of main axle. O—Tubs. P—The small planks. Q—Small upright axles. R—Enlarged part of one. S—Their paddles. T—Their drums which are made of rundles. V—Small horizontal axle set into the end of the main axle. X—Its toothed drums. Y—Three sluices. Z—Their small axles. AA—Spokes. BB—Paddles.

Seven methods of washing are in common use for the ores of many metals; for they are washed either in a simple buddle, or in a divided buddle, or in an ordinary strake, or in a large tank, or in a short strake, or in a canvas strake, or in a jigging sieve. Other methods of washing are either peculiar to some particular metal, or are combined with the method of crushing wet ore by stamps.

A simple buddle is made in the following way. In the first place, the head is higher than the rest of the buddle, and is three feet long and a foot and a half broad; this head is made of planks laid upon a timber and fastened, and on both sides, side-boards are set up so as to hold the water, which flows in through a pipe or trough, so that it shall fall straight down. The middle of the head is somewhat depressed in order that the broken rock and the larger metallic particles may settle into it. The buddle is sunk into the earth to a depth of three-quarters of a foot below the head, and is twelve feet long and a foot and a half wide and deep; the bottom and each side are lined with planks to prevent the earth, when it is softened by the water, from falling in or from absorbing the metallic particles. The lower end of the buddle is obstructed by a board, which is not as high as the sides. To this straight buddle there is joined a second transverse buddle, six feet long and a foot and a half wide and deep, similarly lined with planks; at the lower

"silver into a piece of holland and presse it out upon the metall, which goes forth like a dewe, "alwaies turning and stirring the metall, to the end it may be well incorporate. Before the "invention of these furnaces of fire, they did often mingle their metall with quicksilver in "great troughes, letting it settle some daies, and did then mix it and stirre it againe, until "they thought all the quicksilver were well incorporate with the silver, the which continued "twentie daies and more, and at least nine daies." Frequent mention of the different methods of silver amalgamation is made by the Spanish writers subsequent to this time, the best account being that of Alonso Barba, a priest. Barba was a native of Lepe, in Andalusia, and followed his calling at various places in Peru from about 1600 to about 1630, and at one time held the Curacy of St. Bernard at Potosi. In 1640 he published at Madrid his *Arte de los Metales*, etc., in five books. The first two books of this work were translated into English by the Earl of Sandwich, and published in London in 1674, under the title "The First Book of the Art of Metals." This translation is equally wretched with those in French and German, as might be expected from the translators' total lack of technical understanding. Among the methods of silver amalgamation described by Barba is one which, upon later "discovery" at Virginia City, is now known as the "Washoe Process." None of the Spanish writers, so far as we know, make reference to Biringuccio's account, and the question arises whether the Patio Process was an importation from Europe or whether it was re-invented in Mexico. While there is no direct evidence on the point, the presumption is in favour of the former.

The general introduction of the amalgamation of silver ores into Central Europe seems to have been very slow, and over 200 years elapsed after its adoption in Peru and Mexico before it received serious attention by the German Metallurgists. Ignaz Elder v. Born was the first to establish the process effectually in Europe, he having in 1784 erected a "quick-mill" at Glasshutte, near Shemnitz. He published an elaborate account of a process which he claimed as his own, under the title *Ueber das Anquicken der Gold und Silberhält igen Erze*, Vienna, 1786. The only thing new in his process seems to have been mechanical agitation. According to Born, a Spaniard named Don Juan de Corduba, in the year 1588, applied to the Court at Vienna offering to extract silver from ores with mercury. Various tests were carried out under the celebrated Lazarus Erckern, and although it appears that some vitriol and salt were used, the trials apparently failed, for Erckern concluded his report with the advice: "That their Lordships should not suffer any more expense to be thrown "away upon this experiment." Born's work was translated into English by R. E. Raspe, under the title—"Baron Inigo Born's New Process of Amalgamation, etc.," London, 1791. Some interest attaches to Raspe, in that he was not only the author of "Baron Munchausen," but was also the villain in Scott's "Antiquary." Raspe was a German Professor at Cassel, who fled to England to avoid arrest for theft. He worked at various mines in Cornwall, and in 1791 involved Sir John Sinclair in a fruitless mine, but disappeared before that was known. The incident was finally used by Sir Walter Scott in this novel.

end it is closed up with a board, also lower than the sides of the buddle so that the water can flow away ; this water falls into a launder and is carried outside the building. In this simple buddle is washed the metallic material which has passed on to the floor of the works through the five large sieves. When this has been gathered into a heap, the washer throws it into the head of the buddle, and water is poured upon it through the pipe or small trough, and the portion which sinks and settles in the middle of the head compartment he stirs with a wooden scrubber,—this is what we will henceforth call the implement made of a stick to which is fixed a piece of wood a foot long and a palm broad. The water is made turbid by this stirring, and carries the mud and sand and small particles of metal into the buddle below. Together with the broken rock, the larger metallic particles remain in the head compartment, and when these have been removed, boys throw them upon the platform of a washing tank or the short strake, and separate them from the broken rock. When the buddle is full of mud and sand, the washer closes the pipe through which the water flows into the head; very soon the water which remains in the buddle flows away, and when this has taken

A—Head of buddle. B—Pipe. C—Buddle. D—Board. E—Transverse buddle. F—Shovel. G—Scrubber.

place, he removes with a shovel the mud and sand which are mixed with minute particles of metal, and washes them on a canvas strake. Sometimes before the buddles have been filled full, the boys throw the material into a bowl and carry it to the strakes and wash it.

Pulverized ore is washed in the head of this kind of a buddle; but usually when tin-stone is washed in it, interlacing fir boughs are put into the buddle, in the same manner as in the sluice when wet ore is crushed with stamps. The larger tin-stone particles, which sink in the upper part of the buddle, are washed separately in a strake; those particles which are of medium size, and settle in the middle part, are washed separately in the same way; and the mud mixed with minute particles of tin-stone, which has settled in the lowest part of the buddle below the fir boughs, is washed separately on the canvas strakes.

The divided buddle differs from the last one by having several cross-boards, which, being placed inside it, divide it off like steps; if the buddle is twelve feet long, four of them are placed within; if nine feet long, three. The nearer each one is to the head, the greater is its height; the further from the head, the lower it is; and so when the highest is a foot and a palm high,

A—Pipe. B—Cross launder. C—Small troughs. D—Head of the buddle. E—Wooden scrubber. F—Dividing boards. G—Short strake.

the second is usually a foot and three digits high, the third a foot and two digits, and the lowest a foot and one digit. In this buddle is generally washed that metalliferous material which has been sifted through the large sieve into the tub containing water. This material is continuously thrown with an iron shovel into the head of the buddle, and the water which has been let in is stirred up by a wooden scrubber, until the buddle is full, then the cross-boards are taken out by the washer, and the water is drained off; next the metalliferous material which has settled in the compartments is again washed, either on a short strake or on the canvas strakes or in the jigging sieves. Since a short strake is often united with the upper part of this buddle, a pipe in the first place carries the water into a cross launder, from which it flows down through one little launder into the buddle, and through another into the short strake.

An ordinary strake, so far as the planks are concerned, is not unlike the last two. The head of this, as of the others, is first made of earth stamped down, then covered with planks; and where it is necessary, earth is thrown in and beaten down a second time, so that no crevice may remain through which water carrying the particles of metal can escape. The water ought to fall straight down into the strake, which has a length of eight feet

A—Head B—Strake. C—Trowel. D—Scrubber. E—Canvas. F—Rod by which the canvas is made smooth.

and a breadth of a foot and a half; it is connected with a transverse launder, which then extends to a settling pit outside the building. A boy with a shovel or a ladle takes the impure concentrates or impure tin-stone from a heap, and throws them into the head of the strake or spreads them over it. A washer with a wooden scrubber then agitates them in the strake, whereby the mud mixed with water flows away into the transverse launder, and the concentrates or the tin-stone settle on the strake. Since sometimes the concentrates or fine tin-stone flow down together with the mud into the transverse launder, a second washer closes it, after a distance of about six feet, with a cross-board and frequently stirs the mud with a shovel, in order that when mixed with water it may flow out into the settling-pit; and there remains in the launder only the concentrates or tin-stone. The tin-stuff of Schlackenwald and Erbisdroff is washed in this kind of a strake once or twice; those of Altenberg three or four times; those of Geyer often seven times; for in the ore at Schlackenwald and Erbisdorff the tin-stone particles are of a fair size, and are crushed with stamps; at Altenberg they are of much smaller size, and in the broken ore at Geyer only a few particles of tin-stone can be seen occasionally.

This method of washing was first devised by the miners who treated tin ore, whence it passed on from the works of the tin workers to those of the silver workers and others; this system is even more reliable than washing in jigging-sieves. Near this ordinary strake there is generally a canvas strake.

In modern times two ordinary strakes, similarly made, are generally joined together; the head of one is three feet distant from that of the other, while the bodies are four feet distant from each other, and there is only one cross launder under the two strakes. One boy shovels, from the heap into the head of each, the concentrates or tin-stone mixed with mud. There are two washers, one of whom sits at the right side of one strake, and the other at the left of the other strake, and each pursues his task, using the following sort of implement. Under each strake is a sill, from a socket in which a round pole rises, and is held by half an iron ring in a beam of the building, so that it may revolve; this pole is nine feet long and a palm thick. Penetrating the pole is a small round piece of wood, three palms long and as many digits thick, to which is affixed a small board two feet long and five digits wide, in an opening of which one end of a small axle revolves, and to this axle is fixed the handle of a little scrubber. The other end of this axle turns in an opening of a second board, which is likewise fixed to a small round piece of wood; this round piece, like the first one, is three palms long and as many digits thick, and is used by the washer as a handle. The little scrubber is made of a stick three feet long, to the end of which is fixed a small tablet of wood a foot long, six digits broad, and a digit and a half thick. The washer constantly moves the handle of this implement with one hand; in this way the little scrubber stirs the concentrates or the fine tin-stone mixed with mud in the head of the strake, and the mud, on being stirred, flows on to the strake. In the other hand he holds a second

A—Upper cross launder. B—Small launders. C—Heads of strakes. D—Strakes. E—Lower transverse launder. F—Settling pit. G—Socket in the sill. H—Halved iron rings fixed to beam. I—Pole. K—Its little scrubber. L—Second small scrubber.

little scrubber, which has a handle of half the length, and with this he ceaselessly stirs the concentrates or tin-stone which have settled in the upper part of the strake; in this way the mud and water flow down into the transverse launder, and from it into the settling-pit which is outside the building.

Before the short strake and the jigging-sieve had been invented, metalliferous ores, especially tin, were crushed dry with stamps and washed in a large trough hollowed out of one or two tree trunks; and at the head of this trough was a platform, on which the ore was thrown after being completely crushed. The washer pulled it down into the trough with a wooden scrubber which had a long handle, and when the water had been let into the trough, he stirred the ore with the same scrubber.

A—Trough. B—Platform. C—Wooden scrubber.

The short strake is narrow in the upper part where the water flows down into it through the little launder; in fact it is only two feet wide; at the lower end it is wider, being three feet and as many palms. At the sides, which are six feet long, are fixed boards two palms high. In other respects the head resembles the head of the simple buddle, except that it is not depressed in the middle. Beneath is a cross launder closed by a low board. In this short strake not only is ore agitated and washed with a wooden scrubber, but boys also separate the concentrates from the broken rock in them and collect them in tubs. The short strake is now rarely employed by miners, owing to the carelessness of the boys, which has been frequently detected; for this reason, the jigging-sieve has taken its place. The mud which settles in the launder, if the ore is rich, is taken up and washed in a jigging-sieve or on a canvas strake.

A—Short strake. B—Small launder. C—Transverse launder. D—Wooden scrubber.

A canvas strake is made in the following way. Two beams, eighteen feet long and half a foot broad and three palms thick, are placed on a slope; one half of each of these beams is partially cut away lengthwise, to allow the ends of planks to be fastened in them, for the bottom is covered by planks three feet long, set crosswise and laid close together. One half of each supporting beam is left intact and rises a palm above the planks, in order that the water that is running down may not escape at the sides, but shall flow straight down. The head of the strake is higher than the rest of the body, and slopes so as to enable the water to flow away. The whole strake is covered by six stretched pieces of canvas, smoothed with a stick. The first of them occupies the lowest division, and the second is so laid as to slightly overlap it; on

A—Beams. B—Canvas. C—Head of strake. D—Small launder. E—Settling pit or tank. F—Wooden scrubber. G—Tubs.

the second division, the third is similarly laid, and so on, one on the other. If they are laid in the opposite way, the water flowing down carries the concentrates or particles of tin-stone under the canvas, and a useless task is attempted. Boys or men throw the concentrates or tin-stuff mixed with mud into the head of the strake, after the canvas has been thus stretched, and having opened the small launder they let the water flow in; then they stir the concentrates or tin-stone with a wooden scrubber till the water carries them all on to the canvas; next they gently sweep the linen with the wooden scrubber until the mud flows into the settling-pit or into the transverse launder. As soon as there is little or no mud on the canvas, but only concentrates or tin-stone, they carry the canvas away and wash it in a tub placed close by. The tin-stone settles in the tub, and the men return immediately to the same task. Finally, they pour the water out of the tub, and collect the concentrates or tin-stone. However, if either concentrates or tin-stone have washed down from the canvas and settled in the settling-pit or in the transverse launder, they wash the mud again.

Some neither remove the canvas nor wash it in the tubs, but place over it on each edge narrow strips, of no great thickness, and fix them to the beams with nails. They agitate the metalliferous material with wooden scrubbers and wash it in a similar way. As soon as little or no mud remains on the canvas, but only concentrates or fine tin-stone, they lift one beam so that the whole strake rests on the other, and dash it with water, which has been drawn with buckets out of the small tank, and in this way all the sediment which clings to the canvas falls into the trough placed underneath. This trough is hewn out of a tree and placed in a ditch dug in the ground; the interior of the trough is a foot wide at the top, but narrower in the bottom, because it is rounded out. In the middle of this trough they put a cross-board, in order that the fairly large particles of concentrates or fairly large-sized tin-stone may remain in the forepart into which they have fallen, and the fine concentrates or fine tin-stone in the lower part, for the water flows from one into the other, and at last flows down through an opening into the pit. As for the fairly large-sized concentrates or tin-stone which have been removed from the trough, they are washed again on the ordinary strake.

A—Canvas strake. B—Man dashing water on the canvas. C—Bucket. D—Bucket of another kind. E—Man removing concentrates or tin-stone from the trough.

The fine concentrates and fine tin-stone are washed again on this canvas strake. By this method, the canvas lasts longer because it remains fixed, and nearly double the work is done by one washer as quickly as can be done by two washers by the other method.

The jigging sieve has recently come into use by miners. The metalliferous material is thrown into it and sifted in a tub nearly full of water. The sieve is shaken up and down, and by this movement all the material below the size of a pea passes through into the tub, and the rest remains on the bottom of the sieve. This residue is of two kinds, the metallic particles, which occupy the lower place, and the particles of rock and earth, which take the higher place, because the heavy substance always settles, and the light is borne upward by the force of the water. This light material is taken away with a limp, which is a thin tablet of wood almost semicircular in shape, three-quarters of a foot long, and half a foot wide. Before the lighter portion is taken away the contents of the sieve are generally divided crosswise with a limp, to enable the water to penetrate into it more quickly. Afterward fresh material is again thrown into the sieve and shaken up and down, and when a great quantity of metallic particles have settled in the sieve, they are taken out and put into a tray close by. But since there fall into the tub with the mud, not only particles of gold or silver, but also of sand, pyrites, *cadmia*, galena, quartz, and other substances, and since the water cannot separate these from the metallic particles because they are all heavy, this muddy mixture is washed a second time, and the part which is useless is thrown away. To prevent the sieve passing this sand again too quickly, the washer lays small stones or gravel in the bottom of the sieve. However, if the sieve is not shaken straight up and down, but is tilted to one side, the small stones or broken ore move from one part to another, and the metallic material again falls into the tub, and the operation is frustrated. The miners of our country have made an even finer sieve, which does not fail even with unskilled washers; in washing with this sieve they have no need for the bottom to be strewn with small stones. By this method the mud settles in the tub with the very fine metallic particles, and the larger sizes of metal remain in the sieve and are covered with the valueless sand, and this is taken away with a limp. The concentrates which have been collected are smelted together with other things. The mud mixed with the very fine metallic particles is washed for a third time and in the finest sieve, whose bottom is woven of hair. If the ore is rich in metal, all the material which has been removed by the limp is washed on the canvas strakes, or if the ore is poor it is thrown away.

I have explained the methods of washing which are used in common for the ores of many metals. I now come to another method of crushing ore, for I ought to speak of this before describing those methods of washing which are peculiar to ores of particular metals.

In the year 1512, George, the illustrious Duke of Saxony[14], gave the over-

[14]George, Duke of Saxony, surnamed "The Bearded," was born 1471, and died 1539. He was chiefly known for his bitter opposition to the Reformation.

A—Fine sieves. B—Limp. C—Finer sieve. D—Finest sieve

lordship of all the dumps ejected from the mines in Meissen to the noble and wise Sigismund Maltitz, father of John, Bishop of Meissen. Rejecting the dry stamps, the large sieve, and the stone mills of Dippoldswalde and Altenberg, in which places are dug the small black stones from which tin is smelted, he invented a machine which could crush the ore wet under iron-shod stamps. That is called "wet ore" which is softened by water which flows into the mortar box, and they are sometimes called "wet stamps" because they are drenched by the same water; and on the other hand, the other kinds are called "dry stamps" or "dry ore," because no water is used to soften the ore when the stamps are crushing. But to return to our subject. This machine is not dissimilar to the one which crushes the ore with dry iron-shod stamps, but the heads of the wet stamps are larger by half than the heads of the others. The mortar-box, which is made of oak or beech timber, is set up in the space between the upright posts; it does not open in front, but at one end, and it is three feet long, three-quarters of a foot wide, and one foot and six digits deep. If it has no bottom, it is set up in the same way over a slab of hard, smooth rock placed in the ground, which has been dug down a little. The joints are stopped up all round with moss or cloth rags. If the mortar has a bottom, then an iron sole-plate, three feet long, three-quarters of a foot wide, and a palm thick, is placed in it. In the opening in the end of the mortar there is fixed an iron plate full of holes, in such a way that there is a space of two digits between it and the shoe of the nearest stamp, and the same distance between this screen and the upright post, in an opening through which runs a small but fairly long launder. The crushed particles of silver ore flow through this launder with the water into a settling-pit, while the material which settles in the launder is removed with an iron shovel to the nearest planked floor; that material which has settled in the pit is removed with an iron shovel on to another floor. Most people make two launders, in order that while the workman empties one of them of the accumulation which has settled in it, a fresh deposit may be settling in the other. The water flows in through a small launder at the other end of the mortar that is near the water-wheel which turns the machine. The workman throws the ore to be crushed into the mortar in such a way that the pieces, when they are thrown in among the stamps, do not impede the work. By this method a silver or gold ore is crushed very fine by the stamps.

When tin ore is crushed by this kind of iron-shod stamps, as soon as crushing begins, the launder which extends from the screen discharges the water carrying the fine tin-stone and fine sand into a transverse trough, from which the water flows down through the spouts, which pierce the side of the trough, into the one or other of the large buddles set underneath. The reason why there are two is that, while the washer empties the one which is filled with fine tin-stone and sand, the material may flow into the other. Each buddle is twelve feet long, one cubit deep, and a foot and a half broad. The tin-stone which settles in the upper part of the buddles is called the large size; these are frequently stirred with a shovel, in order that the medium sized particles of tin-stone, and the mud mixed with the very fine

A—Mortar. B—Open end of mortar. C—Slab of rock. D—Iron sole plates. E—Screen. F—Launder. G—Wooden shovel. H—Settling pit. I—Iron shovel. K—Heap of material which has settled. L—Ore which requires crushing. M—Small launder.

particles of the stones may flow away. The particles of medium size generally settle in the middle part of the buddle, where they are arrested by interwoven fir twigs. The mud which flows down with the water settles between the twigs and the board which closes the lower end of the buddle. The tin-stone of large size is removed separately from the buddle with a shovel; those of medium size are also removed separately, and likewise the mud is removed separately, for they are separately washed on the canvas strakes and on the ordinary strake, and separately roasted and smelted. The tin-stone which has settled in the middle part of the buddle, is also always washed separately on the canvas strakes; but if the particles are nearly equal in size to those which have settled in the upper part of the buddle, they are washed with them in the ordinary strake and are roasted and smelted with them. However, the mud is never washed with the others, either on the canvas strakes or on the ordinary strake, but separately, and the fine tin-stone which is obtained from it is roasted and smelted separately. The two large buddles discharge into a cross trough, and it again empties through a launder into a settling-pit which is outside the building.

A—Launder reaching to the screen. B—Transverse trough. C—Spouts. D—Large buddles. E—Shovel. F—Interwoven twigs. G—Boards closing the buddles. H—Cross trough.

This method of washing has lately undergone a considerable change ; for the launder which carries the water, mixed with the crushed tin-stone and fine sand which flow from the openings of the screen, does not reach to a transverse trough which is inside the same room, but runs straight through a partition into a small settling-pit. A boy draws a three-toothed rake through the material which has settled in the portion of the launder outside the room, by which means the larger sized particles of tin-stone settle at the bottom, and these the washer takes out with the wooden shovel and carries into the room ; this material is thrown into an ordinary strake and swept with a wooden scrubber and washed. As for those tin-stone particles which the water carries off from the strake, after they have been brought back on to the strake, he washes them again until they are clean.

The remaining tin-stone, mixed with sand, flows into the small settling-pit which is within the building, and this discharges into two large buddles. The tin-stone of moderate size, mixed with those of fairly large size, settle in the upper part, and the small size in the lower part ; but both are impure, and for this reason they are taken out separately and the former is washed twice,

A—FIRST LAUNDER. B—THREE-TOOTHED RAKE. C—SMALL SETTLING PIT. D—LARGE BUDDLE. E—BUDDLE RESEMBLING THE SIMPLE BUDDLE. F—SMALL ROLLER. G—BOARDS. H—THEIR HOLES. I—SHOVEL. K—BUILDING. L—STOVE. (THIS PICTURE DOES NOT ENTIRELY AGREE WITH THE TEXT).

first in a buddle like the simple buddle, and afterward on an ordinary strake. Likewise the latter is washed twice, first on a canvas strake and afterward on an ordinary strake. This buddle, which is like the simple buddle, differs from it in the head, the whole of which in this case is sloping, while in the case of the other it is depressed in the centre. In order that the boy may be able to rest the shovel with which he cleanses the tin-stone, this sluice has a small wooden roller which turns in holes in two thick boards fixed to the sides of the buddle ; if he did not do this, he would become over-exhausted by his task, for he spends whole days standing over these labours. The large buddle, the one like the simple buddle, the ordinary strake, and the canvas strakes, are erected within a special building. In this building there is a stove that gives out heat through the earthen tiles or iron plates of which it is composed, in order that the washers can pursue their labours even in winter, if the rivers are not completely frozen over.

On the canvas strakes are washed the very fine tin-stone mixed with mud which has settled in the lower end of the large buddle, as well as in the lower end of the simple buddle and of the ordinary strake. The canvas is cleaned in a trough hewn out of one tree trunk and partitioned off with two boards, so that three compartments are made. The first and second pieces of canvas are washed in the first compartment, the third and fourth in the second compartment, the fifth and sixth in the third compartment. Since among the very fine tin-stone there are usually some grains of stone, rock, or marble, the master cleanses them on the ordinary strake, lightly brushing the top of the material with a broom, the twigs of which do not all run the same way, but some straight and some crosswise. In this way the water carries off these impurities from the strake into the settling-pit because they are lighter, and leaves the tin-stone on the table because it is heavier.

Below all buddles or strakes, both inside and outside the building, there are placed either settling-pits or cross-troughs into which they discharge, in order that the water may carry on down into the stream but very few of the most minute particles of tin-stone. The large settling-pit which is outside the building is generally made of joined flooring, and is eight feet in length, breadth and depth. When a large quantity of mud, mixed with very fine tin-stone, has settled in it, first of all the water is let out by withdrawing a plug, then the mud which is taken out is washed outside the house on the canvas strakes, and afterward the concentrates are washed on the strake which is inside the building. By these methods the very finest tin-stone is made clean.

The mud mixed with the very fine tin-stone, which has neither settled in the large settling-pit nor in the transverse launder which is outside the room and below the canvas strakes, flows away and settles in the bed of the stream or river. In order to recover even a portion of the fine tin-stone, many miners erect weirs in the bed of the stream or river, very much like those that are made above the mills, to deflect the current into the races through which it flows to the water-wheels. At one side of each weir there is an area dug out to a depth of five or six or seven feet, and if the nature of

A—Launder from the screen of the mortar-box. B—Three-toothed rake. C—Small settling-pit. D—Canvas. E—Strakes. F—Brooms.

the place will permit, extending in every direction more than sixty feet. Thus, when the water of the river or stream in autumn and winter inundates the land, the gates of the weir are closed, by which means the current carries the mud mixed with fine tin-stone into the area. In spring and summer this mud is washed on the canvas strakes or on the ordinary strake, and even the finest black-tin is collected. Within a distance of four thousand fathoms along the bed of the stream or river below the buildings in which the tin-stuff is washed, the miners do not make such weirs, but put inclined fences in the meadows, and in front of each fence they dig a ditch of the same length, so that the mud mixed with the fine tin-stone, carried along by the stream or river when in flood, may settle in the ditch and cling to the fence. When this mud is collected, it is likewise washed on canvas strakes and on the ordinary strake, in order that the fine tin-stone may be separated from it. Indeed we may see many such areas and fences collecting mud of this kind in Meissen below Altenberg in the river Moglitz,—which is always of a reddish colour when the rock containing the black tin is being crushed under the stamps.

A—River. B—Weir. C—Gate. D—Area. E—Meadow. F—Fence. G—Ditch.

But to return to the stamping machines. Some usually set up four machines of this kind in one place, that is to say, two above and the same number below. By this plan it is necessary that the current which has been diverted should fall down from a greater height upon the upper water-wheels, because these turn axles whose cams raise heavier stamps. The stamp-stems of the upper machines should be nearly twice as long as the stems of the lower ones, because all the mortar-boxes are placed on the same level. These stamps have their tappets near their upper ends, not as in the case of the lower stamps, which are placed just above the bottom. The water flowing down from the two upper water-wheels is caught in two broad races, from which it falls on to the two lower water-wheels. Since all these machines have the stamps very close together, the stems should be somewhat cut away, to prevent the iron shoes from rubbing each other at the point where they are set into the stems. Where so many machines cannot be constructed, by reason of the narrowness of the valley, the mountain is excavated and levelled in two places, one of which is higher than the other, and in this case two machines are constructed and generally placed in one building. A broad race receives in the same way the water which flows down from the upper water-wheel, and similarly lets it fall on the lower water-wheel. The mortar-boxes are not then placed on one level, but each on the level which is appropriate to its own machine, and for this reason, two workmen are then required to throw ore into the mortar-boxes. When no stream can be diverted which will fall from a higher place upon the top of the water-wheel, one is diverted which will turn the foot of the wheel; a great quantity of water from the stream is collected in one pool capable of holding it, and from this place, when the gates are raised, the water is discharged against the wheel which turns in the race. The buckets of a water-wheel of this kind are deeper and bent back, projecting upward; those of the former are shallower and bent forward, inclining downward.

Further, in the Julian and Rhaetian Alps[15] and in the Carpathian Mountains, gold or even silver ore is now put under stamps, which are sometimes placed more than twenty in a row, and crushed wet in a long mortar-box. The mortar has two plates full of holes through which the ore, after being crushed, flows out with the water into the transverse launder placed underneath, and from there it is carried down by two spouts into the heads of the canvas strakes. Each head is made of a thick broad plank, which can be raised and set upright, and to which on each side are fixed pieces projecting upward. In this plank there are many cup-like depressions equal in size and similar in shape, in each of which an egg could be placed. Right down in these depressions are small crevices which can retain the concentrates of gold or silver, and when the hollows are nearly filled with these materials, the plank is raised on one side so that the concentrates will fall into a large bowl. The cup-like depressions are washed out by dashing them with water. These

[15]The Julian Alps are a section east of the Carnic Alps and lie north of Trieste. The term Rhaetian Alps is applied to that section along the Swiss Italian Boundary, about north of Lake Como.

A—First machine. B—Its stamps. C—Its mortar-box. D—Second machine. E—Its stamps. F—Its mortar-box. G—Third machine. H—Its stamps. I—Its mortar-box. K—Fourth machine. L—Its stamps. M—Its mortar-box.

concentrates are washed separately in different bowls from those which have settled on the canvas. This bowl is smooth and two digits wide and deep, being in shape very similar to a small boat; it is broad in the fore part, narrow in the back, and in the middle of it there is a cross groove, in which the particles of pure gold or silver settle, while the grains of sand, since they are lighter, flow out of it.

In some parts of Moravia, gold ore, which consists of quartz mixed with gold, is placed under the stamps and crushed wet. When crushed fine it flows out through a launder into a trough, is there stirred by a wooden scrubber, and the minute particles of gold which settle in the upper end of the trough are washed in a black bowl.

A—Stamps. B—Mortar. C—Plates full of holes. D—Transverse launder. E—Planks full of cup-like depressions. F—Spout. G—Bowl into which the concentrates fall. H—Canvas strake. I—Bowls shaped like a small boat. K—Settling-pit under the canvas strake.

So far I have spoken of machines which crush wet ore with iron-shod stamps. I will now explain the methods of washing which are in a measure peculiar to the ore of certain metals, beginning with gold. The ore which contains particles of this metal, and the sand of streams and rivers which contains grains of it, are washed in frames or bowls; the sands especially are also washed in troughs. More than one method is employed for washing on frames, for these frames either pass or retain the particles or concentrates of gold; they pass them if they have holes, and retain them if they have no holes. But either the frame itself has holes, or a box is substituted for it; if the frame itself is perforated it passes the particles or concentrates of gold into a trough; if the box has them, it passes the gold material into the long sluice. I will first speak of these two methods of washing. The frame is made of two planks joined together, and is twelve feet long and three feet wide, and is full of holes large enough for a pea to pass. To prevent the ore or sand with which the gold is mixed from falling out at the sides, small projecting edge-boards are fixed to it. This frame is set upon two stools, the first of which is higher than the second, in order that the gravel and small stones can roll down it. The washer throws the ore or sand into the head of the frame, which is higher, and opening the small launder, lets the water into it, and then agitates it with a wooden scrubber. In this way, the gravel and small stones roll down the frame on to the ground, while the

A—Head of frame. B—Frame. C—Holes. D—Edge-boards. E—Stools F—Scrubber. G—Trough. H—Launder. I—Bowl.

particles or concentrates of gold, together with the sand, pass through the holes into the trough which is placed under the frame, and after being collected are washed in the bowl.

A box which has a bottom made of a plate full of holes, is placed over the upper end of a sluice, which is fairly long but of moderate width. The gold material to be washed is thrown into this box, and a great quantity of water is let in. The lumps, if ore is being washed, are mashed with an iron shovel. The fine portions fall through the bottom of the box into the sluice, but the coarse pieces remain in the box, and these are removed with a scraper through an opening which is nearly in the middle of one side. Since a large amount of water is necessarily let into the box, in order to prevent it from sweeping away any particles of gold which have fallen into the sluice, the sluice is divided off by ten, or if it is as long again, by fifteen riffles. These riffles are placed equidistant from one another, and each is higher than the one next toward the lower end of the sluice. The little compartments which are thus made are filled with the material and the water which flows through

A—Sluice. B—Box. C—Bottom of inverted box. D—Open part of it. E—Iron hoe. F—Riffles. G—Small launder. H—Bowl with which settlings are taken away. I—Black bowl in which they are washed.

the box ; as soon as these compartments are full and the water has begun to flow over clear, the little launder through which this water enters into the box is closed, and the water is turned in another direction. Then the lowest riffle is removed from the sluice, and the sediment which has accumulated flows out with the water and is caught in a bowl. The riffles are removed one by one and the sediment from each is taken into a separate bowl, and each is separately washed and cleansed in a bowl. The larger particles of gold concentrates settle in the higher compartments, the smaller size, in the lower compartments. This bowl is shallow and smooth, and smeared with oil or some other slippery substance, so that the tiny particles of gold may not cling to it, and it is painted black, that the gold may be more easily discernible ; on the exterior, on both sides and in the middle, it is slightly hollowed out in order that it may be grasped and held firmly in the hands when shaken. By this method the particles or concentrates of gold settle in the back part of the bowl ; for if the back part of the bowl is tapped or shaken with one hand, as is usual, the contents move toward the fore part. In this way the Moravians, especially, wash gold ore.

The gold particles are also caught on frames which are either bare or covered. If bare, the particles are caught in pockets ; if covered, they

A—Plank. B—Side-boards. C—Iron wire. D—Handles.

cling to the coverings. Pockets are made in various ways, either with iron wire or small cross-boards fixed to the frame, or by holes which are sunk into the sluice itself or into its head, but which do not quite go through. These holes are round or square, or are grooves running crosswise. The frames are either covered with skins, pieces of cloth, or turf, which I will deal with one by one in turn.

In order to prevent the sand which contains the particles of gold from spilling out, the washer fixes side-boards to the edges of a plank which is six feet long and one and a quarter wide. He then lays crosswise many iron wires a digit apart, and where they join he fixes them to the bottom plank with iron nails. Then he makes the head of the frame higher, and into this he throws the sand which needs washing, and taking in his hands the handles which are at the head of the frame, he draws it backward and forward several times in the river or stream. In this way the small stones and gravel flow down along the frame, and the sand mixed with particles of gold remains in the pockets between the strips. When the contents of the pockets have been shaken out and collected in one place, he washes them in a bowl and thus cleans the gold dust.

Other people, among whom are the Lusitanians[16], fix to the sides of a sluice, which is about six feet long and a foot and a half broad, many cross-strips or riffles, which project backward and are a digit apart. The washer or his wife lets the water into the head of the sluice, where he throws the sand which contains the particles of gold. As it flows down he agitates it with a wooden scrubber, which he moves transversely to the riffles. He constantly removes with a pointed wooden stick the sediment which settles in the pockets between the riffles, and in this way the particles of gold settle in them, while the sand and other valueless materials are carried by the water into a tub placed below the sluice. He removes the particles of metal with a small wooden shovel into a wooden bowl. This bowl does not exceed a foot and a quarter in breadth, and by moving it up and down in the stream he cleanses the gold dust, for the remaining sand flows out of the dish, and the gold dust settles in the middle of it, where there is a cup-like depression. Some make use of a bowl which is grooved inside like a shell, but with a smooth lip where the water flows out. This smooth place, however, is narrower where the grooves run into it, and broader where the water flows out.

[16]Ancient Lusitania comprised Portugal and some neighbouring portions of Spain.

A—Head of the sluice. B—Riffles. C—Wooden scrubber. D—Pointed stick. E—Dish. F—Its cup-like depression. G—Grooved dish.

The cup-like pockets and grooves are cut or burned at the same time into the bottom of the sluice ; the bottom is composed of three planks ten feet long, and is about four feet wide ; but the lower end, through which the water is discharged, is narrower. This sluice, which likewise has side-boards fixed to its edges, is full of rounded pockets and of grooves which lead to them, there being two grooves to one pocket, in order that the water mixed with sand may flow into each pocket through the upper groove, and that after the sand has partly settled, the water may again flow out through the lower groove. The sluice is set in the river or stream or on the bank, and placed on two stools, of which the first is higher than the second in order that the gravel and small stones may roll down the sluice. The washer throws sand into the head with a shovel, and opening the launder, lets in the water, which carries the particles of metal with a little sand down into the pockets, while the gravel and small stones with the rest of the sand falls into a tub placed below the sluice. As soon as the pockets are filled, he brushes out the concentrates and washes them in a bowl. He washes again and again through this sluice.

A—Head of the sluice. B—Side-boards. C—Lower end of the sluice. D—Pockets. E—Grooves. F—Stools. G—Shovel. H—Tub set below. I—Launder.

Some people cut a number of cross-grooves, one palm distant from each other, in a sluice similarly composed of three planks eight feet long. The upper edge of these grooves is sloping, that the particles of gold may slip into them when the washer stirs the sand with a wooden shovel ; but their lower edge is vertical so that the gold particles may thus be unable to slide out of them. As soon as these grooves are full of gold particles mixed with fine sand, the sluice is removed from the stools and raised up on its head. The head in this case is nothing but the upper end of the planks of which the sluice is composed. In this way the metallic particles, being turned over backward, fall into another tub, for the small stones and gravel have rolled down the sluice. Some people place large bowls under the sluice instead of tubs, and as in the other cases, the unclean concentrates are washed in the small bowl.

The Thuringians cut rounded pockets, a digit in diameter and depth, in the head of the sluice, and at the same time they cut grooves reaching from one to another. The sluice itself they cover with canvas. The sand which

A—Cross grooves. B—Tub set under the sluice. C—Another tub.

is to be washed, is thrown into the head and stirred with a wooden scrubber; in this way the water carries the light particles of gold on to the canvas, and the heavy ones sink in the pockets, and when these hollows are full, the head is removed and turned over a tub, and the concentrates are collected and washed in a bowl. Some people make use of a sluice which has square pockets with short vertical recesses which hold the particles of gold. Other workers use a sluice made of planks, which are rough by reason of the very small shavings which still cling to them; these sluices are used instead of those with coverings, of which this sluice is bare, and when the sand is washed, the particles of gold cling no less to these shavings than to canvas, or skins, or cloths, or turf. The washer sweeps the sluice upward with a broom, and when he has washed as much of the sand as he wishes, he lets a more abundant supply of water into the sluice again to wash out the concentrates, which he collects in a tub set below the sluice, and then washes again in a bowl. Just as Thuringians cover the sluice with canvas, so some people cover it with the skins of oxen or horses. They push the auriferous sand upward with a wooden scrubber, and by this system the light material flows away with the water, while the particles of gold settle among the hairs; the skins are afterward washed in a tub; and the concentrates are colleced in a bowl.

A—Sluice covered with canvas. B—Its head full of pockets and grooves. C—Head removed and washed in a tub. D—Sluice which has square pockets. E—Sluice to whose planks small shavings cling. F—Broom. G—Skins of oxen. H—Wooden scrubber.

The Colchians[17] placed the skins of animals in the pools of springs; and since many particles of gold had clung to them when they were removed,

A—Spring. B—Skin. C—Argonauts.

the poets invented the "golden fleece" of the Colchians. In like manner, it can be contrived by the methods of miners that skins should take up, not only particles of gold, but also of silver and gems.

[17]Colchis, the traditional land of the Golden Fleece, lay between the Caucasus on the north, Armenia on the south, and the Black Sea on the west. If Agricola's account of the metallurgical purpose of the fleece is correct, then Jason must have had real cause for complaint as to the tangible results of his expedition. The fact that we hear nothing of the fleece after the day it was taken from the dragon would thus support Agricola's theory. Tons of ink have been expended during the past thirty centuries in explanations of what the fleece really was. These explanations range through the supernatural and metallurgical, but more recent writers have endeavoured to construct the journey of the Argonauts into an epic of the development of the Greek trade in gold with the Euxine. We will not attempt to traverse them from a metallurgical point of view further than to maintain that Agricola's explanation is as probable and equally as ingenious as any other, although Strabo (xi, 2, 19.) gives much the same view long before.

Alluvial mining—gold washing—being as old as the first glimmer of civilisation, it is referred to, directly or indirectly, by a great majority of ancient writers, poets, historians, geographers, and naturalists. Early Egyptian inscriptions often refer to this industry, but from the point of view of technical methods the description by Pliny is practically the only one of interest, and in Pliny's chapter on the subject, alluvial is badly con-

Many people cover the frame with a green cloth as long and wide as the frame itself, and fasten it with iron nails in such a way that they can easily

A—HEAD OF FRAME. B—FRAME. C—CLOTH. D—SMALL LAUNDER. E—TUB SET BELOW THE FRAME. F—TUB IN WHICH CLOTH IS WASHED.

draw them out and remove the cloth. When the cloth appears to be golden because of the particles which adhere to it, it is washed in a special tub and the particles are collected in a bowl. The remainder which has run down into the tub is again washed on the frame.

fused with vein mining. This passage (XXXIII, 21) is as follows: "Gold is found in "the world in three ways, to say nothing of that found in India by the ants, and in "Scythia by the Griffins. The first is as gold dust found in streams, as, for instance, in the "Tagus in Spain, in the Padus in Italy, in the Hebrus in Thracia, in the Pactolus in Asia, "and in the Ganges in India; indeed, there is no gold found more perfect than this, as the "current polishes it thoroughly by attrition. . . . Others by equal labour and greater "expense bring rivers from the mountain heights, often a hundred miles, for the purpose of "washing this debris. The ditches thus made are called *corrugi*, from our word *corrivatio*, I "suppose; and these entail a thousand fresh labours. The fall must be steep, that the "water may rush down from very high places, rather than flow gently. The ditches "across the valleys are joined by aqueducts, and in other places, impassable rocks have to be "cut away and forced to make room for troughs of hollowed-out logs. Those who cut the "rocks are suspended by ropes, so that to those who watch them from a distance, the "workmen seem not so much beasts as birds. Hanging thus, they take the levels and trace "the lines which the ditch is to take; and thus, where there is no place for man's footstep, "streams are dragged by men. The water is vitiated for washing if the current of the

Some people, in place of a green cloth, use a cloth of tightly woven horsehair, which has a rough knotty surface. Since these knots stand out

A—CLOTH FULL OF SMALL KNOTS, SPREAD OUT. B—SMALL KNOTS MORE CONSPICUOUSLY SHOWN. C—TUB IN WHICH CLOTH IS WASHED.

and the cloth is rough, even the very small particles of gold adhere to it; these cloths are likewise washed in a tub with water.

"stream carries mud with it. This kind of earth is called *urium*, hence these ditches are "laid out to carry the water over beds of pebbles to avoid this *urium*. When they have "reached the head of the fall, at the top of the mountain, reservoirs are excavated a couple "of hundred feet long and wide, and about ten feet deep. In these reservoirs there are "generally five gates left, about three feet square, so that when the reservoir is full, the gates "are opened, and the torrent bursts forth with such violence that the rocks are hurled along. "When they have reached the plain there is yet more labour. Trenches called *agogae* are "dug for the flow of the water. The bottoms of these are spread at regular intervals with *ulex* "to catch the gold. This *ulex* is similar to rosemary, rough and prickly. The sides, too, "are closed in with planks and are suspended when crossing precipitous spots. The earth "is carried to the sea and thus the shattered mountain is washed away and scattered; and "this deposition of the earth in the sea has extended the shore of Spain. . . . The gold "procured from *arrugiae* does not require to be melted, but is already pure gold. It is found "in lumps, in shafts as well, sometimes even exceeding ten *librae* in weight. These lumps "are called *palagae* and *palacurnae*, while the small grains are called *baluce*. The Ulex is "dried and burnt and the ashes are washed on a bed of grassy turf in order that the gold "may settle thereon."

Some people construct a frame not unlike the one covered with canvas, but shorter. In place of the canvas they set pieces of turf in rows. They

A—Head of frame. B—Small launder through which water flows into head of frame. C—Pieces of turf. D—Trough placed under frame. E—Tub in which pieces of turf are washed.

wash the sand, which has been thrown into the head of the frame, by letting in water. In this way the particles of gold settle in the turf, the mud and sand, together with the water, are carried down into the settling-pit or trough below, which is opened when the work is finished. After all the water has passed out of the settling-pit, the sand and mud are carried away and washed over again in the same manner. The particles which have clung to the turf are afterward washed down into the settling-pit or trough by a stronger current of the water, which is let into the frame through a small launder. The concentrates are finally collected and washed in a bowl. Pliny was not ignorant of this method of washing gold. "The ulex," he says, "after being dried, is burnt, and its ashes are washed over a grassy turf, that the gold may settle on it."

A—Tray. B—Bowl-like depression. C—Handles.

Sand mixed with particles of gold is also washed in a tray, or in a trough or bowl. The tray is open at the further end, is either hewn out of a squared trunk of a tree or made out of a thick plank to which side-boards are fixed, and is three feet long, a foot and a half wide, and three digits deep. The bottom is hollowed out into the shape of an elongated bowl whose narrow end is turned toward the head, and it has two long handles, by which it is drawn backward and forward in the river. In this way the fine sand is washed, whether it contains particles of gold or the little black stones from which tin is made.

The Italians who come to the German mountains seeking gold, in order to wash the river sand which contains gold-dust and garnets,[19] use a fairly long shallow trough hewn out of a tree, rounded within and without, open at one end and closed at the other, which they turn in the bed of the stream in such a way that the water does not dash into it, but flows in gently. They stir the sand, which they throw into it, with a wooden hoe, also rounded. To prevent the particles of gold or garnets from running out with the light sand, they close the end with a board similarly rounded, but lower than the sides of the trough. The concentrates of gold or garnets which,

[19]*Carbunculus Carchedonius* — Carthaginian carbuncle. The German is given by Agricola in the *Interpretatio* as *granat, i.e.*, garnet.

A—Trough. B—Its open end. C—End that may be closed. D—Stream. E—Hoe. F—End-board. G—Bag.

with a small quantity of heavy sand, have settled in the trough, they wash in a bowl and collect in bags and carry away with them.

Some people wash this kind of sand in a large bowl which can easily be shaken, the bowl being suspended by two ropes from a beam in a building. The sand is thrown into it, water is poured in, then the bowl is shaken, and the muddy water is poured out and clear water is again poured in, this being done again and again. In this way, the gold particles settle in the back part of the bowl because they are heavy, and the sand in the front part because it is light ; the latter is thrown away, the former kept for smelting. The one who does the washing then returns immediately to his task. This method of washing is rarely used by miners, but frequently by coiners and goldsmiths when they wash gold, silver, or copper. The bowl they employ has only three handles, one of which they grasp in their hands when they shake the bowl, and in the other two is fastened a rope by which the bowl is hung from a beam, or from a cross-piece which is upheld by the forks of two upright posts fixed in the ground. Miners frequently wash ore in a small bowl to test

A—Large bowl. B—Ropes. C—Beam. D—Other large bowl which coiners use. E—Small bowl.

it. This bowl, when shaken, is held in one hand and thumped with the other hand. In other respects this method of washing does not differ from the last.

I have spoken of the various methods of washing sand which contains grains of gold ; I will now speak of the methods of washing the material in which are mixed the small black stones from which tin is made[20]. Eight such methods are in use, and of these two have been invented lately. Such metalliferous material is usually found torn away from veins and stringers and scattered far and wide by the impetus of water, although sometimes *venae dilatatae* are composed of it. The miners dig out the latter material with a broad mattock, while they dig the former with a pick. But they dig out the little stones, which are not rare in this kind of ore, with an instrument like the bill of a duck. In districts which contain this material, if there is an abundant supply of water, and if there are valleys or gentle slopes and hollows, so that rivers can be diverted into them, the washers in summer-

[20]As the concentration of crushed tin ore has been exhaustively treated of already, the descriptions from here on probably refer entirely to alluvial tin.

A—Stream. B—Ditch. C—Mattock. D—Pieces of turf. E—Seven-pronged fork. F—Iron shovel. G—Trough. H—Another trough below it. I—Small wooden trowel.

time first of all dig a long ditch sloping so that the water will run through it rapidly. Into the ditch is thrown the metallic material, together with the surface material, which is six feet thick, more or less, and often contains moss, roots of plants, shrubs, trees, and earth; they are all thrown in with a broad mattock, and the water flows through the ditch. The sand and tin-stone, as they are heavy, sink to the bottom of the ditch, while the moss and roots, as they are light, are carried away by the water which flows through the ditch. The bottom of the ditch is obstructed with turf and stones in order to prevent the water from carrying away the tin-stone at the same time. The washers, whose feet are covered with high boots made of hide, though not of rawhide, themselves stand in the ditch and throw out of it the roots of the trees, shrubs, and grass with seven-pronged wooden forks, and push back the tin-stone toward the head of the ditch. After four weeks, in which they have devoted much work and labour, they raise the tin-stone in the following way; the sand with which it is mixed is repeatedly lifted from the ditch

A—Trough. B—Wooden shovel. C—Tub. D—Launder. E—Wooden trowel. F—Transverse trough. G—Plug. H—Falling water. I—Ditch. K—Barrow conveying material to be washed. L—Pick like the beak of a duck with which the miner digs out the material from which the small stones are obtained.

with an iron shovel and agitated hither and thither in the water, until the sand flows away and only the tin-stone remains on the shovel. The tin-stone is all collected together and washed again in a trough by pushing it up and turning it over with a wooden trowel, in order that the remaining sand may separate from it. Afterward they return to their task, which they continue until the metalliferous material is exhausted, or until the water can no longer be diverted into the ditches.

The trough which I mentioned is hewn out of the trunk of a tree and the interior is five feet long, three-quarters of a foot deep, and six digits wide. It is placed on an incline and under it is put a tub which contains interwoven fir twigs, or else another trough is put under it, the interior of which is three feet long and one foot wide and deep ; the fine tin-stone, which has run out with the water, settles in the bottom. Some people, in place of a trough, put a square launder underneath, and in like manner they wash the tin-stone in this by agitating it up and down and turning it over with a small wooden trowel. A transverse trough is put under the launder, which is either open on one end and drains off into a tub or settling-pit, or else is closed and perforated through the bottom ; in this case, it drains into a ditch beneath, where the water falls when the plug has been partly removed. The nature of this ditch I will now describe.

If the locality does not supply an abundance of water, the washers dig a ditch thirty or thirty-six feet long, and cover the bottom, the full length, with logs joined together and hewn on the side which lies flat on the ground. On each side of the ditch, and at its head also, they place four logs, one above the other, all hewn smooth on the inside. But since the logs are laid obliquely along the sides, the upper end of the ditch is made four feet wide and the tail end, two feet. The water has a high drop from a launder and first of all it falls into interlaced fir twigs, in order that it shall fall straight down for the most part in an unbroken stream and thus break up the lumps by its weight. Some do not place these twigs under the end of the launder, but put a plug in its mouth, which, since it does not entirely close the launder, nor altogether prevent the discharge from it, nor yet allow the water to spout far afield, makes it drop straight down. The workman brings in a wheelbarrow the material to be washed, and throws it into the ditch. The washer standing in the upper end of the ditch breaks the lumps with a seven-pronged fork, and throws out the roots of trees, shrubs, and grass with the same instrument, and thereby the small black stones settle down. When a large quantity of the tin-stone has accumulated, which generally happens when the washer has spent a day at this work, to prevent it from being washed away he places it upon the bank, and other material having been again thrown into the upper end of the ditch, he continues the task of washing. A boy stands at the lower end of the ditch, and with a thin pointed hoe stirs up the sediment which has settled at the lower end, to prevent the washed tin-stone from being carried further, which occurs when the sediment has accumulated to such an extent that the fir branches at the outlet of the ditch are covered.

A—LAUNDER. B—INTERLACING FIR TWIGS. C—LOGS; THREE ON ONE SIDE, FOR THE FOURTH CANNOT BE SEEN BECAUSE THE DITCH IS SO FULL WITH MATERIAL NOW BEING WASHED. D—LOGS AT THE HEAD OF THE DITCH. E—BARROW. F—SEVEN-PRONGED FORK. G—HOE

The third method of washing materials of this kind follows. Two strakes are made, each of which is twelve feet long and a foot and a half wide and deep. A tank is set at their head, into which the water flows through a little launder. A boy throws the ore into one strake; if it is of poor quality he puts in a large amount of it, if it is rich he puts in less. The water is let in by removing the plug, the ore is stirred with a wooden shovel, and in this way the tin-stone, mixed with the heavier material, settles in the bottom of the strake, and the water carries the light material into the launder, through which it flows on to a canvas strake. The very fine tin-stone, carried by the water, settles on to the canvas and is cleansed. A low cross-board is placed in the strake near the head, in order that the largest sized tin-stone may settle there. As soon as the strake is filled with the material which has been washed, he closes the mouth of the tank and continues washing in the other strake, and then the plug is withdrawn and the water and tin-stone flow down into a tank below. Then he pounds the sides

A—STRAKES. B—TANK. C—LAUNDER. D—PLUG. E—WOODEN SHOVEL. F—WOODEN MALLET. G—WOODEN SHOVEL WITH SHORT HANDLE. H—THE PLUG IN THE STRAKE. I—TANK PLACED UNDER THE PLUG.

of the loaded strake with a wooden mallet, in order that the tin-stone clinging to the sides may fall off; all that has settled in it, he throws out with a wooden shovel which has a short handle. Silver slags which have been crushed under the stamps, also fragments of silver-lead alloy and of cakes melted from pyrites, are washed in a strake of this kind.

Material of this kind is also washed while wet, in a sieve whose bottom is made of woven iron wire, and this is the fourth method of washing. The sieve is immersed in the water which is contained in a tub, and is violently shaken. The bottom of this tub has an opening of such size that as much water, together with tailings from the sieve, can flow continuously out of it as water flows into it. The material which settles in the strake, a boy either digs over with a three-toothed iron rake or sweeps with a wooden scrubber; in this way the water carries off a great part of both sand and mud. The tin-stone or metalliferous concentrates settle in the strake and are afterward washed in another strake.

These are ancient methods of washing material which contains tin-stone; there follow two modern methods. If the tin-stone mixed with

A—Sieve. B—Tub. C—Water flowing out of the bottom of it. D—Strake. E—Three-toothed rake. F—Wooden scrubber.

earth or sand is found on the slopes of mountains or hills, or in the level fields which are either devoid of streams or into which a stream cannot be diverted, miners have lately begun to employ the following method of washing, even in the winter months. An open box is constructed of planks, about six feet long, three feet wide, and two feet and one palm deep. At the upper end on the inside, an iron plate three feet long and wide is fixed, at a depth of one foot and a half from the top; this plate is very full of holes, through which tin-stone about the size of a pea can fall. A trough hewn from a tree is placed under the box, and this trough is about twenty-four feet long and three-quarters of a foot wide and deep; very often three cross-boards are placed in it, dividing it off into compartments, each one of which is lower than the next. The turbid waters discharge into a settling-pit.

The metalliferous material is sometimes found not very deep beneath the surface of the earth, but sometimes so deep that it is necessary to drive tunnels and sink shafts. It is transported to the washing-box in wheel-barrows, and when the washers are about to begin they lay a small launder,

A—Box. B—Perforated plate. C—Trough. D—Cross-boards. E—Pool. F—Launder. G—Shovel. H—Rake.

through which there flows on to the iron plate so much water as is necessary for this washing. Next, a boy throws the metalliferous material on to the iron plate with an iron shovel and breaks the small lumps, stirring them this way and that with the same implement. Then the water and sand penetrating the holes of the plate, fall into the box, while all the coarse gravel remains on the plate, and this he throws into a wheelbarrow with the same shovel. Meantime, a younger boy continually stirs the sand under the plate with a wooden scrubber nearly as wide as the box, and drives it to the upper end of the box ; the lighter material, as well as a small amount of tin-stone, is carried by the water down into the underlying trough. The boys carry on this labour without intermission until they have filled four wheelbarrows with the coarse and worthless residues, which they carry off and throw away, or three wheelbarrows if the material is rich in black tin. Then the foreman has the plank removed which was in front of the iron plate, and on which the boy stood. The sand, mixed with the tin-stone, is frequently pushed backward and forward with a scrubber, and the same sand, because it is lighter, takes the upper place, and is removed as soon as it appears ; that which takes the lower place is turned over with a spade, in order that any that is light can flow away ; when all the tin-stone is heaped together, he shovels it out of the box and carries it away. While the foreman does this, one boy with an iron hoe stirs the sand mixed with fine tin-stone, which has run out of the box and has settled in the trough and pushes it back to the uppermost part of the trough, and this material, since it contains a very great amount of tin-stone, is thrown on to the plate and washed again. The material which has settled in the lowest part of the trough is taken out separately and piled in a heap, and is washed on the ordinary strake ; that which has settled in the pool is washed on the canvas strake. In the summer-time this fruitful labour is repeated more often, in fact ten or eleven times. The tin-stone which the foreman removes from the box, is afterward washed in a jigging sieve, and lastly in a tub, where at length all the sand is separated out. Finally, any material in which are mixed particles of other metals, can be washed by all these methods, whether it has been disintegrated from veins or stringers, or whether it originated from *venae dilatatae,* or from streams and rivers.

The sixth method of washing material of this kind is even more modern and more useful than the last. Two boxes are constructed, into each of which water flows through spouts from a cross trough into which it has been discharged through a pipe or launder. When the material has been agitated and broken up with iron shovels by two boys, part of it runs down and falls through the iron plates full of holes, or through the iron grating, and flows out of the box over a sloping surface into another cross trough, and from this into a strake seven feet long and two and a half feet wide. Then the foreman again stirs it with a wooden scrubber that it may become clean. As for the material which has flowed down with the water and settled in the third cross trough, or in the launder which leads from it, a third boy rakes it with a two-toothed rake ; in this way the fine tin-stone settles down

A—Launder. B—Cross trough. C—Two spouts. D—Boxes. E—Plate. F—Grating. G—Shovels. H—Second cross trough. I—Strake. K—Wooden scrubber. L—Third cross trough. M—Launder. N—Three-toothed rake.

and the water carries off the valueless sand into the creek. This method of washing is most advantageous, for four men can do the work of washing in two boxes, while the last method, if doubled, requires six men, for it requires two boys to throw the material to be washed on to the plate and to stir it with iron shovels; two more are required with wooden scrubbers to keep stirring the sand, mixed with the tin-stone, under the plate, and to push it toward the upper end of the box; further, two foremen are required to clean the tin-stone in the way I have described. In the place of a plate full of holes, they now fix in the boxes a grating made of iron wire as thick as the stalks of rye; that these may not be depressed by the weight and become bent, three iron bars support them, being laid crosswise underneath. To prevent the grating from being broken by the iron shovels with which the material is stirred in washing, five or six iron rods are placed on top in cross lines, and are fixed to the box so that the shovels may rub them instead of the grating; for this reason the grating lasts longer than the plates, because it remains intact, while the rods, when worn by rubbing, can easily be replaced by others.

Miners use the seventh method of washing when there is no stream of water in the part of the mountain which contains the black tin, or particles of gold, or of other metals. In this case they frequently dig more than fifty ditches on the slope below, or make the same number of pits, six feet long, three feet wide, and three-quarters of a foot deep, not any great distance from each other. At the season when a torrent rises from storms of great violence or long duration, and rushes down the mountain, some of the miners dig the metalliferous material in the woods with broad hoes and

A—Pits. B—Torrent. C—Seven-pronged fork. D—Shovel.

drag it to the torrent. Other miners divert the torrent into the ditches or pits, and others throw the roots of trees, shrubs, and grass out of the ditches or pits with seven-pronged wooden forks. When the torrent has run down, they remove with shovels the uncleansed tin-stone or particles of metal which have settled in the ditches or pits, and cleanse it.

The eighth method is also employed in the regions which the Lusitanians hold in their power and sway, and is not dissimilar to the last. They drive a great number of deep ditches in rows in the gullies, slopes, and hollows of the mountains. Into these ditches the water, whether flowing down from snow melted by the heat of the sun or from rain, collects and carries together with earth and sand, sometimes tin-stone, or, in the case of the Lusitanians, the particles of gold loosened from veins and stringers. As soon as the waters of the torrent have all run away, the miners throw the material out of the ditches with iron shovels, and wash it in a common sluice box.

A—Gully. B—Ditch. C—Torrent. D—Sluice box employed by the Lusitanians.

The Poles wash the impure lead from *venae dilatatae* in a trough ten feet long, three feet wide, and one and one-quarter feet deep. It is mixed with moist earth and is covered by a wet and sandy clay, and so first of all the clay, and afterward the ore, is dug out. The ore is carried to a stream or river, and thrown into a trough into which water is admitted by a little launder, and the washer standing at the lower end of the trough drags the ore out with a narrow and nearly pointed hoe, whose wooden handle is nearly ten feet long. It is washed over again once or twice in the same way and thus made pure. Afterward when it has been dried in the sun they throw it into a copper sieve, and separate the very small pieces which pass through the sieve from the larger ones ; of these the former are smelted in a faggot pile and the latter in the furnace. Of such a number then are the methods of washing.

A—Trough. B—Launder. C—Hoe. D—Sieve.

One method of burning is principally employed, and two of roasting. The black tin is burned by a hot fire in a furnace similar to an oven[21] ; it is burned if it is a dark-blue colour, or if pyrites and the stone from which iron is made are mixed with it, for the dark blue colour if not burnt, consumes the tin. If pyrites and the other stone are not volatilised into fumes in a furnace of this kind, the tin which is made from the tin-stone is impure. The tin-stone is thrown either into the back part of the furnace, or into one side of it ; but in the former case the wood is placed in front, in the latter case alongside, in such a manner, however, that neither firebrands nor coals may fall upon the tin-stone itself or touch it. The fuel is manipulated by a poker made of wood. The tin-stone is now stirred with a rake with two

[21]From a metallurgical point of view all of these operations are roasting. Even to-day, however, the expression " burning " tin is in use in some parts of Cornwall, and in former times it was general.

teeth, and now again levelled down with a hoe, both of which are made of iron. The very fine tin-stone requires to be burned less than that of moderate size, and this again less than that of the largest size. While the tin-stone is being thus burned, it frequently happens that some of the material runs together.

A—Furnace. B—Its mouth. C—Poker. D—Rake with two teeth. E—Hoe.

The burned tin-stone should then be washed again on the strake, for in this way the material which has been run together is carried away by the water into the cross-trough, where it is gathered up and worked over, and again washed on the strake. By this method the metal is separated from that which is devoid of metal.

Cakes from pyrites, or *cadmia*, or cupriferous stones, are roasted in quadrangular pits, of which the front and top are open, and these pits are generally twelve feet long, eight feet wide, and three feet deep. The cakes of melted pyrites are usually roasted twice over, and those of *cadmia* once. These latter are first rolled in mud moistened with vinegar, to prevent the fire from consuming too much of the copper with the bitumen, or sulphur, or orpiment, or realgar. The cakes of pyrites are first roasted in a slow fire and afterward in a fierce one, and in both cases, during the whole following night, water is let in, in order that, if there is in the cakes any alum or vitriol or saltpetre capable of injuring the metals, although it rarely does injure them, the water may remove it and make the cakes soft. The solidified juices are nearly all harmful to the metal, when cakes or ore of this kind are smelted. The cakes which are to be roasted are placed on wood piled up in the form of a crate, and this pile is fired[22].

A—PITS. B—WOOD. C—CAKES. D—LAUNDER.

The cakes which are made of copper smelted from schist are first thrown upon the ground and broken, and then placed in the furnace on bundles of faggots, and these are lighted. These cakes are generally roasted seven times and occasionally nine times. While this is being done, if they are

[22]There can be no doubt that these are mattes, as will develop in Chapter IX. The German term in the Glossary for *panes ex pyrite* is *stein*, the same as the modern German for matte. Orpiment and realgar are the yellow and red arsenical sulphides. The *cadmia* was no doubt the cobalt-arsenic minerals (see note on p. 112). The "solidified juices" were generally anything that could be expelled short of smelting, *i.e.*, roasted off or leached out, as shown in note 4, p. 1; they embrace the sulphates, salts, sulphur, bitumen, and arsenical sulphides, etc. For further information on leaching out the sulphates, alum, etc., see note 10, p. 564.

bituminous, then the bitumen burns and can be smelled. These furnaces have a structure like the structure of the furnaces in which ore is smelted, except that they are open in front; they are six feet high and four feet wide. As for this kind of furnace, three of them are required for one of those in which the cakes are melted. First of all they are roasted in the first furnace, then when they are cooled, they are transferred into the second furnace and again roasted; later they are carried to the third, and afterward back to the first, and this order is preserved until they have been roasted seven or nine times.

A—Cakes. B—Bundles of faggots. C—Furnaces.

END OF BOOK VIII.

BOOK IX.[1]

INCE I have written of the varied work of preparing the ores, I will now write of the various methods of smelting them. Although those who burn, roast and calcine[2] the ore, take from it something which is mixed or combined with the metals; and those who crush it with stamps take away much; and those who wash, screen and sort it, take away still more; yet they cannot remove all which conceals the metal from the eye and renders it crude and unformed. Wherefore smelting is necessary, for by this means earths, solidified juices, and stones are separated from the metals so that they obtain their proper colour and become pure, and may be of great use to mankind in many ways. When the ore is smelted, those things which were mixed with the metal before it was melted are driven forth, because the metal is perfected by fire in this manner. Since metalliferous ores differ greatly amongst themselves, first as to the metals which they contain, then as to the quantity of the metal which is in them, and then by the fact that some are rapidly melted by fire and others slowly, there are, therefore, many methods of smelting. Constant practice has taught the

[1]The history of the fusion of ores and of metals is the history of individual processes, and such information as we have been able to discover upon the individual methods previous to Agricola we give on the pages where such processes are discussed. In general the records of the beginnings of metallurgy are so nebular that, if one wishes to shirk the task, he can adopt the explanation of William Pryce one hundred and fifty years ago: "It is very "probable that the nature and use of Metals were not revealed to Adam in his state of "innocence: the toil and labour necessary to procure and use those implements of the iron "age could not be known, till they made part of the curse incurred by his fall: 'In the sweat "'of thy face shalt thou eat bread, till thou return unto the ground; in sorrow shalt thou "'eat of it all the days of thy life' (Genesis). That they were very early discovered, "however, is manifest from the Mosaick account of Tubal Cain, who was the first instructor "of every artificer in Brass [*sic*] and Iron" (*Mineralogia Cornubiensis*, p. 2).

It is conceivable that gold could be found in large enough pieces to have had general use in pre-historic times, without fusion; but copper, which was also in use, must have been smelted, and therefore we must assume a considerable development of human knowledge on the subject prior to any human record. Such incidental mention as exists after record begins does not, of course, extend to the beginning of any particular branch of the art—in fact, special arts obviously existed long before such mention, and down to the complete survey of the state of the art by Agricola our dates are necessarily "prior to" some first mention in literature, or "prior to" the known period of existing remains of metallurgical operations. The scant Egyptian records, the Scriptures, and the Shoo King give a little insight prior to 1000 B.C. The more extensive Greek literature of about the 5th to the 3rd centuries B.C., together with the remains of Greek mines, furnish another datum point of view, and the Roman and Greek writers at the beginning of the Christian era give a still larger view. After them our next step is to the Monk Theophilus and the Alchemists, from the 12th to the 14th centuries. Finally, the awakening of learning at the end of the 15th and the beginning of the 16th centuries, enables us for the first time to see practically all that was known. The wealth of literature which exists subsequent to this latter time makes history thereafter a matter of some precision, but it is not included in this undertaking. Considering the great part that the metals have played in civilization, it is astonishing what a minute amount of information is available on metallurgy. Either the ancient metallurgists were secretive as to their art, or the ancient authors despised such common things, or, as is equally probable, the very partial preservation of ancient literature, by painful transcription over a score of centuries, served only for those works of more general interest. In any event, if all the direct or indirect material on metallurgy prior to the 15th century were compiled, it would not fill 40 pages such as these.

[2]See footnote 2, p. 267, on verbs used for roasting.

smelters by which of these methods they can obtain the most metal from any one ore. Moreover, while sometimes there are many methods of smelting the same ore, by which an equal weight of metal is melted out, yet one is done at a greater cost and labour than the others. Ore is either melted with a furnace or without one; if smelted with a furnace the tap-hole is either temporarily closed or always open, and if smelted without a furnace, it is done either in pots or in trenches. But in order to make this matter clearer, I will describe each in detail, beginning with the buildings and the furnaces.

It may be of service to give a tabular summary indicating approximately the time when evidence of particular operations appear on the historical horizon:

Gold washed from alluvial	Prior to recorded civilization
Copper reduced from ores by smelting	Prior to recorded civilization
Bitumen mined and used	Prior to recorded civilization
Tin reduced from ores by smelting	Prior to 3500 B.C.
Bronze made	Prior to 3500 B.C.
Iron reduced from ores by smelting	Prior to 3500 B.C.
Soda mined and used	Prior to 3500 B.C.
Gold reduced from ores by concentration	Prior to 2500 B.C.
Silver reduced from ores by smelting	Prior to 2000 B.C.
Lead reduced from ores by smelting	Prior to 2000 B.C. (perhaps prior to 3500 B.C.)
Silver parted from lead by cupellation	Prior to 2000 B.C.
Bellows used in furnaces	Prior to 1500 B.C.
Steel produced	Prior to 1000 B.C.
Base metals separated from ores by water concentration	Prior to 500 B.C.
Gold refined by cupellation	Prior to 500 B.C.
Sulphide ores smelted for lead	Prior to 500 B.C.
Mercury reduced from ores by..(?)	Prior to 400 B.C.
White-lead made with vinegar	Prior to 300 B.C.
Touchstone known for determining gold and silver fineness	Prior to 300 B.C.
Quicksilver reduced from ore by distillation	Prior to Christian Era
Silver parted from gold by cementation with salt	Prior to ,,
Brass made by cementation of copper and calamine	Prior to ,,
Zinc oxides obtained from furnace fumes by construction of dust chambers	Prior to ,,
Antimony reduced from ores by smelting (accidental)	Prior to ,,
Gold recovered by amalgamation	Prior to ,,
Refining of copper by repeated fusion	Prior to ,,
Sulphide ores smelted for copper	Prior to ,,
Vitriol (blue and green) made	Prior to ,,
Alum made	Prior to ,,
Copper refined by oxidation and poling	Prior to 1200 A.D.
Gold parted from copper by cupelling with lead	Prior to 1200 A.D.
Gold parted from silver by fusion with sulphur	Prior to 1200 A.D.
Manufacture of nitric acid and *aqua regia*	Prior to 1400 A.D.
Gold parted from silver by nitric acid	Prior to 1400 A.D.
Gold parted from silver with antimony sulphide	Prior to 1500 A.D.
Gold parted from copper with sulphur	Prior to 1500 A.D.
Silver parted from iron with antimony sulphide	Prior to 1500 A.D.
First text book on assaying	Prior to 1500 A.D.
Silver recovered from ores by amalgamation	Prior to 1500 A.D.
Separation of silver from copper by liquation	Prior to 1540 A.D.
Cobalt and manganese used for pigments	Prior to 1540 A.D.
Roasting copper ores prior to smelting	Prior to 1550 A.D.
Stamp-mill used	Prior to 1550 A.D.
Bismuth reduced from ore	Prior to 1550 A.D.
Zinc reduced from ore (accidental)	Prior to 1550 A.D.

Further, we believe it desirable to sketch at the outset the development of metallurgical appliances as a whole, leaving the details to special footnotes; otherwise a comprehensive view of the development of such devices is difficult to grasp.

We can outline the character of metallurgical appliances at various periods in a few words. It is possible to set up a description of the imaginary beginning of the

A wall which will be called the "second wall" is constructed of brick or stone, two feet and as many palms thick, in order that it may be strong enough to bear the weight. It is built fifteen feet high, and its length depends on the number of furnaces which are put in the works; there are usually six furnaces, rarely more, and often less. There are three furnace walls, a back one which is against the "second" wall, and two side ones, of which I will speak later. These should be made of natural stone, as this is more serviceable than burnt bricks, because bricks soon become defective and crumble away, when the smelter or his deputy chips off the accretions which adhere to the walls when the ore is smelted. Natural stone resists injury by the fire and lasts a long time, especially that which is soft and devoid of cracks; but, on the contrary, that which is hard and has many cracks is burst asunder by the fire and destroyed. For this reason, furnaces which are made of the latter are easily weakened by the fire, and when the accretions are chipped off they crumble to pieces. The front furnace wall should be made of brick, and there should be in the lower part a mouth three palms wide and one and a half feet high, when the hearth is completed. A hole slanting upward, three palms long, is made through the back furnace wall, at the height of a cubit, before the hearth has been prepared; through this hole and a hole one foot long in the "second" wall—as the back of this wall has an arch—is inserted a pipe of iron or bronze, in which are fixed the nozzles

"bronze age" prior to recorded civilization, starting with the savage who accidentally built a fire on top of some easily reducible ore, and discovered metal in the ashes, etc.; but as this method has been pursued times out of number to no particular purpose, we will confine ourselves to a summary of such facts as we can assemble. "Founders' hoards" of the bronze age are scattered over Western Europe, and indicate that smelting was done in shallow pits with charcoal. With the Egyptians we find occasional inscriptions showing small furnaces with forced draught, in early cases with a blow-pipe, but later—about 1500 B.C.—with bellows also. The crucible was apparently used by the Egyptians in secondary melting, such remains at Mt. Sinai probably dating before 2000 B.C. With the advent of the Prophets, and the first Greek literature—9th to 7th century B.C.—we find frequent references to bellows. The remains of smelting appliances at Mt. Laurion (500–300 B.C.) do not indicate much advance over the primitive hearth; however, at this locality we do find evidence of the ability to separate minerals by specific gravity, by washing crushed ore over inclined surfaces with a sort of buddle attachment. Stone grinding-mills were used to crush ore from the earliest times of Mt. Laurion down to the Middle Ages. About the beginning of the Christian era the writings of Diodorus, Strabo, Dioscorides, and Pliny indicate considerable advance in appliances. Strabo describes high stacks to carry off lead fumes; Dioscorides explains a furnace with a dust-chamber to catch *pompholyx* (zinc oxide); Pliny refers to the upper and lower crucibles (a forehearth) and to the pillars and arches of the furnaces. From all of their descriptions we may conclude that the furnaces had then reached some size, and were, of course, equipped with bellows. At this time sulphide copper and lead ores were smelted; but as to fluxes, except lead for silver, and lead and soda for gold, we have practically no mention. Charcoal was the universal fuel for smelting down to the 18th century. Both Dioscorides and Pliny describe a distillation apparatus used to recover quicksilver. A formidable list of mineral products and metal alloys in use, indicate in themselves considerable apparatus, of the details of which we have no indication; in the main these products were lead sulphide, sulphate, and oxide (red-lead and litharge); zinc oxide; iron sulphide, oxide and sulphate; arsenic and antimony sulphides; mercury sulphide, sulphur, bitumen, soda, alum and potash,; and of the alloys, bronze, brass, pewter, electrum and steel.

From this period to the period of the awakening of learning our only light is an occasional gleam from Theophilus and the Alchemists. The former gave a more detailed description of metallurgical appliances than had been done before, but there is little vital change apparent from the apparatus of Roman times. The Alchemists gave a great stimulus to industrial chemistry in the discovery of the mineral acids, and described distillation apparatus of approximately modern form.

The next period—the Renaissance—is one in which our descriptions are for the first time satisfactory, and a discussion would be but a review of *De Re Metallica*.

of the bellows. The whole of the front furnace wall is not more than five feet high, so that the ore may be conveniently put into the furnace, together with those things which the master needs for his work of smelting. Both the side walls of the furnace are six feet high, and the back one seven feet, and they are three palms thick. The interior of the furnace is five palms wide, six palms and a digit long, the width being measured by the space which lies between the two side walls, and the length by the space between the front and the back walls ; however, the upper part of the furnace widens out somewhat.

There are two doors in the second wall if there are six furnaces, one of the doors being between the second and third furnaces and the other between the fourth and fifth furnaces. They are a cubit wide and six feet high, in order that the smelters may not have mishaps in coming and going. It is necessary to have a door to the right of the first furnace, and similarly one to the left of the last, whether the wall is longer or not. The second wall is carried further when the rooms for the cupellation furnaces, or any other building, adjoin the rooms for the blast furnaces, these buildings being only divided by a partition. The smelter, and the ones who attend to the first and the last furnaces, if they wish to look at the bellows or to do anything else, go out through the doors at the end of the wall, and the other people go through the other doors, which are the common ones. The furnaces are placed at a distance of six feet from one another, in order that the smelters and their assistants may more easily sustain the fierceness of the heat. Inasmuch as the interior of each furnace is five palms wide and each is six feet distant from the other, and inasmuch as there is a space of four feet three palms at the right side of the first furnace and as much at the left side of the last furnace, and there are to be six furnaces in one building, then it is necessary to make the second wall fifty-two feet long ; because the total of the widths of all of the furnaces is seven and a half feet, the total of the spaces between the furnaces is thirty feet, the space on the outer sides of the first and last furnaces is nine feet and two palms, and the thickness of the two transverse walls is five feet, which make a total measurement of fifty-two feet.[3]

Outside each furnace hearth there is a small pit full of powder which is compressed by ramming, and in this manner is made the forehearth which receives the metal flowing from the furnaces. Of this I will speak later.

Buried about a cubit under the forehearth and the hearth of the furnace is a transverse water-tank, three feet long, three palms wide and a cubit deep. It is made of stone or brick, with a stone cover, for if it were not covered, the heat would draw the moisture from below and the vapour might be blown into the hearth of the furnace as well as into the forehearth, and would dampen the blast. The moisture would vitiate the blast, and part of the metal would be absorbed and part would be mixed with the slags, and in this manner the melting would be greatly damaged. From each water-tank is built a walled vent, to the same depth as the tank, but six digits wide ;

[3]Agricola has here either forgotten to take into account his three-palm-thick furnace walls, which will make the length of this long wall sixty-one feet, or else he has included this foot and a half in each case in the six-foot distance between the furnaces, so that the actual clear space is only four and a half feet between the furnace with four feet on the ends.

A—FURNACES. B—FOREHEARTHS.

this vent slopes upward, and sooner or later penetrates through to the other side of the wall, against which the furnace is built. At the end of this vent there is an opening where the steam, into which the water has been converted, is exhausted through a copper or iron tube or pipe. This method of making the tank and the vent is much the best. Another kind has a similar vent but a different tank, for it does not lie transversely under the forehearth, but lengthwise ; it is two feet and a palm long, and a foot and three palms wide, and a foot and a palm deep. This method of making tanks is not condemned by us, as is the construction of those tanks without a vent ; the latter, which have no opening into the air through which the vapour may discharge freely, are indeed to be condemned.

A—Furnaces. B—Forehearth. C—Door. D—Water tank. E—Stone which covers it. F—Material of the vent walls. G—Stone which covers it. H—Pipe exhaling the vapour.

Fifteen feet behind the second wall is constructed the first wall, thirteen feet high. In both of these are fixed roof beams[4], which are a foot wide and

[4]The paucity of terms in Latin for describing structural members, and the consequent repetition of "beam" (*trabs*), "timber" (*tignum*), "billet" (*tigillum*), "pole" (*asser*), with such modifications as small, large, and transverse, and with long explanatory clauses showing their location, renders the original very difficult to follow. We have, therefore, introduced such terms as "posts," "tie-beams," "sweeps," "levers," "rafters," "sills," "moulding," "braces," "cleats," "supports," etc., as the context demands.

thick, and nineteen feet and a palm long; these are placed three feet distant from one another. As the second wall is two feet higher than the first wall, recesses are cut in the back of it two feet high, one foot wide, and a palm deep, and in these recesses, as it were in mortises, are placed one end of each of the beams. Into these ends are mortised the bottoms of just as many posts; these posts are twenty-four feet high, three palms wide and thick, and from the tops of the posts the same number of rafters stretch downward to the ends of the beams superimposed on the first wall; the upper ends of the rafters are mortised into the posts and the lower ends are mortised into the ends of the beams laid on the first wall; the rafters support the roof, which consists of burnt tiles. Each separate rafter is propped up by a separate timber, which is a cross-beam, and is joined to its post. Planks close together are affixed to the posts above the furnaces; these planks are about two digits thick and a palm wide, and they, together with the wicker work interposed between the timbers, are covered with lute so that there may be no risk of fire to the timbers and wicker-work. In this practical manner is constructed the back part of the works, which contains the bellows, their frames, the mechanism for compressing the bellows, and the instrument for distending them, of all of which I will speak hereafter.

In front of the furnaces is constructed the third long wall and likewise the fourth. Both are nine feet high, but of the same length and thickness as the other two, the fourth being nine feet distant from the third; the third is twenty-one and a half feet from the second. At a distance of twelve feet from the second wall, four posts seven and a half feet high, a cubit wide and thick, are set upon rock laid underneath. Into the tops of the posts the roof beam is mortised; this roof beam is two feet and as many palms longer than the distance between the second and the fifth transverse walls, in order that its ends may rest on the transverse walls. If there should not be so long a beam at hand, two are substituted for it. As the length of the long beam is as above, and as the posts are equidistant, it is necessary that the posts should be a distance of nine feet, one palm, two and two-fifths digits from each other, and the end ones this distance from the transverse walls. On this longitudinal beam and to the third and fourth walls are fixed twelve secondary beams twenty-four feet long, one foot wide, three palms thick, and distant from each other three feet, one palm, and two digits. In these secondary beams, where they rest on the longitudinal beams, are mortised the ends of the same number of rafters as there are posts which stand on the second wall. The ends of the rafters do not reach to the tops of the posts, but are two feet away from them, that through this opening, which is like the open part of a forge, the furnaces can emit their fumes. In order that the rafters should not fall down, they are supported partly by iron rods, which extend from each rafter to the opposite post, and partly supported by a few tie-beams, which in the same manner extend from some rafters to the posts opposite, and give them stability. To these tie-beams, as well as to the rafters which face the posts, a number of boards, about two digits thick and a palm wide, are fixed at a distance of a palm from each other, and are covered with lute so that they do not catch fire. In the secondary beams, where they are laid on the fourth wall, are mortised the lower ends of the same number of rafters as those in a set of rafters[5] opposite them. From the third long wall these rafters are joined and tied to the ends of the opposite rafters, so that they may not slip, and besides they are strengthened with substructures which are made of cross and oblique timbers. The rafters support the roof.

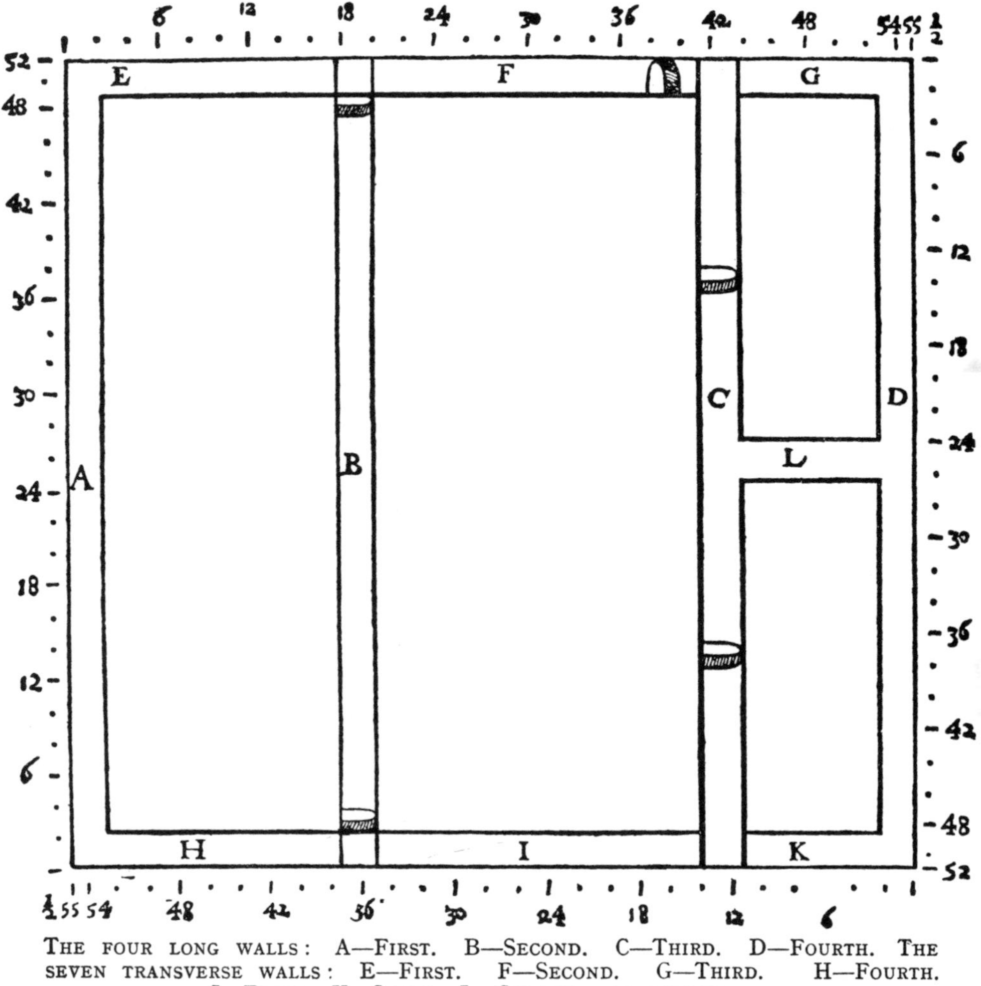

THE FOUR LONG WALLS: A—FIRST. B—SECOND. C—THIRD. D—FOURTH. THE SEVEN TRANSVERSE WALLS: E—FIRST. F—SECOND. G—THIRD. H—FOURTH. I—FIFTH. K—SIXTH. L—SEVENTH, OR MIDDLE.

In this manner the front part of the building is made, and is divided into three parts; the first part is twelve feet wide and is under the hood, which consists of two walls, one vertical and one inclined. The second part is the same number of feet wide and is for the reception of the ore to be smelted, the fluxes, the charcoal, and other things which are needed by the smelter. The third part is nine feet wide and contains two separate rooms of equal size, in one of which is the assay furnace, while the other contains the metal to be melted in the cupellation furnaces. It is thus necessary that in the

[5]This set of rafters appears to start from the longitudinal beam.

building there should be, besides the four long walls, seven transverse walls, of which the first is constructed from the upper end of the first long wall to the upper end of the second long wall; the second proceeds from the end of this to the end of the third long wall; the third likewise from this end of the last extends to the end of the fourth long wall; the fourth leads from the lower end of the first long wall to the lower end of the second long wall; the fifth extends from the end of this to the end of the third long wall; the sixth extends from this last end to the end of the fourth long wall; the seventh divides into two parts the space between the third and fourth long walls.

To return to the back part of the building, in which, as I said, are the bellows[6], their frames, the machinery for compressing them, and the instrument for distending them. Each bellows consists of a body and a head. The body is composed of two "boards," two bows, and two hides. The upper board is a palm thick, five feet and three palms long, and two and a half feet wide at the back part, where each of the sides is a little curved, and it is a cubit wide at the front part near the head. The whole of the body of the bellows tapers toward the head. That which we now call the "board" consists of two pieces of pine, joined and glued together, and of two strips of linden wood which bind the edges of the board, these being seven digits wide at the back, and in front near the head of the bellows one and a half digits wide. These strips are glued to the boards, so that there shall be less damage from the iron nails driven through the hide. There are some people who do not surround the boards with strips, but use boards only, which are very thick. The upper board has an aperture and a handle; the aperture is in the middle of the board and is one foot three palms distant from where the board joins the head of the bellows, and is six digits long and four wide. The lid for this aperture is two palms and a digit long and wide, and three digits thick; toward the back of the lid is a little notch cut into the surface so that it may be caught by the hand; a groove is cut out of the top of the front and sides, so that it may engage in mouldings a palm wide and three digits thick, which are also cut out in a similar manner under the edges. Now, when the lid is drawn forward the hole is closed, and when drawn back it is opened; the smelter opens the aperture a little so that the air may escape from the bellows through it, if he fears the hides might be burst when the bellows are too vigorously and quickly inflated; he, however, closes the aperture if the hides are ruptured and the air escapes. Others perforate the upper board with two or three round holes in the same place as the rectangular one, and they insert plugs in them which they draw out

[6]Devices for creating an air current must be of very old invention, for it is impossible to conceive of anything but the crudest melting of a few simple ores without some forced draft. Wilkinson (The Ancient Egyptians, II, p. 316) gives a copy of an illustration of a foot-bellows from a tomb of the time of Thothmes III. (1500 B.C.). The rest of the world therefore, probably obtained them from the Egyptians. They are mentioned frequently in the Bible, the most pointed reference to metallurgical purposes being Jeremiah (VI, 29): "The bellows are burned, the lead is consumed in the fire; the founder melteth in vain; for "the wicked are not plucked away." Strabo (VII, 3) states that Ephorus ascribed the invention of bellows to Anacharsis—a Thracian prince of about 600 B.C.

when it is necessary. The wooden handle is seven palms long, or even longer, in order that it may extend outside ; one-half of this handle, two palms wide and one thick, is glued to the end of the board and fastened with pegs covered with glue ; the other half projects beyond the board, and is rounded and seven digits thick. Besides this, to the handle and to the board is fixed a cleat two feet long, as many palms wide and one palm thick, and to the under side of the same board, at a distance of three palms from the end, is fixed another cleat two feet long, in order that the board may sustain the force of distension and compression ; these two cleats are glued to the board, and are fastened to it with pegs covered with glue.

The lower bellows-board, like the upper, is made of two pieces of pine and of two strips of linden wood, all glued together ; it is of the same width and thickness as the upper board, but is a cubit longer, this extension being part of the head of which I have more to say a little later. This lower bellows-board has an air-hole and an iron ring. The air-hole is about a cubit distant from the posterior end, and it is midway between the sides of the bellows-board, and is a foot long and three palms wide ; it is divided into equal parts by a small rib which forms part of the board, and is not cut from it ; this rib is a palm long and one-third of a digit wide. The flap of the air-hole is a foot and three digits long, three palms and as many digits wide ; it is a thin board covered with goat skin, the hairy part of which is turned toward the ground. There is fixed to one end of the flap, with small iron nails, one-half of a doubled piece of leather a palm wide and as long as the flap is wide ; the other half of the leather, which is behind the flap, is twice perforated, as is also the bellows-board, and these perforations are seven digits apart. Passing through these a string is tied on the under side of the board ; and thus the flap when tied to the board does not fall away. In this manner are made the flap and the air-hole, so when the bellows are distended the flap opens, when compressed it closes. At a distance of about a foot beyond the air-hole a slightly elliptical iron ring, two palms long and one wide, is fastened by means of an iron staple to the under part of the bellows-board ; it is at a distance of three palms from the back of the bellows. In order that the lower bellows-board may remain stationary, a wooden bolt is driven into the ring, after it penetrates through the hole in the transverse supporting plank which forms part of the frame for the bellows. There are some who dispense with the ring and fasten the bellows-board to the frame with two iron screws something like nails.

The bows are placed between the two boards and are of the same length as the upper board. They are both made of four pieces of linden wood three digits thick, of which the two long ones are seven digits wide at the back and two and a half at the front ; the third piece, which is at the back, is two palms wide. The ends of the bows are a little more than a digit thick, and are mortised to the long pieces, and both having been bored through, wooden pegs covered with glue are fixed in the holes ; they are thus joined and glued to the long pieces. Each of the ends is bowed (*arcuatur*) to meet the end of the long part of the bow, whence its name "bow" originated. The fourth piece keeps the ends of the bow distended, and is placed a cubit distant from the head of the bellows; the ends of this piece are mortised into the ends of the bow and are joined and glued to them; its length without the tenons is a foot, and its width a palm and two digits. There are, besides, two other very small pieces glued to the head of the bellows and to the lower board, and fastened to them by wooden pegs covered with glue, and they are three palms and two digits long, one palm high, and a digit thick, one half being slightly cut away. These pieces keep the ends of the bow away from the hole in the bellows-head, for if they were not there, the ends, forced inward by the great and frequent movement, would be broken.

The leather is of ox-hide or horse-hide, but that of the ox is far preferable to that of the horse. Each of these hides, for there are two, is three and a half feet wide where they are joined at the back part of the bellows. A long leathern thong is laid along each of the bellows-boards and each of the bows, and fastened by T-shaped iron nails five digits long; each of the horns of the nails is two and a half digits long and half a digit wide. The hide is attached to the bellows-boards by means of these nails, so that a horn of one nail almost touches the horn of the next; but it is different with the bows, for the hide is fastened to the back piece of the bow by only two nails, and to the two long pieces by four nails. In this practical manner they put ten nails in one bow and the same number in the other. Sometimes when the smelter is afraid that the vigorous motion of the bellows may pull or tear the hide from the bows, he also fastens it with little strips of pine by means of another kind of nail, but these strips cannot be fastened to the back pieces of the bow, because these are somewhat bent. Some people do not fix the hide to the bellows-boards and bows by iron nails, but by iron screws, screwed at the same time through strips laid over the hide. This method of fastening the hide is less used than the other, although there is no doubt that it surpasses it in excellence.

Lastly, the head of the bellows, like the rest of the body, consists of two boards, and of a nozzle besides. The upper board is one cubit long, one and a half palms thick. The lower board is part of the whole of the lower bellows-board; it is of the same length as the upper piece, but a palm and a digit thick. From these two glued together is made the head, into which, when it has been perforated, the nozzle is fixed. The back part of the head, where it is attached to the rest of the bellows-body, is a cubit wide, but three palms forward it becomes two digits narrower. Afterward it is somewhat cut away so that the front end may be rounded, until it is two palms and as many digits in diameter, at which point it is bound with an iron ring three digits wide.

The nozzle is a pipe made of a thin plate of iron; the diameter in front is three digits, while at the back, where it is encased in the head of the bellows, it is a palm high and two palms wide. It thus gradually widens out, especially at the back, in order that a copious wind can penetrate into it; the whole nozzle is three feet long.

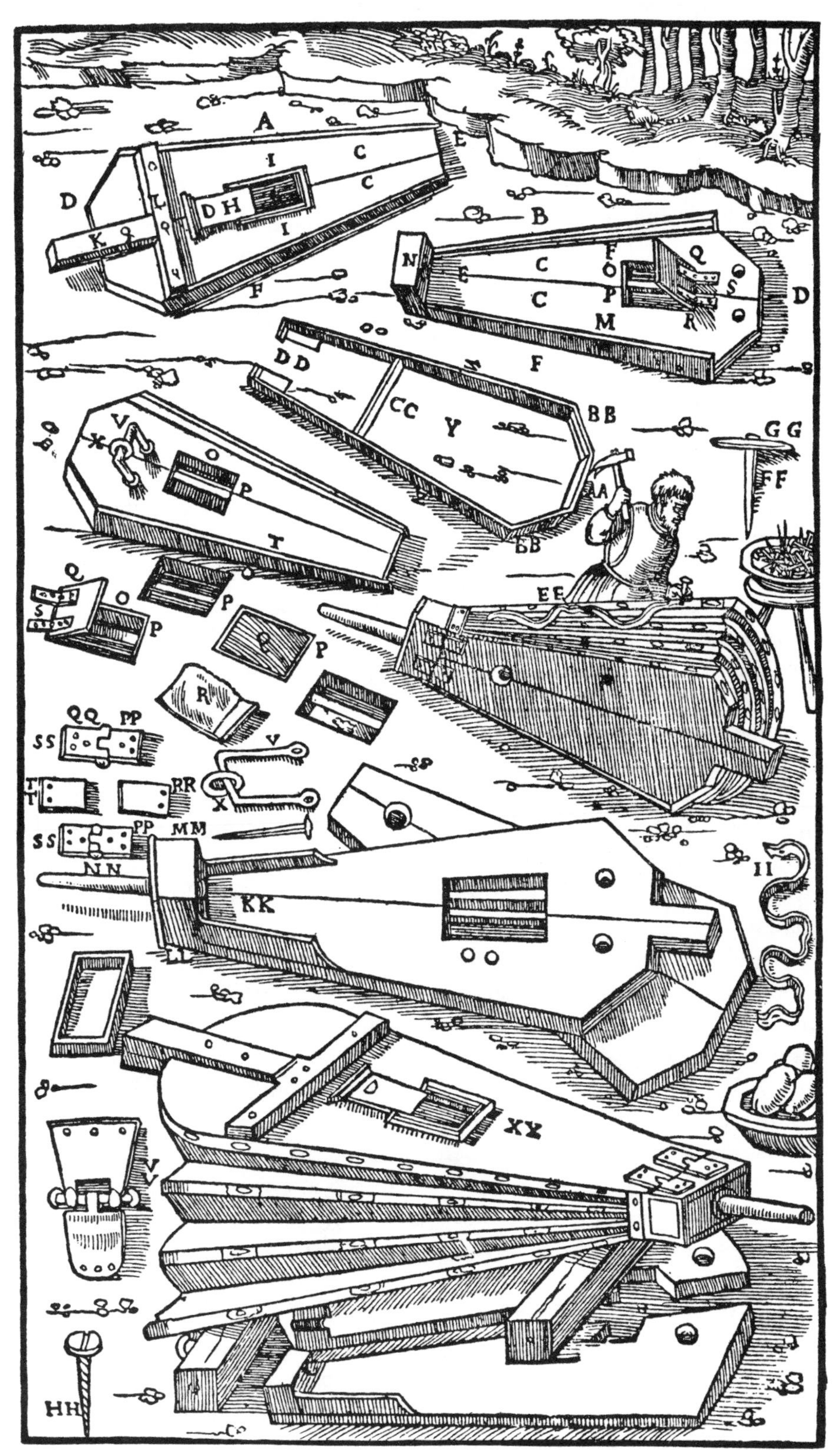

A—Upper bellows-board. B—Lower bellows-board. C—The two pieces of wood of which each consists. D—Posterior arched part of each. E—Tapered front part of each. F—Pieces of linden wood. G—Aperture in the upper board. H—Lid. I—Little mouldings of wood. K—Handle. L—Cleat on the outside. The cleat inside I am not able to depict. M—Interior of the lower bellows-board. N—Part of the head. O—Air-hole. P—Supporting bar. Q—Flap. R—Hide. S—Thong. T—Exterior of the lower board. V—Staple. X—Ring. Y—Bow. Z—Its long pieces. AA—Back piece of the bow. BB—The bowed ends. CC—Crossbar distending the bow. DD—The two little pieces. EE—Hide. FF—Nail. GG—Horn of the nail. HH—A screw. II—Long thong. KK—Head. LL—Its lower board. MM—Its upper board. NN—Nozzle. OO—The whole of the lower bellows-board. PP—The two exterior plates of the head hinges. QQ—Their curved piece. RR—Middle plate of the head. SS—The two outer plates of the upper bellows-board. TT—Its middle plate. VV—Little axle. XX—Whole bellows.

The upper bellows-board is joined to the head of the bellows in the following way. An iron plate[7], a palm wide and one and a half palms long, is first fastened to the head at a distance of three digits from the end ; from this plate there projects a piece three digits long and two wide, curved in a small circle. The other side has a similar plate. Then in the same part of the upper board are fixed two other iron plates, distant two digits from the edge, each of which are six digits wide and seven long ; in each of these plates the middle part is cut away for a little more than three digits in length and for two in depth, so that the curved part of the plates on the head corresponding to them may fit into this cut out part. From both sides of each plate there project pieces, three digits long and two digits wide, similarly curved into small circles. A little iron pin is passed through these curved pieces of the plates, like a little axle, so that the upper board of the bellows may turn upon it. The little axle is six digits long and a little more than a digit thick, and a small groove is cut out of the upper board, where the plates are fastened to it, in such a manner that the little axle when fixed to the plates may not fall out. Both plates fastened to the bellows-board are affixed by four iron nails, of which the heads are on the inner part of the board, whereas the points, clinched at the top, are transformed into heads, so to speak. Each of the other plates is fastened to the head of the bellows by means of a nail with a wide head, and by two other nails of which the heads are on the edge of the bellows-head. Midway between the two plates on the bellows-board there remains a space two palms wide, which is covered by an iron plate fastened to the board by little nails ; and another plate corresponding to this is fastened to the head between the other two plates ; they are two palms and the same number of digits wide.

The hide is common to the head as to all the other parts of the body ; the plates are covered with it, as well as the front part of the upper bellows-board, and both the bows and the back of the head of the bellows, so that the wind may not escape from that part of the bellows. It is three palms and as many digits wide, and long enough to extend from one of the sides of the lower board over the back of the upper ; it is fastened by many T-headed nails on one side to the upper board, and on the other side to the head of the bellows, and both ends are fastened to the lower bellows-board.

In the above manner the bellows is made. As two are required for each furnace, it is necessary to have twelve bellows, if there are to be six furnaces in one works.

Now it is time to describe their framework. First, two sills a little shorter than the furnace wall are placed on the ground. The front one of these is three palms wide and thick, and the back one three palms and two digits. The front one is two feet distant from the back wall of the furnace, and the back one is six feet three palms distant from the front one. They are set into the earth, that they may remain firm ; there are some who accomplish this by means of pegs which, through several holes, penetrate deeply into the ground.

[7]This whole arrangement could be summarized by the word " hinge."

Then twelve short posts are erected, whose lower ends are mortised into the sill that is near the back of the furnace wall ; these posts are two feet high, exclusive of the tenons, and are three palms and the same number of digits wide, and two palms thick. A slot one and a half palms wide is cut through them, beginning two palms from the bottom and extending for a height of three palms. All the posts are not placed at the same intervals, the first being at a distance of three feet five digits from the second, and likewise the third from the fourth, but the second is two feet one palm and three digits from the third ; the intervals between the other posts are arranged in the same manner, equal and unequal, of which each four pertain to two furnaces. The upper ends of these posts are mortised into a transverse beam which is twelve feet, two palms, and three digits long, and projects five digits beyond the first post and to the same distance beyond the fourth ; it is two palms and the same number of digits wide, and two palms thick. Since each separate transverse beam supports four bellows, it is necessary to have three of them.

Behind the twelve short posts the same number of higher posts are erected, of which each has the middle part of the lower end cut out, so that its two resulting lower ends are mortised into the back sill ; these posts, exclusive of the tenons, are twelve feet and two palms high, and are five palms wide and two palms thick. They are cut out from the bottom upward, the slot being four feet and five digits high and six digits wide. The upper ends of these posts are mortised into a long beam imposed upon them ; this long beam is placed close under the timbers which extend from the wall at the back of the furnace to the first long wall ; the beam is three palms wide and two palms thick, and forty-three feet long. If such a long one is not at hand, two or three may be substituted for it, which when joined together make up that length. These higher posts are not placed at equal distances, but the first is at a distance of two feet three palms one digit from the second, and the third is at the same distance from the fourth ; while the second is at a distance of one foot three palms and the same number of digits from the third, and in the same manner the rest of the posts are arranged at equal and unequal intervals. Moreover, there is in every post, where it faces the shorter post, a mortise at a foot and a digit above the slot ; in these mortises of the four posts is tenoned a timber which itself has four mortises. Tenons are enclosed in mortises in order that they may be better joined, and they are transfixed with wooden pins. This timber is thirteen feet three palms one digit long, and it projects beyond the first post a distance of two palms and two digits, and to the same number of palms and digits beyond the fourth post. It is two palms and as many digits wide, and also two palms thick. As there are twelve posts it is necessary to have three timbers of this kind.

On each of these timbers, and on each of the cross-beams which are laid upon the shorter posts, are placed four planks, each nine feet long, two palms three digits wide, and two palms one digit thick. The first plank is five feet one palm one digit distant from the second, at the front as well as at the back.

for each separate plank is placed outside of the posts. The third is at the same distance from the fourth, but the second is one foot and three digits distant from the third. In the same manner the rest of the eight planks are arranged at intervals, the fifth from the sixth and the seventh from the eighth are at the same distances as the first from the second and the third from the fourth; the sixth is at the same distance from the seventh as the second from the third.

Two planks support one transverse plank six feet long, one foot wide, one palm thick, placed at a distance of three feet and two palms from the back posts. When there are six of these supporting planks, on each separate one are placed two bellows; the lower bellows-boards project a palm beyond them. From each of the bellows-boards an iron ring descends through a hole in its supporting plank, and a wooden peg is driven into the ring, so that the bellows-board may remain stationary, as I stated above.

The two bellows communicate, each by its own plank, to the back of a copper pipe in which are set both of the nozzles, and their ends are tightly

A—Front sill. B—Back sill. C—Front posts. D—Their slots. E—Beam imposed upon them. F—Higher posts. G—Their slots. H—Beam imposed upon them. I—Timber joined in the mortises of the posts. K—Planks. L—Transverse supporting planks. M—The holes in them. N—Pipe. O—Its front end. P—Its rear end.

fastened in it. The pipe is made of a rolled copper or iron plate, a foot and two palms and the same number of digits long ; the plate is half a digit thick, but a digit thick at the back. The interior of the pipe is three digits wide, and two and a half digits high in the front, for it is not absolutely round ; and at the back it is a foot and two palms and three digits in diameter. The plate from which the pipe is made is not entirely joined up, but at the front there is left a crack half a digit wide, increasing at the back to three digits. This pipe is placed in the hole in the furnace, which, as I said, was in the middle of the wall and the arch. The nozzles of the bellows, placed in this pipe, are a distance of five digits from its front end.

The levers are of the same number as the bellows, and when depressed by the cams of the long axle they compress the bellows. These levers are eight feet three palms long, one palm wide and thick, and the ends are inserted in the slots of the posts ; they project beyond the front posts to a distance of two palms, and the same distance beyond the back posts in order that each may have its end depressed by its two cams on the axle. The cams not only penetrate into the slots of the back posts, but project three digits beyond them. An iron pin is set in round holes made through both sides of the slot of each front post, at three palms and as many digits from the bottom ; the pin penetrates the lever, which turns about it when depressed or raised. The back of the lever for the length of a cubit is a palm and a digit wider than the rest, and is perforated ; in this hole is engaged a bar six feet and two palms long, three digits wide, and about one and one-half digits thick ; it is somewhat hooked at the upper end, and approaches the handle of the bellows. Under the lever there is a nail, which penetrates through a hole in the bar, so that the lever and bar may move together. The bar is perforated in the upper end at a distance of six digits from the top ; this hole is two palms long and a digit wide, and in it is engaged the hook of an iron implement which is a digit thick. At the upper part this implement has either a round or square opening, like a link, and at the lower end is hooked ; the link is two digits high and wide and the hook is three digits long ; the middle part between the link and the hook is three palms and two digits long. The link of this implement engages either the handle of the bellows, or else a large ring which does engage it. This iron ring is a digit thick, two palms wide on the inside of the upper part, and two digits in the lower part, and this iron ring, not unlike the first one, engages the handle of the bellows. The iron ring either has its narrower part turned upward, and in it is engaged the ring of another iron implement, similar to the first, whose hook, extending upward, grips the rope fastened to the iron ring holding the end of the second lever, of which I will speak presently ; or else the iron ring grips this lever, and then in its hook is engaged the ring of the other implement whose ring engages the handle of the bellows, and in this case the rope is dispensed with.

Resting on beams fixed in the two walls is a longitudinal beam, at a distance of four and a half feet from the back posts ; it is two palms wide,

A—Lever which when depressed by means of a cam compresses the bellows. B—Slots through the posts. C—Bar. D—Iron implement with a rectangular link. E—Iron instrument with round ring. F—Handle of bellows. G—Upper post. H—Upper lever. I—Box with equal sides. K—Box narrow at the bottom. L—Pegs driven into the upper lever.

one and a half palms thick. There are mortised into this longitudinal beam the lower ends of upper posts three palms wide and two thick, which are six feet two palms high, exclusive of their tenons. The upper ends of these posts are mortised into an upper longitudinal beam, which lies close under the rafters of the building; this upper longitudinal beam is two palms wide and one thick. The upper posts have a slot cut out upward from a point two feet from the bottom, and the slot is two feet high and six digits wide. Through these upper posts a round hole is bored from one side to the other at a point three feet one palm from the bottom, and a small iron axle penetrates through the hole and is fastened there. Around this small iron axle turns the second lever when it is depressed and raised. This lever is eight feet long, and its other end is three digits wider than the rest of the lever; at this widest point is a hole two digits wide and three high, in which is fixed an iron ring, to which is tied the rope I have mentioned; it is five palms long, its upper loop is two palms and as many digits wide, and the lower one is one palm one digit wide. This half of the second lever, the end of which I have just mentioned, is three palms high and one wide ; it projects three feet beyond the slot of the post on which it turns ; the other end, which faces the back wall of the furnaces, is one foot and a palm high and a foot wide.

On this part of the lever stands and is fixed a box three and a half feet long, one foot and one palm wide, and half a foot deep ; but these measurements vary ; sometimes the bottom of this box is narrower, sometimes equal in width to the top. In either case, it is filled with stones and earth to make it heavy, but the smelters have to be on their guard and make provision against the stones falling out, owing to the constant motion ; this is prevented by means of an iron band which is placed over the top, both ends being wedge-shaped and driven into the lever so that the stones can be held in. Some people, in place of the box, drive four or more pegs into the lever and put mud between them, the required amount being added to the weight or taken away from it.

There remains to be considered the method of using this machine. The lower lever, being depressed by the cams, compresses the bellows, and the compression drives the air through the nozzle. Then the weight of the box on the other end of the upper lever raises the upper bellows-board, and the air is drawn in, entering through the air-hole.

The machine whose cams depress the lower lever is made as follows. First there is an axle, on whose end outside the building is a water-wheel ; at the other end, which is inside the building, is a drum made of rundles. This drum is composed of two double hubs, a foot apart, which are five digits thick, the radius all round being a foot and two digits ; but they are double, because each hub is composed of two discs, equally thick, fastened together with wooden pegs glued in. These hubs are sometimes covered above and around by iron plates. The rundles are thirty in number, a foot and two palms and the same number of digits long, with each end fastened into a hub ; they are rounded, three digits in diameter, and the same number of digits apart. In this practical manner is made the drum composed of rundles.

There is a toothed wheel, two palms and a digit thick, on the end of another axle ; this wheel is composed of a double disc[8]. The inner disc is composed of four segments a palm thick, everywhere two palms and a digit wide. The outer disc, like the inner, is made of four segments, and is a palm and a digit thick ; it is not equally wide, but where the head of the spokes are inserted it is a foot and a palm and digit wide, while on each side of the spokes it becomes a little narrower, until the narrowest part is only two palms and the same number of digits wide. The outer segments are joined to the inner ones in such a manner that, on the one hand, an outer segment ends in the middle of an inner one, and, on the other hand, the ends of the inner segments are joined in the middle of the outer ones ; there is no doubt that by this kind of joining the wheel is made stronger. The outer segments are fastened to the inner by means of a large number of wooden pegs. Each

[8]The rim of this wheel is obviously made of segments fixed in two layers ; the " disc " meaning the aggregate of segments on either side of the wheel.

A—Axle. B—Water-wheel. C—Drum composed of rundles. D—Other axle. E—Toothed wheel. F—Its spokes. G—Its segments. H—Its teeth. I—Cams of the axle.

segment, measured over its round back, is four feet and three palms long. There are four spokes, each two palms wide and a palm and a digit thick; their length, excluding the tenons, being two feet and three digits. One end of the spoke is mortised into the axle, where it is firmly fastened with pegs; the wide part of the other end, in the shape of a triangle, is mortised into the outer segment opposite it, keeping the shape of the same as far as the segment ascends. They also are joined together with wooden pegs glued in, and these pegs are driven into the spokes under the inner disc. The parts of the spokes in the shape of the triangle are on the inside; the outer part is simple. This triangle has two sides equal, the erect ones as is evident, which are a palm long; the lower side is not of the same length, but is five digits long, and a mortise of the same shape is cut out of the segments. The wheel has sixty teeth, since it is necessary that the rundle drum should revolve twice while the toothed wheel revolves once. The teeth are a foot long, and project one palm from the inner disc of the wheel, and three digits from the outer disc;

they are a palm wide and two and a half digits thick, and it is necessary that they should be three digits apart, as were the rundles.

The axle should have a thickness in proportion to the spokes and the segments. As it has two cams to depress each of the levers, it is necessary that it should have twenty-four cams, which project beyond it a foot and a palm and a digit. The cams are of almost semicircular shape, of which the widest part is three palms and a digit wide, and they are a palm thick; they are distributed according to the four sides of the axle, on the upper, the lower and the two lateral sides. The axle has twelve holes, of which the first penetrates through from the upper side to the lower, the second from one lateral side to the other; the first hole is four feet two palms distant from the second; each alternate one of these holes is made in the same direction, and they are arranged at equal intervals. Each single cam must be opposite another; the first is inserted into the upper part of the first hole, the second into the lower part of the same hole, and so fixed by pegs that they do not fall out; the third cam is inserted into that part of the second hole which is on the right side, and the fourth into that part on the left. In like manner all the cams are inserted into the consecutive holes, for which reason it happens that the cams depress the levers of the

A—Charcoal. B—Mortar-box. C—Stamps.

bellows in rotation. Finally we must not omit to state that this is only one of many such axles having cams and a water-wheel.

I have arrived thus far with many words, and yet it is not unseasonable that I have in this place pursued the subject minutely, since the smelting of all the metals, to which I am about to proceed, could not be undertaken without it.

The ores of gold, silver, copper, and lead, are smelted in a furnace by four different methods. The first method is for the rich ores of gold or silver, the second for the mediocre ores, the third for the poor ores, and the fourth method is for those ores which contain copper or lead, whether they contain precious metals or are wanting in them. The smelting of the first ores is performed in the furnace of which the tap-hole is intermittently closed ; the other three ores are melted in furnaces of which the tap-holes are always open.

First, I will speak of the manner in which the furnaces are prepared for the smelting of the ores, and of the first method of smelting. The powder from which the hearth and forehearth should be made is composed of charcoal and earth (clay ?). The charcoal is crushed by the stamps in a mortar-box, the front of which is closed by a board at the top, while the charcoal,

A—Tub. B—Sieve. C—Rods. D—Bench-frame.

crushed to powder, is removed through the open part below ; the stamps are not shod with iron, but are made entirely of wood, although at the lower part they are bound round at the wide part by an iron band.

The powder into which the charcoal is crushed is thrown on to a sieve whose bottom consists of interwoven withes of wood. The sieve is drawn backward and forward over two wooden or iron rods placed in a triangular position on a tub, or over a bench-frame set on the floor of the building ; the powder which falls into the tub or on to the floor is of suitable size, but the pieces of small charcoal which remain in the sieve are emptied out and thrown back under the stamps.

When the earth is dug up it is first exposed to the sun that it may dry. Later on it is thrown with a shovel on to a screen—set up obliquely and supported by poles,—made of thick, loosely woven hazel withes, and in this way the fine earth and its small lumps pass through the holes of the screen, but the clods and stones do not pass through, but run down to the ground. The earth which passes through the screen is conveyed in a two-wheeled cart to the works and there sifted. This sieve, which is not dissimilar to the one

A—Screen. B—Poles. C—Shovel. D—Two-wheeled cart. E—Hand-sieve. F—Narrow boards. G—Box. H—Covered pit.

described above, is drawn backward and forward upon narrow boards of equal length placed over a long box ; the powder which falls through the sieve into the box is suitable for the mixture ; the lumps that remain in the sieve are thrown away by some people, but by others they are placed under the stamps. This powdered earth is mixed with powdered charcoal, moistened, and thrown into a pit, and in order that it may remain good for a long time, the pit is covered up with boards so that the mixture may not become contaminated.

They take two parts of pulverised charcoal and one part of powdered earth, and mix them well together with a rake ; the mixture is moistened by pouring water over it so that it may easily be made into shapes resembling snowballs ; if the powder be light it is moistened with more water, if heavy with less. The interior of the new furnace is lined with lute, so that the cracks in the walls, if there are any, may be filled up, but especially in order to preserve the rock from injury by fire. In old furnaces in which ore has been melted, as soon as the rocks have cooled the assistant chips away, with a spatula, the accretions which adhere to the walls, and then breaks them up with an iron hoe or a rake with five teeth. The cracks of the furnace are first filled in with fragments of rock or brick, which he does by passing his hand into the furnace through its mouth, or else, having placed a ladder against it, he mounts by the rungs to the upper open part of the furnace. To the upper part of the ladder a board is fastened that he may lean and recline against it. Then standing on the same ladder, with a wooden spatula, he smears the furnace walls over with lute ; this spatula is four feet long, a digit thick, and for a foot upward from the bottom it is a palm wide, or even wider, generally two and a half digits. He spreads the lute equally over the inner walls of the furnace. The mouth of the copper pipe[9] should not protrude from the lute, lest sows[10] form round about it and thus impede the melting, for the furnace bellows could not force a blast through them. Then the same assistant throws a little powdered charcoal into the pit of the forehearth and sprinkles it with pulverised earth. Afterward, with a bucket he pours water into it and sweeps this all over the forehearth pit, and with the broom drives the turbid water into the furnace hearth and likewise sweeps it out. Next he throws the mixed and moistened powder into the furnace, and then a second time mounting the steps of the ladder, he introduces the rammer into the furnace and pounds the powder so that the hearth is made solid. The rammer is rounded and three palms long ; at the bottom it is five digits in diameter, at the top three and a half, therefore it is made in the form of a truncated cone ; the handle of the rammer is round and five feet long and

[9]It has not been considered necessary to introduce the modern term *twyer* in these descriptions, as the literal rendering is sufficiently clear.

[10]*Ferruminata.* These accretions are practically always near the hearth, and would correspond to English "sows," and therefore that term has been adopted. It will be noted that, like most modern metallurgists, Agricola offers no method for treating them. Pliny (XXXIV, 37) describes a "sow," and uses the verb *ferruminare* (to weld or solder) : "Some "say that in the furnace there are certain masses of stone which become soldered together, "and that the copper fuses around it, the mass not becoming liquid unless it is transferred "to another furnace ; it thus forms a sort of knot, as it were, of the metal."

A—FURNACE. B—LADDER. C—BOARD FIXED TO IT. D—HOE. E—FIVE-TOOTHED RAKE. F—WOODEN SPATULA. G—BROOM. H—RAMMER. I—RAMMER, SAME DIAMETER. K—TWO WOODEN SPATULAS. L—CURVED BLADE. M—BRONZE RAMMER. N—ANOTHER BRONZE RAMMER. O—WIDE SPATULA. P—ROD. Q—WICKER BASKET. R—TWO BUCKETS OF LEATHER IN WHICH WATER IS CARRIED FOR PUTTING OUT A CONFLAGRATION, SHOULD THE *officina* CATCH FIRE. S—BRASS PUMP WITH WHICH THE WATER IS SQUIRTED OUT. T—TWO HOOKS. V—RAKE. X—WORKMAN BEATING THE CLAY WITH AN IRON IMPLEMENT.

two and a half digits thick; the upper part of the rammer, where the handle is inserted, is bound with an iron band two digits wide. There are some who, instead, use two rounded rammers three and a half digits in diameter, the same at the bottom as at the top. Some people prefer two wooden spatulas, or a rammer spatula.

In a similar manner, mixed and moistened powder is thrown and pounded with a rammer in the forehearth pit, which is outside the furnace. When this is nearly completed, powder is again put in, and pushed with the rammer up toward the protruding copper pipe, so that from a point a digit under the mouth of the copper pipe the hearth slopes down into the crucible of the forehearth,[11] and the metal can run down. The same is repeated until the

[11]What are known in English as "crucible," "furnace well," "forehearth," "dipping-pot," "tapping-pot," "receiving-pot," etc., are in the text all *catinus*, *i.e.*, crucible. For easier reading, however, we have assigned the names indicated in the context.

forehearth pit is full, then afterward this is hollowed out with a curved blade ; this blade is of iron, two palms and as many digits long, three digits wide, blunt at the top and sharp at the bottom. The crucible of the forehearth must be round, a foot in diameter and two palms deep if it has to contain a *centumpondium* of lead, or if only seventy *librae*, then three palms in diameter and two palms deep like the other. When the forehearth has been hollowed out it is pounded with a round bronze rammer. This is five digits high and the same in diameter, having a curved round handle one and a half digits thick ; or else another bronze rammer is used, which is fashioned in the shape of a cone, truncated at the top, on which is imposed another cut away at the bottom, so that the middle part of the rammer may be grasped by the hand ; this is six digits high, and five digits in diameter at the lower end and four at the top. Some use in its place a wooden spatula two and a half palms wide at the lower end and one palm thick.

The assistant, having prepared the forehearth, returns to the furnace and besmears both sides as well as the top of the mouth with simple lute. In the lower part of the mouth he places lute that has been dipped in charcoal dust, to guard against the risk of the lute attracting to itself the powder of the hearth and vitiating it. Next he lays in the mouth of the furnace a straight round rod three quarters of a foot long and three digits in diameter. Afterward he places a piece of charcoal on the lute, of the same length and width as the mouth, so that it is entirely closed up ; if there be not at hand one piece of charcoal so large, he takes two instead. When the mouth is thus closed up, he throws into the furnace a wicker basket full of charcoal, and in order that the piece of charcoal with which the mouth of the furnace is closed should not then fall out, the master holds it in with his hand. The pieces of charcoal which are thrown into the furnace should be of medium size, for if they are large they impede the blast of the bellows and prevent it from blowing through the tap-hole of the furnace into the forehearth to heat it. Then the master covers over the charcoal, placed at the mouth of the furnace, with lute and extracts the wooden rod, and thus the furnace is prepared. Afterward the assistant throws four or five larger baskets full of charcoal into the furnace, filling it right up ; he also throws a little charcoal into the forehearth, and places glowing coals upon it in order that it may be kindled, but in order that the flames of this fire should not enter through the tap-hole of the furnace and fire the charcoal inside, he covers the tap-hole with lute or closes it with fragments of pottery. Some do not warm the forehearth the same evening, but place large charcoals round the edge of it, one leaning on the other ; those who follow the first method sweep out the forehearth in the morning, and clean out the little pieces of charcoal and cinders, while those who follow the latter method take, early in the morning, burning firebrands, which have been prepared by the watchman of the works, and place them on the charcoal.

At the fourth hour the master begins his work. He first inserts a small piece of glowing coal into the furnace, through the bronze nozzle-pipe of the bellows, and blows up the fire with the bellows ; thus within the space of half an hour the forehearth, as well as the hearth, becomes warmed, and of course more quickly if on the preceding day ores have been smelted in the same furnace, but if not then it warms more slowly. If the hearth and forehearth are not warmed before the ore to be smelted is thrown in, the furnace is injured and the metals lost ; or if the powder from which both are made is damp in summer or frozen in winter, they will be cracked, and, giving out a sound like thunder, they will blow out the metals and other substances with great peril to the workmen. After the furnace has been warmed, the master throws in slags, and these, when melted, flow out through the tap-hole into the forehearth. Then he closes up the tap-hole at once with mixed lute and charcoal dust ; this plug he fastens with his hand to a round wooden rammer that is five digits thick, two palms high, with a handle three feet long. The smelter extracts the slags from the forehearth with a hooked bar ; if the ore to be smelted is rich in gold or silver he puts into the forehearth a *centumpondium* of lead, or half as much if the ore is poor, because the former requires much lead, the latter little ; he immediately throws burning firebrands on to the lead so that it melts. Afterward he performs everything according to the usual manner and order, whereby he first throws into the furnace as many cakes melted from pyrites[12], as he requires to smelt the ore ; then he puts in two wicker baskets full of ore with litharge and hearth-lead[13], and stones which fuse easily by fire of the second order, all mixed together ; then one wicker basket full of charcoal, and lastly the slags. The furnace now being filled with all the things I have mentioned, the ore is slowly smelted ; he does not put too much of it against the back wall of the furnace, lest sows should form around the nozzles of the bellows and the blast be impeded and the fire burn less fiercely.

This, indeed, is the custom of many most excellent smelters, who know how to govern the four elements[14]. They combine in right proportion the ores, which are part earth, placing no more than is suitable in the furnaces ; they pour in the needful quantity of water ; they moderate with skill the air from the bellows ; they throw the ore into that part of the fire which burns fiercely. The master sprinkles water into each part of the furnace to dampen the charcoal slightly, so that the minute parts of ore may adhere to it, which otherwise the blast of the bellows and the force of the fire would agitate and blow away with the fumes. But as the nature of the ores to be smelted varies, the smelters have to arrange the hearth now high, now low, and to place the pipe in which the nozzles of the bellows are inserted sometimes on a great and sometimes at a slight angle, so that the blast of the bellows may

[12]*Panes ex pyrite conflati.* While the term *matte* would cover most cases where this expression appears, and in many cases would be more expressive to the modern reader, yet there are instances where the expression as it stands indicates its particular origin, and it has been, therefore, considered advisable to adhere to the literal rendering.

[13]*Molybdaena.* See note 37, p. 476. It was the saturated furnace bottoms from cupellation.

[14]The four elements were earth, air, fire, and water.

blow into the furnace in either a mild or a vigorous manner. For those ores which heat and fuse easily, a low hearth is necessary for the work of the smelters, and the pipe must be placed at a gentle angle to produce a mild blast from the bellows. On the contrary, those ores that heat and fuse slowly must have a high hearth, and the pipe must be placed at a steep incline in order to blow a strong blast of the bellows, and it is necessary, for this kind of ore, to have a very hot furnace in which slags, or cakes melted from pyrites, or stones which melt easily in the fire[15], are first melted, so that the ore should not settle in the hearth of the furnace and obstruct and choke up the tap-hole, as the minute metallic particles that have been washed from the ores are wont to do. Large bellows have wide nozzles, for if they were narrow the copious and strong blast would be too much compressed and too acutely blown into the furnace, and then the melted material would be chilled, and would form sows around the nozzle, and thus obstruct the opening into the furnace, which would cause great damage to the proprietors' property. If the ores agglomerate and do not fuse, the smelter, mounting on the ladder placed against the side of the furnace, divides the charge with a pointed or hooked bar, which he also pushes down into the pipe in

[15]" Stones which easily melt in the fire." Nowhere in *De Re Metallica* does the author explain these substances. However in the *Interpretatio* (p. 465) he gives three genera or orders with their German equivalents, as follows :—" *Lapides qui igni liquescunt primi generis,—Schöne flüsse ; secundi,—flüsse zum schmeltzen flock quertze* ; *tertii,—quertze oder kiselstein.*" We confess our inability to make certain of most of the substances comprised in the first and second orders. We consider they were in part fluor-spar, and in any event the third order embraced varieties of quartz, flint, and silicious material generally. As the matter is of importance from a metallurgical point of view, we reproduce at some length Agricola's own statements on the subject from *Bermannus* and *De Natura Fossilium.* In the latter (p. 268) he states : " Finally there now remain those stones which I call ' stones which easily melt in " the fire,' because when thrown into hot furnaces they flow (*fluunt*). There are three orders " (*genera*) of these. The first resembles the transparent gems ; the second is not similar, " and is generally not translucent ; it is translucent in some part, and in rare instances " altogether translucent. The first is sparingly found in silver and other mines ; the second " abounds in veins of its own. The third genus is the material from which glass is made, " although it can also be made out of the other two. The stones of the first order are not " only transparent, but are also resplendent, and have the colours of gems, for some resemble " crystal, others emerald, heliotrope, lapis lazuli, amethyst, sapphire, ruby, *chrysolithus, morion* " (cairngorm ?), and other gems, but they differ from them in hardness. . . . To the " first genus belongs the *lapis alabandicus* (modern albandite ?), if indeed it was different " from the alabandic carbuncle. It can be melted, according to Pliny, in the fire, and fused " for the preparation of glass. It is black, but verging upon purple. It comes from " Caria, near Alabanda, and from Miletus in the same province. The second order of stones " does not show a great variety of colours, and seldom beautiful ones, for it is generally white, " whitish, greyish, or yellowish. Because these (stones) very readily melt in the fire, they are " added to the ores from which the metals are smelted. The small stones found in veins, " veinlets, and the spaces between the veins, of the highest peaks of the Sudetic range (*Sudi-* " *torum montium*), belong partly to this genus and partly to the first. They differ in size, " being large and small ; and in shape, some being round or angular or pointed ; in colour they " are black or ash-grey, or yellow, or purple, or violet, or iron colour. All of these are lacking " in metals. Neither do the little stones contain any metals which are usually found in the " streams where gold dust is collected by washing. . . . In the rivers where are collected " the small stones from which tin is smelted, there are three genera of small stones to be found, " all somewhat rounded and of very light weight, and devoid of all metals. The largest are " black, both on the outside and inside, smooth and brilliant like a mirror ; the medium-sized " are either bluish black or ash-grey ; the smallest are of a yellowish colour, somewhat like a " silkworm. But because both the former and the latter stones are devoid of metals, and fly " to pieces under the blows of the hammer, we classify them as sand or gravel. Glass is made " from the stones of the third order, and particularly from sand. For when this is thrown " into the heated furnace it is melted by the fire. . . . This kind of stone is either found

which the nozzle of the bellows is placed, and by a downward movement dislodges the ore and the sows from around it.

After a quarter of an hour, when the lead which the assistant has placed in the forehearth is melted, the master opens the tap-hole of the furnace with a tapping-bar. This bar is made of iron, is three and a half feet long, the forward end pointed and a little curved, and the back end hollow so that into it may be inserted a wooden handle, which is three feet long and thick enough to be well grasped by the hand. The slag first flows from the furnace into the forehearth, and in it are stones mixed with metal or with the metal adhering to them partly altered, the slag also containing earth and solidified juices. After this the material from the melted pyrites flows out, and then the molten lead contained in the forehearth absorbs the gold and silver. When that which has run out has stood for some time in the forehearth, in order to be able to separate one from the other, the master first either skims off the slags with the hooked bar or else lifts them off with an iron fork; the slags, as they are very light, float on the top. He next draws off the cakes of melted pyrites, which as they are of medium weight hold the middle place; he leaves in the forehearth the alloy of gold or silver with the lead, for these being the heaviest, sink to the bottom. As, however, there is a difference

"in its own veins, which are occasionally very wide, or else scattered through the mines. It "is less hard than flint, on account of which no fire can be struck from it. It is not trans-"parent, but it is of many colours—that is to say, white, yellowish, ash-grey, brown, black, "green, blue, reddish or red. This genus of stones occurs here and there in mountainous "regions, on banks of rivers, and in the fields. Those which are black right through to the "interior, and not merely on the surface, are more rare; and very frequently one coloured "vein is intersected by another of a different colour—for instance, a white one by a red one; "the green is often spotted with white, the ash-grey with black, the white with crimson. "Fragments of these stones are frequently found on the surface of the earth, and in the "running water they become polished by rubbing against stones of their own or of another "genus. In this way, likewise, fragments of rocks are not infrequently shaped into spherical "forms. . . . This stone is put to many uses; the streets are paved with it, whatever its "colour; the blue variety is added to the ash of pines for making those other ashes which are "used by wool-dyers. The white variety is burned, ground, and sifted, and from this they "make the sand out of which glass is made. The whiter the sand is, the more useful it is."

Perusal of the following from *Bermannus* (p. 458) can leave little doubt as to the first or second order being in part fluor-spar. Agricola derived the name *fluores* from *fluo* "to flow," and we in turn obtain "fluorite," or "fluorspar," from Agricola. "*Bermannus*. These stones are "similar to gems, but less hard. Allow me to explain word for word. Our miners call them "*fluores*, not inappropriately to my mind, for by the heat of fire, like ice in the sun, they "liquefy and flow away. They are of varied and bright colours. *Naevius*.—Theophrastus "says of them that they are made by a conflux in the earth. These red *fluores*, to employ "the words just used by you, are the ruby silver which you showed us before. *Bermannus*.—At "the first glance it appears so, although it is not infrequently translucent. *Naevius*.—Then "they are rubies? *Bermannus*.—Not that either. *Naevius*.—In what way, then, can they be "distinguished from rubies? *Bermannus*.—Chiefly by this sign, that they glitter more "feebly when translucent. Those which are not translucent may be distinguished from "rubies. Moreover, *fluores* of all kinds melt when they are subject to the first fire; rubies "do not melt in fire. *Naevius*.—You distinguish well. *Bermannus*.—You see the other "kind, of a paler purple colour? *Naevius*.—They appear to be an inferior kind of amethyst, "such as are found in many places in Bohemia. *Bermannus*.—Indeed, they are not very dis-"similar, therefore the common people who do not know amethysts well, set them in rings "for gems, and they are easily sold. The third kind, as you see here, is white. *Naevius*.—I "should have thought it a crystal. *Bermannus*.—A fourth is a yellow colour, a fifth ash colour, "a sixth blackish. Some are violet, some green, others gold-coloured. *Anton*.—What is the "use of *fluores*? *Bermannus*.—They are wont to be made use of when metals are smelted, "as they cause the material in the fire to be much more fluid, exactly like a kind of stone "which we said is made from pyrites (matte); it is, indeed, made not far from here, at Breiten-"brunn, which is near Schwarzenberg. Moreover, from *fluores* they can make colours which "artists use."

in slags, the uppermost containing little metal, the middle more, and the lowest much, he puts these away separately, each in its own place, in order that to each heap, when it is re-smelted, he may add the proper fluxes, and can put in as much lead as is demanded for the metal in the slag ; when the slag is re-melted, if it emits much odour, there is some metal in it ; if it emits no odour, then it contains none. He puts the cakes of melted pyrites away separately, as they were nearest in the forehearth to the metal, and contain a little more of it than the slags ; from all these cakes a conical mound is built up, by always placing the widest of them at the bottom. The hooked bar has a hook on the end, hence its name ; otherwise it is similar to other bars.

Afterward the master closes up the tap-hole and fills the furnace with the same materials I described above, and again, the ores having been melted, he opens the tap-hole, and with a hooked bar extracts the slags and the cakes melted from pyrites, which have run down into the forehearth. He repeats the same operation until a certain and definite part of the ore has been smelted, and the day's work is at an end ; if the ore was rich the work is finished in eight hours ; if poor, it takes a longer time. But if the ore was so rich as to be smelted in less than eight hours, another operation is in the meanwhile combined with the first, and both are performed in the space of ten hours. When all the ore has been smelted, he throws into the furnace a basket full of litharge or hearth-lead, so that the metal which has remained in the accretions may run out with these when melted. When he has finally drawn out of the forehearth the slags and the cakes melted from pyrites, he takes out, with a ladle, the lead alloyed with gold or silver and pours it into little iron or copper pans, three palms wide and as many digits deep, but first lined on the inside with lute and dried by warming, lest the glowing molten substances should break through. The iron ladle is two palms wide, and in other respects it is similar to the others, all of which have a sufficiently long iron shaft, so that the fire should not burn the wooden part of the handle. When the alloy has been poured out of the forehearth, the smelter foreman and the mine captain weigh the cakes.

Then the master breaks out the whole of the mouth of the furnace with a crowbar, and with that other hooked bar, the rabble and the five-toothed rake, he extracts the accretions and the charcoal. This crowbar is not unlike the other hooked one, but larger and wider ; the handle of the rabble is six feet long and is half of iron and half of wood. The furnace having cooled, the master chips off the accretions clinging to the walls with a rectangular spatula six digits long, a palm broad, and sharp on the front edge ; it has a round handle four feet long, half of it being of iron and half of wood. This is the first method of smelting ores.

Because they generally consist of unequal constituents, some of which melt rapidly and others slowly, the ores rich in gold and silver cannot be smelted as rapidly or as easily by the other methods as they can by the first method, for three important reasons. The first reason is that, as often as the closed tap-hole of the furnace is opened with a tapping-bar, so often can the

A, B, C—Three furnaces. At the first stands the smelter, who with a ladle pours the alloy out of the forehearth into the moulds. D—Forehearth. E—Ladle. F—Moulds. G—Round wooden rammer. H—Tapping-bar. At the second furnace stands the smelter, who opens the tap-hole with his tapping-bar. The assistant, standing on steps placed against the third furnace which has been broken open, chips off the accretions. I—Steps. K—Spatula. L—The other hooked bar. M—Mine captain carrying a cake, in which he has stuck the pick, to the scales to be weighed. N—Another mine captain opens a chest in which his things are kept.

smelter observe whether the ore is melting too quickly or too slowly, or whether it is flaming in scattered bits, and not uniting in one mass; in the first case the ore is smelting too slowly and not without great expense; in the second case the metal mixes with the slag which flows out of the furnace into the forehearth, wherefore there is the expense of melting it again; in the third case, the metal is consumed by the violence of the fire. Each of these evils has its remedy; if the ore melts slowly or does not come together, it is necessary to add some amount of fluxes which melt the ore; or if they melt too readily, to decrease the amount.

The second reason is that each time that the furnace is opened with a tapping-bar, it flows out into the forehearth, and the smelter is able to test the alloy of gold and lead or of silver with lead, which is called *stannum*[16]. When the tap-hole is opened the second or third time, this test shows us whether the alloy of gold or silver has become richer, or whether the lead is too debilitated and wanting in strength to absorb any more gold or silver. If it has become richer, some portion of lead added to it should renew its strength; if it has not become richer, it is poured out of the forehearth that it may be replaced with fresh lead.

The third reason is that if the tap-hole of the furnace is always open when the ore and other things are being smelted, the fluxes, which are easily melted, run out of the furnace before the rich gold and silver ores, for these are sometimes of a kind that oppose and resist melting by the fire for a longer period. It follows in this case, that some part of the ore is either consumed or is mixed with the accretions, and as a result little lumps of ore not yet melted are now and then found in the accretions. Therefore when these ores are being smelted, the tap-hole of the furnace should be closed for a time, as it is necessary to heat and mix the ore and the fluxes at the same time; since the fluxes fuse more rapidly than the ore, when the molten fluxes are held in the furnace, they thus melt the ore which does not readily fuse or mix with the lead. The lead absorbs the gold or silver, just as tin or lead when melted in the forehearth absorbs the other unmelted metal which has been thrown into it. But if the molten matter is poured upon that which is not molten, it runs off on all sides and consequently does not melt it. It follows from all this that ores rich in gold or silver, when put into a furnace with its tap-hole always open, cannot for that reason be smelted so successfully as in one where the tap-hole is closed for a time, so that during this time the ore may be melted by the molten fluxes. Afterward, when the tap-hole has been opened, they flow into the forehearth and mix there with the molten lead. This method of smelting the ores is used by us and by the Bohemians.

The three remaining methods of smelting ores are similar to each other in that the tap-holes of the furnaces always remain open, so that the molten metals may continually run out. They differ greatly from each other,

[16]*Stannum* (*Interpretatio,—werck*, modern *werk*). This term has been rendered throughout as "silver-lead" or "silver-lead alloy." It was the argentiferous lead suitable for cupellation. Agricola, in using it in this sense, was no doubt following his interpretation of its use by Pliny. Further remarks upon this subject will be found in note 33, p. 473.

A, B—Two furnaces. C—Forehearths. D—Dipping-pot. The smelter standing by the first furnace draws off the slags with a hooked bar. E—Hooked bar. F—Slags. G—The assistant drawing a bucket of water which he pours over the glowing slags to quench them. H—Basket made of twigs of wood intertwined. I—Rabble. K—Ore to be smelted. The master stands at the other furnace and prepares the forehearth by ramming it with two rammers. M—Crowbar.

however, for the tap-hole of the first of this kind is deeper in the furnace and narrower than that of the third, and besides it is invisible and concealed. It easily discharges into the forehearth, which is one and a half feet higher than the floor of the building, in order that below it to the left a dipping-pot can be made. When the forehearth is nearly full of the slags, which well up from the invisible tap-hole of the furnace, they are skimmed off from the top with a hooked bar ; then the alloy of gold or silver with lead and the melted pyrites, being uncovered, flow into the dipping-pot, and the latter are made into cakes ; these cakes are broken and thrown back into the furnace so that all their metal may be smelted out. The alloy is poured into little iron moulds.

The smelter, besides lead and cognate things, uses fluxes which combine with the ore, of which I gave a sufficient account in Book VII. The metals which are melted from ores that fuse readily in the fire, are profitable because they are smelted in a short time, while those which are difficult to fuse are not as profitable, because they take a long time. When fluxes remain in the furnace and do not melt, they are not suitable ; for this reason, accretions and slags are the most convenient for smelting, because they melt quickly. It is necessary to have an industrious and experienced smelter, who in the first place takes care not to put into the furnace more ores mixed with fluxes than it can accommodate.

The powder out of which this furnace hearth and the adjoining forehearth and the dipping-pot are usually made, consists mostly of equal proportions of charcoal dust and of earth, or of equal parts of the same and of ashes. When the hearth of the furnace is prepared, a rod that will reach to the forehearth is put into it, higher up if the ore to be smelted readily fuses, and lower down if it fuses with difficulty. When the dipping-pot and forehearth are finished, the rod is drawn out of the furnace so that the tap-hole is open, and through it the molten material flows continuously into the forehearth, which should be very near the furnace in order that it may keep very hot and the alloy thus be made purer. If the ore to be smelted does not melt easily, the hearth of the furnace must not be made too sloping, lest the molten fluxes should run down into the forehearth before the ore is smelted, and the metal thus remain in the accretions on the sides of the furnace. The smelter must not ram the hearth so much that it becomes too hard, nor make the mistake of ramming the lower part of the mouth to make it hard, for it could not breathe[17], nor could the molten matter flow freely out of the furnace. The ore which does not readily melt is thrown as much as possible to the back of the furnace, and toward that part where the fire burns very fiercely, so that it may be smelted longer. In this way the smelter may direct it whither he wills. Only when it glows at the part near the bellows' nozzle does it signify that all the ore is smelted which has been thrown to the side of the furnace in which the nozzles are placed. If the ore is easily melted, one or two wicker baskets full are thrown into the front part of the furnace so that the fire, being driven back by it, may also smelt the ore and the sows that

[17]*Expirare*,—to exhale or blow out.

A, B—Two furnaces. C—Forehearth. D—Dipping-pots. The master stands at the one furnace and draws away the slags with an iron fork. E—Iron fork. F—Wooden hoe with which the cakes of melted pyrites are drawn out. G—The forehearth crucible: one-half inside is to be seen open in the other furnace. H—The half outside the furnace. I—The assistant prepares the forehearth, which is separated from the furnace that it may be seen. K—Bar. L—Wooden rammer. M—Ladder. N—Ladle.

form round about the nozzles of the bellows. This process of smelting is very ancient among the Tyrolese[18], but not so old among the Bohemians.

The second method of smelting ores stands in a measure midway between that one performed in a furnace of which the tap-hole is closed intermittently, and the first of the methods performed in a furnace where the tap-hole is always open. In this manner are smelted the ores of gold and silver that are neither very rich nor very poor, but mediocre, which fuse easily and are readily absorbed by the lead. It was found that in this way a large quantity of ore could be smelted at one operation without much labour or great expense, and could thus be alloyed with lead. This furnace has two crucibles, one of which is half inside the furnace and half outside, so that the lead being put into this crucible, the part of the lead which is in the furnace absorbs the metals of the ores which easily fuse; the other crucible is lower, and the alloy and the molten pyrites run into it. Those who make use of this method of smelting, tap the alloy of gold or silver with lead from the upper crucible once or twice if need be, and throw in other lead or litharge, and each absorbs that flux which is nearest. This method of smelting is in use in Styria[19].

The furnace in the third method of smelting ores has the tap-hole likewise open, but the furnace is higher and wider than the others, and its bellows are larger; for these reasons a larger charge of the ore can be thrown into it. When the mines yield a great abundance of ore for the smelter, they smelt in the same furnace continuously for three days and three nights, providing there be no defect either in the hearth or in the forehearth. In this kind of a furnace almost every kind of accretion will be found. The forehearth of the furnace is not unlike the forehearth of the first furnace of all, except that it has a tap-hole. However, because large charges of ore are smelted uninterruptedly, and the melted material runs out and the slags are skimmed off, there is need for a second forehearth crucible, into which the molten material runs through an opened tap-hole when the first is full. When a smelter has spent twelve hours' labour on this work, another always takes his place. The ores of copper and lead and the poorest ores of gold and silver are smelted by this method, because they cannot be smelted by the other three methods on account of the greater expense occasioned. Yet by this method a *centumpondium* of ore containing only one or two *drachmae* of gold, or only a half to one *uncia* of silver,[20] can be smelted; because there is a large amount of ore in each charge, smelting is continuous, and without expensive fluxes such as lead, litharge, and hearth-lead. In this method of smelting we must use only cupriferous pyrites which easily melt in the fire, in truth the cakes melted out from this, if they no longer absorb

[18]*Rhetos*. The ancient Rhaetia comprised not only the greater part of Tyrol, but also parts of Switzerland and Lombardy. The mining section was, however, in Tyrol.

[19]*Noricum* was a region south of the Danube, embracing not only modern Styria, but also parts of Austria, Salzberg, and Carinthia.

[20]One *drachma* of gold to a *centumpondium* would be (if we assume these were Roman weights) 3 ozs. 1 dwt. Troy per short ton. One-half *uncia* of silver would be 12 ozs. 3 dwts. per short ton.

A, B—Two furnaces. C—Tap-holes of furnaces. D—Forehearths. E—Their tap-holes. F—Dipping-pots. G—At the one furnace stands the smelter carrying a wicker basket full of charcoal. At the other furnace stands a smelter who with the third hooked bar breaks away the material which has frozen the tap-hole of the furnace. H—Hooked bar. I—Heap of charcoal. K—Barrow on which is a box made of wicker work in which the coals are measured. L—Iron spade.

much gold or silver, are replenished again from crude pyrites alone. If from this poor ore, with melted pyrites alone, material for cakes cannot be made, there are added other fluxes which have not previously been melted. These fluxes are, namely, lead ore, stones easily fused by fire of the second order and sand made from them, limestone, *tophus*, white schist, and iron stone[21].

Although this method of smelting ores is rough and might not seem to be of great use, yet it is clever and useful; for a great weight of ores, in which the gold, silver, or copper are in small quantities, may be reduced into a few cakes containing all the metal. If on being first melted they are too crude to be suitable for the second melting, in which the lead absorbs the precious metals that are in the cakes, or in which the copper is melted out of them, yet they can be made suitable if they are repeatedly roasted, sometimes as often as seven or eight times, as I have explained in the last book. Smelters of this kind are so clever and expert, that in smelting they take out all the gold and silver which the assayer in assaying the ores has stated to be contained in them, because if during the first operation, when he makes the cakes, there is a *drachma* of gold or half an *uncia* of silver lost from the ores, the smelter obtains it from the slags by the second smelting. This method of smelting ores is old and very common to most of those who use other methods.

Although lead ores are usually smelted in the third furnace—whose tap-hole is always open,—yet not a few people melt them in special furnaces by a method which I will briefly explain. The *Carni*[22] first burn such lead ores, and afterward break and crush them with large round mallets. Between the two low walls of a hearth, which is inside a furnace made of and vaulted with a rock that resists injury by the fire and does not burn into chalk, they place green wood with a layer of dry wood on the top of it; then they throw the ore on to this, and when the wood is kindled the lead drips down and runs on to the underlying sloping hearth[23]. This hearth is made of pulverised

[21]For discussion of these fluxes see note page 232.

[22]*Carni*. Probably the people of modern Austrian Carniola, which lies south of Styria and west of Croatia.

[23]HISTORICAL NOTE ON SMELTING LEAD AND SILVER.—The history of lead and silver smelting is by no means a sequent array of exact facts. With one possible exception, lead does not appear upon the historical horizon until long after silver, and yet their metallurgy is so inextricably mixed that neither can be considered wholly by itself. As silver does not occur native in any such quantities as would have supplied the amounts possessed by the Ancients, we must, therefore, assume its reduction by either (1) intricate chemical processes, (2) amalgamation, (3) reduction with copper, (4) reduction with lead. It is impossible to conceive of the first with the ancient knowledge of chemistry; the second (see note 12, p. 297) does not appear to have been known until after Roman times; in any event, quicksilver appears only at about 400 B.C. The third was impossible, as the parting of silver from copper without lead involves metallurgy only possible during the last century. Therefore, one is driven to the conclusion that the fourth case obtained, and that the lead must have been known practically contemporaneously with silver. There is a leaden figure exhibited in the British Museum among the articles recovered from the Temple of Osiris at Abydos, and considered to be of the Archaic period—prior to 3800 B.C. The earliest known Egyptian silver appears to be a necklace of beads, supposed to be of the XII. Dynasty (2400 B.C.), which is described in the 17th Memoir, Egyptian Exploration Fund (London, 1898, p. 22). With this exception of the above-mentioned lead specimen, silver articles antedate positive evidence of lead by nearly a millennium, and if we assume lead as a necessary factor in silver production, we must conclude it was known long prior to any direct (except the above solitary possibility) evidence of lead itself. Further, if we are to conclude its necessary association with silver, we must assume a knowledge of cupellation for the parting of the two metals. Lead is mentioned in 1500 B.C.

charcoal and earth, as is also a large crucible, one-half of which lies under the furnace and the other half outside it, into which runs the lead. The smelter, having first skimmed off the slags and other things with a hoe, pours the lead with a ladle into moulds, taking out the cakes after they have cooled. At the back of the furnace is a rectangular hole, so that the fire may be allowed more draught, and so that the smelter can crawl through it into the furnace if necessity demands.

The Saxons who inhabit Gittelde, when smelting lead ore in a furnace not unlike a baking oven, put the wood in through a hole at the back of the furnace, and when it begins to burn vigorously the lead trickles out of the ore into a forehearth. When this is full, the smelting being accomplished, the tap-hole is opened with a bar, and in this way the lead, together with the slags, runs into the dipping-pots below. Afterward the cakes of lead, when they are cold, are taken from the moulds.

In Westphalia they heap up ten wagon-loads of charcoal on some hill-side which adjoins a level place, and the top of the heap being made flat, straw is thrown upon it to the thickness of three or four digits. On the top of

among the spoil captured by Thotmes III. Leaden objects have frequently been found in Egyptian tombs as early as Rameses III. (1200 B.C.). The statement is made by Pulsifer (Notes for a History of Lead, New York 1888, p. 146) that Egyptian pottery was glazed with lead. We have been unable to find any confirmation of this. It may be noted, incidentally, that lead is not included in the metals of the "Tribute of Yü" in the Shoo King (The Chinese Classics, 2500 B.C. ?), although silver is so included.

After 1200 or 1300 B.C. evidences of the use of lead become frequent. Moses (Numbers xxxi, 22–23) directs the Israelites with regard to their plunder from the Midianites (1300 B.C.): "Only the gold and the silver, the brass [*sic*], the iron, the tin, and the lead. Everything "that may abide the fire, ye shall make it go through the fire, and it shall be clean; neverthe-"less, it shall be purified with the water of separation, and all that abideth not the fire ye shall "make go through the water." Numerous other references occur in the Scriptures (Psalms xii, 6; Proverbs xvii, 3; xxv. 4, etc.), one of the most pointed from a metallurgical point of view being that of Jeremiah (600 B.C.), who says (vi, 29–30): "The bellows "are burned, the lead is consumed of the fire; the founder melteth in vain; for the wicked "are not plucked away. Reprobate silver shall men call them because the Lord hath rejected "them." From the number of his metaphors in metallurgical terms we may well conclude that Jeremiah was of considerable metallurgical experience, which may account for his critical tenor of mind. These Biblical references all point to a knowledge of separating silver and lead. Homer mentions lead (Iliad XXIV, 109), and it has been found in the remains of ancient Troy and Mycenae (H. Schliemann, "Troy and its Remains," London, 1875, and "Mycenae," New York, 1877). Both Herodotus (I, 186) and Diodorus (II, 1) speak of the lead used to fix iron clamps in the stone bridge of Nitocris (600 B.C.) at Babylon.

Our best evidence of ancient lead-silver metallurgy is the result of the studies at Mt. Laurion by Edouard Ardaillon (*Mines du Laurion dans l'Antiquité*, Paris, 1897). Here the very extensive old workings and the slag heaps testify to the greatest activity. The re-opening of the mines in recent years by a French Company has well demonstrated their technical character, and the frequent mention in Greek History easily determines their date. These deposits of argentiferous galena were extensively worked before 500 B.C., and while the evidence of concentration methods is ample, there is but little remaining of the ancient smelters. Enough, however, remains to demonstrate that the galena was smelted in small furnaces at low heat, with forced draught, and that it was subsequently cupelled. In order to reduce the sulphides the ancient smelters apparently depended upon partial roasting in the furnace at a preliminary period in reduction, or else upon the ferruginous character of the ore, or upon both. See notes p. 27 and p. 265. Theognis (6th century B.C.) and Hippocrates (5th century B.C.) are frequently referred to as mentioning the refining of gold with lead; an inspection of the passages fails to corroborate the importance which has been laid upon them. Among literary evidences upon lead metallurgy of later date, Theophrastus (300 B.C.) describes the making of white-lead with lead plates and vinegar. Diodorus Siculus (1st century B.C.), in his well-known quotation from Agatharchides (2nd century B.C.) with regard to gold mining and treatment in Egypt, describes the refining of gold with lead. (See note 8 p. 279.) Strabo (63 B.C.—24 A.D.) says (III, 2, 8): "The furnaces for

this is laid as much pure lead ore as the heap can bear ; then the charcoal is kindled, and when the wind blows, it fans the fire so that the ore is smelted. In this wise the lead, trickling down from the heap, flows on to the level and forms broad thin slabs. A few hundred pounds of lead ore are kept at hand, which, if things go well, are scattered over the heap. These broad slabs are impure and are laid upon dry wood which in turn is placed on green wood laid over a large crucible, and the former having been kindled, the lead is re-melted.

The Poles use a hearth of bricks four feet high, sloping on both sides and plastered with lute. On the upper level part of the hearth large pieces of wood are piled, and on these is placed small wood with lute put in between ; over the top are laid wood shavings, and upon these again pure lead ore covered with large pieces of wood. When these are kindled, the ore melts and

"silver are constructed lofty in order that the vapour, which is dense and pestilent, "may be raised and carried off." And again (III, 2, 10), in quoting from Polybius (204–125 B.C.): "Polybius, speaking of the silver mines of New Carthage, tells us that they "are extremely large, distant from the city about 20 stadia, and occupy a circuit of 400 "stadia ; that there are 40,000 men regularly engaged in them, and that they yield daily "to the Roman people (a revenue of) 25,000 drachmae. The rest of the process I pass over, "as it is too long ; but as for the silver ore collected, he tells us that it is broken up and "sifted through sieves over water ; that what remains is to be again broken, and the water "having been strained off it is to be sifted and broken a third time. The dregs which remain "after the fifth time are to be melted, and the lead being poured off, the silver is obtained "pure. These silver mines still exist ; however, they are no longer the property of the State, "neither these nor those elsewhere, but are possessed by private individuals. The gold "mines, on the contrary, nearly all belong to the State. Both at Castlon and other places there "are singular lead mines worked. They contain a small proportion of silver, but not sufficient "to pay for the expense of refining" (Hamilton's Trans.). Dioscorides (1st century A.D.), among his medicines, describes several varieties of litharge, their origin, and the manner of making white-lead (see on pp. 465, 440), but he gives no very tangible information on lead smelting. Pliny, at the same period in speaking of silver, (XXXIII, 31), says: "After this "we speak of silver, the next folly. Silver is only found in shafts, there being no indications "like shining particles as in the case of gold. This earth is sometimes red, sometimes of an "ashy colour. It is impossible to melt it except with lead ore (*vena plumbi*), called *galena*, "which is generally found next to silver veins. And this the same agency of fire separates "part into lead, which floats on the silver like oil on water." (We have transferred lead and silver in this last sentence, otherwise it means nothing.) Also (XXXIV, 47) he says: "There "are two different sources of lead, it being smelted from its own ore, whence it comes without "the admixture of any other substance, or else from an ore which contains it in common "with silver. The metal, which flows liquid at the first melting in the furnace, is called "*stannum* that at the second melting is silver ; that which remains in the furnace is *galena*, "which is added to a third part of the ore. This being again melted, produces lead with "a deduction of two-ninths." We have, despite some grammatical objections, rendered this passage quite differently from other translators, none of whom have apparently had any knowledge of metallurgy ; and we will not, therefore, take the several pages of space necessary to refute their extraordinary and unnecessary hypotheses. From a metallurgical point of view, two facts must be kept in mind,—first, that *galena* in this instance was the same substance as *molybdaena*, and they were both either a variety of litharge or of lead carbonates ; second, that the *stannum* of the Ancients was silver-lead alloy. Therefore, the metallurgy of this paragraph becomes a simple melting of an argentiferous lead ore, its subsequent cupellation, with a return of the litharge to the furnace. Pliny goes into considerable detail as to varieties of litharge, for further notes upon which see p. 466. The Romans were most active lead-silver miners, not only in Spain, but also in Britain. There are scores of lead pigs of the Roman era in various English museums, many marked "*ex argent.*" Bruce (The Roman Wall, London, 1852, p. 432) describes some Roman lead furnaces in Cumberland where the draught was secured by driving a tapering tunnel into the hills. The Roman lead slag ran high in metal, and formed a basis for quite an industry in England in the early 18th century (Hunt, British Mining, London, 1887, p. 26, etc.). There is nothing in mediæval literature which carries us further with lead metallurgy than the knowledge displayed by Pliny, until we arrive at Agricola's period. The history of cupellation is specially dealt with in note on p. 465.

A—Furnace of the carni. B—Low wall. C—Wood. D—Ore dripping lead. E—Large crucible. F—Moulds. G—Ladle. H—Slabs of lead. I—Rectangular hole at the back of the furnace. K—Saxon furnace. L—Opening in the back of the furnace. M—Wood. N—Upper crucible. O—Dipping-pot. P—Westphalian method of melting. Q—Heaps of charcoal. R—Straw. S—Wide slabs. T—Crucibles. V—Polish hearth.

runs down on to the lower layer of wood; and when this is consumed by the fire, the metal is collected. If necessity demand, it is melted over and over again in the same manner, but it is finally melted by means of wood laid over the large crucible, the slabs of lead being placed upon it.

The concentrates from washing are smelted together with slags (fluxes ?) in a third furnace, of which the tap-hole is always open.

It is worth while to build vaulted dust-chambers over the furnaces, especially over those in which the precious ores are to be smelted, in order that the thicker part of the fumes, in which metals are not wanting, may be caught and saved. In this way two or more furnaces are combined under the same vaulted ceiling, which is supported by the wall, against which the furnaces are built, and by four columns. Under this the smelters of the ore perform their work. There are two openings through which the fumes rise from the furnaces into the wide vaulted chamber, and the wider this is the more fumes it collects; in the middle of this chamber over the arch is an opening three palms high and two wide. This catches the fumes of both furnaces, which have risen up from both sides of the vaulted chamber to its arch, and have fallen again because they could not force their way out; and they thus pass out through the opening mentioned, into the chimney which the Greeks call καπνοδόχη, the name being taken from the object. The chimney has thin iron plates fastened into the walls, to which the thinner metallic substances adhere when ascending with the fumes. The thicker metallic substances, or *cadmia*,[25] adhere to the vaulted chamber, and often harden into stalactites. On one side of the chamber is a window in which are set panes of glass, so that the light may be transmitted, but the fumes kept in; on the other side is a door, which is kept entirely closed while the ores are being smelted in the furnaces, so that none of the fumes may escape. It is opened in order that the workman, passing through it, may be enabled to enter the chamber and remove the soot and *pompholyx*[26] and chip off

[25]*Cadmia*. In the German Translation this is given as *kobelt*. It would be of uncertain character, but no doubt partially furnace calamine. (See note on p. 112.)

[26]*Pompholyx* (*Interpretatio* gives the German as *Weisser hütten rauch als ober dem garherde und ober dem kupfer ofen*). This was the impure protoxide of zinc deposited in the furnace outlets, and is modern "tutty." The ancient products, no doubt, contained arsenical oxides as well. It was well known to the Ancients, and used extensively for medicinal purposes, they dividing it into two species—*pompholyx* and *spodos*. The first adequate description is by Dioscorides (v, 46): "*Pompholyx* differs from *spodos* in species, not in genus. "For *spodos* is blacker, and is often heavier, full of straws and hairs, like the refuse that is "swept from the floors of copper smelters. But *pompholyx* is fatty, unctuous, white and light "enough to fly in the air. Of this there are two kinds—the one inclines to sky blue and is "unctuous; the other is exceedingly white, and is extremely light. White *pompholyx* is "made every time that the artificer, in the preparation and perfecting of copper (brass ?) "sprinkles powdered *cadmia* upon it to make it more perfect, for the soot which rises being "very fine becomes *pompholyx*. Other *pompholyx* is made, not only in working copper "(brass ?), but is also made from *cadmia* by continually blowing with bellows. The manner "of doing it is as follows:—The furnace is constructed in a two-storied building, and there "is a medium-sized aperture opening to the upper chamber; the building wall nearest the "furnace is pierced with a small opening to admit the nozzle of the bellows. The building must "have a fair-sized door for the artificer to pass in and out. Another small building must "adjoin this, in which are the bellows and the man who works them. Then the charcoal "in the furnace is lighted, and the artificer continually throws broken bits of *cadmia* from "the place above the furnace, whilst his assistant, who is below, throws in charcoals, until "all of the *cadmia* inside is consumed. By this means the finest and lightest part of the

A—Furnaces. B—Vaulted roof. C—Columns. D—Dust-chamber. E—Opening. F—Chimney. G—Window. H—Door. I—Chute.

the *cadmia*; this sweeping is done twice a year. The soot mixed with *pompholyx* and the *cadmia*, being chipped off, is thrown down through a long chute made of four boards joined in the shape of a rectangle, that they should not fly away. They fall on to the floor, and are sprinkled with salt water, and are again smelted with ore and litharge, and become an emolument to the proprietors. Such chambers, which catch the metallic substances that rise with the fumes, are profitable for all metalliferous ores; but especially for the minute metallic particles collected by washing crushed ores and rock, because these usually fly out with the fire of the furnaces.

I have explained the four general methods of smelting ores; now I will state how the ores of each metal are smelted, or how the metal is obtained from the ore. I will begin with gold. Its sand, the concentrates from washing, or the gold dust collected in any other manner, should very often not be smelted, but should be mixed with quicksilver and washed with tepid water, so that all the impurities may be eliminated. This method I explained in Book VII. Or they are placed in the *aqua* which separates gold from silver, for this also separates its impurities. In this method we see the gold sink in the glass ampulla, and after all the *aqua* has been drained from the particles, it frequently remains as a gold-coloured residue at the bottom; this powder, when it has been moistened with oil made from argol[27], is then dried and placed in a crucible, where it is melted with borax or with saltpetre and salt; or the same very fine dust is thrown into molten silver, which absorbs it, and from this it is again parted by *aqua valens*[28].

It is necessary to smelt gold ore either outside the blast furnace in a crucible, or inside the blast furnace; in the former case a small charge of ore is used, in the latter a large charge of it. *Rudis* gold, of whatever colour it is, is crushed with a *libra* each of sulphur and salt, a third of a *libra* of copper,

"stuff flies up with the smoke to the upper chamber, and adheres to the walls of the roof. "The substance which is thus formed has at first the appearance of bubbles on water, after-"ward increasing in size, it looks like skeins of wool. The heaviest parts settle in the bottom, "while some fall over and around the furnaces, and some lie on the floor of the building. "This latter part is considered inferior, as it contains a lot of earth and becomes full of dirt."

Pliny (XXXIV, 33) appears somewhat confused as to the difference between the two species: "That which is called *pompholyx* and *spodos* is found in the copper-smelting "furnaces, the difference between them being that *pompholyx* is separated by washing, while "*spodos* is not washed. Some have called that which is white and very light *pompholyx*, and "it is the soot of copper and *cadmia*; whereas *spodos* is darker and heavier. It is scraped "from the walls of the furnace, and is mixed with particles of metal, and sometimes with "charcoal." (XXXIV, 34.) "The Cyprian *spodos* is the best. It is formed by fusing "*cadmia* with copper ore. This being the lightest part of the metal, it flies up in the fumes "from the furnace, and adheres to the roof, being distinguished from the soot by its whiteness. "That which is less white is immature from the furnace, and it is this which some call '*pom-"pholyx*.'" Agricola (*De Natura Fossilium*, p. 350) traverses much the same ground as the authors previously quoted, and especially recommends the *pompholyx* produced when making brass by melting alternate layers of copper and calamine (*cadmia fossilis*).

[27]*Oleo, ex fece vini sicca confecto*. This oil, made from argol, is probably the same substance mentioned a few lines further on as "wine," distilled by heating argol in a retort. Still further on, salt made from argol is mentioned. It must be borne in mind that this argol was crude tartrates from wine vats, and probably contained a good deal of organic matter. Heating argol sufficiently would form potash, but that the distillation product could be anything effective it is difficult to see.

[28]*Aqua valens*. No doubt mainly nitric acid, the preparation of which is explained at length in Book X, p. 439).

and a quarter of a *libra* of argol; they should be melted in a crucible on a slow fire for three hours, then the alloy is put into molten silver that it may melt more rapidly. Or a *libra* of the same crude gold, crushed up, is mixed together with half a *libra* of *stibium* likewise crushed, and put into a crucible with half an *uncia* of copper filings, and heated until they melt, then a sixth part of granulated lead is thrown into the same crucible. As soon as the mixture emits an odour, iron-filings are added to it, or if these are not at hand, iron hammer-scales, for both of these break the strength of the *stibium*. When the fire consumes it, not alone with it is some strength of the *stibium* consumed, but some particles of gold and also of silver, if it be mixed with the gold[29]. When the button has been taken out of the crucible and cooled, it is melted in a cupel, first until the antimony is exhaled, and thereafter until the lead is separated from it.

Crushed pyrites which contains gold is smelted in the same way; it and the *stibium* should be of equal weight and in truth the gold may be made from them in a number of different ways[30]. One part of crushed material is mixed with six parts of copper, one part of sulphur, half a part of salt, and they are all placed in a pot and over them is poured wine distilled by heating liquid argol in an ampulla. The pot is covered and smeared over with lute and is put in a hot place, so that the mixture moistened with wine may dry for the space of six days, then it is heated for three hours over a gentle fire that it may combine more rapidly with the lead. Finally it is put into a cupel and the gold is separated from the lead[31].

Or else one *libra* of the concentrates from washing pyrites, or other stones to which gold adheres, is mixed with half a *libra* of salt, half a *libra* of argol, a third of a *libra* of glass-galls, a sixth of a *libra* of gold or silver slags, and a *sicilicus* of copper. The crucible into which these are put, after it has been covered with a lid, is sealed with lute and placed in a small furnace that is provided with small holes through which the air is drawn in, and then it is heated until it turns red and the substances put in have alloyed; this should take place within four or five hours. The alloy having cooled, it is again crushed to powder and a pound of litharge is added to it; then it is heated again in another crucible until it melts. The button is taken out, purged of slag, and placed in a cupel, where the gold is separated from the lead.

[29]*Quod cum ignis consumit non modo una cum eo, quae ipsius stibii vis est, aliqua auri particula, sed etiam argenti, si cum auro fuerit permistum, consumitur.* The meaning is by no means clear. On p. 451 is set out the old method of parting silver from gold with antimony sulphide, of which this may be a variation. The silver combines with sulphur, and the reduced antimony forms an alloy with the gold. The added iron and copper would also combine with the sulphur from the antimony sulphide, and no doubt assist by increasing the amount of free collecting agent and by increasing the volume of the matte. (See note 17, p. 451.)

[30]There follow eight different methods of treating crude bullion or rich concentrates. In a general way three methods are involved,—1st, reduction with lead or antimony, and cupellation; 2nd, reduction with silver, and separation with nitric acid; 3rd, reduction with lead and silver, followed by cupellation and parting with nitric acid. The use of sulphur or antimony sulphide would tend to part out a certain amount of silver, and thus obtain fairly pure bullion upon cupellation. But the introduction of copper could only result deleteriously, except that it is usually accompanied by sulphur in some form, and would thus probably pass off harmlessly as a matte carrying silver. (See note 33 below.)

[31]It is not very clear where this lead comes from. Should it be antimony? The German translation gives this as "silver."

Or to a *libra* of the powder prepared from such metalliferous concentrates, is added a *libra* each of salt, of saltpetre, of argol, and of glass-galls, and it is heated until it melts. When cooled and crushed, it is washed, then to it is added a *libra* of silver, a third of copper filings, a sixth of litharge, and it is likewise heated again until it melts. After the button has been purged of slag, it is put into the cupel, and the gold and silver are separated from the lead; the gold is parted from the silver with *aqua valens*. Or else a *libra* of the powder prepared from such metalliferous concentrates, a quarter of a *libra* of copper filings, and two *librae* of that second powder[32] which fuses ores, are heated until they melt. The mixture when cooled is again reduced to powder, roasted and washed, and in this manner a blue powder is obtained. Of this, and silver, and that second powder which fuses ores, a *libra* each are taken, together with three *librae* of lead, and a quarter of a *libra* of copper, and they are heated together until they melt; then the button is treated as before. Or else a *libra* of the powder prepared from such metalliferous concentrates, half a *libra* of saltpetre, and a quarter of a *libra* of salt are heated until they melt. The alloy when cooled is again crushed to powder, one *libra* of which is absorbed by four pounds of molten silver. Or else a *libra* of the powder made from that kind of concentrates, together with a *libra* of sulphur, a *libra* and a half of salt, a third of a *libra* of salt made from argol, and a third of a *libra* of copper resolved into powder with sulphur, are heated until they melt. Afterward the lead is re-melted, and the gold is separated from the other metals. Or else a *libra* of the powder of this kind of concentrates, together with two *librae* of salt, half a *libra* of sulphur, and one *libra* of litharge, are heated, and from these the gold is melted out. By these and similar methods concentrates containing gold, if there be a small quantity of them or if they are very rich, can be smelted outside the blast furnace.

If there be much of them and they are poor, then they are smelted in the blast furnace, especially the ore which is not crushed to powder, and particularly when the gold mines yield an abundance of it[33]. The gold concentrates mixed with litharge and hearth-lead, to which are added iron-scales, are smelted in the blast furnace whose tap-hole is intermittently closed, or else in the first or the second furnaces in which the tap-hole is always open. In this manner an

[32]These powders are described in Book VII., p. 236. It is difficult to say which the second really is. There are numbers of such recipes in the *Probierbüchlein* (see Appendix B), with which a portion of these are identical.

[33]A variety of methods are involved in this paragraph: 1st, crude gold ore is smelted direct; 2nd, gold concentrates are smelted in a lead bath with some addition of iron—which would simply matte off—the lead bullion being cupelled; 3rd, roasted and unroasted pyrites and *cadmia* (probably blende, cobalt, arsenic, etc.) are melted into a matte; this matte is repeatedly roasted, and then re-melted in a lead bath; 4th, if the material "flies out of the furnace" it is briquetted with iron ore and lime, and the briquettes smelted with copper matte. Three products result: (*a*) slag; (*b*) matte; (*c*) copper-gold-silver alloy. The matte is roasted, re-smelted with lead, and no doubt a button obtained, and further matte. The process from this point is not clear. It appears that the copper bullion is melted with lead, and normally this product would be taken to the liquation furnace, but from the text it would appear that the lead-copper bullion was melted again with iron ore and pyrites, in which case some of the copper would be turned into the matte, and the lead alloy would be richer in gold and silver.

alloy of gold and lead is obtained which is put into the cupellation furnace. Two parts of roasted pyrites or *cadmia* which contain gold, are put with one part of unroasted, and are smelted together in the third furnace whose tap-hole is always open, and are made into cakes. When these cakes have been repeatedly roasted, they are re-smelted in the furnace whose tap-hole is temporarily closed, or in one of the two others whose tap-holes are always open. In this manner the lead absorbs the gold, whether pure or argentiferous or cupriferous, and the alloy is taken to the cupellation furnace. Pyrites, or other gold ore which is mixed with much material that is consumed by fire and flies out of the furnace, is melted with stone from which iron is melted, if this is at hand. Six parts of such pyrites, or of gold ore reduced to powder and sifted, four of stone from which iron is made, likewise crushed, and three of slaked lime, are mixed together and moistened with water; to these are added two and a half parts of the cakes which contain some copper, together with one and a half parts of slag. A basketful of fragments of the cakes is thrown into the furnace, then the mixture of other things, and then the slag. Now when the middle part of the forehearth is filled with the molten material which runs down from the furnace, the slags are first skimmed off, and then the cakes made of pyrites; afterward the alloy of copper, gold and silver, which settles at the bottom, is taken out. The cakes are gently roasted and re-smelted with lead, and made into cakes, which are carried to other works. The alloy of copper, gold, and silver is not roasted, but is re-melted again in a crucible with an equal portion of lead. Cakes are also made much richer in copper and gold than those I spoke of. In order that the alloy of gold and silver may be

Historical Note on Gold.—There is ample evidence of gold being used for ornamental purposes prior to any human record. The occurrence of large quantities of gold in native form, and the possibility of working it cold, did not necessitate any particular metallurgical ingenuity. The earliest indications of metallurgical work are, of course, among the Egyptians, the method of washing being figured as early as the monuments of the IV Dynasty (prior to 3800 B.C.). There are in the British Museum two stelae of the XII Dynasty (2400 B.C.) (144 Bay 1 and 145 Bay 6) relating to officers who had to do with gold mining in Nubia, and upon one there are references to working what appears to be ore. If this be true, it is the earliest reference to this subject. The Papyrus map (1500 B.C.) of a gold mine, in the Turin Museum (see note 16, p. 129), probably refers to a quartz mine. Of literary evidences there is frequent mention of refining gold and passing it through the fire in the Books of Moses, arts no doubt learned from the Egyptians. As to working gold ore as distinguished from alluvial, we have nothing very tangible, unless it be the stelae above, until the description of Egyptian gold mining by Agatharcides (see note 8, p. 279). This geographer, of about the 2nd century B.C., describes very clearly indeed the mining, crushing, and concentration of ore and the refining of the concentrates in crucibles with lead, salt, and barley bran. We may mention in passing that Theognis (6th Century B.C.) is often quoted as mentioning the refining of gold with lead, but we do not believe that the passage in question (1101): "But having been put to the test and being rubbed beside (or against) lead as being refined "gold, you will be fair," etc.; or much the same statement again (418) will stand much metallurgical interpretation. In any event, the myriads of metaphorical references to fining and purity of gold in the earliest shreds of literature do not carry us much further than do those of Shakespeare or Milton. Vitruvius and Pliny mention the recovery or refining of gold with mercury (see note 12, p. 297 on Amalgamation); and it appears to us that gold was parted from silver by cementation with salt prior to the Christian era. We first find mention of parting with sulphur in the 12th century, with nitric acid prior to the 14th century, by antimony sulphide prior to the 15th century, and by cementation with nitre by Agricola. (See historical note on parting gold and silver, p. 458.) The first mention of parting gold from copper occurs in the early 16th century (see note 24, p. 462). The first comprehensive description of gold metallurgy in all its branches is in *De Re Metallica.*

made richer, to eighteen *librae* of it are added forty-eight *librae* of crude ore, three *librae* of the stone from which iron is made, and three-quarters of a *libra* of the cakes made from pyrites, and mixed with lead, all are heated together in the crucible until they melt. When the slag and the cakes melted from pyrites have been skimmed off, the alloy is carried to other furnaces.

There now follows silver, of which the native silver or the lumps of *rudis* silver[34] obtained from the mines are not smelted in the blast furnaces, but in small iron pans, of which I will speak at the proper place; these lumps are heated and thrown into molten silver-lead alloy in the cupellation furnace when the silver is being separated from the lead, and refined. The tiny flakes or tiny lumps of silver adhering to stones or marble or rocks, or again the same little lumps mixed with earth, or silver not pure enough, should be smelted in the furnace of which the tap-hole is only closed for a short time, together with cakes melted from pyrites, with silver slags, and with stones which easily fuse in fire of the second order.

In order that particles of silver should not fly away[35] from the lumps of ore consisting of minute threads of pure silver and twigs of native silver, they are enclosed in a pot, and are placed in the same furnace where the rest of the silver ores are being smelted. Some people smelt lumps of native silver not sufficiently pure, in pots or triangular crucibles, whose lids are sealed with lute. They do not place these pots in the blast furnace, but arrange them in the assay furnace into which the draught of the air blows through small holes. To one part of the native silver they add three parts of powdered litharge, as many parts of hearth-lead, half a part of galena[36], and a small quantity of salt and iron-scales. The alloy which settles at the bottom of the other substances in the pot is carried to the cupellation furnace, and the slags are re-melted with the other silver slags. They crush under the stamps and wash the pots or crucibles to which silver-lead alloy or slags adhere, and having collected the concentrates they smelt them together with the slags. This method of smelting *rudis* silver, if there is a small quantity of it, is the best, because the smallest portion of silver does not fly out of the pot or the crucible, and get lost.

If bismuth ore or antimony ore or lead ore[37] contains silver, it is smelted with the other ores of silver; likewise galena or pyrites, if there is a small amount of it. If there be much galena, whether it contain a large or a small amount of silver, it is smelted separately from the others; which process I will explain a little further on.

[34]*Rudis* silver comprised all fairly pure silver ores, such as silver sulphides, chlorides, arsenides, etc. This is more fully discussed in note 6, p. 108.

[35]*Evolent*,—volatilize ?

[36]*Lapidis plumbarii facile liquescentis*. The German Translation gives *glantz*, *i.e.*, Galena, and the *Interpretatio* also gives *glantz* for *lapis plumbarius*. We are, however, uncertain whether this "easily melting" material is galena or some other lead ore.

[37]*Molybdaena* is usually hearth-lead in *De Re Metallica*, but the German translation in this instance uses *pleyertz*, lead ore. From the context it would not appear to mean hearth-lead—saturated bottoms of cupellation furnaces—for such material would not contain appreciable silver. Agricola does confuse what are obviously lead carbonates with his other *molybdaena* (see note 37, p. 476).

Because lead and copper ores and their metals have much in common with silver ores, it is fitting that I should say a great deal concerning them, both now and later on. Also in the same manner, pyrites are smelted separately if there be much of them. To three parts of roasted lead or copper ore and one part of crude ore, are added concentrates if they were made by washing the same ore, together with slags, and all are put in the third furnace whose tap-hole is always open. Cakes are made from this charge, which, when they have been quenched with water, are roasted. Of these roasted cakes generally four parts are again mixed with one part of crude pyrites and re-melted in the same furnace. Cakes are again made from this charge, and if there is a large amount of copper in these cakes, copper is made immediately after they have been roasted and re-melted; if there is little copper in the cakes they are also roasted, but they are re-smelted with a little soft slag. In this method the molten lead in the forehearth absorbs the silver. From the pyritic material which floats on the top of the forehearth are made cakes for the third time, and from them when they have been roasted and re-smelted is made copper. Similarly, three parts of roasted *cadmia*[38] in which there is silver, are mixed with one part of crude pyrites, together with slag, and this charge is smelted and cakes are made from it; these cakes having been roasted are re-smelted in the same furnace. By this method the lead contained in the forehearth absorbs the silver, and the silver-lead is taken to the cupellation furnace. Crude quartz and stones which easily fuse in fire of the third order, together with other ores in which there is a small amount of silver, ought to be mixed with crude roasted pyrites or *cadmia*, because the roasted cakes of pyrites or *cadmia* cannot be profitably smelted separately. In a similar manner earths which contain little silver are mixed with the same; but if pyrites and *cadmia* are not available to the smelter, he smelts such silver ores and earths with litharge, hearth-lead, slags, and stones which easily melt in the fire. The concentrates[39] originating from the washing of *rudis* silver, after first being roasted[40] until they melt, are smelted with mixed litharge and hearth-lead, or else, after being moistened with water, they are smelted with cakes made from pyrites and *cadmia*. By neither of these methods do (the concentrates) fall back in the furnace, or fly out of it, driven by the blast of the bellows and the agitation of the fire. If the concentrates originated from galena they are smelted with it after having been roasted; and if from pyrites, then with pyrites.

Pure copper ore, whether it is its own colour or is tinged with chrysocolla or azure, and copper glance, or grey or black *rudis* copper, is smelted in a furnace of which the tap-hole is closed for a very short time, or else is always

[38]The term *cadmia* is used in this paragraph without the usual definition. Whether it was *cadmia fornacis* (furnace accretions) or *cadmia metallica* (cobalt-arsenic-blende mixture) is uncertain. We believe it to be the former.

[39]*Ramentum si lotura ex argento rudi.* This expression is generally used by the author to indicate concentrates, but it is possible that in this sentence it means the tailings after washing rich silver minerals, because the treatment of the *rudis* silver has been already discussed above.

[40]*Ustum.* This might be rendered "burnt." In any event, it seems that the material is sintered.

open[41]. If there is a large amount of silver in the ore it is run into the forehearth, and the greater part of the silver is absorbed by the molten lead, and the remainder is sold with the copper to the proprietor of the works in which silver is parted from copper[42]. If there is a small amount of silver in the ore, no lead is put into the forehearth to absorb the silver, and the above-

[41]*Aes purum sive proprius ei color insederit, sive chrysocolla vel caeruleo fuerit tinctum, et rude plumbei coloris, aut fusci, aut nigri.* There are six copper minerals mentioned in this sentence, and from our study of Agricola's *De Natura Fossilium* we hazard the following :—*Proprius ei color insederit,*—"its own colour,"—probably cuprite or "ruby copper." *Tinctum chrysocolla*—partly the modern mineral of that name and partly malachite. *Tinctum caeruleo,* partly azurite and partly other blue copper minerals. *Rude plumbei coloris,*—"lead coloured,"—was certainly chalcocite (copper glance). We are uncertain of *fusci aut nigri,* but they were probably alteration products. For further discussion see note on p. 109.

[42]Historical Note on Copper Smelting.—The discoverer of the reduction of copper by fusion, and his method, like the discoverer of tin and iron, will never be known, because he lived long before humanity began to make records of its discoveries and doings. Moreover, as different races passed independently and at different times through the so-called "Bronze Age," there may have been several independent discoverers. Upon the metallurgy of pre-historic man we have some evidence in the many "founders' hoards" or "smelters' hoards" of the Bronze Age which have been found, and they indicate a simple shallow pit in the ground into which the ore was placed, underlaid with charcoal. Rude round copper cakes eight to ten inches in diameter resulted from the cooling of the metal in the bottom of the pit. Analyses of such Bronze Age copper by Professor Gowland and others show a small percentage of sulphur, and this is possible only by smelting oxidized ores. Copper objects appear in the pre-historic remains in Egypt, are common throughout the first three dynasties, and bronze articles have been found as early as the IV Dynasty (from 3800 to 4700 B.C., according to the authority adopted). The question of the origin of this bronze, whether from ores containing copper and tin or by alloying the two metals, is one of wide difference of opinion, and we further discuss the question in note 53, p. 411, under Tin. It is also interesting to note that the crucible is the emblem of copper in the hieroglyphics. The earliest source of Egyptian copper was probably the Sinai Peninsula, where there are reliefs as early as Seneferu (about 3700 B.C.), indicating that he worked the copper mines. Various other evidences exist of active copper mining prior to 2500 B.C. (Petrie, Researches in Sinai, London, 1906, p. 51, etc.). The finding of crucibles here would indicate some form of refining. Our knowledge of Egyptian copper metallurgy is limited to deductions from their products, to a few pictures of crude furnaces and bellows, and to the minor remains on the Sinai Peninsula ; none of the pictures were, so far as we are aware, prior to 2300 B.C., but they indicate a considerable advance over the crude hearth, for they depict small furnaces with forced draught—first a blow-pipe, and in the XVIII Dynasty (about 1500 B.C.) the bellows appear. Many copper articles have been found scattered over the Eastern Mediterranean and Asia Minor of pre-Mycenaean Age, some probably as early as 3000 B.C. This metal is mentioned in the "Tribute of Yü" in the Shoo King (2500 B.C. ?) ; but even less is known of early Chinese metallurgy than of the Egyptian. The remains of Mycenaean, Phoenician, Babylonian, and Assyrian civilizations, stretching over the period from 1800 to 500 B.C., have yielded endless copper and bronze objects, the former of considerable purity, and the latter a fairly constant proportion of from 10% to 14% tin. The copper supply of the pre-Roman world seems to have come largely, first from Sinai, and later from Cyprus, and from the latter comes our word copper, by way of the Romans shortening *aes cyprium* (Cyprian copper) to *cuprum.* Research in this island shows that it produced copper from 3000 B.C., and largely because of its copper it passed successively under the domination of the Egyptians, Assyrians, Phoenicians, Greeks, Persians, and Romans. The bronze objects found in Cyprus show 2% to 10% of tin, although tin does not, so far as modern research goes, occur on that island. There can be no doubt that the Greeks obtained their metallurgy from the Egyptians, either direct or second-hand—possibly through Mycenae or Phoenicia. Their metallurgical gods and the tradition of Cadmus indicate this much.

By way of literary evidences, the following lines from Homer (Iliad, XVIII.) have interest as being the first preserved description in any language of a metallurgical work. Hephaestus was much interrupted by Thetis, who came to secure a shield for Achilles, and whose general conversation we therefore largely omit. We adopt Pope's translation :—

There the lame architect the goddess found
Obscure in smoke, his forges flaming round,
While bathed in sweat from fire to fire he flew ;

mentioned proprietors buy it in with the copper; if there be no silver, copper is made direct. If such copper ore contains some minerals which do not easily melt, as pyrites or *cadmia metallica fossilis*[43], or stone from which iron is melted, then crude pyrites which easily fuse are added to it, together with slag. From this charge, when smelted, they make cakes; and from

And puffing loud the roaring bellows blew.
* * * * * *
In moulds prepared, the glowing ore (metal ?) he pours.
* * * * * *
" Vouchsafe, oh Thetis! at our board to share
The genial rites and hospitable fare;
While I the labours of the forge forego,
And bid the roaring bellows cease to blow."
Then from his anvil the lame artist rose;
Wide with distorted legs oblique he goes,
And stills the bellows, and (in order laid)
Locks in their chests his instruments of trade;
Then with a sponge, the sooty workman dress'd
His brawny arms embrown'd and hairy breast.
* * * * * *
Thus having said, the father of the fires
To the black labours of his forge retires.
Soon as he bade them blow the bellows turn'd
Their iron mouths; and where the furnace burn'd
Resounding breathed: at once the blast expires,
And twenty forges catch at once the fires;
Just as the God directs, now loud, now low,
They raise a tempest, or they gently blow;
In hissing flames huge silver bars are roll'd,
And stubborn brass (copper ?) and tin, and solid gold;
Before, deep fixed, the eternal anvils stand.
The ponderous hammer loads his better hand;
His left with tongs turns the vex'd metal round.
And thick, strong strokes, the doubling vaults rebound
Then first he formed the immense and solid shield;

Even if we place the siege of Troy at any of the various dates from 1350 to 1100 B.C., it does not follow that the epic received its final form for many centuries later, probably 900–800 B.C.; and the experience of the race in metallurgy at a much later period than Troy may have been drawn upon to fill in details. It is possible to fill a volume with indirect allusion to metallurgical facts and to the origins of the art, from Greek mythology, from Greek poetry, from the works of the grammarians, and from the Bible. But they are of no more technical value than the metaphors from our own tongue. Greek literature in general is singularly lacking in metallurgical description of technical value, and it is not until Dioscorides (1st Century A.D.) that anything of much importance can be adduced. Aristotle, however, does make an interesting reference to what may be brass (see note on p. 410), and there can be no doubt that if we had the lost work of Aristotle's successor, Theophrastus (372–288 B.C.), on metals we should be in possession of the first adequate work on metallurgy. As it is, we find the green and blue copper minerals from Cyprus mentioned in his "Stones." And this is the first mention of any particular copper ore. He also mentions (XIX.) pyrites "which melt," but whether it was a copper variety cannot be determined. Theophrastus further describes the making of verdigris (see note 4, p. 440). From Dioscorides we get a good deal of light on copper treatment, but as his objective was to describe medicinal preparations, the information is very indirect. He states (V, 100) that "pyrites is a stone from which copper is made." He mentions *chalcitis* (copper sulphide, see note on, p. 573); while his *misy*, *sory*, *melanteria*, *caeruleum*, and *chrysocolla* were all oxidation copper or iron minerals. (See notes on p. 573.) In giving a method of securing *pompholyx* (zinc oxide), "the soot flies up when the copper refiners sprinkle powdered *cadmia* over the molten metal" (see note 26, p. 394); he indirectly gives us the first definite indication of making brass, and further gives some details as to the furnaces there employed, which embraced bellows and dust chambers. In describing the making of flowers of copper (see note 26, p. 538) he states that in refining copper, when the "molten metal flows through its tube into a receptacle, the work-

[43]*Cadmia metallica fossilis* (see note on p. 112). This was undoubtedly the complex cobalt-arsenic-zinc minerals found in Saxony. In the German translation, however, this is given as *Kalmey*, calamine, which is unlikely from the association with pyrites.

these, when they have been roasted as much as is necessary and re-smelted, the copper is made. But if there be some silver in the cakes, for which an outlay of lead has to be made, then it is first run into the forehearth, and the molten lead absorbs the silver.

Indeed, *rudis* copper ore of inferior quality, whether ash-coloured or purple, blackish and occasionally in parts blue, is smelted in the first furnace whose tap-hole is always open. This is the method of the Tyrolese. To as much *rudis* copper ore as will fill eighteen vessels, each of which holds

" men pour cold water on it, the copper spits and throws off the flowers." He gives the first description of vitriol (see note 11, p. 572), and describes the pieces as " shaped like dice which stick together in bunches like grapes." Altogether, from Dioscorides we learn for the first time of copper made from sulphide ores, and of the recovery of zinc oxides from furnace fumes; and he gives us the first certain description of making brass, and finally the first notice of blue vitriol.

The next author we have who gives any technical detail of copper work is Pliny (23–79 A.D.), and while his statements carry us a little further than Dioscorides, they are not as complete as the same number of words could have afforded had he ever had practical contact with the subject, and one is driven to the conclusion that he was not himself much of a metallurgist. Pliny indicates that copper ores were obtained from veins by underground mining. He gives the same minerals as Dioscorides, but is a good deal confused over *chrysocolla* and *chalcitis*. He gives no description of the shapes of furnaces, but frequently mentions the bellows, and speaks of the *cadmia* and *pompholyx* which adhered to the walls and arches of the furnaces. He has nothing to say as to whether fluxes are used or not. As to fuel, he says (XXXIII, 30) that " for smelting copper and iron pine wood is the best." The following (XXXIV, 20) is of the greatest interest on the subject :—" Cyprian copper is known as *coronarium* and " *regulare*; both are ductile. . . . In other mines are made that known as *regulare* and " *caldarium*. These differ, because the *caldarium* is only melted, and is brittle to the hammer; " whereas the *regulare* is malleable or ductile. All Cyprian copper is this latter kind. But " in other mines with care the difference can be eliminated from *caldarium*, the impurities being " carefully purged away by smelting with fire, it is made into *regulare*. Among the remaining " kinds of copper the best is that of Campania, which is most esteemed for vessels and utensils. " This kind is made in several ways. At Capua it is melted with wood, not with charcoal, " after which it is sprinkled with water and washed through an oak sieve. After it is melted " a number of times Spanish *plumbum argentum* (probably pewter) is added to it in proportion of ten pounds of the lead to one hundred pounds of copper, and thereby it is " made pliable and assumes that pleasing colour which in other kinds of copper is effected " by oil and the sun. In many parts of the Italian provinces they make a similar kind " of metal; but there they add eight pounds of lead, and it is re-melted over charcoal " because of the scarcity of wood. Very different is the method carried on in Gaul, particularly where the ore is smelted between red hot stones, for this burns the metal and renders " it black and brittle. Moreover, it is re-melted only a single time, whereas the oftener this " operation is repeated the better the quality becomes. It is well to remark that all copper " fuses best when the weather is intensely cold." The red hot stones in Gaul were probably as much figments of imagination as was the assumption of one commentator that they were a reverberatory furnace. Apart from the above, Pliny says nothing very direct on refining copper. It is obvious that more than one melting was practised, but that anything was known of the nature of oxidation by a blast and reduction by poling is uncertain. We produce the three following statements in connection with some bye-products used for medicinal purposes, which at least indicate operations subsequent to the original melting. As to whether they represent this species of refining or not, we leave it to the metallurgical profession (XXXIV, 24) :—" The flowers of copper are used in medicine; they are made by fusing copper and moving " it to another furnace, where the rapid blast separates it into a thousand particles, which " are called flowers. These scales are also made when the copper cakes are cooled in water " (XXXIV, 35). *Smega* is prepared in the copper works; when the metal is melted and " thoroughly smelted charcoal is added to it and gradually kindled; after this, being blown " upon by a powerful bellows, it spits out, as it were, copper chaff (XXXIV, 37). There is " another product of these works easily distinguished from *smega*, which the Greeks call " *diphrygum*. This substance has three different origins. . . . A third way of making it " is from the residues which fall to the bottom in copper furnaces. The difference between " the different substances (in the furnace) is that the copper itself flows into a receiver; the " slag makes its escape from the furnace; the flowers float on the top (of the copper ?), and " the *diphrygum* remains behind. Some say that in the furnace there are certain masses of " stone which, being smelted, become soldered together, and that the copper fuses around it, " the mass not becoming liquid unless it is transferred to another furnace. It thus forms a " sort of knot, as it were, in the metal."

almost as much as seven Roman *moduli*[44], the first smelter—for there are three—adds three cartloads of lead slags, one cartload of schist, one fifth of a *centumpondium* of stones which easily fuse in the fire, besides a small quantity of concentrates collected from copper slag and accretions, all of which he smelts for the space of twelve hours, and from which he makes six *centumpondia* of primary cakes and one-half of a *centumpondium* of alloy. One half of the latter consists of copper and silver, and it settles to the bottom of the forehearth. In every *centumpondium* of the cakes there is half a *libra* of silver and sometimes half an *uncia* besides ; in the half of a *centumpondium*

Pliny is a good deal confused over the copper alloys, failing to recognise *aurichalcum* as the same product as that made by mixing *cadmia* and molten copper. Further, there is always the difficulty in translation arising from the fact that the Latin *aes* was indiscriminately copper, brass, and bronze. He does not, except in one instance (XXXIV., 2), directly describe the mixture of *cadmia* and copper. "Next to Livian (copper) this kind (*corduban*, from "Spain) most readily absorbs *cadmia*, and becomes almost as excellent as *aurichalcum* for "making *sesterces*." As to bronze, there is no very definite statement ; but the *argentatium* given in the quotation above from XXXIV, 20, is stated in XXXIV, 48, to be a mixture of tin and lead. The Romans carried on most extensive copper mining in various parts of their empire ; these activities extended from Egypt through Cyprus, Central Europe, the Spanish Peninsula, and Britain. The activity of such works is abundantly evidenced in the mines, but very little remains upon the surface to indicate the equipment ; thus, while mining methods are clear enough, the metallurgy receives little help from these sources. At Rio Tinto there still remain enormous slag heaps from the Romans, and the Phoenician miners before them. Professor W. A. Carlyle informs us that the ore worked must have been almost exclusively sulphides, as only negligible quantities of carbonates exist in the deposits ; they probably mixed basic and siliceous ores. There is some evidence of roasting, and the slags run from .2 to .6%. They must have run down mattes, but as to how they ultimately arrived at metallic copper there is no evidence to show.

The special processes for separating other metals from copper by liquation and matting, or of refining by poling, etc., are none of them clearly indicated in records or remains until we reach the 12th century. Here we find very adequate descriptions of copper smelting and refining by the Monk Theophilus (see Appendix B). We reproduce two paragraphs of interest from Hendrie's excellent translation (p. 305 and 313) : "Copper is engendered in the earth. "When a vein of which is found, it is acquired with the greatest labour by digging and break- "ing. It is a stone of a green colour and most hard, and naturally mixed with lead. This "stone, dug up in abundance, is placed upon a pile and burned after the manner of chalk, "nor does it change colour, but yet loses its hardness, so that it can be broken up. Then, "being bruised small, it is placed in the furnace ; coals and the bellows being applied, it is "incessantly forged by day and night. This should be done carefully and with caution ; "that is, at first coals are placed in, then small pieces of stone are distributed over them, "and again coals, and then stone anew, and it is thus arranged until it is sufficient for the "size of the furnace. And when the stone has commenced to liquefy, the lead flows out "through some small cavities, and the copper remains within. (313) Of the purification of cop- "per. Take an iron dish of the size you wish, and line it inside and out with clay strongly "beaten and mixed, and it is carefully dried. Then place it before a forge upon the coals, "so that when the bellows act upon it the wind may issue partly within and partly above it, "and not below it. And very small coals being placed round it, place copper in it equally, "and add over it a heap of coals. When, by blowing a long time, this has become melted, "uncover it and cast immediately fine ashes of coals over it, and stir it with a thin and dry "piece of wood as if mixing it, and you will directly see the burnt lead adhere to these ashes "like a glue. Which being cast out again superpose coals, and blowing for a long time, as "at first, again uncover it, and then do as you did before. You do this until at length, by "cooking it, you can withdraw the lead entirely. Then pour it over the mould which you have "prepared for this, and you will thus prove if it be pure. Hold it with pincers, glowing as it "is, before it has become cold, and strike it with a large hammer strongly over the anvil, and "if it be broken or split you must liquefy it anew as before."

The next writer of importance was Biringuccio, who was contemporaneous with Agricola, but whose book precedes *De Re Metallica* by 15 years. That author (III, 2) is the first to describe particularly the furnace used in Saxony and the roasting prior to smelting, and the first to mention fluxes in detail. He, however, describes nothing of matte smelting ; in copper refining he gives the whole process of poling, but omits the pole. It is not until we reach *De Re Metallica* that we find adequate descriptions of the copper minerals, roasting, matte smelting, liquation, and refining, with a wealth of detail which eliminates the necessity for a large amount of conjecture regarding technical methods of the time.

[44]The Roman *modius* (*modulus* ?) held about 550 cubic inches, the English peck holding 535 cubic inches. Then, perhaps, his seven *moduli* would be roughly, 1 bushel 3 pecks, and 18 vessels full would be about 31 bushels—say, roughly, 5,400 lbs. of ore.

of the alloy there is a *bes* or three-quarters of silver. In this way every week, if the work is for six days, thirty-six *centumpondia* of cakes are made and three *centumpondia* of alloy, in all of which there is often almost twenty-four *librae* of silver. The second smelter separates from the primary cakes the greater part of the silver by absorbing it in lead. To eighteen *centumpondia* of cakes made from crude copper ore, he adds twelve *centumpondia* of hearth-lead and litharge, three *centumpondia* of stones from which lead is smelted, five *centumpondia* of hard cakes rich in silver, and two *centumpondia* of exhausted liquation cakes[45]; he adds besides, some of the slags resulting from smelting crude copper, together with a small quantity of concentrates made from accretions, all of which he melts for the space of twelve hours, and makes eighteen *centumpondia* of secondary cakes, and twelve *centumpondia* of copper-lead-silver alloy; in each *centumpondium* of the latter there is half a *libra* of silver. After he has taken off the cakes with a hooked bar, he pours the alloy out into copper or iron moulds; by this method they make four cakes of alloy, which are carried to the works in which silver is parted from copper. On the following day, the same smelter, taking eighteen *centumpondia* of the secondary cakes, again adds twelve *centumpondia* of hearth-lead and litharge, three *centumpondia* of stones from which lead is smelted, five *centumpondia* of hard cakes rich in silver, together with slags from the smelting of the primary cakes, and with concentrates washed from the accretions which are usually made at that time. This charge is likewise smelted for the space of twelve hours, and he makes as many as thirteen *centumpondia* of tertiary cakes and eleven *centumpondia* of copper-lead-silver alloy, each *centumpondium* of which contains one-third of a *libra* and half an *uncia* of silver. When he has skimmed off the tertiary cakes with a hooked bar, the alloy is poured into copper moulds, and by this method four cakes of alloy are made, which, like the preceding four cakes of alloy, are carried to the works in which silver is parted from copper. By this method the second smelter makes primary cakes on alternate days and secondary cakes on the intermediate days. The third smelter takes eleven cartloads of the tertiary cakes and adds to them three cartloads of hard cakes poor in silver, together with the slag from smelting the secondary cakes, and the concentrates from the accretions which are usually made at that time. From this charge when smelted, he makes twenty *centumpondia* of quaternary cakes, which are called "hard cakes," and also fifteen *centumpondia* of those "hard cakes rich in silver," each *centumpondium* of which contains a third of a *libra* of silver. These latter cakes the second smelter, as I said before, adds to the primary and secondary cakes when he re-melts them. In the same way, from eleven cartloads of quaternary cakes thrice roasted, he makes the "final" cakes, of which one *centumpondinm* contains only half an *uncia* of silver. In this operation he also makes fifteen *centumpondia* of "hard cakes poor in silver," in each *centumpondium* of which is a sixth of a *libra* of silver. These hard cakes the

[45]Exhausted liquation cakes (*panes aerei fathiscentes*). This is the copper sponge resulting from the first liquation of lead, and still contains a considerable amount of lead. The liquation process is discussed in great detail in Book XI.

third smelter, as I have said, adds to the tertiary cakes when he re-smelts them, while from the "final" cakes, thrice roasted and re-smelted, is made black copper[46].

The *rudis* copper from which pure copper is made, if it contains little silver or if it does not easily melt, is first smelted in the third furnace of which the tap-hole is always open; and from this are made cakes, which after being seven times roasted are re-smelted, and from these copper is melted out; the cakes of copper are carried to a furnace of another kind, in which they are melted for the third time, in order that in the copper "bottoms" there may be more silver, while in the "tops" there may be less, which process is explained in Book XI.

[46]The method of this paragraph involves two main objectives—first, the gradual enrichment of matte to blister copper; and, second, the creation of large cakes of copper-lead-silver alloy of suitable size and ratio of metals for liquation. This latter process is described in detail in Book XI. The following groupings show the circuit of the various products, the "lbs." being Roman *librae* :—

	Charge.		Products.	
1st	Crude ore	5,400 lbs.	Primary matte (1) ..	600 lbs.
	Lead slags	3 cartloads	Silver-copper alloy (A)	50 „
	Schist	1 cartload	Slags (B)	
	Flux	20 lbs.		
	Concentrates from slags & accretions	Small quantity		
2nd	Primary matte (1)	1,800 lbs.	Secondary matte (2)	1,800 lbs.
	Hearth-lead & litharge	1,200 „	Silver-copper-lead alloy (liquation cakes) (A_2) ..	1,200 „
	Lead ore ..	300 „	Slags (B_2)	
	Rich hard cakes (A_4)	500 „		
	Liquated cakes ..	200 „		
	Slags (B)			
	Concentrates from accretions			
3rd	Secondary matte (2)	1,800 lbs.	Tertiary matte (3)	1,300 lbs.
	Hearth-lead & litharge	1,200 „	Silver-copper-lead alloy (liquation cakes) (A_3) ..	1,100 „
	Lead ore	300 „	Slags (B_3)	
	Rich hard cakes (A_4)	500 „		
	Slags (B_2)			
	Concentrates from accretions			
4th	Tertiary matte (3) ..	11 cartloads	Quaternary hard cakes matte (4)	2,000 lbs.
	Poor hard cakes (A_5)	3 „	Rich hard cakes of matte (A_4) ..	1,500 „
	Slags (B_3)			
	Concentrates from accretions			
5th	Roasted quartz		Poor hard cakes of matte (A_5) ..	1,500 lbs.
	Matte (4) (three times roasted)	11 cartloads	Final cakes of matte (5)	

6th Final matte three times roasted is smelted to blister copper.

The following would be a rough approximation of the value of the various products :—

(1.)	Primary matte ..	=	158 ounces troy per short ton.
(2.)	Secondary matte ..	=	85 „ „ „
(3.)	Tertiary matte ..	=	60 „ „ „
(4.)	Quaternary matte	=	Indeterminate.
A.	Copper-silver alloy	=	388 ounces Troy per short ton.
A_2	Copper-silver-lead alloy	=	145 „ „ „
A_3	„ „ „	=	109 „ „ „
A_4	Rich hard cakes	=	97 „ „ „
A_5	Poor hard cakes	=	Indeterminate.
	Final blister copper	=	12 ozs. Troy per short ton.

Pyrites, when they contain not only copper, but also silver, are smelted in the manner I described when I treated of ores of silver. But if they are poor in silver, and if the copper which is melted out of them cannot easily be treated, they are smelted according to the method which I last explained.

Finally, the copper schists containing bitumen or sulphur are roasted, and then smelted with stones which easily fuse in a fire of the second order, and are made into cakes, on the top of which the slags float. From these cakes, usually roasted seven times and re-melted, are melted out slags and two kinds of cakes; one kind is of copper and occupies the bottom of the crucible, and these are sold to the proprietors of the works in which silver is parted from copper; the other kind of cakes are usually re-melted with primary cakes. If the schist contains but a small amount of copper, it is burned, crushed under the stamps, washed and sieved, and the concentrates obtained from it are melted down; from this are made cakes from which, when roasted, copper is made. If either chrysocolla or azure, or yellow or black earth containing copper and silver, adheres to the schist, it is not washed, but is crushed and smelted with stones which easily fuse in fire of the second order.

Lead ore, whether it be *molybdaena*[47], pyrites, (galena ?) or stone from which it is melted, is often smelted in a special furnace, of which I have spoken above, but no less often in the third furnace of which the tap-hole is always open. The hearth and forehearth are made from powder containing a small portion of iron hammer-scales; iron slag forms the principal flux for such ores; both of these the expert smelters consider useful and to the owner's advantage, because it is the nature of iron to attract lead. If it is *molybdaena* or the stone from which lead is smelted, then the lead runs down from the furnace into the forehearth, and when the slags have been skimmed off, the lead is poured out with a ladle. If pyrites are smelted, the first to flow from the furnace into the forehearth, as may be seen at Goslar, is a white molten substance, injurious and noxious to silver, for it consumes it. For this reason the slags which float on the top having been skimmed off, this substance is poured out; or if it hardens, then it is taken out with a hooked bar; and the walls of the furnace exude the same substance[48].

[47]This expression is usually used for hearth-lead, but in this case the author is apparently confining himself to lead ore, and apparently refers to lead carbonates. The German Translation gives *pleyschweiss*. The pyrites mentioned in this paragraph may mean galena, as pyrites was to Agricola a sort of genera.

[48](*Excoquitur*) . . . "*si verò pyrites, primò è fornace, ut Goselariae videre licet, in* "*catinum defluit liquor quidam candidus, argento inimicus et nocivus; id enim comburit:* "*quo circa recrementis, quae supernatant, detractis effunditur: vel induratus conto uncinato* "*extrahitur: eundem liquorem parietes fornacis exudant.*" In the Glossary the following statement appears: "*Liquor candidus primo è fornace defluens cum Goselariae excoquitur* "*pyrites,—kobelt; quem parietes fornacis exudant,—conterfei.*" In this latter statement Agricola apparently recognised that there were two different substances, *i.e.*, that the substance found in the furnace walls—*conterfei*—was not the same substance as that which first flowed from the furnace—*kobelt*. We are at no difficulty in recognizing *conterfei* as metallic zinc; it was long known by that term, and this accidental occurrence is repeatedly mentioned by other authors after Agricola. The substance which first flowed into the forehearth presents greater difficulties; it certainly was not zinc. In *De Natura Fossilium* (p. 347), Agricola says that at Goslar the lead has a certain white slag floating upon it, the "colour derived from the pyrites (*pyriten argenteum*) from which it was produced." *Pyriten argenteum* was either marcasite or mispickel, neither of which offers much suggestion; nor are we able to hazard an explanation of value.

HISTORICAL NOTE ON ZINC. The history of zinc metallurgy falls into two distinct

Then the *stannum* runs out of the furnace into the forehearth ; this is an alloy of lead and silver. From the silver-lead alloy they first skim off the slags, not rarely white, as some pyrites[49] are, and afterward they skim off the cakes of pyrites, if there are any. In these cakes there is usually some copper ; but since there is usually but a very small quantity, and as the forest

lines—first, that of the metal, and second, that of zinc ore, for the latter was known and used to make brass by cementation with copper and to yield oxides by sublimation for medicinal purposes, nearly 2,000 years before the metal became generally known and used in Europe.

There is some reason to believe that metallic zinc was known to the Ancients, for bracelets made of it, found in the ruins of Cameros (prior to 500 B.C.), may have been of that age (Raoul Jagnaux, *Traité de Chimie Générale*, 1887, II, 385) ; and further, a passage in Strabo (63 B.C.—24 A.D.) is of much interest. He states : (XIII, I, 56) "There is found at "Andeira a stone which when burnt becomes iron. It is then put into a furnace, together "with some kind of earth, when it distils a mock silver (*pseudargyrum*), or with the addition "of copper it becomes the compound called *orichalcum*. There is found a mock silver near "Tismolu also." (Hamilton's Trans., II, p. 381). About the Christian era the terms *orichalcum* or *aurichalcum* undoubtedly refer to brass, but whether these terms as used by earlier Greek writers do not refer to bronze only, is a matter of considerable doubt. Beyond these slight references we are without information until the 16th Century. If the metal was known to the Ancients it must have been locally, for by its greater adaptability to brass-making it would probably have supplanted the crude melting of copper with zinc minerals.

It appears that the metal may have been known in the Far East prior to such knowledge in Europe ; metallic zinc was imported in considerable quantities from the East as early as the 16th and 17th centuries under such terms as *tuteneque*, *tuttanego*, *calaëm*, and *spiauter*—the latter, of course, being the progenitor of our term spelter. The localities of Eastern production have never been adequately investigated. W. Hommel (Engineering and Mining Journal, June 15, 1912) gives a very satisfactory review of the Eastern literature upon the subject, and considers that the origin of manufacture was in India, although the most of the 16th and 17th Century product came from China. The earliest certain description seems to be some recipes for manufacture quoted by Praphulla Chandra Ray (A History of Hindu Chemistry, London, 1902, p. 39) dating from the 11th to the 14th Centuries. There does not appear to be any satisfactory description of the Chinese method until that of Sir George Staunton (Journal Asiatique, Paris, 1835, p. 141.) We may add that spelter was produced in India by crude distillation of calamine in clay pots in the early part of the 19th Century (Brooke, Jour. Asiatic Soc. of Bengal, vol. XIX, 1850, p. 212), and the remains of such smelting in Rajputana are supposed to be very ancient.

The discovery of zinc in Europe seems to have been quite independent of the East, but precisely where and when is clouded with much uncertainty. The *marchasita aurea* of Albertus Magnus has been called upon to serve as metallic zinc, but such belief requires a hypothesis based upon a great deal of assumption. Further, the statement is frequently made that zinc is mentioned in Basil Valentine's Triumphant Chariot of Antimony (the only one of the works attributed to this author which may date prior to the 17th Century), but we have been unable to find any such reference. The first certain mention of metallic zinc is generally accredited to Paracelsus (1493–1541), who states (*Liber Mineralium* II.) : "More-"over there is another metal generally unknown called *zinken*. It is of peculiar nature and "origin ; many other metals adulterate it. It can be melted, for it is generated from three "fluid principles ; it is not malleable. Its colour is different from other metals and does not "resemble others in its growth. Its ultimate matter (*ultima materia*) is not to me yet fully "known. It admits of no mixture and does not permit of the *fabricationes* of other metals. "It stands alone entirely to itself." We do not believe that this book was published until after Agricola's works. Agricola introduced the following statements into his revised edition of *Bermannus* (p. 431), published in 1558 : "It (a variety of pyrites) is almost the colour "of galena, but of entirely different components. From it there is made gold and silver, and "a great quantity is dug in Reichenstein, which is in Silesia, as was recently reported to me. "Much more is found at Raurici, which they call *zincum*, which species differs from pyrites, "for the latter contains more silver than gold, the former only gold or hardly any silver." In *De Natura Fossilium* (p. 368) : "For this *cadmia* is put, in the same way as quicksilver, "in a suitable vessel so that the heat of the fire will cause it to sublime, and from it is made "a black or brown or grey body which the Alchemists call *cadmia sublimata*. This "possesses corrosive properties to the highest degree. Cognate with this *cadmia* and pyrites "is a compound which the Noricans and Rhetians call *zincum*." We leave it to readers to decide how near this comes to metallic zinc ; in any event, he apparently did not

---

[49] " . . . *non raro, ut nonnulli pyritae sunt, candida* . . . ." This is apparently the unknown substance mentioned above.

charcoal is not abundant, no copper is made from them. From the silver-lead poured into iron moulds they likewise make cakes; when these cakes have been melted in the cupellation furnace, the silver is parted from the lead, because part of the lead is transformed into litharge and part into hearth-lead, from which in the blast furnace on re-melting they make

recognise his *conterfei* from the furnaces as the same substance as the *zincum* from Silesia. The first correlation of these substances was apparently by Lohneys, in 1617, who says (*Vom Bergwerk*, p. 83–4): "When the people in the smelting works are smelting, there is "made under the furnace and in the cracks in the walls among the badly plastered stones, a "metal which is called *zinc* or *counterfeht*, and when the wall is scraped it falls into a vessel "placed to receive it. This metal greatly resembles tin, but it is harder and less malleable. ". . . . The Alchemists have a great desire for this *zinc* or bismuth." That this metal originated from blende or calamine was not recognised until long after, and Libavis (*Alchymia*, Frankfort, 1606), in describing specimens which came from the East, did not so identify it, this office being performed by Glauber, who says (*De Prosperitate Germanias*, Amsterdam, 1656): "Zink is a volatile mineral or half-ripe metal when it is "extracted from its ore. It is more brilliant than tin and not so fusible or malleable . . . "it turns (copper) into brass, as does *lapis calaminaris*, for indeed this stone is nothing but "infusible zinc, and this zinc might be called a fusible *lapis calaminaris*, inasmuch as both "of them partake of the same nature. . . . It sublimates itself up into the cracks of the "furnace, whereupon the smelters frequently break it out." The systematic distillation of zinc from calamine was not discovered in Europe until the 18th Century. Henkel is generally accredited with the first statement to that effect. In a contribution published as an Appendix to his other works, of which we have had access only to a French translation (*Pyritologie*, Paris, 1760, p. 494), he concludes that zinc is a half-metal of which the best ore is calamine, but believes it is always associated with lead, and mentions that an Englishman lately arrived from Bristol had seen it being obtained from calamine in his own country. He further mentions that it can be obtained by heating calamine and lead ore mixed with coal in a thick earthen vessel. The Bristol works were apparently those of John Champion, established about 1740. The art of distillation was probably learned in the East.

Definite information as to the zinc minerals goes back to but a little before the Christian Era, unless we accept nebular references to *aurichalcum* by the poets, or what is possibly zinc ore in the "earth" mentioned by Aristotle (*De Mirabilibus*, 62): "Men say "that the copper of the Mossynoeci is very brilliant and white, no tin being mixed with it; "but there is a kind of earth there which is melted with it." This might quite well be an arsenical mineral. But whether we can accept the poets or Aristotle or the remark of Strabo given above, as sufficient evidence or not, there is no difficulty with the description of *cadmia* and *pompholyx* and *spodos* of Dioscorides (1st Century), parts of which we reproduce in note 26, p. 394. His *cadmia* is described as rising from the copper furnaces and clinging to the iron bars, but he continues: "*Cadmia* is also prepared by burning the "stone called pyrites, which is found near Mt. Soloi in Cyprus. . . . Some say that "*cadmia* may also be found in stone quarries, but they are deceived by stones having a "resemblance to *cadmia*." *Pompholyx* and *spodos* are evidently furnace calamine. From reading the quotation given on p. 394, there can be no doubt that these materials, natural or artificial, were used to make brass, for he states (v, 46): "White *pompholyx* is made every "time that the artificer in the working and perfecting of the copper sprinkles powdered "*cadmia* upon it to make it more perfect, the soot arising from this . . . . is *pompholyx*." Pliny is confused between the mineral *cadmia* and furnace *calamine*, and none of his statements are very direct on the subject of brass making. His most pointed statement is (XXXIV, 2): ". . . . Next to Livian (copper) this kind best absorbs *cadmia*, and is almost as good as *aurichalcum* for making sesterces and double asses." As stated above, there can be little doubt that the *aurichalcum* of the Christian Era was brass, and further, we do know of brass sesterces of this period. Other Roman writers of this and later periods refer to earth used with copper for making brass. Apart from these evidences, however, there is the evidence of analyses of coins and objects, the earliest of which appears to be a large brass of the Cassia family of 20 B.C., analyzed by Phillips, who found 17.3% zinc (Records of Mining and Metallurgy, London, 1857, p. 13). Numerous analyses of coins and other objects dating during the following century corroborate the general use of brass. Professor Gowland (Presidential Address, Inst. of Metals, 1912) rightly considers the Romans were the first to make brass, and at about the above period, for there appears to be no certainty of any earlier production. The first adequate technical description of brass making is in about 1200 A.D., being that of Theophilus, who describes (Hendrie's Trans., p. 307) calcining *calamina* and mixing it with finely divided copper in glowing crucibles. The process was repeated by adding more calamine and copper until the pots were full of molten metal. This method is repeatedly described with minor variations by Biringuccio, Agricola (*De Nat. Fos.*), and others, down to the 18th Century. For discussion of the zinc minerals see note on p. 112.

de-silverized lead, for in this lead each *centumpondium* contains only a *drachma* of silver, when before the silver was parted from it each *centumpondium* contained more or less than three *unciae* of silver[50].

The little black stones[51] and others from which tin is made, are smelted in their own kind of furnace, which should be narrower than the other furnaces, that there may be only the small fire which is necessary for this ore. These furnaces are higher, that the height may compensate for the narrowness and make them of almost the same capacity as the other furnaces. At the top, in front, they are closed and on the other side they are open, where there are steps, because they cannot have the steps in front on account of the forehearth; the smelters ascend by these steps to put the tin-stone into the furnace. The hearth of the furnace is not made of powdered earth and charcoal, but on the floor of the works are placed sandstones which are not too hard; these are set on a slight slope, and are two and three-quarters feet long, the same number of feet wide, and two feet thick, for the thicker they are the longer they last in the fire. Around them is constructed a rectangular furnace eight or nine feet high, of broad sandstones, or of those common substances which by nature are composed of diverse materials[52]. On the inside the furnace is everywhere evenly covered with lute. The upper part of the interior is two feet long and one foot wide, but below it is not so long and wide. Above it are two hood-walls, between which the fumes ascend from the furnace into the dust chamber, and through this they escape by a narrow opening in the roof. The sandstones are sloped at the bed of the furnace, so that the tin melted from the tin-stone may flow through the tap-hole of the furnace into the forehearth.[53]

[50]One *drachma* is about 3 ounces Troy per short ton. Three *unciae* are about 72 ounces 6 dwts. Troy per short ton.

[51]In this section, which treats of the metallurgy of *plumbum candidum*, "tin," the word *candidum* is very often omitted in the Latin, leaving only *plumbum*, which is confusing at times with lead. The black tin-stone, *lapilli nigri* has been treated in a similar manner, *lapilli* (small stones) constantly occurring alone in the Latin. This has been rendered as "tin-stone" throughout, and the material prior to extraction of the *lapilli nigri* has been rendered "tin-stuff," after the Cornish.

[52]" . . . *ex saxis vilibus, quae natura de diversa materia composuit.*" The Glossary gives *grindstein.* Granite (?).

[53]HISTORICAL NOTES ON TIN METALLURGY. The first appearance of tin lies in the ancient bronzes. And while much is written upon the "Bronze Age" by archæologists, we seriously doubt whether or not a large part of so-called bronze is not copper. In any event, this period varied with each race, and for instance, in Britain may have been much later than Egyptian historic times. The bronze articles of the IV Dynasty (from 3800 to 4700 B.C. depending on the authority) place us on certain ground of antiquity. Professor Gowland (Presidential Address, Inst. of Metals, London, 1912) maintains that the early bronzes were the result of direct smelting of stanniferous copper ores, and while this may be partially true for Western Europe, the distribution and nature of the copper deposits do not warrant this assumption for the earlier scenes of human activity—Asia Minor, Egypt, and India. Further, the lumps of rough tin and also of copper found by Borlase (Tin Mining in Spain, Past and Present, London, 1897, p. 25) in Cornwall, mixed with bronze celts under conditions certainly indicating the Bronze Age, is in itself of considerable evidence of independent melting. To our mind the vast majority of ancient bronzes must have been made from copper and tin mined and smelted independently. As to the source of supply of ancient tin, we are on clear ground only with the advent of the Phœnicians, 1500–1000 B.C., who, as is well known, distributed to the ancient world a supply from Spain and Britain. What the source may have been prior to this time has been subject to much discussion, and while some

As there is no need for the smelters to have a fierce fire, it is not necessary to place the nozzles of the bellows in bronze or iron pipes, but only through a hole in the furnace wall. They place the bellows higher at the back so that the blast from the nozzles may blow straight toward the tap-hole of the furnace. That it may not be too fierce, the nozzles are wide, for if the fire were fiercer, tin could not be melted out from the tin-stone, as it would be consumed and turned into ashes. Near the steps is a hollowed stone, in which is placed the tin-stone to be smelted ; as often as the smelter throws into the furnace an iron shovel-ful of this tin-stone, he puts on charcoal that was first put into a vat and washed with water to be cleansed from the grit and small stones which adhere to it, lest they melt at the same time as the tin-stone and obstruct the tap-hole and impede the flow of tin from the furnace. The tap-hole of the furnace is always open ; in front of it is a forehearth a little more than half a foot deep, three-quarters of two feet long and one foot wide ; this is lined with lute, and the tin from the tap-hole flows into it. On one side of the forehearth is a low wall, three-quarters of a foot wider and one foot longer than the forehearth, on which lies charcoal powder. On the other side the floor of the building slopes, so that the slags may conveniently run down and be carried away. As soon as the tin begins to run from the tap-hole of the furnace into the forehearth, the smelter scrapes

slender threads indicate the East, we believe that a more local supply to Egypt, etc., is not impossible. The discovery of large tin fields in Central Africa and the native-made tin ornaments in circulation among the negroes, made possible the entrance of the metal into Egypt along the trade routes. Further, we see no reason why alluvial tin may not have existed within easy reach and have become exhausted. How quickly such a source of metal supply can be forgotten and no evidence remain, is indicated by the seldom remembered alluvial gold supply from Ireland. However, be these conjectures as they may, the East has long been the scene of tin production and of transportation activity. Among the slender evidences that point in this direction is that the Sanskrit term for tin is *kastira*, a term also employed by the Chaldeans, and represented in Arabic by *kasdir*, and it may have been the progenitor of the Greek *cassiteros*. There can be no doubt that the Phœnicians also traded with Malacca, etc, but beyond these threads there is little to prove the pre-western source. The strained argument of Beckmann (Hist. of Inventions, vol. II., p. 207) that the *cassiteros* of Homer and the *bedil* of the Hebrews was possibly not tin, and that tin was unknown at this time, falls to the ground in the face of the vast amount of tin which must have been in circulation to account for the bronze used over a period 2,000 years prior to those peoples. Tin is early mentioned in the Scriptures (Numbers XXXI, 22), being enumerated among the spoil of the Midianites (1200 B.C. ?), also Ezekiel (600 B.C., XXVII, 12) speaks of tin from Tarshish (the Phœnician settlement on the coast of Spain). According to Homer tin played considerable part in Vulcan's metallurgical stores. Even approximately at what period the Phœnicians began their distribution from Spain and Britain cannot be determined. They apparently established their settlements at Gades (Cadiz) in Tarshish, beyond Gibraltar, about 1100 B.C. The remains of tin mining in the Spanish peninsula prior to the Christian Era indicate most extensive production by the Phœnicians, but there is little evidence as to either mining or smelting methods. Generally as to the technical methods of mining and smelting tin, we are practically without any satisfactory statement down to Agricola. However, such scraps of information as are available are those in Homer (see note on p. 402), Diodorus, and Pliny.

Diodorus says (V, 2) regarding tin in Spain : " They dig it up, and melt it down in the " same way as they do gold and silver ; " and again, speaking of the tin in Britain, he says : " These people make tin, which they dig up with a great deal of care and labour ; being " rocky, the metal is mixed with earth, out of which they melt the metal, and then refine " it." Pliny (XXXIV, 47), in the well-known and much-disputed passage : " Next to be " considered are the characteristics of lead, which is of two kinds, black and white. The " most valuable is the white ; the Greeks called it *cassiteros*, and there is a fabulous story of its " being searched for and carried from the islands of Atlantis in barks covered with hides. " Certainly it is obtained in Lusitania and Gallaecia on the surface of the earth from black-" coloured sand. It is discovered by its great weight, and it is mixed with small pebbles in

down some of the powdered charcoal into it from the wall, so that the slags may be separated from the hot metal, and so that it may be covered, lest any part of it, being very hot, should fly away with the fumes. If after the slag has been skimmed off, the powder does not cover up the whole of the tin, the smelter draws a little more charcoal off the wall with a scraper. After he has opened the tap-hole of the forehearth with a tapping-bar, in order that the tin can flow into the tapping-pot, likewise smeared with lute, he again closes the tap-hole with pure lute or lute mixed with powdered charcoal. The smelter, if he be diligent and experienced, has brooms at hand with which he sweeps down the walls above the furnace ; to these walls and to the dust chamber minute tin-stones sometimes adhere with part of the fumes. If he be not sufficiently experienced in these matters and has melted at the same time all of the tin-stone,—which is commonly of three sizes, large, medium, and very small,—not a little waste of the proprietor's tin results ; because, before the large or the medium sizes have melted, the small have either been burnt up in the furnace, or else, flying up from it, they not only adhere to the walls but also fall in the dust chamber. The owner of the works has the sweepings by right from the owner of the ore. For the above reasons the most experienced smelter melts them down separately ; indeed, he melts the very small size in a wider furnace, the medium in a medium-sized furnace, and the largest size in the narrowest furnace. When he melts down the small size he uses a gentle blast from the bellows, with the medium-sized a moderate one, with the large size a violent blast ; and when he smelts the first size he needs a slow fire, for the second a medium one, and for the third a fierce one ; yet he uses a much less fierce fire than when he smelts the ores of gold, silver, or copper. When the workmen have spent three consecutive days and nights in this work, as is usual, they have finished their labours ; in this time they are able to melt out a large weight of small

" the dried beds of torrents. The miners wash these sands, and that which settles they heat " in the furnace. It is also found in gold mines, which are called *alutiae*. A stream of water " passing through detaches small black pebbles variegated with white spots, the weight of " which is the same as gold. Hence it is that they remain in the baskets of the gold collectors " with the gold ; afterward, they are separated in a *camillum* and when melted become white " lead."

There is practically no reference to the methods of Cornish tin-working over the whole period of 2,000 years that mining operations were carried on there prior to the Norman occupation. From then until Agricola's time, a period of some four centuries, there are occasional references in Stannary Court proceedings, Charters, and such-like official documents which give little metallurgical insight. From a letter of William de Wrotham, Lord Warden of the Stannaries, in 1198, setting out the regulations for the impost on tin, it is evident that the black tin was smelted once at the mines and that a second smelting or refining was carried out in specified towns under the observation of the Crown Officials. In many other official documents there are repeated references to the right to dig turfs and cut wood for smelting the tin. Under note 8, p. 282, we give some further information on tin concentration, and the relation of Cornish and German tin miners. Biringuccio (1540) gives very little information on tin metallurgy, and we are brought to *De Re Metallica* for the first clear exposition.

As to the description on these pages it must be remembered that the tin-stone has been already roasted, thus removing some volatile impurities and oxidizing others, as described on page 348. The furnaces and the methods of working the tin, here described, are almost identical with those in use in Saxony to-day. In general, since Agricola's time tin has not seen the mechanical and metallurgical development of the other metals. The comparatively small quantities to be dealt with ; the necessity of maintaining a strong reducing atmosphere, and consequently a mild cold blast ; and the comparatively low temperature demanded, gave little impetus to other than crude appliances until very modern times.

sized tin-stone which melts quickly, but less of the large ones which melt slowly, and a moderate quantity of the medium-sized which holds the middle course. Those who do not smelt the tin-stone in furnaces made sometimes wide, sometimes medium, or sometimes narrow, in order that great loss should not be occasioned, throw in first the smallest size, then the medium, then the large size, and finally those which are not quite pure ; and the blast of the bellows is altered as required. In order that the tin-stone thrown into the furnace should not roll off from the large charcoal into the forehearth before the tin is melted out of it, the smelter uses small charcoal ; first some of this moistened with water is placed in the furnace, and then he frequently repeats this succession of charcoal and tin-stone.

The tin-stone, collected from material which during the summer was washed in a ditch through which a stream was diverted, and during the winter was screened on a perforated iron plate, is smelted in a furnace a palm wider than that in which the fine tin-stone dug out of the earth is smelted. For the smelting of these, a more vigorous blast of the bellows and a fiercer fire is needed than for the smelting of the large tin-stone. Whichever kind of tin-stone is being smelted, if the tin first flows from the furnace, much of it is made, and if slags first flow from the furnace, then only a little. It happens that the tin-stone is mixed with the slags when it is either less pure or ferruginous—that is, not enough roasted—and is imperfect when put into the furnace, or when it has been put in in a larger quantity than was necessary ; then, although it may be pure and melt easily, the ore either runs out of the furnace at the same time, mixed with the slags, or else it settles so firmly at the bottom of the furnace that the operation of smelting being necessarily interrupted, the furnace freezes up.

The tap-hole of the forehearth is opened and the tin is diverted into the dipping-pot, and as often as the slags flow down the sloping floor of the building they are skimmed off with a rabble ; as soon as the tin has run out of the forehearth, the tap-hole is again closed up with lute mixed with powdered charcoal. Glowing coals are put in the dipping-pot so that the tin, after it has run out, should not get chilled. If the metal is so impure that nothing can be made from it, the material which has run out is made into cakes to be re-smelted in the hearth, of which I shall have something to say later ; if the metal is pure, it is poured immediately upon thick copper plates, at first in straight lines and then transversely over these to make a lattice. Each of these lattice bars is impressed with an iron die ; if the tin was melted out of ore excavated from mines, then one stamp only, namely, that of the Magistrate, is usually imprinted, but if it is made from tin-stone collected on the ground after washing, then it is impressed with two seals, one the Magistrate's and the other a fork which the washers use. Generally, three of this kind of lattice bars are beaten and amalgamated into one mass with a wooden mallet.

The slags that are skimmed off are afterward thrown with an iron shovel into a small trough hollowed from a tree, and are cleansed from charcoal

A—FURNACE. B—ITS TAP-HOLE. C—FOREHEARTH. D—ITS TAP-HOLE. E—SLAGS. F—SCRAPER. G—DIPPING-POT. H—WALLS OF THE CHIMNEY. I—BROOM. K—COPPER PLATE. L—LATTICEWORK BARS. M—IRON SEAL OR DIE. N—HAMMER.

by agitation ; when taken out they are broken up with a square iron mallet, and then they are re-melted with the fine tin-stone next smelted. There are some who crush the slags three times under wet stamps and re-melt them three times ; if a large quantity of this be smelted while still wet, little tin is melted from it, because the slag, soon melted again, flows from the furnace into the forehearth. Under the wet stamps are also crushed the lute and broken rock with which such furnaces are lined, and also the accretions, which often contain fine tin-stone, either not melted or half-melted, and also prills of tin. The tin-stone not yet melted runs out through the screen into a trough, and is washed in the same way as tin-stone, while the partly melted and the prills of tin are taken from the mortar-box and washed in the sieve on which not very minute particles remain, and thence to the canvas strake. The soot which adheres to that part of the chimney which emits the smoke, also often contains very fine tin-stone which flies from the furnace with the fumes, and this is washed in the strake which I have just mentioned, and in other sluices. The prills of tin and the partly melted tin-stone that are contained in the lute and broken rock with which the furnace is lined, and in the remnants of the tin from the forehearth and the dipping-pot, are smelted together with the tin-stone.

When tin-stone has been smelted for three days and as many nights in a furnace prepared as I have said above, some little particles of the rock from which the furnace is constructed become loosened by the fire and fall down ; and then the bellows being taken away, the furnace is broken through at the back, and the accretions are first chipped off with hammers, and afterward the whole of the interior of the furnace is re-fitted with the prepared sandstone, and again evenly lined with lute. The sandstone placed on the bed of the furnace, if it has become faulty, is taken out, and another is laid down in its place ; those rocks which are too large the smelter chips off and fits with a sharp pick.

Some build two furnaces against the wall just like those I have described, and above them build a vaulted ceiling supported by the wall and by four pillars. Through holes in the vaulted ceiling the fumes from the furnaces ascend into a dust chamber, similar to the one described before, except that there is a window on each side and there is no door. The smelters, when they have to clear away the flue-dust, mount by the steps at the side of the furnaces, and climb by ladders into the dust chamber through the apertures in the vaulted ceilings over the furnaces. They then remove the flue-dust from everywhere and collect it in baskets, which are passed from one to the other and emptied. This dust chamber differs from the other described, in the fact that the chimneys, of which it has two, are not dissimilar to those of a house ; they receive the fumes which, being unable to escape through the upper part of the chamber, are turned back and re-ascend and release the tin ; thus the tin set free by the fire and turned to ash, and the little tin-stones which fly up with the fumes, remain in the dust chamber or else adhere to copper plates in the chimney.

A—Furnaces. B—Forehearths. C—Their tap-holes. D—Dipping-pots. E—Pillars. F—Dust-chamber. G—Window. H—Chimneys. I—Tub in which the coals are washed.

If the tin is so impure that it cracks when struck with the hammer, it is not immediately made into lattice-like bars, but into the cakes which I have spoken of before, and these are refined by melting again on a hearth. This hearth consists of sandstones, which slope toward the centre and a little toward a dipping-pot ; at their joints they are covered with lute. Dry logs are arranged on each side, alternately upright and lengthwise, and more closely in the middle ; on this wood are placed five or six cakes of tin which all together weigh about six *centumpondia* ; the wood having been kindled,

A—Hearths. B—Dipping-pots. C—Wood. D—Cakes. E—Ladle. F—Copper plate. G—Lattice-shaped bars. H—Iron dies. I—Wooden mallet. K—Mass of tin bars. L—Shovel.

the tin drips down and flows continuously into the dipping-pot which is on the floor. The impure tin sinks to the bottom of this dipping-pot and the pure tin floats on the top ; then both are ladled out by the master, who first takes out the pure tin, and by pouring it over thick plates of copper makes lattice-like bars. Afterward he takes out the impure tin from which he makes cakes ; he discriminates between them, when he ladles and pours, by the ease or difficulty of the flow. One *centumpondium* of the lattice-like bare sells for more than a *centumpondium* of cakes, for the price of the former exceeds the price of the latter by a gold coin[54]. These lattice-like bars are lighter than the others, and when five of them are pounded and amalgamated with a wooden mallet, a mass is made which is stamped with an iron die. There are some who do not make a dipping-pot on the floor for the tin to run into, but in the hearth itself; out of this the master, having removed the charcoal, ladles the tin and pours it over the copper-plate. The dross which adheres to the wood and the charcoal, having been collected, is re-smelted in the furnace.

A—Furnace. B—Bellows. C—Iron Disc. D—Nozzle. E—Wooden Disc. F—Blow-hole. G—Handle. H—Haft. I—Hoops. K—Masses of tin.

Some of the Lusitanians melt tin from tin-stone in small furnaces. They use round bellows made of leather, of which the fore end is a round iron disc and the rear end a disc of wood; in a hole in the former is fixed the nozzle, in the middle of the latter the blow-hole. Above this is the handle or haft, which draws open the round bellows and lets in the air, or compresses it and drives the air out. Between the discs are several iron hoops to which the leather is fastened, making such folds as are to be seen in paper lanterns that

[54]*Aureo nummo*. German Translation gives *reinschen gülden*, which was the equivalent of about $1.66, or 6.9 shillings. The purchasing power of money was, however, several times as great as at present.

are folded together. Since this kind of bellows does not give a vigorous blast, because they are drawn apart and compressed slowly, the smelter is not able during a whole day to smelt much more than half a *centumpondium* of tin.

Very good iron ore is smelted[55] in a furnace almost like the cupellation furnace. The hearth is three and a half feet high, and five feet long and wide; in the centre of it is a crucible a foot deep and one and a half feet wide, but it may be deeper or shallower, wider or narrower, according to whether more or less ore is to be made into iron. A certain quantity of iron ore is given to the master, out of which he may smelt either much or little iron. He being about to expend his skill and labour on this matter, first throws charcoal into the crucible, and sprinkles over it an iron shovel-ful of crushed iron ore mixed with unslaked lime. Then he repeatedly throws on charcoal and sprinkles it with ore, and continues this until he has slowly built up a heap; it melts when the charcoal has been kindled and the fire violently stimulated by the blast of the bellows, which are skilfully fixed in a pipe.

[55]In the following descriptions of iron-smelting, we have three processes described; the first being the direct reduction of malleable iron from ore, the second the transition stage then in progress from the direct to indirect method by way of cast-iron; and the third a method of making steel by cementation. The first method is that of primitive iron-workers of all times and all races, and requires little comment. A pasty mass was produced, which was subsequently hammered to make it exude the slag, the hammered mass being the ancient "bloom." The second process is of considerable interest, for it marks one of the earliest descriptions of working iron in "a furnace similar to a blast furnace, but much wider "and higher." This original German *Stückofen* or high bloomery furnace was used for making "masses" of wrought-iron under essentially the same conditions as its progenitor the forge—only upon a larger scale. With high temperatures, however, such a furnace would, if desired, yield molten metal, and thus the step to cast-iron as a preliminary to wrought-iron became very easy and natural, in fact Agricola mentions above that if the iron is left to settle in the furnace it becomes hard. The making of malleable iron by subsequent treatment of the cast-iron—the indirect method—originated in about Agricola's time, and marks the beginning of one of those subtle economic currents destined to have the widest bearing upon civilization. It is to us uncertain whether he really understood the double treatment or not. In the above paragraph he says from ore "once or twice smelted they make iron," etc., and in *De Natura Fossilium* (p. 339) some reference is made to pouring melted iron, all of which would appear to be cast-iron. He does not, however, describe the 16th Century method of converting cast into wrought iron by way of in effect roasting the pig iron to eliminate carbon by oxidation, with subsequent melting into a "ball" or "mass." It must be borne in mind that puddling for this purpose did not come into use until the end of the 18th Century. A great deal of discussion has arisen as to where and at what time cast-iron was made systematically, but without satisfactory answer; in any event, it seems to have been in about the end of the 14th Century, as cast cannon began to appear about that time. It is our impression that the whole of this discussion on iron in *De Re Metallica* is an abstract from Biringuccio, who wrote 15 years earlier, as it is in so nearly identical terms. Those interested will find a translation of Biringuccio's statement with regard to steel in Percy's Metallurgy of Iron and Steel, London, 1864, p. 807.

HISTORICAL NOTE ON IRON SMELTING. The archæologists' division of the history of racial development into the Stone, Bronze, and Iron Ages, based upon objects found in tumuli, burial places, etc., would on the face of it indicate the prior discovery of copper metallurgy over iron, and it is generally so maintained by those scientists. The metallurgists have not hesitated to protest that while this distinction of "Ages" may serve the archæologists, and no doubt represents the sequence in which the metal objects are found, yet it by no means follows that this was the order of their discovery or use, but that iron by its rapidity of oxidation has simply not been preserved. The arguments which may be advanced from our side are in the main these. Iron ore is of more frequent occurrence than copper ores, and the necessary reduction of copper oxides (as most surface ores must have been) to fluid metal requires a temperature very much higher than does the reduction of iron oxides to wrought-iron blooms, which do not necessitate fusion. The comparatively greater simplicity of iron metallurgy under primitive conditions is well exemplified by the hill tribes of Northern Nigeria, where in village forges the negroes reduce iron

He is able to complete this work sometimes in eight hours, sometimes in ten, and again sometimes in twelve. In order that the heat of the fire should not burn his face, he covers it entirely with a cap, in which, however, there are holes through which he may see and breathe. At the side of the hearth is a bar which he raises as often as is necessary, when the bellows blow too violent a blast, or when he adds more ore and charcoal. He also uses the bar to draw off the slags, or to open or close the gates of the sluice, through which the waters flow down on to the wheel which turns the axle that compresses the bellows. In this sensible way, iron is melted out and a mass weighing two or three *centumpondia* may be made, providing the iron ore was rich. When this is done the master opens the slag-vent with the tapping-bar, and when all has run out he allows the iron mass to cool. Afterward he and his assistant stir the iron with the bar, and then in order to chip off the slags which had until then adhered to it, and to condense and flatten it, they take it down from the furnace to the floor, and beat it with large wooden mallets having slender handles five feet long. Thereupon it is immediately

sufficient for their needs, from hematite. Copper alone would not be a very serviceable metal to primitive man, and he early made the advance to bronze ; this latter metal requires three metallurgical operations, and presents immeasurably greater difficulties than iron. It is, as Professor Gowland has demonstrated (Presidential Address, Inst. of Metals, London, 1912) quite possible to make bronze from melting stanniferous copper ores, yet such combined occurrence at the surface is rare, and, so far as known, the copper sources from which Asia Minor and Egypt obtained their supply do not contain tin. It seems to us, therefore, that in most cases the separate fusions of different ores and their subsequent re-melting were required to make bronze. The arguments advanced by the archæologists bear mostly upon the fact that, had iron been known, its superiority would have caused the primitive races to adopt it, and we should not find such an abundance of bronze tools. As to this, it may be said that bronze weapons and tools are plentiful enough in Egyptian, Mycenæan, and early Greek remains, long after iron was demonstrably well known. There has been a good deal pronounced by etymologists on the history of iron and copper, for instance, by Max Müller, (Lectures on the Science of Language, Vol. II, p. 255, London, 1864), and many others, but the amazing lack of metallurgical knowledge nullifies practically all their conclusions. The oldest Egyptian texts extant, dating 3500 B.C., refer to iron, and there is in the British Museum a piece of iron found in the Pyramid of Kephron (3700 B.C.) under conditions indicating its co-incident origin. There is exhibited also a fragment of oxidized iron lately found by Professor Petrie and placed as of the VI Dynasty (B.C. 3200). Despite this evidence of an early knowledge of iron, there is almost a total absence of Egyptian iron objects for a long period subsequent to that time, which in a measure confirms the view of its disappearance rather than that of ignorance of it. Many writers have assumed that the Ancients must have had some superior art of hardening copper or bronze, because the cutting of the gigantic stone-work of the time could not have been done with that alloy as we know it ; no such hardening appears among the bronze tools found, and it seems to us that the argument is stronger that the oldest Egyptian stoneworkers employed mostly iron tools, and that these have oxidized out of existence. The reasons for preferring copper alloys to iron for decorative objects were equally strong in ancient times as in the present day, and accounts sufficiently for these articles, and, therefore, iron would be devoted to more humble objects less likely to be preserved. Further, the Egyptians at a later date had some prejudices against iron for sacred purposes, and the media of preservation of most metal objects were not open to iron. We know practically nothing of very early Egyptian metallurgy, but in the time of Thotmes III. (1500 B.C.) bellows were used upon the forge.

Of literary evidences the earliest is in the Shoo King among the Tribute of Yü (2500 B.C. ?). Iron is frequently mentioned in the Bible, but it is doubtful if any of the early references apply to steel. There is scarcely a Greek or Latin author who does not mention iron in some connection, and of the earliest, none are so suggestive from a metallurgical point of view as Homer, by whom "laboured" mass (wrought-iron ?) is often referred to. As, for instance, in the Odyssey (I., 234) Pallas in the guise of Mentes, says according to Pope :

"Freighted with iron from my native land
"I steer my voyage to the Brutian strand,
"To gain by commerce for the laboured mass
"A just proportion of refulgent brass."

A—Hearth. B—Heap. C—Slag-vent. D—Iron mass. E—Wooden mallets. F—Hammer. G—Anvil.

placed on the anvil, and repeatedly beaten by the large iron hammer that is raised by the cams of an axle turned by a water-wheel. Not long afterward it is taken up with tongs and placed under the same hammer, and cut up with a sharp iron into four, five, or six pieces, according to whether it is large or small. These pieces, after they have been re-heated in the blacksmith's forge and again placed on the anvil, are shaped by the smith into square bars or into ploughshares or tyres, but mainly into bars. Four, six, or eight of these bars weigh one-fifth of a *centumpondium*, and from these they make various implements. During the blows from the hammer by which it is shaped by the smith, a youth pours water with a ladle on to the glowing iron, and this is why the blows make such a loud sound that they may be heard a long distance from the works. The masses, if they remain and settle in the crucible of the furnace in which the iron is smelted, become hard iron which can only be hammered with difficulty, and from these they make the iron-shod heads for the stamps, and such-like very hard articles.

But to iron ore which is cupriferous, or which when heated[56] melts with difficulty, it is necessary for us to give a fiercer fire and more labour; because not only must we separate the parts of it in which there is metal from those in which there is no metal, and break it up by dry stamps, but we must also roast it, so that the other metals and noxious juices may be exhaled; and we must wash it, so that the lighter parts may be separated from it. Such ores are smelted in a furnace similar to the blast furnace, but much wider and higher, so that it may hold a great quantity of ore and much charcoal; mounting the stairs at the side of the furnace, the smelters fill it partly with fragments of ore not larger than nuts, and partly with charcoal; and from this kind of ore once or twice smelted they make iron which is suitable for re-heating in the blacksmith's forge, after it is flattened out with the large iron hammer and cut into pieces with the sharp iron.

By skill with fire and fluxes is made that kind of iron from which steel is made, which the Greeks call στόμωμα. Iron should be selected which is easy to melt, is hard and malleable. Now although iron may be smelted from ore which contains other metals, yet it is then either soft or brittle; such (iron) must be broken up into small pieces when it is

(Brass is modern poetic licence for copper or bronze). Also, in the Odyssey (IX, 465) when Homer describes how Ulysses plunged the stake into Cyclop's eye, we have the first positive evidence of steel, although hard iron mentioned in the Tribute of Yü, above referred to, is sometimes given as steel:

> "And as when armourers temper in the ford
> "The keen-edg'd pole-axe, or the shining sword,
> "The red-hot metal hisses in the lake."

No doubt early wrought-iron was made in the same manner as Agricola describes. We are, however, not so clear as to the methods of making steel. Under primitive methods of making wrought-iron it is quite possible to carburize the iron sufficiently to make steel direct from ore. The primitive method of India and Japan was to enclose lumps of wrought-iron in sealed crucibles with charcoal and sawdust, and heat them over a long period. Neither Pliny nor any of the other authors of the period previous to the Christian Era give us much help on steel metallurgy, although certain obscure expressions of Aristotle have been called upon (for instance, St. John V. Day, Prehistoric Use of Iron and Steel, London, 1877, p. 134) to prove its manufacture by immersing wrought-iron in molten cast-iron.

[56]*Quae vel aerosa est, vel cocta.* It is by no means certain that *cocta*, "cooked" is rightly translated, for the author has not hitherto used this expression for heated. This may be residues from roasting and leaching pyrites for vitriol, etc.

A—Furnace. B—Stairs. C—Ore. D—Charcoal.

A—Forge. B—Bellows. C—Tongs. D—Hammer. E—Cold stream.

hot, and then mixed with crushed stone which melts. Then a crucible is made in the hearth of the smith's furnace, from the same moistened powder from which are made the forehearths in front of the furnaces in which ores of gold or silver are smelted; the width of this crucible is about one and a half feet and the depth one foot. The bellows are so placed that the blast may be blown through the nozzle into the middle of the crucible. Then the whole of the crucible is filled with the best charcoal, and it is surrounded by fragments of rock to hold in place the pieces of iron and the superimposed charcoal. As soon as all the charcoal is kindled and the crucible is glowing, a blast is blown from the bellows and the master pours in gradually as much of the mixture of iron and flux as he wishes. Into the middle of this, when it is melted, he puts four iron masses each weighing thirty pounds, and heats them for five or six hours in a fierce fire; he frequently stirs the melted iron with a bar, so that the small pores in each mass absorb the minute particles, and these particles by their own strength consume and expand the thick particles of the masses, which they render soft and similar to dough. Afterward the master, aided by his assistant, takes out a mass with the tongs and places it on the anvil, where it is pounded by the hammer which is alternately raised and dropped by means of the water-wheel; then, without delay, while it is still hot, he throws it into water and tempers it; when it is tempered, he places it again on the anvil, and breaks it with a blow from the same hammer. Then at once examining the fragments, he decides whether the iron in some part or other, or as a whole, appears to be dense and changed into steel; if so, he seizes one mass after another with the tongs, and taking them out he breaks them into pieces. Afterward he heats the mixture up again, and adds a portion afresh to take the place of that which has been absorbed by the masses. This restores the energy of that which is left, and the pieces of the masses are again put back into the crucible and made purer. Each of these, after having been heated, is seized with the tongs, put under the hammer and shaped into a bar. While they are still glowing, he at once throws them into the very coldest nearby running water, and in this manner, being suddenly condensed, they are changed into pure steel, which is much harder and whiter than iron.

The ores of the other metals are not smelted in furnaces. Quicksilver ores and also antimony are melted in pots, and bismuth in troughs.

I will first speak of quicksilver. This is collected when found in pools formed from the outpourings of the veins and stringers; it is cleansed with vinegar and salt, and then it is poured into canvas or soft leather, through which, when squeezed and compressed, the quicksilver runs out into a pot or pan. The ore of quicksilver is reduced in double or single pots. If in double pots, then the upper one is of a shape not very dissimilar to the glass ampullas used by doctors, but they taper downward toward the bottom, and the lower ones are little pots similar to those in which men and women make cheese, but both are larger than these; it is necessary to sink the lower pots up to the rims in earth, sand, or ashes. The ore, broken up into small pieces is put into the upper pots; these having been entirely closed up with moss, are placed upside down in the openings of the lower pots, where they are joined with lute, lest the quicksilver which takes refuge in them should be exhaled. There are some who, after the pots have been buried, do not fear to leave them uncemented, and who boast that they are able to produce no less weight of quicksilver than those who do cement them, but nevertheless cementing with lute is the greatest protection against exhalation. In this manner seven hundred pairs of pots are set together in the ground or on a hearth. They must be surrounded on all sides with a mixture consisting of crushed earth and charcoal, in such a way that the upper pots protrude to a height of a palm above it. On both sides of the hearth rocks are first laid, and upon them poles, across which the workmen place other poles transversely ; these poles do not touch the pots, nevertheless the fire heats the quicksilver, which fleeing from the heat is forced to run down through the moss into the lower pots. If the ore is being reduced in the upper pots, it flees from them, wherever there is an exit, into the lower pots, but if the ore on the contrary is put in the lower pots the quicksilver rises into the upper pot or into the operculum, which, together with the gourd-shaped vessels, are cemented to the upper pots.

A—HEARTH. B—POLES. C—HEARTH WITHOUT FIRE IN WHICH THE POTS ARE PLACED. D—ROCKS. E—ROWS OF POTS. F—UPPER POTS. G—LOWER POTS.

The pots, lest they should become defective, are moulded from the best potters' clay, for if there are defects the quicksilver flies out in the fumes. If the fumes give out a very sweet odour it indicates that the quicksilver is being lost, and since this loosens the teeth, the smelters and others standing by, warned of the evil, turn their backs to the wind, which drives the fumes in the opposite direction ; for this reason, the building should be open around the front and the sides, and exposed to the wind. If these pots are made of cast copper they last a long time in the fire. This process for reducing the ores of quicksilver is used by most people.

In a similar manner the antimony ore,[57] if free from other metals, is reduced in upper pots which are twice as large as the lower ones. Their size, however, depends on the cakes, which have not the same weight everywhere ; for in some places they are made to weigh six *librae*, in other places ten, and elsewhere twenty. When the smelter has concluded his operation, he extinguishes the fire with water, removes the lids from the pots, throws earth mixed with ash around and over them, and when they have cooled, takes out the cakes from the pots.

[57]Agricola draws no sharp line of distinction between antimony the metal, and its sulphide. He uses the Roman term *stibi* or *stibium* (*Interpretatio,—Spiesglas*) throughout this book, and evidently in most cases means the sulphide, but in others, particularly in parting gold and silver, metallic antimony would be reduced out. We have been in much doubt as to the term to introduce into the text, as the English " stibnite " carries too much precision of meaning. Originally the " antimony " of trade was the sulphide. Later, with the application of that term to the metal, the sulphide was termed " grey antimony," and we have either used *stibium* for lack of better alternative, or adopted " grey antimony." The method described by Agricola for treating antimony sulphide is still used in the Harz, in Bohemia, and elsewhere. The stibnite is liquated out at a low heat and drips from the upper to the lower pot. The resulting purified antimony sulphide is the modern commercial " crude antimony " or " grey antimony."

HISTORICAL NOTE ON THE METALLURGY OF ANTIMONY. The Egyptologists have adopted the term " antimony " for certain cosmetics found in Egyptian tombs from a very early period. We have, however, failed to find any reliable analyses which warrant this assumption, and we believe that it is based on the knowledge that antimony was used as a base for eye ointments in Greek and Roman times, and not upon proper chemical investigation. It may be that the ideograph which is interpreted as antimony may really mean that substance, but we only protest that the chemist should have been called in long since. In St. Jerome's translation of the Bible, the cosmetic used by Jezebel (II. Kings IX, 30) and by the lady mentioned by Ezekiel (XXIII, 40), " who didst wash thyself and paintedst thine eyes " is specifically given as *stibio*. Our modern translation carries no hint of the composition of the cosmetic, and whether some of the Greek or Hebrew MSS. do furnish a basis for such translation we cannot say. The Hebrew term for this mineral was *kohl*, which subsequently passed into " alcool " and " alkohol " in other languages, and appears in the Spanish Bible in the above passage in Ezekiel as *alcoholaste*. The term *antimonium* seems to have been first used in Latin editions of Geber published in the latter part of the 15th Century. In any event, the metal is clearly mentioned by Dioscorides (1st Century), who calls it *stimmi*, and Pliny, who termed it *stibium*, and they leave no doubt that it was used as a cosmetic for painting the eyebrows and dilating the eyes. Dioscorides (V, 59) says : " The best *stimmi* " is very brilliant and radiant. When broken it divides into layers with no part earthy or " dirty ; it is brittle. Some call it *stimmi*, others *platyophthalmon* (wide eyed) ; others " *larbason*, others *gynaekeion* (feminine). . . . It is roasted in a ball of dough with " charcoal until it becomes a cinder. . . . It is also roasted by putting it on live charcoal " and blowing it. If it is roasted too much it becomes lead." Pliny states (XXXIII, 33 and 34) : " In the same mines in which silver is found, properly speaking there is a stone froth. " It is white and shining, not transparent ; is called *stimmi*, or *stibi*, or *alabastrum*, and *larbasis*. " There are two kinds of it, the male and the female. The most approved is the female, the " male being more uneven, rougher, less heavy, not so radiant, and more gritty. The female " kind is bright and friable, laminar and not globular. It is astringent and refrigerative, " and its principal use is for the eyes. . . . It is burned in manure in a furnace, is " quenched with milk, ground with rain water in a mortar, and while thus turbid it is poured " into a copper vessel and purified with nitrum . . . . above all in roasting it care

Other methods for reducing quicksilver are given below. Big-bellied pots, having been placed in the upper rectangular open part of a furnace, are filled with the crushed ore. Each of these pots is covered with a lid with a long nozzle—commonly called a *campana*—in the shape of a bell, and they are cemented. Each of the small earthenware vessels shaped like a gourd receives two of these nozzles, and these are likewise cemented. Dried

A—POTS. B—OPERCULA. C—NOZZLES. D—GOURD-SHAPED EARTHENWARE VESSELS.

wood having been placed in the lower part of the furnace and kindled, the ore is heated until all the quicksilver has risen into the operculum which is over the pot; it then flows from the nozzle and is caught in the earthenware gourd-shaped vessel.

" should be taken that it does not turn to lead." There can be little doubt from Dioscorides' statement of its turning to lead that he had seen the metal antimony, although he thought it a species of lead. Of further interest in connection with the ancient knowledge of the metal is the Chaldean vase made of antimony described by Berthelot (*Comptes Rendus*, 1887, CIV, 265). It is possible that Agricola knew the metal, although he gives no details as to de-sulphurizing it or for recovering the metal itself. In *De Natura Fossilium* (p. 181) he makes a statement which would indicate the metal, " *Stibium* when melted in the crucible and " refined has as much right to be regarded as a metal as is accorded to lead by most writers. " If when smelted a certain portion be added to tin, a printer's alloy is made from which " type is cast that is used by those who print books." Basil Valentine, in his " Triumphal " Chariot of Antimony," gives a great deal that is new with regard to this metal, even if we can accredit the work with no earlier origin than its publication—about 1600; it seems

Others build a hollow vaulted chamber, of which the paved floor is made concave toward the centre. Inside the thick walls of the chamber are the furnaces. The doors through which the wood is put are in the outer part of the same wall. They place the pots in the furnaces and fill them with crushed ore, then they cement the pots and the furnaces on all sides with lute, so that none of the vapour may escape from them, and there is no entrance to the

A—ENCLOSED CHAMBER. B—DOOR. C—LITTLE WINDOWS. D—MOUTHS THROUGH THE WALLS. E—FURNACE IN THE ENCLOSED CHAMBER. F—POTS.

furnaces except through their mouths. Between the dome and the paved floor they arrange green trees, then they close the door and the little windows, and cover them on all sides with moss and lute, so that none of the quicksilver can exhale from the chamber. After the wood has been kindled the

possible however, that it was written late in the 15th Century (see Appendix B). He describes the preparation of the metal from the crude ore, both by roasting and reduction from the oxide with argol and saltpetre, and also by fusing with metallic iron. While the first description of these methods is usually attributed to Valentine, it may be pointed out that in the *Probierbüchlein* (1500) as well as in Agricola the separation of silver from iron by antimony sulphide implies the same reaction, and the separation of silver and gold with antimony sulphide, often attributed to Valentine, is repeatedly set out in the *Probierbüchlein* and in *De Re Metallica*. Biringuccio (1540) has nothing of importance to say as to the treatment of antimonial ores, but mentions it as an alloy for bell-metal, which would imply the metal.

ore is heated, and exudes the quicksilver ; whereupon, impatient with the heat, and liking the cold, it escapes to the leaves of the trees, which have a cooling power. When the operation is completed the smelter extinguishes the fire, and when all gets cool he opens the door and the windows, and collects the quicksilver, most of which, being heavy, falls of its own accord from the trees, and flows into the concave part of the floor ; if all should not have fallen from the trees, they are shaken to make it fall.

The following is the fourth method of reducing ores of quicksilver. A larger pot standing on a tripod is filled with crushed ore, and over the ore is put sand or ashes to a thickness of two digits, and tamped ; then in the mouth of this pot is inserted the mouth of another smaller pot and cemented with lute, lest the vapours are emitted. The ore heated by the fire exhales the quicksilver, which, penetrating through the sand or the ashes, takes refuge in the upper pot, where condensing into drops it falls back into the sand or the ashes, from which the quicksilver is washed and collected.

A—Larger pot. B—Smaller. C—Tripod. D—Tub in which the sand is washed.

The fifth method is not very unlike the fourth. In the place of these pots are set other pots, likewise of earthenware, having a narrow bottom and a wide mouth. These are nearly filled with crushed ore, which is likewise covered with ashes to a depth of two digits and tamped in. The pots are covered with lids a digit thick, and they are smeared over on the inside with liquid litharge, and on the lid are placed heavy stones. The pots are set on the furnace, and the ore is heated and similarly exhales quicksilver, which fleeing from the heat takes refuge in the lid; on congealing there, it falls back into the ashes, from which, when washed, the quicksilver is collected.

A—Pots. B—Lids. C—Stones. D—Furnace.

By these five methods quicksilver may be made, and of these not one is to be despised or repudiated; nevertheless, if the mine supplies a great abundance of ore, the first is the most expeditious and practical, because a large quantity of ore can be reduced at the same time without great expense.[58]

[58]Historical Note on the Metallurgy of Quicksilver. The earliest mention of quicksilver appears to have been by Aristotle (*Meteorologica* IV, 8, 11), who speaks of it as fluid silver (*argyros chytos*). Theophrastus (105) states: "Such is the production of "quicksilver, which has its uses. This is obtained from cinnabar rubbed with vinegar in a "brass mortar with a brass pestle." (Hill's Trans., p. 139). Theophrastus also (103) mentions cinnabar from Spain and elsewhere. Dioscorides (v, 70) appears to be the first to describe the recovery of quicksilver by distillation: "Quicksilver (*hydrargyros*, *i.e.*, liquid silver) is made from *ammion*, which is called *cinnabari*. An iron bowl containing *cinnabari* "is put into an earthen vessel and covered over with a cup-shaped lid smeared with clay. "Then it is set on a fire of coals and the soot which sticks to the cover when wiped off and "cooled is quicksilver. Quicksilver is also found in drops falling from the walls of the silver "mines. Some say there are quicksilver mines. It can be kept only in vessels of glass, lead, "tin (?), or silver, for if put in vessels of any other substances it consumes them and flows

Bismuth[59] ore, free from every kind of silver, is smelted by various methods. First a small pit is dug in the dry ground; into this pulverised charcoal is thrown and tamped in, and then it is dried with burning charcoal. Afterward, thick dry pieces of beech wood are placed over the pit, and the bismuth ore is thrown on it. As soon as the kindled wood burns, the heated ore drips with bismuth, which runs down into the pit, from which when cooled the cakes are removed. Because pieces of burnt wood, or often charcoal and occasionally slag, drop into the bismuth which collects in the pit, and make it impure, it is put back into another kind of crucible to be melted, so that pure cakes may be made. There are some who, bearing these things in mind, dig a pit on a sloping place and below it put a forehearth, into which the bismuth continually flows, and thus remains clean; then they take it out with ladles and pour it into iron pans lined inside with lute, and make cakes of it. They cover such pits with flat stones, whose joints are besmeared with a lute of mixed dust and crushed charcoal, lest the joints should absorb the molten bismuth. Another method is to put the ore in troughs made of fir-wood and placed on sloping ground; they place small firewood over it, kindling it when a gentle wind blows, and thus the ore is heated. In this manner the bismuth melts and runs down from the troughs into a pit below, while there remains slag, or stones, which are of a yellow colour, as is also the wood laid across the pit. These are also sold.

"through." Pliny (XXXIII, 41): "There has been discovered a way of extracting *hydrargyros* "from the inferior *minium* as a substitute for quicksilver, as mentioned. There are two "methods: either by pounding *minium* and vinegar in a brass mortar with a brass pestle, "or else by putting *minium* into a flat earthen dish covered with a lid, well luted with potter's "clay. This is set in an iron pan and a fire is then lighted under the pan, and continually "blown by a bellows. The perspiration collects on the lid and is wiped off and is like silver "in colour and as liquid as water." Pliny is somewhat confused over the *minium*—or the text is corrupt, for this should be the genuine *minium* of Roman times. The methods of condensation on the leaves of branches placed in a chamber, of condensing in ashes placed over the mouth of the lower pot, and of distilling in a retort, are referred to by Biringuccio (A.D. 1540), but with no detail.

[59]Most of these methods depend upon simple liquation of native bismuth. The sulphides, oxides, etc., could not be obtained without fusing in a furnace with appropriate de-sulphurizing or reducing agents, to which Agricola dimly refers. In *Bermannus* (p. 439), he says: "*Bermannus.*—I will show you another kind of mineral which is numbered "amongst metals, but appears to me to have been unknown to the Ancients; we call it "*bisemutum*. *Naevius.*—Then in your opinion there are more kinds of metals than the "seven commonly believed? *Bermannus.*—More, I consider; for this which just now I "said we called *bisemutum*, cannot correctly be called *plumbum candidum* (tin) nor *nigrum* "(lead), but is different from both, and is a third one. *Plumbum candidum* is whiter and "*plumbum nigrum* is darker, as you see. *Naevius.*—We see that this is of the colour of "*galena*. *Ancon.*—How then can *bisemutum*, as you call it, be distinguished from *galena*? "*Bermannus.*—Easily; when you take it in your hands it stains them with black unless it "is quite hard. The hard kind is not friable like *galena*, but can be cut. It is blacker than "the kind of crude silver which we say is almost the colour of lead, and thus is different "from both. Indeed, it not rarely contains some silver. It generally shows that there is "silver beneath the place where it is found, and because of this our miners are accustomed "to call it the 'roof of silver.' They are wont to roast this mineral, and from the better "part they make metal; from the poorer part they make a pigment of a kind not to be "despised." This pigment was cobalt blue (see note on p. 112), indicating a considerable confusion of these minerals. This quotation is the first description of bismuth, and the above text the first description of bismuth treatment. There is, however, bare mention of the mineral earlier, in the following single line from the *Probierbüchlein* (p. 1): "Jupiter (con-"trols) the ores of tin and *wismundt*." And it is noted in the *Nütliche Bergbüchlein* in association with silver (see Appendix B).

A—Pit across which wood is placed. B—Forehearth. C—Ladle. D—Iron mould. E—Cakes. F—Empty pot lined with stones in layers. G—Troughs. H—Pits dug at the foot of the troughs. I—Small wood laid over the troughs. K—Wind.

Others reduce the ore in iron pans as next described. They lay small pieces of dry wood alternately straight and transversely upon bricks, one and a half feet apart, and set fire to it. Near it they put small iron pans lined on the inside with lute, and full of broken ore; then when the wind blows the flame of the fierce fire over the pans, the bismuth drips out of the ore; wherefore, in order that it may run, the ore is stirred with the tongs; but when they decide that all the bismuth is exuded, they seize the pans with the tongs and remove them, and pour out the bismuth into empty pans, and by turning many into one they make cakes. Others reduce the ore, when it is not mixed with *cadmia*,[60] in a furnace similar to the iron furnace. In this case they make a pit and a crucible of crushed earth mixed with pulverised

A—Wood. B—Bricks. C—Pans. D—Furnace. E—Crucible. F—Pipe. G—Dipping-pot.

charcoal, and into it they put the broken ore, or the concentrates from washing, from which they make more bismuth. If they put in ore, they reduce it with charcoal and small dried wood mixed, and if concentrates, they use charcoal only; they blow both materials with a gentle blast from

[60]This *cadmia* is given in the German translation as *kobelt*. It is probably the cobalt-arsenic-bismuth minerals common in Saxony. A large portion of the world's supply of bismuth to-day comes from the cobalt treatment works near Schneeberg. For further discussion of *cadmia* see note on p. 112.

a bellows. From the crucible is a small pipe through which the molten bismuth runs down into a dipping-pot, and from this cakes are made.

On a dump thrown up from the mines, other people construct a hearth exposed to the wind, a foot high, three feet wide, and four and a half feet long. It is held together by four boards, and the whole is thickly coated at the top with lute. On this hearth they first put small dried sticks of fir wood, then over them they throw broken ore ; then they lay more wood over it, and when the wind blows they kindle it. In this manner the bismuth drips out of the ore, and afterward the ashes of the wood consumed by the fire and the charcoals are swept away. The drops of bismuth which fall down into the hearth are congealed by the cold, and they are taken away with the tongs and thrown into a basket. From the melted bismuth they make cakes in iron pans.

A—Hearth in which ore is melted. B—Hearth on which lie drops of bismuth. C—Tongs. D—Basket. E—Wind.

Others again make a box eight feet long, four feet wide, and two feet high, which they fill almost full of sand and cover with bricks, thus making the hearth. The box has in the centre a wooden pivot, which turns in a hole in two beams laid transversely one upon the other ; these beams are hard and thick, are sunk into the ground, both ends are perforated, and through these holes wedge-shaped pegs are driven, in order that the beams may remain fixed, and that the box may turn round, and may be turned toward the wind from whichever quarter of the sky it may blow. In such a hearth they put

A—Box. B—Pivot. C—Transverse wood beams. D—Grate. E—Its feet. F—Burning wood. G—Stick. H—Pans in which the bismuth is melted. I—Pans for moulds. K—Cakes. L—Fork. M—Brush.

an iron grate, as long and wide as the box and threequarters of a foot high ; it has six feet, and there are so many transverse bars that they almost touch one another. On the grate they lay pine-wood and over it broken ore, and over this they again lay pine-wood. When it has been kindled the ore melts, out of which the bismuth drips down ; since very little wood is burned, this is the most profitable method of smelting the bismuth. The bismuth drips through the grate on to the hearth, while the other things remain upon the grate with the charcoal. When the work is finished, the workman takes a stick from the hearth and overturns the grate, and the things which have accumulated on it ; with a brush he sweeps up the bismuth and collects it in a basket, and then he melts it in an iron pan and makes cakes. As soon as possible after it is cool, he turns the pans over, so that the cakes may fall out, using for this purpose a two-pronged fork of which one prong is again forked. And immediately afterward he returns to his labours.

END OF BOOK IX.

BOOK X.

UESTIONS as to the methods of smelting ores and of obtaining metals I discussed in Book IX. Following this, I should explain in what manner the precious metals are parted from the base metals, or on the other hand the base metals from the precious[1]. Frequently two metals, occasionally more than two, are melted out of one ore, because in nature generally there is some amount of gold in silver and in copper, and some silver in gold, copper, lead, and iron; likewise some copper in gold, silver, lead, and iron, and some lead in silver; and lastly, some iron in copper[2]. But I will begin with gold.

Gold is parted from silver, or likewise the latter from the former, whether it be mixed by nature or by art, by means of *aqua valens*[3], and by powders which consist of almost the same things as this *aqua*. In order to preserve the sequence, I will first speak of the ingredients of which this *aqua* is made, then of the method of making it, then of the manner in which gold is parted from silver or silver from gold. Almost all these ingredients contain vitriol or alum, which, by themselves, but much more when joined with saltpetre, are powerful to part silver from gold. As to the other things that are added to them, they cannot individually by their own strength and nature separate those metals, but joined they are very powerful. Since there are many combinations, I will set out a few. In the first, the use of which is common and general, there is one *libra* of vitriol and as much salt, added to a third of a *libra* of spring water. The second contains two *librae* of vitriol, one of saltpetre, and as much spring or river water by weight as will pass away whilst the vitriol is being reduced to powder by the fire. The third consists of four *librae* of vitriol, two and a half *librae* of saltpetre, half a *libra* of alum, and one and a half *librae* of spring water. The fourth consists of two *librae* of vitriol, as many *librae* of saltpetre, one quarter of a *libra* of alum, and three-quarters of a *libra* of spring water. The fifth is composed of one *libra* of saltpetre,

[1] *Vile a precioso.*

[2] The reagents mentioned in this Book are much the same as those of Book VII, where (p. 220) a table is given showing the Latin and Old German terms. Footnotes in explanation of our views as to these substances may be most easily consulted through the index.

[3] *Aqua valens*, literally strong, potent, or powerful water. It will appear later, from the method of manufacture, that hydrochloric, nitric, and sulphuric acids and *aqua regia* were more or less all produced and all included in this term. We have, therefore, used either the term *aqua valens* or simply *aqua* as it occurs in the text. The terms *aqua fortis* and *aqua regia* had come into use prior to Agricola, but he does not use them; the Alchemists used various terms, often *aqua dissolvia*. It is apparent from the uses to which this reagent was put in separating gold and silver, from the method of clarifying it with silver and from the red fumes, that Agricola could have had practical contact only with nitric acid. It is probable that he has copied part of the recipes for the compounds to be distilled from the Alchemists and from such works as the *Probierbüchlein*. In any event he could not have had experience with them all, for in some cases the necessary ingredients for making nitric acid are not all present, and therefore could be of no use for gold and silver separation. The essential ingredients for the production of this acid by distillation, were saltpetre, water, and either vitriol or alum. The other substances mentioned were unnecessary, and any speculation as to the combinations which would result, forms a useful exercise in chemistry, but of little purpose here. The first recipe would no doubt produce hydrochloric acid.

three *librae* of alum, half a *libra* of brick dust, and three-quarters of a *libra* of spring water. The sixth consists of four *librae* of vitriol, three *librae* of saltpetre, one of alum, one *libra* likewise of stones which when thrown into a fierce furnace are easily liquefied by fire of the third order, and one and a half *librae* of spring water. The seventh is made of two *librae* of vitriol, one and a half *librae* of saltpetre, half a *libra* of alum, and one *libra* of stones which when thrown into a glowing furnace are easily liquefied by fire of the third order, and five-sixths of a *libra* of spring water. The eighth is made of two *librae* of vitriol, the same number of *librae* of saltpetre, one and a half *librae* of alum, one *libra* of the lees of the *aqua* which parts gold from silver; and to each separate *libra* a sixth of urine is poured over it. The ninth contains two *librae* of powder of baked bricks, one *libra* of vitriol, likewise one *libra* of saltpetre, a handful of salt, and three-quarters of a *libra* of spring water. Only the tenth lacks vitriol and alum, but it contains three *librae* of saltpetre, two *librae* of stones which when thrown into a hot furnace are easily liquefied by fire of the third order, half a *libra* each of verdigris[4], of *stibium*, of iron scales and filings, and of asbestos[5], and one and one-sixth *librae* of spring water.

All the vitriol from which the *aqua* is usually made is first reduced to powder in the following way. It is thrown into an earthen crucible lined on the inside with litharge, and heated until it melts ; then it is stirred with a copper wire, and after it has cooled it is pounded to powder. In the same manner saltpetre melted by the fire is pounded to powder when it has cooled. Some indeed place alum upon an iron plate, roast it, and make it into powder.

Although all these *aquae* cleanse gold concentrates or dust from impurities, yet there are certain compositions which possess singular power.

[4]Agricola, in the *Interpretatio*, gives the German equivalent for the Latin *aerugo* as *Spanschgrün*—"because it was first brought to Germany from Spain ; foreigners call it "*viride aeris* (copper green)." The English "verdigris" is a corruption of *vert de grice*. Both verdigris and white lead were very ancient products, and they naturally find mention together among the ancient authors. The earliest description of the method of making is from the 3rd Century B.C., by Theophrastus, who says (101–2) : "But these are "works of art, as is also Ceruse (*psimythion*) to make which, lead is placed in earthen vessels "over sharp vinegar, and after it has acquired some thickness of a kind of rust, which it "commonly does in about ten days, they open the vessels and scrape off, as it were, a kind "of foulness ; they then place the lead over the vinegar again, repeating over and over "again the same method of scraping it till it is wholly dissolved ; what has been scraped off "they then beat to powder and boil for a long time ; and what at last subsides to the bottom of "the vessel is the white lead. . . . Also in a manner somewhat resembling this, verdigris "(*ios*) is made, for copper is placed over lees of wine (grape refuse ?), and the rust which it "acquires by this means is taken off for use. And it is by this means that the rust which "appears is produced." (Based on Hill's translation.) Vitruvius (VII, 12), Dioscorides (V, 51), and Pliny (XXXIV, 26 and 54), all describe the method of making somewhat more elaborately.

[5]*Amiantus* (*Interpretatio* gives *federwis*, *pliant*, *salamanderhar*). From Agricola's elaborate description in *De Natura Fossilium* (p. 252) there can be no doubt that he means asbestos. This mineral was well-known to the Ancients, and is probably earliest referred to (3rd Century B.C.) by Theophrastus in the following passage (29) : "There is also found in "the mines of Scaptesylae a stone, in its external appearance somewhat resembling wood, "on which, if oil be poured, it burns ; but when the oil is burnt away, the burning of the "stone ceases, as if it were in itself not liable to such accidents." There can be no doubt that Strabo (X, 1) describes the mineral : "At Carystus there is found in the earth a stone, "which is combed like wool, and woven, so that napkins are made of this substance, which, "when soiled, are thrown into the fire and cleaned, as in the washing of linen." It is also described by Dioscorides (V, 113) and Pliny (XIX, 4). Asbestos cloth has been found in Pre-Augustinian Roman tombs.

The first of these consists of one *libra* of verdigris and three-quarters of a *libra* of vitriol. For each *libra* there is poured over it one-sixth of a *libra* of spring or river water, as to which, since this pertains to all these compounds, it is sufficient to have mentioned once for all. The second composition is made from one *libra* of each of the following, artificial orpiment, vitriol, lime, alum, ash which the dyers of wool use, one quarter of a *libra* of verdigris, and one and a half *unciae* of *stibium*. The third consists of three *librae* of vitriol, one of saltpetre, half a *libra* of asbestos, and half a *libra* of baked bricks. The fourth consists of one *libra* of saltpetre, one *libra* of alum, and half a *libra* of sal-ammoniac.[6]

The furnace in which *aqua valens* is made[7] is built of bricks, rectangular, two feet long and wide, and as many feet high and a half besides. It is covered with iron plates supported with iron rods ; these plates are smeared on the top with lute, and they have in the centre a round hole, large enough to hold the earthen vessel in which the glass ampulla is placed, and on each side of the centre hole are two small round air-holes. The lower part of the furnace, in order to hold the burning charcoal, has iron plates at the height of a palm, likewise supported by iron rods. In the middle of the front there is the mouth, made for the purpose of putting the fire into the furnace ; this mouth is half a foot high and wide, and rounded at the top, and under it is the draught opening. Into the earthen vessel set over the hole is placed clean sand a digit deep, and in it the glass ampulla is set as deeply as it is smeared with lute. The lower quarter is smeared eight or ten times with nearly liquid lute, each time to the thickness of a blade, and each time it is dried again, until it has become as thick as the thumb ; this kind of lute is well beaten with an iron rod, and is thoroughly mixed with hair or cotton thread, or with wool and salt, that it should not crackle. The many things of which the compounds are made must not fill the ampulla completely, lest when boiling they rise into the operculum. The operculum is likewise made of glass, and is closely joined to the ampulla with linen, cemented with wheat flour and white of egg moistened with water, and then lute free from salt is spread over that part of it. In a similar way the spout of the operculum is joined by linen covered with lute to another glass ampulla which receives the distilled *aqua*. A kind of thin iron nail or small wooden peg, a little thicker than a needle, is fixed in this joint, in order that when air seems necessary to the artificer distilling by this process he can pull it out ; this is necessary when too much of the vapour has been driven into the upper part. The four air-holes which, as I have said, are on the top of the furnace beside the large hole on which the ampulla is placed, are likewise covered with lute.

[6]This list of four recipes is even more obscure than the previous list. If they were distilled, the first and second mixtures would not produce nitric acid, although possibly some sulphuric would result. The third might yield nitric, and the fourth *aqua regia*. In view of the water, they were certainly not used as cements, and the first and second are deficient in the vital ingredients.

[7]*Distillation*, at least in crude form, is very old. Aristotle (*Meteorologica*, IV.) states that sweet water can be made by evaporating salt-water and condensing the steam. Dioscorides and Pliny both describe the production of mercury by distillation (note 58, p. 432). The Alchemists of the Alexandrian School, from the 1st to the 6th Centuries, mention forms of imperfect apparatus—an ample discussion of which may be found in Kopp, *Beiträge zur Geschichte der Chemie*, Braunschweig, 1869, p. 217).

A—Furnace. B—Its round hole. C—Air-holes. D—Mouth of the furnace. E—Draught opening under it. F—Earthenware crucible. G—Ampulla. H—Operculum. I—Its spout. K—Other ampulla. L—Basket in which this is usually placed lest it should be broken.

All this preparation having been accomplished in order, and the ingredients placed in the ampulla, they are gradually heated over burning charcoal until they begin to exhale vapour and the ampulla is seen to trickle with moisture. But when this, on account of the rising of the vapour, turns red, and the *aqua* distils through the spout of the operculum, then one must work with the utmost care, lest the drops should fall at a quicker rate than one for every five movements of the clock or the striking of its bell, and not slower than one for every ten; for if it falls faster the glasses will be broken, and if it drops more slowly the work begun cannot be completed within the definite time, that is within the space of twenty-four hours. To prevent the first accident, part of the coals are extracted by means of an iron implement similar to pincers; and in order to prevent the second happening, small dry pieces of oak are placed upon the coals, and the substances in the ampulla are heated with a sharper fire, and the air-holes on the furnace are re-opened if need arise. As soon as the drops are being distilled, the glass ampulla which receives them is covered with a piece of linen moistened with water, in order that the powerful vapour which arises may be repelled. When the ingredients have been heated and the ampulla in which they were placed is whitened with moisture, it is heated by a fiercer fire until all the drops have been distilled[8]. After the furnace has cooled, the *aqua* is filtered and poured into a small glass ampulla, and into the same is put half a *drachma* of silver[9], which when dissolved makes the turbid *aqua* clear. This is poured into the ampulla containing all the rest of the *aqua*, and as soon as the lees have sunk to the bottom the *aqua* is poured off, removed, and reserved for use.

Gold is parted from silver by the following method[10]. The alloy, with lead added to it, is first heated in a cupel until all the lead is exhaled, and eight

[8]It is desirable to note the contents of the residues in the retort, for it is our belief that these are the materials to which the author refers as "lees of the water which separates gold from silver," in many places in Book VII. They would be strange mixtures of sodium, potassium, aluminium sulphates, with silica, brickdust, asbestos, and various proportions of undigested vitriol, salt, saltpetre, alum, iron oxides, etc. Their effect must have been uncertain. Many old German metallurgies also refer to the *Todenkopf der Scheidwasser*, among them the *Probierbüchlein* before Agricola, and after him Lazarus Ercker (*Beschreibung Allerfürnemsten*, etc., Prague, 1574). See also note 16, p. 234.

[9]This use of silver could apply to one purpose only, that is, the elimination of minor amounts of hydrochloric from the nitric acid, the former originating no doubt from the use of salt among the ingredients. The silver was thus converted into a chloride and precipitated. This use of a small amount of silver to purify the nitric acid was made by metallurgists down to fairly recent times. Biringuccio (IV, 2) and Lazarus Ercker (p. 71) both recommend that the silver be dissolved first in a small amount of acid, and the solution poured into the newly-manufactured supply. They both recommend preserving this precipitate and its cupellation after melting with lead—which Agricola apparently overlooked.

[10]In this description of parting by nitric acid, the author digresses from his main theme on pages 444 and 445, to explain a method apparently for small quantities where the silver was precipitated by copper, and to describe another cryptic method of precipitation. These subjects are referred to in notes 11 and 12 below. The method of parting set out here falls into six stages : *a*-cupellation, *b*-granulation, *c*-solution in acid, *d*-treatment of the gold residues, *e*-evaporation of the solution, *f*-reduction of the silver nitrate. For nitric acid parting, bullion must be free from impurities, which cupellation would ensure ; if copper were left in, it would have the effect he mentions if we understand "the silver "separated from the gold soon unites with it again," to mean that the silver unites with the copper, for the copper would go into solution and come down with the silver on evaporation. Agricola does not specifically mention the necessity of an excess of silver in this description, although he does so elsewhere, and states that the ratio must be at least three parts silver to one part gold. The first description of the solution of the silver is clear enough, but that on p. 445 is somewhat difficult to follow, for the author states that the bullion is placed in a retort with the acid, and that distillation is carried on between each additional charge of acid. So far as the arrangement of a receiver might relate to the saving of any acid that came over accidentally in the boiling, it can be understood, but to distill off much acid would soon result in the crystallization of the silver nitrate, which would greatly impede the action of subsequent acid additions, and finally the gold could not be separated from such nitrate in the way described. The explanation may be (apart from incidental evaporation when heating) that the acids used were very weak, and that by the evaporation of a certain amount of water, not only was the acid concentrated, but room was provided for the further charges. The acid in the gold wash-water, mentioned in the following paragraph, was apparently thus concentrated. The "glass" mentioned as being melted with litharge, argols, nitre, etc., was no doubt the silver nitrate. The precipitation of the silver from the solution as a chloride, by the use of salt, so generally used during the 18th and 19th Centuries, was known in Agricola's time, although he does not mention it. It is mentioned in Geber and the *Probierbüchlein.* The clarity of the latter on the subject is of some interest (p. 34a) : "How to pulverise silver "and again make it into silver. Take the silver and dissolve it in water with the *starcken-"wasser, aqua fort,* and when that is done, take the silver water and pour it into warm salty "water, and immediately the silver settles to the bottom and becomes powder. Let it stand "awhile until it has well settled, then pour away the water from it and dry the settlings, "which will become a powder like ashes. Afterward one can again make it into silver. "Take the powder and put it on a *test*, and add thereto the powder from the settlings from "which the *aqua forte* has been made, and add lead. Then if there is a great deal, blow on

ounces of the alloy contain only five *drachmae* of copper or at most six, for if there is more copper in it, the silver separated from the gold soon unites with it again. Such molten silver containing gold is formed into granules, being stirred by means of a rod split at the lower end, or else is poured into an iron mould, and when cooled is made into thin leaves. As the process of making granules from argentiferous gold demands greater care and diligence than making them from any other metals, I will now explain the method briefly. The alloy is first placed in a crucible, which is then covered with a lid and placed in another earthen crucible containing a few ashes. Then they are placed in the furnace, and after they are surrounded by charcoal, the fire is blown by the blast of a bellows, and lest the charcoal fall away it is surrounded by stones or bricks. Soon afterward charcoal is thrown over the upper crucible and covered with live coals ; these again are covered with charcoal, so that the crucible is surrounded and covered on all sides with it. It is necessary to heat the crucibles with charcoal for the space of half an hour or a little longer, and to provide that there is no deficiency of charcoal, lest the alloy become chilled ; after this the air is blown in through the nozzle of the bellows, that the gold may begin to melt. Soon afterward it is turned round, and a test is quickly taken to see whether it be melted, and if it is melted, fluxes are thrown into it ; it is advisable to cover up the crucible again closely that the contents may not be exhaled. The contents are heated together for as long as it would take to walk fifteen paces, and then the crucible is seized with tongs and the gold is emptied into an oblong vessel containing very cold water, by pouring it slowly from a height so that the granules will not be too big ; in proportion as they are lighter, more fine and more irregular, the better they are, therefore the water is frequently stirred with a rod split into four parts from the lower end to the middle.

The leaves are cut into small pieces, and they or the silver granules are put into a glass ampulla, and the *aqua* is poured over them to a height of a digit above the silver. The ampulla is covered with a bladder or with waxed linen, lest the contents exhale. Then it is heated until the silver is dissolved, the indication of which is the bubbling of the *aqua*. The gold remains in the bottom, of a blackish colour, and the silver mixed with the *aqua* floats above. Some pour the latter into a copper bowl and pour into it cold water, which immediately congeals the silver ; this they take out and dry, having poured off the *aqua*[11]. They heat the dried silver in an earthenware crucible until it melts, and when it is melted they pour it into an iron mould.

The gold which remains in the ampulla they wash with warm water, filter, dry, and heat in a crucible with a little *chrysocolla* which is called borax, and when it is melted they likewise pour it into an iron mould.

" it until the lead has incorporated itself . . . blow it until it *plickt* (*blickens*). Then " you will have as much silver as before."

[11]The silver is apparently precipitated by the copper of the bowl. It would seem that this method was in considerable use for small amounts of silver nitrate in the 16th Century. Lazarus Ercker gives elaborate directions for this method (*Beschreibung Allerfürnemsten*, etc., Prague, 1574, p. 77).

Some workers, into an ampulla which contains gold and silver and the *aqua* which separates them, pour two or three times as much of this *aqua valens* warmed, and into the same ampulla or into a dish into which all is poured, throw fine leaves of black lead and copper ; by this means the gold adheres to the lead and the silver to the copper, and separately the lead from the gold, and separately the copper from the silver, are parted in a cupel. But no method is approved by us which loses the *aqua* used to part gold from silver, for it might be used again[12].

A glass ampulla, which bulges up inside at the bottom like a cone, is covered on the lower part of the outside with lute in the way explained above, and into it is put silver bullion weighing three and a half Roman *librae*. The *aqua* which parts the one from the other is poured into it, and the ampulla is placed in sand contained in an earthen vessel, or in a box, that it may be warmed with a gentle fire. Lest the *aqua* should be exhaled, the top of the ampulla is plastered on all sides with lute, and it is covered with a glass operculum, under whose spout is placed another ampulla which receives the distilled drops ; this receiver is likewise arranged in a box containing sand. When the contents are heated it reddens, but when the redness no longer appears to increase, it is taken out of the vessel or box and shaken ; by this motion the *aqua* becomes heated again and grows red ; if this is done two or three times before other *aqua* is added to it, the operation is sooner concluded, and much less *aqua* is consumed. When the first charge has all been distilled, as much silver as at first is again put into the ampulla, for if too much were put in at once, the gold would be parted from it with difficulty. Then the second *aqua* is poured in, but it is warmed in order that it and the ampulla may be of equal temperature, so that the latter may not be cracked by the cold ; also if a cold wind blows on it, it is apt to crack. Then the third *aqua* is poured in, and also if circumstances require it, the fourth, that is to say more *aqua* and again more is poured in until the gold assumes the colour of burned brick. The artificer keeps in hand two *aquae*, one of which is stronger than the other ; the stronger is used at first, then the less strong, then at the last again the stronger. When the gold becomes of a reddish yellow colour, spring water is poured in and heated until it boils. The gold is washed four times and then heated in the crucible until it melts. The water with which it was washed is put back, for there is a little silver in it ; for this reason it is poured into an ampulla and heated, and the drops first distilled are received by one ampulla, while those which come later, that is to say when the operculum begins to get red, fall into another. This latter *aqua* is useful for testing the gold, the former for washing it ; the former may also be poured over the ingredients from which the *aqua valens* is made.

The *aqua* that was first distilled, which contains the silver, is poured into an ampulla wide at the base, the top of which is also smeared with lute and covered by an operculum, and is then boiled as before in order that it may be separated from the silver. If there be so much *aqua* that (when boiled) it

[12]We confess to a lack of understanding of this operation with leaves of lead and copper.

A—Ampullae arranged in the vessels. B—An ampulla standing upright between iron rods. C—Ampullae placed in the sand which is contained in a box, the spouts of which reach from the opercula into ampullae placed under them. D—Ampullae likewise placed in sand which is contained in a box, of which the spout from the opercula extends crosswise into ampullae placed under them. E—Other ampullae receiving the distilled *aqua* and likewise arranged in sand contained in the lower boxes. F—Iron tripod, in which the ampulla is usually placed when there are not many particles of gold to be parted from the silver. G—Vessel.

rises into the operculum, there is put into the ampulla one lozenge or two; these are made of soap, cut into small pieces and mixed together with powdered argol, and then heated in a pot over a gentle fire; or else the contents are stirred with a hazel twig split at the bottom, and in both cases the *aqua* effervesces, and soon after again settles. When the powerful vapour appears, the *aqua* gives off a kind of oil, and the operculum becomes red. But, lest the vapours should escape from the ampulla and the operculum in that part where their mouths communicate, they are entirely sealed all round. The *aqua* is boiled continually over a fiercer fire, and enough charcoal must be put into the furnace so that the live coals touch the vessel. The ampulla is taken out as soon as all the *aqua* has been distilled, and the silver, which is dried by the heat of the fire, alone remains in it; the silver is shaken out and put in an earthenware crucible, and heated until it melts. The molten glass is extracted with an iron rod curved at the lower end, and the silver is made into cakes. The glass extracted from the crucible is ground to powder, and to this are added litharge, argol, glass-galls, and saltpetre, and they are melted in an earthen crucible. The button that settles is transferred to the cupel and re-melted.

If the silver was not sufficiently dried by the heat of the fire, that which is contained in the upper part of the ampulla will appear black; this when melted will be consumed. When the lute, which was smeared round the lower part of the ampulla, has been removed, it is placed in the crucible and is re-melted, until at last there is no more appearance of black[13].

If to the first *aqua* the other which contains silver is to be added, it must be poured in before the powerful vapours appear, and the *aqua* gives off the oily substance, and the operculum becomes red; for he who pours in the *aqua* after the vapour appears causes a loss, because the *aqua* generally spurts out and the glass breaks. If the ampulla breaks when the gold is being parted from the silver or the silver from the *aqua*, the *aqua* will be absorbed by the sand or the lute or the bricks, whereupon, without any delay, the red hot coals should be taken out of the furnace and the fire extinguished. The sand and bricks after being crushed should be thrown into a copper vessel, warm water should be poured over them, and they should be put aside for the space of twelve hours; afterward the water should be strained through a canvas, and the canvas, since it contains silver, should be dried by the heat of the sun or the fire, and then placed in an earthen crucible and heated until the silver melts, this being poured out into an iron mould. The strained water should be poured into an ampulla and separated from the silver, of which it contains a minute portion; the sand should be mixed with litharge, glass-galls, argol, saltpetre, and salt, and heated in an earthen crucible. The button which settles at the bottom should be transferred to a cupel, and should be re-melted, in order that the lead may be separated from the silver. The lute, with lead added, should be heated in an earthen crucible, then re-melted in a cupel.

We also separate silver from gold by the same method when we assay them. For this purpose the alloy is first rubbed against a touchstone, in order to learn what proportion of silver there is in it; then as much silver as is necessary is added to the argentiferous gold, in a *bes* of which there must be less than a *semi-uncia* or a *semi-uncia* and a *sicilicus*[14] of copper. After lead has been added, it is melted in a cupel until the lead and the copper have exhaled, then the alloy of gold with silver is flattened out, and little tubes are made of the leaves; these are put into a glass ampulla, and strong *aqua* is poured over them two or three times. The tubes after this are absolutely pure, with the exception of only a quarter of a *siliqua*, which is silver; for only this much silver remains in eight *unciae* of gold[15].

[13]We do not understand this "appearance of black." If the nitrate came into contact with organic matter it would, of course, turn black by reduction of the silver, and sunlight would have the same effect.

[14]This would be equal to from 62 to 94 parts of copper in 1,000.

[15]As 144 *siliquae* are 1 *uncia*, then $\frac{1}{4}$ *siliqua* in 8 *unciae* would equal one part silver in 4,608 parts gold, or about 999.8 fine.

As great expense is incurred in parting the metals by the methods that I have explained, as night vigils are necessary when *aqua valens* is made, and as generally much labour and great pains have to be expended on this matter, other methods for parting have been invented by clever men, which are less costly, less laborious, and in which there is less loss if through carelessness an error is made. There are three methods, the first performed with sulphur, the second with antimony, the third by means of some compound which consists of these or other ingredients.

In the first method,[16] the silver containing some gold is melted in a crucible and made into granules. For every *libra* of granules, there is taken a sixth of a *libra* and a *sicilicus* of sulphur (not exposed to the fire) ; this, when crushed, is sprinkled over the moistened granules, and then they are put into a new earthen pot of the capacity of four *sextarii*, or into several of them if there is an abundance of granules. The pot, having been filled, is covered with an earthen lid and smeared over, and placed within a circle of fire set one and a half feet distant from the pot on all sides, in order that the sulphur added to the silver should not be distilled when melted. The pot is opened,

[16]The object of this treatment with sulphur and copper is to separate a considerable portion of silver from low-grade bullion (*i.e.*, silver containing some gold), in preparation for final treatment of the richer gold-silver alloy with nitric acid. Silver sulphide is created by adding sulphur, and is drawn off in a silver-copper regulus. After the first sentence, the author uses silver alone where he obviously means silver " containing some gold," and further he speaks of the " gold lump " (*massula*) where he likewise means a button containing a great deal of silver. For clarity we introduced the term " regulus " for the Latin *mistura*. The operation falls into six stages : *a*, granulation ; *b*, sulphurization of the granulated bullion ; *c*, melting to form a combination of the silver sulphide with copper into a regulus, an alloy of gold and silver settling out ; *d*, repetition of the treatment to abstract further silver from the " lump ; " *e*, refining the " lump " with nitric acid ; *f*, recovery of the silver from the regulus by addition of lead, liquation and cupellation.

The use of a " circle of fire " secures a low temperature that would neither volatilize the sulphur nor melt the bullion. The amount of sulphur given is equal to a ratio of 48 parts bullion and 9 parts sulphur. We are not certain about the translation of the paragraph in relation to the proportion of copper added to the granulated bullion ; because in giving definite quantities of copper to be added in the contingencies of various original copper contents in the bullion, it would be expected that they were intended to produce some positive ratio of copper and silver. However, the ratio as we understand the text in various cases works out to irregular amounts, *i.e.*, 48 parts of silver to 16, 12.6, 24, 20.5, 20.8, 17.8, or 18 parts of copper. In order to obtain complete separation there should be sufficient sulphur to have formed a sulphide of the copper as well as of the silver, or else some of the copper and silver would come down metallic with the " lump ". The above ratio of copper added to the sulphurized silver, in the first instance would give about 18 parts of copper and 9 parts of sulphur to 48 parts of silver. The copper would require 4.5 parts of sulphur to convert it into sulphide, and the silver about 7 parts, or a total of 11.5 parts required against 9 parts furnished. It is plain, therefore, that insufficient sulphur is given. Further, the litharge would probably take up some sulphur and throw down metallic lead into the " lump ". However, it is necessary that there should be some free metallics to collect the gold, and, therefore, the separation could not be complete in one operation. In any event, on the above ratios the " gold lump " from the first operation was pretty coppery, and contained some lead and probably a good deal of silver, because the copper would tend to desulphurize the latter. The " powder " of glass-galls, salt, and litharge would render the mass more liquid and assist the " gold lump " to separate out.

The Roman silver *sesterce*, worth about 2⅛ pence or 4.2 American cents, was no doubt used by Agricola merely to indicate an infinitesimal quantity. The test to be applied to the regulus by way of cupellation and parting of a sample with nitric acid, requires no explanation. The truth of the description as to determining whether the gold had settled out, by using a chalked iron rod, can only be tested by actual experiment. It is probable, however, that the sulphur in the regulus would attack the iron and make it black. The re-melting of the regulus, if some gold remains in it, with copper and " powder " without more sulphur, would provide again free metallics to gather the remaining gold, and by desulphurizing some silver this button would probably not be very pure.

A—Pot. B—Circular fire. C—Crucibles. D—Their lids. E—Lid of the pot. F—Furnace. G—Iron rod.

the black-coloured granules are taken out, and afterward thirty-three *librae* of these granules are placed in an earthen crucible, if it has such capacity. For every *libra* of silver granules, weighed before they were sprinkled with

From the necessity for some free metallics besides the gold in the first treatment, it will be seen that a repetition of the sulphur addition and re-melting is essential gradually to enrich the "lump". Why more copper is added is not clear. In the second melting, the ratio is 48 parts of the "gold lump", 12 parts of sulphur and 12 parts copper. In this case the added copper would require about 3 parts sulphur, and if we consider the deficiency of sulphur in the first operations pertained entirely to the copper, then about 2.5 parts would be required to make good the shortage, or in other words the second addition of sulphur is sufficient. In the final parting of the "lump" it will be noticed that the author states that the silver ratio must be arranged as three of silver to one of gold. As to the recovery of the silver from the regulus, he states that 66 *librae* of silver give 132 *librae* of *regulus*. To this, 500 *librae* of lead are added, and it is melted in the "second" furnace, and the litharge and hearth-lead made are re-melted in the "first" furnace, the cakes made being again treated in the "third" furnace to separate the copper and lead. The "first" is usually the blast furnace, the "second" furnace is the cupellation furnace, and the "third" the liquation furnace. It is difficult to understand this procedure. The charge sent to the cupellation furnace would contain between 3% and 5% copper, and between 3% and 5% sulphur. However, possibly the sulphur and copper could be largely abstracted in the skimmings from the cupellation furnace, these being subsequently liquated in the "third" furnace. It may be noted that two whole lines from this paragraph are omitted in the editions of *De Re Metallica* after 1600. For historical note on sulphur separation see page 461.

sulphur, there is weighed out also a sixth of a *libra* and a *sicilicus* of copper, if each *libra* consists either of three-quarters of a *libra* of silver and a quarter of a *libra* of copper, or of three-quarters of a *libra* and a *semi-uncia* of silver and a sixth of a *libra* and a *semi-uncia* of copper. If, however, the silver contains five-sixths of a *libra* of silver and a sixth of a *libra* of copper, or five-sixths of a *libra* and a *semi-uncia* of silver and an *uncia* and a half of copper, then there are weighed out a quarter of a *libra* of copper granules. If a *libra* contains eleven-twelfths of a *libra* of silver and one *uncia* of copper, or eleven-twelfths and a *semi-uncia* of silver and a *semi-uncia* of copper, then are weighed out a quarter of a *libra* and a *semi-uncia* and a *sicilicus* of copper granules. Lastly, if there is only pure silver, then as much as a third of a *libra* and a *semi-uncia* of copper granules are added. Half of these copper granules are added soon afterward to the black-coloured silver granules. The crucible should be tightly covered and smeared over with lute, and placed in a furnace, into which the air is drawn through the draught-holes. As soon as the silver is melted, the crucible is opened, and there is placed in it a heaped ladleful more of granulated copper, and also a heaped ladleful of a powder which consists of equal parts of litharge, of granulated lead, of salt, and of glass-galls ; then the crucible is again covered with the lid. When the copper granules are melted, more are put in, together with the powder, until all have been put in.

A little of the regulus is taken from the crucible, but not from the gold lump which has settled at the bottom, and a *drachma* of it is put into each of the cupels, which contain an *uncia* of molten lead ; there should be many of these cupels. In this way half a *drachma* of silver is made. As soon as the lead and copper have been separated from the silver, a third of it is thrown into a glass ampulla, and *aqua valens* is poured over it. By this method is shown whether the sulphur has parted all the gold from the silver, or not. If one wishes to know the size of the gold lump which has settled at the bottom of the crucible, an iron rod moistened with water is covered with chalk, and when the rod is dry it is pushed down straight into the crucible, and the rod remains bright to the height of the gold lump ; the remaining part of the rod is coloured black by the regulus, which adheres to the rod if it is not quickly removed.

If when the rod has been extracted the gold is observed to be satisfactorily parted from the silver, the regulus is poured out, the gold button is taken out of the crucible, and in some clean place the regulus is chipped off from it, although it usually flies apart. The lump itself is reduced to granules, and for every *libra* of this gold they weigh out a quarter of a *libra* each of crushed sulphur and of granular copper, and all are placed together in an earthen crucible, not into a pot. When they are melted, in order that the gold may more quickly settle at the bottom, the powder which I have mentioned is added.

Although minute particles of gold appear to scintillate in the regulus of copper and silver, yet if all that are in a *libra* do not weigh as much as a single sesterce, then the sulphur has satisfactorily parted the gold from the silver; but if it should weigh a sesterce or more, then the regulus is thrown back again into the earthen crucible, and it is not advantageous to add sulphur, but only a little copper and powder, by which method a gold lump is again made to settle at the bottom; and this one is added to the other button which is not rich in gold.

When gold is parted from sixty-six *librae* of silver, the silver, copper, and sulphur regulus weighs one hundred and thirty-two *librae*. To separate the copper from the silver we require five hundred *librae* of lead, more or less, with which the regulus is melted in the second furnace. In this manner litharge and hearth-lead are made, which are re-smelted in the first furnace. The cakes that are made from these are placed in the third furnace, so that the lead may be separated from the copper and used again, for it contains very little silver. The crucibles and their covers are crushed, washed, and the sediment is melted together with litharge and hearth-lead.

Those who wish to separate all the silver from the gold by this method leave one part of gold to three of silver, and then reduce the alloy to granules. Then they place it in an ampulla, and by pouring *aqua valens* over it, part the gold from the silver, which process I explained in Book VII.

If sulphur from the lye with which *sal artificiosus* is made, is strong enough to float an egg thrown into it, and is boiled until it no longer emits fumes, and melts when placed upon glowing coals, then, if such sulphur is thrown into the melted silver, it parts the gold from it.

Silver is also parted from gold by means of *stibium*[17]. If in a *bes* of gold there are seven, or six, or five double *sextulae* of silver, then three parts of *stibium* are added to one part of gold; but in order that the *stibium* should not consume the gold, it is melted with copper in a red hot earthern crucible. If the gold contains some portion of copper, then to eight *unciae* of *stibium*

[17]There can be no doubt that in most instances Agricola's *stibium* is antimony sulphide, but it does not follow that it was the mineral *stibnite*, nor have we considered it desirable to introduce the precision of either of these modern terms, and have therefore retained the Latin term where the sulphide is apparently intended. The use of antimony sulphide to part silver from gold is based upon the greater affinity of silver than antimony for sulphur. Thus the silver, as in the last process, is converted into a sulphide, and is absorbed in the regulus, while the metallic antimony alloys with the gold and settles to the bottom of the pot. This process has several advantages over the sulphurization with crude sulphur; antimony is a more convenient vehicle of sulphur, for it saves the preliminary sulphurization with its attendant difficulties of volatilization of the sulphur; it also saves the granulation necessary in the former method; and the treatment of the subsequent products is simpler. However, it is possible that the sulphur-copper process was better adapted to bullion where the proportion of gold was low, because the fineness of the bullion mentioned in connection with the antimonial process was apparently much higher than the previous process. For instance, a *bes* of gold, containing 5, 6, or 7 double *sextulae* of silver would be .792, .750. or .708 fine. The antimonial method would have an advantage over nitric acid separation, in that high-grade bullion could be treated direct without artificial decrease of fineness required by inquartation to about .250 fine, with the consequent incidental losses of silver involved.

The process in this description falls into six operations: *a*, sulphurization of the silver by melting with antimony sulphide; *b*, separation of the gold "lump" (*massula*) by jogging; *c*, re-melting the regulus (*mistura*) three or four times for recovery of further "lumps"; *d*, re-melting of the "lump" four times, with further additions of antimony sulphide; *e*, cupellation of the regulus to recover the silver; *f*, cupellation of the antimony from the "lump" to recover the gold. Percy seems to think it difficult to understand the insistence upon the addition of copper. Biringuccio (IV, 6) states, among other things, that copper makes the ingredients more liquid. The later metallurgists, however, such as Ercker, Lohneys, and Schlüter, do not mention this addition; they do mention the "swelling and

a *sicilicus* of copper is added; and if it contains no copper, then half an *uncia*, because copper must be added to *stibium* in order to part gold from silver. The gold is first placed in a red hot earthen crucible, and when melted it swells, and a little *stibium* is added to it lest it run over; in a short space of time, when this has melted, it likewise again swells, and when this occurs it is advisable to put in all the remainder of the *stibium*, and to cover the crucible with a lid, and then to heat the mixture for the time required to walk thirty-five paces. Then it is at once poured out into an iron pot, wide at the top and narrow at the bottom, which was first heated and smeared over with tallow or wax, and set on an iron or wooden block. It is shaken violently, and by this agitation the gold lump settles to the bottom, and when the pot has cooled it is tapped loose, and is again melted four times in the same way. But each time a less weight of *stibium* is added to the gold, until finally only twice as much *stibium* is added as there is gold, or a little more; then the gold lump is melted in a cupel. The *stibium* is melted again three or four times in an earthen crucible, and each time a gold lump settles, so that there are three or four gold lumps, and these are all melted together in a cupel.

To two *librae* and a half of such *stibium* are added two *librae* of argol and one *libra* of glass-galls, and they are melted in an earthen crucible, where a lump likewise settles at the bottom; this lump is melted in the cupel. Finally, the *stibium* with a little lead added, is melted in the cupel, in which, after all the rest has been consumed by the fire, the silver alone remains. If the *stibium* is not first melted in an earthen crucible with argol and glass-galls, before it is melted in the cupel, part of the silver is consumed, and is absorbed by the ash and powder of which the cupel is made.

The crucible in which the gold and silver alloy are melted with *stibium*, and also the cupel, are placed in a furnace, which is usually of the kind

frothing," and recommend that the crucible should be only partly filled. As to the copper, we suggest that it would desulphurize part of the antimony and thus free some of that metal to collect the gold. If we assume bullion of the medium fineness mentioned and containing no copper, then the proportions in the first charge would be about 36 parts gold, 12 parts silver, 41 parts sulphur, 103 parts antimony, and 9 parts copper. The silver and copper would take up 4.25 parts of sulphur, and thus free about 10.6 parts of antimony as metallics. It would thus appear that the amount of metallics provided to assist the collection of the gold was little enough, and that the copper in freeing 5.6 parts of the antimony was useful. It appears to have been necessary to have a large excess of antimony sulphide; for even with the great surplus in the first charge, the reaction was only partial, as is indicated by the necessity for repeated melting with further antimony.

The later metallurgists all describe the separation of the metallic antimony from the gold as being carried out by oxidation of the antimony, induced by a jet of air into the crucible, this being continued until the mass appears limpid and no cloud forms in the surface in cooling. Agricola describes the separation of the silver from the regulus by preliminary melting with argols, glass-gall, and some lead, and subsequent cupellation of the lead-silver alloy. The statement that unless this preliminary melting is done, the cupel will absorb silver, might be consonant with an attempt at cupellation of sulphides, and it is difficult to see that much desulphurizing could take place with the above fluxes. In fact, in the later descriptions of the process, iron is used in this melting, and we are under the impression that Agricola had omitted this item for a desulphurizing reagent. At the Dresden Mint, in the methods described by Percy (Metallurgy Silver and Gold, p. 373) the gold lumps were tested for fineness, and from this the amount of gold retained in the regulus was computed. It is not clear from Agricola's account whether the test with nitric acid was applied to the regulus or to the "lumps". For historical notes see p. 461.

A—Furnace in which the air is drawn in through holes. B—Goldsmith's forge. C—Earthen crucibles. D—Iron pots. E—Block.

in which the air is drawn in through holes; or else they are placed in a goldsmith's forge.

Just as *aqua valens* poured over silver, from which the sulphur has parted the gold, shows us whether all has been separated or whether particles of gold remain in the silver; so do certain ingredients, if placed in the pot or crucible "alternately" with the gold, from which the silver has been parted by *stibium*, and heated, show us whether all have been separated or not.

We use cements[18] when, without *stibium*, we part silver or copper or both so ingeniously and admirably from gold. There are various cements. Some

[18]As will be shown in the historical note, this process of separating gold and silver is of great antiquity—in all probability the only process known prior to the Middle Ages, and in any event, the first one used. In general the process was performed by "cementing" the disintegrated bullion with a paste and subjecting the mass to long-continued heat at a temperature under the melting point of the bullion. The cement (*compositio*) is of two different species; in the first species saltpetre and vitriol and some aluminous or silicious medium are the essential ingredients, and through them the silver is converted into nitrate and absorbed by the mass; in the second species, common salt and the same sort of medium are the essentials, and in this case the silver is converted into a chloride. Agricola does not distinguish between these two species, for, as shown by the text, his ingredients are badly mixed.

consist of half a *libra* of brick dust, a quarter of a *libra* of salt, an *uncia* of saltpetre, half an *uncia* of sal-ammoniac, and half an *uncia* of rock salt. The bricks or tiles from which the dust is made must be composed of fatty clays, free from sand, grit, and small stones, and must be moderately burnt and very old.

Another cement is made of a *bes* of brick dust, a third of rock salt, an *uncia* of saltpetre, and half an *uncia* of refined salt. Another cement is made of a *bes* of brick dust, a quarter of refined salt, one and a half *unciae* of saltpetre, an *uncia* of sal-ammoniac, and half an *uncia* of rock salt. Another has one *libra* of brick dust, and half a *libra* of rock salt, to which some add a sixth of a *libra* and a *sicilicus* of vitriol. Another is made of half a *libra* of brick dust, a third of a *libra* of rock salt, an *uncia* and a half of vitriol, and one *uncia* of saltpetre. Another consists of a *bes* of brick dust, a third of refined salt, a sixth of white vitriol[19], half an *uncia* of verdigris, and likewise half an *uncia* of saltpetre. Another is made of one and a third *librae* of brick dust, a *bes* of rock salt, a sixth of a *libra* and half an *uncia* of sal-ammoniac, a sixth and half an *uncia* of vitriol, and a sixth of saltpetre. Another contains a *libra* of brick dust, a third of refined salt, and one and a half *unciae* of vitriol.

The process as here described falls into five operations : *a*, granulation of the bullion or preparation of leaves ; *b*, heating alternate layers of cement and bullion in pots ; *c*, washing the gold to free it of cement ; *d*, melting the gold with borax or soda ; *e*, treatment of the cement by way of melting with lead and cupellation to recover the silver. Investigation by Boussingalt (*Ann. De Chimie*, 1833, p. 253–6), D'Elhuyar (*Bergbaukunde*, Leipzig, 1790, Vol. II, p. 200), and Percy (Metallurgy of Silver and Gold, p. 395), of the action of common salt upon silver under cementation conditions, fairly well demonstrated the reactions involved in the use of this species of cement. Certain factors are essential besides salt : *a*, the admission of air, which is possible through the porous pots used ; *b*, the presence of some moisture to furnish hydrogen ; *c*, the addition of alumina or silica. The first would be provided by Agricola in the use of new pots, the second possibly by use of wood fuel in a closed furnace, the third by the inclusion of brickdust. The alumina or silica at high temperatures decomposes the salt, setting free hydrochloric acid and probably also free chlorine. The result of the addition of vitriol in Agricola's ingredients is not discussed by those investigators, but inasmuch as vitriol decomposes into sulphuric acid under high temperatures, this acid would react upon the salt to free hydrochloric acid, and thus assist to overcome deficiencies in the other factors. It is possible also that sulphuric acid under such conditions would react directly upon the silver to form silver sulphates, which would be absorbed into the cement. As nitric acid is formed by vitriol and saltpetre at high temperatures, the use of these two substances as a cementing compound would produce nitric acid, which would at once attack the silver to form silver nitrate, which would be absorbed into the melted cement. In this case the brickdust probably acted merely as a vehicle for the absorption, and to lower the melting point of the mass and prevent fusion of the metal. While nitric acid will only part gold and silver when the latter is in great excess, yet when applied as fumes under cementation conditions it appears to react upon a minor ratio of silver. While the reactions of the two above species of compounds can be accounted for in a general way, the problem furnished by Agricola's statements is by no means simple, for only two of his compounds are simply salt cements, the others being salt and nitre mixtures. An inspection of these compounds produces at once a sense of confusion. Salt is present in every compound, saltpetre in all but two, vitriol in all but three. Lewis (*Traité Singulier de Métallique*, Paris, 1743, II, pp. 48–60), in discussing these processes, states that salt and saltpetre must never be used together, as he asserts that in this case *aqua regia* would be formed and the gold dissolved. Agricola, however, apparently found no such difficulty. As to the other ingredients, apart from nitre, salt, vitriol, and brickdust, they can have been of no use. Agricola himself points out that ingredients of " metallic origin " corrupt the gold and that brickdust and common salt are sufficient. In a description of this process in the *Probierbüchlein* (p. 58), no nitre is mentioned. This booklet does mention the recovery of the silver from the cement by amalgamation with mercury—the earliest mention of silver amalgamation.

[19]While a substance which we now know to be natural zinc sulphate was known to Agricola (see note 11, p. 572), it is hardly possible that it is referred to here. If green vitriol be dehydrated and powdered, it is white.

Those ingredients above are peculiar to each cement, but what follows is common to all. Each of the ingredients is first separately crushed to powder ; the bricks are placed on a hard rock or marble, and crushed with an iron implement ; the other things are crushed in a mortar with a pestle ; each is separately passed through a sieve. Then they are all mixed together, and are moistened with vinegar in which a little sal-ammoniac has been dissolved, if the cement does not contain any. But some workers, however, prefer to moisten the gold granules or gold-leaf instead.

The cement should be placed, alternately with the gold, in new and clean pots in which no water has ever been poured. In the bottom the cement is levelled with an iron implement, and afterward the gold granules or leaves are placed one against the other, so that they may touch it on all sides ; then, again, a handful of the cement, or more if the pots are large, is thrown in and levelled with an iron implement ; the granules and leaves are laid over this in the same manner, and this is repeated until the pot is filled. Then it is covered with a lid, and the place where they join is smeared over with artificial lute, and when this is dry the pots are placed in the furnace.

The furnace has three chambers, the lowest of which is a foot high ; into this lowest chamber the air penetrates through an opening, and into it the

A—FURNACE. B—POT. C—LID. D—AIR-HOLES.

ashes fall from the burnt wood, which is supported by iron rods, arranged to form a grating. The middle chamber is two feet high, and the wood is pushed in through its mouth. The wood ought to be oak, holmoak, or turkey-oak, for from these the slow and lasting fire is made which is necessary for this operation. The upper chamber is open at the top so that the pots, for which it has the depth, may be put into it; the floor of this chamber consists of iron rods, so strong that they may bear the weight of the pots and the heat of the fire; they are sufficiently far apart that the fire may penetrate well and may heat the pots. The pots are narrow at the bottom, so that the fire entering into the space between them may heat them; at the top the pots are wide, so that they may touch and hold back the heat of the fire. The upper part of the furnace is closed in with bricks not very thick, or with tiles and lute, and two or three air-holes are left, through which the fumes and flames may escape.

The gold granules or leaves and the cement, alternately placed in the pots, are heated by a gentle fire, gradually increasing for twenty-four hours, if the furnace was heated for two hours before the full pots were stood in it, and if this was not done, then for twenty-six hours. The fire should be increased in such a manner that the pieces of gold and the cement, in which is the potency to separate the silver and copper from the gold, may not melt, for in this case the labour and cost will be spent in vain; therefore, it is ample to have the fire hot enough that the pots always remain red. After so many hours all the burning wood should be drawn out of the furnace. Then the refractory bricks or tiles are removed from the top of the furnace, and the glowing pots are taken out with the tongs. The lids are removed, and if there is time it is well to allow the gold to cool by itself, for then there is less loss; but if time cannot be spared for that operation, the pieces of gold are immediately placed separately into a wooden or bronze vessel of water and gradually quenched, lest the cement which absorbs the silver should exhale it. The pieces of gold, and the cement adhering to them, when cooled or quenched, are rolled with a little mallet so as to crush the lumps and free the gold from the cement. Then they are sifted by a fine sieve, which is placed over a bronze vessel; in this manner the cement containing the silver or the copper or both, falls from the sieve into the bronze vessel, and the gold granules or leaves remain on it. The gold is placed in a vessel and again rolled with the little mallet, so that it may be cleansed from the cement which absorbs silver and copper.

The particles of cement, which have dropped through the holes of the sieve into the bronze vessel, are washed in a bowl, over a wooden tub, being shaken about with the hands, so that the minute particles of gold which have fallen through the sieve may be separated. These are again washed in a little vessel, with warm water, and scrubbed with a piece of wood or a twig broom, that the moistened cement may be detached. Afterward all the gold is again washed with warm water, and collected with a bristle brush, and should be washed in a copper full of holes, under which is placed a little vessel. Then it is necessary to put the gold on an iron plate, under which is a vessel, and to wash it with warm water. Finally, it is placed in a bowl, and, when dry, the granules or leaves are rubbed against a touchstone at the same time as a touch-needle, and considered carefully as to whether they be pure or alloyed. If they are not pure enough, the granules or the leaves, together with the cement which attracts silver and copper, are arranged alternately in layers in the same manner, and again heated; this is done as often as is necessary, but the last time it is heated as many hours as are required to cleanse the gold.

Some people add another cement to the granules or leaves. This cement lacks the ingredients of metalliferous origin, such as verdigris and vitriol, for if these are in the cement, the gold usually takes up a little of the base metal; or if it does not do this, it is stained by them. For this reason some very rightly never make use of cements containing these things, because brick dust and salt alone, especially rock salt, are able to extract all the silver and copper from the gold and to attract it to themselves.

It is not necessary for coiners to make absolutely pure gold, but to heat it only until such a fineness is obtained as is needed for the gold money which they are coining.

The gold is heated, and when it shows the necessary golden yellow colour and is wholly pure, it is melted and made into bars, in which case they are either prepared by the coiners with *chrysocolla*, which is called by the Moors borax, or are prepared with salt of lye made from the ashes of ivy or of other salty herbs.

The cement which has absorbed silver or copper, after water has been poured over it, is dried and crushed, and when mixed with hearth-lead and de-silverized lead, is smelted in the blast furnace. The alloy of silver and lead, or of silver and copper and lead, which flows out, is again melted in the cupellation furnace, in order that the lead and copper may be separated from the silver. The silver is finally thoroughly purified in the refining furnace, and in this practical manner there is no silver lost, or only a minute quantity.

There are besides this, certain other cements[20] which part gold from silver, composed of sulphur, *stibium* and other ingredients. One of these compounds consists of half an *uncia* of vitriol dried by the heat of the fire and reduced to powder, a sixth of refined salt, a third of *stibium*, half a *libra*

[20]The processes involved by these "other" compounds are difficult to understand, because of the lack of information given as to the method of operation. It might be thought that these were five additional recipes for cementing pastes, but an inspection of their internal composition soon dissipates any such assumption, because, apart from the lack of brickdust or some other similar necessary ingredient, they all contain more or less sulphur. After describing a preliminary treatment of the bullion by cupellation, the author says: "Then the silver is sprinkled with two *unciae* of that powdered compound and is "stirred. Afterward it is poured into another crucible . . . . and violently shaken. "The rest is performed according to the process I have already explained." As he has already explained four or five parting processes, it is not very clear to which one this refers. In fact, the whole of this discussion reads as if he were reporting hearsay, for it lacks in every respect the infinite detail of his usual descriptions. In any event, if the powder was introduced into the molten bullion, the effect would be to form some silver sulphides in a regulus of different composition depending upon the varied ingredients of different compounds. The enriched bullion was settled out in a "lump" and treated "as I have explained," which is not clear.

of prepared sulphur (not exposed to the fire), one *sicilicus* of glass, likewise one *sicilicus* of saltpetre, and a *drachma* of sal-ammoniac.[21] The sulphur is prepared as follows: it is first crushed to powder, then it is heated for six hours in sharp vinegar, and finally poured into a vessel and washed with warm water; then that which settles at the bottom of the vessel is dried. To refine the salt it is placed in river water and boiled, and again evaporated. The second compound contains one *libra* of sulphur (not exposed to fire) and two *librae* of refined salt. The third compound is made from one

[21]HISTORICAL NOTE ON PARTING GOLD AND SILVER. Although the earlier Classics contain innumerable references to refining gold and silver, there is little that is tangible in them, upon which to hinge the metallurgy of parting the precious metals. It appears to us, however, that some ability to part the metals is implied in the use of the touchstone, for we fail to see what use a knowledge of the ratio of gold and silver in bullion could have been without the power to separate them. The touchstone was known to the Greeks at least as early as the 5th Century B.C. (see note 37, p. 252), and a part of Theophrastus' statement (LXXVIII.) on this subject bears repetition in this connection: "The nature of the stone which tries gold "is also very wonderful, as it seems to have the same power as fire; which is also a test of "that metal. . . . The trial by fire is by the colour and the quantity lost by it, but "that of the stone is made only by rubbing," etc. This trial by fire certainly implies a parting of the metals. It has been argued from the common use of *electrum*—a gold-silver alloy—by the Ancients, that they did not know how to part the two metals or they would not have wasted gold in such a manner, but it seems to us that the very fact that *electrum* was a positive alloy (20% gold, 80% silver), and that it was deliberately made (Pliny XXXIII, 23) and held of value for its supposed superior brilliancy to silver and the belief that goblets made of it detected poison, is sufficient answer to this.

To arrive by a process of elimination, we may say that in the Middle Ages, between 1100 and 1500 A.D., there were known four methods of parting these metals: *a*, parting by solution in nitric acid; *b*, sulphurization of the silver in finely-divided bullion by heating it with sulphur, and the subsequent removal of the silver sulphide in a regulus by melting with copper, iron, or lead; *c*, melting with an excess of antimony sulphide, and the direct conversion of the silver to sulphide and its removal in a regulus; *d*, cementation of the finely-divided bullion with salt, and certain necessary collateral re-agents, and the separation of the silver by absorption into the cement as silver chloride. Inasmuch as it can be clearly established that mineral acids were unknown to the Ancients, we can eliminate that method. Further, we may say at once that there is not, so far as has yet been found, even a remote statement that could be applied to the sulphide processes. As to cementation with salt, however, we have some data at about the beginning of the Christian Era.

Before entering into a more detailed discussion of the history of various processes, it may be useful, in a word, to fix in the mind of the reader our view of the first authority on various processes, and his period.

(1) Separation by cementation with salt, Strabo (?) 63 B.C.–24 A.D.; Pliny 23–79 A.D.
(2) Separation by sulphur, Theophilus, 1150–1200 A.D.
(3) Separation by nitric acid, Geber, prior to 14th Century.
(4) Separation by antimony sulphide, Basil Valentine, end 14th Century, or *Probierbüchlein*, beginning 15th Century.
(5) Separation by antimony sulphide and copper, or sulphur and copper, *Probierbüchlein*, beginning 15th Century.
(6) Separation by cementation with saltpetre, Agricola, 1556.
(7) Separation by sulphur and iron, Schlüter, 1738.
(8) Separation by sulphuric acid, D'Arcet, 1802.
(9) Separation by chloride gas, Thompson, 1833.
(10) Separation electrolytically, latter part 19th Century.

PARTING BY CEMENTATION. The following passage from Strabo is of prime interest as the first definite statement on parting of any kind (III, 2, 8): "That when they have "melted the gold and purified it by means of a kind of aluminous earth, the residue left is "*electrum*. This, which contains a mixture of silver and gold, being again subjected to the "fire, the silver is separated and the gold left (pure); for this metal is easily dissipated and "fat, and on this account gold is most easily molten by straw, the flame of which is soft, and "bearing a similarity (to the gold) causes it easily to dissolve, whereas coal, besides wasting a "great deal, melts it too much, by reason of its vehemence, and carries it off (in vapour)." This statement has provoked the liveliest discussion, not only on account of the metallurgical

libra of sulphur (not exposed to the fire), half a *libra* of refined salt, a quarter of a *libra* of sal-ammoniac, and one *uncia* of red-lead. The fourth compound consists of one *libra* each of refined salt, sulphur (not exposed to the fire) and argol, and half a *libra* of *chrysocolla* which the Moors call borax. The fifth compound has equal proportions of sulphur (not exposed to the fire), sal-ammoniac, saltpetre, and verdigris.

The silver which contains some portion of gold is first melted with lead in an earthen crucible, and they are heated together until the silver exhales the lead. If there was a *libra* of silver, there must be six *drachmae* of lead. Then the silver is sprinkled with two *unciae* of that powdered com-

interest and obscurity, but also because of differences of view as to its translation; we have given that of Mr. H. C. Hamilton (London, 1903). A review of this discussion will be found in Percy's Metallurgy of Gold and Silver, p. 399. That it refers to cementation at all hangs by a slender thread, but it seems more nearly this than anything else.

Pliny (XXXIII, 25) is a little more ample : "(The gold) is heated with double its "weight of salt and thrice its weight of *misy*, and again with two portions of salt and one of a "stone which they call *schistos*. The *virus* is drawn out when these things are burnt together "in an earthen crucible, itself remaining pure and incorrupt, the remaining ash being "preserved in an earthen pot and mixed with water as a lotion for *lichen* (ring-worm) on the "face." Percy, (Metallurgy Silver and Gold, p. 398) rightly considers that this undoubtedly refers to the parting of silver and gold by cementation with common salt. Especially as Pliny further on states that with regard to *misy*, "In purifying gold they mix it with this "substance." There can be no doubt from the explanations of Pliny and Dioscorides that *misy* was an oxidized pyrite, mostly iron sulphate. Assuming the latter case, then all of the necessary elements of cementation, *i.e.*, vitriol, salt, and an aluminous or silicious element, are present.

The first entirely satisfactory evidence on parting is to be found in Theophilus (12th Century), and we quote the following from Hendrie's translation (p. 245) : "Of Heating the "Gold. Take gold, of whatsoever sort it may be, and beat it until thin leaves are made in "breadth three fingers, and as long as you can. Then cut out pieces that are equally long "and wide and join them together equally, and perforate through all with a fine cutting "iron. Afterwards take two earthen pots proved in the fire, of such size that the gold can "lie flat in them, and break a tile very small, or clay of the furnace burned and red, weigh "it, powdered, into two equal parts, and add to it a third part salt for the same weight; "which things being slightly sprinkled with urine, are mixed together so that they may not "adhere together, but are scarcely wetted, and put a little of it upon a pot about the "breadth of the gold, then a piece of the gold itself, and again the composition, and "again the gold, which in the digestion is thus always covered, that gold may not be in "contact with gold; and thus fill the pot to the top and cover it above with another pot, "which you carefully lute round with clay, mixed and beaten, and you place it over the fire, "that it may be dried. In the meantime compose a furnace from stones and clay, two feet "in height, and a foot and a half in breadth, wide at the bottom, but narrow at the top, "where there is an opening in the middle, in which project three long and hard stones, which "may be able to sustain the flame for a long time, upon which you place the pots with the "gold, and cover them with other tiles in abundance. Then supply fire and wood, and take "care that a copious fire is not wanting for the space of a day and night. In the morning "taking out the gold, again melt, beat and place it in the furnace as before. Again also, "after a day and night, take it away and mixing a little copper with it, melt it as before, and "replace it upon the furnace. And when you have taken it away a third time, wash and dry "it carefully, and so weighing it, see how much is wanting, then fold it up and keep it."

The next mention is by Geber, of whose date and authenticity there is great doubt, but, in any event, the work bearing his name is generally considered to be prior to the 14th, although he has been placed as early as the 8th Century. We quote from Russell's translation, pp. 17 and 224, which we have checked with the Latin edition of 1542 : "Sol, or gold, is beaten into thin plates "and with them and common salt very well prepared lay upon lay in a vessel of "calcination which set into the furnace and calcine well for three days until the whole is "subtily calcined. Then take it out, grind well and wash it with vinegar, and dry it in the "sun. Afterwards grind it well with half its weight of cleansed *sal-armoniac*; then set it "to be dissolved until the whole be dissolved into most clear water." Further on : "Now "we will declare the way of cementing. Seeing it is known to us that cement is very necessary "in the examen of perfection, we say it is compounded of inflammable things. Of this

pound and is stirred; afterward it is poured into another crucible, first warmed and lined with tallow, and then violently shaken. The rest is performed according to the process I have already explained.

Gold may be parted without injury from silver goblets and from other gilt vessels and articles[22], by means of a powder, which consists of one part of sal-ammoniac and half a part of sulphur. The gilt goblet or other article is smeared with oil, and the powder is dusted on; the article is seized in the hand, or with tongs, and is carried to the fire and sharply tapped, and by this means the gold falls into water in vessels placed underneath, while the goblet remains uninjured.

"kind are, all blackening, flying, penetrating, and burned things; as is vitriol, *sal-armoniac*, "*flos aeris* (copper oxide scales) and the ancient *fictile* stone (earthen pots), and a very small "quantity, or nothing, of sulphur, and urine with like acute and penetrating things. All "these are impasted with urine and spread upon thin plates of that body which you intend "shall be examined by this way of probation. Then the said plates must be laid upon a "grate of iron included in an earthen vessel, yet so as one touch not the other that the virtue "of the fire may have free and equal access to them. Thus the whole must be kept in fire "in a strong earthen vessel for the space of three days. But here great caution is required "that the plates may be kept but not melt."

Albertus Magnus (1205–1280) *De Mineralibus et Rebus Metallicis*, Lib. IV, describes the process as follows:—"But when gold is to be purified an earthen vessel is made like a "cucurbit or dish, and upon it is placed a similar vessel; and they are luted together with "the tenacious lute called by alchemists the lute of wisdom. In the upper vessel there are "numerous holes by which vapour and smoke may escape; afterwards the gold in the form "of short thin leaves is arranged in the vessel, the leaves being covered consecutively with a "mixture obtained by mixing together soot, salt, and brick dust; and the whole is strongly "heated until the gold becomes perfectly pure and the base substances with which it was "mixed are consumed." It will be noted that salt is the basis of all these cement compounds. We may also add that those of Biringuccio and all other writers prior to Agricola were of the same kind, our author being the first to mention those with nitre.

PARTING WITH NITRIC ACID. The first mention of nitric acid is in connection with this purpose, and, therefore, the early history of this reagent becomes the history of the process. Mineral acids of any kind were unknown to the Greeks or Romans. The works of the Alchemists and others from the 12th to the 15th Centuries, have been well searched by chemical historians for indications of knowledge of the mineral acids, and many of such suspected indications are of very doubtful order. In any event, study of the Alchemists for the roots of chemistry is fraught with the greatest difficulty, for not only is there the large ratio of fraud which characterised their operations, but there is even the much larger field of fraud which characterised the authorship and dates of writing attributed to various members of the cult. The mention of saltpetre by Roger Bacon (1214—94), and Albertus Magnus (1205–80), have caused some strain to read a knowledge of mineral acids into their works, but with doubtful result. Further, the Monk Theophilus (1150–1200) is supposed to have mentioned products which would be mineral acids, but by the most careful scrutiny of that work we have found nothing to justify such an assertion, and it is of importance to note that as Theophilus was a most accomplished gold and silver worker, his failure to mention it is at least evidence that the process was not generally known. The transcribed manuscripts and later editions of such authors are often altered to bring them "up-to-date." The first mention is in the work attributed to Geber, as stated above, of date prior to the 14th Century. The following passage from his *De Inventione Veritatis* (Nuremberg edition, 1545, p. 182) is of interest:—"First take one *librá* of vitriol of Cyprus and one-half *libra* "of saltpetre and one-quarter of alum of Jameni, extract the *aqua* with the redness of the "alembic—for it is very solvative—and use as in the foregoing chapters. This can be made "acute if in it you dissolve a quarter of sal-ammoniac, which dissolves gold, sulphur, and "silver." Distilling vitriol, saltpetre and alum would produce nitric acid. The addition of sal-ammoniac would make *aqua regia;* Geber used this solvent water—probably without being made "more acute"—to dissolve silver, and he crystallized out silver nitrate. It

[22]There were three methods of gilding practised in the Middle Ages—the first by hammering on gold leaf; the second by laying a thin plate of gold on a thicker plate of silver, expanding both together, and fabricating the articles out of the sheets thus prepared; and the third by coating over the article with gold amalgam, and subsequently driving off the mercury by heat. Copper and iron objects were silver-plated by immersing them in molten silver after coating with sal-ammoniac or borax. Tinning was done in the same way.

Gold is also parted from silver on gilt articles by means of quicksilver. This is poured into an earthen crucible, and so warmed by the fire that the finger can bear the heat when dipped into it ; the silver-gilt objects are placed in it, and when the quicksilver adheres to them they are taken out and placed on a dish, into which, when cooled, the gold falls, together with the quicksilver. Again and frequently the same silver-gilt object is placed in heated quicksilver, and the same process is continued until at last no more gold is visible on the object ; then the object is placed in the fire, and the quicksilver which adheres to it is exhaled. Then the artificer takes a hare's foot, and brushes up into a dish the quicksilver and the gold which have

would not be surprising to find all the Alchemists subsequent to Geber mentioning acids. It will thus be seen that even the approximate time at which the mineral acids were first made cannot be determined, but it was sometime previous to the 15th Century, probably not earlier than the 12th Century. Beckmann (Hist. of Inventions II, p. 508) states that it appears to have been an old tradition that acid for separating the precious metals was first used at Venice by some Germans ; that they chiefly separated the gold from Spanish silver and by this means acquired great riches. Beckmann considers that the first specific description of the process seems to be in the work of William Budaeus (*De Asse*, 1516, III, p. 101), who speaks of it as new at this time. He describes the operation of one, Le Conte, at Paris, who also acquired a fortune through the method. Beckmann and others have, however, entirely overlooked the early *Probierbüchlein*. If our conclusions are correct that the first of these began to appear at about 1510, then they give the first description of inquartation. This book (see appendix) is made up of recipes, like a cook-book, and four or five different recipes are given for this purpose ; of these we give one, which sufficiently indicates a knowledge of the art (p. 39) : " If you would part them do it this way : " Beat the silver which you suppose to contain gold, as thin as possible ; cut it in small " pieces and place it in 'strong' water (*starkwasser*). Put it on a mild fire till it becomes " warm and throws up blisters or bubbles. Then take it and pour off the water into a copper- " bowl ; let it stand and cool. Then the silver settles itself round the copper bowl ; let the " silver dry in the copper bowl, then pour the water off and melt the silver in a crucible. " Then take the gold also out of the glass *kolken* and melt it together." Biringuccio (1540, Book VI.) describes the method, but with much less detail than Agricola. He made his acid from alum and saltpetre and calls it *lacque forti.*

Parting with Sulphur. This process first appears in Theophilus (1150-1200), and in form is somewhat different from that mentioned by Agricola. We quote from Hendrie's Translation, p. 317, " How gold is separated from silver. When you have scraped the gold " from silver, place this scraping in a small cup in which gold or silver is accustomed to be " melted, and press a small linen cloth upon it, that nothing may by chance be abstracted " from it by the wind of the bellows, and placing it before the furnace, melt it ; and directly " lay fragments of sulphur in it, according to the quantity of the scraping, and carefully " stir it with a thin piece of charcoal until its fumes cease ; and immediately pour it into " an iron mould. Then gently beat it upon the anvil lest by chance some of that black may " fly from it which the sulphur has burnt, because it is itself silver. For the sulphur con- " sumes nothing of the gold, but the silver only, which it thus separates from the gold, and " which you will carefully keep. Again melt this gold in the same small cup as before, and " add sulphur. This being stirred and poured out, break what has become black and keep " it, and do thus until the gold appear pure. Then gather together all that black, which you " have carefully kept, upon the cup made from the bone and ash, and add lead, and so burn it " that you may recover the silver. But if you wish to keep it for the service of niello, before " you burn it add to it copper and lead, according to the measure mentioned above, and " mix with sulphur." This process appears in the *Probierbüchlein* in many forms, different recipes containing other ingredients besides sulphur, such as salt, saltpetre, sal-ammoniac, and other things more or less effective. In fact, a series of hybrid methods between absolute melting with sulphur and cementation with salt, were in use, much like those mentioned by Agricola on p. 458.

Parting with Antimony Sulphide. The first mention of this process lies either in Basil Valentine's " Triumphant Chariot of Antimony " or in the first *Probierbüchlein*. The date to be assigned to the former is a matter of great doubt. It was probably written about the end of the 15th Century, but apparently published considerably later. The date of the *Probierbüchlein* we have referred to above. The statement in the " Triumphal Chariot " is as follows (Waite's Translation, p. 117-118) : " The elixir prepared in this way has the " same power of penetrating and pervading the body with its purifying properties that " antimony has of penetrating and purifying gold. . . . This much, however, I have " proved beyond a possibility of doubt, that antimony not only purifies gold and frees it

fallen together from the silver article, and puts them into a cloth made of woven cotton or into a soft leather; the quicksilver is squeezed through one or the other into another dish.[23] The gold remains in the cloth or the leather, and when collected is placed in a piece of charcoal hollowed out, and is heated until it melts, and a little button is made from it. This button is heated with a little *stibium* in an earthen crucible and poured out into another little vessel, by which method the gold settles at the bottom, and the *stibium* is seen to be on the top; then the work is completed. Finally, the gold button is put in a hollowed-out brick and placed in the fire, and by this method the gold is made pure. By means of the above methods gold is parted from silver and also silver from gold.

Now I will explain the methods used to separate copper from gold[24].

"from foreign matter, but it also ameliorates all other metals, but it does the same for animal "bodies." There are most specific descriptions of this process in the other works attributed to Valentine, but their authenticity is so very doubtful that we do not quote. The *Probierbüchlein* gives several recipes for this process, all to the same metallurgical effect, of which we quote two: "How to separate silver from gold. Take 1 part of golden silver, 1 "part of *spiesglass*, 1 part copper, 1 part lead; melt them together in a crucible. "When melted pour into the crucible pounded sulphur and directly you have poured it in "cover it up with soft lime so that the fumes cannot escape, and let it get cold and you will "find your gold in a button. Put that same in a pot and blow on it." "How to part gold "and silver by melting or fire. Take as much gold-silver as you please and granulate it; "take 1 *mark* of these grains, 1 *mark* of powder; put them together in a crucible. Cover it "with a small cover, put it in the fire, and let it slowly heat; blow on it gently until it melts; "stir it all well together with a stick, pour it out into a mould, strike the mould gently with "a knife so that the button may settle better, let it cool, then turn the mould over, strike off "the button and twice as much *spiesglas* as the button weighs, put them in a "crucible, blow on it till it melts, then pour it again into a mould and break away the button "as at first. If you want the gold to be good always add to the button twice as much "*spiesglass*. It is usually good gold in three meltings. Afterward take the button, place "it on a cupel, blow on it till it melts. And if it should happen that the gold is covered "with a membrane, then add a very little lead, then it shines (*plickt*) and becomes "clearer." Biringuccio (1540) also gives a fairly clear exposition of this method. All the old refiners varied the process by using mixtures of salt, antimony sulphide, and sulphur, in different proportions, with and without lead or copper; the net effect was the same. Later than Agricola these methods of parting bullion by converting the silver into a sulphide and carrying it off in a regulus took other forms. For instance, Schlüter (*Hütte-Werken*, Braunschweig, 1738) describes a method by which, after the granulated bullion had been sulphurized by cementation with sulphur in pots, it was melted with metallic iron. Lampadius (*Grundriss Einer Allgemeinen Hüttenkunde*, Göttingen, 1827) describes a treatment of the bullion, sulphurized as above, with litharge, thus creating a lead-silver regulus and a lead-silver-gold bullion which had to be repeatedly put through the same cycle. The principal object of these processes was to reduce silver bullion running low in gold to a ratio acceptable for nitric acid treatment.

Before closing the note on the separation of gold and silver, we may add that with regard to the three processes largely used to-day, the separation by solution of the silver from the bullion by concentrated sulphuric acid where silver sulphate is formed, was first described by D'Arcet, Paris, in 1802; the separation by introducing chlorine gas into the molten bullion and thus forming silver chlorides was first described by Lewis Thompson in a communication to the Society of Arts, 1833, and was first applied on a large scale by F. B. Miller at the Sydney Mint in 1867–70; we do not propose to enter into the discussion as to who is the inventor of electrolytic separation.

[23]See note 12, p. 297, for complete discussion of amalgamation.

[24]These nine methods of separating gold from copper are based fundamentally upon the sulphur introduced in each case, whereby the copper is converted into sulphides and separated off as a matte. The various methods are much befogged by the introduction of extraneous ingredients, some of which serve as fluxes, while others would provide metallics in the shape of lead or antimony for collection of the gold, but others would be of no effect, except to increase the matte or slag. Inspection will show that the amount of sulphur introduced in many instances is in so large ratio that unless a good deal of volatilization took place there would be insufficient metallics to collect the gold, if it happened to be in small quantities. In a general way the auriferous button is gradually impoverished in copper

The salt which we call *sal-artificiosus*,[25] is made from a *libra* each of vitriol, alum, saltpetre, and sulphur not exposed to the fire, and half a *libra* of sal-ammoniac; these ingredients when crushed are heated with one part of lye made from the ashes used by wool dyers, one part of unslaked lime, and four parts of beech ashes. The ingredients are boiled in the lye until the whole has been dissolved. Then it is immediately dried and kept in a hot place, lest it turn into oil; and afterward when crushed, a *libra* of lead-ash is mixed with it. With each *libra* of this powdered compound one and a half *unciae* of the copper is gradually sprinkled into a hot crucible, and it is stirred rapidly and frequently with an iron rod. When the crucible has cooled and been broken up, the button of gold is found.

The second method for parting is the following. Two *librae* of sulphur not exposed to the fire, and four *librae* of refined salt are crushed and mixed; a sixth of a *libra* and half an *uncia* of this powder is added to a *bes* of granules made of lead, and twice as much copper containing gold; they are heated together in an earthen crucible until they melt. When cooled, the button is taken out and purged of slag. From this button they again make granules, to a third of a *libra* of which is added half a *libra* of that powder of which I have spoken, and they are placed in alternate layers in the crucible; it is well to cover the crucible and to seal it up, and afterward it is heated over a gentle fire until the granules melt. Soon afterward, the crucible is taken off the fire, and when it is cool the button is extracted. From this, when purified and again melted down, the third granules are made, to which, if they weigh a sixth of a *libra*, is added one half an *uncia* and a *sicilicus* of the powder, and they are heated in the same manner, and the button of gold settles at the bottom of the crucible.

The third method is as follows. From time to time small pieces of sulphur, enveloped in or mixed with wax, are dropped into six *librae* of the molten copper, and consumed; the sulphur weighs half an *uncia* and a *sicilicus*. Then one and a half *sicilici* of powdered saltpetre are dropped into the same copper and likewise consumed; then again half an *uncia* and a *sicilicus* of sulphur enveloped in wax; afterward one and a half *sicilici* of lead-ash enveloped in wax, or of minium made from red-lead. Then immediately the copper is taken out, and to the gold button, which is now mixed with only a little copper, they add *stibium* to double the amount of the button; these are heated together until the *stibium* is driven off; then the button, together with lead of half the weight of the button, are heated in a cupel.

until it is fit for cupellation with lead, except in one case where the final stage is accomplished by amalgamation. The lore of the old refiners was much after the order of that of modern cooks—they treasured and handed down various efficacious recipes, and of those given here most can be found in identical terms in the *Probierbüchlein*, some editions of which, as mentioned before, were possibly fifty years before *De Re Metallica*. This knowledge, no doubt, accumulated over long experience; but, so far as we are aware, there is no description of sulphurizing copper for this purpose prior to the publication mentioned.

[25] *Sal artificiosus.* The compound given under this name is of quite different ingredients from the stock fluxes given in Book VII under the same term. The method of preparation, no doubt, dehydrated this one; it would, however, be quite effective for its purpose of sulphurizing the copper. There is a compound given in the *Probierbüchlein* identical with this, and it was probably Agricola's source of information.

Finally, the gold is taken out of this and quenched, and if there is a blackish colour settled in it, it is melted with a little of the *chrysocolla* which the Moors call borax; if too pale, it is melted with *stibium*, and acquires its own golden-yellow colour. There are some who take out the molten copper with an iron ladle and pour it into another crucible, whose aperture is sealed up with lute, and they place it over glowing charcoal, and when they have thrown in the powders of which I have spoken, they stir the whole mass rapidly with an iron rod, and thus separate the gold from the copper; the former settles at the bottom of the crucible, the latter floats on the top. Then the aperture of the crucible is opened with the red-hot tongs, and the copper runs out. The gold which remains is re-heated with *stibium*, and when this is exhaled the gold is heated for the third time in a cupel with a fourth part of lead, and then quenched.

The fourth method is to melt one and a third *librae* of the copper with a sixth of a *libra* of lead, and to pour it into another crucible smeared on the inside with tallow or gypsum; and to this is added a powder consisting of half an *uncia* each of prepared sulphur, verdigris, and saltpetre, and an *uncia* and a half of *sal coctus*. The fifth method consists of placing in a crucible one *libra* of the copper and two *librae* of granulated lead, with one and a half *unciae* of *sal-artificiosus*; they are at first heated over a gentle fire and then over a fiercer one. The sixth method consists in heating together a *bes* of the copper and one-sixth of a *libra* each of sulphur, salt, and *stibium*. The seventh method consists of heating together a *bes* of the copper and one-sixth each of iron scales and filings, salt, *stibium*, and glass-galls. The eighth method consists of heating together one *libra* of the copper, one and a half *librae* of sulphur, half a *libra* of verdigris, and a *libra* of refined salt. The ninth method consists of placing in one *libra* of the molten copper as much pounded sulphur, not exposed to the fire, and of stirring it rapidly with an iron rod; the lump is ground to powder, into which quicksilver is poured, and this attracts to itself the gold.

Gilded copper articles are moistened with water and placed on the fire, and when they are glowing they are quenched with cold water, and the gold is scraped off with a brass rod. By these practical methods gold is separated from copper.

Either copper or lead is separated from silver by the methods which I will now explain.[26] This is carried on in a building near by the works, or in the works in which the gold or silver ores or alloys are smelted. The middle wall of such a building is twenty-one feet long and fifteen feet high, and from this a front wall is distant fifteen feet toward the river; the rear wall

[26]Throughout the book the cupellation furnace is styled the *secunda fornax* (Glossary, *Treibeherd*). Except in one or two cases, where there is some doubt as to whether the author may not refer to the second variety of blast furnace, we have used "cupellation furnace." Agricola's description of the actual operation of the old German cupellation is less detailed than that of such authors as Schlüter (*Hütte-Werken*, Braunschweig, 1738) or Winkler (*Beschreibung der Freyberger Schmelz Huttenprozesse*, Freyberg, 1837). The operation falls into four periods. In the first period, or a short time after melting, the first scum—the *abzug*—arises. This material contains most of the copper, iron, zinc, or sulphur impurities in the lead. In the second period, at a higher temperature, and with the blast turned on, a second scum

is nineteen feet distant, and both these walls are thirty-six feet long and fourteen feet high ; a transverse wall extends from the end of the front wall to the end of the rear wall ; then fifteen feet back a second transverse wall is built out from the front wall to the end of the middle wall. In that space which is between those two transverse walls are set up the stamps, by means of which the ores and the necessary ingredients for smelting are broken up. From the further end of the front wall, a third transverse wall leads to the other end of the middle wall, and from the same to the end of the rear wall. The space between the second and third transverse walls, and between the rear and middle long walls, contains the cupellation furnace, in which lead

arises—the *abstrich*. This material contains most of the antimony and arsenical impurities. In the third stage the litharge comes over. At the end of this stage the silver brightens—"*blicken*"—due to insufficient litharge to cover the entire surface. Winkler gives the following average proportion of the various products from a charge of 100 *centners* :—

Abzug	2	*centners*, containing	64% lead
Abstrich	5½	,, ,,	73% ,,
Herdtplei	21½	,, ,,	60% ,,
Impure litharge ..	18	,, ,,	85% ,,
Litharge	66	,, ,,	89% ,,
Total	113 *centners*		

He estimates the lead loss at from 8% to 15%, and gives the average silver contents of *blicksilber* as about 90%. Many analyses of the various products may be found in Percy (Metallurgy of Lead, pp. 198–201), Schnabel and Lewis (Metallurgy, Vol. I, p. 581) ; but as they must vary with every charge, a repetition of them here is of little purpose.

Historical Note on Cupellation. The cupellation process is of great antiquity, and the separation of silver from lead in this manner very probably antedates the separation of gold and silver. We can be certain that the process has been used continuously for at least 2,300 years, and was only supplanted in part by Pattinson's crystallization process in 1833, and further invaded by Parks' zinc method in 1850, and during the last fifteen years further supplanted in some works by electrolytic methods. However, it yet survives as an important process. It seems to us that there is no explanation possible of the recovery of the large amounts of silver possessed from the earliest times, without assuming reduction of that metal with lead, and this necessitates cupellation. If this be the case, then cupellation was practised in 2500 B.C. The subject has been further discussed on p. 389. The first direct evidence of the process, however, is from the remains at Mt. Laurion (note 6, p. 27), where the period of greatest activity was at 500 B.C., and it was probably in use long before that time. Of literary evidences, there are the many metaphorical references to "fining silver" and "separating dross" in the Bible, such as Job (XXVIII, 1), Psalms (XII, 6, LXVI, 10), Proverbs (XVII, 3). The most certain, however, is Jeremiah (VI, 28–30) : "They are all brass [*sic*] and iron ; they "are corrupters. The bellows are burned, the lead is consumed in the fire, the founder "melteth in vain ; for the wicked are not plucked away. Reprobate silver shall men call "them." Jeremiah lived about 600 B.C. His contemporary Ezekiel (XXII, 18) also makes remark : "All they are brass and tin and iron and lead in the midst of the furnace ; "they are even the dross of the silver." Among Greek authors Theognis (6th century B.C.) and Hippocrates (5th century B.C.) are often cited as mentioning the refining of gold with lead, but we do not believe their statements will stand this construction without strain. Aristotle (Problems XXIV, 9) makes the following remark, which has been construed not only as cupellation, but also as the refining of silver in "tests." "What is the reason that boiling "water does not leap out of the vessel . . . . silver also does this when it is purified. "Hence those whose office it is in the silversmiths' shops to purify silver, derive gain by "appropriation to themselves of the sweepings of silver which leap out of the melting-pot."

The quotation of Diodorus Siculus from Agatharcides (2nd century B.C.) on gold refining with lead and salt in Egypt we give in note 8, p. 279. The methods quoted by Strabo (63 B.C.–24 A.D.) from Polybius (204–125 B.C.) for treating silver, which appear to involve cupellation, are given in note 8, p. 281. It is not, however, until the beginning of the Christian era that we get definite literary information, especially with regard to litharge, in Dioscorides and Pliny. The former describes many substances under the terms *scoria, molybdaena, scoria argyros* and *lithargyros*, which are all varieties of litharge. Under the latter term he says (V, 62) : "One kind is produced from a lead sand (concentrates ?), which has been heated in the furnaces "until completely fused ; another (is made) out of silver ; another from lead. The best is is separated from gold or silver. The vertical wall of its chimney is erected upon the middle wall, and the sloping chimney-wall rests on the beams which extend from the second transverse wall to the third; these are so located that they are at a distance of thirteen feet from the middle long wall and four from the rear wall, and they are two feet wide and thick. From the ground up to the roof-beams is twelve feet, and lest the sloping chimney-wall should fall down, it is partly supported by means of many iron rods, and partly by means of a few tie-beams covered with lute, which extend from the small beams of the sloping chimney-wall to the beams of the vertical chimney-wall. The rear roof is arranged in the same way as the roof

"from Attica, the second (best) from Spain; after that the kinds made in Puteoli, in Campania, "and at Baia in Sicily, for in these places it is mostly produced by burning lead plates. The "best of all is that which is a bright golden colour, called *chrysitis*, that from Sicily (is called) "*argyritis*, that made from silver is called *lauritis*." Pliny refers in several passages to litharge (*spuma argenti*) and to what is evidently cupellation, (XXXIII, 31): "And this the "same agency of fire separates part into lead, which floats on the silver like oil on water" (XXXIV, 47). "The metal which flows liquid at the first melting is called *stannum*, the second "melting is silver; that which remains in the furnace is *galena*, which is added to a third part "of the ore. This being again melted, produced lead with a deduction of two-ninths." Assuming *stannum* to be silver-lead alloy, and *galena* to be *molybdaena*, and therefore litharge, this becomes a fairly clear statement of cupellation (see note 23, p. 392). He further states (XXXIII, 35): "There is made in the same mines what is called *spuma argenti* (litharge). "There are three varieties of it; the best, known as *chrysitis*; the second best, which is called "*argyritis*; and a third kind, which is called *molybditis*. And generally all these colours "are to be found in the same tubes (see p. 480). The most approved kind is that of Attica; "the next, that which comes from Spain. *Chrysitis* is the product from the ore itself; "*argyritis* is made from the silver, and *molybditis* is the result of smelting of lead, which is "done at Puteoli, and from this has its name. All three are made as the material when "smelted flows from an upper crucible into a lower one. From this last it is raised with an "iron bar, and is then twirled round in the flames in order to make it less heavy (made in "tubes). Thus, as may be easily perceived from the name, it is in reality the *spuma* of a "boiling substance—of the future metal, in fact. It differs from slag in the same way that "the scum of a liquid differs from the lees, the one being purged from the material while "purifying itself, the other an excretion of the metal when purified."

The works of either Theophilus (1150–1200 A.D.) or Geber (prior to the 14th century) are the first where adequate description of the cupel itself can be found. The uncertainty of dates renders it difficult to say which is earliest. Theophilus (Hendrie's Trans., p. 317) says: "How gold is separated from copper: But if at any time you have broken copper "or silver-gilt vessels, or any other work, you can in this manner separate the gold. Take "the bones of whatever animal you please, which (bones) you may have found in the street, "and burn them, being cold, grind them finely, and mix with them a third part of beech-"wood ashes, and make cups as we have mentioned above in the purification of silver; you "will dry these at the fire or in the sun. Then you carefully scrape the gold from the copper, "and you will fold this scraping in lead beaten thin, and one of these cups being placed in "the embers before the furnace, and now become warm, you place in this fold of lead with the "scraping, and coals being heaped upon it you will blow it. And when it has become "melted, in the same manner as silver is accustomed to be purified, sometimes by removing "the embers and by adding lead, sometimes by re-cooking and warily blowing, you burn it "until, the copper being entirely absorbed, the gold may appear pure."

We quote Geber from the Nuremberg edition of 1545, p. 152: "Now we describe the "method of this. Take sifted ashes or *calx*, or the powder of the burned bones of animals, "or all of them mixed, or some of them; moisten with water, and press it with your hand to "make the mixture firm and solid, and in the middle of this bed make a round solid crucible "and sprinkle a quantity of crushed glass. Then permit it to dry. When it is dry, place "into the crucible that which we have mentioned which you intend to test. On it kindle "a strong fire, and blow upon the surface of the body that is being tested until it melts, which, "when melted, piece after piece of lead is thrown upon it, and blow over it a strong flame. "When you see it agitated and moved with strong shaking motion it is not pure. Then wait "until all of the lead is exhaled. If it vanishes and does not cease its motion it is not purified. "Then again throw lead and blow again until the lead separates. If it does not become quiet "again, throw in lead and blow on it until it is quiet and you see it bright and clear on the "surface."

Cupellation is mentioned by most of the alchemists, but as a metallurgical operation on a large scale the first description is by Biringuccio in 1540.

of the works in which ore is smelted. In the space between the middle and the front long walls and between the second[27] and the third transverse walls are the bellows, the machinery for depressing and the instrument for raising them. A drum on the axle of a water-wheel has rundles which turn the toothed drum of an axle, whose long cams depress the levers of the bellows, and also another toothed drum on an axle, whose cams raise the tappets of the stamps, but in the opposite direction. So that if the cams which depress the levers of the bellows turn from north to south, the cams of the stamps turn from south to north.

Lead is separated from gold or silver in a cupellation furnace, of which the structure consists of rectangular stones, of two interior walls of which the one intersects the other transversely, of a round sole, and of a dome. Its crucible is made from powder of earth and ash ; but I will first speak of the structure and also of the rectangular stones. A circular wall is built four feet and three palms high, and one foot thick ; from the height of two feet and three palms from the bottom, the upper part of the interior is cut away to the width of one palm, so that the stone sole may rest upon it. There are usually as many as fourteen stones ; on the outside they are a foot and a palm wide, and on the inside narrower, because the inner circle is much smaller than the outer ; if the stones are wider, fewer are required, if narrower more ; they are sunk into the earth to a depth of a foot and a palm. At the top each one is joined to the next by an iron staple, the points of which are embedded in holes, and into each hole is poured molten lead. This stone structure has six air-holes near the ground, at a height of a foot above the ground ; they are two feet and a palm from the bottom of the stones ; each of these air-holes is in two stones, and is two palms high, and a palm and three digits wide. One of them is on the right side, between the wall which protects the main wall from the fire, and the channel through which the litharge flows out of the furnace crucible ; the other five air-holes are distributed all round at equal distances apart ; through these escapes the moisture which the earth exhales when heated, and if it were not for these openings the crucible would absorb the moisture and be damaged. In such a case a lump would be raised, like that which a mole throws up from the earth, and the ash would float on the top, and the crucible would absorb the silver-lead alloy ; there are some who, because of this, make the rear part of the structure entirely open. The two inner walls, of which one intersects the other, are built of bricks, and are a brick in thickness. There are four air-holes in these, one in each part, which are about one digit's breadth higher and wider than the others. Into the four compartments is thrown a wheelbarrowful of slag, and over this is placed a large wicker basket full of charcoal dust. These walls extend a cubit above the ground, and on these, and on the ledge cut in the rectangular stones, is placed the stone sole ; this sole is a palm and three digits thick, and on all sides touches the rectangular stones ; if there are any cracks in it they are filled up with fragments of stone or brick. The front part of the sole is sloped so that a channel can be made, through which

[27]In Agricolas' text this is " first,"—obviously an error.

the litharge flows out. Copper plates are placed on this part of the sole-stone so that the silver-lead or other alloy may be more rapidly heated.

A dome which has the shape of half a sphere covers the crucible. It consists of iron bands and of bars and of a lid. There are three bands, each about a palm wide and a digit thick ; the lowest is at a distance of one foot from the middle one, and the middle one a distance of two feet from the upper one. Under them are eighteen iron bars fixed by iron rivets ; these bars are of the same width and thickness as the bands, and they are of such a length, that curving, they reach from the lower band to the upper, that is two feet and three palms long, while the dome is only one foot and three palms high. All the bars and bands of the dome have iron plates fastened on the underside with iron wire. In addition, the dome has four apertures ; the rear one, which is situated opposite the channel through which the litharge flows out, is two feet wide at the bottom ; toward the top, since it slopes gently, it is narrower, being a foot, three palms, and a digit wide ; there is no bar at this place, for the aperture extends from the upper band to the middle one, but not to the lower one. The second aperture is situated above the

A—Rectangular stones. B—Sole-stone. C—Air-holes. D—Internal walls. E—Dome. F—Crucible. G—Bands. H—Bars. I—Apertures in the dome. K—Lid of the dome. L—Rings. M—Pipes. N—Valves. O—Chains.

channel, is two and a half feet wide at the bottom, and two feet and a palm at the top; and there is likewise no bar at this point; indeed, not only does the bar not extend to the lower band, but the lower band itself does not extend over this part, in order that the master can draw the litharge out of the crucible. There are besides, in the wall which protects the principal wall against the heat, near where the nozzles of the bellows are situated, two apertures, three palms wide and about a foot high, in the middle of which two rods descend, fastened on the inside with plates. Near these apertures are placed the nozzles of the bellows, and through the apertures extend the pipes in which the nozzles of the bellows are set. These pipes are made of iron plates rolled up; they are two palms three digits long, and their inside diameter is three and a half digits; into these two pipes the nozzles of the bellows penetrate a distance of three digits from their valves. The lid of the dome consists of an iron band at the bottom, two digits wide, and of three curved iron bars, which extend from one point on the band to the point opposite; they cross each other at the top, where they are fixed by means of iron rivets. On the under side of the bars there are likewise plates fastened by rivets; each of the plates has small holes the size of a finger, so that the lute will adhere when the interior is lined. The dome has three iron rings engaged in wide holes in the heads of iron claves, which fasten the bars to the middle band at these points. Into these rings are fastened the hooks of the chains with which the dome is raised, when the master is preparing the crucible.

On the sole and the copper plates and the rock of the furnace, lute mixed with straw is placed to a depth of three digits, and it is pounded with a wooden rammer until it is compressed to a depth of one digit only. The rammer-head is round and three palms high, two palms wide at the bottom, and tapering upward; its handle is three feet long, and where it is set into the rammer-head it is bound around with an iron band. The top of the stonework in which the dome rests is also covered with lute, likewise mixed with straw, to the thickness of a palm. All this, as soon as it becomes loosened, must be repaired.

The artificer who undertakes the work of parting the metals, distributes the operation into two shifts of two days. On the one morning he sprinkles a little ash into the lute, and when he has poured some water over it he brushes it over with a broom. Then he throws in sifted ashes and dampens them with water, so that they could be moulded into balls like snow. The ashes are those from which lye has been made by letting water percolate through them, for other ashes which are fatty would have to be burnt again in order to make them less fat. When he has made the ashes smooth by pressing them with his hands, he makes the crucible slope down toward the middle; then he tamps it, as I have described, with a rammer. He afterward, with two small wooden rammers, one held in each hand, forms the channel through which the litharge flows out. The heads of these small rammers are each a palm wide, two digits thick, and one foot high; the handle of each is somewhat rounded, is a digit and a half less in

A—An artificer tamping the crucible with a rammer. B—Large rammer. C—Broom. D—Two smaller rammers. E—Curved iron plates. F—Part of a wooden strip. G—Sieve. H—Ashes. I—Iron shovel. K—Iron plate. L—Block of wood. M—Rock. N—Basket made of woven twigs. O—Hooked bar. P—Second hooked bar. Q—Old linen rag. R—Bucket. S—Doeskin. T—Bundles of straw. V—Wood. X—Cakes of lead alloy. Y—Fork. Z—Another workman covers the outside of the furnace with lute where the dome fits on it. AA—Basket full of ashes. BB—Lid of the dome. CC—The assistant standing on the steps pours charcoal into the crucible through the hole at the top of the dome. DD—Iron implement with which the lute is beaten. EE—Lute. FF—Ladle with which the workman or master takes a sample. GG—Rabble with which the scum of impure lead is drawn off. HH—Iron wedge with which the silver mass is raised.

diameter than the rammer-head, and is three feet in length; the rammer-head as well as the handle is made of one piece of wood. Then with shoes on, he descends into the crucible and stamps it in every direction with his feet, in which manner it is packed and made sloping. Then he again tamps it with a large rammer, and removing his shoe from his right foot he draws a circle around the crucible with it, and cuts out the circle thus drawn with an iron plate. This plate is curved at both ends, is three palms long, as many digits wide, and has wooden handles a palm and two digits long, and two digits thick; the iron plate is curved back at the top and ends, which penetrate into handles. There are some who use in the place of the plate a strip of wood, like the rim of a sieve; this is three digits wide, and is cut out at both ends that it may be held in the hands. Afterward he tamps the channel through which the litharge discharges. Lest the ashes should fall out, he blocks up the aperture with a stone shaped to fit it, against which he places a board, and lest this fall, he props it with a stick. Then he pours in a basketful of ashes and tamps them with the large rammer; then again and again he pours in ashes and tamps them with the rammer. When the channel has been made, he throws dry ashes all over the crucible with a sieve, and smooths and rubs it with his hands. Then he throws three basketsful of damp ashes on the margin all round the edge of the crucible, and lets down the dome. Soon after, climbing upon the crucible, he builds up ashes all around it, lest the molten alloy should flow out. Then, having raised the lid of the dome, he throws a basketful of charcoal into the crucible, together with an iron shovelful of glowing coals, and he also throws some of the latter through the apertures in the sides of the dome, and he spreads them with the same shovel. This work and labour is finished in the space of two hours.

An iron plate is set in the ground under the channel, and upon this is placed a wooden block, three feet and a palm long, a foot and two palms and as many digits wide at the back, and two palms and as many digits wide in front; on the block of wood is placed a stone, and over it an iron plate similar to the bottom one, and upon this he puts a basketful of charcoal, and also an iron shovelful of burning charcoals. The crucible is heated in an hour, and then, with the hooked bar with which the litharge is drawn off, he stirs the remainder of the charcoal about. This hook is a palm long and three digits wide, has the form of a double triangle, and has an iron handle four feet long, into which is set a wooden one six feet long. There are some who use instead a simple hooked bar. After about an hour's time, he stirs the charcoal again with the bar, and with the shovel throws into the crucible the burning charcoals lying in the channel; then again, after the space of an hour, he stirs the burning charcoals with the same bar. If he did not thus stir them about, some blackness would remain in the crucible and that part would be damaged, because it would not be sufficiently dried. Therefore the assistant stirs and turns the burning charcoal that it may be entirely burnt up, and so that the crucible may be well heated, which takes three hours; then the crucible is left quiet for the remaining two hours.

When the hour of eleven has struck, he sweeps up the charcoal ashes with a broom and throws them out of the crucible. Then he climbs on to the dome, and passing his hand in through its opening, and dipping an old linen rag in a bucket of water mixed with ashes, he moistens the whole of the crucible and sweeps it. In this way he uses two bucketsful of the mixture, each holding five Roman *sextarii*,[28] and he does this lest the crucible, when the metals are being parted, should break open; after this he rubs the crucible with a doe skin, and fills in the cracks. Then he places at the left side of the channel, two fragments of hearth-lead, laid one on the top of the other, so that when partly melted they remain fixed and form an obstacle, that the litharge will not be blown about by the wind from the bellows, but remain in its place. It is expedient, however, to use a brick in the place of the hearth-lead, for as this gets much hotter, therefore it causes the litharge to form more rapidly. The crucible in its middle part is made two palms and as many digits deeper.[29]

There are some who having thus prepared the crucible, smear it over with incense[30], ground to powder and dissolved in white of egg, soaking it up in a sponge and then squeezing it out again; there are others who smear over it a liquid consisting of white of egg and double the amount of bullock's blood or marrow. Some throw lime into the crucible through a sieve.

Afterward the master of the works weighs the lead with which the gold or silver or both are mixed, and he sometimes puts a hundred *centumpondia*[31] into the crucible, but frequently only sixty, or fifty, or much less. After it has been weighed, he strews about in the crucible three small bundles of straw, lest the lead by its weight should break the surface. Then he places in the channel several cakes of lead alloy, and through the aperture at the rear of the dome he places some along the sides; then, ascending to the opening at the top of the dome, he arranges in the crucible round about the dome the cakes which his assistant hands to him, and after ascending again and passing his hands through the same aperture, he likewise places other cakes inside the crucible. On the second day those which remain he, with an iron fork, places on the wood through the rear aperture of the dome.

When the cakes have been thus arranged through the hole at the top of the dome, he throws in charcoal with a basket woven of wooden twigs. Then he places the lid over the dome, and the assistant covers over the joints with lute. The master himself throws half a basketful of charcoal into the crucible through the aperture next to the nozzle pipe, and prepares the bellows, in order to be able to begin the second operation on the morning of the following day. It takes the space of one hour to carry out such a piece of work, and

[28]The Roman *sextarius* was about a pint.

[29]This sentence continues, *Ipsa vero media pars praeterea digito*, to which we are unable to attribute any meaning.

[30]*Thus*, or *tus*—"incense."

[31]One *centumpondium*, Roman, equals about 70.6 lbs. avoirdupois; one *centner*, old German, equals about 114.2 lbs. avoirdupois. Therefore, if German weights are meant, the maximum charge would be about 5.7 short tons; if Roman weights, about 3.5 short tons.

at twelve all is prepared. These hours all reckoned up make a sum of eight hours.

Now it is time that we should come to the second operation. In the morning the workman takes up two shovelsful of live charcoals and throws them into the crucible through the aperture next to the pipes of the nozzles; then through the same hole he lays upon them small pieces of fir-wood or of pitch pine, such as are generally used to cook fish. After this the water-gates are opened, in order that the machine may be turned which depresses the levers of the bellows. In the space of one hour the lead alloy is melted; and when this has been done, he places four sticks of wood, twelve feet long, through the hole in the back of the dome, and as many through the channel; these sticks, lest they should damage the crucible, are both weighted on the ends and supported by trestles; these trestles are made of a beam, three feet long, two palms and as many digits wide, two palms thick, and have two spreading legs at each end. Against the trestle, in front of the channel, there is placed an iron plate, lest the litharge, when it is extracted from the furnace, should splash the smelter's shoes and injure his feet and legs. With an iron shovel or a fork he places the remainder of the cakes through the aperture at the back of the dome on to the sticks of wood already mentioned.

The native silver, or silver glance, or grey silver, or ruby silver, or any other sort, when it has been flattened out[32], and cut up, and heated in an iron crucible, is poured into the molten lead mixed with silver, in order that impurities may be separated. As I have often said, this molten lead mixed with silver is called *stannum*[33].

When the long sticks of wood are burned up at the fore end, the master, with a hammer, drives into them pointed iron bars, four feet long and two digits wide at the front end, and beyond that one and a half digits wide

[32]See description, p. 269.

[33]*Stannum*, as a term for lead-silver alloys, is a term which Agricola (*De Natura Fossilium*, pp. 341–3) adopted from his views of Pliny. In the *Interpretatio* and the Glossary he gives the German equivalent as *werk*, which would sufficiently identify his meaning were it not obvious from the context. There can be little doubt that Pliny uses the term for lead alloys, but it had come into general use for tin before Agricola's time. The Roman term was *plumbum candidum*, and as a result of Agricola's insistence on using it and *stannum* in what he conceived was their original sense, he managed to give considerable confusion to mineralogic literature for a century or two. The passages from Pliny, upon which he bases his use, are (XXXIV, 47): "The metal which flows liquid at the first melting in the furnace is called *stannum*, " the second melting is silver," etc. (XXXIV, 48): "When copper vessels are coated with " *stannum* they produce a less disagreeable flavour, and it prevents verdigris. It is also " remarkable that the weight is not increased. . . . At the present day a counterfeit " *stannum* is made by adding one-third of white copper to tin. It is also made in another way, " by mixing together equal parts of tin and lead; this last is called by some *argentarium*. " . . . . There is also a composition called *tertiarium*, a mixture of two parts of lead and " one of tin. Its price is twenty *denarii* per pound, and it is used for soldering pipes. Persons " still more dishonest mix together equal parts of *tertiarium* and tin, and calling the compound " *argentarium*, when it is melted coat articles with it." Although this last passage probably indicates that *stannum* was a tin compound, yet it is not inconsistent with the view that the genuine *stannum* was silver-lead, and that the counterfeits were made as stated by Pliny. At what period the term *stannum* was adopted for tin is uncertain. As shown by Beckmann (Hist. of Inventions II, p. 225), it is used as early as the 6th century in occasions where tin was undoubtedly meant. We may point out that this term appears continuously in the official documents relating to Cornish tin mining, beginning with the report of William de Wrotham in 1198.

and thick; with these he pushes the sticks of wood forward and the bars then rest on the trestles. There are others who, when they separate metals, put two such sticks of wood into the crucible through the aperture which is between the bellows, as many through the holes at the back, and one through the channel; but in this case a larger number of long sticks of wood is necessary, that is, sixty; in the former case, forty long sticks of wood suffice to carry out the operation. When the lead has been heated for two hours, it is stirred with a hooked bar, that the heat may be increased.

If it be difficult to separate the lead from the silver, he throws copper and charcoal dust into the molten silver-lead alloy. If the alloy of argentiferous gold and lead, or the silver-lead alloy, contains impurities from the ore, then he throws in either equal portions of argol and Venetian glass or of sal-ammoniac, or of Venetian glass and of Venetian soap; or else unequal portions, that is, two of argol and one of iron rust; there are some who mix a little saltpetre with each compound. To one *centumpondium* of the alloy is added a *bes* or a *libra* and a third of the powder, according to whether it is more or less impure. The powder certainly separates the impurities from the alloy. Then, with a kind of rabble he draws out through

A—Furnace. B—Sticks of wood. C—Litharge. D—Plate. E—The foreman when hungry eats butter, that the poison which the crucible exhales may not harm him, for this is a special remedy against that poison.

the channel, mixed with charcoal, the scum, as one might say, of the lead; the lead makes this scum when it becomes hot, but that less of it may be made it must be stirred frequently with the bar.

Within the space of a quarter of an hour the crucible absorbs the lead; at the time when it penetrates into the crucible it leaps and bubbles. Then the master takes out a little lead with an iron ladle, which he assays, in order to find what proportion of silver there is in the whole of the alloy; the ladle is five digits wide, the iron part of its handle is three feet long and the wooden part the same. Afterward, when they are heated, he extracts with a bar the litharge which comes from the lead and the copper, if there be any of it in the alloy. Wherefore, it might more rightly be called *spuma* of lead than of silver[34]. There is no injury to the silver, when the lead and copper are separated from it. In truth the lead becomes much purer in the crucible of the other furnace, in which silver is refined. In ancient times, as the author Pliny[35] relates, there was under the channel of the crucible another crucible, and the litharge flowed down from the upper one into the lower one, out of which it was lifted up and rolled round with a stick in order that it might be of moderate weight. For which reason, they formerly made it into small tubes or pipes, but now, since it is not rolled round a stick, they make it into bars.

If there be any danger that the alloy might flow out with the litharge, the foreman keeps on hand a piece of lute, shaped like a cylinder and pointed at both ends; fastening this to a hooked bar he opposes it to the alloy so that it will not flow out.

Now when the colour begins to show in the silver, bright spots appear, some of them being almost white, and a moment afterward it becomes absolutely white. Then the assistant lets down the water-gates, so that, the race being closed, the water-wheel ceases to turn and the bellows are still. Then the master pours several buckets of water on to the silver to cool it; others pour beer over it to make it whiter, but this is of no importance since the silver has yet to be refined. Afterward, the cake of silver is raised with the pointed iron bar, which is three feet long and two digits wide, and has a wooden handle four feet long fixed in its socket. When the cake of silver has been taken from the crucible, it is laid upon a stone, and from part of it the hearth-lead, and from the other part the litharge, is chipped away with a hammer; then it is cleansed with a bundle of brass wire dipped in water. When the lead is separated from the silver, more silver is frequently found than when it was assayed; for instance, if before there were three *unciae* and as many *drachmae* in a *centumpondium*, they now sometimes find three *unciae* and a half[36]. Often the hearth-lead remaining in the crucible is a palm deep; it is taken out with the rest of the ashes and is sifted, and that which remains in the sieve, since it is hearth-lead, is added to the hearth-lead[37].

[34]The Latin term for litharge is *spuma argenti*, spume of silver.

[35]Pliny, XXXIII, 35. This quotation is given in full in the footnote p. 466. Agricola illustrates these "tubes" of litharge on p. 481.

[36]Assuming Roman weights, three *unciae* and three *drachmae* per *centumpondium* would be about 82 ozs., and the second case would equal about 85 ozs. per short ton.

[37]Agricola uses throughout *De Re Metallica* the term *molybdaena* for this substance.

A—Cake. B—Stone. C—Hammer. D—Brass wire. E—Bucket containing water. F—Furnace from which the cake has been taken, which is still smoking. G—Labourer carrying a cake out of the works.

The ashes which pass through the sieve are of the same use as they were at first, for, indeed, from these and pulverised bones they make the cupels. Finally, when much of it has accumulated, the yellow *pompholyx* adhering to the walls of the furnace, and likewise to those rings of the dome near the apertures, is cleared away.

I must also describe the crane with which the dome is raised. When it is made, there is first set up a rectangular upright post twelve feet long, each side of which measures a foot in width. Its lower pinion turns in a bronze socket set in an oak sill ; there are two sills placed crosswise so

It is obvious from the context that he means saturated furnace bottoms—the *herdpley* of the old German metallurgists—and, in fact, he himself gives this equivalent in the *Interpretatio*, and describes it in great detail in *De Natura Fossilium* (p. 353). The derivatives coined one time and another from the Greek *molybdos* for lead, and their applications, have resulted in a stream of wasted ink, to which we also must contribute. Agricola chose the word *molybdaena* in the sense here used from his interpretation of Pliny. The statements in Pliny are a hopeless confusion of *molybdaena* and *galena*. He says (XXXIII, 35) : " There are three varieties of " it (litharge)—the best-known is *chrysitis* ; the second best is called *argyritis* ; and " a third kind is called *molybditis*. . . . . . . *Molybditis* is the result of the smelting of " lead. . . . Some people make two kinds of litharge, which they call *scirerytis* and " *peumene* ; and a third variety being *molybdaena*, will be mentioned with lead." (XXXIV, 53) : " *Molybdaena*, which in another place I have called *galena*, is an ore of mixed silver

that the one fits in a mortise in the middle of the other, and the other likewise fits in the mortise of the first, thus making a kind of a cross; these sills are three feet long and one foot wide and thick. The crane-post is round at its upper end and is cut down to a depth of three palms, and turns in a band fastened at each end to a roof-beam, from which springs the inclined chimney wall. To the crane-post is affixed a frame, which is made in this way: first, at a height of a cubit from the bottom, is mortised into the crane-post a small cross-beam, a cubit and three digits long, except its tenons, and two palms in width and thickness. Then again, at a height of five feet above it, is another small cross-beam of equal length, width, and thickness, mortised into the crane-post. The other ends of these two small cross-beams are mortised into an upright timber, six feet three palms long, and three-quarters wide and thick; the mortise is transfixed by wooden pegs. Above, at a height of three palms from the lower small cross-beam, are two bars, one foot one palm long, not including the tenons, a palm three digits wide, and a palm thick, which are mortised in the other sides of the crane-post. In the same manner, under the upper small cross-beam are two bars of the same size. Also in the upright timber there are mortised the same number of bars, of the same length as the preceding, but three digits thick, a palm two digits wide, the two lower ones being above the lower small cross-beam. From the upright timber near the upper small cross-beam, which at its other end is mortised into the crane-post, are two mortised bars. On the outside of this frame, boards are fixed to the small cross-beams, but the front and back parts of the frame have doors, whose hinges are fastened to the boards which are fixed to the bars that are mortised to the sides of the crane-post.

Then boards are laid upon the lower small cross-beam, and at a height of two palms above these there is a small square iron axle, the sides of which are two digits wide; both ends of it are round and turn in bronze or iron bearings, one of these bearings being fastened in the crane-post, the other in the upright timber. About each end of the small axle is a wooden disc, of three palms and a digit radius and one palm thick, covered on the rim with an iron band; these two discs are distant two palms and as many digits from each

"and lead. It is considered better in quality the nearer it approaches to a golden colour "and the less lead there is in it; it is also friable and moderately heavy. When it is boiled "with oil it becomes liver-coloured, adheres to the gold and silver furnaces, and in this state "it is called *metallica*." From these two passages it would seem that *molybdaena*, a variety of litharge, might quite well be hearth-lead. Further (in XXXIV, 47), he says: "The metal "which flows liquid at the first melting in the furnace is called *stannum*, at the second melting is silver, that which remains in the furnace is *galena*." If we still maintain that *molybdaena* is hearth-lead, and *galena* is its equivalent, then this passage becomes clear enough, the second melting being cupellation. The difficulty with Pliny, however, arises from the passage (XXXIII, 31), where, speaking of silver ore, he says: "It is impossible to melt it except with lead ore, called *galena*, which is generally found next to silver veins." Agricola (*Bermannus*, p. 427, &c.), devotes a great deal of inconclusive discussion to an attempt to reconcile this conflict of Pliny, and also that of Dioscorides. The probable explanation of this conflict arises in the resemblance of cupellation furnace bottoms to lead carbonates, and the native *molybdaena* of Dioscorides; and some of those referred to by Pliny may be this sort of lead ores. In fact, in one or two places in Book IX, Agricola appears to use the term in this sense himself. After Agricola's time the term *molybdaenum* was applied to substances resembling lead, such as graphite, and what we now know as *molybdenite* (MoS_2). Some time in the latter part of the 18th century, an element being separated from the latter, it was dubbed *molybdenum*, and confusion was five times confounded.

other, and are joined with five rundles; these rundles are two and a half digits thick and are placed three digits apart. Thus a drum is made, which is a palm and a digit distant from the upright timber, but further from the crane-post, namely, a palm and three digits. At a height of a foot and a palm above this little axle is a second small square iron axle, the thickness of which is three digits ; this one, like the first one, turns in bronze or iron bearings. Around it is a toothed wheel, composed of two discs a foot three palms in diameter, a palm and two digits thick ; on the rim of this there are twenty-three teeth, a palm wide and two digits thick ; they protrude a palm from the wheel and are three digits apart. And around this same axle, at a distance of two palms and as many digits toward the upright timber, is another disc of the same diameter as the wheel and a palm thick ; this turns in a hollowed-out place in the upright timber. Between this disc and the disc of the toothed wheel another drum is made, having likewise five rundles. There is, in addition to this second axle, at a height of a cubit above it, a small wooden axle, the journals of which are of iron ; the ends are bound round with iron rings so that the journals may remain firmly fixed, and the journals, like the little iron axles, turn in bronze or iron bearings. This third axle is at a distance of about a cubit from the upper small cross-beam ; it has, near the upright timber, a toothed wheel two and a half feet in diameter, on the rim of which are twenty-seven teeth ; the other part of this axle, near the crane-post, is covered with iron plates, lest it should be worn away by the chain which winds around it. The end link of the chain is fixed in an iron pin driven into the little axle ; this chain passes out of the frame and turns over a little pulley set between the beams of the crane-arm.

Above the frame, at a height of a foot and a palm, is the crane-arm. This consists of two beams fifteen feet long, three palms wide, and two thick, mortised into the crane-post, and they protrude a cubit from the back of the crane-post and are fastened together. Moreover, they are fastened by means of a wooden pin which penetrates through them and the crane-post ; this pin has at the one end a broad head, and at the other a hole, through which is driven an iron bolt, so that the beams may be tightly bound into the crane-post. The beams of the crane-arm are supported and stayed by means of two oblique beams, six feet and two palms long, and likewise two palms wide and thick ; these are mortised into the crane-post at their lower ends, and their upper ends are mortised into the beams of the crane-arm at a point about four feet from the crane-post, and they are fastened with iron nails. At the back of the upper end of these oblique beams, toward the crane-post, is an iron staple, fastened into the lower sides of the beams of the crane-arm, in order that it may hold them fast and bind them. The outer end of each beam of the crane-arm is set in a rectangular iron plate, and between these are three rectangular iron plates, fixed in such a manner that the beams of the crane-arm can neither move away from, nor toward, each other. The upper sides of these crane-arm beams are covered with iron plates for a length of six feet, so that a trolley can move on it.

A—Crane-post. B—Socket. C—Oak cross-sills. D—Band. E—Roof-beam. F—Frame. G—Lower small cross-beam. H—Upright timber. I—Bars which come from the sides of the crane-post. K—Bars which come from the sides of the upright timber. L—Rundle drums. M—Toothed wheels. N—Chain. O—Pulley. P—Beams of the crane-arm. Q—Oblique beams supporting the beams of the crane-arm. R—Rectangular iron plates. S—Trolley. T—Dome of the furnace. V—Ring. X—Three chains. Y—Crank. Z—The crane-post of the other contrivance. AA—Crane-arm. BB—Oblique beam. CC—Ring of the crane-arm. DD—The second ring. EE—Lever-bar. FF—Third ring. GG—Hook. HH—Chain of the dome. II—Chain of the lever-bar

The body of the trolley is made of wood from the Ostrya or any other hard tree, and is a cubit long, a foot wide, and three palms thick; on both edges of it the lower side is cut out to a height and width of a palm, so that the remainder may move backward and forward between the two beams of the crane-arm; at the front, in the middle part, it is cut out to a width of two palms and as many digits, that a bronze pulley, around a small iron axle, may turn in it. Near the corners of the trolley are four holes, in which as many small wheels travel on the beams of the crane-arm. Since this trolley, when it travels backward and forward, gives out a sound somewhat similar to the barking of a dog, we have given it this name[38]. It is propelled forward by means of a crank, and is drawn back by means of a chain. There is an iron hook whose ring turns round an iron pin fastened to the right side of the trolley, which hook is held by a sort of clavis, which is fixed in the right beam of the crane-arm.

At the end of the crane-post is a bronze pulley, the iron axle of which is fastened in the beams of the crane-arm, and over which the chain passes as it comes from the frame, and then, penetrating through the hollow in the top of the trolley, it reaches to the little bronze pulley of the trolley, and passing over this it hangs down. A hook on its end engages a ring, in which are fixed the top links of three chains, each six feet long, which pass through the three iron rings fastened in the holes of the claves which are fixed into the middle iron band of the dome, of which I have spoken.

Therefore when the master wishes to lift the dome by means of the crane, the assistant fits over the lower small iron axle an iron crank, which projects from the upright beam a palm and two digits; the end of the little axle is rectangular, and one and a half digits wide and one digit thick; it is set into a similar rectangular hole in the crank, which is two digits long and a little more than a digit wide. The crank is semi-circular, and one foot three palms and two digits long, as many digits wide, and one digit thick. Its handle is straight and round, and three palms long, and one and a half digits thick. There is a hole in the end of the little axle, through which an iron pin is driven so that the crank may not come off. The crane having four drums, two of which are rundle-drums and two toothed-wheels, is more easily moved than another having two drums, one of which has rundles and the other teeth.

Many, however, use only a simple contrivance, the pivots of whose crane-post turn in the same manner, the one in an iron socket, the other in a ring. There is a crane-arm on the crane-post, which is supported by an oblique beam; to the head of the crane-arm a strong iron ring is fixed, which engages a second iron ring. In this iron ring a strong wooden lever-bar is fastened firmly, the head of which is bound by a third iron ring, from which hangs an iron hook, which engages the rings at the ends of the chains from the dome. At the other end of the lever-bar is another chain, which, when it is pulled down, raises the opposite end of the bar and thus the dome; and when it is relaxed the dome is lowered.

[38]Agricola here refers to the German word used in this connection, *i.e.*, *hundt*, a dog.

A—Chamber of the furnace. B—Its bed. C—Passages. D—Rammer. E—Mallet. F—Artificer making tubes from litharge according to the roman method. G—Channel. H—Litharge. I—Lower crucible or hearth. K—Stick. L—Tubes.

In certain places, as at Freiberg in Meissen, the upper part of the cupellation furnace is vaulted almost like an oven. This chamber is four feet high and has either two or three apertures, of which the first, in front, is one and a half feet high and a foot wide, and out of this flows the litharge ; the second aperture and likewise the third, if there be three, are at the sides, and are a foot and a half high and two and a half feet wide, in order that he who prepares the crucible may be able to creep into the furnace. Its circular bed is made of cement, it has two passages two feet high and one foot wide, for letting out the vapour, and these lead directly through from one side to the other, so that the one passage crosses the other at right angles, and thus four openings are to be seen ; these are covered at the top by rocks, wide, but only a palm thick. On these and on the other parts of the interior of the bed made of cement, is placed lute mixed with straw, to a depth of three digits, as it was placed over the sole and the plates of copper and the rocks of that other furnace. This, together with the ashes which are thrown in, the master or the assistant, who, upon his knees, prepares the crucible, tamps down with short wooden rammers and with mallets likewise made of wood.

A—Furnace similar to an oven. B—Passage C—Iron bars. D—Hole through which the litharge is drawn out. E—Crucible which lacks a dome. F—Thick sticks. G—Bellows

The cupellation furnace in Poland and Hungary is likewise vaulted at the top, and is almost similar to an oven, but in the lower part the bed is solid, and there is no opening for the vapours, while on one side of the crucible is a wall, between which and the bed of the crucible is a passage in place of the opening for vapours ; this passage is covered by iron bars or rods extending from the wall to the crucible, and placed a distance of two digits from each other. In the crucible, when it is prepared, they first scatter straw, and then they lay in it cakes of silver-lead alloy, and on the iron bars they lay wood, which when kindled heats the crucible. They melt cakes to the weight of sometimes eighty *centumpondia* and sometimes a hundred *centumpondia*[39]. They stimulate a mild fire by means of a blast from the bellows, and throw on to the bars as much wood as is required to make a flame which will reach into the crucible, and separate the lead from the silver. The litharge is drawn out on the other side through an aperture that is just wide enough for the master to creep through into the crucible. The Moravians and Carni, who very rarely make more than a *bes* or five-sixths of a *libra* of silver, separate the lead from it, neither in a furnace resembling an oven, nor in the crucible covered by a dome, but on a crucible which is without a cover and exposed to the wind ; on this crucible they lay cakes of silver-lead alloy, and over them they place dry wood, and over these again thick green wood. The wood having been kindled, they stimulate the fire by means of a bellows.

I have explained the method of separating lead from gold or silver. Now I will speak of the method of refining silver, for I have already explained the process for refining gold. Silver is refined in a refining furnace, over whose hearth is an arched chamber built of bricks ; this chamber in the front part is three feet high. The hearth itself is five feet long an four wide. The walls are unbroken along the sides and back, but in front one chamber is placed over the other, and above these and the wall is the upright chimney. The hearth has a round pit, a cubit wide and two palms deep, into which are thrown sifted ashes, and in this is placed a prepared earthenware " test," in such a manner that it is surrounded on all sides by ashes to a height equal to its own. The earthenware test is filled with a powder consisting of equal portions of bones ground to powder, and of ashes taken from the crucible in which lead is separated from gold or silver ; others mix crushed brick with the ashes, for by this method the powder attracts no silver to itself. When the powder has been made up and moistened with water, a little is thrown into the earthenware test and tamped with a wooden pestle. This pestle is round, a foot long, and a palm and a digit wide, out of which extend six teeth, each a digit thick, and a digit and a third long and wide, and almost a digit apart ; these six teeth form a circle, and in the centre of them is the seventh tooth, which is round and of the same length as the others, but a digit and a half thick ; this pestle tapers a little from the bottom up, that the upper part of the handle may be round and three digits thick. Some use a round pestle without teeth. Then a

[39]If Agricola means the German *centner*, this charge would be from about 4.6 to 5.7 short tons. If he is using Roman weights, it would be from about 3 to 3.7 short tons.

A—Pestle with teeth. B—Pestle without teeth. C—Dish or tray full of ashes. D—Prepared tests placed on boards or shelves. E—Empty tests. F—Wood. G—Saw.

little powder is again moistened, and thrown into the test, and tamped ; this work is repeated until the test is entirely full of the powder, which the master then cuts out with a knife, sharp on both sides, and turned upward at both ends so that the central part is a palm and a digit long ; therefore it is partly straight and partly curved. The blade is one and a half digits wide, and at each end it turns upward two palms, which ends to the depth of a palm are either not sharpened or they are enclosed in wooden handles. The master holds the knife with one hand and cuts out the powder from the test, so that it is left three digits thick all round ; then he sifts the powder of dried bones over it through a sieve, the bottom of which is made of closely-woven bristles. Afterward a ball made of very hard wood, six digits in diameter, is placed in the test and rolled about with both hands, in order to make the inside even and smooth ; for that matter he may move the ball about with only one hand. The tests[40] are of various capacities, for some of them when prepared

[40]The refining of silver in " tests " (Latin *testa*) is merely a second cupellation, with greater care and under stronger blast. Stirring the mass with an iron rod serves to raise the impurities which either volatilize as litharge or, floating to the edges, are absorbed into the " test." The capacity of the tests, from 15 *librae* to 50 *librae*, would be from about 155 to 515 ozs. Troy.

A—Straight knife having wooden handles. B—Curved knife likewise having wooden handles. C—Curved knife without wooden handles. D—Sieve. E—Balls. F—Iron door which the master lets down when he refines silver, lest the heat of the fire should injure his eyes. G—Iron implement on which the wood is placed when the liquid silver is to be refined. H—Its other part passing through the ring of another iron implement enclosed in the wall of the furnace. I—Tests in which burning charcoal has been thrown.

hold much less than fifteen *librae* of silver, others twenty, some thirty, others forty, and others fifty. All these tests thus prepared are dried in the sun, or set in a warm and covered place ; the more dry and old they are the better. All of them, when used for refining silver, are heated by means of burning charcoal placed in them. Others use instead of these tests an iron ring ; but the test is more useful, for if the powder deteriorates the silver remains in it, while there being no bottom to the ring, it falls out ; besides, it is easier to place in the hearth the test than the iron ring, and furthermore it requires much less powder. In order that the test should not break and damage the silver, some bind it round with an iron band.

In order that they may be more easily broken, the silver cakes are placed upon an iron grate by the refiner, and are heated by burning charcoal placed under them. He has a brass block two palms and two digits long and wide, with a channel in the middle, which he places upon a block of hard wood. Then with a double-headed hammer, he beats the hot cakes of silver placed on the brass block, and breaks them in pieces. The head of this hammer is a foot and two digits long, and a palm wide. Others use for this purpose merely a block of wood channelled in the top. While the fragments of the cake are still hot, he seizes them with the tongs and throws them into a bowl with holes in the bottom, and pours water over them. When the fragments are cooled, he puts them nicely into the test by placing them so that they stand upright and project from the test to a height of two palms, and lest one should fall against the other, he places little pieces of charcoal between them; then he places live charcoal in the test, and soon two twig basketsful of charcoal. Then he blows in air with the bellows. This bellows is double, and four feet two palms long, and two feet and as many palms wide at the back; the other parts are similar to those described in Book VII. The nozzle of the bellows is placed in a bronze pipe a foot long, the aperture in this pipe being a digit in diameter in front and quite round, and at the back two palms wide. The master, because he needs for the operation of refining

A—Grate. B—Brass block. C—Block of wood. D—Cakes of silver. E—Hammer. F—Block of wood channelled in the middle. G—Bowl full of holes. H—Block of wood fastened to an iron implement. I—Fir-wood. K—Iron bar. L—Implement with a hollow end. The implement which has a circular end is shown in the next picture. M—Implement, the extremity of which is bent upwards. N—Implement in the shape of tongs.

silver a fierce fire, and requires on that account a vigorous blast, places the bellows very much inclined, in order that, when the silver has melted, it may blow into the centre of the test. When the silver bubbles, he presses the nozzle down by means of a small block of wood moistened with water and fastened to an iron rod, the outer end of which bends upward. The silver melts when it has been heated in the test for about an hour; when it is melted, he removes the live coals from the test and places over it two billets of fir-wood, a foot and three palms long, a palm two digits wide, one palm thick at the upper part, and three digits at the lower. He joins them together at the lower edges, and into the billets he again throws the coals, for a fierce fire is always necessary in refining silver. It is refined in two or three hours, according to whether it was pure or impure, and if it is impure it is made purer by dropping granulated copper or lead into the test at the same time. In order that the refiner may sustain the great heat from the fire while the silver is being refined, he lets down an iron door, which is three feet long and a foot and three palms high; this door is held on both ends in iron plates, and when the operation is concluded, he raises it again with an iron shovel, so that its edge holds against the iron hook in the arch, and thus the door is held open. When the silver is nearly refined, which may be judged by the space of time, he dips into it an iron bar, three and a half feet long and a digit thick, having a round steel point. The small drops of silver that adhere to the bar he places on the brass block and flattens with a hammer, and from their colour he decides whether the silver is sufficiently refined or not. If it is thoroughly purified it is very white, and in a *bes* there is only a *drachma* of impurities. Some ladle up the silver with a hollow iron implement. Of each *bes* of silver one *sicilicus* is consumed, or occasionally when very impure, three *drachmae* or half an *uncia*[41].

The refiner governs the fire and stirs the molten silver with an iron implement, nine feet long, a digit thick, and at the end first curved toward the right, then curved back in order to form a circle, the interior of which is a palm in diameter; others use an iron implement, the end of which is bent directly upward. Another iron implement has the shape of tongs, with which, by compressing it with his hands, he seizes the coals and puts them on or takes them off; this is two feet long, one and a half digits wide, and the third of a digit thick.

When the silver is seen to be thoroughly refined, the artificer removes the coals from the test with a shovel. Soon afterward he draws water in a copper ladle, which has a wooden handle four feet long; it has a small hole at a point half-way between the middle of the bowl and the edge, through which a hemp seed just passes. He fills this ladle three times with water, and three times it all flows out through the hole on to the silver, and slowly quenches it; if he suddenly poured much water on it, it would burst asunder and injure those standing near. The artificer has a pointed iron bar, three

[41]A *drachma* of impurities in a *bes*, would be one part in 64, or 984.4 fine. A loss of a *sicilicus* of silver to the *bes*, would be one part in 32, or about 3.1%; three *drachmae* would equal 4.7%, and half an *uncia* 6.2%, or would indicate that the original bullion had a fineness in the various cases of about 950, 933, and 912.

A—Implement with a ring. B—Ladle. C—Its hole. D—Pointed bar. E—Forks. F—Cake of silver laid upon the implement shaped like tongs. G—Tub of water. H—Block of wood, with a cake laid upon it. I—Hammer. K—Silver again placed upon the implement resembling tongs. L—Another tub full of water. M—Brass wires. N—Tripod. O—Another block. P—Chisel. Q—Crucible of the furnace. R—Test still smoking.

feet long, which has a wooden handle as many feet long, and he puts the end of this bar into the test in order to stir it. He also stirs it with a hooked iron bar, of which the hook is two digits wide and a palm deep, and the iron part of its handle is three feet long and the wooden part the same. Then he removes the test from the hearth with a shovel or a fork, and turns it over, and by this means the silver falls to the ground in the shape of half a sphere; then lifting the cake with a shovel he throws it into a tub of water, where it gives out a great sound. Or else, having lifted the cake of silver with a fork, he lays it upon the iron implement similar to tongs, which are placed across a tub full of water; afterward, when cooled, he takes it from the tub again and lays it on the block made of hard wood and beats it with a hammer, in order to break off any of the powder from the test which adheres to it. The cake is then placed on the implement similar to tongs, laid over the tub full of water, and cleaned with a bundle of brass wire dipped into the water; this operation of beating and cleansing is repeated until it is all clean. Afterward he places it on an iron grate or tripod; the tripod is a palm and two digits high, one and a half digits wide, and its span is two palms wide; then he puts burning charcoal under the tripod or grate, in order again to dry the silver that was moistened by the water. Finally, the Royal Inspector[42] in the employment of the King or Prince, or the owner, lays the silver on a block of wood, and with an engraver's chisel he cuts out two

A—Muffle. B—Its little windows. C—Its little bridge. D—Bricks. E—Iron door. F—Its little window. G—Bellows. H—Hammer-chisel. I—Iron ring which some use instead of the test. K—Pestle with which the ashes placed in the ring are pounded.

small pieces, one from the under and the other from the upper side. These are tested by fire, in order to ascertain whether the silver is thoroughly refined or not, and at what price it should be sold to the merchants. Finally he impresses upon it the seal of the King or the Prince or the owner, and, near the same, the amount of the weight.

There are some who refine silver in tests placed under iron or earthenware muffles. They use a furnace, on the hearth of which they place the test containing the fragments of silver, and they place the muffle over it; the

[42]*Praefectus Regis.*

muffle has small windows at the sides, and in front a little bridge. In order to melt the silver, at the sides of the muffle are laid bricks, upon which the charcoal is placed, and burning firebrands are put on the bridge. The furnace has an iron door, which is covered on the side next to the fire with lute in order that it may not be injured. When the door is closed it retains the heat of the fire, but it has a small window, so that the artificers may look into the test and may at times stimulate the fire with the bellows. Although by this method silver is refined more slowly than by the other, nevertheless it is more useful, because less loss is caused, for a gentle fire consumes fewer particles than a fierce fire continually excited by the blast of the bellows. If, on account of its great size, the cake of silver can be carried only with difficulty when it is taken out of the muffle, they cut it up into two or three pieces while it is still hot, with a wedge or a hammer-chisel ; for if they cut it up after it has cooled, little pieces of it frequently fly off and are lost.

END OF BOOK X.

BOOK XI.

IFFERENT methods of parting gold from silver, and, on the other hand, silver from gold, were discussed in the last book; also the separation of copper from the latter, and further, of lead from gold as well as from silver; and, lastly, the methods for refining the two precious metals. Now I will speak of the methods by which silver must be separated from copper, and likewise from iron.[1]

The *officina*, or the building necessary for the purposes and use of those who separate silver from copper, is constructed in this manner. First, four long walls are built, of which the first, which is parallel with the bank of a stream, and the second, are both two hundred and sixty-four feet long. The second, however, stops at one hundred and fifty-one feet, and after, as it were, a break for a length of twenty-four feet, it continues again until it is of a length equal to the first wall. The third wall is one hundred and twenty feet long, starting at a point opposite the sixty-seventh foot of the other walls, and reaching to their one hundred and eighty-sixth foot.

[1]The whole of this Book is devoted to the subject of the separation of silver from copper by liquation, except pages 530-9 on copper refining, and page 544 on the separation of silver from iron. We believe a brief outline of the liquation process here will refresh the mind of the reader, and enable him to peruse the Book with more satisfaction. The fundamental principle of the process is that if a copper-lead alloy, containing a large excess of lead, be heated in a reducing atmosphere, above the melting point of lead but below that of copper, the lead will liquate out and carry with it a large proportion of the silver. As the results are imperfect, the process cannot be carried through in one operation, and a large amount of bye-products is created which must be worked up subsequently. The process, as here described, falls into six stages. 1st, Melting the copper and lead in a blast furnace to form " liquation cakes "—that is, the " leading." If the copper contain too little silver to warrant liquation directly, then the copper is previously enriched by melting and drawing off from a settling pot the less argentiferous " tops " from the metal, liquation cakes being made from the enriched " bottoms." 2nd, Liquation of the argentiferous lead from the copper. This work was carried out in a special furnace, to which the admission of air was prevented as much as possible in order to prevent oxidation. 3rd, " Drying " the residual copper, which retained some lead, in a furnace with a free admission of air. The temperature was raised to a higher degree than in the liquation furnace, and the expelled lead was oxidized. 4th, Cupellation of the argentiferous lead. 5th, Refining of the residual copper from the " drying " furnace by oxidation of impurities and poling in a " refining furnace." 6th, Re-alloy and re-liquation of the bye-products. These consist of: *a*, " slags " from " leading "; *b*, " slags " from " drying "; *c*, " slags " from refining of the copper. All of these " slags " were mainly lead oxides, containing some cuprous oxides and silica from the furnace linings; *d*, " thorns " from liquation; *e*, " thorns " from " drying "; *f*, " thorns " from skimmings during cupellation; these were again largely lead oxides, but contained rather more copper and less silica than the " slags "; *g*, " ash-coloured copper," being scales from the " dried " copper, were cuprous oxides, containing considerable lead oxides; *h*, concentrates from furnace accretions, crushed bricks, &c.

The discussion of detailed features of the process has been reserved to notes attached to the actual text, to which the reader is referred. As to the general result of liquation, Karsten (see below) estimates the losses in the liquation of the equivalent of 100 lbs. of argentiferous copper to amount to 32–35 lbs. of lead and 5 to 6 lbs. of copper. Percy (see below) quotes results at Lautenthal in the Upper Harz for the years 1857–60, showing losses of 25% of the silver, 9.1% of the copper, and 36.37 lbs. of lead to the 100 lbs. of copper, or say, 16% of the lead; and a cost of £8 6s. per ton of copper. The theoretical considerations involved in liquation have not been satisfactorily determined. Those who may wish to pursue the subject will find repeated descriptions and much discussion in the following works, which have been freely consulted in the notes which follow upon particular features of the process. It may be mentioned that Agricola's treatment of the subject is more able than any down to the 18th century. Ercker (*Beschreibung Allerfürnemsten Mineralischen*, etc., Prague, 1574). Lohneys (*Bericht vom Bergwercken*, etc., Zellerfeldt, 1617). Schlüter (*Gründlicher Unterricht*

The fourth wall is one hundred and fifty-one feet long. The height of each of these walls, and likewise of the other two and of the transverse walls, of which I will speak later on, is ten feet, and the thickness two feet and as many palms. The second long wall only is built fifteen feet high, because of the furnaces which must be built against it. The first long wall is distant fifteen feet from the second, and the third is distant the same number of feet from the fourth, but the second is distant thirty-nine feet from the third. Then transverse walls are built, the first of which leads from the beginning of the first long wall to the beginning of the second long wall; and the second transverse wall from the beginning of the second long wall to the beginning of the fourth long wall, for the third long wall does not reach so far. Then from the beginning of the third long wall are built two walls—the one to the sixty-seventh foot of the second long wall, the other to the same point in the fourth long wall. The fifth transverse wall is built at a distance of ten feet from the fourth transverse wall toward the second transverse wall;

von Hütte-Werken, Braunschweig, 1738). *Karsten* (*System der Metallurgie V.* and *Archiv für Bergbau und Hüttenwesen*, 1st series, 1825). Berthier (*Annales des Mines*, 1825, II.). Percy (Metallurgy of Silver and Gold, London, 1880).

NOMENCLATURE.—This process held a very prominent position in German metallurgy for over four centuries, and came to have a well-defined nomenclature of its own, which has never found complete equivalents in English, our metallurgical writers to the present day adopting more or less of the German terms. Agricola apparently found no little difficulty in adapting Latin words to his purpose, but stubbornly adhered to his practice of using no German at the expense of long explanatory clauses. The following table, prepared for convenience in translation, is reproduced. The German terms are spelled after the manner used in most English metallurgies, some of them appear in Agricola's Glossary to *De Re Metallica*.

English.	Latin.	German.
Blast furnace	*Prima fornax*	*Schmeltzofen*
Liquation furnace	*Fornax in qua argentum et plumbum ab aere secernuntur*	*Saigernofen*
Drying furnace	*Fornax in qua aerei panes fathiscentes torrentur*	*Darrofen*
Refining hearth	*Fornax in qua panes aerei torrefacti coquuntur*	*Gaarherd*
Cupellation furnace	*Secunda fornax*, or *fornax in qua plumbum ab argento separatur*	*Treibherd*
Leading	*Mistura*	*Frischen*
Liquating	*Stillare*, or *distillare*	*Saigern*
"Drying"	*Torrere*	*Darren*
Refining	*Aes ex panibus torrefactis conficere*	*Gaarmachen*
Liquation cakes	*Panes ex aere ac plumbo misti*	*Saigerstock*
Exhausted liquation cakes	*Panes fathiscentes*	*Kiehnstock*, or *Kinstocke*
"Dried" cakes	*Panes torrefacti*	*Darrlinge*
Slags:		
from leading	*Recrementa* (with explanatory phrases)	*Frischschlacke*
,, drying	,, ,, ,,	*Darrost*
,, refining	,, ,, ,,	*Gaarschlacke*
Liquation thorns	*Spinae* (with explanatory phrases)	*Saigerdörner*, or *Röstdörner*
Thorns from "drying"	,, ,, ,,	*Darrsöhle*
,, ,, cupellation	,, ,, ,,	*Abstrich*
Silver-lead or liquated silver-lead	*Stannum*	*Saigerwerk* or *saigerblei*
Ash-coloured copper	*Aes cinereum*	*Pickschiefer* or *schifer*
Furnace accretions or "accretions"	*Cadmiae*	*Offenbrüche*

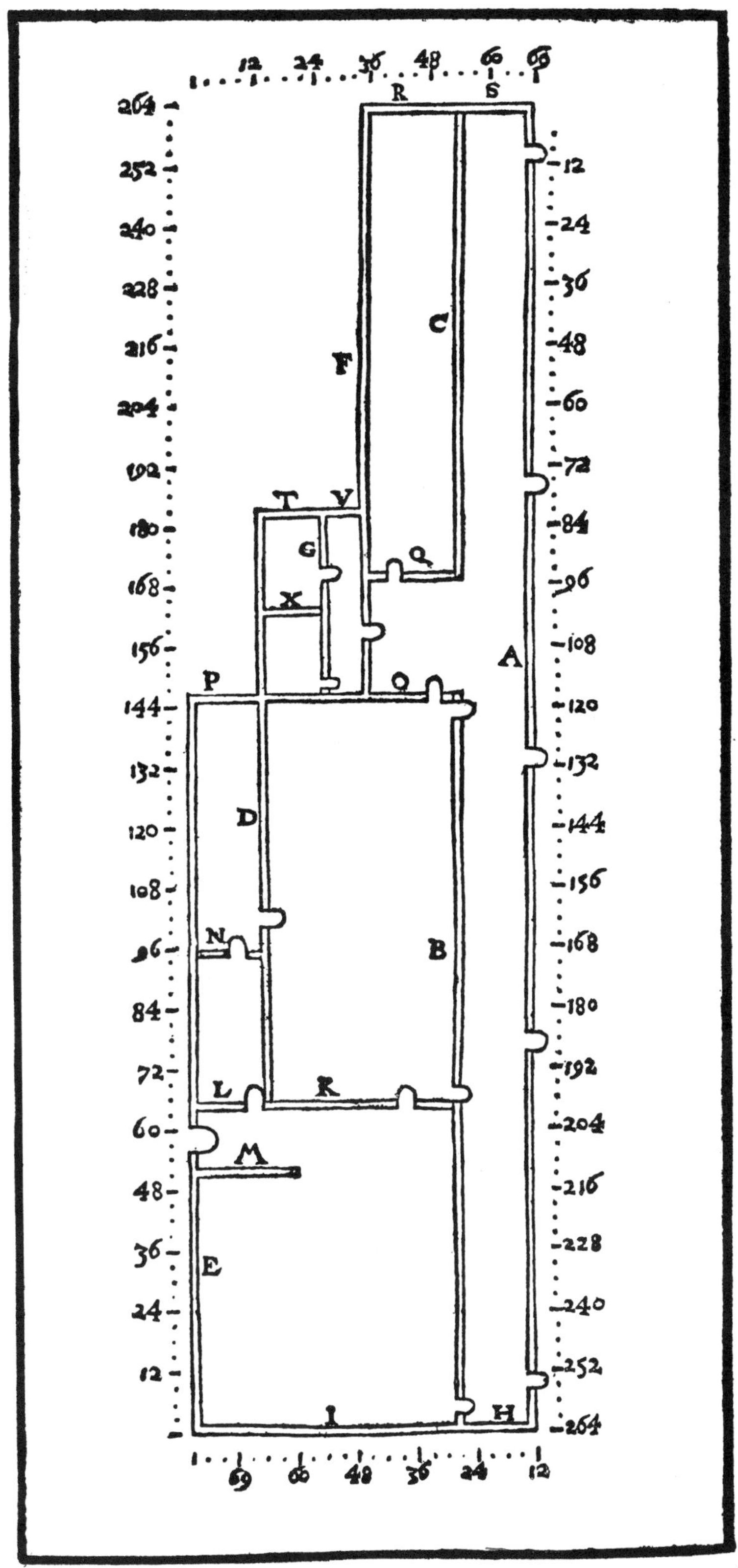

SIX LONG WALLS: A—THE FIRST. B—THE FIRST PART OF THE SECOND. C—THE FURTHER PART OF THE SECOND. D—THE THIRD. E—THE FOURTH. F—THE FIFTH. G—THE SIXTH. FOURTEEN TRANSVERSE WALLS: H—THE FIRST. I—THE SECOND. K—THE THIRD. L—THE FOURTH. M—THE FIFTH. N—THE SIXTH. O—THE SEVENTH. P—THE EIGHTH. Q—THE NINTH. R—THE TENTH. S—THE ELEVENTH. T—THE TWELFTH. V—THE THIRTEENTH. X—THE FOURTEENTH.

it is twenty feet long, and starts from the fourth long wall. The sixth transverse wall is built also from the fourth long wall, at a point distant thirty feet from the fourth transverse wall, and it extends as far as the back of the third long wall. The seventh transverse wall is constructed from the second long wall, where this first leaves off, to the third long wall; and from the back of the third long wall the eighth transverse wall is built, extending to the end of the fourth long wall. Then the fifth long wall is built from the seventh transverse wall, starting at a point nineteen feet from the second long wall; it is one hundred and nine feet in length; and at a point twenty-four feet along it, the ninth transverse wall is carried to the third end of the second long wall, where that begins again. The tenth transverse wall is built from the end of the fifth long wall, and leads to the further end of the second long wall; and from there the eleventh transverse wall leads to the further end of the first long wall. Behind the fifth long wall, and five feet toward the third long wall, the sixth long wall is built, leading from the seventh transverse wall; its length is thirty-five feet, and from its further end the twelfth transverse wall is built to the third long wall, and from it the thirteenth transverse wall is built to the fifth long wall. The fourteenth transverse wall divides into equal parts the space which lies between the seventh transverse wall and the twelfth.

The length, height, breadth, and position of the walls are as above. Their archways, doors, and openings are made at the same time that the walls are built. The size of these and the way they are made will be much better understood hereafter. I will now speak of the furnace hoods and of the roofs. The first side[2] of the hood stands on the second long wall, and is similar in every respect to those whose structure I explained in Book IX, when I described the works in whose furnaces are smelted the ores of gold, silver, and copper. From this side of the hood a roof, which consists of burnt tiles, extends to the first long wall; and this part of the building contains the bellows, the machinery for compressing them, and the instruments for inflating them. In the middle space, which is situated between the second and third transverse walls, an upright post eight feet high and two feet thick

HISTORICAL NOTE.—So far as we are aware, this is the first complete discussion of this process, although it is briefly mentioned by one writer before Agricola—that is, by Biringuccio (III, 5, 8), who wrote ten years before this work was sent to the printer. His account is very incomplete, for he describes only the bare liquation, and states that the copper is re-melted with lead and re-liquated until the silver is sufficiently abstracted. He neither mentions "drying" nor any of the bye-products. In his directions the silver-lead alloy was cupelled and the copper ultimately refined, obviously by oxidation and poling, although he omits the pole. In A.D. 1150 Theophilus (p. 305, Hendrie's Trans.) describes melting lead out of copper ore, which would be a form of liquation so far as separation of these two metals is concerned, but obviously not a process for separating silver from copper. This passage is quoted in the note on copper smelting (Note on p. 405). A process of such well-developed and complicated a character must have come from a period long before Agricola; but further than such a surmise, there appears little to be recorded. Liquation has been during the last fifty years displaced by other methods, because it was not only tedious and expensive, but the losses of metal were considerable.

[2]*Paries*,—"Partition" or "wall." The author uses this term throughout in distinction to *murus*, usually applying the latter to the walls of the building and the former to furnace walls, chimney walls, etc. In order to gain clarity, we have introduced the term "hood" in distinction to "chimney," and so far as possible refer to the *paries* of these constructions and furnaces as "side of the furnace," "side of the hood," etc.

and wide, is erected on a rock foundation, and is distant thirteen feet from the second long wall. On that upright post, and in the second transverse wall, which has at that point a square hole two feet high and wide, is placed a beam thirty-four feet and a palm long. Another beam, of the same length, width, and thickness, is fixed on the same upright post and in the third transverse wall. The heads of those two beams, where they meet, are joined together with iron staples. In a similar manner another post is erected, at a distance of ten feet from the first upright post in the direction of the fourth wall, and two beams are laid upon it and into the same walls in a similar way to those I have just now described. On these two beams and on the fourth long wall are fixed seventeen cross-beams, forty-three feet and three palms long, a foot wide, and three palms thick; the first of these is laid upon the second transverse wall, the last lies along the third and fourth transverse walls; the rest are set in the space between them. These cross-beams are three feet apart one from the other.

In the ends of these cross-beams, facing the second long wall, are mortised the ends of the same number of rafters reaching to those timbers which stand upright on the second long wall, and in this manner is made the inclined side of the hood in a similar way to the one described in Book IX. To prevent this from falling toward the vertical wall of the hood, there are iron rods securing it, but only a few, because the four brick chimneys which have to be built in that space partly support it. Twelve feet back are likewise mortised into the cross-beams, which lie upon the two longitudinal beams and the fourth long wall, the lower ends of as many rafters, whose upper ends are mortised into the upper ends of an equal number of similar rafters, whose lower ends are mortised to the ends of the beams at the fourth long wall. From the first set of rafters[4] to the second set of rafters is a distance of twelve feet, in order that a gutter may be well placed in the middle space. Between these two are again erected two sets of rafters, the lower ends of which are likewise mortised into the beams, which lie on the two longitudinal beams and the fourth long wall, and are interdistant a cubit. The upper ends of the ones fifteen feet long rest on the backs of the rafters of the first set; the ends of the others, which are eighteen feet long, rest on the backs of the rafters of the second set, which are longer; in this manner, in the middle of the rafters, is a sub-structure. Upon each alternate cross-beam which is placed upon the two longitudinal beams and the fourth long wall is erected an upright post, and that it may be sufficiently firm it is strengthened by means of a slanting timber. Upon these posts is laid a long beam, upon which rests one set of middle rafters. In a similar manner the other set of middle rafters rests on a long beam which is placed upon other posts. Besides this, two feet above every cross-beam, which is placed on the two longitudinal beams and the

[4]From this point on, the construction of the roofs, in the absence of illustration, is hopeless of intelligent translation. The constant repetition of "*tignum*," "*tigillum*," "*trabs*," for at least fifteen different construction members becomes most hopelessly involved, especially as the author attempts to distinguish between them in a sort of "House-that-Jack-built" arrangement of explanatory clauses.

fourth long wall, is placed a tie-beam which reaches from the first set of middle rafters to the second set of middle rafters; upon the tie-beams is placed a gutter hollowed out from a tree. Then from the back of each of the first set of middle rafters a beam six feet long reaches almost to the gutter; to the lower end of this beam is attached a piece of wood two feet long; this is repeated with each rafter of the first set of middle rafters. Similarly from the back of each rafter of the second set of middle rafters a little beam, seven feet long, reaches almost to the gutter; to the lower end of it is likewise attached a short piece of wood; this is repeated on each rafter of the second set of middle rafters. Then in the upper part, to the first and second sets of principal rafters are fastened long boards, upon which are fixed the burnt tiles; and in the same manner, in the middle part, they are fastened to the first and second sets of middle rafters, and at the lower part to the little beams which reach from each rafter of the first and second set of middle rafters almost to the gutter; and, finally, to the little boards fastened to the short pieces of wood are fixed shingles of pinewood extending into the gutter, so that the violent rain or melted snow may not penetrate into the building. The substructures in the interior which support the second set of rafters, and those on the opposite side which support the third, being not unusual, I need not explain.

In that part of the building against the second long wall are the furnaces, in which exhausted liquation cakes which have already been "dried" are smelted, that they may recover once again the appearance and colour of copper, inasmuch as they really are copper. The remainder of the room is occupied by the passage which leads from the door to the furnaces, together with two other furnaces, in one of which the whole cakes of copper are heated, and in the other the exhausted liquation cakes are "dried" by the heat of the fire.

Likewise, in the room between the third and seventh[5] transverse walls, two posts are erected on rock foundation; both of them are eight feet high and two feet wide and thick. The one is at a distance of thirteen feet from the second long wall; the other at the same distance from the third long wall; there is a distance of thirteen feet between them. Upon these two posts and upon the third transverse wall are laid two longitudinal beams, forty-one feet and one palm long, and two feet wide and thick. Two other beams of the same length, width, and thickness are laid upon the upright posts and upon the seventh transverse wall, and the heads of the two long beams, where they meet, are joined with iron staples. On these longitudinal beams are again placed twenty-one transverse beams, thirteen feet long, a foot wide, and three palms thick, of which the first is set on the third transverse wall, and the last on the seventh transverse wall; the rest are laid in the space between these two, and they are distant from one another three feet. Into the ends of the transverse beams which face the second long wall, are mortised the ends of the same number of rafters erected toward the upright posts which are placed upon the second long wall, and in this manner is made

[5]In the original text this is given as the "fifth," a manifest impossibility.

the second inclined side wall of the hood. Into the ends of the transverse beams facing the third long wall, are mortised the ends of the same number of rafters rising toward the rafters of the first inclined side of the second hood, and in this manner is made the other inclined side of the second hood. But to prevent this from falling in upon the opposite inclined side of the hood, and that again upon the opposite vertical one, there are many iron rods reaching from some of the rafters to those opposite them ; and this is also prevented in part by means of a few tie-beams, extending from the back of the rafters to the back of those which are behind them. These tie-beams are two palms thick and wide, and have holes made through them at each end ; each of the rafters is bound round with iron bands three digits wide and half a digit thick, which hold together the ends of the tie-beams of which I have spoken ; and so that the joints may be firm, an iron nail, passing through the plate on both sides, is driven through the holes in the ends of the beams. Since one weight counter-balances another, the rafters on the opposite hoods cannot fall. The tie-beams and middle posts which have to support the gutters and the roof, are made in every particular as I stated above, except only that the second set of middle rafters are not longer than the first set of middle rafters, and that the little beams which reach from the back of each rafter of the second set of middle rafters nearly to the gutter are not longer than the little beams which reach from the back of each rafter of the first set of middle rafters almost to the gutter. In this part of the building, against the second long wall, are the furnaces in which copper is alloyed with lead, and in which "slags" are re-smelted. Against the third long wall are the furnaces in which silver and lead are liquated from copper. The interior is also occupied by two cranes, of which one deposits on the ground the cakes of copper lifted out of the moulding pans ; the other lifts them from the ground into the second furnace.

On the third and the fourth long walls are set twenty-one beams eighteen feet and three palms long. In mortises in them, two feet behind the third long wall, are set the ends of the same number of rafters erected opposite to the rafters of the other inclined wall of the second furnace hood, and in this manner is made the third inclined wall, exactly similar to the others. The ends of as many rafters are mortised into these beams where they are fixed in the fourth long wall ; these rafters are erected obliquely, and rest against the backs of the preceding ones and support the roof, which consists entirely of burnt tiles and has the usual substructures. In this part of the building there are two rooms, in the first of which the cakes of copper, and in the other the cakes of lead, are stored.

In the space enclosed between the ninth and tenth transverse walls and the second and fifth long walls, a post twelve feet high and two feet wide and thick is erected on a rock foundation ; it is distant thirteen feet from the second long wall, and six from the fifth long wall. Upon this post and upon the ninth transverse wall is laid a beam thirty-three feet and three palms long, and two palms wide and thick. Another beam, also of the same length, width and thickness, is laid upon the same post and upon the tenth transverse wall, and the ends of these two beams where they meet are joined by means of iron staples. On these beams and on the fifth long wall are placed ten cross-beams, eight feet and three palms long, the first of which is placed on the ninth transverse wall, the last on the tenth, the remainder in the space between them; they are distant from one another three feet. Into the ends of the cross-beams facing the second long wall, are mortised the ends of the same number of rafters inclined toward the posts which stand vertically upon the second long wall. This, again, is the manner in which the inclined side of the furnace hood is made, just as with the others; at the top where the fumes are emitted it is two feet distant from the vertical side. The ends of the same number of rafters are mortised into the cross-beams, where they are set in the fifth long wall; each of them is set up obliquely and rests against the back of one of the preceding set; they support the roof, made of burnt tiles. In this part of the building, against the second long wall, are four furnaces in which lead is separated from silver, together with the cranes by means of which the domes are lifted from the crucibles.

In that part of the building which lies between the first long wall and the break in the second long wall, is the stamp with which the copper cakes are crushed, and the four stamps with which the accretions that are chipped off the walls of the furnace are broken up and crushed to powder, and likewise the bricks on which the exhausted liquation cakes of copper are stood to be "dried." This room has the usual roof, as also has the space between the seventh transverse wall and the twelfth and thirteenth transverse walls.

At the sides of these rooms are the fifth, the sixth, and the third long walls. This part of the building is divided into two parts, in the first of which stand the little furnaces in which the artificer assays metals; and the bone ash, together with the other powders, are kept here. In the other room is prepared the powder from which the hearths and the crucibles of the furnaces are made. Outside the building, at the back of the fourth long wall, near the door to the left as you enter, is a hearth in which smaller masses of lead are melted from large ones, that they may be the more easily weighed; because the masses of lead, just as much as the cakes of copper, ought to be first prepared so that they can be weighed, and a definite weight can be melted and alloyed in the furnaces. To begin with, the hearth in which the masses of lead are liquefied is six feet long and five wide; it is protected on both sides by rocks partly sunk into the earth, but a palm higher than the hearth, and it is lined in the inside with lute. It slopes toward the middle and toward the front, in order that the molten lead may run down and flow out into the dipping-pot. There is a wall at the back of the hearth which protects the fourth long wall from damage by the heat; this wall, which is made of bricks and lute, is four feet high, three palms thick, and five feet long at the bottom, and at the top three feet and two palms long; therefore it narrows gradually, and in the upper part are laid seven bricks, the middle ones of which are set upright, and the end ones inclined; they are all thickly coated with lute. In front of the hearth is a dipping-pot, whose pit is a foot deep, and a foot and three palms wide at the top, and gradually narrows.

A—Hearth. B—Rocks sunk into the ground. C—Walls which protect the fourth long wall from damage by fire. D—Dipping-pot. E—Masses of lead. F—Trolley. G—Its wheels. H—Crane. I—Tongs. K—Wood. L—Moulds. M—Ladle. N—Pick. O—Cakes.

When the masses of lead are to be melted, the workman first places the wood in the hearth so that one end of each billet faces the wall, and the other end the dipping-pot. Then, assisted by other workmen, he pushes the mass of lead forward with crowbars on to a low trolley, and draws it to the crane. The trolley consists of planks fastened together, is two and one-half feet wide and five feet long, and has two small iron axles, around which at each end revolve small iron wheels, two palms in diameter and as many digits wide. The trolley has a tongue, and attached to this is a rope, by which it is drawn to the crane. The crane is exactly similar to those in the second part of the works, except that the crane-arm is not so long. The tongs in whose jaws[6] the masses of lead are seized, are two feet a palm and two digits long; both of the jaws, when struck with a hammer, impinge upon the mass and are driven into it. The upper part of both handles of the tongs are curved back, the one to the right, the other to the left, and each handle is engaged in one of the lowest links of two short chains, which are three links long. The upper links are engaged in a large round ring, in which is fixed the hook of a chain let down from the pulley of the crane-arm. When the crank of the crane is turned, the mass is lifted and is carried by the crane-arm to the hearth and placed on the wood. The workmen wheel up one mass after another and place them in a similar manner on the wood of the hearth; masses which weigh a total of about a hundred and sixty *centumpondia*[7] are usually placed upon the wood and melted at one time. Then a workman throws charcoal on the masses, and all are made ready in the evening. If he fears that it may rain, he covers it up with a cover, which may be moved here and there; at the back this cover has two legs, so that the rain which it collects may flow down the slope on to the open ground. Early in the morning of the following day, he throws live coals on the charcoal with a shovel, and by this method the masses of lead melt, and from time to time charcoal is added. The lead, as soon as it begins to run into the dipping-pot, is ladled out with an iron ladle into copper moulds such as the refiners generally use. If it does not cool immediately he pours water over it, and then sticks the pointed pick into it and pulls it out. The pointed end of the pick is three palms long and the round end is two digits long. It is necessary to smear the moulds with a wash of lute, in order that, when they have been turned upside down and struck with the broad round end of the pick, the cakes of lead may fall out easily. If the moulds are not washed over with the lute, there is a risk that they may be melted by the lead and let it through. Others take hold of a billet of wood with their left hand, and with the heavy lower end of it they pound the mould, and with the right hand they stick the point of the pick into the cake of lead, and thus pull it out. Then immediately the workman pours other lead into the empty moulds, and this he does until the work of melting the lead is finished. When the lead is melted, something similar to litharge is produced; but it is no wonder that it should be possible to make

[6]*Chelae*,—" claws."

[7]If Roman weights, this would be 5.6 short tons, and 7.5 tons if German *centner* is meant.

it in this case, when it used formerly to be produced at Puteoli from lead alone when melted by a fierce fire in the cupellation furnace.[8] Afterward these cakes of lead are carried into the lead store-room.

The cakes of copper, put into wheelbarrows, are carried into the third part of the building, where each is laid upon a saddle, and is broken up by the impact of successive blows from the iron-shod stamp. This machine is made by placing upon the ground a block of oak, five feet long and three feet

A—Block of wood. B—Upright posts. C—Transverse beams. D—Head of the stamp. E—Its tooth. F—The hole in the stamp-stem. G—Iron bar. H—Masses of lead. I—The bronze saddle. K—Axle. L—Its arms. M—Little iron axle. N—Bronze pipe.

wide and thick ; it is cut out in the middle for a length of two feet and two palms, a width of two feet, and a depth of three palms and two digits, and is open in front; the higher part of it is at the back, and the wide part lies flat in the block. In the middle of it is placed a bronze saddle. Its base is a palm and two digits wide, and is planted between two masses of lead, and extends under them to a depth of a palm on both sides. The whole saddle is three palms and two digits wide, a foot long, and

[8]This is, no doubt, a reference to Pliny's statement (xxxiii, 35) regarding litharge at Puteoli. This passage from Pliny is given in the footnote on p. 466. Puteoli was situated on the Bay of Naples.

two palms thick. Upon each end of the block stands a post, a cubit wide and thick, the upper end of which is somewhat cut away and is mortised into the beams of the building. At a height of four feet and two digits above the block there are joined to the posts two transverse beams, each of which is three palms wide and thick; their ends are mortised into the upright posts, and holes are bored through them; in the holes are driven iron claves, horned in front and so driven into the post that one of the horns of each points upward and the other downward; the other end of each clavis is perforated, and a wide iron wedge is inserted and driven into the holes, and thus holds the transverse beams in place. These transverse beams have in the middle a square opening three palms and half a digit wide in each direction, through which the iron-shod stamp passes. At a height of three feet and two palms above these transverse beams there are again two beams of the same kind, having also a square opening and holding the same stamp. This stamp is square, eleven feet long, three palms wide and thick; its iron shoe is a foot and a palm long; its head is two palms long and wide, a palm two digits thick at the top, and at the bottom the same number of digits, for it gradually narrows. But the tail is three palms long; where the head begins is two palms wide and thick, and the further it departs from the same the narrower it becomes. The upper part is enclosed in the stamp-stem, and it is perforated so that an iron bolt may be driven into it; it is bound by three rectangular iron bands, the lowest of which, a palm wide, is between the iron shoe and the head of the stamp; the middle band, three digits wide, follows next and binds round the head of the stamp, and two digits above is the upper one, which is the same number of digits wide. At a distance of two feet and as many digits above the lowest part of the iron shoe, is a rectangular tooth, projecting from the stamp for a distance of a foot and a palm; it is two palms thick, and when it has extended to a distance of six digits from the stamp it is made two digits narrower. At a height of three palms upward from the tooth there is a round hole in the middle of the stamp-stem, into which can be thrust a round iron bar two feet long and a digit and a half in diameter; in its hollow end is fixed a wooden handle two palms and the same number of digits long. The bar rests on the lower transverse beam, and holds up the stamp when it is not in use. The axle which raises the stamp has on each side two arms, which are two palms and three digits distant from each other, and which project from the axle a foot, a palm and two digits; penetrating through them are bolts, driven in firmly; the arms are each a palm and two digits wide and thick, and their round heads, for a foot downward on either side, are covered with iron plates of the same width as the arms and fastened by iron nails. The head of each arm has a round hole, into which is inserted an iron pin, passing through a bronze pipe; this little axle has at the one end a wide head, and at the other end a perforation through which is driven an iron nail, lest this little axle should fall out of the arms. The bronze pipe is two palms long and one in diameter; the little iron axle penetrates through its round interior, which is two digits in diameter. The bronze pipe not only revolves round the little iron axle, but it also rotates with it; therefore, when the axle revolves, the little axle and the bronze tube in their turn raise the tooth and the stamp. When the little iron axle and the bronze pipe have been taken out of the arms, the tooth of the stamps is not raised, and other stamps may be raised without this one. Further on, a drum with spindles fixed around the axle of a water-wheel moves the axle of a toothed drum, which depresses the sweeps of the bellows in the adjacent fourth part of the building; but it turns in the contrary direction; for the axis of the drum which raises the stamps turns toward the north, while that one which depresses the sweeps of the bellows turns toward the south.

Those cakes which are too thick to be rapidly broken by blows from the iron-shod stamp, such as are generally those which have settled in the bottom of the crucible,[9] are carried into the first part of the building. They are there heated in a furnace, which is twenty-eight feet distant from the second long wall and twelve feet from the second transverse wall. The three sides of this furnace are built of rectangular rocks, upon which bricks are laid; the back furnace wall is three feet and a palm high, and the rear of the side walls is the same; the side walls are sloping, and where the furnace is open in front they are only two feet and three palms high; all the walls are a foot and a palm thick. Upon these walls stand upright posts not less thick, in order that they may bear the heavy weight placed upon them, and they are covered with lute; these posts support the sloping chimney and penetrate through the roof. Moreover, not only the ribs of the chimney, but also the rafters, are covered thickly with lute. The hearth of the furnace is six feet long on each side, is sloping, and is paved with bricks. The cakes of copper are placed in the furnace and heated in the following way. They are first of all placed in the furnace in rows, with as many small stones the size of an egg between, so that the heat of the fire can penetrate through the spaces between them; indeed, those cakes which are placed at the bottom of the crucible are each raised upon half a brick for the same reason. But lest the last row, which lies against the mouth of the furnace, should fall out, against the mouth are placed iron plates, or the copper cakes which are the first taken from the crucible when copper is made, and against them are laid exhausted liquation cakes or rocks. Then charcoal is thrown on the cakes, and then live coals; at first the cakes are heated by a gentle fire, and afterward more charcoal is added to them until it is at times three-quarters of a foot deep. A fiercer fire is certainly required to heat the hard cakes of copper than the fragile ones. When the cakes have been sufficiently heated, which usually occurs within the space of about two hours, the exhausted liquation cakes or the rocks and the iron plate are removed from the mouth of the furnace. Then the hot cakes are taken out row after row with a two-pronged rabble, such as the one which is used by those who "dry" the exhausted liquation cakes. Then the first cake is laid upon the exhausted liquation cakes, and beaten by two workmen with hammers until it breaks; the hotter the cakes are, the

[9]By this expression is apparently meant the "bottoms" produced in enriching copper, as described on p. 510.

sooner they are broken up; the less hot, the longer it takes, for now and then they bend into the shape of copper basins. When the first cake has been broken, the second is put on to the other fragments and beaten until it breaks into pieces, and the rest of the cakes are broken up in the same manner in due order. The head of the hammer is three palms long and one wide, and sharpened at both ends, and its handle is of wood three feet long. When they have been broken by the stamp, if cold, or with hammers if hot, the fragments of copper or the cakes are carried into the store-room for copper.

A—BACK WALL. B—WALLS AT THE SIDES. C—UPRIGHT POSTS. D—CHIMNEY. E—THE CAKES ARRANGED. F—IRON PLATES. G—ROCKS. H—RABBLE WITH TWO PRONGS. I—HAMMERS.

The foreman of the works, according to the different proportions of silver in each *centumpondium* of copper, alloys it with lead, without which he could not separate the silver from the copper.[10] If there be a moderate

[10]The details of the preparation of liquation cakes—" leading "—were matters of great concern to the old metallurgists. The size of the cakes, the proportion of silver in the original copper and in the liquated lead, the proportion of lead and silver left in the residual cakes, all had to be reached by a series of compromises among militant forces. The cakes were generally two and one-half to three and one-half inches thick and about two feet in diameter, and

amount of silver in the copper, he alloys it fourfold; for instance, if in three-quarters of a *centumpondium* of copper there is less than the following proportions, *i.e.*: half a *libra* of silver, or half a *libra* and a *sicilicus*, or half a *libra* and a *semi-uncia*, or half a *libra* and *semi-uncia* and a *sicilicus*, then rich lead—that is, that from which the silver has not yet been separated—is added, to the amount of half a *centumpondium* or a whole *centumpondium*, or a whole and a half, in such a way that there may be in the copper-lead alloy some one of the proportions of silver which I have just mentioned, which is the first alloy. To this "first" alloy is added such a weight of de-silverized lead or litharge as is required to make out of all of these a single liquation cake that will contain approximately two *centumpondia* of lead; but as usually from one hundred and thirty *librae* of litharge only one hundred *librae* of lead are made, a greater proportion of litharge than of de-silverized lead is added as a supplement. Since four cakes of this kind are placed at the same time into the furnace in which the silver and lead is liquated from copper, there will be in all the cakes three *centumpondia* of copper and eight *centumpondia* of lead. When the lead has been liquated from the copper, it weighs six *centumpondia*, in each *centumpondium* of which there is a quarter of a *libra* and almost a *sicilicus* of silver. Only seven *unciae* of the silver remain in the exhausted liquation cakes and in that copper-lead alloy which we call "liquation thorns"; they are not called by this name so much because they have sharp points as because they are base. If in three-quarters of a *centumpondium* of copper there are less than seven *uncia* and a *semi-uncia* or a *bes* of silver, then so much rich lead must be added as to make in the copper and lead alloy one of the proportions of silver which I have already mentioned. This is the "second" alloy. To this is again to be added as great a weight

weighed 225 to 375lbs. This size was wonderfully persistent from Agricola down to modern times; and was, no doubt, based on sound experience. If the cakes were too small, they required proportionately more fuel and labour; whilst if too large, the copper began to melt before the maximum lead was liquated. The ratio of the copper and lead was regulated by the necessity of enough copper to leave a substantial sponge mass the shape of the original cake, and not so large a proportion as to imprison the lead. That is, if the copper be in too small proportion the cakes break down; and if in too large, then insufficient lead liquates out, and the extraction of silver decreases. Ercker (p. 106–9) insists on the equivalent of about 3 copper to 9.5 lead; Lohneys (p. 99), 3 copper to 9 or 10 lead. Schlüter (p. 479, etc.) insists on a ration of 3 copper to about 11 lead. Kerl (*Handbuch Der Metallurgischen Hütten kunde*, 1855; Vol. III., p. 116) gives 3 copper to 6 to 7 parts lead. Agricola gives variable amounts of 3 parts copper to from 8 to 12 parts lead. As to the ratio of silver in the copper, or to the cakes, there does not, except the limit of payability, seem to have been any difficulty on the minimum side. On the other hand, Ercker, Lohneys, Schlüter, and Karsten all contend that if the silver ran above a certain proportion, the copper would retain considerable silver. These authors give the outside ratio of silver permissible for good results in one liquation at what would be equivalent to 45 to 65 ozs. per ton of cakes, or about 190 to 250 ozs. per ton on the original copper. It will be seen, however, that Agricola's cakes greatly exceed these values. A difficulty did arise when the copper ran low in silver, in that the liquated lead was too poor to cupel, and in such case the lead was used over again, until it became rich enough for this purpose. According to Karsten, copper containing less than an equivalent of 80 to 90 ozs. per ton could not be liquated profitably, although the Upper Harz copper, according to Kerl, containing the equivalent of about 50 ozs. per ton, was liquated at a profit. In such a case the cakes would run only 12 to 14 ozs. per ton. It will be noticed that in the eight cases given by Agricola the copper ran from 97 to over 580 ozs. per ton, and in the description of enrichment of copper "bottoms" the original copper runs 85 ozs., and "it cannot be separated easily"; as a result, it is raised to 110 ozs. per ton before treatment. In addition to the following tabulation of the proportions here given by Agricola, the reader should refer to footnotes 15 and 17, where four more combinations are tabulated. It will be observed from

of de-silverized lead, or of litharge, as will make it possible to obtain from that alloy a liquation cake containing two and a quarter *centumpondia* of lead, in which manner in four of these cakes there will be three *centumpondia* of copper and nine *centumpondia* of lead. The lead which liquates from these cakes weighs seven *centumpondia*, in each *centumpondium* of which there is a quarter of a *libra* of silver and a little more than a *sicilicus*. About seven *unciae* of silver remain in the exhausted liquation cakes and in the liquation thorns, if we may be allowed to make common the old name (*spinae*=thorns) and bestow it upon a new substance. If in three-quarters of a *centumpondium* of copper there is less than three-quarters of a *libra* of silver, or three-quarters and a *semi-uncia*, then as much rich lead must be added as will produce one of the proportions of silver in the copper-lead alloy above mentioned; this is the " third " alloy. To this is added such an amount of de-silverized lead or of litharge, that a liquation cake made from it contains in all two and three-quarters *centumpondia* of lead. In this manner four such cakes will contain three *centumpondia* of copper and eleven *centumpondia* of lead. The lead which these cakes liquate, when they are melted in the furnace, weighs about nine *centumpondia*, in each *centumpondium* of which there is a quarter of a *libra* and more than a *sicilicus* of silver; and seven *unciae* of silver remain in the exhausted liquation cakes and in the liquation thorns. If, however, in three-quarters of a *centumpondium* of copper there is less than ten-twelfths of a *libra* or ten-twelfths of a *libra* and a *semi-uncia* of silver, then such a proportion of rich lead is added as will produce in the copper-lead alloy one of the proportions of silver which I mentioned above; this is the " fourth " alloy. To this is added such a weight of de-silverized lead or of litharge, that a liquation cake made from it contains three *centumpondia* of

this table that with the increasing richness of copper an increased proportion of lead was added, so that the products were of similar value. It has been assumed (see footnote 13 p. 509), that Roman weights are intended. It is not to be expected that metallurgical results of this period will " tie up " with the exactness of the modern operator's, and it has not been considered necessary to calculate beyond the nearest pennyweight. Where two or more values are given by the author the average has been taken.

	1st Charge.	2nd Charge.	3rd Charge.	4th Charge.
Amount of argentiferous copper ..	211.8 lbs.	211.8 lbs.	211.8 lbs.	211.8 lbs.
Amount of lead ..	564.8 ,,	635.4 ,,	776.6 ,,	847.2 ,,
Weight of each cake..	193.5 ,,	211.5 ,,	247.1 ,,	264.75 ,,
Average value of charge	56 ozs. 3dwts.	62 ozs. 4dwts.	64 ozs. 4dwts.	66 ozs. 7dwts.
Per cent. of copper..	27.2%	25%	21.4%	20%
Average value of original copper per ton	207 ozs. 4dwts.	251 ozs. 3dwts.	299 ozs. 15dwts.	332 ozs. 3dwts.
Weight of argentiferous lead liquated out	423.6 lbs.	494.2 lbs.	635.4 lbs.	706 lbs.
Average value of liquated lead per ton..	79 ozs.	79 ozs.	79 ozs.	85 ozs.
Weight of residues (residual copper and thorns)	353 lbs.	353 lbs.	353 lbs.	353 lbs.
Average value of residues per ton ..	34 ozs.	34 ozs.	34 ozs.	34 ozs. to 38 ozs.
Extraction of silver into the argentiferous lead ..	76.5%	73.4%	79%	85.3%

lead, and in four cakes of this kind there are three *centumpondia* of copper and twelve *centumpondia* of lead. The lead which is liquated therefrom weighs about ten *centumpondia*, in each *centumpondium* of which there is a quarter of a *libra* and more than a *semi-uncia* of silver, or seven *unciae* ; a *bes*, or seven *unciae* and a *semi-uncia*, of silver remain in the exhausted liquation cakes and in the liquation thorns.

Against the second long wall in the second part of the building, whose area is eighty feet long by thirty-nine feet wide, are four furnaces in which the copper is alloyed with lead, and six furnaces in which "slags" are re-smelted. The interior of the first kind of furnace is a foot and three palms wide, two feet three digits long ; and of the second is a foot and a palm wide and a foot three palms and a digit long. The side walls of these furnaces are the same height as the furnaces in which gold or silver ores are smelted. As the whole room is divided into two parts by upright posts, the front part must have, first, two furnaces in which "slags" are re-melted ; second, two furnaces in which copper is alloyed with lead ; and third, one furnace in which "slags" are re-melted. The back part of the room has first, one furnace in which "slags" are re-melted ; next, two furnaces in which copper is alloyed with lead ; and third, two furnaces in which "slags" are re-melted. Each of these is six feet distant from the next ; on the right side of the first is a space of three feet and two palms, and on the left side of the last one of seven feet. Each pair of furnaces has a common door, six feet high and a cubit wide, but the first and the tenth furnace each has one of its own. Each of the furnaces is set in an arch of its own in the back wall, and in front has a forehearth pit ; this is filled with a powder compound rammed down and compressed in order to make a crucible. Under each furnace is a hidden receptacle for the moisture,[11] from which a vent is made through the back wall toward the right, which allows the vapour to escape. Finally, to the right, in front, is the copper mould into which the copper-lead alloy is poured from the forehearth, in order that liquation cakes of equal weight may be made. This copper mould is a digit thick, its interior is two feet in diameter and six digits deep. Behind the second long wall are ten pairs of bellows, two machines for compressing them, and twenty instruments for inflating them. The way in which these should be made may be understood from Book IX.

The smelter, when he alloys copper with lead, with his hand throws into the heated furnace, first the large fragments of copper, then a basketful of charcoal, then the smaller fragments of copper. When the copper is melted and begins to run out of the tap-hole into the forehearth, he throws litharge into the furnace, and, lest part of it should fly away, he first throws charcoal over it, and lastly lead. As soon as he has thrown into the furnace the copper and the lead, from which alloy the first liquation cake is made, he again throws in a basket of charcoal, and then fragments of copper are thrown over them, from which the second cake may be made. Afterward with a rabble he skims the "slag" from the copper and lead as they flow into the forehearth. Such a rabble is a board into which an iron bar is fixed ; the

[11]See p. 356.

board is made of elder-wood or willow, and is ten digits long, six wide, and one and a half digits thick; the iron bar is three feet long, and the wooden handle inserted into it is two and a half feet long. While he purges the alloy and pours it out with a ladle into the copper mould, the fragments of copper from which he is to make the second cake are melting. As soon as this begins to run down he again throws in litharge, and when he has put on more charcoal he adds the lead. This operation he repeats until thirty liquation cakes have been made, on which work he expends nine hours, or at most ten; if more than thirty cakes must be made, then he is paid for another shift when he has made an extra thirty.

At the same time that he pours the copper-lead alloy into the copper mould, he also pours water slowly into the top of the mould. Then, with a cleft stick, he takes a hook and puts its straight stem into the molten cake. The hook itself is a digit and a half thick; its straight stem is two palms long and two digits wide and thick. Afterward he pours more water over the cakes. When they are cold he places an iron ring in the hook of the chain

A—Furnace in which "slags" are re-smelted. B—Furnace in which copper is alloyed with lead. C—Door. D—Fore-hearths on the ground. E—Copper moulds. F—Rabble. G—Hook. H—Cleft stick. I—Arm of the crane. K—The hook of its chain.

let down from the pulley of the crane arm; the inside diameter of this ring is six digits, and it is about a digit and a half thick; the ring is then engaged in the hook whose straight stem is in the cake, and thus the cake is raised from the mould and put into its place.

The copper and lead, when thus melted, yield a small amount of "slag"[12] and much litharge. The litharge does not cohere, but falls to pieces like the residues from malt from which beer is made. *Pompholyx* adheres to the walls in white ashes, and to the sides of the furnace adheres *spodos*.

In this practical manner lead is alloyed with copper in which there is but a moderate portion of silver. If, however, there is much silver in it, as, for instance, two *librae*, or two *librae* and a *bes*, to the *centumpondium*,—which weighs one hundred and thirty-three and a third *librae*, or one hundred and forty-six *librae* and a *bes*,[13]—then the foreman of the works adds to a *centumpondium* of such copper three *centumpondia* of lead, in each *centumpondium* of which there is a third of a *libra* of silver, or a third of a *libra* and a *semi-uncia*. In this manner three liquation cakes are made, which contain altogether three *centumpondia* of copper and nine *centumpondia* of lead.[14] The lead, when it has been liquated from the copper, weighs seven *centumpondia*; and in each *centumpondium*—if the *centumpondium* of copper contain two *librae* of silver, and the lead contain a third of a *libra*—there will be a *libra* and a sixth and more than a *semi-uncia* of silver; while in the exhausted liquation cakes, and in the liquation thorns, there remains a third of a *libra*.

[12]An analysis of this "slag" by Karsten (*Archiv.* 1st Series IX, p. 24) showed 63.2% lead oxide, 5.1% cuprous oxide, 20.1% silica (from the fuel and furnace linings), together with some iron alumina, etc. The *pompholyx* and *spodos* were largely zinc oxide (see note, p. 394).

[13]This description of a *centumpondium* which weighed either 133⅓ *librae*, or 146¾ *librae*, adds confusion to an already much mixed subject (see Appendix C.). Assuming the German *pfundt* to weigh 7,219 troy grains, and the Roman *libra* 4,946 grains, then a *centner* would weigh 145.95 *librae*, which checks up fairly well with the second case; but under what circumstances a *centner* can weigh 133⅓ *librae* we are unable to record. At first sight it might appear from this statement that where Agricola uses the word *centumpondium* he means the German *centner*. On the other hand, in the previous five or six pages the expressions one-third, five-sixths, ten-twelfths of a *libra* are used, which are even divisions of the Roman 12 *unciae* to one *libra*, and are used where they manifestly mean divisions of 12 units. If Agricola had in mind the German scale, and were using the *libra* for a *pfundt* of 16 *untzen*, these divisions would amount to fractions, and would not total the *sicilicus* and *drachma* quantities given, nor would they total any of the possibly synonymous divisions of the German *untzen* (see also page 254).

[14]If we assume Roman weights, the charge in the first case can be tabulated as follows, and for convenience will be called the fifth charge:—

	5TH CHARGE (3 cakes).
Amount of copper	211.8 lbs.
Amount of lead	635.4 lbs.
Weight of each cake	282.4 lbs.
Average value of charge	218 ozs. 18 dwts.
Per cent. of copper	25%
Average value of original copper per ton	583 ozs. 6 dwts. 16 grs.
Weight of argentiferous lead liquated out	494.2 lbs.
Average value of liquated lead per ton	352 ozs. 8 dwts.
Weight of residues	353 lbs.
Average value of residues per ton	20 ozs. (about).
Extraction of silver into the argentiferous lead	94%

The results given in the second case where the copper contains 2 *librae* and a *bes* per *centumpondium* do not tie together at all, for each liquation cake should contain 3 *librae* 9½ *unciae*, instead of 1½ *librae* and ½ *uncia* of silver.

If a *centumpondium* of copper contains two *librae* and a *bes* of silver, and the lead a third of a *libra* and a *semi-uncia*, there will be in each liquation cake one and a half *librae* and a *semi-uncia*, and a little more than a *sicilicus* of silver. In the exhausted liquation cakes there remain a third of a *libra* and a *semi-uncia* of silver.

If there be in the copper only a minute proportion of silver, it cannot be separated easily until it has been re-melted in other furnaces, so that in the "bottoms" there remains more silver and in the "tops" less.[15] This

A—FURNACE. B—FOREHEARTH. C—DIPPING-POT. D—CAKES.

furnace, vaulted with unbaked bricks, is similar to an oven, and also to the cupellation furnace, in which the lead is separated from silver, which I described in the last book. The crucible is made of ashes, in the same manner as

[15]In this enrichment of copper by the "settling" of the silver in the molten mass the original copper ran, in the two cases given, 60 ozs. 15 dwts and 85 ozs. 1 dwt. per ton. The whole charge weighed 2,685 lbs., and contained in the second case 114 ozs. Troy, omitting fractions. On melting, 1,060 lbs. were drawn off as "tops," containing 24 ozs. of silver, or running 45 ozs. per ton, and there remained 1,625 lbs. of "bottoms," containing 90 ozs. of silver, or averaging 110 ozs. per ton. It will be noticed later on in the description of making liquation cakes from these copper bottoms, that the author alters the value from one-third *librae*, a *semi-uncia* and a *drachma* per *centumpondium* to one-third of a *libra*, *i.e.*, from 110 ozs. to 97 ozs. 4 dwts. per ton. In the Glossary this furnace is described as a *spleisofen*, *i.e.*, a refining hearth.

in the latter, and in the front of the furnace, three feet above the floor of the building, is the mouth out of which the re-melted copper flows into a forehearth and a dipping-pot. On the left side of the mouth is an aperture, through which beech-wood may be put into the furnace to feed the fire. If in a *centumpondium* of copper there were a sixth of a *libra* and a *semi-uncia* of silver, or a quarter of a *libra*, or a quarter of a *libra* and a *semi-uncia*—there is re-melted at the same time thirty-eight *centumpondia* of it in this furnace, until there remain in each *centumpondium* of the copper " bottoms " a third of a *libra* and a *semi-uncia* of silver. For example, if in each *centumpondium* of copper not yet re-melted, there is a quarter of a *libra* and a *semi-uncia* of silver, then the thirty-eight *centumpondia* that are smelted together must contain a total of eleven *librae* and an *uncia* of silver. Since from fifteen *centumpondia* of re-melted copper there was a total of four and a third *librae* and a *semi-uncia* of silver, there remain only two and a third *librae*. Thus there is left in the " bottoms," weighing twenty-three *centumpondia*, a total of eight and three-quarter *librae* of silver. Therefore, each *centumpondium* of this contains a third of a *libra* and a *semi-uncia*, a *drachma*, and the twenty-third part of a *drachma* of silver ; from such copper it is profitable to separate the silver. In order that the master may be more certain of the number of *centumpondia* of copper in the " bottoms," he weighs the " tops " that have been drawn off from it ; the " tops " were first drawn off into the dipping-pot, and cakes were made from them. Fourteen hours are expended on the work of thus dividing the copper. The " bottoms," when a certain weight of lead has been added to them, of which alloy I shall soon speak, are melted in the blast furnace ; liquation cakes are then made, and the silver is afterward separated from the copper. The " tops " are subsequently melted in the blast furnace, and re-melted in the refining furnace, in order that red copper shall be made[16] ; and the " tops " from this are again smelted in the blast furnace, and then again in the refining furnace, that therefrom

[16]The latter part of this paragraph presents great difficulties. The term " refining furnace " is given in the Latin as the " second furnace," an expression usually applied to the cupellation furnace. The whole question of refining is exhaustively discussed on pages 530 to 539. Exactly what material is meant by the term red (*rubrum*), yellow (*fulvum*) and *caldarium* copper is somewhat uncertain. They are given in the German text simply as *rot*, *geel*, and *lebeter kupfer*, and apparently all were " coarse " copper of different characters destined for the refinery. The author states in *De Natura Fossilium* (p. 334) : " Copper has a " red colour peculiar to itself ; this colour in smelted copper is considered the most excellent. " It, however, varies. In some it is red, as in the copper smelted at Neusohl. . . . . " Other copper is prepared in the smelters where silver is separated from copper, which is " called yellow copper (*luteum*), and is *regulare*. In the same place a dark yellow copper is " made which is called *caldarium*, taking its name among the Germans from a caldron. " . . . . *Regulare* differs from *caldarium* in that the former is not only fusible, but " also malleable ; while the latter is, indeed, fusible, but is not ductile, for it breaks when " struck with the hammer." Later on in *De Re Metallica* (p. 542) he describes yellow copper as made from " baser " liquation thorns and from exhausted liquation cakes made from thorns. These products were necessarily impure, as they contained, among other things, the concentrates from furnace accretions. Therefore, there was ample source for zinc, arsenic or other metallics which would lighten the colour. *Caldarium* copper is described by Pliny (see note, p. 404), and was, no doubt, " coarse " copper, and apparently Agricola adopted this term from that source, as we have found it used nowhere else. On page 542 the author describes making *caldarium* copper from a mixture of yellow copper and a peculiar *cadmia*, which he describes as the " slags " from refining copper. These " slags," which are the result of oxidation and poling, would contain almost any of the metallic impurities of the original ore, antimony, lead, arsenic, zinc, cobalt, etc. Coming from these two sources the *caldarium* must have been, indeed, impure.

shall be made *caldarium* copper. But when the copper, yellow or red or *caldarium* is re-smelted in the refining furnace, forty *centumpondia* are placed in it, and from it they make at least twenty, and at most thirty-five, *centumpondia*. About twenty-two *centumpondia* of exhausted liquation cakes and ten of yellow copper and eight of red, are simultaneously placed in this latter furnace and smelted, in order that they may be made into refined copper.

The copper "bottoms" are alloyed in three different ways with lead.[17] First, five-eighths of a *centumpondium* of copper and two and three-quarters *centumpondia* of lead are taken; and since one liquation cake is made from this, therefore two and a half *centumpondia* of copper and eleven *centumpondia* of lead make four liquation cakes. Inasmuch as in each *centumpondium* of copper there is a third of a *libra* of silver, there would be in the whole of the copper ten-twelfths of a *libra* of silver; to these are added four *centumpondia* of lead re-melted from "slags," each *centumpondium* of which contains a *sicilicus* and a *drachma* of silver, which weights make up a total of an *uncia* and a half of silver. There is also added seven *centumpondia* of de-silverized lead, in each *centumpondium* of which there is a *drachma* of silver; therefore in the four cakes of copper-lead alloy there is a total of a *libra*, a *sicilicus* and a *drachma* of silver. In each single *centumpondium* of lead, after it has been liquated from the copper, there is an *uncia* and a *drachma* of silver, which alloy we call "poor" argentiferous lead, because it contains but little silver. But as five cakes of that kind are placed together in the furnace, they liquate from them usually as much as nine and three-quarters *centumpondia* of poor

[17]The liquation of these low-grade copper "bottoms" required that the liquated lead should be re-used again to make up fresh liquation cakes, in order that it might eventually become rich enough to warrant cupellation. In the following table the "poor" silver-lead is designated (A) the "medium" (B) and the "rich" (C). The three charges here given are designated sixth, seventh, and eighth for purposes of reference. It will be seen that the data is insufficient to complete the ninth and tenth. Moreover, while the author gives directions for making four cakes, he says the charge consists of five, and it has, therefore, been necessary to reduce the volume of products given to this basis.

	6th Charge.	7th Charge.	8th Charge.
Amount of copper bottoms	176.5 lbs.	176.5 lbs.	176.5 lbs.
Amount of lead	282.4 lbs. (slags)	564.8 lbs. or (A)	635.4 lbs. of (B)
Amount of de-silverized lead	494.2 lbs.	211.8 lbs.	141.2 lbs. (A)
Weight of each cake ..	238.3 lbs.	238.3 lbs.	238.3 lbs
Average value of charge per ton	22 ozs. 5dwts.	35 ozs. 15dwts.	50 ozs. 5dwts.
Per cent. of copper ..	18.5%	18.5%	18.5%
Average value per ton original copper ..	97 ozs. 4dwts.	97 ozs. 4dwts.	97 ozs. 4dwts.
Average value per ton of..	90 ozs. 2dwts. (slags)	28 ozs. 5dwts. (A)	28 ozs. 5dwts. (A)
Average value per ton of..	3 ozs. 1dwt. (lead)	3 ozs. 1dwt. (lead)	42 ozs. 10dwts (B)
Weight of liquated lead..	550.6 lbs.		
Average value of the liquated lead per ton ..	28 ozs. 5dwts. (A)	42 ozs. 10dwts. (B)	63 ozs. 16dwts. (C)
Weight of exhausted liquation cakes	225.9 lbs.		
Average value of the exhausted liquation cakes per ton	12 ozs. 3dwts.		
Weight of liquation thorns	169.4 lbs.		
Average value of the liquation thorns per ton ..	18 ozs. 4dwts.		
Extraction of silver into the liquated lead ..	71%		

argentiferous lead, in each *centumpondium* of which there is an *uncia* and a *drachma* of silver, or a total of ten *unciae* less four *drachmae*. Of the liquation thorns there remain three *centumpondia*, in each *centumpondium* of which there are three *sicilici* of silver; and there remain four *centumpondia* of exhausted liquation cakes, each *centumpondium* of which contains a *semi-uncia* or four and a half *drachmae*. Inasmuch as in a *centumpondium* of copper "bottoms" there is a third of a *libra* and a *semi-uncia* of silver, in five of those cakes there must be more than one and a half *unciae* and half a *drachma* of silver.

Then, again, from another two and a half *centumpondia* of copper "bottoms," together with eleven *centumpondia* of lead, four liquation cakes are made. If in each *centumpondium* of copper there was a third of a *libra* of silver, there would be in the whole of the *centumpondia* of base metal five-sixths of a *libra* of the precious metal. To this copper is added eight *centumpondia* of poor argentiferous lead, each *centumpondium* of which contains an *uncia* and a *drachma* of silver, or a total of three-quarters of a *libra* of silver. There is also added three *centumpondia* of de-silverized lead, in each *centumpondium* of which there is a *drachma* of silver. Therefore, four liquation cakes contain a total of a *libra*, seven *unciae*, a *sicilicus* and a *drachma* of silver; thus each *centumpondium* of lead, when it has been liquated from the copper, contains an *uncia* and a half and a *sicilicus* of silver, which alloy we call "medium" silver-lead.

Then, again, from another two and a half *centumpondia* of copper "bottoms," together with eleven *centumpondia* of lead, they make four liquation cakes. If in each *centumpondium* of copper there were likewise a third of a *libra* of silver, there will be in all the weight of the base metal five-sixths of a *libra* of the precious metal. To this is added nine *centumpondia* of medium silver-lead, each *centumpondium* of which contains an *uncia* and a half and a *sicilicus* of silver; or a total of a *libra* and a quarter and a *semi-uncia* and a *sicilicus* of silver. And likewise they add two *centumpondia* of poor silver-lead, in each of which there is an *uncia* and a *drachma* of silver. Therefore the four liquation cakes contain two and a third *librae* of silver. Each *centumpondium* of lead, when it has been liquated from the copper, contains a sixth of a *libra* and a *semi-uncia* and a *drachma* of silver. This alloy we call "rich" silver-lead; it is carried to the cupellation furnace, in which lead is separated from silver. I have now mentioned in how many ways copper containing various proportions of silver is alloyed with lead, and how they are melted together in the furnace and run into the casting pan.

Now I will speak of the method by which lead is liquated from copper simultaneously with the silver. The liquation cakes are raised from the ground with the crane, and placed on the copper plates of the furnaces. The hook of the chain let down from the arm of the crane, is inserted in a ring of the tongs, one jaw of which has a tooth; a ring is engaged in each of the handles of the tongs, and these two rings are engaged in a third, in which the hook of the chain is inserted. The tooth on the one jaw of the tongs is struck by a hammer, and driven into the hole in the cake, at the point where the straight end of the hook was driven into it when it was lifted out of the copper mould; the other jaw of the tongs, which has no tooth, squeezes the cake, lest the tooth should fall out of it; the tongs are one and a half feet long, each ring is a digit and a half thick, and the inside is a palm and two digits in diameter. Those cranes by which the cakes are lifted out of the copper pans and placed on the ground, and lifted up again from there and placed in the furnaces, are two in number—one in the middle space between the third transverse wall and the two upright posts, and the other in

A—CRANE. B—DRUM CONSISTING OF RUNDLES. C—TOOTHED DRUM. D—TROLLEY AND ITS WHEELS. E—TRIANGULAR BOARD. F—CAKES. G—CHAIN OF THE CRANE. H—ITS HOOK. I—RING. K—THE TONGS.

the middle space between the same posts and the seventh transverse wall. The rectangular crane-post of both of these is two feet wide and thick, and is eighteen feet from the third long wall, and nineteen from the second long wall. There are two drums in the framework of each—one drum consisting of rundles, the other being toothed. The crane-arm of each extends seventeen feet, three palms and as many digits from the post. The trolley of each crane is two feet and as many palms long, a foot and two digits wide, and a palm and two digits thick; but where it runs between the beams of the crane-arm it is three digits wide and a palm thick; it has five notches, in which turn five brass wheels, four of which are small, and the fifth much larger than the rest. The notches in which the small wheels turn are two palms long and as much as a palm wide ; those wheels are a palm wide and a palm and two digits in diameter ; four of the notches are near the four corners of the trolley ; the fifth notch is between the two front ones, and it is two palms back from the front. Its pulley is larger than the rest, and turns in its own notch ; it is three palms in diameter and one palm wide, and grooved on the circumference, so that the iron chain may run in the groove. The trolley has two small axles, to the one in front are fastened three, and to the one at the back, the two wheels ; two wheels run on the one beam of the crane-arm, and two on the other ; the fifth wheel, which is larger than the others, runs between those two beams. Those people who have no cranes place the cakes on a triangular board, to which iron cleats are affixed, so that it will last longer ; the board has three iron chains, which are fixed in an iron ring at the top ; two workmen pass a pole through the ring and carry it on their shoulders, and thus take the cake to the furnace in which silver is separated from copper.

From the vicinity of the furnaces in which copper is mixed with lead and the "slags" are re-melted, to the third long wall, are likewise ten furnaces, in which silver mixed with lead is separated from copper. If this space is eighty feet and two palms long, and the third long wall has in the centre a door three feet and two palms wide, then the spaces remaining at either side of the door will be thirty-eight feet and two palms ; and if each of the furnaces occupies four feet and a palm, then the interval between each furnace and the next one must be a foot and three palms ; thus the width of the five furnaces and four interspaces will be twenty-eight feet and a palm. Therefore, there remain ten feet and a palm, which measurement is so divided that there are five feet and two digits between the first furnace and the transverse wall, and as many feet and digits between the fifth furnace and the door ; similarly in the other part of the space from the door to the sixth furnace, there must be five feet and two digits, and from the tenth furnace to the seventh transverse wall, likewise, five feet and two digits. The door is six feet and two palms high ; through it the foreman of the *officina* and the workmen enter the store-room in which the silver-lead alloy is kept.

Each furnace has a bed, a hearth, a rear wall, two sides and a front, and a receiving-pit. The bed consists of two sole-stones, four rectangular stones, and two copper plates ; the sole-stones are five feet and a palm long, a cubit wide, a foot and a palm thick, and they are sunk into the ground, so that they emerge a palm and two digits ; they are distant from each other about three palms, yet the distance is narrower at the back than the front. Each of the rectangular stones is two feet and as many palms long, a cubit wide, and a cubit thick at the outer edge, and a foot and a palm thick on the inner edge which faces the hearth, thus they form an incline, so that there is a slope to the copper plates which are laid upon them. Two of these rectangular stones are placed on one sole-stone ; a hole is cut in the upper edge of each, and into the holes are placed iron clamps, and lead is poured in ; they are so placed on the sole-stones that they project a palm at the sides, and at the front the sole-stones project to the same extent; if rectangular stones are not available, bricks are laid in their place. The copper plates are four feet two palms and as many digits long, a cubit wide, and a palm thick; each edge has a protuberance, one at the front end, the other at the back; these are a palm and three digits long, and a palm wide and thick. The plates are so laid upon the rectangular stones that their rear ends are three digits from the third long wall; the stones project beyond the plate the same number of digits in front, and a palm and three digits at the sides. When the plates have been joined, the groove which is between the protuberances is a palm and three digits wide, and four feet long, and through it flows the silver-lead which liquates from the cakes. When the plates are corroded either by the fire or by the silver-lead, which often adheres to them in the form of stalactites, and is chipped off, they are exchanged, the right one being placed to the left, and the left one, on the contrary, to the right; but the left side of the plates, which, when the fusion of the copper took place, came into contact with the copper, must lie flat; so that when the exchange of the plates has been carried out, the protuberances, which are thus on the underside, raise the plate from the stones, and they have to be partially chipped off, lest they should prove an impediment to the work; and in each of their places is laid a piece of iron, three palms long, a digit thick at both ends, and a palm thick in the centre for the length of a palm and three digits.

The passage under the plates between the rectangular stones is a foot wide at the back, and a foot and a palm wide at the front, for it gradually widens out. The hearth, which is between the sole-stones, is covered with a bed of hearth-lead, taken from the crucible in which lead is separated from silver. The rear end is the highest, and should be so high that it reaches to within six digits of the plates, from which point it slopes down evenly to the front end, so that the argentiferous lead alloy which liquates from the cakes can flow into the receiving-pit. The wall built against the third long wall in order to protect it from injury by fire, is constructed of bricks joined together with lute, and stands on the copper plates; this wall is two feet, a palm and two digits high, two palms thick, and three feet, a palm and three digits wide at the bottom, for it reaches across both of them; at the top it is three feet wide, for it rises up obliquely on each side. At each side of this wall, at a height of a palm and two digits above the top of it, there is inserted in a hole in the third long wall a hooked iron rod, fastened in with molten lead; the rod projects two palms from the wall, and is two digits wide and one digit thick; it has two hooks, the one at the side, the other at the end. Both of these hooks open toward the wall, and both are a digit thick, and both are inserted in the last, or the adjacent, links of a short iron chain. This chain consists of four links, each of which is a palm and a digit long and half a digit thick; the first link is engaged in the first hole in a long iron rod, and one or other of the remaining three links engages the hook of the hooked rod. The two long rods are three feet and as many palms and digits long, two digits wide, and one digit thick; both ends of both of these rods have holes,

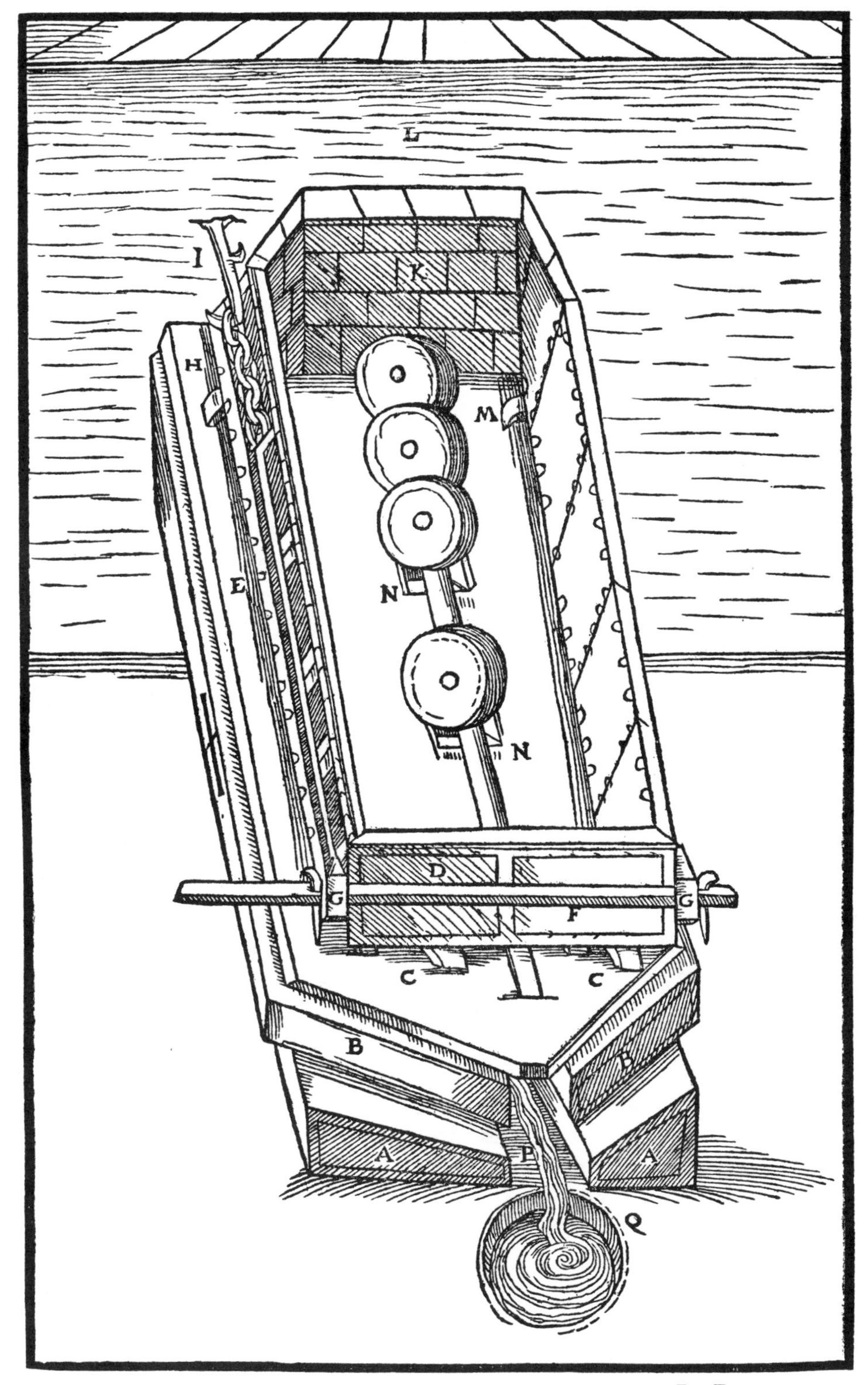

A—Sole-stones. B—Rectangular stones. C—Copper plates. D—Front panel. E—Side panels. F—Bar. G—Front end of the long iron rods. H—Short chain. I—Hooked rod. K—Wall which protects the third long wall from injury by fire. L—Third long wall. M—Feet of the panels. N—Iron blocks. O—Cakes. P—Hearth. Q—Receiving-pit.

the back one of which is round and a digit in diameter, and in this is engaged the first link of the chain as I have stated ; the hole at the front end is two digits and a half long and a digit and a half wide. This end of each rod is made three digits wide, while for the rest of its length it is only two digits, and at the back it is two and a half digits. Into the front hole of each rod is driven an iron bar, which is three feet and two palms long, two digits wide and one thick ; in the end of this bar are five small square holes, two-thirds of a digit square ; each hole is distant from the other half a digit, the first being at a distance of about a digit from the end. Into one of these holes the refiner drives an iron pin ; if he should desire to make the furnace narrower, then he drives it into the last hole ; if he should desire to widen it, then into the first hole ; if he should desire to contract it moderately, then into one of the middle holes. For the same reason, therefore, the hook is sometimes inserted into the last link of the chain, and sometimes into the third or the second. The furnace is widened when many cakes are put into it, and contracted when there are but few, but to put in more than five is neither usual nor possible ; indeed, it is because of thin cakes that the walls are contracted. The bar has a hump, which projects a digit on each side at the back, of the same width and thickness as itself. These humps project, lest the bar should slip through the hole of the right-hand rod, in which it remains fixed when it, together with the rods, is not pressing upon the furnace walls.

There are three panels to the furnace—two at the sides, one in front and another at the back. Those which are at the sides are three feet and as many palms and two digits long, and two feet high ; the front one is two feet and a palm and three digits long, and, like the side ones, two feet high. Each consists of iron bars, of feet, and of iron plates. Those which are at the side have seven bars, the lower and upper of which are of the same length as the panels ; the former holds up the upright bars ; the latter is placed upon them ; the uprights are five in number, and have the same height as the panels ; the middle ones are inserted into holes in the upper and lower bars ; the outer ones are made of one and the same bar as the lower and upper ones. They are two digits wide and one thick. The front panel has five bars ; the lower one holds similar uprights, but there are three of them only ; the upper bar is placed on them. Each of these panels has two feet fixed at each end of the lower bar, and these are two palms long, one wide, and a digit thick. The iron plates are fastened to the inner side of the bars with iron wire, and they are covered with lute, so that they may last longer and may be uninjured by the fire. There are, besides, iron blocks three palms long, one wide, and a digit and a half thick ; the upper surface of these is somewhat hollowed out, so that the cakes may stand in them ; these iron blocks are dipped into a vessel in which there is clay mixed with water, and they are used only for placing under the cakes of copper and lead alloy made in the furnaces. There is more silver in these than in those which are made of liquation thorns, or furnace accretions, or re-melted "slags." Two iron blocks are placed under each cake, in order that, by raising it up, the fire may bring more force to bear upon it ; the one is put on the right bed-plate,

A — Furnace in which the operation of liquation is being performed. B—Furnace in which it is not being performed. C—Receiving-pit. D—Moulds. E—Cakes. F—Liquation thorns.

the other on the left. Finally, outside the hearth is the receiving-pit, which is a foot wide and three palms deep; when this is worn away it is restored with lute alone, which easily retains the lead alloy.

If four liquation cakes are placed on the plates of each furnace, then the iron blocks are laid under them; but if the cakes are made from copper "bottoms," or from liquation thorns, or from the accretions or "slags," of which I have partly written above and will further describe a little later, there are five of them, and because they are not so large and heavy, no blocks are placed under them. Pieces of charcoal six digits long are laid between the cakes, lest they should fall one against the other, or lest the last one should fall against the wall which protects the third long wall from injury by fire. In the middle empty spaces, long and large pieces of charcoal are likewise laid. Then when the panels have been set up, and the bar has been closed, the furnace is filled with small charcoal, and a wicker basket full of charcoal is thrown into the receiving-pit, and over that are thrown live coals; soon afterward the burning coal, lifted up in a shovel, is spread over all parts of the furnace, so that the charcoal in it may be kindled; any charcoal which remains in the receiving-pit is thrown into the passage, so that it may likewise be heated. If this has not been done, the silver-lead alloy liquated from the cakes is frozen by the coldness of the passage, and does not run down into the receiving-pit.

After a quarter of an hour the cakes begin to drip silver-lead alloy,[18] which runs down through the openings between the copper plates into the passage. When the long pieces of charcoal have burned up, if the cakes lean toward the wall, they are placed upright again with a hooked bar, but if they lean toward the front bar they are propped up by charcoal; moreover, if some cakes shrink more than the rest, charcoal is added to the former and not to the others. The silver drips together with the lead, for both melt more rapidly than copper. The liquation thorns do not flow away, but remain in the passage, and should be turned over frequently with a hooked bar, in order that the silver-lead may liquate away from them and flow down into the receiving pit; that which remains is again melted in the blast furnace, while that which flows into the receiving pit is at once carried with the remain-

[18]For the liquation it was necessary to maintain a reducing atmosphere, otherwise the lead would oxidize; this was secured by keeping the cakes well covered with charcoal and by preventing the entrance of air as much as possible. Moreover, it was necessary to preserve a fairly even temperature. The proportions of copper and lead in the three liquation products vary considerably, depending upon the method of conducting the process and the original proportions. From the authors consulted (see note p. 492) an average would be about as follows:—The residual copper—exhausted liquation cakes—ran from 25 to 33% lead; the liquated lead from 2 to 3 % copper; and the liquation thorns, which were largely oxidized, contained about 15% copper oxides, 80% lead oxides, together with impurities, such as antimony, arsenic, etc. The proportions of the various products would obviously depend upon the care in conducting the operation; too high temperature and the admission of air would increase the copper melted and oxidize more lead, and thus increase the liquation thorns. There are insufficient data in Agricola to adduce conclusions as to the actual ratios produced. The results given for the 6th charge (note 17, p. 512) would indicate about 30% lead in the residual copper, and would indicate that the original charge was divided into about 24% of residual copper, 18% of liquation thorns, and 57% of liquated lead. This, however, was an unusually large proportion of liquation thorns some of the authors giving instances of as low as 5%.

ing products to the cupellation furnace, where the lead is separated from the silver. The hooked bar has an iron handle two feet long, in which is set a wooden one four feet long. The silver-lead which runs out into the receiving-pit is poured out by the refiner with a bronze ladle into eight copper moulds, which are two palms and three digits in diameter; these are first smeared with a lute wash so that the cakes of silver-lead may more easily fall out when they are turned over. If the supply of moulds fails because the silver-lead flows down too rapidly into the receiving-pit, then water is poured on them, in order that the cakes may cool and be taken out of them more rapidly; thus the same moulds may be used again immediately; if no such necessity urges the refiner, he washes over the empty moulds with a lute wash. The ladle is exactly similar to that which is used in pouring out the metals that are melted in the blast furnace. When all the silver-lead has run down from the passage into the receiving-pit, and has been poured out into copper moulds, the thorns are drawn out of the passage into the receiving-pit with a rabble; afterward they are raked on to the ground from the receiving-pit, thrown with a shovel into a wheelbarrow, and, having been conveyed away to a heap, are melted once again. The blade of the rabble is two palms and as many digits long, two palms and a digit wide, and joined to its back is an iron handle three feet long; into the iron handle is inserted a wooden one as many feet in length.

The residue cakes, after the silver-lead has been liquated from the copper, are called "exhausted liquation cakes" (*fathiscentes*), because when thus smelted they appear to be dried up. By placing a crowbar under the cakes they are raised up, seized with tongs, and placed in the wheelbarrow; they are then conveyed away to the furnace in which they are "dried." The crowbar is somewhat similar to those generally used to chip off the accretions that adhere to the walls of the blast furnace. The tongs are two and a half feet long. With the same crowbar the stalactites are chipped off from the copper plates from which they hang, and with the same instrument the iron blocks are struck off the exhausted liquation cakes to which they adhere. The refiner has performed his day's task when he has liquated the silver-lead from sixteen of the large cakes and twenty of the smaller ones; if he liquates more than this, he is paid separately for it at the price for extraordinary work.

Silver, or lead mixed with silver, which we call *stannum*, is separated by the above method from copper. This silver-lead is carried to the cupellation furnace, in which lead is separated from silver; of these methods I will mention only one, because in the previous book I have explained them in detail. Amongst us some years ago only forty-four *centumpondia* of silver-lead and one of copper were melted together in the cupellation furnaces, but now they melt forty-six *centumpondia* of silver-lead and one and a half *centumpondia* of copper; in other places, usually a hundred and twenty *centumpondia* of silver-lead alloy and six of copper are melted, in which manner they make about one hundred and ten *centumpondia* more or less of litharge and thirty of hearth-lead. But in all these methods the silver which is in the copper is mixed with the remainder of silver; the copper itself, equally with the lead, will be changed partly into litharge and partly into hearth-lead.[19] The silver-lead alloy which does not melt is taken from the margin of the crucible with a hooked bar.

The work of "drying" is distributed into four operations, which are performed in four days. On the first—as likewise on the other three days—the master begins at the fourth hour of the morning, and with his assistant chips

A—CAKES. B—HAMMER.

off the stalactites from the exhausted liquation cakes. They then carry the cakes to the furnace, and put the stalactites upon the heap of liquation thorns. The head of the chipping hammer is three palms and as many digits

[19]The first instance given, of 44 *centumpondia* (3,109 lbs.) lead and one *centumpondium* (70.6 lbs.) copper, would indicate that the liquated lead contained 2.2% copper. The second, of 46 *centumpondia* (3,250 lbs.) lead and 1½ *centumpondia* copper (106 lbs.), would indicate 3% copper; and in the third, 120 *centumpondia* (8,478 lbs.) lead and six copper (424 lbs.) would show 4.76% copper. This charge of 120 *centumpondia* in the cupellation furnace would normally make more than 110 *centumpondia* of litharge and 30 of hearth-lead, *i.e.*, saturated furnace bottoms. The copper would be largely found in the silver-lead "which does not melt," at the margin of the crucible. These skimmings are afterward referred to as "thorns." It is difficult to understand what is meant by the expression that the silver which is in the copper is mixed with the remaining (*reliquo*) silver. The coppery skimmings from the cupellation furnace are referred to again in Note 28, p. 539.

long ; its sharp edge is a palm wide ; the round end is three digits thick ; the wooden handle is four feet long.

The master throws pulverised earth into a small vessel, sprinkles water over it, and mixes it ; this he pours over the whole hearth, and sprinkles charcoal dust over it to the thickness of a digit. If he should neglect this, the copper, settling in the passages, would adhere to the copper bed-plates, from which it can be chipped off only with difficulty ; or else it would adhere to the bricks, if the hearth was covered with them, and when the copper is chipped off these they are easily broken. On the second day, at the same time, the master arranges bricks in ten rows ; in this manner twelve passages are made. The first two rows of bricks are between the first and the second openings on the right of the furnace ; the next three rows are between the second and third openings, the following three rows are between the third and the fourth openings, and the last two rows between the fourth and fifth openings. These bricks are a foot and a palm long, two palms and a digit wide, and a palm and two digits thick ; there are seven of these thick bricks in a row, so there are seventy all together. Then on the first three rows of bricks they lay exhausted liquation cakes and a layer five digits thick of large charcoal ; then in a similar way more exhausted liquation cakes are laid upon the other bricks, and charcoal is thrown upon them ; in this manner seventy *centumpondia* of cakes are put on the hearth of the furnace. But if half of this weight, or a little more, is to be "dried," then four rows of bricks will suffice. Those who dry exhausted liquation cakes[20] made from copper "bottoms" place ninety or a hundred *centumpondia*[21] into the furnace at the same time. A place is left in the front part of the furnace for the topmost cakes removed from the forehearth in which copper is made, these being more suitable for supporting the exhausted liquation cakes than are iron plates ; indeed, if the former cakes drip copper from the heat, this can be taken back with the liquation thorns to the first furnace, but melted iron is of no use to us in these matters. When the cakes of this kind have been placed in front of the exhausted liquation cakes, the workman inserts the iron bar into the holes on the inside of the wall, which are at a height of three palms and two digits above the hearth ; the hole to the left penetrates through into the wall, so that the bar may be pushed back

[20]A further amount of lead could be obtained in the first liquation, but a higher temperature is necessary, which was more economical to secure in the "drying" furnace. Therefore, the "drying" was really an extension of liquation ; but as air was admitted the lead and copper melted out were oxidized. The products were the final residual copper, called by Agricola the "dried" copper, together with lead and copper oxides, called by him the "slags," and the scale of copper and lead oxides termed by him the "ash-coloured copper." The German metallurgists distinguished two kinds of slag : the first and principal one, the *darrost*, and the second the *darrsöhle*, this latter differing only in that it contained more impurities from the floor of the furnace, and remained behind until the furnace cooled. Agricola possibly refers to these as "more liquation thorns," because in describing the treatment of the bye-products he refers to thorns from the process, whereas in the description of "drying" he usually refers to "slags." A number of analyses of these products, given by Karsten, show the "dried" copper to contain from 82.7 to 90.6% copper, and from 9.4 to 17.3% lead ; the "slag" to contain 76.5 to 85.1% lead oxide, and from 4.1 to 7.8% cuprous oxide, with 9 to 13% silica from the furnace bottoms, together with some other

[21]If Roman weights, this would equal from 6,360 lbs. to 7,066 lbs.

and forth. This bar is round, eight feet long and two digits in diameter; on the right side it has a haft made of iron, which is about a foot from the right end; the aperture in this haft is a palm wide, two digits high, and a digit thick. The bar holds the exhausted liquation cakes opposite, lest they should fall down. When the operation of "drying" is completed, a workman draws out this bar with a crook which he inserts into the haft, as I will explain hereafter.

In order that one should understand those things of which I have spoken, and concerning which I am about to speak, it is necessary for me to give some information beforehand about the furnace and how it is to be made. It stands nine feet from the fourth long wall, and as far from the wall which is between the second and fourth transverse walls. It consists of walls, an arch, a chimney, an interior wall, and a hearth; the two walls are at the sides; and they are eleven feet three palms and two digits long, and where they support the chimney they are eight feet and a palm high. At the front of the arch they are only seven feet high; they are two feet three palms and two digits thick, and are made either of rock or of bricks; the distance between them is eight feet, a palm and two digits. There are two of the arches, for the space at the rear between the walls is also arched from the ground, in order that it may be able to support the chimney; the foundations of these arches are the walls of the furnace; the span of the arch has the same length as the space between the walls; the top of the arch is five feet, a palm and two digits high. In the rear arch there is a wall made of bricks joined with lime; this wall at a height of a foot and three palms from the ground has five vent-holes, which are two palms and a digit high, a palm and a digit wide, of which the first is near the right interior wall, and the last near the left interior wall, the remaining three in the intervening space; these vent-holes penetrate through the interior of the wall which is in the arch. Half-bricks can be placed over the vent-holes, lest too much air should be drawn into the furnace, and they can be taken out at times, in order that he who is "drying" the exhausted liquation cakes may inspect the passages, as they are called, to see whether the cakes are being properly "dried." The front arch is three feet two palms distant from the rear one; this arch is the same thickness as that of the rear arch, but the span is six feet wide;

impurities; the "ash-coloured copper" to contain about 60% cuprous oxide and 30% lead oxide, with some metallic copper and minor impurities. An average of proportions given by various authors shows, roughly, that out of 100 *centners* of "exhausted" liquation cakes, containing about 70% copper and 30% lead, there were about 63 *centners* of "dried" copper, 38 *centners* of "slag," and 6½ *centners* of "ash-coloured copper." According to Karsten, the process fell into stages; first, at low temperature some metallic lead appeared; second, during an increasing temperature for over 14 to 15 hours the slags ran out; third, there was a period of four hours of lower temperature to allow time for the lead to diffuse from the interior of the cakes; and fourth, during a period of eight hours the temperature was again increased. In fact, the latter portion of the process ended with the economic limit between leaving some lead in the copper and driving too much copper into the "slags." Agricola gives the silver contents of the "dried" copper as 3 *drachmae* to 1 *centumpondium*, or equal to about 9 ozs. per ton; and assuming that the copper finally recovered from the bye-products ran no higher, then the first four charges (see note on p. 506) would show a reduction in the silver values of from 95 to 97%; the 7th and 8th charges (note on p. 512) of about 90%.

A—Side walls. B—Front arch. C—Rear arch. D—Wall in the rear arch. E—Inner wall. F—Vent holes. G—Chimney. H—Hearth. I—Tank. K—Pipe. L—Plug. M—Iron door. N—Transverse bars. O—Upright bars. P—Plates. Q—Rings of the bars. R—Chains. S—Rows of bricks. T—Bar. V—Its haft. X—Copper bed-plates.

the interior of the aroh itself is of the same height as the walls. A chimney is built upon the arches and the walls, and is made of bricks joined together with lime; it is thirty-six feet high and penetrates through the roof. The interior wall is built against the rear arch and both the side walls, from which it juts out a foot; it is three feet and the same number of palms high, three palms thick, and is made of bricks joined together with lute and smeared thickly with lute, sloping up to the height of a foot above it. This wall is a kind of shield, for it protects the exterior walls from the heat of the fire, which is apt to injure them; the latter cannot be easily re-made, while the former can be repaired with little work.

The hearth is made of lute, and is covered either with copper plates, such as those of the furnaces in which silver is liquated from copper, although they have no protuberances, or it may be covered with bricks, if the owners are unwilling to incur the expense of copper plates. The wider part of the hearth is made sloping in such a manner that the rear end reaches as high as the five vent-holes, and the front end of the hearth is so low that the back of the front arch is four feet, three palms and as many digits above it, and the front five feet, three palms and as many digits. The hearth beyond the furnaces is paved with bricks for a distance of six feet. Near the furnace, against the fourth long wall, is a tank thirteen feet and a palm long, four feet wide, and a foot and three palms deep. It is lined on all sides with planks, lest the earth should fall into it; on one side the water flows in through pipes, and on the other, if the plug be pulled out, it soaks into the earth; into this tank of water are thrown the cakes of copper from which the silver and lead have been separated. The fore part of the front furnace arch should be partly closed with an iron door; the bottom of this door is six feet and two digits wide; the upper part is somewhat rounded, and at the highest point, which is in the middle, it is three feet and two palms high. It is made of iron bars, with plates fastened to them with iron wire, there being seven bars—three transverse and four upright—each of which is two digits wide and half a digit thick. The lowest transverse bar is six feet and two palms long; the middle one has the same length; the upper one is curved and higher at the centre, and thus longer than the other two. The upright bars are two feet distant from one another; both the outer ones are two feet and as many palms high; but the centre ones are three feet and two palms. They project from the upper curved transverse bar and have holes, in which are inserted the hooks of small chains two feet long; the topmost links of these chains are engaged in the ring of a third chain, which, when extended, reaches to one end of a beam which is somewhat cut out. The chain then turns around the beam, and again hanging down, the hook in the other end is fastened in one of the links. This beam is eleven feet long, a palm and two digits wide, a palm thick, and turns on an iron axle fixed in a near-by timber; the rear end of the beam has an iron pin, which is three palms and a digit long, and which penetrates through it where it lies under a timber, and projects from it a palm and two digits on one side, and three digits on the other side. At this point the pin is perforated, in order that a ring may be fixed in it and hold it, lest it should fall out of the beam; that end is hardly a digit thick, while the other round end is thicker than a digit. When the door is to be shut, this pin lies under the timber and holds the door so that it cannot fall; the pin likewise prevents the rectangular iron band which encircles the end of the beam, and into which is inserted the ring of a long hook, from falling from the end. The lowest link of an iron chain, which is six feet long, is inserted in the ring of a staple driven into the right wall of the furnace, and fixed firmly by filling in with molten lead. The hook suspended at the top from the ring should be inserted in one of these lower links, when the door is to be raised; when the door is to be let down, the hook is taken out of that link and put into one of the upper links.

On the third day the master sets about the principal operation. First he throws a basketful of charcoals on to the ground in front of the hearth, and kindles them by adding live coals, and having thrown live coals on to the cakes placed within, he spreads them equally all over with an iron shovel. The blade of the shovel is three palms and a digit long, and three palms wide; its iron handle is two palms long, and the wooden one ten feet long, so that it can reach to the rear wall of the furnace. The exhausted liquation cakes become incandescent in an hour and a half, if the copper was good and hard,

A—The door let down. B—Bar. C—Exhausted liquation cakes. D—Bricks. E—Tongs.

or after two hours, if it was soft and fragile. The workman adds charcoal to them where he sees it is needed, throwing it into the furnace through the openings on both sides between the side walls and the closed door. This opening is a foot and a palm wide. He lets down the door, and when the " slags " begin to flow he opens the passages with a bar; this should take place after five hours; the door is let down over the upper open part of the arch for two feet and as many digits, so that the master can bear the violence of the heat. When the cakes shrink, charcoal should not be added to them lest they should melt. If the cakes made from poor and fragile copper are " dried " with cakes made from good hard copper, very often the copper so settles into the passages that a bar thrust into them cannot penetrate them. This bar is of iron, six feet and two palms long, into which a wooden handle five feet long is inserted. The refiner draws off the " slags " with a rabble from the right side of the hearth. The blade of the rabble is made of an iron plate a foot and a palm wide, gradually narrowing toward the handle; the blade is two palms high, its iron handle is two feet long, and the wooden handle set into it is ten feet long.

When the exhausted liquation cakes have been " dried," the master

A—The door raised. B—Hooked bar. C—Two-pronged rake. D—Tongs. E—Tank.

raises the door in the manner I have described, and with a long iron hook inserted into the haft of the bar he draws it through the hole in the left wall from the hole in the right wall; afterward he pushes it back and replaces it. The master then takes out the exhausted liquation cakes nearest to him with the iron hook; then he pulls out the cakes from the bricks. This hook is two palms high, as many digits wide, and one thick; its iron handle is two feet long, and the wooden handle eleven feet long. There is also a two-pronged rake with which the "dried" cakes are drawn over to the left side so that they may be seized with tongs; the prongs of the rake are pointed, and are two palms long, as many digits wide, and one digit thick; the iron part of the handle is a foot long, the wooden part nine feet long. The "dried" cakes, taken out of the hearth by the master and his assistants, are seized with other tongs and thrown into the rectangular tank, which is almost filled with water. These tongs are two feet and three palms long, both the handles are round and more than a digit thick, and the ends are bent for a palm and two digits; both the jaws are a digit and a half wide in front and sharpened; at the back they are a digit thick, and then gradually taper, and when closed, the interior is two palms and as many digits wide.

The "dried" cakes which are dripping copper are not immediately dipped into the tank, because, if so, they burst in fragments and give out a sound like thunder. The cakes are afterward taken out of the tank with the tongs, and laid upon the two transverse planks on which the workmen stand; the sooner they are taken out the easier it is to chip off the copper that has become ash-coloured. Finally, the master, with a spade, raises up the bricks a little from the hearth, while they are still warm. The blade of the spade is a palm and two digits long, the lower edge is sharp, and is a palm and a digit wide, the upper end a palm wide; its handle is round, the iron part being two feet long, and the wooden part seven and a half feet long.

On the fourth day the master draws out the liquation thorns which have settled in the passages; they are much richer in silver than those that are made when the silver-lead is liquated from copper in the liquation furnace. The "dried" cakes drip but little copper, but nearly all their remaining silver-lead and the thorns consist of it, for, indeed, in one *centumpondium* of "dried" copper there should remain only half an *uncia* of silver, and there sometimes remain only three *drachmae*.[22] Some smelters chip off the metal adhering to the bricks with a hammer, in order that it may be melted again; others, however, crush the bricks under the stamps and wash them, and the copper and lead thus collected is melted again. The master, when he has taken these things away and put them in their places, has finished his day's work.

The assistants take the "dried" cakes out of the tank on the next day, place them on an oak block, and first pound them with rounded hammers in order that the ash-coloured copper may fall away from them,

[22]One half *uncia* or three *drachmae* of silver would equal either 12 ozs. or 9 ozs. per ton. If we assume the values given for residual copper in the first four charges (note p. 506) of 34 ozs., this would mean an extraction of, roughly, 65% of the silver from the exhausted liquation cakes.

and then they dig out with pointed picks the holes in the cakes, which contain the same kind of copper. The head of the round hammer is three palms and a digit long; one end of the head is round and two digits long and thick; the other end is chisel-shaped, and is two digits and a half long. The sharp pointed hammer is the same length as the round hammer, but one end is pointed, the other end is square, and gradually tapers to a point.

A—TANK. B—BOARD. C—TONGS. D—"DRIED" CAKES TAKEN OUT OF THE TANKS. E—BLOCK. F—ROUNDED HAMMER. G—POINTED HAMMER.

The nature of copper is such that when it is "dried" it becomes ash coloured, and since this copper contains silver, it is smelted again in the blast furnaces.[23]

I have described sufficiently the method by which exhausted liquation cakes are "dried"; now I will speak of the method by which they are made into copper after they have been "dried." These cakes, in order that they may recover the appearance of copper which they have to some extent lost, are melted in four furnaces, which are placed against the second long wall in the part of the building between the second and third transverse walls. This space is sixty-three feet and two palms long, and since each of

[23]See note 29, p. 540.

these furnaces occupies thirteen feet, the space which is on the right side of the first furnace, and on the left of the fourth, are each three feet and three palms wide, and the distance between the second and third furnace is six feet. In the middle of each of these three spaces is a door, a foot and a half wide and six feet high, and the middle one is common to the master of each of the furnaces. Each furnace has its own chimney, which rises between the two long walls mentioned above, and is supported by two arches and a partition wall. The partition wall is between the two furnaces, and is five feet long, ten feet high, and two feet thick; in front of it is a pillar belonging in common to the front arches of the furnace on either side, which is two feet and as many palms thick, three feet and a half wide. The front arch reaches from this common pillar to another pillar that is common to the side arch of the same furnace; this arch on the right spans from the second long wall to the same pillar, which is two feet and as many palms wide and thick at the bottom. The interior of the front arch is nine feet and a palm wide, and eight feet high at its highest point; the interior of the arch which is on the right side, is five feet and a palm wide, and of equal height to the other, and both the arches are built of the same height as the partition wall. Imposed upon these arches and the partition wall are the walls of the chimney; these slope upward, and thus contract, so that at the upper part, where the fumes are emitted, the opening is eight feet in length, one foot and three palms in width. The fourth wall of the chimney is built vertically upon the second long wall. As the partition wall is common to the two furnaces, so its superstructure is common to the two chimneys. In this sensible manner the chimney is built. At the front each furnace is six feet two palms long, and three feet two palms wide, and a cubit high; the back of each furnace is against the second long wall, the front being open. The first furnace is open and sloping at the right side, so that the slags may be drawn out; the left side is against the partition wall, and has a little wall built of bricks cemented together with lute; this little wall protects the partition wall from injury by the fire. On the contrary, the second furnace has the left side open and the right side is against the partition wall, where also it has its own little wall which protects the partition wall from the fire. The front of each furnace is built of rectangular rocks; the interior of it is filled up with earth. Then in each of the furnaces at the rear, against the second long wall, is an aperture through an arch at the back, and in these are fixed the copper pipes. Each furnace has a round pit, two feet and as many palms wide, built three feet away from the partition wall. Finally, under the pit of the furnace, at a depth of a cubit, is the hidden receptacle for moisture, similar to the others, whose vent penetrates through the second long wall and slopes upward to the right from the first furnace, and to the left from the second. If copper is to be made the next day, then the master cuts out the crucible with a spatula, the blade of which is three digits wide and as many palms long, the iron handle being two feet long and one and a half digits in diameter; the wooden handle inserted into it is round, five feet long and two digits in diameter. Then, with another cutting spatula, he makes the crucible smooth ; the blade of this spatula is a palm wide and two palms long ; its handle, partly of iron, partly of wood, is similar in every respect to the first one. Afterward he throws pulverised clay and charcoal into the crucible, pours water over it, and sweeps it over with a broom into which a stick is fixed. Then immediately he throws into the crucible a powder, made of two wheelbarrowsful of sifted charcoal dust, as many wheelbarrowsful of

A—Hearth of the furnace. B—Chimney. C—Common pillar. D—Other pillars. The partition wall is behind the common pillar and not to be seen. E—Arches. F—Little walls which protect the partition wall from injury by the fire. G—Crucibles. H—Second long wall. I—Door. K—Spatula. L—The other spatula. M—The broom in which is inserted a stick. N—Pestles. O—Wooden mallet. P—Plate. Q—Stones. R—Iron rod.

pulverised clay likewise sifted, and six basketsful of river sand which has passed through a very fine sieve. This powder, like that used by smelters, is sprinkled with water and moistened before it is put into the crucible, so that it may be fashioned by the hands into shapes similar to snowballs. When it has been put in, the master first kneads it and makes it smooth with his hands, and then pounds it with two wooden pestles, each of which is a cubit long ; each pestle has a round head at each end, but one of these is a palm in diameter, the other three digits ; both are thinner in the middle, so that they may be held in the hand. Then he again throws moistened powder into the crucible, and again makes it smooth with his hands, and kneads it with his fists and with the pestles; then, pushing upward and pressing with his fingers, he makes the edge of the crucible smooth. After the crucible has been made smooth, he sprinkles in dry charcoal dust, and again pounds it with the same pestles, at first with the narrow heads, and afterward with the wider ones. Then he pounds the crucible with a wooden mallet two feet long, both heads of which are round and three digits in diameter; its wooden handle is two palms long, and one and a half digits in diameter. Finally, he throws into the crucible as much pure sifted ashes as both hands can hold, and pours water into it, and, taking an old linen rag, he smears the crucible over with the wet ashes. The crucible is round and sloping. If copper is to be made from the best quality of "dried" cakes, it is made two feet wide and one deep, but if from other cakes, it is made a cubit wide and two palms deep. The master also has an iron band curved at both ends, two palms long and as many digits wide, and with this he cuts off the edges of the crucible if they are higher than is necessary. The copper pipe is inclined, and projects three digits from the wall, and has its upper end and both sides smeared thick with lute, that it may not be burned; but the underside of the pipe is smeared thinly with lute, for this side reaches almost to the edge of the crucible, and when the crucible is full the molten copper touches it. The wall above the pipe is smeared over with lute, lest that should be damaged. He does the same to the other side of an iron plate, which is a foot and three palms long and a foot high; this stands on stones near the crucible at the side where the hearth slopes, in order that the slag may run out under it. Others do not place the plates upon stones, but cut out of the plate underneath a small piece, three digits long and three digits wide; lest the plate should fall, it is supported by an iron rod fixed in the wall at a height of two palms and the same number of digits, and it projects from the wall three palms.

Then with an iron shovel, whose wooden handle is six feet long, he throws live charcoal into the crucible; or else charcoal, kindled by means of a few live coals, is added to them. Over the live charcoal he lays "dried" cakes, which, if they were of copper of the first quality, weigh all together three *centumpondia*, or three and a half *centumpondia*; but if they were of copper of the second quality, then two and a half *centumpondia*; if they were of the third quality, then two *centumpondia* only; but if they were of copper of very superior quality, then they place upon it six *centumpondia*, and in this case they make the crucible wider and deeper.[24] The lowest "dried" cake is placed at a distance of two palms from the pipe, the rest at a greater distance, and when the lower ones are melted the upper ones fall down and get nearer to the pipe; if they do not fall down they must be pushed with a shovel. The blade of the shovel is a foot long, three palms and two digits wide, the iron part of the handle is two palms long, the

[24]Assuming Roman weights:

2	*centumpondia*	= 141.3 lbs.
2½	"	= 176.6 "
3	"	= 211.9 "
3½	"	= 248.2 "
6	"	= 423.9 "

wooden part nine feet. Round about the "dried" cakes are placed large long pieces of charcoal, and in the pipe are placed medium-sized pieces. When all these things have been arranged in this manner, the fire must be more violently excited by the blast from the bellows. When the copper is melting and the coals blaze, the master pushes an iron bar into the middle of them in order that they may receive the air, and that the flame can force its way out. This pointed bar is two and a half feet long, and its wooden handle four feet long. When the cakes are partly melted, the master, passing out through the door, inspects the crucible through the bronze pipe, and if he should find that too much of the "slag" is adhering to the mouth of the pipe, and thus impeding the blast of the bellows, he inserts the hooked iron bar into the pipe through the nozzle of the bellows, and, turning this about the mouth of the pipe, he removes the "slags" from it. The hook on this bar is two digits high ; the iron part of the handle is three feet long ; the wooden part is the same number of palms long. Now it is time to insert the bar under the iron plate, in order that the "slags" may flow out. When the cakes, being all melted, have run into the crucible, he takes out a sample of copper with the third round bar, which is made wholly of iron, and is three feet long, a digit thick, and has a steel point lest its pores should absorb the copper.

A—Pointed bar. B—Thin copper layer. C—Anvil. D—Hammer.

When he has compressed the bellows, he introduces this bar as quickly as possible into the crucible through the pipe between the two nozzles, and takes out samples two, three, or four times, until he finds that the copper is perfectly refined. If the copper is good it adheres easily to the bar, and two samples suffice; if it is not good, then many are required. It is necessary to smelt it in the crucible until the copper adhering to the bar is seen to be of a brassy colour, and if the upper as well as the lower part of the thin layer of copper may be easily broken, it signifies that the copper is perfectly melted; he places the point of the bar on a small iron anvil, and chips off the thin layer of copper from it with a hammer.[25]

If the copper is not good, the master draws off the "slags" twice, or three times if necessary—the first time when some of the cakes have been melted, the second when all have melted, the third time when the copper has been heated for some time. If the copper was of good quality, the "slags" are not drawn off before the operation is finished, but at the time they are to be drawn off, he depresses the bar over both bellows, and places over both a stick, a cubit long and a palm wide, half cut away at the upper part, so that it may pass under the iron pin fixed at the back in the perforated wood. This he does likewise when the copper has been completely melted. Then the assistant removes the iron plate with the tongs; these tongs are four feet three palms long, their jaws are about a foot in length, and their straight part measures two palms and three digits, and the curved a palm and a digit. The same assistant, with the iron shovel, throws and heaps up the larger pieces of charcoal into that part of the hearth which is against the little wall which protects the other wall from injury by fire, and partly extinguishes them by pouring water over them. The master, with a hazel stick inserted

[25]This description of refining copper in an open hearth by oxidation with a blast and "poling"—the *gaarmachen* of the Germans—is so accurate, and the process is so little changed in some parts of Saxony, that it might have been written in the 20th century instead of the 16th. The best account of the old practice in Saxony after Agricola is to be found in Schlüter's *Hütte Werken* (Braunschweig, 1738, Chap. CXVIII.). The process has largely been displaced by electrolytic methods, but is still in use in most refineries as a step in electrolytic work. It may be unnecessary to repeat that the process is one of subjecting the molten mass of impure metal to a strong and continuous blast, and as a result, not only are the impurities to a considerable extent directly oxidized and taken off as a slag, but also a considerable amount of copper is turned into cuprous oxide. This cuprous oxide mostly melts and diffuses through the metallic copper, and readily parting with its oxygen to the impurities further facilitates their complete oxidation. The blast is continued until the impurities are practically eliminated, and at this stage the molten metal contains a great deal of dissolved cuprous oxide, which must be reduced. This is done by introducing a billet of green wood ("poling"), the dry distillation of which generates large quantities of gases, which reduce the oxide. The state of the metal is even to-day in some localities tested by dipping into it the point of an iron rod; if it be at the proper state the adhering copper has a net-like appearance, should be easily loosened from the rod by dipping in water, is of a reddish-copper colour and should be quite pliable; if the metal is not yet refined, the sample is thick, smooth, and detachable with difficulty; if over-refined, it is thick and brittle. By allowing water to run on to the surface of the molten metal, thin cakes are successively formed and taken off. These cakes were the article known to commerce over several centuries as "rosetta copper." The first few cakes are discarded as containing impurities or slag, and if the metal be of good quality the cakes are thin and of a red colour. Their colour and thinness, therefore, become a criterion of purity. The cover of charcoal or charcoal dust maintained upon the surface of the metal tended to retard oxidation, but prevented volatilization and helped to secure the impurities as a slag instead. Karsten (*Archiv.*, 1st series, p. 46) gives several analyses of the

into the crucible, stirs it twice. Afterward he draws off the slags with a rabble, which consists of an iron blade, wide and sharp, and of alder-wood; the blade is a digit and a half in width and three feet long; the wooden handle inserted in its hollow part is the same number of feet long, and the alder-wood in which the blade is fixed must have the figure of a rhombus; it must be three palms and a digit long, a palm and two digits wide, and a palm thick. Subsequently he takes a broom and sweeps the charcoal dust and small coal over the whole of the crucible, lest the copper should cool before it flows together; then, with a third rabble, he cuts off the slags which may adhere to the edge of the crucible. The blade of this rabble is two palms long and a palm and one digit wide, the iron part of the handle is a foot and three palms long, the wooden part six feet. Afterward he again draws off the slags from the crucible, which the assistant does not quench by pouring water upon them, as the other slags are usually quenched, but he sprinkles over them a little water and allows them to cool. If the copper should bubble, he presses down the bubbles with the rabble. Then he pours water on the wall and the pipes, that it may flow down warm into the crucible, for, the copper, if cold water were to be poured over it while still hot, would spatter about. If a stone, or a piece of lute or wood, or a damp coal should then fall into it, the crucible would vomit out all the copper with a loud noise like thunder, and whatever it touches it injures and sets on fire. Subsequently he lays a curved board with a notch in it over the front part of the crucible; it is two feet long, a palm and two digits wide, and a digit thick. Then the copper in the crucible should be divided into cakes with an iron wedge-shaped bar; this is three feet long, two digits wide, and steeled on the end for the distance of two digits, and its wooden handle is three feet long. He places this bar on the notched board, and, driving it into the copper, moves

slag from refining "dried" copper, showing it to contain from 51.7 to 67.4% lead oxide, 6.2 to 19.2% cuprous oxide, and 21.4 to 23.9 silica (from the furnace bottoms), with minor quantities of iron, antimony, etc. The "bubbles" referred to by Agricola were apparently the shower of copper globules which takes place upon the evolution of sulphur dioxide, due to the reaction of the cuprous oxide upon any remaining sulphide of copper when the mass begins to cool.

HISTORICAL NOTE.—It is impossible to say how the Ancients refined copper, beyond the fact that they often re-smelted it. Such notes as we can find are set out in the note on copper smelting (note 42, p. 402). The first authentic reference to poling is in Theophilus (1150 to 1200 A.D., Hendrie's translation, p. 313), which shows a very good understanding of this method of refining copper:—"Of the Purification of Copper. Take an iron dish of "the size you wish, and line it inside and out with clay strongly beaten and mixed, and it is "carefully dried. Then place it before a forge upon the coals, so that when the bellows act "upon it the wind may issue partly within and partly above it, and not below it. And very "small coals being placed round it, place the copper in it equally, and add over it a heap of "coals. When by blowing a long time this has become melted, uncover it and cast immedi-"ately fine ashes of coals over it, and stir it with a thin and dry piece of wood as if mixing it, "and you will directly see the burnt lead adhere to these ashes like a glue, which being cast "out again superpose coals, and blowing for a long time, as at first, again uncover it, and "then do as you did before. You do this until at length by cooking it you can withdraw the "lead entirely. Then pour it over the mould which you have prepared for this, and you will "thus prove if it be pure. Hold it with the pincers, glowing as it is, before it has become "cold, and strike it with a large hammer strongly over the anvil, and if it be broken or split "you must liquefy it anew as before. If, however, it should remain sound, you will cool it in "water, and you cook other (copper) in the same manner." Biringuccio (III, 8) in 1540 describes the process briefly, but omits the poling, an essential in the production of malleable copper.

A—Crucible. B—Board. C—Wedge-shaped bar. D—Cakes of copper made by separating them with the wedge-shaped bar. E—Tongs. F—Tub.

it forward and back, and by this means the water flows into the vacant space in the copper, and he separates the cake from the rest of the mass. If the copper is not perfectly smelted the cakes will be too thick, and cannot be taken out of the crucible easily. Each cake is afterward seized by the assistant with the tongs and plunged into the water in the tub; the first one is placed aside so that the master may re-melt it again immediately, for, since some "slags" adhere to it, it is not as perfect as the subsequent ones; indeed, if the copper is not of good quality, he places the first two cakes aside. Then, again pouring water over the wall and the pipes, he separates out the second cake, which the assistant likewise immerses in water and places on the ground together with the others separated out in the same way, which he piles upon them. These, if the copper was of good quality, should be thirteen or more in number; if it was not of good quality, then fewer. If the copper was of good quality, this part of the operation, which indeed is distributed into four parts, is accomplished by the master in two hours; if of mediocre quality, in two and a half hours; if of bad quality, in three. The "dried" cakes are re-melted, first in the first crucible and then in the second. The assistant must, as quickly as possible, quench all the cakes with water, after they have been cut out of the second crucible. Afterward with the tongs he replaces in its proper place the iron plate which was in front of the furnace, and throws the charcoal back into the crucible with a shovel. Meanwhile the master, continuing his work, removes the wooden stick from the bars of the bellows, so that in re-melting the other cakes he may accomplish the third part of his process; this must be carefully done, for if a particle from any iron implement should by chance fall into the crucible, or should be thrown in by any malevolent person, the copper could not be made until the iron had been consumed, and therefore double labour would have to be expended upon it. Finally, the assistant extinguishes all the glowing coals, and chips off the dry lute from the mouth of the copper pipe with a hammer; one end of this hammer is pointed, the other round, and it has a wooden handle five feet long. Because there is danger that the copper would be scattered if the *pompholyx* and *spodos*, which adhere to the walls and the hood erected upon them, should fall into the crucible, he cleans them off in the meantime. Every week he takes the copper flowers out of the tub, after having poured off the water, for these fall into it from the cakes when they are quenched.[26]

[26]*Pompholyx* and *spodos* were impure zinc oxides (see note 26, p. 394).

The copper flowers were no doubt cupric oxide. They were used by the Ancients for medicinal purposes. Dioscorides (V, 48) says: "Of flowers of copper, which some call the "scrapings of old nails, the best is friable; it is gold-coloured when rubbed, is like millet in "shape and size, is moderately bright, and somewhat astringent. It should not be mixed "with copper filings, with which it is often adulterated. But this deception is easily detected, "for when bitten in the teeth the filings are malleable. It (the flowers) is made when the "copper fused in a furnace has run into the receptacle through the spout pertaining to it, "for then the workmen engaged in this trade cleanse it from dirt and pour clear water over it "in order to cool it; from this sudden condensation the copper spits and throws out the "aforesaid flowers." Pliny (XXXIV, 24) says: "The flower, too, of copper (*æris flos*) is "used in medicine. This is made by fusing copper, and then removing it to another furnace, "where the repeated blast makes the metal separate into small scales like millet, known as "flowers. These scales also fall off when the cakes of metal are cooled in water; they become "red, too, like the scales of copper known as '*lepis*,' by use of which the flowers of copper are "adulterated, it being also sold for it. These are made when hammering the nails that are

The bellows which this master uses differ in size from the others, for the boards are seven and a half feet long; the back part is three feet wide; the front, where the head is joined on is a foot, two palms and as many digits. The head is a cubit and a digit long; the back part of it is a cubit and a palm wide, and then becomes gradually narrower. The nozzles of the bellows are bound together by means of an iron chain, controlled by a thick bar, one end of which penetrates into the ground against the back of the long wall, and the other end passes under the beam which is laid upon the foremost perforated beams. These nozzles are so placed in a copper pipe that they are at a distance of a palm from the mouth; the mouth should be made three digits in diameter, that the air may be violently expelled through this narrow aperture.

There now remain the liquation thorns, the ash-coloured copper, the "slags," and the *cadmia*.[27] Liquation cakes are made from thorns in the following manner.[28] There are taken three-quarters of a *centumpondium* of thorns, which have their origin from the cakes of copper-lead alloy when lead-silver is liquated, and as many parts of a *centumpondium* of the thorns derived from cakes made from once re-melted thorns by the same method, and to them are added a *centumpondium* of de-silverized lead and half a *centumpondium* of hearth-lead. If there is in the works plenty of litharge, it is substituted for the de-silverized lead. One and a half *centumpondia* of litharge and hearth-lead is added to the same weight of primary thorns, and half a *centumpondium* of thorns which have their origin from liquation cakes composed of thorns twice re-melted by the same method (tertiary thorns), and a fourth part of a *centumpondium* of thorns which are pro-

"made from the cakes of copper. All these methods are carried on in the works of Cyprus; "the difference between these substances is that the *squamae* (copper scales) are detached from "hammering the cakes, while the flower falls off spontaneously." Agricola (*De Nat. Fos.*, p. 352) notes that "flowers of copper (*flos æris*) have the same properties as 'roasted "copper.'"

[27]It seems scarcely necessary to discuss in detail the complicated "flow scheme" of the various minor bye-products. They are all re-introduced into the liquation circuit, and thereby are created other bye-products of the same kind *ad infinitum*. Further notes are given on:—

Liquation thorns	Note 28.
Slags	" 30.
Ash-coloured copper ..	" 29.
Concentrates	" 33.
Cadmia	" 32.

There are no data given, either by Agricola or the later authors, which allow satisfactory calculation of the relative quantities of these products. A rough estimate from the data given in previous notes would indicate that in one liquation only about 70% of the original copper came out as refined copper, and that about 70% of the original lead would go to the cupellation furnace, *i.e.*, about 30% of the original metal sent to the blast furnace would go into the "thorns," "slags," and "ash-coloured copper." The ultimate losses were very great, as given before (p. 491), they probably amounted to 25% of the silver, 9% copper, and 16% of the lead.

[28]There were the following classes of thorns:—

1st. From liquation.
2nd. From drying.
3rd. From cupellation.

In a general way, according to the later authors, they were largely lead oxide, and contained from 5% to 20% cuprous oxide. If a calculation be made backward from the products given as the result of the charge described, it would appear that in this case they must have contained at least one-fifth copper. The silver in these liquation cakes would run about 24 ozs. per ton, in the liquated lead about 36 ozs. per ton, and in the liquation thorns 24 ozs. per ton. The extraction into the liquated lead would be about 80% of the silver.

duced when the exhausted liquation cakes are "dried." By both methods one single liquation cake is made from three *centumpondia*. In this manner the smelter makes every day fifteen liquation cakes, more or less; he takes great care that the metallic substances, from which the first liquation cake is made, flow down properly and in due order into the fore-hearth, before the material of which the subsequent cake is to be made. Five of these liquation cakes are put simultaneously into the furnace in which silver-lead is liquated from copper, they weigh almost fourteen *centumpondia*, and the "slags" made therefrom usually weigh quite a *centumpondium*. In all the liquation cakes together there is usually one *libra* and nearly two *unciae* of silver, and in the silver-lead which drips from those cakes, and weighs seven and a half *centumpondia*, there is in each an *uncia* and a half of silver. In each of the three *centumpondia* of liquation thorns there is almost an *uncia* of silver, and in the two *centumpondia* and a quarter of exhausted liquation cakes there is altogether one and a half *unciae;* yet this varies greatly for each variety of thorns, for in the thorns produced from primary liquation cakes made of copper and lead when silver-lead is liquated from the copper, and those produced in "drying" the exhausted liquation cakes, there are almost two *unciae* of silver; in the others not quite an *uncia*. There are other thorns besides, of which I will speak a little further on.

Those in the Carpathian Mountains who make liquation cakes from the copper "bottoms" which remain after the upper part of the copper is divided from the lower, in the furnace similar to an oven, produce thorns when the poor or mediocre silver-lead is liquated from the copper. These, together with those made of cakes of re-melted thorns, or made with re-melted litharge, are placed in a heap by themselves; but those that are made from cakes melted from hearth-lead are placed in a heap separate from the first, and likewise those produced from "drying" the exhausted liquation cakes are placed separately; from these thorns liquation cakes are made. From the first heap they take the fourth part of a *centumpondium*, from the second the same amount, from the third a *centumpondium*,—to which thorns are added one and a half *centumpondia* of litharge and half a *centumpondium* of hearth-lead, and from these, melted in the blast furnace, a liquation cake is made; each workman makes twenty such cakes every day. But of theirs enough has been said for the present; I will return to ours.

The ash-coloured copper [29] which is chipped off, as I have stated, from the "dried" cakes, used some years ago to be mixed with the thorns produced from liquation of the copper-lead alloy, and contained in themselves, equally with the first, two *unciae* of silver; but now it is mixed with the concentrates washed from the accretions and the other material. The inhabitants of the Carpathian Mountains melt this kind of copper in furnaces in which are re-melted the "slags" which flow out when the copper is refined; but as this soon melts and flows down out of the furnace, two workmen are required for

[29] The "ash-coloured copper" is a cuprous oxide, containing some 3% lead oxide; and if Agricola means they contained two *unciae* of silver to the *centumpondium*, then they ran about 48 ozs. per ton, and would contain much more silver than the mass.

the work of smelting, one of whom smelts, while the other takes out the thick cakes from the forehearth. These cakes are only "dried," and from the "dried" cakes copper is again made.

The "slags"[30] are melted continually day and night, whether they have been drawn off from the alloyed metals with a rabble, or whether they adhered to the forehearth to the thickness of a digit and made it smaller and were taken off with spatulas. In this manner two or three liquation cakes are made, and afterward much or little of the "slag," skimmed from the molten alloy of copper and lead, is re-melted. Such liquation cakes should weigh up to three *centumpondia*, in each of which there is half an *uncia* of silver. Five cakes are placed at the same time in the furnace in which argentiferous lead is liquated from copper, and from these are made lead which contains half an *uncia* of silver to the *centumpondium*. The exhausted liquation cakes are laid upon the other baser exhausted liquation cakes, from both of which yellow copper is made. The base thorns thus obtained are re-melted with a few baser "slags," after having been sprinkled with concentrates from furnace accretions and other material, and in this manner six or seven liquation cakes are made, each of which weighs some two *centumpondia*. Five of these are placed at the same time in the furnace in which silver-lead is liquated from copper; these drip three *centumpondia* of lead, each of which contains half an *uncia* of silver. The basest thorns thus produced should be re-melted with only a little "slag." The copper alloyed with lead, which flows down from the furnace into the forehearth, is poured out with a ladle into oblong copper moulds; these cakes are "dried" with base exhausted liquation cakes. The thorns they produce are added to the base thorns, and they are made into cakes according to the method I have described. From the "dried" cakes they make copper, of which some add a small portion to the best "dried" cakes when copper is made from them, in order that by mixing the base copper with the good it may be sold without loss. The "slags," if they are utilisable, are re-melted a second and a third time, the cakes made from them are "dried," and from the "dried" cakes is made copper, which is mixed with the good copper. The "slags," drawn off by the master who makes copper out of "dried" cakes, are sifted, and those which fall through the sieve into a vessel placed underneath are washed; those which remain in it are emptied into a wheelbarrow and wheeled away to the blast furnaces, and they are re-melted together with other "slags," over which are sprinkled the concentrates from washing the slags or furnace accretions made at this time. The copper which flows out

[30]There are three principal "slags" mentioned—

1st. Slag from "leading."
2nd. Slag from "drying."
3rd. Slag from refining the copper.

From the analyses quoted by various authors these ran from 52% to 85% lead oxide, 5% to 20% cuprous oxide, and considerable silica from the furnace bottoms. They were reduced in the main into liquation cakes, although Agricola mentions instances of the metal reduced from "slags" being taken directly to the "drying" furnace. Such liquation cakes would run very low in silver, and at the values given only averaged 12 ozs. per ton; therefore the liquated lead running the same value as the cakes, or less than half that of the "poor" lead mentioned in Note 17, p. 512, could not have been cupelled directly.

of the furnace into the forehearth, is likewise dipped out with a ladle into oblong copper moulds ; in this way nine or ten cakes are made, which are "dried," together with bad exhausted liquation cakes, and from these "dried" cakes yellow[31] copper is made.

The *cadmia*,[32] as it is called by us, is made from the "slags" which the master, who makes copper from "dried" cakes, draws off together with other re-melted base "slags" ; for, indeed, if the copper cakes made from such "slags" are broken, the fragments are called *cadmia* ; from this and yellow copper is made *caldarium* copper in two ways. For either two parts of *cadmia* are mixed with one of yellow copper in the blast furnaces, and melted ; or, on the contrary, two parts of yellow copper with one of *cadmia*, so that the *cadmia* and yellow copper may be well mixed ; and the copper which flows down from the furnace into the forehearth is poured out with a ladle into oblong copper moulds heated beforehand. These moulds are sprinkled over with charcoal dust before the *caldarium* copper is to be poured into them, and the same dust is sprinkled over the copper when it is poured in, lest the *cadmia* and yellow copper should freeze before they have become well mixed. With a piece of wood the assistant cleanses each cake from the dust, when it is turned out of the mould. Then he throws it into the tub containing hot water, for the *caldarium* copper is finer if quenched in hot water. But as I have so often made mention of the oblong copper moulds, I must now speak of them a little ; they are a foot and a palm long, the inside is three palms and a digit wide at the top, and they are rounded at the bottom.

The concentrates are of two kinds—precious and base.[33] The first are obtained from the accretions of the blast furnace, when liquation cakes are made from copper and lead, or from precious liquation thorns, or from the better quality "slags," or from the best grade of concentrates, or from the sweepings and bricks of the furnaces in which exhausted liquation cakes are "dried"; all of these things are crushed and washed, as I explained in Book VIII. The base concentrates are made from accretions formed when cakes are cast from base thorns or from the worst quality of slags. The smelter who makes liquation cakes from the precious concentrates, adds to them three wheelbarrowsful of litharge and four barrowsful of hearth-lead and one of ash-coloured copper, from all of which nine or ten liquation cakes are melted out, of which five at a time are placed in the furnace in which silver-lead is liquated from copper ; a *centumpondium* of the lead which drips from these cakes contains one *uncia* of silver. The liquation thorns are

[31]See Note 16, p. 511, for discussion of yellow and *caldarium* copper.

[32]This *cadmia* is given in the Glossary and the German translation as *kobelt*. A discussion of this substance is given in the note on p. 112 ; and it is sufficient to state here that in Agricola's time the metal cobalt was unknown, and the substances designated *cadmia* and *cobaltum* were arsenical-cobalt-zinc minerals. A metal made from "slag" from refining, together with "base" thorns, would be very impure ; for the latter, according to the paragraph on concentrates a little later on, would contain the furnace accretions, and would thus be undoubtedly zincky. It is just possible that the term *kobelt* was used by the German smelters at this time in the sense of an epithet—"black devil" (see Note 21, p. 214).

[33]It is somewhat difficult to see exactly the meaning of base (*vile*) and precious (*preciosum*) in this connection. While "base" could mean impure, "precious" could hardly mean pure, and while "precious" could mean high value in silver, the reverse does not seem entirely *apropos*. It is possible that "bad" and "good" would be more appropriate terms.

A—FURNACE. B—FOREHEARTH. C—OBLONG MOULDS.

placed apart by themselves, of which one basketful is mixed with the precious thorns to be re-melted. The exhausted liquation cakes are "dried" at the same time as other good exhausted liquation cakes.

The thorns which are drawn off from the lead, when it is separated from silver in the cupellation furnace[34], and the hearth-lead which remains in the crucible in the middle part of the furnaces, together with the hearth material which has become defective and has absorbed silver-lead, are all melted together with a little slag in the blast furnaces. The lead, or rather the silver-lead, which flows from the furnace into the fore-hearth, is poured out into copper moulds such as are used by the refiners; a *centumpondium* of such lead contains four *unciae* of silver, or, if the hearth was defective, it contains more. A small portion of this material is added to the copper and lead when liquation cakes are made from them, if more were to be added the alloy would be much richer than it should be, for which reason the wise

[34]The skimmings from the molten lead in the early stages of cupellation have been discussed in Note 28, p. 539. They are probably called thorns here because of the large amount of copper in them. The lead from liquation would contain 2% to 3% of copper, and this would be largely recovered in these skimmings, although there would be some copper in the furnace bottoms—hearth-lead—and the litharge. These "thorns" are apparently fairly rich, four *unciae* to the *centumpondium* being equivalent to about 97 ozs. per ton, and they are only added to low-grade liquation material.

foreman of the works mixes these thorns with other precious thorns. The hearth-lead which remains in the middle of the crucible, and the hearth material which absorbs silver-lead, is mixed with other hearth-lead which remains in the cupellation furnace crucible ; and yet some cakes, made rich in this manner, may be placed again in the cupellation furnaces, together with the rest of the silver-lead cakes which the refiner has made.

The inhabitants of the Carpathian Mountains, if they have an abundance of finely crushed copper[35] or lead either made from "slags," or collected from the furnace in which the exhausted liquation cakes are dried, or litharge, alloy them in various ways. The "first" alloy consists of two *centumpondia* of lead melted out of thorns, litharge, and thorns made from hearth-lead, and of half a *centumpondium* each of lead collected in the furnace in which exhausted liquation cakes are "dried," and of copper *minutum*, and from these are made liquation cakes; the task of the smelter is finished when he has made forty liquation cakes of this kind. The "second" alloy consists of two *centumpondia* of litharge, of one and a quarter *centumpondia* of de-silverized lead or lead from "slags," and of half a *centumpondium* of lead made from thorns, and of as much copper *minutum*. The "third" alloy consists of three *centumpondia* of litharge and of half a *centumpondium* each of de-silverized lead, of lead made from thorns, and of copper *minutum contusum*. Liquation cakes are made from all these alloys ; the task of the smelters is finished when they have made thirty cakes.

The process by which cakes are made among the Tyrolese, from which they separate the silver-lead, I have explained in Book IX.

Silver is separated from iron in the following manner. Equal portions of iron scales and filings and of *stibium* are thrown into an earthenware crucible which, when covered with a lid and sealed, is placed in a furnace, into which air is blown. When this has melted and again cooled, the crucible is broken ; the button that settles in the bottom of it, when taken out, is pounded to powder, and the same weight of lead being added, is mixed and melted in a second crucible ; at last this button is placed in a cupel and the lead is separated from the silver.[36]

There are a great variety of methods by which one metal is separated from other metals, and the manner in which the same are alloyed I have explained partly in the eighth book of *De Natura Fossilium*, and partly I will explain elsewhere. Now I will proceed to the remainder of my subject.

[35]*Particulis aeris tusi*. Unless this be the fine concentrates from crushing the material mentioned, we are unable to explain the expression.

[36]This operation would bring down a button of antimony under an iron matte, by de-sulphurizing the antimony. It would seem scarcely necessary to add lead before cupellation. This process is given in an assay method, in the *Probierbüchlein* (folio 31) 50 years before *De Re Metallica* : "How to separate silver from iron : Take that silver which is "in iron *plechen* (*plachmal*), pulverize it finely, take the same iron or *plec* one part, *spiesglasz* "(antimony sulphide) one part, leave them to melt in a crucible placed in a closed *windtofen*. "When it is melted, let it cool, break the crucible, chip off the button that is in the bottom, "and melt it in a crucible with as much lead. Then break the crucible, and seek from the "button in the cupel, and you will find what silver it contains."

END OF BOOK XI.

BOOK XII.

REVIOUSLY I have dealt with the methods of separating silver from copper. There now remains the portion which treats of solidified juices; and whereas they might be considered as alien to things metallic, nevertheless, the reasons why they should not be separated from it I have explained in the second book.

Solidified juices are either prepared from waters in which nature or art has infused them, or they are produced from the liquid juices themselves, or from stony minerals. Sagacious people, at first observing the waters of some lakes to be naturally full of juices which thickened on being dried up by the heat of the sun and thus became solidified juices, drew such waters into other places, or diverted them into low-lying places adjoining hills, so that the heat of the sun should likewise cause them to condense. Subsequently, because they observed that in this wise the solidified juices could be made only in summer, and then not in all countries, but only in hot and temperate regions in which it seldom rains in summer, they boiled them in vessels over a fire until they began to thicken. In this manner, at all times of the year, in all regions, even the coldest, solidified juices could be obtained from solutions of such juices, whether made by nature or by art. Afterward, when they saw juices drip from some roasted stones, they cooked these in pots in order to obtain solidified juices in this wise also. It is worth the trouble to learn the proportions and the methods by which these are made.

I will therefore begin with salt, which is made from water either salty by nature, or by the labour of man, or else from a solution of salt, or from lye, likewise salty. Water which is salty by nature, is condensed and converted into salt in salt-pits by the heat of the sun, or else by the heat of a fire in pans or pots or trenches. That which is made salty by art, is also condensed by fire and changed into salt. There should be as many salt-pits dug as the circumstance of the place permits, but there should not be more made than can be used, although we ought to make as much salt as we can sell. The depth of salt-pits should be moderate, and the bottom should be level, so that all the water is evaporated from the salt by the heat of the sun. The salt-pits should first be encrusted with salt, so that they may not suck up the water. The method of pouring or leading sea-water into salt-pits is very old, and is still in use in many places. The method is not less old, but less common, to pour well-water into salt-pits, as was done in Babylon, for which Pliny is the authority, and in Cappadocia, where they used not only well-water, but also spring-water. In all hot countries salt-water and lake-water are conducted, poured or carried into salt-pits, and, being dried by the heat of the sun, are converted into salt.[1] While the salt-water contained in the salt-pits is being heated by the sun, if they be flooded with great and frequent showers of rain the evaporation is hindered. If this happens rarely, the salt acquires a disagreeable[2] flavour, and in this case the salt-pits have to be filled with other sweet water.

Salt from sea-water is made in the following manner. Near that part of the seashore where there is a quiet pool, and there are wide, level plains which the inundations of the sea do not overflow, three, four, five, or six trenches are dug six feet wide, twelve feet deep, and six hundred feet long, or longer if the level place extends for a longer distance ; they are two hundred feet distant from one another ; between these are three transverse trenches. Then are dug the principal pits, so that when the water has been raised from the pool it can flow into the trenches, and from thence into the salt-pits, of which there are numbers on the level ground between the trenches. The salt-pits are basins dug to a moderate depth ; these are banked round with the earth which was dug in sinking them or in cleansing them, so that between the basins, earth walls are made a foot high, which retain the water let into them. The trenches have openings, through which the first basins receive the water ; these basins also have openings, through which the water flows again from one into the other. There should be a slight fall, so that the water may flow from one basin into the other, and can thus be replenished. All these things having been done rightly and in order, the gate is raised that opens the mouth of the pool which contains sea-water mixed with rain-water or river-water ; and thus all of the trenches are filled. Then the gates of the first basins are opened, and thus the remaining basins are filled with the water from the first ; when this salt-water condenses, all these basins are incrusted, and thus made clean from earthy matter. Then again the first basins are filled up from the nearest trench with the same kind of water, and left until much of the thin liquid is converted into vapour by the heat of the sun and dissipated, and the remainder is considerably thickened. Then their gates being opened, the water passes into the second basins ; and when it has remained there for a certain space of time the gates are opened, so that it flows into the third basins, where it is all condensed into salt. After the salt has been taken out, the basins are filled again and again with sea-water. The salt is raked up with wooden rakes and thrown out with shovels.

Salt-water is also boiled in pans, placed in sheds near the wells from which it is drawn. Each shed is usually named from some animal or other thing which is pictured on a tablet nailed to it. The walls of these sheds are made either from baked earth or from wicker work covered with thick

[1]The history of salt-making in salt-pans, from sea-water or salt springs, goes further back than human records. From an historical point of view the real interest attached to salt lies in the bearing which localities rich in either natural salt or salt springs, have had upon the movements of the human race. Many ancient trade routes have been due to them, and innumerable battles have been fought for their possession. Salt has at times served for currency, and during many centuries in nearly every country has served as a basis of taxation. These subjects do not, however, come within the scope of this text. For the quotation from Pliny referred to, see Note 14 below, on bitumen.

[2]The first edition gives *graviorem*, the latter editions *gratiorem*, which latter would have quite the reverse meaning from the above.

A—Sea. B—Pool. C—Gate. D—Trenches. E—Salt basins. F—Rake. G—Shovel.

mud, although some may be made of stones or bricks. When of brick they are often sixteen feet high, and if the roof rises twenty-four feet high, then the walls which are at the ends must be made forty feet high, as likewise the interior partition walls. The roof consists of large shingles four feet long, one foot wide, and two digits thick; these are fixed on long narrow planks placed on the rafters, which are joined at the upper end and slope in opposite directions. The whole of the under side is plastered one digit thick with straw mixed with lute; likewise the roof on the outside is plastered one and a half feet thick with straw mixed with lute, in order that the shed should not run any risk of fire, and that it should be proof against rain, and be able to retain the heat necessary for drying the lumps of salt. Each shed is divided into three parts, in the first of which the firewood and straw are placed; in the middle room, separated from the first room by a partition, is the fireplace on which is placed the caldron. To the right of the caldron is a tub, into which is emptied the brine brought into the shed by the porters; to the left is a bench, on which there is room to lay thirty pieces of salt. In the third room, which is in the back part of the house, there is made a pile of clay or ashes eight feet higher than the floor, being the same height as the bench. The master and his assistants, when they carry away the lumps of salt from the caldrons, go from the former to the latter. They ascend from the right side of the caldron, not by steps, but by a slope of earth. At the top of the end wall are two small windows, and a third is in the roof, through which the smoke escapes. This smoke, emitted from both the back and the front of the furnace, finds outlet through a hood through which it makes its way up to the windows; this hood consists of boards projecting one beyond the other, which are supported by two small beams of the roof. Opposite the fireplace the middle partition has an open door eight feet high and four feet wide, through which there is a gentle draught which drives the smoke into the last room; the front wall also has a door of the same height and width. Both of these doors are large enough to permit the firewood or straw or the brine to be carried in, and the lumps of salt to be carried out; these doors must be closed when the wind blows, so that the boiling will not be hindered. Indeed, glass panes which exclude the wind but transmit the light, should be inserted in the windows in the walls.

They construct the greater part of the fireplace of rock-salt and of clay mixed with salt and moistened with brine, for such walls are greatly hardened by the fire. These fireplaces are made eight and a half feet long, seven and three quarters feet wide, and, if wood is burned in them, nearly four feet high; but if straw is burned in them, they are six feet high. An iron rod, about four feet long, is engaged in a hole in an iron foot, which stands on the base of the middle of the furnace mouth. This mouth is three feet in width, and has a door which opens inward; through it they throw in the straw.

The caldrons are rectangular, eight feet long, seven feet wide, and half a foot high, and are made of sheets of iron or lead, three feet long and of the same width, all but two digits. These plates are not very thick, so that the

A—Shed. B—Painted signs. C—First room. D—Middle room. E—Third room. F—Two little windows in the end wall. G—Third little window in the roof. H—Well. I—Well of another kind. K—Cask. L—Pole. M—Forked sticks in which the porters rest the pole when they are tired.

water is heated more quickly by the fire, and is boiled away rapidly. The more salty the water is, the sooner it is condensed into salt. To prevent the brine from leaking out at the points where the metal plates are fastened with rivets, the caldrons are smeared over with a cement made of ox-liver and ox-blood mixed with ashes. On each side of the middle of the furnace two rectangular posts, three feet long, and half a foot thick and wide are set into the ground, so that they are distant from each other only one and a half feet. Each of them rises one and a half feet above the caldron. After the caldron has been placed on the walls of the furnace, two beams of the same width and thickness as the posts, but four feet long, are laid on these posts, and are mortised in so that they shall not fall. There rest transversely upon these beams three bars, three feet long, three digits wide, and two digits thick, distant from one another one foot. On each of these hang three iron hooks, two beyond the beams and one in the middle ; these are a foot long, and are hooked at both ends, one hook turning to the right, the other to the left. The bottom hook catches in the eye of a staple, whose ends are fixed in the bottom of the caldron, and the eye projects from it. There are besides, two longer bars six feet long, one palm wide, and three digits thick, which pass under the front beam and rest upon the rear beam. At the rear end of each of the bars there is an iron hook two feet and three digits long, the lower end of which is bent so as to support the caldron. The rear end of the caldron does not rest on the two rear corners of the fireplace, but is distant from the fireplace two thirds of a foot, so that the flame and smoke can escape ; this rear end of the fireplace is half a foot thick and half a foot higher than the caldron. This is also the thickness and height of the wall between the caldron and the third room of the shed, to which it is adjacent. This back wall is made of clay and ashes, unlike the others which are made of rock-salt. The caldron rests on the two front corners and sides of the fireplace, and is cemented with ashes, so that the flames shall not escape. If a dipperful of brine poured into the caldron should flow into all the corners, the caldron is rightly set upon the fireplace.

The wooden dipper holds ten Roman *sextarii*, and the cask holds eight dippers full[3]. The brine drawn up from the well is poured into such casks and carried by porters, as I have said before, into the shed and poured into a tub, and in those places where the brine is very strong it is at once transferred with the dippers into the caldron. That brine which is less strong is thrown into a small tub with a deep ladle, the spoon and handle of which are hewn out of one piece of wood. In this tub rock-salt is placed in order

[3]The following are approximately the English equivalents :—

		Pints.	Quarts.	Gallons.
1 *Cyathus*	.. ..	.08 ..		..
3 *Cyathi*	= 1 *Quartarius* ..	.24 ..		..
4 *Quartarii*	= 1 *Sextarius*	.99 ..		..
6 *Sextarii*	= 1 *Congius*	5.94 ..	2.97 ..	
16 *Sextarii*	= 1 *Modius*	15.85 ..	7.93 ..	1.98
8 *Congii*	= 1 *Amphora*	47.57 ..	23.78 ..	5.94

The dipper mentioned would thus hold about one and one quarter gallons, and the cask ten gallons.

A—Fireplace. B—Mouth of fireplace. C—Caldron. D—Posts sunk into the ground. E—Cross-beams. F—Shorter bars. G—Iron hooks. H—Staples. I—Longer bars. K—Iron rod bent to support the caldron.

that the water should be made more salty, and it is then run off through a launder which leads into the caldron. From thirty-seven dippersful of brine the master or his deputy, at Halle in Saxony,[4] makes two cone-shaped pieces of salt. Each master has a helper, or in the place of a helper his wife assists him in his work, and, in addition, a youth who throws wood or straw under the caldron. He, on account of the great heat of the workshop, wears a straw cap on his head and a breech cloth, being otherwise quite naked. As soon as the master has poured the first dipperful of brine into the caldron the youth sets fire to the wood and straw laid under it. If the firewood is bundles of faggots or brushwood, the salt will be white, but if straw is burned, then it is not infrequently blackish, for the sparks, which are drawn up with the smoke into the hood, fall down again into the water and colour it black.

In order to accelerate the condensation of the brine, when the master has poured in two casks and as many dippersful of brine, he adds about a Roman *cyathus* and a half of bullock's blood, or of calf's blood, or buck's blood, or else he mixes it into the nineteenth dipperful of brine, in order that it may be dissolved and distributed into all the corners of the caldron ; in other places the blood is dissolved in beer. When the boiling water seems to be mixed with scum, he skims it with a ladle ; this scum, if he be working with rock-salt, he throws into the opening in the furnace through which the smoke escapes, and it is dried into rock-salt ; if it be not from rock-salt, he pours it on to the floor of the workshop. From the beginning to the boiling and skimming is the work of half-an-hour ; after this it boils down for another quarter-of-an-hour, after which time it begins to condense into salt. When it begins to thicken with the heat, he and his helper stir it assiduously with a wooden spatula, and then he allows it to boil for an hour. After this he pours in a *cyathus* and a half of beer. In order that the wind should not blow into the caldron, the helper covers the front with a board seven and a half feet long and one foot high, and covers each of the sides with boards three and three quarters feet long. In order that the front board may hold more firmly, it is fitted into the caldron itself, and the sideboards are fixed on the front board and upon the transverse beam. Afterward, when the boards have been lifted off, the helper places two baskets, two feet high and as many wide at the top, and a palm wide at the bottom, on the transverse beams, and into them the master throws the salt with a shovel, taking half-an-hour to fill them. Then, replacing the boards on the caldron, he allows the brine to boil for three quarters of an hour. Afterward the salt has again to be removed with a shovel, and when the baskets are full, they pile up the salt in heaps.

In different localities the salt is moulded into different shapes. In the baskets the salt assumes the form of a cone ; it is not moulded in baskets alone, but also in moulds into which they throw the salt, which are made in

[4]The salt industry, founded upon salt springs, is still of importance to this city. It was a salt centre of importance to the Germanic tribes before Charles, the son of Charlemagne, erected a fortress here in 806. Mention of the salt works is made in the charter by Otto I., conveying the place to the Diocese of Magdeburg, in 968.

A—Wooden dipper. B—Cask. C—Tub. D—Master. E—Youth. F—Wife. G—Wooden spade. H—Boards. I—Baskets. K—Hoe. L—Rake. M—Straw. N—Bowl. O—Bucket containing the blood. P—Tankard which contains beer.

the likeness of many objects, as for instance tablets. These tablets and cones are kept in the higher part of the third room of the house, or else on the flat bench of the same height, in order that they may dry better in the warm air. In the manner I have described, a master and his helper continue one after the other, alternately boiling the brine and moulding the salt, day and night, with the exception only of the annual feast days. No caldron is able to stand the fire for more than half a year. The master pours in water and washes it out every week ; when it is washed out he puts straw under it and pounds it ; new caldrons he washes three times in the first two weeks, and afterward twice. In this manner the incrustations fall from the bottom ; if they are not cleared off, the salt would have to be made more slowly over a fiercer fire, which requires more brine and burns the plates of the caldron. If any cracks make their appearance in the caldron they are filled up with cement. The salt made during the first two weeks is not so good, being usually stained by the rust at the bottom where incrustations have not yet adhered.

Although salt made in this manner is prepared only from the brine of

A—Pool. B—Pots. C—Ladle. D—Pans. E—Tongs.

springs and wells, yet it is also possible to use this method in the case of river-, lake-, and sea-water, and also of those waters which are artificially salted. For in places where rock-salt is dug, the impure and the broken pieces are thrown into fresh water, which, when boiled, condenses into salt. Some, indeed, boil sea-salt in fresh water again, and mould the salt into the little cones and other shapes.

Some people make salt by another method, from salt water which flows from hot springs that issue boiling from the earth. They set earthenware pots in a pool of the spring-water, and into them they pour water scooped up with ladles from the hot spring until they are half full. The perpetual heat of the waters of the pool evaporates the salt water just as the heat of the fire does in the caldrons. As soon as it begins to thicken, which happens when it has been reduced by boiling to a third or more, they seize the pots with tongs and pour the contents into small rectangular iron pans, which have also been placed in the pool. The interior of these pans is usually three feet long, two feet wide, and three digits deep, and they stand on four heavy legs, so that the water flows freely all round, but not into them. Since the water flows continuously from the pool through the little canals, and the spring

A—Pots. B—Tripod. C—Deep ladle.

always provides a new and copious supply, always boiling hot, it condenses the thickened water poured into the pans into salt; this is at once taken out with shovels, and then the work begins all over again. If the salty water contains other juices, as is usually the case with hot springs, no salt should be made from them.

Others boil salt water, and especially sea-water, in large iron pots; this salt is blackish, for in most cases they burn straw under them. Some people boil in these pots the brine in which fish is pickled. The salt which they make tastes and smells of fish.

A—Trench. B—Vat into which the salt water flows. C—Ladle. D—Small bucket with pole fastened into it.

Those who make salt by pouring brine over firewood, lay the wood in trenches which are twelve feet long, seven feet wide, and two and one half feet deep, so that the water poured in should not flow out. These trenches are constructed of rock-salt wherever it is to be had, in order that they should not soak up the water, and so that the earth should not fall in on the front, back and sides. As the charcoal is turned into salt at the same time as the

A—Large vat. B—Plug. C—Small tub. D—Deep ladle. E—Small vat. F—Caldron.

salt liquor, the Spaniards think, as Pliny writes[5], that the wood itself turns into salt. Oak is the best wood, as its pure ash yields salt ; elsewhere hazel-wood is lauded. But with whatever wood it be made, this salt is not greatly appreciated, being black and not quite pure ; on that account this method of salt-making is disdained by the Germans and Spaniards.

The solutions from which salt is made are prepared from salty earth or from earth rich in salt and saltpetre. Lye is made from the ashes of reeds and rushes. The solution obtained from salty earth by boiling, makes salt only ; from the other, of which I will speak more a little later, salt and saltpetre are made ; and from ashes is derived lye, from which its own salt is obtained. The ashes, as well as the earth, should first be put into a large vat ; then fresh water should be poured over the ashes or earth, and it should be stirred for about twelve hours with a stick, so that it may dissolve the salt. Then the plug is pulled out of the large vat ; the solution of salt or the lye is drained into a small tub and emptied with ladles into small vats ; finally, such a solution is transferred into iron or lead caldrons and boiled, until the water having evaporated, the juices are condensed into salt. The above are the various methods for making salt. (Illustration p. 557.)

Nitrum[6] is usually made from *nitrous* waters, or from solutions or from lye. In the same manner as sea-water or salt-water is poured into salt-pits and evaporated by the heat of the sun and changed into salt, so the *nitrous* Nile is led into *nitrum* pits and evaporated by the heat of the sun and con-

[5]Pliny XXXI., 39–40. " In the Gallic provinces in Germany they pour salt water " upon burning wood. The Spaniards in a certain place draw the brine from wells, which " they call *Muria*. They indeed think that the wood turns to salt, and that the oak is the " best, being the kind which is itself salty. Elsewhere the hazel is praised. Thus the char- " coal even is turned into salt when it is steeped in brine. Whenever salt is made with wood it " is black."

[6]We have elsewhere in this book used the word " soda " for the Latin term *nitrum*, because we believe as used by Agricola it was always soda, and because some confusion of this term with its modern adaptation for saltpetre (nitre) might arise in the mind of the reader. Fortunately, Agricola usually carefully mentions other alkalis, such as the product from lixiviation of ashes, separately from his *nitrum*. In these paragraphs, however, he has soda and potash hopelessly mixed, wherefore we have here introduced the Latin term. The actual difference between potash and soda—the *nitrum* of the Ancients, and the *alkali* of Geber (and the glossary of Agricola), was not understood for two hundred years after Agricola, when Duhamel made his well-known determinations ; and the isolation of sodium and potassium was, of course, still later by fifty years. If the reeds and rushes described in this paragraph grew near the sea, the salt from lixiviation would be soda, and likewise the Egyptian product was soda, but the lixiviation of wood-ash produces only potash ; as seen above, all are termed *nitrum* except the first.

HISTORICAL NOTES.—The word *nitrum*, *nitron*, *nitri*, *neter*, *nether*, or similar forms, occurs in innumerable ancient writings. Among such references are Jeremiah (II., 22) Proverbs (XXV., 20), Herodotus (II., 86, 87), Aristotle (*Prob.* I., 39, *De Mirab.* 54), Theophrastus (*De Igne* 435 ed. Heinsii, Hist. Plants III., 9), Dioscorides (V., 89), Pliny (XIV., 26, and XXXI., 46). A review of disputations on what salts this term comprised among the Ancients would itself fill a volume, but from the properties named it was no doubt mostly soda, more rarely potash, and sometimes both mixed with common salt. There is every reason to believe from the properties and uses mentioned, that it did not generally comprise nitre (saltpetre)—into which superficial error the nomenclature has led many translators. The preparation by way of burning, and the use of *nitrum* for purposes for which we now use soap, for making glass, for medicines, cosmetics, salves, painting, in baking powder, for preserving food, embalming, etc., and the descriptions of its taste in " nitrous " waters,—all answer for soda and potash, but not for saltpetre. It is possible that the common occurrence of saltpetre as an efflorescence on walls might naturally lead to its use, but in any event its distinguishing characteristics are nowhere mentioned. As sal-ammoniac occurred

A—NILE. B—NITRUM-PITS, SUCH AS I CONJECTURE THEM TO BE.[7]

verted into *nitrum*. Just as the sea, in flowing of its own will over the soil of this same Egypt, is changed into salt, so also the Nile, when it overflows in the dog days, is converted into *nitrum* when it flows into the *nitrum* pits. The solution from which *nitrum* is produced is obtained from fresh water percolating through *nitrous* earth, in the same manner as lye is made from fresh water percolating through ashes of oak or hard oak. Both solutions are taken out of vats and poured into rectangular copper caldrons, and are boiled until at last they condense into *nitrum*.

in the volcanoes in Italy, it also may have been included in the *nitrum* mentioned. *Nitrum* was in the main exported from Egypt, but Theophrastus mentions its production from wood-ash, and Pliny very rightly states that burned lees of wine (argol) had the nature of *nitrum*. Many of the ancient writers understood that it was rendered more caustic by burning, and still more so by treatment with lime. According to Beckmann (Hist. of Inventions II., p. 488), the form of the word *natron* was first introduced into Europe by two travellers in Egypt, Peter Ballon and Prosper Alpinus, about 1550. The word was introduced into mineralogy by Linnaeus in 1736. In the first instance *natron* was applied to

[7]This wondrous illustration of soda-making from Nile water is no doubt founded upon Pliny (XXXI., 46). "It is made in almost the same manner as salt, except that sea-water "is put into salt pans, whereas in the nitrous pans it is water of the Nile; these, with the "subsidence of the Nile during the forty days, are impregnated with *nitrum*."

Native as well as manufactured *nitrum* is mixed in vats with urine and boiled in the same caldrons; the decoction is poured into vats in which are copper wires, and, adhering to them, it hardens and becomes *chrysocolla*, which the Moors call *borax*. Formerly *nitrum* was compounded with Cyprian verdigris, and ground with Cyprian copper in Cyprian mortars, as Pliny writes. Some *chrysocolla* is made of rock-alum and sal-ammoniac.[8]

soda and potash in distinction to *nitre* for saltpetre, and later *natron* was applied solely to soda.

It is desirable to mention here two other forms of soda and potash which are frequently mentioned by Agricola. "Ashes which wool dyers use" (*cineres quo infectores lanarum utuntur*).—There is no indication in any of Agricola's works as to whether this was some special wood-ash or whether it was the calcined residues from wool washing. The "yolk" or "suint" of wool, originating from the perspiration of the animal, has long been a source of crude potash. The water, after washing the wool, is evaporated, and the residue calcined. It contains about 85% K_2CO_3, the remainder being sodium and potassium sulphates. Another reason for assuming that it was not a wood-ash product, is that these products are separately mentioned. In either event, whether obtained from wool residues or from lixiviation of wood-ash, it would be an impure potash. In some methods of wool dyeing, a wash of soda was first given, so that it is barely possible that this substance was sodium carbonate.

"Salt made from the ashes of musk ivy" (*sal ex anthyllidis cinere factus*,—Glossary. *salalkali*). This would be largely potash.

[8]This paragraph displays hopeless ignorance. Borax was known to Agricola and greatly used in his time; it certainly was not made from these compounds, but was imported from Central Asia. Sal-ammoniac was also known in his time, and was used like borax as a soldering agent. The reaction given by Agricola would yield free ammonia. The following historical notes on borax and sal-ammoniac may be of service.

Borax.—The uncertainties of the ancient distinctions in salts involve borax deeply. The word *Baurach* occurs in Geber and the other early Alchemistic writings, but there is nothing to prove that it was modern borax. There cannot be the slightest doubt, however, that the material referred to by Agricola as *borax* was our borax, because of the characteristic qualities incidentally mentioned in Book VII. That he believed it was an artificial product from *nitrum* is evident enough from his usual expression "*chrysocolla* made from *nitrum*, which the Moors call *borax*." Agricola, in *De Natura Fossilium* (p. 206–7), makes the following statements, which could leave no doubt on the subject:—"Native *nitrum* is found "in the earth or on the surface. . . . It is from this variety that the Venetians make "*chrysocolla*, which I call *borax*. . . . The second variety of artificial *nitrum* is made "at the present day from the native *nitrum*, called by the Arabs *tincar*, but I call it usually "by the Greek name *chrysocolla;* it is really the Arabic *borax*. . . . This *nitrum* does not "decrepitate nor fly out of the fire; however, the native variety swells up from within." The application of the word *chrysocolla* (*chrysos*, gold; *colla*, solder) to soldering materials, and at the same time to the copper mineral, is of Greek origin. If any further proof were needed as to the substance meant by Agricola, it lies in the word *tincar*. For a long time the borax of Europe was imported from Central Asia, through Constantinople and Venice, under the name of *tincal* or *tincar*. When this trade began, we do not know; evidently before Agricola's time. The statement here of making borax from alum and sal-ammoniac is identical with the assertion of Biringuccio (II., 9).

Sal-ammoniac.—The early history of this—ammonium chloride—is also under a cloud. Pliny (XXXI., 39) speaks of a *sal-hammoniacum*, and Dioscorides (V., 85) uses much the same word. Pliny describes it as from near the temple of Ammon in Egypt. None of the distinctive characteristics of sal-ammoniac are mentioned, and there is every reason to believe it was either common salt or soda. Herodotus, Strabo, and others mention common salt sent from about the same locality. The first authentic mention is in Geber, who calls it *sal-ammoniacum*, and describes a method of making, and several characteristic reactions. It was known in the Middle Ages under various names, among them *sal-aremonicum*. Agricola (*De Nat. Fos.*, III., p. 206) notes its characteristic quality of volatilization. "Sal-"ammoniac . . in the fire neither crackles nor flies out, but is totally consumed." He also says (p. 208): "Borax is used by goldsmiths to solder gold, likewise silver. The "artificers who make iron needles (tacks ?) similarly use sal-ammoniac when they cover the "heads with tin." The statement from Pliny mentioned in this paragraph is from XXXIII., 29, where he describes the *chrysocolla* used as gold solder as made from verdigris, *nitrum*, and urine in the way quoted. It is quite possible that this solder was sal-ammoniac, though not made in quite this manner. Pliny refers in several places (XXXIII., 26, 27, 28, and 29, XXXV., 28, etc.) to *chrysocolla*, about which he is greatly confused as between gold-solder, the copper mineral, and a green pigment, the latter being of either mineral origin.

A—Vat in which the soda is mixed. B—Caldron. C—Tub in which *chrysocolla* is condensed. D—Copper wires. E—Mortar.

Saltpetre[9] is made from a dry, slightly fatty earth, which, if it be retained for a while in the mouth, has an acrid and salty taste. This earth, together with a powder, are alternately put into a vat in layers a palm deep. The powder consists of two parts of unslaked lime and three parts of ashes of oak, or holmoak, or Italian oak, or Turkey oak, or of some similar kind. Each vat is filled with alternate layers of these to within three-quarters of a foot of the top, and then water is poured in until it is full. As the water percolates through the material it dissolves the saltpetre ; then, the plug being pulled out from the vat, the solution is drained into a tub and ladled out into small

[9]Saltpetre was secured in the Middle Ages in two ways, but mostly from the treatment of calcium nitrate efflorescence on cellar and similar walls, and from so-called saltpetre plantations. In this description of the latter, one of the most essential factors is omitted until the last sentence, *i.e.*, that the nitrous earth was the result of the decay of organic or animal matter over a long period. Such decomposition, in the presence of potassium and calcium carbonates—the lye and lime—form potassium and calcium nitrates, together with some magnesium and sodium nitrates. After lixiviation, the addition of lye converts the calcium and magnesium nitrates into saltpetre, *i.e.*, $Ca\,(NO^3)^2 + K^2CO^3 = Ca\,CO^3 + 2KNO^3$. The carbonates precipitate out, leaving the saltpetre in solution, from which it was evaporated and crystallised out. The addition of alum as mentioned would scarcely improve the situation.

The purification by repeated re-solution and addition of lye, and filtration, would eliminate the remaining other salts. The purification with sulphur, however, is more difficult

vats. If when tested it tastes very salty, and at the same time acrid, it is good; but, if not, then it is condemned, and it must be made to percolate again through the same material or through a fresh lot. Even two or three waters may be made to percolate through the same earth and become full of saltpetre, but the solutions thus obtained must not be mixed together unless all have the same taste, which rarely or never happens. The first of these solutions is poured into the first vat, the next into the second, the third into the third vat; the second and third solutions are used instead of plain water to percolate through fresh material; the first solution is made in this manner from both the second and third. As soon as there is an abundance of this solution it is poured into the rectangular copper caldron and evaporated to one half by boiling; then it is transferred into a vat covered with a lid, in which the earthy matter settles to the bottom. When the solution is clear it is poured back into the same pan, or into another, and re-boiled. When it bubbles and forms a scum, in order that it should not run over and that it may be greatly purified, there is poured into it three or four pounds of lye, made from three parts of oak or similar ash and one of unslaked lime. But in the water, prior to its being poured in, is dissolved rock-alum, in the proportion of one hundred and twenty *librae* of the former to five

to understand. In this case the saltpetre is melted and the sulphur added and set alight. Such an addition to saltpetre would no doubt burn brilliantly. The potassium sulphate formed would possibly settle to the bottom, and if the "greasy matter" were simply organic impurities, they might be burned off. This method of refining appears to have been copied from Biringuccio (x., 1), who states it in almost identical terms.

HISTORICAL NOTE.—As mentioned in Note 6 above, it is quite possible that the Ancients did include efflorescence of walls under *nitrum*; but, so far as we are aware, no specific mention of such an occurrence of *nitrum* is given, and, as stated before, there is every reason to believe that all the substances under that term were soda and potash. Especially the frequent mention of the preparation of *nitrum* by way of burning, argues strongly against saltpetre being included, as they would hardly have failed to notice the decrepitation. Argument has been put forward that Greek fire contained saltpetre, but it amounts to nothing more than argument, for in those receipts preserved, no salt of any kind is mentioned. It is most likely that the leprosy of house-walls of the Mosaic code (Leviticus XIV., 34 to 53) was saltpetre efflorescence. The drastic treatment by way of destruction of such "unclean" walls and houses, however, is sufficient evidence that this salt was not used. The first certain mention of saltpetre (*sal petrae*) is in Geber. As stated before, the date of this work is uncertain; in any event it was probably as early as the 13th Century. He describes the making of "solvative water" with alum and saltpetre, so there can be no doubt as to the substance (see Note on p. 460, on nitric acid). There is also a work by a nebulous Marcus Graecus, where the word *sal petrosum* is used. And it appears that Roger Bacon (died 1294) and Albertus Magnus (died 1280) both had access to that work. Bacon uses the term *sal petrae* frequently enough, and was the first to describe gunpowder (*De Mirabili Potestate Artis et Naturae* 1242). He gives no mention of the method of making his *sal petrae*. Agricola uses throughout the Latin text the term *halinitrum*, a word he appears to have coined himself. However, he gives its German equivalent in the *Interpretatio* as *salpeter*. The only previous description of the method of making saltpetre, of which we are aware, is that of Biringuccio (1540), who mentions the boiling of the excrescences from walls, and also says a good deal about boiling solutions from "nitrous" earth, which may or may not be of "plantation" origin. He also gives this same method of refining with sulphur. In any event, this statement by Agricola is the first clear and complete description of the saltpetre "plantations." Saltpetre was in great demand in the Middle Ages for the manufacture of gunpowder, and the first record of that substance and of explosive weapons necessarily involves the knowledge of saltpetre. However, authentic mention of such weapons only begins early in the 14th Century. Among the earliest is an authority to the Council of Twelve at Florence to appoint persons to make cannon, etc., (1326), references to cannon in the stores of the Tower of London, 1388, &c.

librae of the latter. Shortly afterward the solution will be found to be clear and blue. It is boiled until the waters, which are easily volatile (*subtiles*), are evaporated, and then the greater part of the salt, after it has settled at the bottom of the pan, is taken out with iron ladles. Then the concentrated solution is transferred to the vat in which rods are placed horizontally and vertically, to which it adheres when cold, and if there be much, it is condensed in three or four days into saltpetre. Then the solution which has not congealed, is poured out and put on one side or re-boiled. The saltpetre being cut out and washed with its own solution, is thrown on to boards that it may drain and dry. The yield of saltpetre will be much or little in proportion to whether the solution has absorbed much or little ; when the saltpetre has been obtained from lye, which purifies itself, it is somewhat clear and pure.

The purest and most transparent, because free from salt, is made if it is drawn off at the thickening stage, according to the following method. There

A—Caldron. B—Large vat into which sand is thrown. C– Plug. D—Tub. E—Vat containing the rods.

are poured into the caldron the same number of *amphorae* of the solution as of *congii* of the lye of which I have already spoken, and into the same caldron is thrown as much of the already made saltpetre as the solution and lye will dissolve. As soon as the mixture effervesces and forms scum, it is transferred to a vat, into which on a cloth has been thrown washed sand obtained from a river. Soon afterward the plug is drawn out of the hole at the bottom, and the mixture, having percolated through the sand, escapes into a tub. It is then reduced by boiling in one or another of the caldrons, until the greater part of the solution has evaporated; but as soon as it is well boiled and forms scum, a little lye is poured into it. Then it is transferred to another vat in which there are small rods, to which it adheres and congeals in two days if there is but little of it, or if there is much in three days, or at the most in four days; if it does not condense, it is poured back into the caldron and re-boiled down to half; then it is transferred to the vat to cool. The process must be repeated as often as is necessary.

Others refine saltpetre by another method, for with it they fill a pot made of copper, and, covering it with a copper lid, set it over live coals, where it is heated until it melts. They do not cement down the lid, but it has a handle, and can be lifted for them to see whether or not the melting has taken place. When it has melted, powdered sulphur is sprinkled in, and if the pot set on the fire does not light it, the sulphur kindles, whereby the thick, greasy matter floating on the saltpetre burns up, and when it is consumed the saltpetre is pure. Soon afterward the pot is removed from the fire, and later, when cold, the purest saltpetre is taken out, which has the appearance of white marble, the earthy residue then remains at the bottom. The earths from which the solution was made, together with branches of oak or similar trees, are exposed under the open sky and sprinkled with water containing saltpetre. After remaining thus for five or six years, they are again ready to be made into a solution.

Pure saltpetre which has rested many years in the earth, and that which exudes from the stone walls of wine cellars and dark places, is mixed with the first solution and evaporated by boiling.

Thus far I have described the methods of making *nitrum*, which are not less varied or multifarious than those for making salt. Now I propose to describe the methods of making alum,[10] which are likewise neither all alike, nor simple, because it is made from boiling aluminous water until it condenses to alum, or else from boiling a solution of alum which is obtained from a kind of earth, or from rocks, or from pyrites, or other minerals.

[10]There are three methods of manufacturing alum described by Agricola, the first and third apparently from shales, and the second from alum rock or "alunite." The reasons for assuming that the first process was from shales, are the reference to the "aluminous earth" as ore (*venae*) coming from "veins," and also the mixture of vitriol. In this process the free sulphuric acid formed by the oxidation of pyrites reacts upon the argillaceous material to form aluminium sulphate. The decomposed ore is then placed in tanks and lixiviated. The solution would contain aluminium sulphate, vitriol, and other impurities. By the addition of urine, the aluminium sulphate would be converted into ammonia alum. Agricola is, of course, mistaken as to the effect of the addition, being under the belief that it separated the vitriol from the alum; in fact, this belief was general until the latter part of the 18th Century, when Lavoisier determined that alum must have an alkali base. Nor is it clear

This kind of earth having first been dug up in such quantity as would make three hundred wheelbarrow loads, is thrown into two tanks; then the water is turned into them, and if it (the earth) contains vitriol it must be diluted with urine. The workmen must many times a day stir the ore with long, thick sticks in order that the water and urine may be mixed with it; then the plugs having been taken out of both tanks, the solution is drawn off into a trough, which is carved out of one or two trees. If the locality is supplied with an abundance of such ore, it should not immediately be thrown into the tanks, but first conveyed into open spaces and heaped up, for the longer it is exposed to the air and the rain, the better it is; after some months, during which the ore has been heaped up in open spaces into mounds, there are generated veinlets of far better quality than the ore. Then it is conveyed into six or more tanks, nine feet in length and breadth and five in depth, and afterward water is drawn into them of similar solution. After this, when the water has absorbed the alum, the plugs are pulled out, and the solution escapes into a round reservoir forty feet wide and three feet deep. Then the ore is thrown out of the tanks into other tanks, and water again being run into the latter and the urine added and stirred by means of poles, the plugs are withdrawn and the solution is run off into the same reservoir. A few days afterward, the reservoirs containing the solution are emptied through a small launder, and run into rectangular lead caldrons; it is boiled in them until the

from this description exactly how they were separated. In a condensed solution allowed to cool, the alum would precipitate out as "alum meal," and the vitriol would "float on top"—in solution. The reference to "meal" may represent this phenomenon, and the re-boiling referred to would be the normal method of purification by crystallization. The "asbestos" and gypsum deposited in the caldrons were no doubt feathery and mealy calcium sulphate. The alum produced would, in any event, be mostly ammonia alum.

The second process is certainly the manufacture from "alum rock" or "alunite" (the hydros sulphate of aluminium and potassium), such as that mined at La Tolfa in the Papal States, where the process has been for centuries identical with that here described. The alum there produced is the double basic potassium alum, and crystallizes into cubes instead of octrahedra, *i.e.*, the Roman alum of commerce. The presence of much ferric oxide gives the rose colour referred to by Agricola. This account is almost identical with that of Biringuccio (II., 4), and it appears from similarity of details that Agricola, as stated in his preface, must have "refreshed his mind" from this description; it would also appear from the preface that he had himself visited the locality.

The third process is essentially the same as the first, except that the decomposition of the pyrites was hastened by roasting. The following obscure statement of some interest occurs in Agricola's *De Natura Fossilium*, p. 209:—". . . . alum is made from vitriol, "for when oil is made from the latter, alum is distilled out (*expirat*). This absorbs the clay "which is used in cementing glass, and when the operation is complete the clay is macerated "with pure water, and the alum is soon afterward deposited in the shape of small cubes." Assuming the oil of vitriol to be sulphuric acid and the clay "used in cementing glass" to be kaolin, we have here the first suggestion of a method for producing alum which came into use long after.

"Burnt alum" (*alumen coctum*).—Agricola frequently uses this expression, and on p. 568, describes the operation, and the substance is apparently the same as modern dehydrated alum, often referred to as "burnt alum."

HISTORICAL NOTES.—Whether the Ancients knew of alum in the modern sense is a most vexed question. The Greeks refer to a certain substance as *stypteria*, and the Romans refer to this same substance as *alumen*. There can be no question as to their knowledge and common use of vitriol, nor that substances which they believed were entirely different from vitriol were comprised under the above names. Beckmann (Hist. of Inventions, Vol. I., p. 181) seems to have been the founder of the doctrine that the ancient *alumen* was vitriol, and scores of authorities seem to have adopted his arguments without inquiry, until that belief

greater part of the water has evaporated. The earthy sediment deposited at the bottom of the caldron is composed of fatty and aluminous matter, which usually consists of small incrustations, in which there is not infrequently found a very white and very light powder of asbestos or gypsum. The solution now seems to be full of meal. Some people instead pour the partly evaporated solution into a vat, so that it may become pure and clear; then pouring it back into the caldron, they boil it again until it becomes mealy. By whichever process it has been condensed, it is then poured into a wooden tub sunk into the earth in order to cool it. When it becomes cold it is poured into vats, in which are arranged horizontal and vertical twigs, to which the alum clings when it condenses; and thus are made the small white transparent cubes, which are laid to dry in hot rooms.

If vitriol forms part of the aluminous ore, the material is dissolved in water without being mixed with urine, but it is necessary to pour that into the clear and pure solution when it is to be re-boiled. This separates the vitriol from the alum, for by this method the latter sinks to the bottom of the caldron, while the former floats on the top; both must be poured separately into smaller vessels, and from these into vats to condense. If, however, when the solution was re-boiled they did not separate, then they must be poured from the smaller vessels into larger vessels and covered over; then the vitriol separating from the alum, it condenses. Both are cut out and put to dry in the hot room, and are ready to be sold; the solution which did not congeal in

is now general. One of the strongest reasons put forward was that alum does not occur native in appreciable quantities. Apart from the fact that the weight of this argument has been lost by the discovery that alum does occur in nature to some extent as an aftermath of volcanic action, and as an efflorescence from argillaceous rocks, we see no reason why the Ancients may not have prepared it artificially. One of the earliest mentions of such a substance is by Herodotus (II., 180) of a thousand talents of *stypteria*, sent by Amasis from Egypt as a contribution to the rebuilding of the temple of Delphi. Diodorus (V., 1) mentions the abundance which was secured from the Lipari Islands (Stromboli, etc.), and a small quantity from the Isle of Melos. Dioscorides (V., 82) mentions Egypt, Lipari Islands, Melos, Sardinia, Armenia, etc., "and generally in any other places where one finds red ochre (*rubrica*)." Pliny (XXXV., 52) gives these same localities, and is more explicit as to how it originates—"from an earthy water which exudes from the earth." Of these localities, the Lipari Islands (Stromboli, etc.), and Melos are volcanic enough, and both Lipari and Melos are now known to produce natural alum (Dana. Syst. Min., p. 95; and Tournefort, "*Relation d'un voyage du Levant*," London, 1717, *Lettre* IV., Vol. I.). Further, the hair-like alum of Dioscorides, repeated by Pliny below, was quite conceivably fibrous *kalinite*, native potash alum, which occurs commonly as an efflorescence. Be the question of native alum as it may—and vitriol is not much more common—our own view that the ancient *alumen* was alum, is equally based upon the artificial product. Before entering upon the subject, we consider it desirable to set out the properties of the ancient substance, a complete review of which is given by Pliny (XXXV., 52), he obviously quoting also from Dioscorides, which, therefore, we do not need to reproduce. Pliny says:—

"Not less important, or indeed dissimilar, are the uses made of *alumen;* by which "name is understood a sort of salty earth. Of this, there are several kinds. In Cyprus there "is a white *alumen*, and a darker kind. There is not a great difference in their colour, "though the uses made of them are very dissimilar,—the white *alumen* being employed in a "liquid state for dyeing wool bright colours, and the dark-coloured *alumen*, on the other "hand, for giving wool a sombre tint. Gold is purified with black *alumen*. Every kind of "*alumen* is from a *limus* water which exudes from the earth. The collection of it commences "in winter, and it is dried by the summer sun. That portion of it which first matures is the "whitest. It is obtained in Spain, Egypt, Armenia, Macedonia, Pontus, Africa, and the "islands of Sardinia, Melos, Lipari, and Strongyle; the most esteemed, however, is that of "Egypt, the next best from Melos. Of this last there are two kinds, the liquid *alumen*, and "the solid. Liquid *alumen*, to be good, should be of a limpid and milky appearance; when

A—Tanks. B—Stirring poles. C—Plug. D—Trough. E—Reservoir. F—Launder. G—Lead caldron. H—Wooden tubs sunk into the earth. I—Vats in which twigs are fixed.

the vessels and vats is again poured back into the caldron to be re-boiled. The earth which settled at the bottom of the caldron is carried back to the tanks, and, together with the ore, is again dissolved with water and urine. The earth which remains in the tanks after the solution has been drawn off is emptied in a heap, and daily becomes more and more aluminous in the same way as the earth from which saltpetre was made, but fuller of its juices, wherefore it is again thrown into the tanks and percolated by water.

Aluminous rock is first roasted in a furnace similar to a lime kiln. At the bottom of the kiln a vaulted fireplace is made of the same kind of rock; the remainder of the empty part of the kiln is then entirely filled with the same aluminous rocks. Then they are heated with fire until they are red hot and have exhaled their sulphurous fumes, which occurs, according to their divers nature, within the space of ten, eleven, twelve, or more hours. One thing the master must guard against most of all is not to roast the rock either too much or too little, for on the one hand they would not soften when sprinkled with water, and on the other they either would be too hard or would crumble into ashes; from neither would much alum be obtained, for the strength which they have would be decreased. When the rocks are cooled they are drawn out and conveyed into an open space, where they are piled one upon the other in heaps fifty feet long, eight feet wide, and four feet high, which are sprinkled for forty days with water carried in deep ladles. In spring the sprinkling is done both morning and evening, and in summer at

"rubbed, it should be without roughness, and should give a little heat. This is called "*phorimon*. The mode of detecting whether it has been adulterated is by pomegranate "juice, for, if genuine, the mixture turns black. The other, or solid, is pale and rough "and turns dark with nut-galls; for which reason it is called *paraphoron*. Liquid *alumen* is "naturally astringent, indurative, and corrosive; used in combination with honey, it heals "ulcerations. . . . There is one kind of solid *alumen*, called by the Greeks *schistos*, "which splits into filaments of a whitish colour; for which reason some prefer calling it "*trichitis* (hair like). *Alumen* is produced from the stone *chalcitis*, from which copper is also "made, being a sort of coagulated scum from that stone. This kind of *alumen* is less "astringent than the others, and is less useful as a check upon bad humours of the body. . . "The mode of preparing it is to cook it in a pan until it has ceased being a liquid. There "is another variety of *alumen* also, of a less active nature, called *strongyle*. It is of two kinds. "The fungous, which easily dissolves, is utterly condemned. The better kind is the pumice-"like kind, full of small holes like a sponge, and is in round pieces, more nearly white in colour, "somewhat greasy, free from grit, friable, and does not stain black. This last kind is cooked "by itself upon charcoal until it is reduced to pure ashes. The best kind of all is that called "*melinum*, from the Isle of Melos, as I have said, none being more effectual as an astringent, "for staining black, and for indurating, and none becomes more dry. . . . Above all other "properties of *alumen* is its remarkable astringency, whence its Greek name. . . . It is "injected for dysentry and employed as a gargle." The lines omitted refer entirely to medical matters which have no bearing here. The following paragraph (often overlooked) from Pliny (xxxv., 42) also has an important bearing upon the subject:—"In Egypt they "employ a wonderful method of dyeing. The white cloth, after it is pressed, is stained "in various places, not with dye stuffs, but with substances which absorb colours. These "applications are not apparent on the cloth, but when it is immersed in a caldron of hot "dye it is removed the next moment brightly coloured. The remarkable circumstance "is that although there be only one dye in the caldron yet different colours appear in the "cloth."

It is obvious from Pliny's description above, and also from the making of vitriol (see Note 11, p. 572), that this substance was obtained from liquor resulting from natural or artificial lixiviation of rocks—in the case of vitriols undoubtedly the result of decomposition of pyritiferous rocks (such as *chalcitis*). Such liquors are bound to contain aluminum sulphate if there is any earth or clay about, and whether they contained alum would be a question of an alkali being present. If no alkali were present in this liquor, vitriol would

noon besides. After being moistened for this length of time the rocks begin to fall to pieces like slaked lime, and there originates a certain new material of the future alum, which is soft and similar to the *liquidae medullae* found in the rocks. It is white if the stone was white before it was roasted, and rose-coloured if red was mixed with the white; from the former, white alum is obtained, and from the latter, rose-coloured. A round furnace is made, the lower part of which, in order to be able to endure the force of the heat, is made of rock that neither melts nor crumbles to powder by the fire. It is constructed in the form of a basket, the walls of which are two feet high, made of the same rock. On these walls rests a large round caldron made of copper plates, which is concave at the bottom, where it is eight feet in diameter. In the empty space under the bottom they place the wood to be kindled with fire. Around the edge of the bottom of the caldron, rock is built in cone-shaped, and the diameter of the bottom of the rock structure is seven feet, and of the top ten feet; it is eight feet deep. The inside, after being rubbed over with oil, is covered with cement, so that it may be able to hold boiling water; the cement is composed of fresh lime, of which the lumps are slaked with wine, of iron-scales, and of sea-snails, ground and mixed with the white of eggs and oil. The edges of the caldron are surmounted with a circle of wood a foot thick and half a foot high, on which the workmen rest the wooden shovels with which they cleanse the water of earth and of the undissolved lumps of rock that remain at

crystallize out first, and subsequent condensation would yield aluminum sulphate. If alkali were present, the alum would crystallize out either before or with the vitriol. Pliny's remark, "that portion of it which first matures is whitest", agrees well enough with this hypothesis. No one will doubt that some of the properties mentioned above belong peculiarly to vitriol, but equally convincing are properties and uses that belong to alum alone. The strongly astringent taste, white colour, and injection for dysentry, are more peculiar to alum than to vitriol. But above all other properties is that displayed in dyeing, for certainly if we read this last quotation from Pliny in conjunction with the statement that white *alumen* produces bright colours and the dark kind, sombre colours, we have the exact reactions of alum and vitriol when used as mordants. Therefore, our view is that the ancient salt of this character was a more or less impure mixture ranging from alum to vitriol—" the whiter the better." Further, considering the ancient knowledge of soda (*nitrum*), and the habit of mixing it into almost everything, it does not require much flight of imagination to conceive its admixture to the " water," and the absolute production of alum.

Whatever may have been the confusion between alum and vitriol among the Ancients, it appears that by the time of the works attributed to Geber (12th or 13th Century), the difference was well known. His work (*Investigationes perfectiones*, IV.) refers to *alumen glaciale* and *alumen jameni* as distinguished from vitriol, and gives characteristic reactions which can leave no doubt as to the distinction. We may remark here that the repeated statement apparently arising from Meyer (History of Chemistry, p. 51) that Geber used the term *alum de rocca* is untrue, this term not appearing in the early Latin translations. During the 15th Century alum did come to be known in Europe as *alum de rocca*. Various attempts have been made to explain the origin of this term, ranging from the Italian root, a " rock, " to the town of Rocca in Syria, where alum was supposed to have been produced. In any event, the supply for a long period prior to the middle of the 15th Century came from Turkey, and the origin of the methods of manufacture described by Agricola, and used down to the present day, must have come from the Orient.

In the early part of the 15th Century, a large trade in alum was done between Italy and Asia Minor, and eventually various Italians established themselves near Constantinople and Smyrna for its manufacture (Dudae, *Historia Byzantina Venetia*, 1729, p. 71). The alum was secured by burning the rock, and lixiviation. With the capture of Constantinople by the Turks (1453), great feeling grew up in Italy over the necessity of buying this requisite for their dyeing establishments from the infidel, and considerable exertion was made to find other sources of supply. Some minor works were attempted, but nothing much

the bottom of the caldron. The caldron, being thus prepared, is entirely filled through a launder with water, and this is boiled with a fierce fire until it bubbles. Then little by little eight wheelbarrow loads of the material, composed of roasted rock moistened with water, are gradually emptied into the caldron by four workmen, who, with their shovels which reach to the bottom, keep the material stirred and mixed with water, and by the same means they lift the lumps of undissolved rock out of the caldron. In this manner the material is thrown in, in three or four lots, at intervals of two or three hours more or less; during these intervals, the water, which has been cooled by the rock and material, again begins to boil. The water, when sufficiently purified and ready to congeal, is ladled out and run off with launders into thirty troughs. These troughs are made of oak, holm oak, or Turkey oak; their interior is six feet long, five feet deep, and four feet wide. In these the water congeals and condenses into alum, in the spring in the space of four days, and in summer in six days. Afterward the holes at the bottom of the oak troughs being opened, the water which has not congealed is drawn off into buckets and poured back into the caldron; or it may be preserved in empty troughs, so that the master of the workmen, having seen it, may order his helpers to pour it into the caldron, for the water which is not altogether wanting in alum, is considered better than that which has none at all. Then the alum is hewn out with a knife or a chisel. It is thick and excellent according to the strength of the rock, either white or pink according to the colour of the rock. The earthy powder, which remains three to four digits thick as the residue of the alum at the bottom of the trough is again thrown into the caldron and boiled with fresh aluminous material. Lastly, the alum cut out is washed, and dried, and sold.

Alum is also made from crude pyrites and other aluminous mixtures. It is first roasted in an enclosed area: then, after being exposed for some

eventuated until the appearance of one John de Castro. From the Commentaries of Pope Pius II. (1614, p. 185), it appears that this Italian had been engaged in dyeing cloth in Constantinople, and thus became aware of the methods of making alum. Driven out of that city through its capture by the Turks, he returned to Italy and obtained an office under the Apostolic Chamber. While in this occupation he discovered a rock at Tolfa which appeared to him identical with that used at Constantinople in alum manufacture. After experimental work, he sought the aid of the Pope, which he obtained after much vicissitude. Experts were sent, who after examination "shed tears of joy, they kneeling down three times, worshipped God and praised His kindness in conferring such a gift on their age." Castro was rewarded, and the great papal monopoly was gradually built upon this discovery. The industry firmly established at Tolfa exists to the present day, and is the source of the Roman alum of commerce. The Pope maintained this monopoly strenuously, by fair means and by excommunication, gradually advancing the price until the consumers had greater complaint than against the Turks. The history of the disputes arising over the papal alum monopoly would alone fill a volume.

By the middle of the 15th Century alum was being made in Spain, Holland, and Germany, and later in England. In her efforts to encourage home industries and escape the tribute to the Pope, Elizabeth (see Note on p. 283) invited over "certain foreign chymistes and mineral masters" and gave them special grants to induce them to "settle in these realmes." Among them was Cornelius Devoz, to whom was granted the privilege of "mining and digging in our Realm of England for allom and copperas." What Devoz accomplished is not recorded, but the first alum manufacture on a considerable scale seems to have been in Yorkshire, by one Thomas Chaloner (about 1608), who was supposed to have seduced workmen from the Pope's alum works at Tolfa, for which he was duly cursed with all the weight of the Pope and Church. (Pennant, Tour of Scotland, 1786).

A—Furnace. B—Enclosed space. C—Aluminous rock. D—Deep ladle. E—Caldron. F—Launder. G—Troughs.

months to the air in order to soften it, it is thrown into vats and dissolved. After this the solution is poured into the leaden rectangular pans and boiled until it condenses into alum. The pyrites and other stones which are not mixed with alum alone, but which also contain vitriol, as is most usually the case, are both treated in the manner which I have already described. Finally, if metal is contained in the pyrites and other rock, this material must be dried, and from it either gold, silver, or copper is made in a furnace.

Vitriol[11] can be made by four different methods; by two of these methods

[11]The term for vitriol used by the Roman authors, followed by Agricola, is *atramentum sutorium*, literally shoemaker's blacking, the term no doubt arising from its ancient (and modern) use for blackening leather. The Greek term was *chalcanthon*. The term "vitriol" seems first to appear in Albertus Magnus (*De Mineralibus, Liber* v.), who died in 1280, where he uses the expression "*atramentum viride a quibusdam vitreolum vocatur*." Agricola (*De Nat. Foss.*, p. 213) states, "In recent years the name *vitriolum* has been given to it." The first adequate description of vitriol is by Dioscorides (v., 76), as follows:—"Vitriol "(*chalcanthon*) is of one genus, and is a solidified liquid, but it has three different species. "One is formed from the liquids which trickle down drop by drop and congeal in certain "mines; therefore those who work in the Cyprian mines call it *stalactis*. Petesius calls "this kind *pinarion*. The second kind is that which collects in certain caverns; afterward "it is poured into trenches, where it congeals, whence it derives its name *pēctos*. The "third kind is called *hephthon* and is mostly made in Spain; it has a beautiful colour but is "weak. The manner of preparing it is as follows: dissolving it in water, they boil it, and "then they transfer it to cisterns and leave it to settle. After a certain number of days it "congeals and separates into many small pieces, having the form of dice, which stick "together like grapes. The most valued is blue, heavy, dense, and translucent." Pliny (xxxiv., 32) says:—"By the name which they have given to it, the Greeks indicate "the similar nature of copper and *atramentum sutorium*, for they call it *chalcanthon*. There "is no substance of an equally miraculous nature. It is made in Spain from wells of this kind "of water. This water is boiled with an equal quantity of pure water, and is then poured "into wooden tanks (fish ponds). Across these tanks there are fixed beams, to which hang "cords stretched by little stones. Upon these cords adheres the *limus* (Agricola's 'juice') in "drops of a vitreous appearance, somewhat resembling a bunch of grapes. After removal, it "is dried for thirty days. It is of a blue colour, and of a brilliant lustre, and is very like "glass. Its solution is the blacking used for colouring leather. *Chalcanthon* is made in "many other ways: its kind of earth is sometimes dug from ditches, from the sides of which "exude drops, which solidify by the winter frosts into icicles, called *stalagmia*, and there is "none more pure. When its colour is nearly white, with a slight tinge of violet, it is called "*leukoïon*. It is also made in rock basins, the rain water collecting the *limus* into them, "where it becomes hardened. It is also made in the same way as salt by the intense heat of "the sun. Hence it is that some distinguish two kinds, the mineral and the artificial; the "latter being paler than the former and as much inferior to it in quality as it is in colour."

While Pliny gives prominence to blue vitriol, his solution for colouring leather must have been the iron sulphate. There can be no doubt from the above, however, that both iron and copper sulphates were known to the Ancients. From the methods for making vitriol given here in *De Re Metallica*, it is evident that only the iron sulphate would be produced, for the introduction of iron strips into the vats would effectually precipitate any copper. It is our belief that generally throughout this work, the iron sulphate is meant by the term *atramentum sutorium*. In *De Natura Fossilium* (p. 213–15) Agricola gives three varieties of *atramentum sutorium*,—*viride*, *caeruleum*, and *candidum*, *i.e.*, green, blue, and white. Thus the first mention of white vitriol (zinc sulphate) appears to be due to him, and he states further (p. 213): "A white sort is found, especially at Goslar, in the shape "of icicles, transparent like crystals." And on p. 215: "Since I have explained "the nature of vitriol and its relatives, which are obtained from cupriferous pyrites, "I will next speak of an acrid solidified juice which commonly comes from *cadmia*. It is found "at Annaberg in the tunnel driven to the Saint Otto mine; it is hard and white, and so "acrid that it kills mice, crickets, and every kind of animal. However, that feathery sub- "tance which oozes out from the mountain rocks and the thick substance found hanging "in tunnels and caves from which saltpetre is made, while frequently acrid, does not come "from *cadmia*." Dana (Syst. of Min., p. 939) identifies this as *Goslarite*—native zinc sulphate. It does not appear, however, that artificial zinc vitriol was made in Agricola's time. Schlüter (*Huette-Werken*, Braunschweig 1738, p. 597) states it to have been made for the first time at Rammelsberg about 1570.

from water containing vitriol; by one method from a solution of *melanteria, sory* and *chalcitis;* and by another method from earth or stones mixed with vitriol.

The vitriol water is collected into pools, and if it cannot be drained into them, it must be drawn up and carried to them in buckets by a workman.

It is desirable here to enquire into the nature of the substances given by all of the old mineralogists under the Latinized Greek terms *chalcitis, misy, sory,* and *melanteria.* The first mention of these minerals is in Dioscorides, who (v., 75–77) says: "The best *chalcitis* "is like copper. It is friable, not stony, and is intersected by long brilliant veins. . . . "*Misy* is obtained from Cyprus; it should have the appearance of gold, be hard, and when "pulverised it should have the colour of gold and sparkle like stars. It has the same "properties as *chalcitis.* . . . The best is from Egypt. . . . One kind of *melanteria* "congeals like salt in the entries to copper mines. The other kind is earthy and appears "on the surface of the aforesaid mines. It is found in the mines of Cilicia and other regions. "The best has the colour of sulphur, is smooth, pure, homogenous, and upon contact with "water immediately becomes black. . . . . Those who consider *sory* to be the same "as *melanteria,* err greatly. *Sory* is a species of its own, though it is not dissimilar. The "smell of *sory* is oppressive and provokes nausea. It is found in Egypt and in other regions, "as Libya, Spain, and Cyprus. The best is from Egypt, and when broken is black, porous, "greasy, and astringent." Pliny (XXXIV., 29–31) says:—"That is called *chalcitis* from "which, as well as itself copper (?) is extracted by heat. It differs from *cadmia* in that this "is obtained from rocks near the surface, while that is taken from rocks below the surface. "Also *chalcitis* is immediately friable, being naturally so soft as to appear like compressed "wool. There is also this other distinction; *chalcitis* contains three other substances, "copper, *misy,* and *sory.* Of each of these we shall speak in their appropriate places. "It contains elongated copper veins. The most approved kind is of the colour of honey; "it is streaked with fine sinuous veins and is friable and not stony. It is considered most "valuable when fresh. . . . The *sory* of Egypt is the most esteemed, being much superior "to that of Cyprus, Spain, and Africa; although some prefer the *sory* from Cyprus for affec- "tions of the eyes. But from whatever nation it comes, the best is that which has the "strongest odour, and which, when ground up, becomes greasy, black, and spongy. It is "a substance so unpleasant to the stomach that some persons are nauseated by its smell. "Some say that *misy* is made by the burning of stones in trenches, its fine yellow "powder being mixed with the ashes of pine-wood. The truth is, as I said above, that "though obtained from the stone, it is already made and in solid masses, which require force "to detach them. The best comes from the works of Cyprus, its characteristics being that "when broken it sparkles like gold, and when ground it presents a sandy appearance, but on "the contrary, if heated, it is similar to *chalcitis. Misy* is used in refining gold. . . ."

Agricola's views on the subject appear in *De Natura Fossilium.* He says (p. 212):— "The cupriferous pyrites (*pyrites aerosus*) called *chalcitis* is the mother and cause of *sory* "—which is likewise known as mine *vitriol* (*atramentum metallicum*)—and *melanteria.* "These in turn yield vitriol and such related things. This may be seen especially at Goslar, "where the nodular lumps of dark grey colour are called vitriol stone (*lapis atramenti*). "In the centre of them is found greyish pyrites, almost dissolved, the size of a walnut. It "is enclosed on all sides, sometimes by *sory,* sometimes by *melanteria.* From them start "little veinlets of greenish vitriol which spread all over it, presenting somewhat the appear- "ance of hairs extending in all directions and cohering together. . . . There are five "species of this solidified juice, *melanteria, sory, chalcitis, misy,* and vitriol. Sometimes many "are found in one place, sometimes all of them, for one originates from the other. From "pyrites, which is, as one might say, the root of all these juices, originates the above- "mentioned *sory* and *melanteria.* From *sory, chalcitis,* and *melanteria* originate the various "kinds of vitriol. . . . *Sory, melanteria, chalcitis,* and *misy* are always native; vitriol "alone is either native or artificial. From them vitriol effloresces white, and sometimes "green or blue. *Misy* effloresces not only from *sory, melanteria,* and *chalcitis,* but also from "all the vitriols, artificial as well as natural. . . . *Sory* and *melanteria* differ somewhat "from the others, but they are of the same colours, grey and black; but *chalcitis* is red and "copper-coloured; *misy* is yellow or gold-coloured. All these native varieties have the "odour of lightning (brimstone), but *sory* is the most powerful. The feathery vitriol is soft "and fine and hair-like, and *melanteria* has the appearance of wool and it has a similarity to "salt; all these are rare and light; *sory, chalcitis,* and *misy* have the following relations. "*Sory* because of its density has the hardness of stone, although its texture is very coarse. "*Misy* has a very fine texture. *Chalcitis* is between the two; because of its roughness and "strong odour it differs from *melanteria,* although they do not differ in colour. The vitriols, "whether natural or artificial, are hard and dense . . . as regarding shape, *sory, chalcitis,* "*misy,* and *melanteria* are nodular, but *sory* is occasionally porous, which is peculiar to it.

A—TUNNEL. B—BUCKET. C—PIT.

In hot regions or in summer, it is poured into out-of-door pits which have been dug to a certain depth, or else it is extracted from shafts by pumps and poured into launders, through which it flows into the pits, where it is condensed by the heat of the sun. In cold regions and in winter these vitriol waters are boiled down with equal parts of fresh water in rectangular leaden caldrons; then, when cold, the mixture is poured into vats or into tanks, which Pliny calls wooden fish-tanks. In these tanks light cross-beams are fixed to the upper part, so that they may be stationary, and from them hang ropes stretched with little stones; to these the contents of the thickened solutions congeal and adhere in transparent cubes or seeds of vitriol, like bunches of grapes.

" *Misy* when it effloresces in no great quantity from the others is like a kind of pollen, other-
" wise it is nodular. *Melanteria* sometimes resembles wool, sometimes salt."

The sum and substance, therefore, appears to be that *misy* is a yellowish material, possibly ochre, and *sory* a blackish stone, both impregnated with vitriol. *Chalcitis* is a partially decomposed pyrites; and *melanteria* is no doubt native vitriol. From this last term comes the modern *melanterite*, native hydrous ferrous sulphate. Dana (System of Mineralogy, p. 964) considers *misy* to be in part *copiapite*—basic ferric sulphate—but any such part would not come under Agricola's objection to it as a source of vitriol. The disabilities of this and *chalcitis* may, however, be due to their copper content.

A—Caldron. B—Tank. C—Cross-bars. D—Ropes. E—Little stones.

By the third method vitriol is made out of *melanteria* and *sory*. If the mines give an abundant supply of *melanteria* and *sory*, it is better to reject the *chalcitis*, and especially the *misy*, for from these the vitriol is impure, particularly from the *misy*. These materials having been dug and thrown into the tanks, they are first dissolved with water; then, in order to recover the pyrites from which copper is not rarely smelted and which forms a sediment at the bottom of the tanks, the solution is transferred to other vats, which are nine feet wide and three feet deep. Twigs and wood which float on the surface are lifted out with a broom made of twigs, and afterward all the sediment settles at the bottom of this vat. The solution is poured into a rectangular leaden caldron eight feet long, three feet wide, and the same in depth. In this caldron it is boiled until it becomes thick and viscous, when it is poured into a launder, through which it runs into another leaden caldron of the same size as the one described before. When cold, the solution is drawn off through twelve little launders, out of which it flows into as many wooden tubs four and a half feet deep and three feet wide. Upon these tubs are placed perforated crossbars distant from each other from four to six digits, and from the holes hang thin laths, which reach to the bottom, with pegs or wedges driven into them. The vitriol adheres to these laths, and within the space of a few days congeals into cubes, which are taken away and put into a chamber having a sloping board floor, so that the moisture which drips from the vitriol may flow into a tub beneath. This solution is re-boiled, as is also that solution which was left in the twelve tubs, for, by reason of its having become too thin and liquid, it did not congeal, and was thus not converted into vitriol.

A—Wooden tub. B—Cross-bars. C—Laths. D—Sloping floor of the chamber. E—Tub placed under it.

The fourth method of making vitriol is from vitriolous earth or stones. Such ore is at first carried and heaped up, and is then left for five or six months exposed to the rain of spring and autumn, to the heat of summer, and to the rime and frost of winter. It must be turned over several times with shovels, so that the part at the bottom may be brought to the top, and it is thus ventilated and cooled; by this means the earth crumbles up and loosens, and the stone changes from hard to soft. Then the ore is covered with a roof, or else it is taken away and placed under a roof, and remains in that place six, seven, or eight months. Afterward as large a portion as is required is thrown into a vat, which is half-filled with water; this vat is one hundred feet long, twenty-four feet wide, eight feet deep. It has an opening at the bottom, so that when it is opened the dregs of the ore from which the vitriol comes may be drawn off, and it has, at the height of one foot from the bottom, three or four little holes, so that, when closed, the water may be retained, and when opened the solution flows out. Thus the ore is mixed with water, stirred with poles and left in the tank until the earthy portions sink to the bottom and the water absorbs the juices. Then the little holes are opened, the solution flows out of the vat, and is caught in a vat below it ; this vat is of the same length as the other, but twelve feet wide and four feet deep. If the solution is not sufficiently vitriolous it is mixed with fresh ore ; but if it contains enough vitriol, and yet has not exhausted all of the ore rich in vitriol, it is well to dissolve the ore again with fresh water. As soon as the solution becomes clear, it is poured into the rectangular leaden caldron through launders, and is boiled until the water is evaporated. Afterward as many thin strips of iron as the nature of the solution requires, are thrown in, and then it is boiled again until it is thick enough, when cold, to congeal into vitriol. Then it is poured into tanks or vats, or any other receptacle, in which all of it that is apt to congeal does so within two or three days. The solution which does not congeal is either poured back into the caldron to be boiled again, or

A—Caldron. B—Moulds. C—Cakes

it is put aside for dissolving the new ore, for it is far preferable to fresh water. The solidified vitriol is hewn out, and having once more been thrown into the caldron, is re-heated until it liquefies ; when liquid, it is poured into moulds that it may be made into cakes. If the solution first poured out is not satisfactorily thickened, it is condensed two or three times, and each time liquefied in the caldron and re-poured into the moulds, in which manner pure cakes, beautiful to look at, are made from it.

The vitriolous pyrites, which are to be numbered among the mixtures (*mistura*), are roasted as in the case of alum, and dissolved with water, and the solution is boiled in leaden caldrons until it condenses into vitriol. Both alum and vitriol are often made out of these, and it is no wonder, for these juices are cognate, and only differ in the one point,—that the former is less, the latter more, earthy. That pyrites which contains metal must be smelted in the furnace. In the same manner, from other mixtures of vitriolic and metalliferous material are made vitriol and metal. Indeed, if ores of vitriolous pyrites abound, the miners split small logs down the centre and cut them off in lengths as long as the drifts and tunnels are wide, in which they lay them down transversely ; but, that they may be stable, they are laid on the ground with the wide side down and the round side up, and they touch each other at the bottom, but not at the top. The intermediate space is filled with pyrites, and the same crushed are scattered over the wood, so that, coming in or going out, the road is flat and even. Since the drifts or tunnels drip with water, these pyrites are soaked, and from them are freed the vitriol and cognate things. If the water ceases to drip, these dry and harden, and then they are raised from the shafts, together with the pyrites not yet dissolved in the water, or they are carried out from the tunnels ; then they are thrown into vats or tanks, and boiling water having been poured over them, the vitriol is freed and the pyrites are dissolved. This green solution is transferred to other vats or tanks, that it may be made clear and pure ; it is then boiled in the lead caldrons until it thickens ; afterward it is poured into wooden tubs, where it condenses on rods, or reeds, or twigs, into green vitriol.

Sulphur is made from sulphurous waters, from sulphurous ores, and from sulphurous mixtures. These waters are poured into the leaden caldrons and boiled until they condense into sulphur. From this latter, heated together with iron-scales, and transferred into pots, which are afterward covered with lute and refined sulphur, another sulphur is made, which we call *caballinum*.[12]

The ores[13] which consist mostly of sulphur and of earth, and rarely of other minerals, are melted in big-bellied earthenware pots. The furnaces,

[12]Agricola (*De Nat. Fos.*, 221) says :—" There is a species of artificial sulphur made " from sulphur and iron hammer-scales, melted together and poured into moulds. This, because it heals scabs of horses, is generally called *caballinum*." It is difficult to believe such a combination was other than iron sulphide, but it is equally difficult to understand how it was serviceable for this purpose.

[13]Inasmuch as pyrites is discussed in the next paragraph, the material of the first distillation appears to be native sulphur. Until the receiving pots became heated above the melting point of the sulphur, the product would be " flowers of sulphur," and not the wax-

A—Pots having spouts. B—Pots without spouts. C—Lids.

which hold two of these pots, are divided into three parts; the lowest part is a foot high, and has an opening at the front for the draught; the top of this is covered with iron plates, which are perforated near the edges, and these support iron rods, upon which the firewood is placed. The middle part of the furnace is one and a half feet high, and has a mouth in front, so that the wood may be inserted; the top of this has rods, upon which the bottom of the pots stand. The upper part is about two feet high, and the pots are also two feet high and one digit thick; these have below their mouths a long, slender spout. In order that the mouth of the pot may be covered, an earthenware lid is made which fits into it. For every two of these pots there must be one pot

like product. The equipment described for pyrites in the next paragraph would be obviously useful only for coarse material.

But little can be said on the history of sulphur; it is mentioned often enough in the Bible and also by Homer (Od. xxii., 481). The Greeks apparently knew how to refine it, although neither Dioscorides nor Pliny specifically describes such an operation. Agricola says (*De Nat. Fos.*, 220): "Sulphur is of two kinds; the mineral, which the Latins call *vivum*, and the Greeks *apyron*, which means 'not exposed to the fire' (*ignem non expertum*) as rightly interpreted by Celsius; and the artificial, called by the Greeks *pepyromenon*, that is, 'exposed to the fire.'" In Book X., the expression *sulfur ignem non expertum* frequently appears, no doubt in Agricola's mind for native sulphur, although it is quite possible that the Greek distinction was between "flowers" of sulphur and the "wax-like" variety.

of the same size and shape, and without a spout, but having three holes, two of which are below the mouth and receive the spouts of the two first pots; the third hole is on the opposite side at the bottom, and through it the sulphur flows out. In each furnace are placed two pots with spouts, and the furnace must be covered by plates of iron smeared over with lute two digits thick; it is thus entirely closed in, but for two or three ventholes through which the mouths of the pots project. Outside of the furnace, against one side, is placed the pot without a spout, into the two holes of which the two spouts of the other pots penetrate, and this pot should be built in at both sides to keep it steady. When the sulphur ore has been placed in the pots, and these placed in the furnace, they are closely covered, and it is desirable to smear the joint over with lute, so that the sulphur will not exhale, and for the same reason the pot below is covered with a lid, which is also smeared with lute. The wood having been kindled, the ores are heated until the sulphur is exhaled, and the vapour, arising through the spout, penetrates into the lower pot and thickens into sulphur, which falls to the bottom like melted wax. It then flows out through the hole, which, as I said, is at the bottom of this pot; and the workman makes it into cakes, or thin sticks or thin pieces of wood are dipped in it. Then he takes the burning wood and glowing charcoal from the furnace, and when it has cooled, he opens the two pots, empties the residues, which, if the ores were composed of sulphur and earth, resemble naturally extinguished ashes; but if the ores consisted of sulphur and earth and stone, or sulphur and stone only, they resemble earth completely dried or stones well roasted. Afterward the pots are re-filled with ore, and the whole work is repeated.

The sulphurous mixture, whether it consists of stone and sulphur only, or of stone and sulphur and metal, may be heated in similar pots, but with perforated bottoms. Before the furnace is constructed, against the "second" wall of the works two lateral partitions are built seven feet high, three feet long, one and a half feet thick, and these are distant from each other twenty-seven feet. Between them are seven low brick walls, that measure but two feet and the same number of digits in height, and, like the other walls, are three feet long and one foot thick; these little walls are at equal distances from one another, consequently they will be two and one half feet apart. At the top, iron bars are fixed into them, which sustain iron plates three feet long and wide and one digit thick, so that they can bear not only the weight of the pots, but also the fierceness of the fire. These plates have in the middle a round hole one and a half digits wide; there must not be more than eight of these, and upon them as many pots are placed. These pots are perforated at the bottom, and the same number of whole pots are placed underneath them; the former contain the mixture, and are covered with lids; the latter contain water, and their mouths are under the holes in the plates. After wood has been arranged round the upper pots and ignited, the mixture being heated, red, yellow, or green sulphur drips from it and flows down through the hole, and is caught by the pots placed underneath the plates, and is at once cooled by the water. If the mixture contains metal, it is reserved for smelting, and, if not, it is thrown away.

A—LONG WALL. B—HIGH WALLS. C—LOW WALLS. D—PLATES. E—UPPER POTS. F—LOWER POTS.

The sulphur from such a mixture can best be extracted if the upper pots are placed in a vaulted furnace, like those which I described among other metallurgical subjects in Book VIII., which has no floor, but a grate inside; under this the lower pots are placed in the same manner, but the plates must have larger holes.

Others bury a pot in the ground, and place over it another pot with a hole at the bottom, in which pyrites or *cadmia*, or other sulphurous stones are so enclosed that the sulphur cannot exhale. A fierce fire heats the sulphur, and it drips away and flows down into the lower pot, which contains water. (Illustration p. 582).

Bitumen[14] is made from bituminous waters, from liquid bitumen, and from mixtures of bituminous substances. The water, bituminous as well as

[14]The substances referred to under the names *bitumen, asphalt, maltha, naphtha, petroleum, rock-oil*, etc., have been known and used from most ancient times, and much of our modern nomenclature is of actual Greek and Roman ancestry. These peoples distinguished three related substances,—the Greek *asphaltos* and Roman *bitumen* for the hard material,—Greek *pissasphaltos* and Roman *maltha* for the viscous, pitchy variety—and occasionally the Greek *naphtha* and Roman *naphtha* for petroleum proper, although it is often enough referred to as liquid *bitumen* or liquid *asphaltos*. The term *petroleum* apparently first appears in Agricola's *De Natura Fossilium* (p. 222), where he says the " oil of bitumen . . . now

A—Lower pot. B—Upper pot. C—Lid.

salty, at Babylon, as Pliny writes, was taken from the wells to the salt works and heated by the great heat of the sun, and condensed partly into liquid bitumen and partly into salt. The bitumen being lighter, floats on the top, while the salt being heavier, sinks to the bottom. Liquid bitumen, if there is much floating on springs, streams and rivers, is drawn up in buckets or other vessels; but, if there is little, it is collected with goose wings, pieces

called *petroleum*." Bitumen was used by the Egyptians for embalming from prehistoric times, *i.e.*, prior to 5000 B.C., the term "mummy" arising from the Persian word for bitumen, *mumiai*. It is mentioned in the tribute from Babylonia to Thotmes III., who lived about 1500 B.C. (Wilkinson, Ancient Egyptians I., p. 397). The Egyptians, however, did not need to go further afield than the Sinai Peninsula for abundant supplies. Bitumen is often cited as the real meaning of the "slime" mentioned in Genesis (XI., 3; XIV., 10), and used in building the Tower of Babel. There is no particular reason for this assumption, except the general association of Babel, Babylon, and Bitumen. However, the Hebrew word *sift* for pitch or bitumen does occur as the cement used for Moses's bulrush cradle (Exodus II., 3), and Moses is generally accounted about 1300 B.C. Other attempts to connect Biblical reference to petroleum and bitumen revolve around Job XXIX., 6, Deut. XXXII., 13, Maccabees II., 1, 18, Matthew V., 13, but all require an unnecessary strain on the imagination.

The plentiful occurrence of bitumen throughout Asia Minor, and particularly in the Valley of the Euphrates and in Persia, is the subject of innumerable references by writers from Herodotus (484–424 B.C.) down to the author of the company prospectus of recent months. Herodotus (I., 179) and Diodorus Siculus (I) state that the walls of Babylon were mortared with bitumen—a fact partially corroborated by modern investigation. The follow-

A—Bituminous spring. B—Bucket. C—Pot. D—Lid.

of linen, *ralla*, shreds of reeds, and other things to which it easily adheres, and it is boiled in large brass or iron pots by fire and condensed. As this bitumen is put to divers uses, some mix pitch with the liquid, others old cart-grease, in order to temper its viscosity; these, however long they are

ing statement by Herodotus (VI., 119) is probably the source from which Pliny drew the information which Agricola quotes above. In referring to a well at Ardericca, a place about 40 miles from ancient Susa, in Persia, Herodotus says:—"For from the well they "get bitumen, salt, and oil, procuring it in the way that I will now describe: they draw "with a swipe, and instead of a bucket they make use of the half of a wine-skin; with "this the man dips and, after drawing, pours the liquid into a reservoir, wherefrom it "passes into another, and there takes three different shapes. The salt and bitumen forth-"with collect and harden, while the oil is drawn off into casks. It is called by the Persians "*rhadinace*, is black, and has an unpleasant smell." (Rawlinsons, Trans. III., p. 409). The statement from Pliny (XXXI., 39) here referred to by Agricola, reads:—"It (salt) is "made from water of wells poured into salt-pans. At Babylon the first condensed is a "bituminous liquid like oil which is burned in lamps. When this is taken off, salt is found "beneath. In Cappadocia also the water from both wells and springs is poured into salt-"pans." When petroleum began to be used as an illuminant it is impossible to say. A passage in Aristotle's *De Mirabilibus* (127) is often quoted, but in reality it refers only to a burning spring, a phenomenon noted by many writers, but from which to its practical use is not a great step. The first really definite statement as to the use of petroleum as an

boiled in the pots, cannot be made hard. The mixtures containing bitumen are also treated in the same manner as those containing sulphur, in pots having a hole in the bottom, and it is rare that such bitumen is not highly esteemed.

Since all solidified juices and earths, if abundantly and copiously mixed with the water, are deposited in the beds of springs, streams or rivers, and the stones therein are coated by them, they do not require the heat of the sun or fire to harden them. This having been pondered over by wise men, they discovered methods by which the remainder of these solidified juices and unusual earths can be collected. Such waters, whether flowing from springs or tunnels, are collected in many wooden tubs or tanks arranged in consecutive order, and deposit in them such juices or earths; these being scraped off every year, are collected, as *chrysocolla*[15] in the Carpathians and as ochre in the Harz.

There remains glass, the preparation of which belongs here, for the reason that it is obtained by the power of fire and subtle art from certain solidified juices and from coarse or fine sand. It is transparent, as are certain solidified juices, gems, and stones; and can be melted like fusible stones and metals. First I must speak of the materials from which glass is made; then of the furnaces in which it is melted; then of the methods by which it is produced.

It is made from fusible stones and from solidified juices, or from other juicy substances which are connected by a natural relationship. Stones which are fusible, if they are white and translucent, are more excellent than

illuminant is Strabo's quotation (XVI., I, 15) from Posidonius: " Asphaltus is found in " great abundance in Babylonia. Eratosthenes describes it as follows:—The liquid *asphaltus*, " which is called *naphtha*, is found in Susa; the dry kind, which can be made solid, in " Babylonia. There is a spring of it near the Euphrates. . . . Others say that the liquid " kind is also found in Babylonia. . . . The liquid kind, called *naphtha*, is of a singular " nature. When it is brought near the fire, the fire catches it. . . . Posidonius says " that there are springs of *naphtha* in Babylonia, some of which produce white, others black " *naphtha*; the first of these, I mean white *naphtha*, which attracts flame, is liquid sulphur; " the second or black *naphtha* is liquid *asphaltus*, and is burnt in lamps instead of oil." (Hamilton's Translation, Vol. III., p. 151). Eratosthenes lived about 200 B.C., and Posidonius about 100 years later. Dioscorides (I., 83), after discussing the usual sources of bitumen says: " It is found in a liquid state in Agrigentum in Sicily, flowing on streams; they use it " for lights in lanterns in place of oil. Those who call the Sicilian kind oil are under a delusion, " for it is agreed that it is a kind of liquid bitumen." Pliny adds nothing much new to the above quotations, except in regard to these same springs (XXXV., 51) that " The inhabitants " collect it on the panicles of reeds, to which it quickly adheres and they use it for burning " in lamps instead of oil." Agricola (*De Natura Fossilium*, Book IV.) classifies petroleum, coal, jet, and obsidian, camphor, and amber as varieties of bitumen, and devotes much space to the refutation of the claims that the last two are of vegetable origin.

[15]Agricola (*De Natura Fossilium*, p. 215) in discussing substances which originate from copper, gives among them green *chrysocolla* (as distinguished from borax, etc., see Note 8 above), and says: " Native *chrysocolla* originates in veins and veinlets, and is found mostly " by itself like sand, or adhering to metallic substances, and when scraped off from this " appears similar to its own sand. Occasionally it is so thin that very little can be scraped " off. Or else it occurs in waters which, as I have said, wash these minerals, and afterward " it settles as a powder. At Neusohl in the Carpathians, green water flowing from an " ancient tunnel wears away this *chrysocolla* with it. The water is collected in thirty large " reservoirs, where it deposits the *chrysocolla* as a sediment, which they collect every " year and sell,"—as a pigment. This description of its occurrence would apply equally well to modern *chrysocolla* or to malachite. The solution from copper ores would deposit some sort of green incrustation, of carbonates mostly.

A—Mouth of the tunnel. B—Trough. C—Tanks. D—Little trough.

the others, for which reason crystals take the first place. From these, when pounded, the most excellent transparent glass was made in India, with which no other could be compared, as Pliny relates. The second place is accorded to stones which, although not so hard as crystal, are yet just as white and transparent. The third is given to white stones, which are not transparent. It is necessary, however, first of all to heat all these, and afterward they are subjected to the pestle in order to break and crush them into coarse sand, and then they are passed through a sieve. If this kind of coarse or fine sand is found by the glass-makers near the mouth of a river, it saves them much labour in burning and crushing. As regards the solidified juices, the first place is given to soda ; the second to white and translucent rock-salt ; the third to salts which are made from lye, from the ashes of the musk ivy, or from other salty herbs. Yet there are some who give to this latter, and not to the former, the second place. One part of coarse or fine sand made from fusible stones should be mixed with two parts of soda or of rock-salt or of herb salts, to which are added minute particles of *magnes*.[16] It is true that in our

[16]The statement in Pliny (XXXVI., 66) to which Agricola refers is as follows : "Then "as ingenuity was not content with the mixing of *nitrum*, they began the addition of *lapis*

day, as much as in ancient times, there exists the belief in the singular power of the latter to attract to itself the vitreous liquid just as it does iron, and by attracting it to purify and transform green or yellow into white; and afterward fire consumes the *magnes*. When the said juices are not to be had, two parts of the ashes of oak or holmoak, or of hard oak or Turkey oak, or if these be not available, of beech or pine, are mixed with one part of coarse or fine sand, and a small quantity of salt is added, made from salt water or sea-water, and a small particle of *magnes;* but these make a less white and translucent glass. The ashes should be made from old trees, of which the trunk at a height of six feet is hollowed out and fire is put in, and thus the whole tree is consumed and converted into ashes. This is done in winter when the snow lies long, or in summer when it does not rain, for the showers at other times of the year, by mixing the ashes with earth, render them impure; for this reason, at such times, these same trees are cut up into many pieces and burned under cover, and are thus converted into ashes.

Some glass-makers use three furnaces, others two, others only one. Those who use three, melt the material in the first, re-melt it in the second,

"*magnes*, because of the belief that it attracts liquefied glass as well as iron. In a similar "manner many kinds of brilliant stones began to be added to the melting, and then shells "and fossil sand. Authors tell us that the glass of India is made of broken crystal, and "in consequence nothing can compare with it. Light and dry wood is used for fusing, "*cyprium* (copper ?) and *nitrum* being added, particularly *nitrum* from Ophir etc."

A great deal of discussion has arisen over this passage, in connection with what this *lapis magnes* really was. Pliny (XXXVI., 25) describes the lodestone under this term, but also says: "There (in Ethiopia) also is *haematites magnes*, a stone of blood colour, which "shows a red colour if crushed, or of saffron. The *haematites* has not the same property of "attracting iron as *magnes*." Relying upon this sentence for an exception to the ordinary sort of *magnes*, and upon the impossible chemical reaction involved, most commentators have endeavoured to show that lodestone was not the substance meant by Pliny, but manganese, and thus they find here the first knowledge of this mineral. There can be little doubt that Pliny assumed it to be the lodestone, and Agricola also. Whether the latter had any independent knowledge on this point in glass-making or was merely quoting Pliny—which seems probable—we do not know. In any event, Biringuccio, whose work preceded *De Re Metallica* by fifteen years, does definitely mention manganese in this connection. He dismisses this statement of Pliny with the remark (p. 37–38): "The "Ancients wrote about lodestones, as Pliny states, and they mixed it together with *nitrum* "in their first efforts to make glass." The following passage from this author (p. 36–37), however, is not only of interest in this connection, but also as possibly being the first specific mention of manganese under its own name. Moreover, it has been generally overlooked in the many discussions of the subject. "Of a similar nature (to *zaffir*) is also another "mineral called *manganese*, which is found, besides in Germany, at the mountain of "Viterbo in Tuscany . . . it is the colour of *ferrigno scuro* (iron slag ?). In melting it "one cannot obtain any metal . . . but it gives a very fine colour to glass, so that the "glass workers use it in their pigments to secure an azure colour. . . . It also has such "a property that when put into melted glass it cleanses it and makes it white, even if it were "green or yellow. In a hot fire it goes off in a vapour like lead, and turns into ashes."

To enter competently into the discussion of the early history of glass-making would employ more space than can be given, and would lead but to a sterile end. It is certain that the art was pre-Grecian, and that the Egyptians were possessed of some knowledge of making and blowing it in the XI Dynasty (according to Petrie 3,500 B.C.), the wall painting at Beni Hassen, which represents glass-blowing, being attributed to that period. The remains of a glass factory at Tel el Amarna are believed to be of the XVIII Dynasty. (Petrie, 1,500 B.C.). The art reached a very high state of development among the Greeks and Romans. No discussion of this subject omits Pliny's well-known story (XXXVI. 65), which we also add: "The tradition is that a merchant ship laden with "*nitrum* being moored at this place, the merchants were preparing their meal on the beach, "and not having stones to prop up their pots, they used lumps of *nitrum* from the ship, "which fused and mixed with the sands of the shore, and there flowed streams of a new "translucent liquid, and thus was the origin of glass."

A—Lower chamber of the first furnace. B—Upper chamber. C—Vitreous mass.

and in the third they cool the glowing glass vessels and other articles. Of these the first furnace must be vaulted and similar to an oven. In the upper chamber, which is six feet long, four feet wide, and two feet high, the mixed materials are heated by a fierce fire of dry wood until they melt and are converted into a vitreous mass. And if they are not satisfactorily purified from dross, they are taken out and cooled and broken into pieces; and the vitreous pieces are heated in pots in the same furnace.

The second furnace is round, ten feet in diameter and eight feet high, and on the outside, so that it may be stronger, it is encompassed by five arches, one and one half feet thick; it consists in like manner of two chambers, of which the lower one is vaulted and is one and one half feet thick. In front this chamber has a narrow mouth, through which the wood can be put into the hearth, which is on the ground. At the top and in the middle of its vault, there is a large round hole which opens to the upper chamber, so that the flames can penetrate into it. Between the arches in the walls of the upper chamber are eight windows, so large that the big-bellied pots may be placed through them on to the floor of the chamber, around the large hole. The thickness of these pots is about two digits, their height the same number of feet, and the diameter of the belly one and a half feet, and of the mouth and bottom one foot. In the back part of the furnace is a rectangular hole, measuring in height and width a palm, through which the heat penetrates into a third furnace which adjoins it.

This third furnace is rectangular, eight feet long and six feet wide; it also consists of two chambers, of which the lower has a mouth in front, so that firewood may be placed on the hearth which is on the ground. On each side of this opening in the wall of the lower chamber is a recess for oblong earthenware receptacles, which are about four feet long, two feet high, and one and a half feet wide. The upper chamber has two holes, one on the right side, the other on the left, of such height and width that earthenware receptacles may be conveniently placed in them. These latter receptacles are three feet long, one and a half feet high, the lower part one foot wide, and the upper part rounded. In these receptacles the glass articles, which have been blown, are placed so that they may cool in a milder temperature; if they were not cooled slowly they would burst asunder. When the vessels are taken from the upper chamber, they are immediately placed in the receptacles to cool.

A—Arches of the second furnace. B—Mouth of the lower chamber. C—Windows of the upper chamber. D—Big-bellied pots. E—Mouth of the third furnace. F—Recesses for the receptacles. G—Openings in the upper chamber. H—Oblong receptacles.

A—Lower chamber of the other second furnace. B—Middle one. C—Upper one. D—Its opening. E—Round opening. F—Rectangular opening.

Some who use two furnaces partly melt the mixture in the first, and not only re-melt it in the second, but also replace the glass articles there. Others partly melt and re-melt the material in different chambers of the second furnace. Thus the former lack the third furnace, and the latter, the first. But this kind of second furnace differs from the other second furnace, for it is, indeed, round, but the interior is eight feet in diameter and twelve feet high, and it consists of three chambers, of which the lowest is not unlike the lowest of the other second furnace. In the middle chamber wall there are six arched openings, in which are placed the pots to be heated, and the remainder of the small windows are blocked up with lute. In the middle top of the middle chamber is a square opening a palm in length and width. Through this the heat penetrates into the upper chamber, of which the rear part has an opening to receive the oblong earthenware receptacles, in which are placed the glass articles to be slowly cooled. On this side, the ground of the workshop is higher, or else a bench is placed there, so that the glass-makers may stand upon it to stow away their products more conveniently.

Those who lack the first furnace in the evening, when they have accomplished their day's work, place the material in the pots, so that the heat during the night may melt it and turn it into glass. Two boys alternately, during night and day, keep up the fire by throwing dry wood on to the hearth. Those who have but one furnace use the second sort, made with three chambers. Then in the evening they pour the material into the pots, and in the morning, having extracted the fused material, they make the glass objects, which they place in the upper chamber, as do the others.

The second furnace consists either of two or three chambers, the first of which is made of unburnt bricks dried in the sun. These bricks are made of a kind of clay that cannot be easily melted by fire nor resolved into powder; this clay is cleaned of small stones and beaten with rods. The bricks are laid with the same kind of clay instead of lime. From the same clay the potters also make their vessels and pots, which they dry in the shade. These two parts having been completed, there remains the third.

The vitreous mass having been made in the first furnace in the manner I described, is broken up, and the assistant heats the second furnace, in order that the fragments may be re-melted. In the meantime, while they are doing this, the pots are first warmed by a slow fire in the first furnace, so that the vapours may evaporate, and then by a fiercer fire, so that they become red in drying. Afterward the glass-makers open the mouth of the furnace, and, seizing the pots with tongs, if they have not cracked and fallen to pieces, quickly place them in the second furnace, and they fill them up with the fragments of the heated vitreous mass or with glass. Afterward they close up all the windows with lute and bricks, with the exception that in each there are two little windows left free; through one of these they inspect the glass contained in the pot, and take it up by means of a blow-pipe; in the other they rest another blow-pipe, so that it may get warm. Whether it is made of brass, bronze, or iron, the blow-pipe must be three feet long.

A—BLOW-PIPE. B—LITTLE WINDOW. C—MARBLE. D—FORCEPS. E—MOULDS BY MEANS OF WHICH THE SHAPES ARE PRODUCED.

In front of the window is inserted a lip of marble, on which rests the heaped-up clay and the iron shield. The clay holds the blow-pipe when it is put into the furnace, whereas the shield preserves the eyes of the glass-maker from the fire. All this having been carried out in order, the glass-makers bring the work to completion. The broken pieces they re-melt with dry wood, which emits no smoke, but only a flame. The longer they re-melt it, the purer and more transparent it becomes, the fewer spots and blisters there are, and therefore the glass-makers can carry out their work more easily. For this reason those who only melt the material from which glass is made for one night, and then immediately make it up into glass articles, make them less pure and transparent than those who first produce a vitreous mass and then re-melt the broken pieces again for a day and a night. And, again, these make a less pure and transparent glass than do those who melt it again for two days and two nights, for the excellence of the glass does not consist solely in the material from which it is made, but also in the melting. The glass-makers often test the glass by drawing it up with the blowpipes ; as soon as they observe that the fragments have been re-melted and purified satisfactorily, each of them with another blow-pipe which is in the pot, slowly stirs and takes up the glass which sticks to it in the shape of a ball like a glutinous, coagulated gum. He takes up just as much as he needs to complete the article he wishes to make ; then he presses it against the lip of marble and kneads it round and round until it consolidates. When he blows through the pipe he blows as he would if inflating a bubble ; he blows into the blow-pipe as often as it is necessary, removing it from his mouth to re-fill his cheeks, so that his breath does not draw the flames into his mouth. Then, twisting the lifted blow-pipe round his head in a circle, he makes a long glass, or moulds the same in a hollow copper mould, turning it round and round, then warming it again, blowing it and pressing it, he widens it into the shape of a cup or vessel, or of any other object he has in mind. Then he again presses this against the marble to flatten the bottom, which he moulds in the interior with his other blow-pipe. Afterward he cuts out the lip with shears, and, if necessary, adds feet and handles. If it so please him, he gilds it and paints it with various colours. Finally, he lays it in the oblong earthenware receptacle, which is placed in the third furnace, or in the upper chamber of the second furnace, that it may cool. When this receptacle is full of other slowly-cooled articles, he passes a wide iron bar under it, and, carrying it on the left arm, places it in another recess.

The glass-makers make divers things, such as goblets, cups, ewers, flasks, dishes, plates, panes of glass, animals, trees, and ships, all of which excellent and wonderful works I have seen when I spent two whole years in Venice some time ago. Especially at the time of the Feast of the Ascension they were on sale at Morano, where are located the most celebrated glass-works. These I saw on other occasions, and when, for a certain reason, I visited Andrea Naugerio in his house which he had there, and conversed with him and Francisco Asulano.

END OF BOOK XII.

APPENDIX A.

AGRICOLA'S WORKS.

EORGIUS AGRICOLA was not only the author of works on Mining and allied subjects, usually associated with his name, but he also interested himself to some extent in political and religious subjects. For convenience in discussion we may, therefore, divide his writings on the broad lines of (1) works on mining, geology, mineralogy, and allied subjects; (2) works on other subjects, medical, religious, critical, political, and historical. In respect especially to the first division, and partially with regard to the others, we find three principal cases: (*a*) Works which can be authenticated in European libraries to-day; (*b*) references to editions of these in bibliographies, catalogues, etc., which we have been unable to authenticate; and (*c*) references to works either unpublished or lost. The following are the short titles of all of the published works which we have been able to find on the subjects allied to mining, arranged according to their present importance:—*De Re Metallica*, first edition, 1556; *De Natura Fossilium*, first edition, 1546; *De Ortu et Causis Subterraneorum*, first edition, 1546; *Bermannus*, first edition, 1530; *Rerum Metallicarum Interpretatio*, first edition, 1546; *De Mensuris et Ponderibus*, first edition, 1533; *De Precio Metallorum et Monetis*, first edition, 1550; *De Veteribus et Novis Metallis*, first edition, 1546; *De Natura eorum quae Effluunt ex Terra*, first edition, 1546; *De Animantibus Subterraneis*, first edition, 1549.

Of the "lost" or unpublished works, on which there is some evidence, the following are the most important:—*De Metallicis et Machinis, De Ortu Metallorum Defensio ad Jacobum Scheckium, De Jure et Legibus Metallicis, De Varia Temperie sive Constitutione Aeris, De Terrae Motu*, and *Commentariorum, Libri VI.*

The known published works upon other subjects are as follows:—Latin Grammar, first edition, 1520; Two Religious Tracts, first edition, 1522; *Galen* (Joint Revision of Greek Text), first edition, 1525; *De Bello adversus Turcam*, first edition, 1528; *De Peste*, first edition, 1554.

The lost or partially completed works on subjects unrelated to mining, of which some trace has been found, are:—*De Medicatis Fontibus, De Putredine solidas partes*, etc., *Castigationes in Hippocratem, Typographia Mysnae et Toringiae, De Traditionibus Apostolicis, Oratio de rebus gestis Ernesti et Alberti, Ducum Saxoniae.*

REVIEW OF PRINCIPAL WORKS.

Before proceeding with the bibliographical detail, we consider it desirable to review briefly the most important of the author's works on subjects related to mining.

De Natura Fossilium. This is the most important work of Agricola, excepting *De Re Metallica.* It has always been printed in combination with other works, and first appeared at Basel, 1546. This edition was considerably revised by the author, the amended edition being that of 1558, which we have used in giving references. The work comprises ten " books " of a total of 217 folio pages. It is the first attempt at systematic mineralogy, the minerals[1] being classified into (1) " earths " (clay, ochre, etc.), (2) " stones properly so-called " (gems, semi-precious and unusual stones, as distinguished from rocks), (3) " solidified juices " (salt, vitriol, alum, etc.), (4) metals, and (5) " compounds " (homogeneous " mixtures " of simple substances, thus forming such minerals as galena, pyrite, etc.). In this classification Agricola endeavoured to find some fundamental basis, and therefore adopted solubility, fusibility, odour, taste, etc., but any true classification without the atomic theory was, of course, impossible. However, he makes a very creditable performance out of their properties and obvious characteristics. All of the external characteristics which we use to-day in discrimination, such as colour, hardness, lustre, etc., are enumerated, the origin of these being attributed to the proportions of the Peripatetic elements and their binary properties. Dana, in his great work[2], among some fourscore minerals which he identifies as having been described by Agricola and his predecessors, accredits a score to Agricola himself. It is our belief, however, that although in a few cases Agricola has been wrongly credited, there are still more of which priority in description might be assigned to him. While a greater number than fourscore of so-called species are given by Agricola and his predecessors, many of these are, in our modern system, but varieties ; for instance, some eight or ten of the ancient species consist of one form or another of silica.

Book I. is devoted to mineral characteristics—colour, brilliance, taste, shape, hardness, etc., and to the classification of minerals ; Book II., " earths "—clay, Lemnian earth, chalk, ochre, etc. ; Book III., " solidified juices "—salt, *nitrum* (soda and potash), saltpetre, alum, vitriol, chrysocolla, *caeruleum* (part azurite), orpiment, realgar, and sulphur ; Book IV., camphor, bitumen, coal, bituminous shales, amber ; Book V., lodestone, bloodstone, gypsum, talc, asbestos, mica, calamine, various fossils, geodes, emery, touchstones, pumice, fluorspar, and quartz ; Book VI., gems and precious stones ; Book VII., " rocks "—marble, serpentine, onyx, alabaster, limestone, etc. ; Book VIII., metals—gold, silver, quicksilver, copper, lead, tin, antimony, bismuth, iron, and alloys, such as electrum, brass, etc. ; Book IX., various furnace operations, such as making brass, gilding, tinning, and products such as slags, furnace accretions, *pompholyx* (zinc oxide), copper flowers, litharge, hearth-lead, verdigris, white-lead, red-lead, etc. ; Book X., " compounds," embracing the description of a number of recognisable silver, copper, lead, quicksilver, iron, tin, antimony, and zinc minerals, many of which we set out more fully in Note 8, page 108.

De Ortu et Causis Subterraneorum. This work also has always been published in company with others. The first edition was printed at Basel,

[1]See footnote 4, page 1. [2]System of Mineralogy.

1546; the second at Basel, 1558, which, being the edition revised and added to by the author, has been used by us for reference. There are five "books," and in the main they contain Agricola's philosophical views on geologic phenomena. The largest portion of the actual text is occupied with refutations of the ancient philosophers, the alchemists, and the astrologers; and these portions, while they exhibit his ability in observation and in dialectics, make but dull reading. Those sections of the book which contain his own views, however, are of the utmost importance in the history of science, and we reproduce extensively the material relating to ore deposits in the footnotes on pages 43 to 52. Briefly, Book I. is devoted to discussion of the origin and distribution of ground waters and juices. The latter part of this book and a portion of Book II. are devoted to the origin of subterranean heat, which he assumes is in the main due to burning bitumen—a genus which with him embraced coal—and also, in a minor degree, to friction of internal winds and to burning sulphur. The remainder of Book II. is mainly devoted to the discussion of subterranean "air", "vapour", and "exhalations", and he conceives that volcanic eruptions and earthquakes are due to their agency, and in these hypotheses he comes fairly close to the modern theory of eruptions from explosions of steam. "Vapour arises when the internal heat of the "earth or some hidden fire burns earth which is moistened with vapour. "When heat or subterranean fire meets with a great force of vapour which "cold has contracted and encompassed in every direction, then the vapour, "finding no outlet, tries to break through whatever is nearest to it, in order "to give place to the insistent and urgent cold. Heat and cold cannot abide "together in one place, but expel and drive each other out of it by turns".

As he was, we believe, the first to recognise the fundamental agencies of mountain sculpture, we consider it is of sufficient interest to warrant a reproduction of his views on this subject: "Hills and mountains are pro-"duced by two forces, one of which is the power of water, and the other the "strength of the wind. There are three forces which loosen and demolish "the mountains, for in this case, to the power of the water and the strength "of the wind we must add the fire in the interior of the earth. Now we can "plainly see that a great abundance of water produces mountains, for the "torrents first of all wash out the soft earth, next carry away the harder "earth, and then roll down the rocks, and thus in a few years they excavate "the plains or slopes to a considerable depth; this may be noticed in moun-"tainous regions even by unskilled observers. By such excavation to a "great depth through many ages, there rises an immense eminence on each "side. When an eminence has thus arisen, the earth rolls down, loosened by "constant rain and split away by frost, and the rocks, unless they are exceed-"ingly firm, since their seams are similarly softened by the damp, roll down "into the excavations below. This continues until the steep eminence is "changed into a slope. Each side of the excavation is said to be a mountain, "just as the bottom is called a valley. Moreover, streams, and to a far greater "extent rivers, effect the same results by their rushing and washing; for this "reason they are frequently seen flowing either between very high mountains

"which they have created, or close by the shore which borders them. . . .
"Nor did the hollow places which now contain the seas all formerly exist,
"nor yet the mountains which check and break their advance, but in many
"parts there was a level plain, until the force of winds let loose upon it a
"tumultuous sea and a scathing tide. By a similar process the impact of
"water entirely overthrows and flattens out hills and mountains. But
"these changes of local conditions, numerous and important as they are, are
"not noticed by the common people to be taking place at the very moment
"when they are happening, because, through their antiquity, the time, place,
"and manner in which they began is far prior to human memory. The wind
"produces hills and mountains in two ways: either when set loose and free
"from bonds, it violently moves and agitates the sand; or else when, after
"having been driven into the hidden recesses of the earth by cold, as into a
"prison, it struggles with a great effort to burst out. For hills and mountains
"are created in hot countries, whether they are situated by the sea coasts or
"in districts remote from the sea, by the force of winds; these no longer held
"in check by the valleys, but set free, heap up the sand and dust, which they
"gather from all sides, to one spot, and a mass arises and grows together. If
"time and space allow, it grows together and hardens, but if it be not allowed
"(and in truth this is more often the case), the same force again scatters the
"sand far and wide. . . . Then, on the other hand, an earthquake
"either rends and tears away part of a mountain, or engulfs and devours the
"whole mountain in some fearful chasm. In this way it is recorded the
"Cybotus was destroyed, and it is believed that within the memory of man
"an island under the rule of Denmark disappeared. Historians tell us that
"Taygetus suffered a loss in this way, and that Therasia was swallowed up
"with the island of Thera. Thus it is clear that water and the powerful
"winds produce mountains, and also scatter and destroy them. Fire only
"consumes them, and does not produce at all, for part of the mountains—
"usually the inner part—takes fire."

The major portion of Book III. is devoted to the origin of ore channels, which we reproduce at some length on page 47. In the latter part of Book III., and in Books IV. and V., he discusses the principal divisions of the mineral kingdom given in *De Natura Fossilium*, and the origin of their characteristics. It involves a large amount of what now appears fruitless tilting at the Peripatetics and the alchemists; but nevertheless, embracing, as Agricola did, the fundamental Aristotelian elements, he must needs find in these same elements and their subordinate binary combinations cause for every variation in external character.

Bermannus. This, Agricola's first work in relation to mining, was apparently first published at Basel, 1530. The work is in the form of a dialogue between "Bermannus," who is described as a miner, mineralogist, and "a student of mathematics and poetry," and "Nicolaus Ancon" and "Johannes Neavius," both scholars and physicians. Ancon is supposed to be of philosophical turn of mind and a student of Moorish literature, Naevius to be particularly learned in the writings of Dioscorides, Pliny, Galen, etc. "Bermannus" was probably an adaptation by Agricola of the name of his friend Lorenz Berman, a prominent miner. The book is in the main devoted to a correlation of the minerals mentioned by the Ancients with those found in the Saxon mines. This phase is interesting as indicating the natural trend of Agricola's scholastic mind when he first comes into contact with the sciences to which he devoted himself. The book opens with a letter of commendation from Erasmus, of Rotterdam, and with the usual dedication and preface by the author. The three conversationalists are supposed to take walks among the mines and to discuss, incidentally, matters which come to their attention; therefore the book has no systematic or logical arrangement. There are occasional statements bearing on the history, management, titles, and methods used in the mines, and on mining lore generally. The mineralogical part, while of importance from the point of view of giving the first description of several minerals, is immensely improved upon in *De Natura Fossilium*, published 15 years later. It is of interest to find here the first appearance of the names of many minerals which we have since adopted from the German into our own nomenclature. Of importance is the first description of bismuth, although, as pointed out on page 433, the metal had been mentioned before. In the revised collection of collateral works published in 1558, the author makes many important changes and adds some new material, but some of the later editions were made from the unrevised older texts.

Rerum Metallicarum Interpretatio. This list of German equivalents for Latin mineralogical terms was prepared by Agricola himself, and first appears in the 1546 collection of *De Ortu et Causis, De Natura Fossilium*, etc., being repeated in all subsequent publications of these works. It consists of some 500 Latin mineralogical and metallurgical terms, many of which are of Agricola's own coinage. It is of great help in translation and of great value in the study of mineralogic nomenclature.

De Mensuris et Ponderibus. This work is devoted to a discussion of the Greek and Roman weights and measures, with some correlation to those used in Saxony. It is a careful work still much referred to by students of these subjects. The first edition was published at Paris in 1533, and in the 1550 edition at Basel appears, for the first time, *De Precio Metallorum et Monetis*.

De Veteribus et Novis Metallis. This short work comprises 31 folio pages, and first appears in the 1546 collection of collateral works. It consists mainly of historical and geographical references to the occurrence of metals and mines, culled from the Greek and Latin classics, together with some information as to the history of the mines in Central Europe. The latter is the only original material, and unfortunately is not very extensive. We have incorporated some of this information in the footnotes.

De Animantibus Subterraneis. This short work was first printed in Basel, 1549, and consists of one chapter of 23 folio pages. Practically the whole is devoted to the discussion of various animals who at least a portion of their time live underground, such as hibernating, cave-dwelling, and burrowing animals, together with cave-dwelling birds, lizards, crocodiles, serpents, etc. There are only a few lines of remote geological interest as to migration of animals imposed by geologic phenomena, such as earthquakes, floods, etc. This book also discloses an occasional vein of credulity not to be expected from the author's other works, in that he apparently believes Aristotle's story of the flies which were born and lived only in the smelting furnace ; and further, the last paragraph in the book is devoted to underground gnomes. This we reproduce in the footnote on page 217.

De Natura eorum quae Effluunt ex Terra. This work of four books, comprising 83 folio pages, first appears in the 1546 collection. As the title indicates, the discussion is upon the substances which flow from the earth, such as water, bitumen, gases, etc. Altogether it is of microscopic value and wholly uninteresting. The major part refers to colour, taste, temperature, medicinal uses of water, descriptions of rivers, lakes, swamps, and aqueducts.

BIBLIOGRAPHICAL NOTES.

For the following we have mainly to thank Miss Kathleen Schlesinger, who has been employed many months in following up every clue, and although the results display very considerable literary activity on the part of the author, they do not by any means indicate Miss Schlesinger's labours. Agricola's works were many of them published at various times in combination, and therefore to set out the title and the publication of each work separately would involve much repetition of titles, and we consequently give the titles of the various volumes arranged according to dates. For instance, *De Natura Fossilium, De Ortu et Causis, De Veteribus et Novis Metallis, De Natura eorum quae Effluunt ex Terra,* and *Interpretatio* have always been published together, and the Latin and Italian editions of these works always include *Bermannus* as well. Moreover, the Latin *De Re Metallica* of 1657 includes all of these works.

We mark with an asterisk the titles to editions which we have been able to authenticate by various means from actual books. Those unmarked are editions which we are satisfied do exist, but the titles of which are possibly incomplete, as they are taken from library catalogues, etc. Other editions to which we find reference and of which we are not certain are noted separately in the discussion later on.[8]

*1530 (8vo).

Georgii Agricolae Medici, Bermannus sive de re Metallica.

(Froben's mark).

Basileae in aedibus Frobenianis Anno. MDXXX.

Bound with this edition is (p. 131–135), at least occasionally, *Rerum metallicarum appellationes juxta vernaculam Germanorum linguam, autori Plateano.*

Basileae in officina Frobeniana, Anno. MDXXX.

*1533 (8vo):

Georgii Agricolae Medici libri quinque de Mensuris et Ponderibus : in quibus plaeraque à Budaeo et Portio parum animadversa diligenter excutiuntur. Opus nunc primum in lucem aeditum.

(Wechelus's Mark).

Parisiis. Excudebat Christianus Wechelus, in vico Iacobaeo, sub scuto Basileiensi, Anno MDXXXIII.

261 pages and index of 5 pages.

---

[8]The following are the titles of the works referred to in this discussion :—

Petrus Albinus : *Meissnische Land und Berg Chronica In welcher ein wollnstendige description des Landes,* etc., Dresden, 1590 (contains part 1, *Commentatorium de Mysnia). Newe Chronica und Beschreibung des Landes zu Meissen,* pp. 1 to 449, besides preface and index, and Part II. *Meissnische Bergk Chronica,* Dresden, 1590, pp. 1 to 205, besides preface and index.

Adam Daniel Richter : *Unständliche . . . . Chronica der . . . . Stadt Chemnitz nebst beygefügten Urkunden,* 2 pts. 4to, Zittau & Leipzig, 1767.

Ben. G. Weinart : *Versuch einer Litteratur d. Sächsischen Geschichte und Staats kunde,* Leipzig, 1885.

Friedrich August Schmid : *Georg Agrikola's Bermannus : Einleitung in die metallurgischen Schriften desselben,* Freyberg, Craz & Gerlach, 1806, pp. VIII., 1–260.

Franz Ambros Reuss : *Mineralogische Geographie von Böhmen.* 2 vols. 4to, Dresden, 1793–97. (Agricola Vol. 1, p. 2).

Jacob Leupold : *Prodromus Bibliothecae Metallicae,* corrected, continued, and augmented by F. E. Brückmann. Wolfenbüttel, 1732, s.v. Agricola.

Christian Gottlieb Göcher : *Allgemeines Gelehrten-Lexicon,* with continuation and supplements by Adelung, Leipzig, 1750, s.v. Agricola.

John Anton Van der Linden : *De Scriptis medicis, Libri duo,* Amsterdam, 1662, s.v. Georgius Agricola.

Nicolas François Joseph Eloy : *Dictionnaire Historique de la Médecine,* Liége & Francfort (chez J. F. Bassompierre), 1755, 8vo (Agricola p. 28, vol. 1).

Georg Abraham Mercklinus : *Lindenius Renovatus de scriptis medicis continuati . . . . amplificati,* etc., Amsterdam, 1686, s.v. Georgius Agricola.

John Ferguson : *Bibliotheca Chemica :* A catalogue of the Alchemical, Chemical, and Pharmaceutical books in the collection of the late James Young of Kelly & Durris, Esq., LL.D., F.R.S., F.R.S.E. Glasgow, 1906, 4to, 2 vols., s.v. Agricola.

Christoph Wilhelm Gatterer : *Allgemeines Repertorium der mineralogischen, bergwerks und Salz werkswissenschaftlichen Literatur,* Göttingen, 1798, vol. 1.

Dr. Reinhold Hofmann : *Dr. Georg Agricola, Ein Gelehrtenleben aus dem Zeitalter der Reformation,* 8vo, Gotha, 1905.

Georg Heinrich Jacobi : *Der Mineralog Georgius Agricola und sein Verhältnis zur wissenschaft seiner Zeit,* etc., 8vo. Zwickau (1889), (*Dissertation*—Leipzig).

Georg Draud : *Bibliotheca Classica,* Frankfurt-am-Main, 1611.

B. G. Struve : *Bibliotheca Saxonica,* 8vo, Halle, 1736.

*1533 (4to):

Georgii Agricolae Medici Libri quinque. De Mensuris et Ponderibus: *In quibus pleraque à Budaeo et Portio parum animadversa diligenter excutiuntur.*

(Froben's Mark).

Basileae ex Officina Frobeniana Anno MDXXXIII. *Cum gratia et privilegio Caesareo ad sex annos.*

1534 (4to):

Georgii Agricolae. Epistola ad Plateanum, cui sunt adiecta aliquot loca castigata in libris de mensuris et ponderibus nuper editis.

Froben, Basel, 1534.

*1535 (8vo):

Georgii Agricolae Medici libri V. de Mensuris et Ponderibus: *in quibus pleraque à Budaeo et Portio parum animadversa diligenter excutiuntur.*

(Printer's Mark).

At the end of Index: *Venitüs per Joan Anto. de Nicolinis de Sabio, sumptu vero et requisitione Dñi Melchionis Sessae. Anno. Dñi* MDXXXV. *Mense Julii.* 116 folios.

On back of title page is given: *Liber primus de mensuris Romanis, Secundus de mensuris Graecis, Tertius de rerum quas metimur pondere, Quartus de ponderibus Romanis, Quintus de ponderibus Graecis.*

*1541 (8vo):

Georgii Agricolae Medici Bermannus sive de re metallica.

Parisiis. Apud Hieronymum Gormontiú. In Vico Jacobeo sub signotrium coronarum. 1541.

*1546 (8vo):

Georgii Agricolae medici Bermannus, sive de metallica ab accurata autoris recognitione et emendatione nunc primum editus cum nomenclatura rerum metallicarum.

Eorum Lipsiae In officina Valentini Papae Anno. MDXLVI.

*1546 (folio):

Georgii Agricolae De ortu et causis subterraneorum Lib. V. De natura eorum quae effluunt ex terra Lib. IIII. De natura fossilium Lib. X. De veteribus et novis metallis, Lib. II. Bermannus sive De re Metallica dialogus. Interpretatio Germanica vocum rei metallicae addito Indice faecundissimo.

Apud Hieron Frobenium et Nicolaum Episcopium Basileae, MDXLVI. *Cum privilegio Imp. Maiestatis ad quinquennium.*

*1549 (8vo):

Georgii Agricolae de animantibus subterraneis Liber.

Froben, Basel, MDXLIX.

*1550 (8vo):

Di Georgio Agricola De la generatione de le cose, che sotto la terra sono, e de le cause de' loro effetti e natura, Lib. V. De La Natura di quelle cose, che de la terra scorrono Lib. IIII. De La Natura de le cose Fossili, e che sotto la terra si Cavano Lib. X. De Le Minere antiche e moderne Lib. II. Il Bermanno, ò de le cose Metallice Dialogo, Recato tutto hora dal Latino in Buona Lingua volgare.

(Vignette of Sybilla surrounded by the words)—*Qv Al Piv Fermo E Il Mio Foglio È Il Mio Presaggio.*

Col Privilegio del Sommo Pontefice Papa Giulio III. Et del Illustriss. Senato Veneto per anni. XX.

(Colophon). *In Vinegia per Michele Tramezzino,* MDL.

*1550 (folio):

Georgii Agricolae. De Mensuris et ponderibus Rom. atque Graec. lib. V. De externis mensuris et ponderibus Lib. II. Ad ea quae Andreas Alciatus denuo disputavit De Mensuris et Ponderibus brevis defensio Lib. I. De Mensuris quibus intervalla metimur Lib. I. De restituendis ponderibus atque mensuris. Lib. I. De precio metallorum et monetis. Lib. III.

Basileae. Froben. MDL. *Cum privilegio Imp. Maiestatis ad quinquennium.*[4]

*1556 (folio):

Georgii Agricolae De Re Metallica Libri XII. quibus Officia, Instrumenta, Machinae, ac omnia denique ad Metallicam spectantia, non modo luculentissime describuntur, sed et per effigies, suis locis insertas, adjunctis Latinis, Germanicisque appellationibus ita ob oculos ponuntur, ut clarius tradi non possint Eiusdem De Animantibus Subterraneis Liber, ab Autore recognitus: cum Indicibus diversis, quicquid in opere tractatum est, pulchre demonstrantibus.

(Froben's Mark).

Basileae MDLVI. Cum Privilegio Imperatoris in annos V. et Galliarum Regis ad Sexennium.

Folio 538 pages and preface, glossary and index amounting to 86 pages. This is the first edition of *De Re Metallica.* We reproduce this title-page on page XIX.

---

[4]Albinus states (p. 354): *Omnes simul editi Anno.* 1549, *iterum* 1550, *Basileae,* as though two separate editions.

*1557 (folio):

Vom Bergkwerck xii Bücher darinn alle Empter, Instrument, Gezeuge, unnd Alles zu disem Handel gehörig, mitt schönen figuren vorbildet, und Klärlich beschriben seindt erstlich in Lateinischer Sprach durch den Hochgelerten und weittberümpten Herrn Georgium Agricolam, Doctorn und. Bürgermeistern der Churfürstlichen statt Kempnitz, jezundt aber verteüscht durch den Achtparen. unnd Hochgelerten Herrn Philippum Bechium, Philosophen, Artzer und in der Loblichen Universitet zu Basel Professorn.

Gedruckt zu Basel durch Jeronymus Froben Und Niclausen Bischoff im 1557 *Jar mitt Keiserlicher Freyheit.*

*1558 (folio):

Georgii Agricolae De ortu et causis subterraneorum Lib. V. De natura eorum quae effluunt ex terra Lib. IV. De natura fossilium Lib. X. De veteribus et novis metallis Lib. II. Bermannus, sive De Re Metallica Dialogus Liber. Interpretatio Germanica vocum rei metallicae, addito duplici Indice, altero rerum, altero locorum Omnia ab ipso authore, cum haud poenitenda accessione, recens recognita.

Froben, et Episcop. Basileae MDLVIII. *Cum Imp. Maiestatis renovato privilegio ad quinquennium.*

270 pages and index. As the title states, this is a revised edition by the author, and as the changes are very considerable it should be the one used. The Italian translation and the 1612 Wittenberg edition, mentioned below, are taken from the 1546 edition, and are, therefore, very imperfect.

*1561 (folio):

Second edition of *De Re Metallica* including *De Animantibus Subterraneis*, with same title as the first edition except the addition, after the body of the title, of the words *Atque omnibus nunc iterum ad archetypum diligenter restitutis et castigatis* and the year MDLXI. 502 pages and 72 pages of glossary and index.

*1563 (folio):

Opera di Giorgio Agricola de L'arte de Metalli Partita in XII. libri, ne quali si descrivano tutte le sorti, e qualità de gli uffizii, de gli strumenti, delle macchine, e di tutte l'altre cose attenenti a cotal arte, non pure con parole chiare ma eziandio si mettano a luoghi loro le figure di dette cose, ritratte al naturale, con l'aggiunta de nomi di quelle, cotanto chiari, e spediti, che meglio non si puo desiderare, o havere.

Aggiugnesi il libro del medesimo autore, che tratta de gl' Animali di sottoterra da lui stesso corretto et riveduto. Tradotti in lingua Toscana da M. Michelangelo Florio Fiorentino.

Con l'Indice di tutte le cose piu notabili alla fine (Froben's mark) *in Basilea per Hieronimo Frobenio et Nicolao Episcopio,* MDLXIII.

542 pages with 6 pages of index.

*1580 (folio):

Bergwerck Buch: Darinn nicht Allain alle Empte Instrument Gezeug und alles so zu diesem Handel gehörig mit figuren vorgebildet und klärlich beschriben, etc. Durch den Hochgelehrten . . . . Herrn Georgium Agricolam der Artzney Doctorn und Burgermeister der Churfürstlichen Statt Kemnitz erstlich mit grossem fleyss mühe und arbeit in Latein beschriben und in zwölff Bücher abgetheilt: Nachmals aber durch den Achtbarn und auch Hochgelehrten Philippum Bechium Philosophen Artzt und in der Löblichen Universitet zu Basel Professorn mit sonderm fleyss Teutscher Nation zu gut verteutscht und an Tag geben. Allen Berckherrn Gewercken Berckmeistern Geschwornen Schichtmeistern Steigern Berckheuwern Wäschern und Schmeltzern nicht allein nützlich und dienstlich sondern auch zu wissem hochnotwendig.

Mit Römischer Keys. May Freyheit nicht nachzutrucken.

Getruckt in der Keyserlichen Reichsstatt, Franckfort am Mayn, etc. Im Jahr MDLXXX.

*1612 (12mo):

Georgii Agricolae De ortu et causis subterraneorum Lib. V. De natura eorum quae effluunt ex terra, Lib. IV. De natura fossilium Lib. X. De veteribus et novis metallis Lib. II. Bermannus, sive de re metallica Dialogus. Interpretatio Germanica vocum rei metallicae.

Addito Indice faecundissimo, Plurimos jam annos à Germanis, et externarum quoque nationum doctissimis viris, valde desiderati et expetiti.

Nunc vero in rei metallicae studiosorum gratiam recensiti, in certa capita distributi, capitum argumentis, et nonnullis scholiis marginalibus illustrati à Johanne Sigfrido Philos: et Medicinae Doctore et in illustri Julia Professore ordinario.

Accesserunt De metallicis rebus et nominibus observationes variae et eruditae, ex schedis Georgii Fabricii, quibus ea potissimum explicantur, quae Georgius Agricola praeteriit.

Wittebergae Sumptibus Zachariae Schüreri Bibliopolae Typis Andreae Rüdingeri, 1612.

There are 970 pages in the work of Agricola proper, the notes of Fabricius comprising a further 44 pages, and the index 112 pages.

*1614 (8vo):

Georgii Agricolae De Animantibus Subterraneis Liber Hactenus à multis desideratus, nunc vero in gratiam studiosorum seorsim editus, in certa capita divisus, capitum argumentis et nonnullis marginalibus exornatus à Johanne Sigfrido, Phil. & Med. Doctore, etc.

Wittebergae. Typis Meisnerianis: Impensis Zachariae. Schureri Bibliop. Anno. MDCXIV.

*1621 (folio):

Georgii Agricolae Kempnicensis Medici ac Philosophi Clariss. De Re Metallica Libri XII Quibus Officia, Instrumenta, Machinae, ac omnia denique ad metallicam spectantia, non modo Luculentissimè describuntur; sed et per effigies, suis locis insertas adjunctis Latinis, Germanicisque; appellationibus, ita ob oculos ponuntur, ut clarius tradi non possint.

Ejusdem De Animantibus Subterraneis Liber, ab Autore recognitus cum Indicibus diversis quicquid in Opere tractatum est, pulchrè demonstrantibus.

(Vignette of man at assay furnace).

Basileae Helvet. Sumptibus itemque typis chalcographicis Ludovici Regis Anno MDCXXI.

502 pages and 58 pages glossary and indices.

*1621 (folio):

Bergwerck Buch Darinnen nicht allein alle Empter Instrument Gezeug und alles so zu disem Handel gehörig mit Figuren vorgebildet und klärlich beschrieben: . . . . Durch den Hochgelehrten und weitberühmten Herrn Georgium Agricolam, der Artzney Doctorn und Burgermeister der Churfürstlichen Statt Kemnitz Erstlich mit grossem fleiss mühe und arbeit in Latein beschrieben und in zwölff Bücher abgetheilt: Nachmals aber durch den Achtbarn und auch Hochgelehrten Philippum Bechium. Philosophen, Artzt, und in der loblichen Universitet zu Basel Professorn mit sonderm fleiss Teutscher Nation zu gut verteutscht und an Tag geben und nun zum andern mal getruckt.

Allen Bergherrn Gewercken Bergmeistern Geschwornen Schichtmeistern Steigern Berghäwern Wäschern unnd Schmeltzern nicht allein nutzlich und dienstlich sondern auch zu wissen hochnohtwendig.

(Vignette of man at assay furnace).

Getruckt zu Basel inverlegung Ludwig Königs Im Jahr, MDCXXI.

491 pages 5 pages glossary—no index.

*1657 (folio):

Georgii Agricolae Kempnicensis Medici ac Philosophi Clariss. De Re Metallica Libri XII. Quibus Officia, instrumenta, machinae, ac omnia denique ad metallicam spectantia, non modo luculentissimè describuntur: sed et per effigies, suis locis insertas, adjunctis Latinis, Germanicisque appellationibus, ita ob oculos ponuntur, ut clarius tradi non possint. Quibus accesserunt hac ultima editione, Tractatus ejusdem argumenti, ab eodem conscripti, sequentes.

De Animantibus Subterraneis Lib. I., De Ortu et Causis Subterraneorum Lib. V., De Natura eorum quae effluunt ex Terra Lib. IV., De Natura Fossilium Lib. X., De Veteribus et Novis Metallis Lib. II., Bermannus sive de Re Metallica, Dialogus Lib. I.

Cum Indicibus diversis, quicquid in Opere tractatum est, pulchrè demonstrantibus.

(Vignette of assayer and furnace).

Basileae Sumptibus et Typis Emanuelis König. Anno MDCLVII.

Folio, 708 pages and 90 pages of glossary and indices. This is a very serviceable edition of all of Agricola's important works, and so far as we have noticed there are but few typographical errors.

*1778 (8vo):

Gespräch vom Bergwesen, wegen seiner Fürtrefflich keit aus dem Lateinischen in das Deutsche übersetzet, mit nützl. Anmerkungen erläutert. u. mit einem ganz neuen Zusatze von Zlüglicher Anstellung des Bergbaues u. von der Zugutemachung der Erze auf den Hüttenwerken versehen von Johann Gottlieb Stör.

Rotenburg a. d. Fulda, Hermstädt 1778. 180 pages.

*1806 (8vo):

Georg Agricola's Bermannus eine Einleitung in die metallurgischen Schriften desselben, übersetzt und mit Exkursionen herausgegeben von Friedrich August Schmid. Haushalts-und Befahrungs-Protokollist im Churf. vereinigten Bergamte zu St. Annaberg.

Freyberg 1806. *Bey Craz und Gerlach.*

*1807–12 (8vo).

Georg Agrikola's Mineralogische Schriften übersetzt und mit erläuternden Anmerkungen. Begleitet von Ernst Lehmann Bergamts-Assessor, Berg- Gegen- und Receszschreiber in Dem Königl. Sächs. Bergamte Voigtsberg der jenaischen Societät für die gesammte Mineralogie Ehrenmitgliede.

Freyberg, 1807–12. *Bey Craz und Gerlach.*

This German translation consists of four parts: the first being *De Ortu et Causis,* the second *De Natura eorum quae effluunt ex terra,* and the third in two volumes *De Natura Fossilium,* the fourth *De Veteribus et Novis Metallis*; with glossary and index to the four parts.

We give the following notes on other possible prints, as a great many references to the above works occur in various quarters, of date other than the above. Unless otherwise convinced it is our belief that most of these refer to the prints given above, and are due to error in giving titles or dates. It is always possible that such prints do exist and have escaped our search.

De Re Metallica. Leupold, Richter, Schmid, van der Linden, Mercklinus and Eloy give an 8vo edition of *De Re Metallica* without illustrations, Schweinfurt, 1607. We have found no trace of this print. Leupold, van der Linden, Richter, Schmid and Eloy mention an 8vo edition, Wittenberg, 1614. It is our belief that this refers to the 1612 Wittenberg edition of the selected works, which contains a somewhat similar title referring in reality to *Bermannus*, which was and is still continually confused with *De Re Metallica.* Ferguson mentions a German edition, Schweinfurt, 8vo, 1687. We can find no trace of this; it may refer to the 1607 Schweinfurt edition mentioned above.

De Natura Fossilium. Leupold and Gatter refer to a folio edition of 1550. This was probably an error for either the 1546 or the 1558 editions. Watt refers to an edition of 1561 combined with *De Medicatis Fontibus.* We find no trace of such edition, nor even that the latter work was ever actually printed. He also refers to an edition of 1614 and one of 1621, this probably being an error for the 1612 edition of the subsidiary works and the *De Re Metallica* of 1621. Leupold also refers to an edition of 1622, this probably being an error for 1612.

De Ortu et Causis. Albinus, Hofmann, Jacobi, Schmid, Richter, and Reuss mention an edition of 1544. This we believe to be an error in giving the date of the dedication instead of that of the publication (1546). Albinus and Ferguson give an edition of 1555, which date is, we believe, an error for 1558. Ferguson gives an edition of the Italian translation as 1559; we believe this should be 1550. Draud gives an edition of 1621; probably this should be 1612.

Bermannus. Albinus, Schmid, Reuss, Richter, and Weinart give the first edition as 1528. We have been unable to learn of any actual copy of that date, and it is our belief that the date is taken from the dedication instead of from the publication, and should be 1530. Leupold, Schmid, and Reuss give an edition by Froben in 1549; we have been unable to confirm this. Leupold also gives an edition of 1550 (folio), and Jöcher gives an edition of Geneva 1561 (folio); we have also been unable to find this, and believe the latter to be a confusion with the *De Re Metallica* of 1561, as it is unlikely that *Bermannus* would be published by itself in folio. The catalogue of the library at Siena (Vol. III., p. 78) gives *Il Bermanno, Vinegia,* 1550, 8vo. We have found no trace of this edition elsewhere.

De Mensuris et Ponderibus. Albinus and Schmid mention an edition of 1539, and one of 1550. The Biographie Universelle, Paris, gives one of 1553, and Leupold one of 1714, all of which we have been unable to find. An epitome of this work was published at various times, sometimes in connection with editions of Vitruvius; so far as we are aware on the following dates, 1552, 1585, 1586, 1829. There also appear extracts in relation to liquid measures in works entitled *Vocabula rei numariae ponderum et mensurarum,* etc. Paul Eber and Caspar Peucer, *Lipsiae,* 1549, and in same Wittenberg, 1552.

De Veteribus et Novis Metallis. Watt gives an edition, Basel, 1530, and Paris, 1541; we believe this is incorrect and refers to *Bermannus.* Reuss mentions a folio print of Basel, 1550. We consider this very unlikely.

De Natura eorum quae Effluunt ex Terra. Albinus, Hofmann, Schmid, Jacobi, Richter, Reuss, and Weinart give an edition of 1545. We believe this is again the dedication instead of the publication date (1546).

De Animantibus Subterraneis. Van der Linden gives an edition at Schweinfurt, 8vo, 1607. Although we have been unable to find a copy, this slightly confirms the possibility of an octavo edition of *De Re Metallica* of this date, as they were usually published together. Leupold gives assurance that he handled an octavo edition of Wittenberg, 1612, *cum notis Johann Sigfridi.* We think he confused this with *Bermannus sive de re metallica* of that date and place. Schmid, Richter, and Draud all refer to an edition similarly annotated, Leipzig, 1613, 8vo. We have no trace of it otherwise.

UNPUBLISHED WORKS ON SUBJECTS RELATED TO MINING.

Agricola apparently projected a complete series of works covering the whole range of subjects relating to minerals: geology, mineralogy, mining, metallurgy, history of metals, their uses, laws, etc. In a letter[5] from Fabricius to Meurer (March, 1553), the former states that Agricola intended writing about 30 books (chapters) in addition to those already published, and to the twelve books *De Re Metallica* which he was about to publish. Apparently a number of these works were either unfinished or unpublished at Agricola's death, for his friend George Fabricius seems to have made some effort to secure their publication, but did not succeed, through lack of sympathy on the part of Agricola's family. Hofmann[6] states on this matter: "His intentions were frustrated mainly through the lack of support with which "he was met by the heirs of the Mineralogist. These, as he complains to a Councillor of the "Electorate, Christopher von Carlovitz, in 1556, and to Paul Eber in another letter, adopted "a grudging and ungracious tone with regard to his proposal to collect all Agricola's works "left behind, and they only consented to communicate to him as much as they were obliged

[5] *G. Fabricii epistolae ad W. Meurerum et alios aequales,* by Baumgarten-Crusius, Leipzig, 1845, p. 83.

[6] *Dr. Georg Agricola,* Gotha, 1905, pp. 60–61.

"by express command of the Prince. At the Prince's command they showed him a little, "but he supposed that there was much more that they had suppressed or not preserved. "The attempt to purchase some of the works—the Elector had given Fabricius money for "the purpose (30 nummos unciales)—proved unavailing, owing to the disagreeableness of "Agricola's heirs. It is no doubt due to these regrettable circumstances that all the works "of the industrious scholar did not come down to us." The "disagreeableness" was probably due to the refusal of the Protestant townsfolk to allow the burial of Agricola in the Cathedral at Chemnitz. So far as we know the following are the unpublished or lost works.

De Jure et Legibus Metallicis. This work on mining law is mentioned at the end of Book IV. of *De Re Metallica*, and it is referred to by others apparently from that source. We have been unable to find any evidence that it was ever published.

De Varia temperie sive Constitutione Aeris. In a letter[7] to Johann Naevius, Agricola refers to having a work in hand of this title.

De Metallis et Machinis. Hofmann[8] states that a work of this title by Agricola, dated Basel 1543, was sold to someone in America by a Frankfort-on-Main bookseller in 1896. This is apparently the only reference to it that we know of, and it is possibly a confusion of titles or a "separate" of some chapters from *De Re Metallica.*

De Ortu Metallorum Defensio ad Jacobum Scheckium. Referred to by Fabricius in a letter[9] to Meurer. If published was probably only a tract.

De Terrae Motu. In a letter[10] from Agricola to Meurer (Jan. 1, 1544) is some reference which might indicate that he was formulating a work on earthquakes under this title, or perhaps may be only incidental to the portions of *De Ortu et Causis* dealing with this subject.

Commentariorum in quibus utriusque linguae scriptorum locos difficiles de rebus subterraneis explicat, Libri VI. Agricola apparently partially completed a work under some such title as this, which was to embrace chapters entitled *De Methodis* and *De Demonstratione.* The main object seems to have been a commentary on the terms and passages in the classics relating to mining, mineralogy, etc. It is mentioned in the Preface of *De Veteribus et Novis Metallis*, and in a letter[11] from one of Froben's firm to Agricola in 1548, where it is suggested that Agricola should defer sending his new commentaries until the following spring. The work is mentioned by Albinus[12], and in a letter from Georg Fabricius to Meurer on the 2nd Jan. 1548,[13] in another from G. Fabricius, to his brother Andreas on Oct. 28, 1555,[14] and in a third from Fabricius to Melanchthon on December 8th, 1555[15], in which regret is expressed that the work was not completed by Agricola.

---

[7]Albinus, *Landchronik*, pp. 354-5.

[8]*Dr. Georg Agricola*, p. 63.

[9]*Baumgarten-Crusius*, p. 115.

[10]*Virorum Clarorum Saec.* XVI. *et* XVII. *Epistolae Selectae* by Ernst Weber, Leipzig, 1894, p. 2.

[11]Nicholas Episcopius to Georg Agricola, Sept. 17, 1548, published in Schmid's *Bermannus* p. 38. See also Hofmann, op. cit. pp. 62 and 140.

[12]*Meissnische Landchronik*, Dresden, 1589, p. 354.

[13]Printed in Baumgarten-Crusius, pp. 48-49, letter XLVIII.

[14]Printed in Hermann Peter's *Meissner Jahresbericht der Fürstenschule*, 1891, p. 24.

[15]Baumgarten-Crusius. *Georgii Fabricii Chemnicensis Epistolae*, Leipzig, 1845, p. 139.

WRITINGS NOT RELATED TO MINING, INCLUDING LOST OR UNPUBLISHED WORKS.

Latin Grammar. This was probably the first of Agricola's publications, the full title to which is *Georgii Agricolae Glaucii Libellus de prima ac simplici institutione grammatica. Excusum Lipsiae in Officina Melchioris Lottheri. Anno* MDXX. (4to), 24 folios.[16] There is some reason to believe that Agricola also published a Greek grammar, for there is a letter[17] from Agricola dated March 18th, 1522, in which Henicus Camitianus is requested to send a copy to Stephan Roth.

Theological Tracts. There are preserved in the Zwickau Rathsschul Library[18] copies by Stephan Roth of two tracts, the one entitled, *Deum non esse auctorem Peccati*, the other, *Religioso patri Petri Fontano, sacre theologie Doctori eximio Georgius Agricola salutem dicit in Christo.* The former was written from Leipzig in 1522, and the latter, although not dated, is assigned to the same period. Both are printed in *Zwei theologische Abhandlungen des Georg Agricola*, an article by Otto Clemen, *Neuen Archiv für Sächsische Geschichte*, etc., Dresden, 1900. There is some reason (from a letter of Fabricius to Melanchthon, Dec. 8th, 1555) to believe that Agricola had completed a work on the unwritten traditions concerning the Church. There is no further trace of it.

Galen. Agricola appears to have been joint author with Andreas Asulanus and J. B. Opizo of a revision of this well-known Greek work. It was published at Venice in 1525, under the title of *Galeni Librorum*, etc., etc. Agricola's name is mentioned in a prefatory letter to Opizo by Asulanus.

De Bello adversus Turcam. This political tract, directed against the Turks, was written in Latin and first printed by Froben, Basel, 1528. It was translated into German apparently by Agricola's friend Laurenz Berman, and published under the title *Oration Anrede Und Vormanunge . . . . widder den Türcken* by Frederich Peypus, Nuremberg, in 1531 (8vo), and either in 1530 or 1531 by Wolfgang Stöckel, Dresden, 4to. It was again printed in Latin by Froben, Basel, 1538, 4to; by H. Grosius, Leipzig, 1594, 8vo; it was included among other works published on the same subject by Nicholas Reusnerus, Leipzig, 1595; by Michael Lantzenberger, Frankfurt-am-Main, 1597, 4to. Further, there is reference by Watt to an edition at Eisleben, 1603, of which we have no confirmation. There is another work on the subject, or a revision by the author mentioned by Albinus[19] as having been, after Agricola's death, sent to Froben by George Fabricius to be printed; nothing further appears in this matter however.

De Peste. This work on the Plague appears to have been first printed by Froben, Basel, 1554, 8vo. The work was republished at Schweinfurt, 1607, and at Augsburg in 1614, under various editors. It would appear from Albinus[20] that the work was revised by Agricola and in Froben's hands for publication after the author's death.

De Medicatis Fontibus. This work is referred to by Agricola himself in *De Natura Eorum*,[21] in the prefatory letter in *De Veteribus et Novis Metallis*; and Albinus[22] quotes a letter of Agricola to Sebastian Munster on the subject. Albinus states (*Bergchronik*, p. 193) that to his knowledge it had not yet been published. Conrad Gesner, in his work *Excerptorum et observationum de Thermis*, which is reprinted in *De Balneis*, Venice, 1553, after Agricola's *De Natura Eorum*, states [23] concerning Agricola *in libris quos de medicatis fontibus instituerit copiosus se dicturum pollicetur.* Watts mentions it as having been published in 1549, 1561, 1614, and 1621. He, however, apparently confuses it with *De Natura Eorum.* We are unable to state whether it was ever printed or not. A note of inquiry to the principal libraries in Germany gave a negative result.

De Putredine solidas partes humani corporis corrumpente. This work, according to Albinus was received by Fabricius a year after Agricola's death, but whether it was published or not is uncertain.[24]

Castigationes in Hippocratem et Galenum. This work is referred to by Agricola in the preface of *Bermannus*, and Albinus[25] mentions several letters referring to the preparation of the work. There is no evidence of publication.

Typographia Mysnae et Toringiae. It seems from Agricola's letter[26] to Munster that Agricola prepared some sort of a work on the history of Saxony and of the Royal Family

---

[16]There is a copy of this work in the Rathsschul Library at Zwickau.

[17]In the Rathsschul Library at Zwickau.

[18]Contained in Vols. XXXVII. and XL. of Stephan Roth's *Kollectanenbände* Volumes of Transcripts.

[19]*Landchronik*, p. 354.

[20]Op. cit., p. 354.

[21]Book IV.

[22]Op. cit., p. 355.

[23]Page 291.

[24]See Baumgarten-Crusius, p. 114, letter from Georg Fabricius.

[25]Op. cit., p. 354.

[26]Albinus, Op. cit., p. 355.

thereof at the command of the Elector and sent it to him when finished, but it was never published as written by Agricola. Albinus, Hofmann, and Struve give some details of letters in reference to it. Fabricius in a letter[27] dated Nov. 11, 1536 asks Meurer to send Agricola some material for it; in a letter from Fabricius to Meurer dated Oct. 30, 1554, it appears that the Elector had granted Agricola 200 thalers to assist in the work. After Agricola's death the material seems to have been handed over to Fabricius, who made use of it (as he states in the preface) in preparing the work he was commissioned by the Elector to write, the title of which was, *Originum illustrissimae stirpis Saxonicae Libri*, and was published in Leipzig, 1597. It includes on page 880 a fragment of a work entitled *Oratio de rebus Gestis Ernesti et Alberti Ducum Saxoniae*, by Agricola.

WORKS WRONGLY ATTRIBUTED TO GEORGIUS AGRICOLA.

The following works have been at one time or another wrongly attributed to Georgius Agricola :—

Galerazeya sive Revelator Secretorum De Lapide Philosophorum, Cologne, 1531 and 1534, by one Daniel Agricola, which is merely a controversial book with a catch-title, used by Catholics for converting heretics.

Rechter Gebrauch der Alchimey, a book of miscellaneous receipts which treats very slightly of transmutation.[28]

Chronik der Stadt Freiberg by a Georg Agricola (died 1630), a preacher at Freiberg.

Dominatores Saxonici, by the same author.

Breviarum de Asse by Guillaume Bude.

De Inventione Dialectica by Rudolph Agricola.

[27]Baumgarten-Crusius, p. 2.

[28]See Ferguson, *Bibliotheca Chemica*, s.v. Daniel Agricola.

APPENDIX B.

ANCIENT AUTHORS.

We give the following brief notes on early works containing some reference to mineralogy, mining, or metallurgy, to indicate the literature available to Agricola and for historical notes bearing upon the subject. References to these works in the footnotes may be most easily consulted through the personal index.

GREEK AUTHORS.—Only a very limited Greek literature upon subjects allied to mining or natural science survives. The whole of the material of technical interest could be reproduced on less than twenty of these pages. Those of most importance are : Aristotle (384–322 B.C.), Theophrastus (371–288 B.C.), Diodorus Siculus (1st Century B.C.), Strabo (64 B.C.—25 A.D.), and Dioscorides (1st Century A.D.).

Aristotle, apart from occasional mineralogical or metallurgical references in *De Mirabilibus*, is mostly of interest as the author of the Peripatetic theory of the elements and the relation of these to the origin of stones and metals. Agricola was, to a considerable measure, a follower of this school, and their views colour much of his writings. We, however, discuss elsewhere[1] at what point he departed from them. Especially in *De Ortu et Causis* does he quote largely from Aristotle's *Meteorologica*, *Physica*, and *De Coelo* on these subjects. There is a spurious work on stones attributed to Aristotle of some interest to mineralogists. It was probably the work of some Arab early in the Middle Ages.

Theophrastus, the principal disciple of Aristotle, appears to have written at least two works relating to our subject—one "On Stones", and the other on metals, mining or metallurgy, but the latter is not extant. The work "On Stones" was first printed in Venice in 1498, and the Greek text, together with a fair English translation by Sir John Hill, was published in London in 1746 under the title "Theophrastus on Stones"; the translation is, however, somewhat coloured with Hill's views on mineralogy. The work comprises 120 short paragraphs, and would, if reproduced, cover but about four of these pages. In the first paragraphs are the Peripatetic view of the origin of stones and minerals, and upon the foundation of Aristotle he makes some modifications. The principal interest in Theophrastus' work is the description of minerals ; the information given is, however, such as might be possessed by any ordinary workman, and betrays no particular abilities for natural philosophy. He enumerates various exterior characteristics, such as colour, tenacity, hardness, smoothness, density, fusibility, lustre, and transparence, and their quality of reproduction, and then proceeds to describe various substances, but usually omits his enumerated characteristics. Apart from the then known metals and certain "earths" (ochre, marls, clay, etc.), it is possible to identify from his descriptions the following rocks and minerals :—marble, pumice, onyx, gypsum, pyrites, coal, bitumen, amber, azurite, chrysocolla, realgar, orpiment, cinnabar, quartz in various forms, lapis lazuli, emerald, sapphire, diamond, and ruby. Altogether there are some sixteen distinct mineral species. He also describes the touchstone and its uses, the making of white-lead and verdigris, and of quicksilver from cinnabar.

Diodorus Siculus was a Greek native of Sicily. His "historical library" consisted of some 40 books, of which parts of 15 are extant. The first print was in Latin, 1472, and in Greek in 1539 ; the first translation into English was by Thomas Stocker, London, 1568, and later by G. Booth, 1700. We have relied upon Booth's translation, but with some amendments by friends, to gain more literal statement. Diodorus, so far as relates to our subject, gives merely the occasional note of a traveller. The most interesting paragraphs are his quotation from Agatharchides on Egyptian mining and upon British tin.

Strabo was also a geographer. His work consists of 17 books, and practically all survive. We have relied upon the most excellent translation of Hamilton and Falconer, London, 1903, the only one in English. Mines and minerals did not escape such an acute geographer, and the matters of greatest interest are those with relation to Spanish mines.

Dioscorides was a Greek physician who wrote entirely from the standpoint of materia medica, most of his work being devoted to herbs; but Book V. is devoted to minerals and rocks, and their preparation for medicinal purposes. The work has never been translated into English, and we have relied upon the Latin translation of Matthioli, Venice, 1565, and notes upon the Greek text prepared for us by Mr. C. Katopodes. In addition to most of the substances known before, he, so far as can be identified, adds schist, *cadmia* (blende or calamine), *chalcitis* (copper sulphide), *misy*, *melanteria*, *sory* (copper or iron sulphide oxidation minerals). He describes the making of certain artificial products, such as copper oxides, vitriol, litharge, *pompholyx*, and *spodos* (zinc and / or arsenical oxides). His principal interest for us, however, lies in the processes set out for making his medicines.

Occasional scraps of information relating to the metals or mines in some connection are to be found in many other Greek writers, and in quotations by them from others which are not now extant, such as Polybius, Posidonius, etc. The poets occasionally throw a gleam

[1]See pages 44 and 46

of light on ancient metallurgy, as for instance in Homer's description of Vulcan's foundry; while the historians, philosophers, statesmen, and physicians, among them Herodotus, Xenophon, Demosthenes, Galen, and many others, have left some incidental references to the metals and mining, helpful to gleaners from a field, which has been almost exhausted by time. Even Archimedes made pumps, and Hero surveying instruments for mines.

ROMAN AUTHORS.—Pre-eminent among all ancient writers on these subjects is, of course, Pliny, and in fact, except some few lines by Vitruvius, there is practically little else in extant Roman literature of technical interest, for the metallurgical metaphors of the poets and orators were threadbare by this time, and do not excite so much interest as upon their first appearance among the Greeks and Hebrews.

Pliny (Caius Plinius Secundus) was born 23 A.D., and was killed by eruption of Vesuvius 79 A.D. His Natural History should be more properly called an encyclopædia, the whole comprising 37 books; but only portions of the last four books relate to our subject, and over one-half of the material there is upon precious stones. To give some rough idea of the small quantity of even this, the most voluminous of ancient works upon our subject, we have made an estimate that the portions of metallurgical character would cover, say, three pages of this text, on mining two pages, on building and precious stones about ten pages. Pliny and Dioscorides were contemporaries, and while Pliny nowhere refers to the Greek, internal evidence is most convincing, either that they drew from the same source, or that Pliny drew from Dioscorides. We have, therefore, throughout the text given precedence in time to the Greek author in matters of historical interest. The works of Pliny were first printed at Venice in 1469. They have passed dozens of editions in various languages, and have been twice translated into English. The first translation by Philemon Holland, London, 1601, is quite impossible. The second translation, by Bostock and Riley, London, 1855, was a great advance, and the notes are most valuable, but in general the work has suffered from a freedom justifiable in the translation of poetry, but not in science. We have relied upon the Latin edition of Janus, Leipzig, 1870. The frequent quotations in our footnotes are sufficient indication of the character of Pliny's work. In general it should be remembered that he was himself but a compiler of information from others, and, so far as our subjects are concerned, of no other experience than most travellers. When one considers the reliability of such authors to-day on technical subjects, respect for Pliny is much enhanced. Further, the text is no doubt much corrupted through the generations of transcription before it was set in type. So far as can be identified with any assurance, Pliny adds but few distinct minerals to those enumerated by Theophrastus and Dioscorides. For his metallurgical and mining information we refer to the footnotes, and in general it may be said that while those skilled in metallurgy can dimly see in his statements many metallurgical operations, there is little that does not require much deduction to arrive at a conclusion. On geology he offers no new philosophical deductions of consequence; the remote connection of building stones is practically all that can be enumerated, lest one build some assumption of a knowledge of ore-deposits on the use of the word "vein". One point of great interest to this work is that in his search for Latin terms for technical purposes Agricola relied almost wholly upon Pliny, and by some devotion to the latter we have been able to disentangle some very puzzling matters of nomenclature in *De Re Metallica*, of which the term *molybdaena* may be cited as a case in point.

Vitruvius was a Roman architect of note of the 1st Century B.C. His work of ten books contains some very minor references to pumps and machinery, building stones, and the preparation of pigments, the latter involving operations from which metallurgical deductions can occasionally be safely made. His works were apparently first printed in Rome in 1496. There are many editions in various languages, the first English translation being from the French in 1692. We have relied upon the translation of Joseph Gwilt, London, 1875, with such alterations as we have considered necessary.

MEDIÆVAL AUTHORS. For convenience we group under this heading the writers of interest from Roman times to the awakening of learning in the early 16th Century. Apart from Theophilus, they are mostly alchemists; but, nevertheless, some are of great importance in the history of metallurgy and chemistry. Omitting a horde of lesser lights upon whom we have given some data under the author's preface, the works principally concerned are those ascribed to Avicenna, Theophilus, Geber, Albertus Magnus, Roger Bacon, and Basil Valentine. Judging from the Preface to *De Re Metallica*, and from quotations in his subsidiary works, Agricola must have been not only familiar with a wide range of alchemistic material, but also with a good deal of the Arabic literature, which had been translated into Latin. The Arabs were, of course, the only race which kept the light of science burning during the Dark Ages, and their works were in considerable vogue at Agricola's time.

Avicenna (980–1037) was an Arabian physician of great note, a translator of the Greek classics into Arabic, and a follower of Aristotle to the extent of attempting to reconcile the Peripatetic elements with those of the alchemists. He is chiefly known to the world through the works which he compiled on medicine, mostly from the Greek and Latin authors. These works for centuries dominated the medical world, and were used in certain European Universities until the 17th century. A great many works are attributed to him, and he is copiously quoted by Agricola, principally in his *De Ortu et Causis*, apparently for the purpose of exposure.

Theophilus was a Monk and the author of a most illuminating work, largely upon working metal and its decoration for ecclesiastical purposes. An excellent translation, with the Latin text, was published by Robert Hendrie, London, 1847, under the title "An Essay upon various Arts, in three books, by Theophilus, called also Rugerus, Priest and Monk." Hendrie, for many sufficient reasons, places the period of Theophilus as the latter half of the 11th century. The work is mainly devoted to preparing pigments, making glass, and working metals, and their conversion into ecclesiastical paraphernalia, such as mural decoration, pictures, windows, chalices, censers, bells, organs, etc. However, he incidentally describes the making of metallurgical furnaces, cupellation, parting gold and silver by cementation with salt, and by melting with sulphur, the smelting of copper, liquating lead from it, and the refining of copper under a blast with poling.

Geber was until recent years considered to be an Arab alchemist of a period somewhere between the 7th and 12th centuries. A mere bibliography of the very considerable literature which exists in discussion of who, where, and at what time the author was, would fill pages. Those who are interested may obtain a start upon such references from Hermann Kopp's *Beiträge zur Geschichte der Chemie*, Braunschweig, 1875, and in John Ferguson's *Bibliotheca Chemica*, Glasgow, 1906. Berthelot, in his *Chimie au Moyen Age*, Paris, 1893, considers the works under the name of Geber were not in the main of Arabic origin, but composed by some Latin scholar in the 13th century. In any event, certain works were, under this name, printed in Latin as early as 1470–80, and have passed innumerable editions since. They were first translated into English by Richard Russell, London, 1678, and we have relied upon this and the Nuremberg edition in Latin of 1541. This work, even assuming Berthelot's view, is one of the most important in the history of chemistry and metallurgy, and is characterised by a directness of statement unique among alchemists. The making of the mineral acids—certainly nitric and *aqua regia*, and perhaps hydrochloric and sulphuric—are here first described. The author was familiar with saltpetre, sal-ammoniac, and alkali, and with the acids he prepared many salts for the first time. He was familiar with amalgamation, cupellation, the separation of gold and silver by cementation with salt and by nitric acid. His views on the primary composition of bodies dominated the alchemistic world for centuries. He contended that all metals were composed of "spiritual" sulphur (or arsenic, which he seems to consider a special form of sulphur) and quicksilver, varying proportions and qualities yielding different metals. The more the quicksilver, the more "perfect" the metal.

Albertus Magnus (Albert von Bollstadt) was a Dominican Monk, afterwards Bishop, born about 1205, and died about 1280. He was rated the most learned man of his time, and evidence of his literary activities lies in the complete edition of his works issued by Pierre Jammy, Lyons, 1651, which comprises 21 folio volumes. However, there is little doubt that a great number of works attributed to him, especially upon alchemy, are spurious. He covered a wide range of theology, logic, alchemy, and natural science, and of the latter the following works which concern our subject are considered genuine:—*De Rebus Metallicis et Mineralibus*, *De Generatione et Corruptione*, and *De Meteoris*. They are little more than compilations and expositions of the classics muddled with the writings of the Arabs, and in general an attempt to conciliate the Peripatetic and Alchemistic schools. His position in the history of science has been greatly over-estimated. However, his mineralogy is, except for books on gems, the only writing of any consequence at all on the subject between Pliny and Agricola, and while there are but two or three minerals mentioned which are not to be found in the ancient authors, this work, nevertheless, deserves some place in the history of science, especially as some attempt at classification is made. Agricola devotes some thousands of words to the refutation of his "errors."

Roger Bacon (1214–1294) was a Franciscan Friar, a lecturer at Oxford, and a man of considerable scientific attainments for his time. He was the author of a large number of mathematical, philosophical, and alchemistic treatises. The latter are of some importance in the history of chemistry, but have only minute bearing upon metallurgy, and this chiefly as being one of the earliest to mention saltpetre.

Basil Valentine is the reputed author of a number of alchemistic works, of which none appeared in print until early in the 17th century. Internal evidence seems to indicate that the "Triumphant Chariot of Antimony" is the only one which may possibly be authentic, and could not have been written prior to the end of the 15th or early 16th century, although it has been variously placed as early as 1350. To this work has been accredited the first mention of sulphuric and hydrochloric acid, the separation of gold and silver by the use of antimony (sulphide), the reduction of the antimony sulphide to the metal, the extraction of copper by the precipitation of the sulphate with iron, and the discovery of various antimonial salts. At the time of the publication of works ascribed to Valentine practically all these things were well known, and had been previously described. We are, therefore, in much doubt as to whether this author really deserves any notice in the history of metallurgy.

EARLY 16TH CENTURY WORKS. During the 16th century, and prior to *De Re Metallica*, there are only three works of importance from the point of view of mining technology—the *Nützlich Bergbüchlin*, the *Probierbüchlein*, and Biringuccio's *De La Pirotechnia*. There are also some minor works by the alchemists of some interest for isolated statements, particularly those of Paracelsus. The three works mentioned, however, represent such a stride of advance over anything previous, that they merit careful consideration.

Eyn Nützlich Bergbüchlin. Under this title we frequently refer to a little booklet on veins and ores, published at the beginning of the 16th century. The title page of our copy is as below :—

Ein nützlich Berg büchlin von allen Metal-len/als Golt/Silber/Zcyn/Kupfer ertz/Eiſen ſtein/Bleyertz/vnd vom Queckſilber.

This book is small 8vo, comprises 24 folios without pagination, and has no typographical indications upon the title page, but the last line in the book reads : *Gedruckt zu Erffurd durch Johan Loersfelt*, 1527. Another edition in our possession, that of " Frankfurt am Meyn", 1533, by Christian Egenolph, is entitled *Bergwerk und Probierbüchlin*, etc., and contains, besides the above, an extract and plates from the *Probierbüchlein* (referred to later on),and a few recipes for assay tests. All of these booklets, of which we find mention, comprise instructions from Daniel, a skilled miner, to Knappius, " his mining boy". Although the little books of this title are all anonymous, we are convinced, largely from the statement in the Preface of *De Re Metallica*, that one Calbus of Freiberg was the original author of this work. Agricola says : " Two books have been written in our tongue : the one on the assaying of mineral sub- " stances and metals, somewhat confused, whose author is unknown ; the other ' On Veins', " of which Pandulfus Anglus is also said to have written, *although the German book was written " by Calbus of Freiberg, a well-known doctor ; but neither of them accomplished the task he had " begun.*" He again refers to Calbus at the end of Book III.[2] of *De Re Metallica*, and gives an almost verbatim quotation from the *Nützlich Bergbüchlin.* Jacobi[3] says : " Calbus " Fribergius, so called by Agricola himself, is certainly no other than the Freiberg doctor, " Rühlein von C(K)albe." There are also certain internal evidences that support Agricola's statement, for the work was evidently written in Meissen, and the statement of Agricola that the book was unfinished is borne out by a short dialogue at the end of the earlier editions, designed to introduce further discussion. Calbus (or Dr. Ulrich Rühlein von Kalbe) was a very active citizen of Freiberg, having been a town councillor in 1509, burgomaster in 1514, a mathematician, mining surveyor, founder of a school of liberal arts, and in general a physician. He died in 1523.[4] The book possesses great literary interest, as it is, so far as we are aware,

[2]Page 75.
[3]*Der Mineralog Georgius Agricola*, Zwickau, 1889, p. 46.
[4]Andreas Möller, *Theatrum Freibergense Chronicum*, etc., Freiberg, 1653.

undoubtedly the first work on mining geology, and in consequence we have spent some effort in endeavour to find the date of its first appearance. Through the courtesy of M. Polain, who has carefully examined for us the *Nützlich Bergbüchlein* described in Marie Pellechet's *Catalogue Général des Incunables des Bibliothèques Publiques de France*,[5] we have ascertained that it is similar as regards text and woodcuts to the Erfurt edition, 1527. This copy in the Bibliothèque Nationale is without typographical indications, and M. Polain considers it very possible that it is the original edition printed at the end of the fifteenth or begininng of the sixteenth centuries. Mr. Bennett Brough,[6] quoting Hans von Dechen,[7] states that the first edition was printed at Augsburg in 1505, no copy of which seems to be extant. The Librarian at the School of Mines at Freiberg has kindly furnished us with the following notes as to the titles of the copies in that Institution :—(1) *Eyn Wolgeordent und Nützlich Bergbüchlein*, etc., Worms, 1512[8] and 1518[9] (the place and date are written in) ; (2) the same as ours (1527) ; (3) the same, Heinrich Steyner, Augsburg, 1534 ; (4) the same, 1539. On comparing these various editions (to which may be added one probably published in Nürnberg by Friedrich Peypus in 1532[10]) we find that they fall into two very distinct groups, characterised by their contents and by two entirely different sets of woodcuts.

GROUP I.

(*a*) *Eyn Nützlich Bergbüchlein* (in *Bibl. Nat.*, Paris) before 1500 (?).
(*b*) Ditto, Erfurt, 1527.

GROUP II.

(*c*) *Wolgeordent Nützlich Bergbüchlein*, Worms, Peter Schöfern, 1512.
(*d*) *Wolgeordent Nützlich Bergbüchlein*, Worms, Peter Schöfern, 1518.
(*e*) *Bergbüchlin von Erkantnus der Berckwerck*, Nürnberg, undated, 1532 (?).
(*f*) *Bergwerckbuch & Probirbuch*, Christian Egenolph, Frankfurt-am-Meyn, 1533.
(*g*) *Wolgeordent Nützlich Bergbüchlein*, Augsburg, Heinrich Steyner, 1534.
(*h*) *Wolgeordent Nützlich Bergbüchlein*, Augsburg, Heinrich Steyner, 1539.

There are also others of later date toward the end of the sixteenth century.

The *Büchlein* of Group I. terminate after the short dialogue between Daniel and Knappius with the words : *Mitt welchen das kleinspeissig ertz geschmeltzt soll werden ;* whereas in those of Group II. these words are followed by a short explanation of the signs used in the woodcuts, and by directions for colouring the woodcuts, and in some cases by several pages containing definitions of some 92 mining terms. In the editions of Group I. the woodcut on the title page represents a miner hewing ore in a vein and two others working a windlass. In those of Group II. the woodcut on the title page represents one miner hewing on the surface, another to the right carting away ore in a handcart, and two others carrying between them a heavy timber. In our opinion Group I. represents the older and original work of Calbus ; but as we have not seen the copy in the *Bibliothèque Nationale*, and the Augsburg edition of 1505 has only so far been traced to Veith's catalogue,[11] the question of the first edition cannot be considered settled at present. In any event, it appears that the material grafted on in the second group was later, and by various authors.

The earliest books comprise ten chapters, in which Daniel delivers about 6,000 words of instruction. The first four chapters are devoted to the description of veins and the origin of the metals, of the remaining six chapters one each to silver, gold, tin, copper, iron, lead, and quicksilver. Among the mining terms are explained the meaning of country rock (*zechstein*), hanging and footwalls (*hangends* and *liegends*), the strike (*streichen*), dip (*fallen*), and outcrop (*ausgehen*). Of the latter two varieties are given, one of the "whole vein," the other of the *gesteins*, which may be the ore-shoot. Various veins are illustrated, and also for the first time a mining compass. The account of the origin of the metals is a muddle of the Peripatetics, the alchemists, and the astrologers, for which acknowledgment to Albertus Magnus is given. They are represented to originate from quicksilver and sulphur through heat, cold, dampness, and dryness, and are drawn out as exhalations through the veins, each metal owing its origin to the special influence of some planet ; the Moon for silver, Saturn for lead, etc. Two types of veins are mentioned, "standing" (*stehendergang*) and flat (*flachgang*). Stringers are given the same characteristics as veins, but divided into hanging, footwall, and other varieties. Prominence is also given to the *geschick* (selvage seams or joints ?).

[5]Paris, 1897, Vol. I. p. 501.

[6]Cantor Lectures, London, April 1892.

[7]Hans von Dechen, *Das älteste deutsche Bergwerksbuch*, reprint from *Zts. für Bergrecht Bd.* XXVI., Bonn, 1885.

[8]Panzer's *Annalen*, Nürnberg, 1782, p. 422, gives an edition Worms *bei* Peter Schöfern, 1512.

[9]The Royal Library at Dresden and the State Library at Munich have each a copy, dated 1518, Worms.

[10]Hans von Decken *op. cit.*, p. 48–49.

[11]*Annales typographiae augustanae ab ejus origine*, MCCCLXVI. *usque ad. an.* M.D.XXX. *Accedit dom Franc. Ant. Veith. Diatribe de origine . . . artis typographicae in urbe augusta vindelica edidit . . .* Georgius G. Zapf., Augsburg, 1778, X. p. 23.

The importance of the bearing of the junctions of veins and stringers on enrichment is elaborated upon, and veins of east-west strike lying upon a south slope are considered the best. From the following notes it will be seen that two or three other types of deposits besides veins are referred to.

In describing silver veins, of peculiar interest is the mention of the association of bismuth (*wismuth*), this being, we believe, the first mention of that metal, galena (*glantz*), quartz (*quertz*), spar (*spar*), hornstone (*hornstein*), ironstone and pyrites (*kies*), are mentioned as gangue materials, "according to the mingling of the various vapours." The term *glasertz* is used, but it is difficult to say if silver glance is meant; if so, it is the first mention of this mineral. So far as we know, this is the first use of any of the terms in print. Gold alluvial is described, part of the gold being assumed as generated in the gravel. The best alluvial is in streams running east and west. The association of gold with pyrites is mentioned, and the pyrites is found "in some places as a complete stratum carried through horizontally, and is called a *schwebender gang*." This sort of occurrence is not considered very good "because the work of the heavens can be but little completed on account of the unsuitability of the position." Gold pyrites that comes in veins is better. Tin is mentioned as found in alluvial, and also in veins, the latter being better or worse, according to the amount of pyrites, although the latter can be burned off. Tin-stone is found in masses, copper ore in schist and in veins sometimes with pyrites. The ore from veins is better than schist. Iron ore is found in masses, and sometimes in veins; the latter is the best. "The iron veins with good hanging- and foot-"walls are not to be despised, especially if their strike be from east to west, their dip to the "south, the foot-wall and outcrop to the north, then if the ironstone is followed down, the "vein usually reveals gold or other valuable ore". Lead ore is found in *schwebenden gang* and *stehenden gang*. Quicksilver, like other ore, is sometimes found in brown earth, and sometimes, again, in caves where it has run out like water. The classification of veins is the same as in *De Re Metallica*.[12] The book generally, however, seems to have raised Agricola's opposition, for the quotations are given in order to be demolished.

Probierbüchlein. Agricola refers in the Preface of *De Re Metallica* to a work in German on assaying and refining metals, and it is our belief that it was to some one of the little assay books published early in the 16th century. There are several of them, seemingly revised editions of each other; in the early ones no author's name appears, although among the later editions various names appear on the title page. An examination of these little books discloses the fact that their main contents are identical, for they are really collections of recipes after the order of cookery books, and intended rather to refresh the memory of those

Probier büch-
lein/auff Gold/Silber/kupffer/
vnd Bley/Auch allerlay Metall
wie man die zů nutz arbayten vñ
Probieren soll.

Alle Müntzmaystern/Wardeyn/Gold
werckern/Bergkleüten/vñ kauffleütē
der Metall zů nutz mit grossem fleyß zů
samen gebracht.

[12]See p. 44

already skilled than to instruct the novice. The books appear to have grown by accretions from many sources, for a large number of methods are given over and over again in the same book with slight variations. We reproduce the title page of our earliest copy.

The following is a list of these booklets so far as we have been able to discover actual copies :—

Date.	*Place.*	*Publisher.*	*Title (Short).*	*Author.*
Unknown	Unknown	Unknown	*Probierbüchlein*	Anon.
	(Undated ; but catalogue of British Museum suggests Augsburg, 1510.)			
1524	Magdeburg		*Probirbüchleyn tzu Gotteslob*	Anon.
1531	Augsburg	Unknown	*Probierbuch aller Sachsischer Ertze*	Anon.
1533	Frankfurt a. Meyn		*Bergwerck und Probierbüchlein*	Anon.
1534	Augsburg	Heinrich Steyner, 8vo.	*Probirbüchlein*	Anon.
1546	Augsburg	Ditto, ditto	*Probirbüchlein*	Anon.
1549	Augsburg	Ditto, ditto	*Probirbüchlein*	Anon.
1564	Augsburg	Math. Francke, 4to	*Probirbüchlein*	Zach. Lochner
1573	Augsburg	8vo.	*Probirbuch*	Sam. Zimmermann
1574	Franckfurt a. Meyn		*Probierbüchlein*	Anon.
1578	Ditto		*Probierbüchlein Fremde und subtile Kunst*	Cyriacus Schreittmann
1580	Ditto		*Probierbüchlein*	Anon.
1595	Ditto		*Probierbüchlein darinn gründlicher Bericht*	Modestin Fachs
1607	Dresden	4to	*Metallische Probier Kunst Bericht vom Ursprung und Erkenntniss der Metallischen erze*	C. C. Schindler
1669	Amsterdam		*Probierbüchlein darinn gründlicher Bericht*	Modestin Fachs
1678	Leipzig		*Probierbüchlein darinn gründlicher Bericht*	Modestin Fachs
1689	Leipzig		*Probierbüchlein darinn gründlicher Bericht*	Modestin Fachs
1695	Nürnberg	12mo.	*Deutliche Vorstellung der Probier Kunst*	Anon.
1744	Lübeck	8vo.	*Neu-eröffnete Probier Buch*	Anon.
1755	Frankfurt and Leipzig	8vo.	*Scheid-Künstler . . . alle Ertz und Metalle . . . probiren*	Anon.
1782	Rotenburg an der Fulde	8vo.	*Probierbuch aus Erfahrung aufgesetzt*	K. A. Scheidt

As mentioned under the *Nützlich Bergbüchlein*, our copy of that work, printed in 1533, contains only a portion of the *Probierbüchlein.* Ferguson[13] mentions an edition of 1608, and the Freiberg School of Mines Catalogue gives also Frankfort, 1608, and Nürnberg, 1706. The British Museum copy of earliest date, like the title page reproduced, contains no date. The title page woodcut, however, in the Museum copy is referred from that above, possibly indicating an earlier date of the Museum copy.

The booklets enumerated above vary a great deal in contents, the successive prints representing a sort of growth by accretion. The first portion of our earliest edition is devoted to weights, in which the system of " lesser weights " (the principle of the " assay ton ") is explained. Following this are exhaustive lists of touch-needles of various composition. Directions are given with regard to assay furnaces, cupels, muffles, scorifiers, and crucibles, granulated and leaf metals, for washing, roasting, and the preparation of assay charges. Various reagents, including glass-gall, litharge, salt, iron filings, lead, " alkali ", talc, argol, saltpetre, sal-ammoniac, alum, vitriol, lime, sulphur, antimony, *aqua fortis*, or *scheidwasser*, etc., are made use of. Various assays are described and directions given for crucible, scorification, and cupellation tests. The latter part of the book is devoted to the refining and parting of precious metals. Instructions are given for the separation of silver from iron, from lead, and from antimony ; of gold from silver with antimony (sulphide) and sulphur, or with sulphur alone, with " *scheidwasser*," and by cementation with salt ; of gold from copper with sulphur and with lead. The amalgamation of gold and silver is mentioned.

[13]*Bibliotheca Chemica.*

The book is diffuse and confused, and without arrangement or system, yet a little consideration enables one of experience to understand most statements. There are over 120 recipes, with, as said before, much repetition; for instance, the parting of gold and silver by use of sulphur is given eight times in different places. The final line of the book is: "Take this in good part, dear reader, after it, please God, there will be a better." In truth, however, there are books on assaying four centuries younger that are worse. This is, without doubt, the first written word on assaying, and it displays that art already full grown, so far as concerns gold and silver, and to some extent copper and lead; for if we eliminate the words dependent on the atomic theory from modern works on dry assaying, there has been but very minor progress. The art could not, however, have reached this advanced stage but by slow accretion, and no doubt this collection of recipes had been handed from father to son long before the 16th century. It is of wider interest that these booklets represent the first milestone on the road to quantitative analysis, and in this light they have been largely ignored by the historians of chemistry. Internal evidence in Book VII. of *De Re Metallica*, together with the reference in the Preface, leave little doubt that Agricola was familiar with these booklets. His work, however, is arranged more systematically, each operation stated more clearly, with more detail and fresh items; and further, he gives methods of determining copper and lead which are but minutely touched upon in the *Probierbüchlein*, while the directions as to tin, bismuth, quicksilver, and iron are entirely new.

Biringuccio (Vanuccio). We practically know nothing about this author. From the preface to the first edition of his work it appears he was styled a mathematician, but in the text[14] he certainly states that he was most of his time engaged in metallurgical operations, and that in pursuit of such knowledge he had visited Germany. The work was in Italian, published at Venice in 1540, the title page of the first edition as below:—

DE LA PIRO-
TECHNIA.

LIBRI. X. DOVE AMPIAMEN
te ſi tratta non ſolo di ogni ſorte & di-
uerſita di Miniere, ma anchora quan
to ſi ricerca intorno à la prattica di
quelle coſe di quel che ſi appartiene
à l'arte de la fuſione ouer gitto de me
talli come d'ogni altra coſa ſimile à
queſta. Compoſti per il. S. Vanoc-
cio Biringuccio Senneſe.

Con Priuilegio Apoſtolico & de la
Ceſarea Maeſta & del Illuſtriſs. Sena
to Veneto. M D XL.

[14] Book I., Chap. 2.

It comprises ten chapters in 168 folios demi-octavo. Other Italian editions of which we find some record are the second at Venice, 1552; third, Venice, 1558; fourth, Venice, 1559; fifth, Bologna, 1678. A French translation, by Jacques Vincent, was published in Paris, 1556, and this translation was again published at Rouen in 1627. Of the ten chapters the last six are almost wholly devoted to metal working and founding, and it is more largely for this description of the methods of making artillery, munitions of war and bells that the book is celebrated. In any event, with the exception of a quotation which we give on page 297 on silver amalgamation, there is little of interest on our subject in the latter chapters. The first four chapters are undoubtedly of importance in the history of metallurgical literature, and represent the first work on smelting. The descriptions are, however, very diffuse, difficult to follow, and lack arrangement and detail. But like the *Probierbüchlein*, the fact that it was written prior to *De Re Metallica* demands attention for it which it would not otherwise receive. The ores of gold, silver, copper, lead, tin, and iron are described, but much interrupted with denunciations of the alchemists. There is little of geological or mineralogical interest, he too holding to a muddle of the classic elements astrology and alchemy. He has nothing of consequence to say on mining, and dismisses concentration with a few words. Upon assaying his work is not so useful as the *Probierbüchlein*. On ore smelting he describes the reduction of iron and lead ores and cupriferous silver or gold ores with lead. He gives the barest description of a blast furnace, but adds an interesting account of a *reverbero* furnace. He describes liquation as consisting of one operation; the subsequent treatment of the copper by refining with an oxidising blast, but does not mention poling; the cupellation of argentiferous lead and the reduction of the litharge; the manufacture of nitric acid and that method of parting gold and silver. He also gives the method of parting with antimony and sulphur, and by cementation with common salt. Among the side issues, he describes the method of making brass with calamine; of making steel; of distilling quicksilver; of melting out sulphur; of making vitriol and alum. He states that *arsenico* and *orpimento* and *etrisagallio* (realgar) are the same substance, and are used to colour copper white.

In general, Biringuccio should be accredited with the first description (as far as we are aware) of silver amalgamation, of a reverberatory furnace, and of liquation, although the description is not complete. Also he is, so far as we are aware, the first to mention cobalt blue (*Zaffre*) and manganese, although he classed them as "half" metals. His descriptions are far inferior to Agricola's; they do not compass anything like the same range of metallurgy, and betray the lack of a logical mind.

Other works. There are several works devoted to mineralogy, dating from the fifteenth and early sixteenth centuries, which were, no doubt, available to Agricola in the compilation of his *De Natura Fossilium*. They are, however, practically all compiled from the jeweller's point of view rather than from that of the miner. Among them we may mention the poem on precious stones by Marbodaeus, an author who lived from 1035 to 1123, but which was first printed at Vienna in 1511; *Speculum Lapidum*, a work on precious stones, by Camilli Leonardi, first printed in Venice in 1502. A work of wider interest to mineralogists is that by Christoph Entzelt (or Enzelius, Encelio, Encelius, as it is variously given), entitled *De Re Metallica*, and first printed in 1551. The work is five years later than *De Natura Fossilium*, but contains much new material and was available to Agricola prior to his revised editions.

APPENDIX C.

WEIGHTS AND MEASURES.

As stated in the preface, the nomenclature to be adopted for weights and measures has presented great difficulty. Agricola uses, throughout, the Roman and the Romanized Greek scales, but in many cases he uses these terms merely as lingual equivalents for the German quantities of his day. Moreover the classic language sometimes failed him, whereupon he coined new Latin terms adapted from the Roman scale, and thus added further confusion. We can, perhaps, make the matter clearer by an illustration of a case in weights. The Roman *centumpondium*, composed of 100 *librae*, the old German *centner* of 100 *pfundt*, and the English hundredweight of 112 pounds can be called lingual equivalents. The first weighs about 494,600 Troy grains, the second 721,900, and the third 784,000. While the divisions of the *centumpondium* and the *centner* are the same, the *libra* is divided into 12 *unciae* and the *pfundt* into 16 *untzen*, and in most places a summation of the units given proves that the author had in mind the Roman ratios. However, on p. 509 he makes the direct statement that the *centumpondium* weighs 146 *librae*, which would be about the correct weight if the *centumpondium* referred to was a *centner*. If we take an example such as "each *centumpondium* of lead contains one *uncia* of silver", and reduce it according to purely lingual equivalents, we should find that it runs 24.3 Troy ounces per short ton, on the basis of Roman values, and 18·25 ounces per short ton, on the basis of old German. If we were to translate these into English lingual equivalents of one ounce per hundredweight, then the value would be 17.9 ounces per short ton.

Several possibilities were open in translation: first, to calculate the values accurately in the English units; second, to adopt the nearest English lingual equivalent; third, to introduce the German scale of the period; or, fourth, to leave the original Latin in the text. The first would lead to an indefinite number of decimals and to constant doubt as to whether the values, upon which calculations were to be based, were Roman or German. The second, that is the substitution of lingual equivalents, is objectionable, not only because it would indicate values not meant by the author, but also because we should have, like Agricola, to coin new terms to accommodate the lapses in the scales, or again to use decimals. In the third case, that is in the use of the old German scale, while it would be easier to adapt than the English, it would be more unfamiliar to most readers than the Latin, and not so expressive in print, and further, in some cases would present the same difficulties of calculation as in using the English scale. Nor does the contemporary German translation of *De Re Metallica* prove of help, for its translator adopted only lingual equivalents, and in consequence the summation of his weights often gives incorrect results. From all these possibilities we have chosen the fourth, that is simply to reproduce the Latin terms for both weights and measures. We have introduced into the footnotes such reductions to the English scale as we considered would interest readers. We have, however, digressed from the rule in two cases, in the adoption of "foot" for the Latin *pes*, and "fathom" for *passus*. Apart from the fact that these were not cases where accuracy is involved, Agricola himself explains (p. 77) that he means the German values for these particular terms, which, fortunately, fairly closely approximate to the English. Further, we have adopted the Anglicized words "digit", "palm", and "cubit", instead of their Latin forms.

For purposes of reference, we reproduce the principal Roman and old German scales, in so far as they are used by Agricola in this work, with their values in English. All students of weights and measures will realize that these values are but approximate, and that this is not an occasion to enter upon a discussion of the variations in different periods or by different authorities. Agricola himself is the author of one of the standard works on Ancient Weights and Measures (see Appendix A), and further gives fairly complete information on contemporary scales of weight and fineness for precious metals in Book VII. p. 262 etc., to which we refer readers.

ROMAN SCALES OF WEIGHTS.

				Troy Grains.
1 *Siliqua*	=	..	..	2.87
6 *Siliquae*	=	1 *Scripulum*	..	17.2
4 *Scripula*	=	1 *Sextula*	..	68.7
6 *Sextulae*	=	1 *Uncia*	..	412.2
12 *Unciae*	=	1 *Libra*	..	4946.4
100 *Librae*	=	1 *Centumpondium*	..	494640.0
		Also		
1 *Scripulum*	=	..	..	17.2
3 *Scripula*	=	1 *Drachma*	..	51.5
2 *Drachmae*	=	1 *Sicilicus*	..	103.0
4 *Sicilici*	=	1 *Uncia*	..	412.2
8 *Unciae*	=	1 *Bes* ..	..	3297.6

SCALE OF FINENESS
(AGRICOLA'S ADAPTATION).

4 *Siliquae*	=	1 Unit of *Siliquae*
3 *Units of Siliquae*	=	1 *Semi-sextula*
4 *Semi-sextulae*	=	1 *Duella*
24 *Duellae*	=	1 *Bes*

OLD GERMAN SCALE OF WEIGHTS.

			Troy Grains.
1 *Pfennig*	=	..	14.1
4 *Pfennige*	=	1 *Quintlein*	56.4
4 *Quintlein*	=	1 *Loth*	225.6
2 *Loth*	=	1 *Untzen*	451.2
8 *Untzen*	=	1 *Mark*	3609.6
2 *Mark*	=	1 *Pfundt*	7219.2
100 *Pfundt*	=	1 *Centner*	721920.0

SCALE OF FINENESS.

3 *Grenlin*	=	1 *Gran*
4 *Gran*	=	1 *Krat*
24 *Krat*	=	1 *Mark*

ROMAN LONG MEASURE.

			Inches.
1 *Digitus*	=	..	.726
4 *Digiti*	=	1 *Palmus*	2.90
4 *Palmi*	=	1 *Pes*	11.61
1½ *Pedes*	=	1 *Cubitus*	17.41
5 *Pedes*	=	1 *Passus*	58.1

Also

1 Roman *Uncia*	=	..	.97
12 *Unciae*	=	*Pes*	11.61

GREEK LONG MEASURE.

1 *Dactylos*	=	..	.758
4 *Dactyloi*	=	1 *Palaiste*	3.03
4 *Palaistai*	=	1 *Pous*	12.135
1½ *Pous*		1 *Pechus*	18.20
6 *Pous*	=	1 *Oryguia*	72.81

OLD GERMAN LONG MEASURE.

			Inches.
1 *Querfinger*	=	..	.703
16 *Querfinger*	=	1 *Werckschuh*	11.247
2 *Werckschuh*	=	1 *Elle*	22.494
3 *Elle*	=	1 *Lachter*	67.518

Also

1 *Zoll*	=	..	.85
12 *Zoll*	=	1 *Werkschuh*	

ROMAN LIQUID MEASURE.

			Cubic inches.	Pints.
1 *Quartarius*	=		8.6	.247
4 *Quartarii*	=	1 *Sextarius*	31.4	.991
6 *Sextarii*	=	1 *Congius*	206.4	5.947
16 *Sextarii*	=	1 *Modius*	550.4	15.867
8 *Congii*	=	1 *Amphora*	1650.0	47.577

(Agricola nowhere uses the Saxon liquid measures, nor do they fall into units comparable with the Roman).

GENERAL INDEX.

NOTE.—The numbers in **heavy type** refer to the Text; those in plain type to the Footnotes, Appendices, etc.

INDEX TO PERSONS AND AUTHORITIES.

NOTE.—The numbers in heavy type refer to the Text; those in plain type to the Footnotes, Appendices, etc.

INDEX TO ILLUSTRATIONS.